世界の昆虫 英名辞典
vol. 1 A-L

矢野　宏二　編

A dictionary of English names of the world insects

Edited by Koji Yano

櫂歌書房

A dictionary of English names of the world insects

Date of publication 12. May. 2018
ISBN 978-4-434-24028-7
Edited by Koji Yano
Published by Touka Shobo
Printed and distributed by Touka Shobo
Sarayama 4-14-2, Minami-ku, Fukuoka-shi, 811-1365, Japan
Tel:+81-92-511-8111 e-mail: e@touka.com

はじめに

　昆虫の英名は一般社会で自由自在に使用されている。人や地域、一見してわかる形態、摂食植物や寄主動物などの名が入るものは、その由来は分かりやすいが、何故その英名なのかというものも多い。それよりも、英名から種を特定することが困難である。例えば red ant といえば樹上に巣を作る著名なツムギアリだが（red tree ant, weaver ant などの英名もあり、学名は *Oecophilla smaragdina* Fabricius）、キイロクシケアリ、ケズネアカヤマアリ、シワクシケアリ他も red ant といわれる。北米のチャイロイチモンジとカバイチモンジはともに viceroy という英名をもつ。また、mango fly は Oriental fruit fly ともいわれるミカンコミバエと、別のミバエ科の一種、そしてアジアのオオヨコバイ科の一種の英名であり、所属する目も違う3種の英名である。このような同英名はたくさんある。これらの場合、英名だけではどの種をさしているのか分からない。

　一種で多くの英名をもつ種、前記したように複数の種が同じ英名をもつ場合、動物や植物の英名がそのまま昆虫の英名になっているなど、英名はさまざまである。drinker や widow, typewriter など思わずその由来に思いをはせるものも多い。

　専門の研究や学術論文と離れた一般社会では、一般名（和名、英名など）がその生物を示す手段となっているが、国際動物命名規約により厳密に規定されている学名と異なり、前記のように一般名は不明確な側面が多い。

　本書は世界の昆虫の英名をアルファベット順に掲載し、和名（ある場合。同種で複数の和名がある場合は原則として1つ）、その種あるいは属以上のタクサ（分類単位）の学名、分類上の所属、分布などを記した。基本的に分類群の英名を対象とし、生理、生態、形態分野の英名や術語は除外した。

　日本産昆虫の英名リスト（矢野、2004）に外国種も参考までに少し付記したが、本書はそれを世界に広げたものである。掲載した英名は印刷物で学名と併記されていたものであり、英名だけとか口頭の使用例は含まない。IT上の英名は吟味の上一部を採録した。掲載項目数は約36,000だが、別種同英名とか、同一タクサで複数の英名をもつものが多く、タクサ数ではない。使用程度の多さを念頭に採録したが、紙数の関係で未採録の英名も多くある。

　英名の多くは英語圏で使用、定着してきたものであり、英語を母国語としない著者には、その由来が不明なものが多い。さらに、個人的な感覚、思い入れでつけられた名は、他人には推測するしかない。したがって、本書では個々の英名の由来は一部の種に限った。

　多様な印刷物から採録した英名なので、誤記、誤用のまま定着したと思われるものを含め、その不明確さを追求することは本書の主旨ではない。非英語圏で作られた英語として不適切な英名もそのまま掲載した。本書は実態を示したもので、その点を大前提として利用されることを願いたい。忻・夏 1978、Xiao, G. ed. 1997 など種の実態が分かりにくい文献は除外した。本書で掲載した以外にも英名があるが、紙数の関係で除外した。

　同種であるが別学名（シノニム）で別英名をもつ場合、整理不十分で掲載されていること

があることをお断りしておきたい。

　アメリカ昆虫学会で害虫など主要種に対して英名を一つにする努力がなされているが、本書は辞典として関係資料を提示することを目的とするので、選別はしていない。望ましい英名に収斂するための一助となることを望みたい。分類順のリストをつけると同種の別英名が分かり易いが紙数の関係で断念した。

謝辞

　本書は九州大学名誉教授平嶋義宏先生により企画されたもので、準備段階で種々ご配慮をいただいた。ここに厚くお礼を申し上げる。文献やPC用ファイルの整理などで多大なご協力を得た竹松葉子山口大学教授はじめ、多田内修九州大学名誉教授、笹川満廣京都府立大学名誉教授、吉安　裕京都府立大学名誉教授、緒方一夫九州大学教授，林　正美埼玉大学名誉教授、持田　作博士には文献や有益なご意見をいただいた。感謝申し上げる。

協力者

　九州大学名誉教授矢田　脩博士はチョウの採録と原稿確認で多大なご協力をいただいた。厚くお礼を申し上げる。

2018年　編者

著者略歴

矢野　宏二（やの こうじ）

1932 年生まれ
大阪府立大学農学部卒業
九州大学大学院農学研究科博士課程単位取得退学。ハワイ、ビショップ博物館に 1 年滞在
九州大学農学部助手、助教授を経て山口大学農学部教授と鳥取大学大学院連合農学研究科教授を併任。山口大学名誉教授、農学博士。
昆虫分類学 (チョウ目ミノガ科、スカシバガ科、トリバガ科、ハエ目ヤチバエ科、ユスリカ科、アタマアブ科) から水田昆虫の生活史、熱帯昆虫学の研究を進めてきた。東南アジアからインド、パキスタン各地で野外調査を実施。

主な著書

熱帯昆虫学。編著。1999. 九州大学出版会
水田の昆虫誌ーイネをめぐる多様な昆虫たち。2002、東海大学出版会
日本産昆虫の英名リスト。2004. 東海大学出版会
農薬の陰謀. 訳書。1984. 社会思想社
日本原色虫えい図鑑。共著。1996。全国農村教育協会 他

留学生たちと（札幌 1986）

主な論文

Taxonomic and biological studies of Pterophoridae of Japan (Lepidoptera). Pacific Insects 5: 65-209. 1963

Lepidoptera: Pterophoridae. Insects of Micronesia 9(3). Micronesia 28(2): 131-151. 共著。1996

Faunal and biological studies on the insects of paddy fields in Asia. Part I. Introduction and Sciomyzidae from Asia (Diptera). Esakia (11):1-27. 1978 他

目次

はじめに iii
著者略歴 v
凡例 ix
参考文献 xiii
世界の昆虫英名
    A 1
    M ....【別冊 vol.2】 669

和名索引 ....【別冊 索引 vol.3 】 1

学名索引 ....【別冊 索引 vol.3 】 75

凡例

1. 英名のアルファベット順に掲載したが、ハイフンの有無や語の結合があっても原則として同一タクサ（分類単位）は同じ所に掲載した。

2. 英名が亜種で異なる場合は原則として別にし、原名亜種の英名を参照と記して関係が分かるようにした。

3. 属名や種小名が英名に使われる場合、大文字か小文字かで統一せず、使用実態のままとした。固有名詞以外の英名は小文字で記したが、levant と Levant のように両方ある語もある。

4. 日本以外の分布は動物地理区で記したが、全ての分布区を記したわけではなく、当該英名が使用される地域を示した方が多い。日本には分布しないが汎世界的に分布する種も汎世界と記した。

5. 同じ英名が複数ある場合、下記のように整理してある。
   例。 beet leaf miner (1) *Pegomya betae* (Curtis)（ハエ目、ハナバエ科）。分布。
   　　 beet leafminer (2) *Liriomyza chenopodii* (Watt)（ハエ目、ハモグリバエ科）。分布。
   　　 beet leafminer (3) テンサイモグリハナバエ *Pegomya cunicularia* (Rondani)
   　　　　　　　　　　（ハエ目、ハナバエ科）。分布。
   　　 beet leafminer　　 beet fly を見よ。

   beet leafminer は上記のように4つあり、4種の英名であることが分かる。4つ目の beet fly を見ると、
   　　 beet fly　　*Pegomya hyoscyami* (Panzer)（ハエ目、ハナバエ科）。分布。
   　　 beet fly　　beet leaf miner (1) を見よ。

   の2つあり、学名などを記述してある上段の beet fly が beet leafminer の別英名でもあることを示す。下段の beet fly は beet leaf miner (1) の種が beet fly ともいわれることを示す。

   beet leafminer (1)(2)(3) は、他の英名の項で beet leafminer を見よ、と記す場合、(1)(2)(3) の3つのどれを見るかを区別するためである。例えば mangold fly を見ると　beet leaf miner (1) を見よ、と記している。

   このように、（XX を見よ）と記したのは、単に参照せよとの意味ではなく、XX の種の英名でもあることを示す。

また、同種で複数英名がある場合、例えば

  American cotton bollworm corn earworm (1)(2) を見よ。

と記しているが、American cotton bollworm は corn earworm (1) と同 (2) の 2 種の英名でもあることを示す。

6．同種で複数の異なる英名がある場合、分類順に掲載していないので一個所ではわからない。例えば *Agrius convolvuli* (Linnaeus) エビガラスズメは

  A．sweetpotato hawk-moth
  B．sweetpotato horn worm
  C．convolvulus hawkmoth
  D．morning-glory sphinx

の 4 英名がある。このうち、使用例が多く、定着していると思われる A をえらび、和名、学名、分類上の所属と分布、若干の解説を記した。そして B, C, D の英名の項では、これらの記述を省き、A を見よ、と記した。

A を選ぶ基準は上記したが、複数英名間に明らかな使用差がなかったり、使用地域により異なる場合もあり、多くの場合は編者の判断によった。

7．6 の例にあげたエビガラスズメの英名の A と B、シャクガの *Nepytia canosaria* の英名の false hemlock moth と false hemlock looper などは成虫と幼虫の英名であるが、いずれも種の英名として成育期に関係なく使用され、幼虫名の方が定着している場合も多い。従って成虫は XX というとか、幼虫は XX というとかと記した場合もあるが、同様である。

8．同種で複数の英名がある場合、語尾などが少し異なる場合は同じところで記したが、かなり違う場合は別記した。例えば half-yellow moth のところに half-yellow ともいう、と記したが、yellow-cloaked midget は別記した。分類順（学名）のリストを用意すればよいが、紙数の関係で断念した。

9．種の解説は紙数の関係で最小限とし、日本産の主要害虫を主に記述した。幼虫が加害する場合が多いが、単に X の害虫と記した。また、加害植物全てを記さず、代表的な植物に限った。

10．解説の部分で、日本亜種の学名、和名を示した場合がある。例えば

  buff footman *Eilema deplana* (Esper)（チョウ目、ヒトリガ科）。旧北区。
  日本亜種は *E. d. puvescens* (Butler) ムジホソバ。

とあるのは、日本亜種の英名使用例はないが、旧北区の原名亜種は buff footman の英名であることを示す。

11. 英名に含まれるハイフンは採録した文献に準拠したので leaf miner, leaf-miner, leafminer のように統一していない。検索する場合は留意をお願いしたい。例えば armyworm と army worm、blow fly と blowfly、many-dotted appleworm moth と manydotted apple worm のように出典に準拠して掲載してある。これが英名であり、本書では採録文献の使用例に準拠したが、別の使用例（ハイフンの有無）もあるので注意をお願いしたい。このため、アルファベット順の掲載も leaf miner, leaf-miner, leafminer のあとに leafhopper がくるので注意いただきたい。また、通常はされない単語の結合もみられる。

12. 属以上のタクサの英名は下記のように複数形で記されるのが慣例である。
    stink bugs　　カメムシ科 Pentatomidae（カメムシ目）の昆虫の総称。
    cockchafers　　*Melolontha* (コウチュウ目、コガネムシ科）の昆虫の総称。
    これはカメムシ科の種は stink bug、*Melolontha* 属の種は cockchafer といわれることを示すが、これらタクサの英名として使用されている。

13. 最近の分類学的研究の結果、掲載種の所属科名の変更が多く生じているので留意いただきたい（ジャノメチョウ、ドクチョウなどを含むタテハチョウ科やメイガ科など）。

14. 採録文献の英名に併記された学名は、現行学名に変更した場合がある。

参考文献

Alford, D. V. 1991. A colour atlas of pests of ornamental trees, shrubs and flowers. Wolfe Pub. Ltd., London. 448 pp.

青木　淳一・奥谷　喬司・松浦　啓一（編）2002.　虫の名，貝の名、魚の名。和名にまつわる話題。東海大学出版会、東京 . xiv+242 pp.

Arnett, R. H. Jr. 1985. American insects. A handbook of the insects of America north of Mexico. Van Nostrand Reinhold Co., New York. xiii+850 pp. (2nd ed. 2000, 960 pp.)

Barrett, C. and A. No. Burns 1951. Butterflies of Australia and New Guinea. N. H. Seward Pty. Ltd., Melbourne. 187 pp.

Beirne, B. P. 1954. British pyralid and plume moths. Frederick Warne & Co., Ltd., London and N. Y. 208 pp.

Borror, D. J. and D. M. DeLong 1964. An introduction to the study of insects. Rev. ed. Holt, Rinehart and Winston, N. Y. xi+819 pp.

Borror, D. J. and R. E. White 1970. A field guide to the insects of America north of Mexico. Houghton Mifflin Co., 404 pp.

Bosik, J. J. 1997. Common names of insects and related organisms. Ent. Soc. Amer., Maryland. 232 pp.

Brues, C. T. , A. L. Melandar and F. M. Carpenter 1954. Classification of insects. Harvard Univ., Mus. Compar. Zool. Bull. 108: 1-917.

Chinery, M. 1993. Collins field guide. Insects of Britain and northern Europe. (3rd ed.). Harper Collins Pub., 320 pp.

中国科学院動物研究所（編）1983. 拉英汉昆虫名称 , 科学出版社 , 北京 ,v+911pp.

Common, I. F. B. and D. F. Waterhouse 1972. Butterflies of Australia. Angus & Robertson Pub., Sydney. 498 pp.

Corbet, A. S. and H. M. Pendlebury 1992. The butterflies of the Malay Peninsula. Malayan Nature Soc., Malaysia, Kuala Lumpur. 595 pp., 69 pls.

Curran, C H. 1965. The families and genera of North American Diptera. Henry Tripp, N. Y. 515 pp.

Dammerman, K. W. 1929. The agricultural zoology of the Malay Archipelago. J. H. de Bussy Ltd., Amsterdam. xi+473 pp.

DeBach, P. 1964. Biological control of insect pests and weeds. Chapman and Hall Ltd., London. xxiv+844 pp.

Easton, E. R. and W-W. Pun 1996. New records of moths from Macau, Southeast China. Trop. Lep. 7(2): 113-118.

Ebeling, W. 1950. Subtropical entomology. Lithotype Process Co., San Francisco. 747 pp.

Emanoli, M. (ed.) 1994. Encyclopedia of endangered species. Gale Res. Inst. pp. 1014-1049.

Essig, E. O. 1942. College entomology. McMillan Co., N. Y. vii+900 pp.

Ferro, D. N. 1977. Standard names for common insects of New Zealand. Bull. Ent. Soc. NZ. 2: 1-42.

Frohlich, G. and W. Rodeward 1969. Pests and diseases of tropical crops and their control. Pergamon Press, Oxford, ix+371 pp.

Gardiner, B. O. C. (ed.) 1995. An index to the modern names for use with J. W. Tutt's practical hints for the field lepidopterist. Amateur Ent. 23A: 1-58.

Gaskin, D. E. 1966 The butterflies and common moths of New Zealand. Whitcombe and Tombs Ltd. 219 pp.

Gay, T., I. D. Kehimkar and J. C. Punetha 1992. Common butterflies of India. Oxford Univ. Press., Oxford. 67 pp.

Gibbons, B. 1995. Field guide to the insects of Britain and northern Europe. Crowood Press, Ltd. Wilshire. 320 pp.

Gilbert, P. and C. J. Hamilton 1983. Entomology. A guide to information sources. Mansell Pub. Ltd., London. vi+237 pp.

Gordh, G. and D. Headrick 2010. A dictionary of entomology. 2nd ed., CABI Pub., Wallingford. ix+1526 pp.

Gozmany, L. 1979. Vocabularium animalium europae septem linguis redactum. Akademiai Kiado, Budapest. Vol. 1, 42+1171 pp., Vol. II, 1015 pp.

Greiff, M. 1985. Spanish-English-Spanish lexicon of entomological and related forms with indexes of Spanish common names of arthropods and their Latin and English equivalents. CIE, London. 101+115+43 pp.

Grist, D. H. and R. J. A. W. Lever 1959. Pests of rice. Longmans, Green and Co., Ltd. London. xi+520 pp..

Gurney, A. B. 1953. An appeal for a clear understanding of principles concerning the use of common names. Jour. econ. Ent. 46: 207-211.

Hawaiian Ent. Soc. (ed.) 1975 (rev. ed.). List of common names of some insects and organisms. Haw. Ent. Soc. 29 pp. (mimeog.).

Heinrich, E. A. (ed.) 1994. Biology and management of rice insects. John Willey & Sons, IRRI, x+779 pp.

Heppner, J. B. 1998. Revised family list for Lepidoptera. Lep. News 3: 57-62.

Higgins, L. G. and N. D. Riley 1993. A field guide to the butterflies of Britain and Europe. Harper Collins Pub., London. 384 pp.

Hill, D. S. 1983 (3rd ed.) Agricultural insect pests of the tropics and their control. Cambridge

Univ. Press, Cambridge. xii+746 pp.

平嶋義宏　1999. 新版蝶の学名。その語源と解説。九州大学出版会、福岡. ix+714 pp., 8 pls.

平嶋義宏　2007. 生物学名辞典。東京大学出版会、東京. xxii+1292 pp.

市河　三喜（編）1960. 新英和大辞典。研究社、東京. xlvi+2204 pp.

Imms, A. D. 1957. A general textbook of entomology. 9th ed. Revised by Richards, O. W. and R. G. Davies. x+886 pp. Methuen & Co. Ltd., London. x+886 pp.

石原　保　1949. 日本及び近隣産蝶類の外国名。宝塚昆虫館報 54: 1-10.

Jean, L. L. 1960a. Common names of insects approved by the Entomological Society of America. Bull. Ent. Soc. Amer. 6: 175-210.

Jean, L. L. 1960b. Proposed changes in and addition to the list of common names of insects and other invertebrates approved by the Entomological Society of America. Bull. Ent. Soc. Amer. 6: 211.

Jean, L. L. 1963. Proposed changes in and addition to the list of common names of insects approved by the Entomological Society of America. Bull. Ent. Soc. Amer. 8: 84.

Johnston, G. and B. Johnston 1980. This is Hong Kong : Butterflies. Hong Kong Government Pub., Hong Kong. 224 pp.

Kirby, W. F. 1901. Familiar butteflies and moths. Cassell and Co., Ltd., London. 144 pp.

Klots , A. B. 1973. A field guide to the butterflies of North America, East of the Great Plains. Houghton Mifflin Co., Boston. 349 pp.

Larsen, T. B. 1991. The butterflies of Kenya and their natural history. Oxford Univ. Press, Oxford. 490 pp.

Maxwell-Lefroy, H. 1910. List of common names used in India for common insects. Agr. Res. Inst., Pusa, Bull. 19. 49+xvii.

McCall, W. A. 1994. British & European butterfly vernacular names including forms, subspecies & abberations. G. Toth, Hungary. 77 pp.

McCall, W. A. 2000. British, European & Asian butterfly vernacular names. Gergely Toth, Hungary. 51 pp.

Merino-Rodoriguez, M. 1964. Elsevier's lexicon of parasites and diseases in livestock. Elsevier, Amsterdam. 125 pp.

Metcalf, C. L. 1942. Common names of insect. Jour. econ. Ent. 35: 795-797.

Metcalf, C. L. and W. P. Flint 1962. Destructive and useful insects. Their habits and control. 4th ed. McGraw Hill Book Co., Inc. 1087 pp.

Moucha, J. 1973. A color guide to familiar butterflies, caterpillars and chrysalides. Oxtopus Books, Czechoslovkia. 191 pp.

Nauman, I. 1993 (6th ed.) CSIRO Handbook of Australian insect names. CSIRO, Melbourne.

193 pp.

日本応用動物昆虫学会（編）2006. 農林有害動物・昆虫名鑑。増補改訂版。日本応用動物昆虫学会. 東京 v+387 pp.

日本特殊農薬株式会社（編）1966. 日本有用植物病害虫名彙 (付雑草)。日本特殊農薬株式会社、東京. 591 pp.

緒方　一喜（編）1962. 日本産衛生害虫の学名・普通名表。衛生動物 13: 134-152.

Pawar, A. D. 1975. Common names (including scientific names and important synonyms) and distribution of major insect pests of rice of the world. Rice Ent. Newsl. (2): 7-13 (Palampur, India).

Powel, J. A. and . L. Hogue 1979. California insects. Univ. California Press, Berkeley. x+388 pp.

Scott, J. A. 1986. The butterflies of North America. Stanford Univ. Press, Stanford. 583 pp.

瀬能　宏 2002. 標準和名の安定化に向けて。青木他編 2002. 192-225 pp.

Seymour, P. 1980. Invertebrates of economic importance in Britain: common and scientific names. HMSO, London. viii+132 pp.

素木　得一 1954. 昆虫の分類。北隆館、東京. xiii+961 pp.

素木　得一（編）1962. 昆虫学辞典。北隆館、東京. 1098+114+41 pp., 52 pls.

Slater, J. A. 1964. Proposed changes in and addition to the list of common names of insects approved by the Entomological Society of America. Bull. Ent. Soc. Amer. 8: 198-199.

Smith, C. 1989. Butterflies of Nepal (Central Himalaya). T. C. Majupuria and Craftsman Press, Bangkok. 352 pp.

South, R. 1961. The moths of the British Isles. 2 (new edition). Frederick Warne, London. 417 pp.

Sutherland, D. W. S. (ed.) 1978. Common names of insects and related organisms. (1978 ed.). Ent. Soc. Amer., Maryland. 132 pp.

丁　慧琳・王　雯珊（編）1986. 昆虫学名詞辞典。名山出版社、台北. 396 pp.

Thomas, I., H. W. Janson and A. D. Aitken, 1968 (2nd ed.). Common names of British insect and other pests. Min. Agr. Fisheries and Food, Tech. Bull. 6: 1-72.

van den Bosch, R. and P. S. Messenger 1973 Biological control. Intext Educational Pub. xii+180 pp.

Wheeler, W. M. 1960. Ants. Their structure, development and behavior. Columbia Univ. Press, New York. 663 pp.

Williams, J. G. 1973. A field guide to the butterflies of Africa. Collins, Lonodon. 238 pp.

Wrobel, M. 2000. Elsevier's dictionary of butterflies and moths in Latin, English, German, French and Italian. Elsevier, Amsterdam. 278 pp.

Wrobel, M. 2003. Elsevier's dictionary of entomology in Latin, English, German, Frech and

Italian. Elsevier, Amsterdam. 374 pp.

蕭　剛柔（編）1997. 拉漢英昆虫蜱蟎蜘蛛線虫名称。中国林業出版社、北京. 894 pp (Xiang Gangrou ed. Latin-Chinese-English names of insects, ticks, mites, spiders and nematodes)

忻　介六、夏　松云（編）1978. 英漢昆虫俗名詞滙。湖南人民出版社、湖南省. 441 pp. (Xin, Jie-lu and Xing Song Yun. Glossary of common names of insects in English and Chinese)

矢野宏二（編）2004. 日本産昆虫の英名リスト　附　主要外国種の英名。東海大学出版会、秦野. xii+171 pp.

横浜植物防疫所 1977（追補 1985）。植物検疫重要病害虫解説。害虫。植物防疫資料第 4 号。

横浜植物防疫所（監）1985。侵入を警戒する病害虫と早期発見の手引。日本植物防疫協会、東京. 126 pp.

Zahradnik, J. 1991. Bees, wasps and ants. Hamlyn Pub. Group Ltd.192 pp.

Zhang, Bin-Cheng 1994. Index of economically important Lepidoptera. CAB Int., UK. 599 pp.

世界の昆虫
英名辞典
vol. 1 A-L

英名	和名	学名	所属、分布、ほか
A			
Aaron's skipper			Saffron skipper を見よ
Abbott's bagworm moth			live oak bagworm を見よ
Abbott's pine sawfly			white pine sawfly を見よ
Abbott's pine sphinx		*Lapara coniferarum* (J. E. Smith)	（チョウ目、スズメガ科）新北区
Abbott's skimmer		*Orthetrum abbotti* Calvert	（トンボ目、トンボ科）エチオピア区
Abbott's sphinx		*Sphecodina abbotti* (Swainson)	（チョウ目、スズメガ科）新北区。Abbott's sphinx moth ともいう
abbreviated button slug moth		*Tortrcidia flexuosa* (Grote)	（チョウ目、イラガ科）新北区
abbreviated underwing moth		*Catocala abbreviatella* Grote	（チョウ目、ヤガ科）新北区
abbreviated wireworm		*Hypolithus abbreviatus* (Say)	（コウチュウ目、コメツキムシ科）新北区
Abdera Acraea		*Acraea abdera* Hewitson	（チョウ目、タテハチョウ科）エチオピア区
aberrant buprestids		Schizopinae	（コウチュウ目、タマムシ科）の昆虫の総称
aberrant bushblue			aberrant oakblue を見よ
aberrant long-horned beetles		*Parandra*	（コウチュウ目、カミキリムシ科）の昆虫の総称
aberrant long-horns	ニセクワガタカミキリ亜科	Parandrinae	（コウチュウ目、カミキリムシ科）の昆虫の総称
aberrant oakblue		*Arhopala abseus* (Hewitson)	（チョウ目、シジミチョウ科）東洋区
aberrant shieldback		*Idiostatus aberrans* Renz	（バッタ目、キリギリス科）新北区
abies needle gall midge	トドマツノタマバエ	*Paradiplosia manii* (Inouye)	（ハエ目、タマバエ科）日本。トドマツ類の害虫
abies psylla	トドキジラミ	*Psylla abieti* Kuwayama	（カメムシ目、キジラミ科）日本。トドマツ類の害虫
abies torymid	モミモンオナガコバチ	*Megastigmus borriesi* Crosby	（ハチ目、オナガコバチ科）日本、旧北区
abietis aphid	トドワタムシ	*Mindarus japonicus* Takahashi	（カメムシ目、アブラムシ科）日本、旧北区
abietis bark beetle			fir bark beetle (1) を見よ
abnormal silverline		*Cigaritis abnormis* Moore	（チョウ目、シジミチョウ科）東洋区
Abor freak		*Calinaga aborica* Tyler	（チョウ目、タテハチョウ科）東洋区
Abraham's ruby-eye		*Carystoides abrahami* Freeman	（チョウ目、セセリチョウ科）新熱帯区
abraxas			magpie を見よ　スグリシロエダシャクなどの *Abraxas* 属に由来
abrupt brother moth		*Raphia abrupta* Grote	（チョウ目、ヤガ科）新北区
abrupt Epiblema moth		*Epiblema abruptana* (Walsingham)	（チョウ目、ハマキガ科）新北区
Absolon forester		*Bebearia absolon* (Fabricius)	（チョウ目、タテハチョウ科）エチオピア区
abutilon moth			cotton leaf caterpillar (1) を見よ

英名	和名	学名	所属、分布、ほか
Abyssinian admiral		*Vanessa abyssinica* (C. et R. Felder)	(チョウ目、タテハチョウ科) エチオピア区
acacia ants	クシフタフシアリ類	*Pseudomyrmex*	(ハチ目、アリ科) の昆虫の総称
acacia blue		*Surendra vivarna* (Horsfield)	(チョウ目、シジミチョウ科) 東洋区。*S. v. amisena* (Hewitson) も同英名
acacia blues		*Surendra*	(チョウ目、シジミチョウ科) の昆虫の総称
acacia borer		*Zeuzera eucalypti* (Boisduval)	(チョウ目、ボクトウガ科) 豪州区
acacia carpenter moth		*Xyleutes strix* Linnaeus	(チョウ目、ボクトウガ科) 東洋区
acacia carpenter moth			wattle goat moth を見よ
acacia long-horned beetle			Hawaiian black wattle long-horned beetle を見よ
acacia long-horned beetle			Hawaiian koa long-horned beetle を見よ
acacia psyllid		*Psylla uncatoides* (Ferris et Klyver)	(カメムシ目、キジラミ科) 新北区
acacia skipper		*Cogia hippalus* (Edwards)	(チョウ目、セセリチョウ科) 新北区
acacia spotting bug		*Rayieria tumidiceps* (Horvath)	(カメムシ目、カスミカメムシ科) 豪州区
acacia sprite		*Pseudagrion acaciae* Förster	(トンボ目、イトトンボ科) エチオピア区
Acadian hairstreak	ヤナギカラスシジミ	*Satyrium acadica* (Edwards)	(チョウ目、シジミチョウ科) 新北区
acalypterate muscoid flies	無弁翅類	Acalyptrata	(ハエ目) の昆虫の総称
acanaloniid planthoppers		Acanaloniidae	(カメムシ目) の昆虫の総称
acanthopanax bark beetle	ウコギノコキクイムシ	*Ernoporus acanthopanaxi* Niijima	(コウチュウ目、キクイムシ科) 日本
acanthopteroctetid moths			archaic sun moths を見よ
Acapulco Calephelis		*Calephelis acapulcoensis* McAlpine	(チョウ目、シジミタテハ科) 新熱帯区
Acara Acraea		*Acraea acara* Hewitson	(チョウ目、タテハチョウ科) エチオピア区
acartophthalmid flies		Acartophthalmidae	(ハエ目) の昆虫の総称
accent gem		*Ctenoplusia accentifera* (Lefebvre)	(チョウ目、ヤガ科) エチオピア区
accented Nephele		*Nephele accentifera* (Palisot de Beauvois)	(チョウ目、スズメガ科) エチオピア区
accused metalmark		*Symmachia accusatrix* Westwood	(チョウ目、シジミタテハ科) 新熱帯区
acer bark beetle	ベニイタヤノキクイムシ	*Dryocoetes picipennis* Eggers	(コウチュウ目、キクイムシ科) 日本
acer scale	トウカエデフクロカイガラムシ	*Eriococcus tokaedae* Kuwana	(カメムシ目、フクロカイガラムシ科) 日本
acerenntomids	クシカマアシムシ科	Acerentomidae	(カマアシムシ目) の昆虫の総称

英名	和名	学名	所属、分布、ほか
aces	ヒメコチャバネセセリ属	*Halpe*	(チョウ目、セセリチョウ科) の昆虫の総称
achemon sphinx		*Eumorpha achemon* (Drury)	(チョウ目、スズメガ科) 新北区。ブドウ害虫。achemon はギリシャ寓話由来。achemon sphinx moth ともいう
achilid planthoppers	コガシラウンカ科	Achilidae	(カメムシ目) の昆虫の総称
achilids			achilid planthoppers を見よ
Achilles blue Morpho		*Morpho achilles* (Linnaeus)	(チョウ目、タテハチョウ科) 新熱帯区
Achilles Morpho			Achilles blue Morpho を見よ
Acidalia leafwing		*Memphis acidalia* Hübner	(チョウ目、タテハチョウ科) 新熱帯区
Acis hairstreak		*Strymon acis* (Drury)	(チョウ目、シジミチョウ科) 新北区
Acisoma			ivory pintail を見よ
aclerdid scales	カタカイガラモドキ科	Aclerdidae	(カメムシ目) の昆虫の総称
Acmon blue		*Aricia acmon* (Westwood)	(チョウ目、シジミチョウ科) 新北区
Acontius firewing		*Catonephele acontius* Linnaeus	(チョウ目、タテハチョウ科) 新熱帯区
acorn and nut weevils			acorn weevils (1) を見よ
acorn cup gall cynipid		*Andricus quercuscalicis* (Burgsdorf)	(ハチ目、タマバチ科) 旧北区。本種がつくる虫えいを knopper gall という
acorn moth		*Valentinia glandulella* Riley	(チョウ目、ネマルハキバガ科) 新北区。幼虫はドングリに食入
acorn moth			Catalina cherry moth を見よ
acorn weevil		*Curculio glandium* Marsham	(コウチュウ目、ゾウムシ科) 旧北区
acorn weevil (1)		*Curculio nucum* Linnaeus	(コウチュウ目、ゾウムシ科) 全北区。ハシバミの実の害虫
acorn weevils (1)	シギゾウムシ亜科	Curculioninae	(コウチュウ目、ゾウムシ科) の昆虫の総称
acorn weevils (2)		*Curculio*	(コウチュウ目、ゾウムシ科) の昆虫の総称
Acraea mimic (1)		*Castilia perilla* (Hewitson)	(チョウ目、タテハチョウ科) 新熱帯区
Acraea mimic	ホソチョウシジミ属	*Mimacraea*	(チョウ目、シジミチョウ科) の昆虫の総称
Acraea moth	キシタゴマダラヒトリ	*Estigmene acraea* (Drury)	(チョウ目、ヒトリガ科) 新北区、新熱帯区。Acrea と記されること多し
Acraea swordtail	ベニモンタイマイ(ホソバタイマイ)	*Graphium ridleyanus* (White)	(チョウ目、アゲハチョウ科) エチオピア区
Acraeas	ホソチョウ属	*Acraea*	(チョウ目、タテハチョウ科) の昆虫の総称　学名を使用した英名
acraeid mimic satyr		*Lymanopoda acraeida* Butler	(チョウ目、タテハチョウ科) 新熱帯区
Acrea moth			Acraea moth を見よ
acridoid grasshoppers			acridoids を見よ
acridoids			grass hoppers (2) を見よ
acrobat ant	アクロバットアリ	*Crematogaster lineolata* (Say)	(ハチ目、アリ科) 新北区。腹部を持ち上げる習性から

英名	和名	学名	所属、分布、ほか
acrolepiid moths			false diamondback moths を見よ
acrolepiids			false diamondback moths を見よ
Actinodaphne aphid	カゴノキハスジアブラモドキ	*Aiceona actinodaphnis* Takahashi	(カメムシ目、アブラムシ科) 日本、東洋区
Actorion owlet			confused owlet を見よ
Actoris metalmark		*Cremna actoris* (Cramer)	(チョウ目、シジミタテハ科) 新熱帯区
aculeate Hymenoptera		Aculeata	(ハチ目) の昆虫の総称
aculeates			aculeate Hymenoptera を見よ
acuminate scale	ハラビロカタカイガラムシ	*Kilifia acuminata* (Signoret)	(カメムシ目、カタカイガラムシ科) 日本、新北区、汎熱帯
acuminate snaketail		*Ophiogomphus acuminatus* Carle	(トンボ目、サナエトンボ科) 新北区
acute-angled fungus beetle	トゲムネキスイ	*Cryptophagus acutangulus* Gyllenhal	(コウチュウ目、キスイムシ科) 日本、旧北区、新熱帯区
acute sunbeam	ウラギンシジミ	*Curetis acuta paracuta* de Nicéville	(チョウ目、ウラギンシジミチョウ科) 日本
Adachi weevil	アダチアナアキゾウ	*Hylobius adachii* Kono	(コウチュウ目、ゾウムシ科) 日本
Adam's gem		*Libellago adami* Fraser	(トンボ目、Chlorocyphidae) 東洋区
Adams leaf beetle	カバノキハムシ	*Syneta adamsi* Baly	(コウチュウ目、ハムシ科) 日本、旧北区
Adam's shadowdamsel		*Drepanosticta adami* (Fraser)	(トンボ目、Platystictidae) 東洋区
Adamson's zebra blue		*Leptotes adamsoni* Collins et Larsen	(チョウ目、シジミチョウ科) エチオピア区
Adana pine tip moth			pine tip moth (2) を見よ　Adana tip moth ともいう
adderbolt			dragonflies (1) を見よ
adderflies			dragonflies (1) を見よ
adelgids			bark aphids を見よ
adelgids			conifer aphids を見よ
adeloid plantlice			conifer aphids を見よ
Aden skipper		*Spialia doris* (Walker)	(チョウ目、セセリチョウ科) 旧北区、エチオピア区
adephagans			carnivorous beetles を見よ
Adiaste fritillary	ムモンギンボシヒョウモン	*Speyeria adiaste* (Edwards)	(チョウ目、タテハチョウ科) 新北区
Adina glasswing		*Dircenna adina xannthophane* Hopffer	(チョウ目、タテハチョウ科) 新熱帯区
Adirondack black fly		*Prosimulium hirtipes* (Fries)	(ハエ目、ブユ科) 新北区。アメリカ東部の Adirondack 山地に由来
adjutants		Aethriamanta	(トンボ目、トンボ科) の昆虫の総称
Adkin's apple ermel			apple ermine moth を見よ
admiral			チョウ類の俗称
admirals	イチモンジチョウ属	Limenitis	(チョウ目、タテハチョウ科) の昆虫の総称
Adonia white-spots		*Osmodes adonia* Evans	(チョウ目、セセリチョウ科) エチオピア区

英名	和名	学名	所属、分布、ほか
Adonis blue	アドニスヒメシジミ	*Polyommatus bellargus* (Rottemburg)	(チョウ目、シジミチョウ科) 旧北区。Adonis blue butterfly ともいう
Adonis ladybird		*Hippodamia variegata* (Goeze)	(コウチュウ目、テントウムシ科) 旧北区
Adonis Morpho		*Morpho marcus* Schaller	(チョウ目、タテハチョウ科) 新熱帯区
Adonis opal		*Chrysoritis adonis* (Pennington)	(チョウ目、シジミチョウ科) エチオピア区
adorable brocade moth		*Platypolia mactata* (Guenée)	(チョウ目、ヤガ科) 新北区
Adytum swamp damselfly		*Megalagrion adytum* (Perkins)	(トンボ目、イトトンボ科) 大洋区
adzuki bean borer	アズキノメイガ	*Ostrinia scapulalis* (Walker)	(チョウ目、メイガ科) 日本。アズキ、ダイズなどの害虫
adzuki bean bug			adzuki squash bug を見よ
adzuki bean podworm	アズキサヤムシガ	*Matsumuraeses azukivora* (Matsumura)	(チョウ目、ハマキガ科) 日本。マメ科牧草の害虫
adzuki bean weevil			adzuki weevil を見よ
adzuki squash bug	アズキヘリカメムシ	*Homoeocerus marginiventris* Dohrn	(カメムシ目、ヘリカメムシ科) 日本、東洋区
adzuki weevil	アズキマメゾウムシ	*Callosobruchus chinensis* (Linnaeus)	(コウチュウ目、ハムシ科) 日本、全北区、大洋区。アズキなどの害虫
Aecas ruby-eye		*Flaccilla aecas* (Stoll)	(チョウ目、セセリチョウ科) 新熱帯区
Aega Morpho			Brazilian Morpho を見よ
aegialitid beetles	チビキカワムシ科	Aegialitidae	(コウチュウ目) の昆虫の総称
aegialitid beetles	イワハムシ亜科	Aegialitinae	(コウチュウ目、チビキカワムシ科) の昆虫の総称
Aegina numberwing		*Callicore lyca* (Doubleday)	(チョウ目、タテハチョウ科) 新熱帯区。*C. l. aegina* (C. et R. Felder) も同英名
Aeluropis skipper		*Mesodina aeluropis* Meyrick	(チョウ目、セセリチョウ科) 豪州区
aeneous small leaf beetle	アオガネヒメサルハムシ	*Nodina chalcosoma* Baly	(コウチュウ目、ハムシ科) 日本、東洋区
aerial brown moth		*Ozarba aeria* (Grote)	(チョウ目、ヤガ科) 新北区
aerial yellowjacket		*Dolichovespula arenaria* (Fabricius)	(ハチ目、スズメバチ科) 新北区
aeroplane			(チョウ目、タテハチョウ科) コミスジ属、キンミスジ属、トガリミスジ属の種
aeroplanes	キンミスジ属	*Pantoporia*	(チョウ目、タテハチョウ科) の昆虫の総称
aeroplanes	トガリミスジ属	*Phaedyma*	(チョウ目、タテハチョウ科) の昆虫の総称
aeroplanes			sailers を見よ 豪州での英名
aeschnoid dragonflies	ヤンマ上科	Aeschnoidea	(トンボ目) の昆虫の総称
aeschnoids			aeschnoid dragonflies を見よ
aetalionid treehoppers		Aetalionidae	(カメムシ目) の昆虫の総称 豪州、アフリカに50種
Aetherie fritillary		*Melitaea aetherie* Hübner	(チョウ目、タテハチョウ科) 旧北区
aethiops groundling		*Xenolechia aethiops* (Hamphreys et Westwood)	(チョウ目、キバガ科) 旧北区

英名	和名	学名	所属、分布、ほか
Affica sister		*Adelpha affica* (C. et R. Felder)	（チョウ目、タテハチョウ科）新熱帯区
affinis skipper		*Arrhenes marmas* (Felder)	（チョウ目、セセリチョウ科）豪州区
afflicted dagger moth		*Acronicta afflicta* Grote	（チョウ目、ヤガ科）新北区
afranius duskywing	クロミヤマセセリ	*Erynnis afranius* (Lintner)	（チョウ目、セセリチョウ科）新北区、新熱帯区
African		*Perigea capensis* (Guenée)	（チョウ目、ヤガ科）エチオピア区、旧北区
African albatross			diverse white を見よ
African apefly			scribbled harvester を見よ
African army ant		*Dorylus gribodi* Emery	（ハチ目、アリ科）エチオピア区
African armyworm	アフリカシロナヨトウ	*Spodoptera exempta* (Walker)	（チョウ目、ヤガ科）日本、東洋区、豪州区、新北区、エチオピア区。イネ害虫
African Babul blue			topaz blue を見よ
African ball-rolling dung beetle		*Kheper nigroaeneus* (Boheman)	（コウチュウ目、コガネムシ科）エチオピア区
African bark mantis		*Tarachodes* sp.	（カマキリ目、カマキリ科）エチオピア区
African beak		*Libythea labdaca* Westwood	（チョウ目、タテハチョウ科）エチオピア区
African black beetle		*Heteronychus arator* (Fabricius)	（コウチュウ目、コガネムシ科）エチオピア区、豪州区
African blue louse	アフリカヤギジラミ	*Linognathus africanus* Kellog et Paine	（シラミ目、ケモノホソジラミ科）日本、汎世界
African blue tiger	アフリカウスコモンマダラ	*Tirumala petiverana* (Doubleday)	（チョウ目、タテハチョウ科）エチオピア区
African bluetail			common bluetail (1) を見よ
African bollworm			corn earworm (2) を見よ
African bush hopper		*Ampittia capenas* (Hewitson)	（チョウ目、セセリチョウ科）エチオピア区
African cabbage white			meadow white を見よ
African caper			African common white を見よ
African carnation tortrix		*Epichoristodes acerbella* (Walker)	（チョウ目、ハマキガ科）旧北区、エチオピア区。African carnation tortrix moth ともいう
African castor	アオカバタテハ	*Ariadne enotrea* (Cramer)	（チョウ目、タテハチョウ科）エチオピア区
African citrus psyllid			citrus psylla (2) を見よ
African clouded yellow	エレクトモンキチョウ	*Colias electo electo* (Linnaeus)	（チョウ目、シロチョウ科）エチオピア区
African coffee thrips		*Diarthrothrips coffeae* Williams	（アザミウマ目、アザミウマ科）エチオピア区
African common white	アフリカシロチョウ	*Belenois creona* (Cramer)	（チョウ目、シロチョウ科）エチオピア区
African copper		*Aloeides talkosama* (Wallengren)	（チョウ目、シジミチョウ科）エチオピア区
African cotton bollworm			corn earworm (2) を見よ
African cotton leafworm			Egyptian cotton leafworm を見よ
African cupid		*Euchrysops osiris* (Hopffer)	（チョウ目、シジミチョウ科）エチオピア区

英名	和名	学名	所属、分布、ほか
African dung beetle			brown dung beetle を見よ
African emerald		*Hemicordulia africana* Dijkstra	(トンボ目、エゾトンボ科) エチオピア区
African emigrant			African migrant を見よ
African giant swallowtail	ドルーリーオオアゲハ	*Papilio antimachus* Drury	(チョウ目、アゲハチョウ科) エチオピア区
African golden Arab	スジグロツマアカシロチョウ	*Colotis aurigineus* (Butler)	(チョウ目、シロチョウ科) エチオピア区
African grass blue		*Zizeeria knysna* (Trimen)	(チョウ目、シジミチョウ科) 旧北区、エチオピア区
African grass mantis		*Oxyothespis dumonti* Chopard	(カマキリ目、カマキリ科) エチオピア区
African humming bird		*Macroglossum trochilus* (Hübner)	(チョウ目、スズメガ科) エチオピア区
African hummingbird hawk-moth			African humming bird を見よ
African leaf	メスジロコノハチョウ	*Kallima rumia* Doubleday	(チョウ目、タテハチョウ科) エチオピア区。African leaf buterfly ともいう
African leafwing		*Kallimoides rumia jadyae* (Fox)	(チョウ目、タテハチョウ科) エチオピア区
African leopard			common leopard を見よ
African line blue		*Pseudonacaduba sichela* (Wallengren)	(チョウ目、シジミチョウ科) エチオピア区
African line blues		*Pseudonacaduba*	(チョウ目、シジミチョウ科) の昆虫の総称
African lined mantis			African mantis (2) を見よ
African lunar moth			African moon moth を見よ
African maiden moths			maiden moths を見よ
African mallow		*Gomalia elma* (Trimen)	(チョウ目、セセリチョウ科) 東洋区、エチオピア区
African mallow skipper			African mallow を見よ
African mantis		*Sphodromantis balachowski* La Greca	(カマキリ目、カマキリ科) エチオピア区
African mantis (1)		*Sphodromantis gastrica* Stål	(カマキリ目、カマキリ科) エチオピア区
African mantis (2)		*Sphodromantis lineola* (Burmeister)	(カマキリ目、カマキリ科) エチオピア区。African praying mantis ともいう
African mantis (3)		*Sphodromantis centralis* Rehm	(カマキリ目、カマキリ科) エチオピア区
African mantis			giant African mantis を見よ
African map butterfly	アフリカイシガケチョウ (ヘリモンイシガケチョウ)	*Cyresties camillus* (Fabricius)	(チョウ目、タテハチョウ科) エチオピア区
African migrant	アフリカウラナミシロチョウ	*Catopsilia florella* (Fabricius)	(チョウ目、シロチョウ科) エチオピア区、東洋区
African migrant white			African migrant を見よ

英名	和名	学名	所属、分布、ほか
African migratory locust	アフリカトビバッタ	*Locusta migratoria migratorioides* Reiche et Fairmaire	(バッタ目、バッタ科) エチオピア区。大発生して移動群飛することで著名
African mole cricket		*Gryllotalpa africana* Palisot et Beauvois	(バッタ目、ケラ科) 全北区。東洋区
African mole cricket			mole cricket (1) を見よ
African monarch	カバマダラ	*Danaus chrysippus* (Linnaeus)	(チョウ目、タテハチョウ科) 日本、東洋区、豪州区、エチオピア区
African monarch			southern milkweed butterfly を見よ
African moon moth		*Argema mimosae* (Boisduval)	(チョウ目、ヤママユガ科) エチオピア区
African moth			African を見よ
African moth butterfly		*Euliphyra mirifica* Holland	(チョウ目、シジミチョウ科) エチオピア区
African mound-building termite		*Macrotermes michaelseni* (Sjöstedt)	(シロアリ目、シロアリ科) エチオピア区。mound-building termite ともいう
African palm weevil		*Rhynchophorus phoenicis* (Fabricius)	(コウチュウ目、オサゾウムシ科) エチオピア区
African palmfly		*Elymniopsis bammakoo* (Westwood)	(チョウ目、タテハチョウ科) エチオピア区
African pansy			commodore (1) を見よ
African pierrot			blue tiger (1) を見よ
African pierrot			common tiger blue を見よ
African pink borer		*Sesamia calamistis* Hampson	(チョウ目、ヤガ科) エチオピア区
African pinstripe mantis		*Miomantis binotata* (Giglio-Tos)	(カマキリ目、カマキリ科) エチオピア区
African praying mantis			African mantis (2) を見よ
African primitive ghost moths		Prototheoridae	(チョウ目) の昆虫の総称
African purple spot		*Precis clelia* Cramer	(チョウ目、タテハチョウ科) エチオピア区
African queen			African monarch を見よ
African red admirals	オナガアカタテハ属	*Antanartia*	(チョウ目、タテハチョウ科) の昆虫の総称
African rhinoceros beetle		*Oryctes boas* (Fabricius)	(コウチュウ目、コガネムシ科) エチオピア区
African rhinoceros beetle		*Oryctes monoceros* (Oliver)	(コウチュウ目、コガネムシ科) エチオピア区
African rice borer		*Chilo zacconius* Bleszynski	(チョウ目、メイガ科) エチオピア区。イネ害虫
African rice bug		*Stenocoris southwoodi* Ahmad	(カメムシ目、ホソヘリカメムシ科) エチオピア区。イネ害虫
African rice gall midge		*Orseolia oryzivora* Harris et Gagné	(ハエ目、タマバエ科) エチオピア区。イネ害虫
African rice hispa		*Trichispa sericea* (Guérin)	(コウチュウ目、ハムシ科) エチオピア区
African ringlet		*Ypthima asterope* (Klug)	(チョウ目、タテハチョウ科) エチオピア区
African scavenger moth		*Praeacedes atomosella* (Walker)	(チョウ目、ヒロズコガ科) 全北区、豪州区、大洋区

英名	和名	学名	所属、分布、ほか
African skipjack		*Tetralobus flabellicornis* (Linnaeus)	（コウチュウ目、コメツキムシ科）エチオピア区。成虫はアリの巣に住み、幼虫は作物を加害
African skipper moths		Apoprogonidae	（チョウ目）の昆虫の総称
African slug caterpillar moths		Chrysopolomidae	（チョウ目）の昆虫の総称
African small white		*Dixeia charina* (Boisduval)	（チョウ目、シロチョウ科）エチオピア区
African small white (1)		*Dixeia doxo* (Godart)	（チョウ目、シロチョウ科）エチオピア区
African straight			straight swift (1) を見よ
African striped borer			African rice borer を見よ
African swallowtail			mocker swallowtail を見よ
African sweet potato weevil		*Cylas puncticollis* Boheman	（コウチュウ目、ミツギリゾウムシ科）エチオピア区
African termite		*Coptotermes sjostedti* Holmgren	（シロアリ目、ミゾガシラシロアリ科）エチオピア区
African twig mantis		*Popa spurca* Stål	（カマキリ目、カマキリ科）エチオピア区
African vagrant			African migrant を見よ
African veined white		*Belenois gidica* (Godart)	（チョウ目、シロチョウ科）エチオピア区
African vine borer		*Melittia oedipus* Oberthür	（チョウ目、スカシバガ科）エチオピア区、大洋区
African violet mealybug			Pritchard ground mealybug を見よ
African wart hog louse		*Haematopinus phacochaeri* Enderlein	（シラミ目、ケモノジラミ科）エチオピア区
African white borer			white rice borer を見よ
African white Judy		*Abisara delicata* Lathy	（チョウ目、シジミタテハ科）エチオピア区
African white rice borer			white rice borer を見よ
African white stemborer			white rice borer を見よ
African white-striped borer		*Parerupa africana* (Aurivillius)	（チョウ目、メイガ科）エチオピア区。イネ害虫
African wood white		*Leptosia alcesta* (Stoll)	（チョウ目、シロチョウ科）エチオピア区
Africanized honey bee		*Apis mellifera scutella* Lepeletier	（ハチ目、ミツバチ科）エチオピア区
Afrotropical filth fly		*Chrysomya chloropyga* (Wiedemann)	（ハエ目、クロバエ科）エチオピア区
agaontid wasps			fig wasps (1) を見よ
agaric clothes		*Morophaga boleti* (Denis et Schiffermüller)	（チョウ目、ヒロズコガ科）旧北区
agaric clothes-moth		*Morophaga choragella* (Denis et Schiffermuller)	（チョウ目、ヒロズコガ科）旧北区
Agaricon skipper		*Pyrrhopygopsis agaricon* Druce	（チョウ目、セセリチョウ科）新熱帯区
agaristids			forester moths を見よ
agate knot-horn		*Nyctegretis lineana* (Scopoli)	（チョウ目、メイガ科）旧北区

英名	和名	学名	所属、分布、ほか
Agatha blue ringlet		*Chloreuptychia agatha* Butler	(チョウ目、タテハチョウ科) 新熱帯区
Agathina emperor	ツマジロアメリカコムラサキ	*Doxocopa agathina* (Cramer)	(チョウ目、タテハチョウ科) 新熱帯区
agave caterpillar			Tequila giant-skipper を見よ
agave metalmark		*Pseudonymphidia agave agave* (Godman et Salvin)	(チョウ目、シジミタテハ科) 新熱帯区
agave weevil		*Scyphophorus acuipunctatus* Gyllenhal	(コウチュウ目、ゾウムシ科) 新北区
agave yellow		*Eurema agave millerorum* Llorente et Luis	(チョウ目、シロチョウ科) 新熱帯区
Agenjo's anomalous blue		*Polyommatus agenjoi* (Petersen)	(チョウ目、シジミチョウ科) 旧北区
aggravating grasshopper		*Euconocephalus nasutus* (Thunberg)	(バッタ目、キリギリス科) 大洋区
aggressive riverhawk		*Onychothemis tonkinensis ceylanica* Ris	(トンボ目、トンボ科) 東洋区
agile ground mantid			minor ground mantid を見よ
agile meadow katydid		*Orchelimum agile* (De Geer)	(バッタ目、キリギリス科) 新北区
aglaonema scale	クサギウスマルカイガラムシ	*Aspidiotus excisus* Green	(カメムシ目、マルカイガラムシ科) 日本、東洋区、大洋区。バナナ害虫
Agnosia glasswing		*Ithomia agnosia* Hewitson	(チョウ目、タテハチョウ科) 新熱帯区
agonoxenid moths			palm moths を見よ
agreeable tiger moth		*Spilosoma congrua* Walker	(チョウ目、ヒトリガ科) 新北区
agricultural ants	フタフシアリ亜科	Myrmicinae	(ハチ目、アリ科) の昆虫の総称
agricultural ants			harvested ants を見よ
agrilus			twig girdlers を見よ
Agrippina underwing moth		*Catocala agrippina* Strecker	(チョウ目、ヤガ科) 新北区。Agrippina underwing ともいう
Agrippina's fantasy		*Pseudaletis agrippina* Druce	(チョウ目、シジミチョウ科) エチオピア区
Ahira sabre-wing		*Parelbella ahira* (Hewitson)	(チョウ目、セセリチョウ科) 新熱帯区
ahogapollo			rose chafer (3) を見よ
Ahola hairstreak			half-blue hairstreak を見よ
aholiba underwing moth		*Catocala aholibah* Strecker	(チョウ目、ヤガ科) 新北区
ailanthus silkmoth			cynthia moth を見よ
ailanthus silkworm			cynthia moth を見よ　ailanthus ニワウルシに由来
ailanthus webworm			ailanthus webworm moth (1) (2) を見よ
ailanthus webworm moth (1)		*Atteva punctella* (Cramer)	(チョウ目、スガ科) 新北区
ailanthus webworm moth (2)		*Atteva aurea* (Fitch)	(チョウ目、スガ科) 新北区。ailanthus webworm ともいう
airplane moths			plume moths を見よ

英名	和名	学名	所属、分布、ほか
airy Apamea moth		*Apamea vultuosa* Grote	（チョウ目、ヤガ科）新北区。airy Apamea ともいう
akebia leaf-like moth			fruit-piercing moth (5) を見よ
akebia psylla	ベニキジラミ	*Cacopsylla coccinea* (Kuwayama)	（カメムシ目、キジラミ科）日本、東洋区、旧北区。アケビにつくことに由来
akebia whitefly	アケビコナジラミ	*Odontaleyrodes akebiae* (Kuwana)	（カメムシ目、コナジラミ科）日本
Alabama skipper		*Euphyes dion alabamae* (Lindsey)	（チョウ目、セセリチョウ科）新北区。sedge skipper を参照
Alabama underwing moth		*Catocala alabamae* Grote	（チョウ目、ヤガ科）新北区。Alabama underwing ともいう
Alakai swamp damselfly			Adytum swamp damselfly を見よ
Alala sister	ツマキアオオビイチモンジ	*Adelpha alala* (Hewitson)	（チョウ目、タテハチョウ科）新熱帯区
Alamo moth		*Condylorrhiza vestigialis* (Guenée)	（チョウ目、メイガ科）新北区、新熱帯区
Alana white skipper		*Heliopetes alana* (Reakirt)	（チョウ目、セセリチョウ科）新熱帯区
alang-alang gall midge		*Orseolia javanica* Kieffer et van Leeuwen-Reijinvaan	（ハエ目、タマバエ科）東洋区
Alaskan swallowtail		*Papilio machaon aliaska* Scudder	（チョウ目、アゲハチョウ科）新北区。old world swallowtail (1) を参照
Albany pitcher plant fly		*Badisis ambulans* McAlpine	（ハエ目、マルズヤセバエ科）豪州区。豪州西部の地名 Albany に由来
albatross			woodland white を見よ
albatross white			sabine albatross を見よ
Alberta alpine fritillary			Alberta fritillary を見よ
Alberta arctic		*Oeneis alberta* Elwes	（チョウ目、タテハチョウ科）新北区
Alberta fritillary		*Clossiana alberta* (Edwards)	（チョウ目、タテハチョウ科）全北区
Alberta grayring			Alberta arctic を見よ　米国での英名
Alberta lutestring moth		*Ceranemota albertae* Clarke	（チョウ目、カギバガ科）新北区。Alberta lutestring ともいう
Albertus metalmark		*Monethe albertus* C. et R. Felder	（チョウ目、シジミタテハ科）新熱帯区
Albin's dwarf		*Elachista unifasciella* (Haworth)	（チョウ目、クサモグリガ科）旧北区
Albinus metalmark		*Ariconias albinus* (C. et R. Felder)	（チョウ目、シジミタテハ科）新熱帯区
Albizia long-horned beetle			Coptops long-horned beetle を見よ
Albizia sailer			common sailer を見よ
Albocaerulean	サツマシジミ	*Udara albocaerulea* (Moore)	（チョウ目、シジミチョウ科）日本、東洋区。*U. a. scharffi* Corbet も同英名
albopicta scale		*Acutaspis albopicta* (Cockerell)	（カメムシ目、マルカイガラムシ科）新北区、新熱帯区
alcalypha scale		*Pseudoparlatoria ostreata* (Cockerell)	（カメムシ目、マルカイガラムシ科）新熱帯区

英名	和名	学名	所属、分布、ほか
alchymist		*Catephria alchymista* (Denis et Schiffermüller)	（チョウ目、ヤガ科）旧北区
alchymist moth			alchymist を見よ
Alcione Bematistes		*Acraea alcione* C. et R. Felder	（チョウ目、タテハチョウ科）エチオピア区。Bematistes は旧属名
Alciope Acraea		*Acraea alciope* Hewitson	（チョウ目、タテハチョウ科）エチオピア区
Alcmena clearwing		*Pteronymia alcmena* (Godman et Salvin)	（チョウ目、タテハチョウ科）新熱帯区
Alcon blue	アルコンゴマシジミ（キタゴマシジミ）	*Maculinea alcon* Denis et Schiffermüller	（チョウ目、シジミチョウ科）旧北区
Alda hairstreak		*Aubergina alda* (Hewitson)	（チョウ目、シジミチョウ科）新熱帯区
Aldania mimic	アサギマダラミスジチョウ	*Aldania imitans* (Oberthür)	（チョウ目、タテハチョウ科）旧北区
alder			alder moth を見よ
alder bark beetle		*Alniphagus aspericollis* (LeConte)	（コウチュウ目、キクイムシ科）新北区
alder bark beetle	ハンノキカバイロキクイムシ	*Alniphagus costatus* (Blandford)	（コウチュウ目、キクイムシ科）日本、旧北区。ハンノキ類の害虫
alder bud moth			raven-feather case を見よ
alder dagger			alder moth を見よ
alder dagger			fingered dagger moth を見よ
alder flea beetle		*Macrohaltica ambiens* (LeConte)	（コウチュウ目、ハムシ科）新北区
alder flea weevil		*Rhynchaenus alni scutellaris* Redtenbacher	（コウチュウ目、ゾウムシ科）旧北区
alderflies (1)	センブリ科	Sialidae	（アミメカゲロウ目）の昆虫の総称
alderflies (2)	ヘビトンボ亜目	Megaloptera	の昆虫の総称　広翅目としてアミメカゲロウ目から分けることもある
alderflies			lacewings (1) を見よ
alder fly		*Sialis lutaria* (Linnaeus)	（アミメカゲロウ目、センブリ科）旧北区
alder froghopper		*Aphrophora alni* (Fallén)	（カメムシ目、アワフキムシ科）旧北区
alder kitten		*Furcula bicuspis* (Borkhausen)	（チョウ目、シャチホコガ科）旧北区
alder lace-bug		*Corythucha pergandei* Heidemann	（カメムシ目、グンバイムシ科）新北区
alder leaf beetle	ハンノキハムシ	*Agelastica coerulea* Baly	（コウチュウ目、ハムシ科）日本、旧北区。ハンノキ類の害虫
alder leaf beetle (1)		*Agelastica alni* (Linnaeus)	（コウチュウ目、ハムシ科）旧北区
alder leafhopper	ハンノヒロズヨコバイ	*Oncopsis alni* (Schrank)	（カメムシ目、オオヨコバイ科）日本、旧北区。ハンノキ類の害虫
alder leafhopper (1)	ハンノナガヨコバイ	*Speudotettix subfusculus* (Fallén)	（カメムシ目、オオヨコバイ科）日本、旧北区。ハンノキ類の害虫
alder leafminer	ハンノハムグリバチ	*Fenusa dohrni* (Tischbein)	（ハチ目、ハバチ科）日本、全北区、エチオピア区。幼虫がハンノキ類の害虫
alder leafminer moth		*Caloptilia alnivorella* (Chambers)	（チョウ目、ホソガ科）新北区。alder leafminer ともいう
alder longicorn			alder longicorn beetle を見よ

英名	和名	学名	所属、分布、ほか
alder longicorn beetle	ハンノカミキリ	*Cagosima sanguinolenta* Thomson	(コウチュウ目、カミキリムシ科) 日本、旧北区。ハンノキ類の害虫
alder moth	ハンノケンモン	*Jocheaera alni* (Linnaeus)	(チョウ目、ヤガ科) 日本、旧北区
alder patch		*Bucculatrix cidarella* (Zeller)	(チョウ目、チビガ科) 旧北区
alder pistol case		*Coleophora binderella* (Kollar)	(チョウ目、ツツミノガ科) 旧北区
alder psylla	ハンノキキジラミ	*Psylla alni* (Linnaeus)	(カメムシ目、キジラミ科) の本、全北区。ハンノキ類の害虫
alder quaker moth		*Homorthodes communis* (Dyar)	(チョウ目、ヤガ科) 新北区
alder red midget		*Phyllonorycter rajella* (Linnaeus)	(チョウ目、ホソガ科) 旧北区
alder sawfly (1)	ミツクリハバチ	*Eriocampa mitsukurii* Rohwer	(ハチ目、ハバチ科) 日本、旧北区、東洋区。ハンノキ類の害虫
alder sawfly (2)	ヒラアシハバチ	*Craesus japonicus* Takeuchi	(ハチ目、ハバチ科) 日本、旧北区。幼虫がハンノキ類の害虫
alder small shoot		*Heliozela resplendella* (Stainton)	(チョウ目、ツヤコガ科) 旧北区
alder spittle bug		*Clastoptera obtusa* Say	(カメムシ目、コガシラアワフキ科) 新北区
alder sucker			alder psylla を見よ
alder tree beetle			alder leaf beetle (1) を見よ
alder tubemaker moth		*Acrobasis rubrifasciella* Packard	(チョウ目、メイガ科) 新北区
alder wood wasp	クビナガキバチ	*Xiphydria camelus* (Linnaeus)	(ハチ目、クビナガキバチ科) 日本、旧北区。ハンノキ、カシ類の害虫
alderman			red admiiral (1) を見よ
Alea hairstreak		*Strymon alea* (Godman et Salvin)	(チョウ目、シジミチョウ科) 新北区
aleppo gall			*Cynips tinctoria* Hartig (ハチ目、タマバチ科) がカシにつくる虫えい。旧北区
Ales' flat-headed citrus borer	アレスミカンナガタマムシ	*Agrilus alesi* Obenberger	(コウチュウ目、タマムシ科) 日本。カンキツ害虫
Alexander's tree cricket		*Oecanthus alexanderi* Walker	(バッタ目、カンタン科) 新北区
Alexandra sulphur			Queen Alexandra's sulphur を見よ
Alexicles moth		*Alexicles aspersa* Grote	(チョウ目、ヒトリガ科) 新北区
Alexina glasswing		*Oleria alexina* (Hewitson)	(チョウ目、タテハチョウ科) 新熱帯区
Alexon crescent		*Anthanassa nebulosa alexon* (Godman et Salvin)	(チョウ目、タテハチョウ科) 新熱帯区
aleyrodids			white flies を見よ
alfalfa aphis			cowpea aphid を見よ
alfalfa blotch leafminer		*Agromyza frontella* (Rondani)	(ハエ目、ハモグリバエ科) 新北区
alfalfa butterfly	オオアメリカモンキチョウ	*Colias eurytheme* Boisduval	(チョウ目、シロチョウ科) 新北区、新熱帯区。幼虫は alfalfa caterpillar
alfalfa chalcid			clover seed chalcid (1) を見よ
alfalfa flower midge			alfalfa leaf midge を見よ 米国での英名

英名	和名	学名	所属、分布、ほか
alfalfa gall midge		*Asphondylia websteri* Felt	（ハエ目、タマバエ科）新北区
alfalfa gall midge			alfalfa seed midge を見よ
alfalfa hopper			three-cornered alfalfa hopper を見よ
alfalfa leaf midge		*Dasineura medicaginis* Rubsaamen	（ハエ目、タマバエ科）全北区
alfalfa leaf weevil		*Phytonomus murinus* Fabricius	（コウチュウ目、ゾウムシ科）旧北区
alfalfa leafcutter bee			alfalfa leafcutting bee (1) を見よ
alfalfa leafcutting bee (1)		*Megachile rotundata* (Fabricius)	（ハチ目、ハキリバチ科）新北区
alfalfa leafcutting bee		*Megachile pacifica* (Panzer)	（ハチ目、ハキリバチ科）新北区
alfalfa leaf-cutting bee			common leaf-cutter bee を見よ
alfalfa leaftier		*Dichomeris ianthes* (Meyrick)	（チョウ目、キバガ科）新北区、大洋区
alfalfa leaftier			clover gelechid を見よ
alfalfa looper moth		*Autographa californica* (Speyer)	（チョウ目、ヤガ科）新北区。alfalfa looper ともいう
alfalfa membracid		*Strictocephala festina* (Say)	（カメムシ目、ツノゼミ科）大洋区
alfalfa plant bug	ヒゲナガカスミカメムシ	*Adelphocoris lineolatus* (Goeze)	（カメムシ目、カスミカメムシ科）日本、全北区
alfalfa seed chalcid			lucerne seed wasp を見よ
alfalfa seed midge		*Asphondylia miki* Wachtl	（ハエ目、タマバエ科）全北区。欧州での英名
alfalfa semilooper			alfalfa looper moth を見よ
alfalfa snout beetle		*Otiorhynchus ligustici* (Linnaeus)	（コウチュウ目、ゾウムシ科）全北区
alfalfa sprout midge		*Dasineura ignorata* Wachtl	（ハエ目、タマバエ科）全北区
alfalfa thrips			western flower thrips (1) を見よ
alfalfa webworm		*Loxostege cereralis* (Zeller)	（チョウ目、メイガ科）新北区。幼虫はアルファルファを食う。alfalfa webworm moth ともいう
alfalfa webworm		*Loxostege commixtalis* (Walker)	（チョウ目、メイガ科）新北区
alfalfa weevil	アルファルファタコゾウムシ	*Hypera postica* (Gyllenhal)	（コウチュウ目、ゾウムシ科）日本、全北区。アルファルファなどの害虫
alfalfa weevil			clover leaf weevil (1)(3) を見よ　米国での英名
alfalfa weevil			lucerne weevil を見よ　北米での英名
alfalfa weevils	タコゾウムシ亜科	Hyperinae	（コウチュウ目、ゾウムシ科）の昆虫の総称
alfalfa worm			black armyworm を見よ
alga beetles			crawling water beetles を見よ
Algea crow			long-branded crow を見よ
Algerian grayling		*Hipparchia ellena* (Oberthür)	（チョウ目、タテハチョウ科）旧北区
Algerian grizzled skipper		*Muschampia leuzeae* (Oberthür)	（チョウ目、セセリチョウ科）旧北区

英名	和名	学名	所属、分布、ほか
Alicia's metalmark		*Esthemopsis alicia alicia* (H. Bates)	(チョウ目、シジミタテハ科) 新熱帯区
alien probole moth		*Probole alienaria* Herrich-Schäffer	(チョウ目、シャクガ科) 新北区
alienicola			複数形は alienicolae。アブラムシ類の単性胎生雌被譲型で seconds ともいう。二次宿主上で発生し、幹母、移住型から異なった形態をもつ
Aliphera longwing		*Eueides aliphera gracilis* Stichel	(チョウ目、タテハチョウ科) 新熱帯区
alkali bee		*Nomia melanderi* Cockerell	(ハチ目、コハナバチ科) 新北区
alkali bite borer moth		*Comadia suaedivora* Brown et Allen	(チョウ目、ボクトウガ科) 新北区
alkali bluet		*Enallagma clausum* Morse	(トンボ目、イトトンボ科) 新北区
alkali fly		*Ephydra hians* Say	(ハエ目、ミギワバエ科) 新北区、新熱帯区。カリフォルニアの塩水のモノ湖に大発生し、成虫は水に潜る
alkali skipper			salt-grass skipper を見よ
all-black snout		*Nemotelus nigrinus* Fallén	(ハエ目、ミズアブ科) 旧北区
Allard's blue		*Plebejus allardi* (Oberthür)	(チョウ目、シジミチョウ科) 旧北区
Allard's ground cricket		*Allonemobius allardi* (Alexander et Thomas)	(バッタ目、コオロギ科) 新北区
Allard's meadow katydid		*Conocephalus allardi* (Caudell)	(バッタ目、キリギリス科) 新北区
Allard's silver-line		*Cigaritis allardi* Oberthür	(チョウ目、シジミチョウ科) 旧北区
alleculid beetles			comb-clawed beetles (1) を見よ
allegheny mound ant		*Formica exsectoides* Forel	(ハチ目、アリ科) 新北区
allegheny river cruiser		*Macromia alleghaniensis* Williamson	(トンボ目、ヤマトンボ科) 新北区
allegheny spruce beetle		*Dendroctonus punctatus* LeConte	(コウチュウ目、キクイムシ科) 新北区
Allen's Tortrix moth		*Aphelia alleniana* (Fernald)	(チョウ目、ハマキガ科) 新北区
allied crane fly			common crane fly (1) を見よ
allied ermel		*Hyponomeuta cognatellus* Hübner	(チョウ目、スガ科) 旧北区
allied fanner (1)		*Glyphipterix cramerella* (Fabricius)	(チョウ目、ホソハマキモドキガ科) 旧北区
allied fanner (2)		*Glyphipterix simpliciella* (Stephens)	(チョウ目、ホソハマキモドキガ科) 旧北区
allied marbled conch		*Phalonidia affinitana* (Douglas)	(チョウ目、ホソハマキガ科) 旧北区
allied shade moth		*Cnephasia incertana* (Treitschke)	(チョウ目、ハマキガ科) 旧北区
allied willow beetle			blue willow beetle を見よ
alligator ticks			giant water bugs (2) を見よ　alligator fleas ともいわれる
alligatorweed flea beetle		*Agasicles hygrophila* Selman et Vogt	(コウチュウ目、ハムシ科) 新北区、豪州区

英名	和名	学名	所属、分布、ほか
alligatorweed moth		*Vogtia malloi* Pastrana	（チョウ目、メイガ科）豪州区
alligatorweed stem borer moth		*Arcola malloi* (Pastrana)	（チョウ目、メイガ科）新北区、新熱帯区
allium leafminer	ネギコガ	*Acrolepiopsis sapporensis* (Matsumura)	（チョウ目、アトヒゲコガ科）日本、旧北区。ネギ害虫
Allyn's Ridens		*Ridens allyni* Freeman	（チョウ目、セセリチョウ科）新熱帯区
Almansor white-lady swordtail	アルマンソールタイマイ	*Graphium almansor* (Honrath)	（チョウ目、アゲハチョウ科）エチオピア区
Almeida copper		*Aloeides almeida* (C. Felder)	（チョウ目、シジミチョウ科）エチオピア区
Almella frass moth		*Ctenomeristis almella* (Meyrick)	（チョウ目、メイガ科）豪州区
Almoda skipper		*Phanes almoda* (Hewitson)	（チョウ目、セセリチョウ科）新熱帯区
almond aphid		*Brachycaudus amygdalinus* (Shouteden)	（カメムシ目、アブラムシ科）旧北区
almond aphid			peach aphid を見よ
almond bark beetle		*Scolytus amygdali* Guérin-Méneville	（コウチュウ目、キクイムシ科）旧北区
almond blossom weevil		*Anthonomus amygdali* Hustache	（コウチュウ芽、ゾウムシ科）旧北区
almond borer		*Capnodis carbonaria* (Klug)	（コウチュウ目、タマムシ科）旧北区
almond-eyed brown			almond-eyed ringlet を見よ
almond-eyed owl			Brazilian owl を見よ
almond-eyed ringlet		*Erebia alberganus* (de Prunner)	（チョウ目、タテハチョウ科）旧北区
almond moth	スジマダラメイガ	*Cadra cautella* (Walker)	（チョウ目、メイガ科）日本、汎世界。貯穀害虫
almond seed chalcid		*Eurytoma amygdali* Enderlein	（ハチ目、カタビロコバチ科）旧北区
almond tree leaf-skeletonizer moth		*Agalope infausta* (Linnaeus)	（チョウ目、マダラガ科）旧北区
alnus ambrosia beetle	ハンノキキクイムシ	*Xylosandrus germanus* (Blandford)	（コウチュウ目、キクイムシ科）日本、東洋区、全北区。果樹類の害虫
alnus bark beetle			Saxena ambrosia beetle を見よ
alnus dagger moth			alder moth を見よ
alnus myzocallis	ハンノヒゲナガアブラムシ	*Hannabura alnicola* Matsumura	（カメムシ目、アブラムシ科）日本、旧北区
alnus spined aphid	ハンノウスブチアブラムシ	*Recticallis alnijaponicae* Matsumura	（カメムシ目、アブラムシ科）日本、旧北区
aloe aphid		*Aloephagus myersi* Essig	（カメムシ目、アブラムシ科）豪州区
Alope sphinx moth		*Erinnyis alope* (Drury)	（チョウ目、スズメガ科）新北区、新熱帯区。Alope sphinx ともいう
alpine Argus		*Albulina orbitulus* (de Prunner)	（チョウ目、シジミチョウ科）旧北区
alpine blue			alpine Argus を見よ
alpine butterfly			Apollo を見よ
alpine case moth		*Lomera caespitosae* (Oke)	（チョウ目、ミノガ科）豪州区

英名	和名	学名	所属、分布、ほか
alpine checkered skipper			northern grizzled skipper を見よ
alpine dark bush cricket		*Pholidoptera aptera* (Fabricius)	(バッタ目、キリギリス科) 旧北区
alpine darner		*Austroaeschna flavomaculata* Tillyard	(トンボ目、ヤンマ科) 豪州区
alpine earwig		*Chelidura aptera* (Megerle)	(ハサミムシ目、クギヌキハサミムシ科) 旧北区
alpine emerald	クモマエゾトンボ	*Somatochlora alpestris* (Selys)	(トンボ目、エゾトンボ科) 日本、旧北区
alpine grassgrub		*Oncopera alpina* Tindale	(チョウ目、コウモリガ科) 豪州区
alpine grasshopper		*Melanoplus alpinus* Scudder	(バッタ目、バッタ科) 新北区。高地に生息
alpine grass-veneer		*Platytes alpinella* Hübner	(チョウ目、メイガ科) 旧北区
alpine grayling		*Oeneis glacialis* (Moll)	(チョウ目、タテハチョウ科) 旧北区
alpine green hairstreak		*Callophrys sheridanii lemberti* Tilden	(チョウ目、シジミチョウ科) 新北区
alpine grey		*Eudonia alpina* (Curtis)	(チョウ目、メイガ科) 旧北区
alpine grizzled skipper	タカネチャマダラセセリ	*Pyrgus andromedae* (Wallengren)	(チョウ目、セセリチョウ科) 旧北区
alpine hairy cicada			hairy cicada を見よ
alpine heath	コゲチャヒメヒカゲ	*Coenonympha gardetta* de Prunner	(チョウ目、タテハチョウ科) 旧北区
alpine redspot		*Austropetalia tonyana* Theischinger	(トンボ目、Austropetaliidae) 豪州区
alpine scree weta			scree weta を見よ
alpine shieldback		*Acrodectes philopagus* Rehn et Hebard	(バッタ目、キリギリス科) 新北区
alpine silver-Y moth		*Syngrapha ottolenguii* (Dyar)	(チョウ目、ヤガ科) 日本
alpine skipper		*Oreisplanus munionga* (Olliff)	(チョウ目、セセリチョウ科) 豪州区
alpine thermocolour grasshopper		*Kosciuscola tristis* Sjöstedt	(バッタ目、バッタ科) 豪州区。山地に生息
alpine zephyr blue		*Plebejus trappi* (Verity)	(チョウ目、シジミチョウ科) 旧北区
alpines			ringlets (1) を見よ
Alstroemer's flat-body		*Agonopterix alstroemeriana* (Clerck)	(チョウ目、マルハキバガ科) 全北区。ツガの害虫
alternate ruby-sye		*Talides alternata* Bell	(チョウ目、セセリチョウ科) 新熱帯区
alternate-barred conch			barred conch を見よ
Althoff's Acraea		*Acraea althoffi* Dewitz	(チョウ目、タテハチョウ科) エチオピア区
Altissima admiral			Andean painted lady を見よ
Alumna grass skipper		*Cymaenes alumna* (Butler)	(チョウ目、セセリチョウ科) 新熱帯区。Alumna skipper ともいう
Amakosa rocksitter		*Durbania amakosa* Trimen	(チョウ目、シジミチョウ科) エチオピア区
Amakusa bark beetle	アマクサコキクイムシ	*Hypothenemus amakusanus* (Murayama)	(コウチュウ目、キクイムシ科) 日本

英名	和名	学名	所属、分布、ほか
Amalda glasswing		*Oleria amalda* (Hewitson)	（チョウ目、タテハチョウ科）新熱帯区
Amanda's blue		*Polyommatus amandus* (Schneider)	（チョウ目、シジミチョウ科）旧北区
Amani buff		*Baliochila amanica* Stempffer et Bennet	（チョウ目、シジミチョウ科）エチオピア区
amaryllis azure	アマリリスヤドリギシジミ	*Ogyris amaryllis* Hewitson	（チョウ目、シジミチョウ科）豪州区
amaryllis mealybug		*Pseudococcus amaryllidis* Bouché	（カメムシ目、コナカイガラムシ科）旧北区
amaryllis weevils			bulb snout beetles を見よ
Amatola malachite		*Chlorolestes apricans* Wilmot	（トンボ目、Synlestidae）エチオピア区
Amazon angel		*Chorinea amazon* Saunders	（チョウ目、シジミタテハ科）新熱帯区
Amazon ant	アマゾンアリ	*Polyergus rufescens* Latreille	（ハチ目、アリ科）旧北区、新熱帯区。米国での英名
Amazon ants		*Polyergus*	（ハチ目、アリ科）の昆虫の総称　新熱帯区。女武者アリの異称あり
Amazon blue cracker		*Hamadryas chloe* (Stoll)	（チョウ目、タテハチョウ科）新熱帯区
Amazon darner		*Anax amazili* (Burmeister)	（トンボ目、ヤンマ科）新北区、新熱帯区
Amazon fly		*Metagonistylum minense* Townsend	（ハエ目、ヤドリバエ科）新熱帯区
Amazon nightfighter		*Dyscophellus porcius* (C. et R. Felder)	（チョウ目、セセリチョウ科）新熱帯区
Amazonian primitive ghost moths		Neotheoridae	（チョウ目）の昆虫の総称
Amazonicus beauty		*Baeotus aeilus* (Stoll)	（チョウ目、タテハチョウ科）新熱帯区。Amazon beauty ともいう
amber daggerwing	ベラニアツルギタテハ	*Marpesia berania* (Hewitson)	（チョウ目、タテハチョウ科）新熱帯区
amber ladybird			Adonis ladybird を見よ
amber-marked birch leafminer			exotic birch-leafmining sawfly を見よ
amber paradise skipper		*Phocides oreides* (Hewitson)	（チョウ目、セセリチョウ科）新熱帯区
amber phantom		*Haetera piera* (Linnaeus)	（チョウ目、タテハチョウ科）新熱帯区
amber-spot		*Libellula semifasciata* Burmeister	（トンボ目、トンボ科）新北区
amber-spotted sailor		*Patsuia sinensium* (Oberthür)	（チョウ目、タテハチョウ科）旧北区
amber-winged glider		*Hydrobasileus croceus* (Brauer)	（トンボ目、トンボ科）東洋区
amber-winged hawker			brown hawker を見よ
amber-winged skimmer			common amberwing を見よ
amber-winged spreadwing		*Lestes eurinus* Say	（トンボ目、アオイトトンボ科）新北区

英名	和名	学名	所属、分布、ほか
Ambigua firetip		*Mysoria barcatus ambigua* (Mabille et Boullet)	（チョウ目、セセリチョウ科）新熱帯区
ambiguous moth		*Lascoria ambigualis* Walker	（チョウ目、ヤガ科）新北区
ambitious ground cricket		*Pictonemobius ambitiosus* (Scudder)	（バッタ目、コオロギ科）新北区
amblycerans			amblycerous lice を見よ
amblycerous lice		*Amblycera*	（ハジラミ目）の昆虫の総称
Ambon onyx			yellow onyx（1）を見よ
Ambrax butterfly	パプアシロオビアゲハ	*Papilio ambrax* Boisduval	（チョウ目、アゲハチョウ科）東洋区、豪州区
Ambrax hairstreak		*Strephonota ambrax* (Westwood)	（チョウ目、シジミチョウ科）新熱帯区
Ambrax swallowtail			Ambrax butterfly を見よ
ambrosia beetle (1)		*Platypus sulcatus* Dejean	（コウチュウ目、ナガキクイムシ科）新熱帯区。カンキツ害虫
ambrosia beetle		*Platypus compositus* Say	（コウチュウ目、ナガキクイムシ科）新熱帯区。カンキツ害虫
ambrosia beetle			oak borer (2) を見よ
ambrosia beetle			shot-hole borer (2) を見よ
ambrosia beetle			striped ambrosia beetle を見よ
ambrosia beetles		*Xyleborus*	（コウチュウ目、キクイムシ科）の昆虫の総称
ambrosia beetles			pinhole beetles を見よ
ambrosia beetles			bark beetles (1) を見よ
ambrosia plume moth		*Adaina ambrosiae* (Murtfeldt)	（チョウ目、トリバガ科）新北区
ambush bug		*Phymata americana* Melin	（カメムシ目、ヒゲブトカメムシ科）新北区。捕食性
ambush bug		*Phymata crassipes* (Fabricius)	（カメムシ目、ヒゲブトサシガメ科）旧北区
ambush bugs	ヒゲブトカメムシ科	Phymatidae	（カメムシ目）の昆虫の総称　捕食性
ambush bugs		*Phymata*	（カメムシ目、ヒゲブトサシガメ科）の昆虫の総称
amelanchier sawfly	シデザクラハバチ	*Pristiphora amelanchieris* (Takeuchi)	（ハチ目、ハバチ科）日本
Amelia's Charaxes	ミズイロフタオチョウ	*Charaxes ameliae amelina* Doumet	（チョウ目、タテハチョウ科）エチオピア区
Amelia's threadtail		*Neoneura amelia* Calvert	（トンボ目、Protoneuridae）新北区、新熱帯区
American angle shades moth		*Euplexia benesimilis* McDunnough	（チョウ目、ヤガ科）新北区。American angle shades ともいう
American Apollo			clodius Apollo を見よ
American armyworm			armyworm (1) を見よ
American aspen beetle		*Gonioctena americana* (Schaeffer)	（コウチュウ目、ハムシ科）新北区
American barred umber moth			barred umber を見よ

英名	和名	学名	所属、分布、ほか
American bird grasshopper		*Schistocerca americana* (Drury)	（バッタ目、バッタ科）新北区
American bird's-wing moth		*Dypterygia rozmani* Berio	（チョウ目、ヤガ科）新北区
American black flour beetle			black flour beetle (1) を見よ
American blight			woolly apple aphid を見よ
American bollworm			corn earworm (2) を見よ
American bumble bee		*Bombus pennsylvanicus* (De Geer)	（ハチ目、ミツバチ科）新北区、大洋区
American burying beetle		*Nicrophorus americanus* Olivier	（コウチュウ目、シデムシ科）新北区。絶滅危惧種
American butterfly moths		Hedylidae	（チョウ目）の昆虫の総称
American carrion beetle		*Silpha americana* Linnaeus	（コウチュウ目、シデムシ科）新北区
American celery webworm			lucerne moth を見よ
American clover seed midge			clover seed midge を見よ
American cockroach	ワモンゴキブリ	*Periplaneta americana* (Linnaeus)	（ゴキブリ目、ゴキブリ科）日本、汎世界。普通種
American cockroaches			cockroaches (1) を見よ　英国での英名
American copper			small copper (1) を見よ　*L. p. americana* Harris の英名
American copper underwing			copper underwing (1) を見よ
American cotton bollworm			corn earworm (1)(2) を見よ
American dagger moth			dagger moth (1) を見よ
American dried-plant moth		*Idaea bonifata* Hulst	（チョウ目、シャクガ科）全北区
American dun-bar moth		*Cosmia calami* (Harvey)	（チョウ目、ヤガ科）新北区
American ear moth		*Amphipoea americana* Speyer	（チョウ目、ヤガ科）新北区
American emerald		*Cordulia shurtleffii* Scudder	（トンボ目、エゾトンボ科）新北区
American ermine moth		*Yponomeuta multipunctella* Clemens	（チョウ目、スガ科）新北区
American false tiger moths			oak moths を見よ
American flower fly		*Metasyrphus americanus* (Wiedemann)	（ハエ目、ハナアブ科）新北区
American gallinipper mosquito			gallinipper を見よ　American gallinipper ともいう
American grape berry moth			grape berry moth (1) を見よ

英名	和名	学名	所属、分布、ほか
American grass bug		*Acetropis americana* Knight	（カメムシ目、カスミカメムシ科）新北区
American grass mealybug		*Trionymus americanus* (Cockerell)	（カメムシ目、コナカイガラムシ科）新北区
American grass thrips	クサキイロアザミウマ	*Anaphothrips obscurus* (Müller)	（アザミウマ目、アザミウマ科）日本、全北区、豪州区、新熱帯区。イネ、ムギなどの害虫
American grasshopper		*Schistocera pallens* (Thunberg)	（バッタ目、バッタ科）新北区、新熱帯区
American grasshopper (1)		*Schistocera americana* (Drury)	（バッタ目、バッタ科）新北区、新熱帯区。バナナ、ワタなど多くの作物害虫
American grayling			large wood nymph を見よ
American ground mealybug		*Rhizoecus americanus* (Hambleton)	（カメムシ目、コナカイガラムシ科）新北区
American holly leaf miner			holly leafminer を見よ
American hornet moth		*Sesia tibialis* (Harris)	（チョウ目、スカシバガ科）新北区。リラ、トネリコ害虫
American horse-fly		*Tabanus americanus* Foerster	（ハエ目、アブ科）新北区
American hover fly		*Eupeodes americanus* Wiedemann	（ハエ目、ハナアブ科）新北区
American hover fly			American flower fly を見よ
American Idia moth		*Idia americalis* (Guenée)	（チョウ目、ヤガ科）新北区。American Idia, American snout ともいう
American juniper aphid			cypress aphid（2）を見よ
American lady			American painted lady を見よ　American lady butterfly ともいう
American lady beetle		*Hippodamia americana* Crotch	（コウチュウ目、テントウムシ科）新北区
American lawn ant			cornfield ant（1）を見よ
American lotus borer moth		*Ostrinia penitalis* (Grote)	（チョウ目、メイガ科）新北区
American moon moth			luna moth を見よ
American mouse flea		*Stenopomia americana* (Baker)	（ノミ目、ヒトノミ科）新北区
American oil beetle		*Meloe americanus* Leach	（コウチュウ目、ツチハンミョウ科）新北区
American painted lady	アメリカヒメアカタテハ	*Vanessa virginiensis* (Drury)	（チョウ目、タテハチョウ科）新北区
American palm cixiid		*Myndus crudus* Van Duzee	（カメムシ目、ヒシウンカ科）新北区
American pelecinid wasp		*Pelecinus polyturator* (Drury)	（ハチ目、Pelecinidae）新北区
American periodical cicada			seventeen-year locust (1) を見よ
American pistol casebearer moth			pistol case bearer (2) を見よ
American plum borer		*Euzophera semifuneralis* (Walker)	（チョウ目、メイガ科）新北区。ペカン害虫。American plume borer moth ともいう

英名	和名	学名	所属、分布、ほか
American plum borer		*Prosoeuzophera impletela* Zeller	（チョウ目、メイガ科）新北区
American puss moths		*Cerura*	（チョウ目、シャチホコガ科）の昆虫の総称
American rice-flour beetle			confused flour beetle を見よ
American rice leaf miner			rice leaafminer (2) を見よ
American rice stalk borer			American rice stem borer を見よ
American rice stem borer		*Chilo plejadellus* (Zincken)	（チョウ目、メイガ科）新北区、新熱帯区。イネ害虫
American rice stink bug			rice seed bug を見よ
American rice water weevil			rice water weevil を見よ
American rose slug			rose slug sawfly を見よ
American ruby-spot		*Hetaerina americana* (Fabricius)	（トンボ目、イトトンボ科）新北区。腹部が赤色
American salmon fly			giant stonefly (2) を見よ
American seed beetle			bean weevil を見よ
American serpentine leaf miner			legume leafminer を見よ
American shieldback		*Atlanticus americanus* (Saussure)	（バッタ目、キリギリス科）新北区
American silkworm moths			apatelodid moths を見よ
American snout		*Libytheana carinenta* (Cramer)	（チョウ目、タテハチョウ科）新北区、新熱帯区
American snout			southern snout を見よ　*L. c. larvata* (Strecker), *L. c. mexicana* Michener, *L. c. streckeri* Austin et Emmel も同英名
American soldier fly			black soldier fly を見よ　豪州での英名
American spanish fly		*Lytta vulnerata* LeConte	（コウチュウ目、ツチハンミョウ科）新北区
American spider beetle			black spider beetle を見よ
American stag beetle		*Lucanus elaphus* Fabricius	（コウチュウ目、クワガタムシ科）新北区
American sugarcane borer			sugarcane borer (1) を見よ
American swallowtail			black swallowtail を見よ
American swallowtail moths		Sematuridae	（チョウ目）の昆虫の総称
American swordgrass moth			red swordgrass moth を見よ
American syrphid		*Syrphus americanus* Wiedemann	（ハエ目、ハナアブ科）新北区
American tarnished plant bug			tarnished plant bug (1) を見よ

英名	和名	学名	所属、分布、ほか
American tent caterpillar		*Malacosoma americana* (Fabricius)	（チョウ目、カレハガ科）新北区。成虫は American tent caterpillar moth
American thuja shoot moth			arborvitae lef miner を見よ
American vine moth			grape berry moth (1) を見よ
American wainscot			armyworm (1) を見よ
American water weevil			rice water weevil を見よ
American wax moth		*Vitula edmandsii* (Packard)	（チョウ目、メイガ科）新北区
American whitebacked rice planthopper			rice delphacid を見よ
amethyst dancer		*Argia pallens* Calvert	（トンボ目、イトトンボ科）新北区
amethyst hairstreak		*Chlorostrymon maesites* (Herrich-Schäffer)	（チョウ目、シジミチョウ科）新北区
amethyst hairstreak			Icilius blue を見よ
amethyst jewel			Elgner's jewel を見よ
ammabium aphid			leaf-curl plum aphid を見よ
amnemus weevil		*Amnemus quadrituberculatus* (Boheman)	（コウチュウ目、ゾウムシ科）豪州区
amole giant skipper		*Aryxna polingi* (Skinner)	（チョウ目、セセリチョウ科）新北区
amole giant skipper			Poling's giant skipper を見よ
amorbia		*Amorbia essigana* Busck	（チョウ目、ハマキガ科）新北区。アボカド害虫
Amorbus bugs		*Amorbus*	（カメムシ目、ヘリカメムシ科）の昆虫の総称
Ampa skipper		*Euphyes ampa* Evans	（チョウ目、セセリチョウ科）新熱帯区
ampelopsis berry gall midge			ampelopsis fruit gall midge を見よ
ampelopsis fruit gall midge	ノブドウミタマバエ	*Asphondylia baca* Monzen	（ハエ目、タマバエ科）日本。ブドウ害虫
amphibiocorid bugs			pond skaters を見よ
Amphiro redring		*Pyrrhogyra amphiro* Bates	（チョウ目、セセリチョウ科）新熱帯区
ampulicid wasps		Ampulicidae	（ハチ目）の昆虫の総称　アナバチ科 Sphecidae の異名
ampulicid wasps			cockroach hunting wasps を見よ
ampulicids			ampulicid wasps を見よ
amur caddice fly	アムールエグリトビケラ	*Asynarchus amurensis* (Ulmer)	（トビケラ目、エグリトビケラ科）日本、旧北区
amur mining moth		*Acrocercops amurensis* Kuznetsov	（チョウ目、ホソガ科）旧北区
Amymone		*Mestra amymone* Ménétriès	（チョウ目、タテハチョウ科）新北区。Amymone butterfly ともいう
Amyntor greenstreak		*Cyanophrys amyntor* (Cramer)	（チョウ目、シジミチョウ科）新北区、大洋区、新熱帯区
anajapygid entotrophs		Anajapygidae	（コムシ目）の昆虫の総称

英名	和名	学名	所属、分布、ほか
ananas scale			brown pineapple scale を見よ
anaxyelid wood wasps			cedar wood wasps を見よ
anchor-faced wasp			red wasp を見よ
anchor stink bug		*Stiretrus atricornis* Van Duzee	（カメムシ目、カメムシ科）新北区
anchor stink bug (1)		*Stiretrus anchorago* (Fabricius)	（カメムシ目、カメムシ科）新北区。捕食性
anchorage bug			anchor stink bug (1) を見よ
ancient barklice		Archipsocidae	（チャタテムシ目）の昆虫の総称
ancient metalmark		*Voltinia danforthi* (Warren et Opler)	（チョウ目、シジミタテハ科）新熱帯区
Ancyra blue			Felder's lineblue (1) を見よ
Andalusian anomalous blue		*Polyommatus violetae* (Bustillo, Hermosa et Borrego)	（チョウ目、シジミチョウ科）旧北区
Andalusian false grayling		*Arethusana boabdii* (Rambur)	（チョウ目、タテハチョウ科）旧北区
Andaman clubtail		*Losaria rhodifer* Butler	（チョウ目、アゲハチョウ科）東洋区
Andaman crow		*Euploea andamanensis* (Atkinson)	（チョウ目、タテハチョウ科）東洋区
Andaman Helen		*Papilio prexaspes andamanicus* Goeze	（チョウ目、アゲハチョウ科）東洋区。blue Helen を参照
Andaman map		*Cyrestis andamanensis* Staudinger	（チョウ目、タテハチョウ科）東洋区
Andaman Mormon		*Papilio mayo* Atkinson	（チョウ目、アゲハチョウ科）東洋区
Andaman oakleaf		*Kallima albofasciata* (Moore)	（チョウ目、タテハチョウ科）東洋区
Andaman palmfly		*Elymnias cottonis* Hewitson	（チョウ目、タテハチョウ科）東洋区
Andaman palmking		*Amathusia andamanensis* Fruhstorfer	（チョウ目、タテハチョウ科）東洋区
Andaman sergeant		*Athyma rufula* de Nicéville	（チョウ目、タテハチョウ科）東洋区
Andaman swordtail		*Graphium epaminondas* Oberthür	（チョウ目、アゲハチョウ科）東洋区
Andaman tailless oakblue		*Arhopala zeta* (Moore)	（チョウ目、シジミチョウ科）東洋区
Andaman viscount		*Tanaecia cibaritis* Hewitson	（チョウ目、タテハチョウ科）東洋区
Andean buckeye		*Junonia vestina* Felder et Felder	（チョウ目、タテハチョウ科）新熱帯区
Andean grass yellow			Salome yellow を見よ
Andean moon moths		Cercophanidae	（チョウ目）の昆虫の総称
Andean painted lady		*Vanessa altissima* (Rosenberg et Talbot)	（チョウ目、タテハチョウ科）新熱帯区
Anderson mealybug		*Spirococcus andersoni* (Coleman)	（カメムシ目、コナカイガラムシ科）新北区

英名	和名	学名	所属、分布、ほか
Anderson's labyrinth		*Lethe andersoni* (Atkinson)	（チョウ目、タテハチョウ科）東洋区
Anderson's Mellana		*Quasimellana andersoni* Burns	（チョウ目、セセリチョウ科）新熱帯区
Anderson's skipper		*Toxidia andersoni* (Kirby)	（チョウ目、セセリチョウ科）豪州区
Andicola blue		*Leptotes andicola* (Godman et Salvin)	（チョウ目、シジミチョウ科）新熱帯区
andrenas		*Andrena*	（ハチ目、ヒメハナバチ科）の昆虫の総称
andrenid bees	ヒメハナバチ科	Andrenidae	（ハチ目）の昆虫の総称　世界に4,000種以上
andrenid wasps			andrenid bees を見よ
Androgeus	フトオビアゲハ	*Papilio androgeus* Cramer	（チョウ目、アゲハチョウ科）新熱帯区
Androgeus swallowtail			Androgeus を見よ
Andromache underwing moth		*Catocala andromache* H. Edwards	（チョウ目、ヤガ科）新北区。Andromache underwing ともいう
Andromeda lace bug	トサカグンバイ	*Stephanitis takeyai* Drake et Maa	（カメムシ目、グンバイムシ科）日本、新北区
Andromeda underwing moth		*Catocala andromedae* Guenée	（チョウ目、ヤガ科）新北区。Andromeda underwing ともいう
Andromeda wood nymph		*Taygetis thamyra* (Cramer)	（チョウ目、タテハチョウ科）新熱帯区
Andromedana moth		*Zomaria andromedana* (Barnes et McDunnough)	（チョウ目、ハマキガ科）新北区
Andromica clearwing		*Greta andromica* (Hewitson)	（チョウ目、タテハチョウ科）新熱帯区
Andromica glasswing			Andromica clearwing を見よ
Andromorph Palla		*Palla publius* Staudinger	（チョウ目、タテハチョウ科）エチオピア区
aner			雄アリに対する英名
angel insects (1)	ジュズヒゲムシ目	Zoraptera	の昆虫の総称　世界に24種
angel insects	ジュズヒゲムシ科	Zorotypidae	（ジュズヒゲムシ目）の昆虫の総称
angel lichen moth		*Cisthene angelus* (Dyar)	（チョウ目、ヒトリガ科）新北区
angel moth		*Oleclostera angelica* (Grote)	（チョウ目、カイコガ科）新北区
angelic Crocidiphora moth		*Crocidophora serratissimalis* Zeller	（チョウ目、メイガ科）新北区
angelic hairstreak		*Electrostrymon angelia* (Hewitson)	（チョウ目、シジミチョウ科）新北区
angelic leafcutter bee		*Megachile angelica* Mitchell	（ハチ目、ハキリバチ科）新北区
angelica aphid	ヤマトフタオアブラムシ	*Cavariella angelicae* (Matsumura)	（カメムシ目、アブラムシ科）日本、旧北区
angelica borer moth		*Papaipema angelica* (Smith)	（チョウ目、ヤガ科）新北区。angelica borer ともいう
angelica flat-body	キガシラヒラタマルハキバガ	*Agonopterix angelicella* (Hübner)	（チョウ目、マルハキバガ科）旧北区
Angelita's playboy		*Pilodeudorix angelita* (Suffert)	（チョウ目、シジミチョウ科）エチオピア区

英名	和名	学名	所属、分布、ほか
angelwings	キタテハ属	*Polygonia*	(チョウ目、タテハチョウ科) の昆虫の総称
angle-band rose pigmy		*Ectoedemia angulifasciella* (Stainton)	(チョウ目、モグリチビガ科) 旧北区
angle-barred bell		*Epinotia immundana* (Fischer von Röslerstamm)	(チョウ目、ハマキガ科) 旧北区
angle-barred coast groundling		*Scrobipalpa salinella* (Zeller)	(チョウ目、キバガ科) 旧北区
angle-barred pug		*Eupithecia innotata* (Hufnagel)	(チョウ目、シャクガ科) 旧北区
angle-lined carpet moth		*Cladara anguilineata* (Grote et Robinson)	(チョウ目、シャクガ科) 新北区
angle-lined prominent moth		*Clostera inclusa* (Hübner)	(チョウ目、シャチホコガ科) 新北区
angle shades		*Phlogophora meticulosa* (Linnaeus)	(チョウ目、ヤガ科) 旧北区
angle shades moth			angle shades を見よ
angle-striped			angle-striped sallow を見よ
angle-striped sallow	ウスシタキリガ	*Enargia paleacea* (Esper)	(チョウ目、ヤガ科) 日本、旧北区
angle-wing katydids			angular-winged katydids を見よ
angle-winged emerald moth		*Chloropteryx tepperaria* (Hulst)	(チョウ目、シャクガ科) 新北区
angle-winged fern moth		*Azelina fortinata* Guenée	(チョウ目、シャクガ科) 豪州区
angle-winged Telesiphe		*Podotricha telesiphe* Hewitson	(チョウ目、タテハチョウ科) 新熱帯区
angled caster	カバタテハ	*Ariadne ariadne pallidior* (Fruhstorfer)	(チョウ目、タテハチョウ科) 日本、東洋区。*A. ariadne* (Linnaeus) の英名
angled cyclops		*Erites angularis* Moore	(チョウ目、タテハチョウ科) 東洋区
angled darkie		*Allotinus fabius* (Distant et Pryer)	(チョウ目、シジミチョウ科) 東洋区
angled gem		*Anomis sabulifera* (Guenée)	(チョウ目、ヤガ科) エチオピア区
angled grass yellow		*Eurema desjardinsii* (Boisduval)	(チョウ目、シロチョウ科) エチオピア区。*E. d. marshalli* (Butler) も同英名
angled leafwing			crinkled leafwing を見よ
angled Metarranthis moth		*Metarranthis angularia* Barnes et McDunnough	(チョウ目、シャクガ科) 新北区
angled pierrot		*Caleta caleta* (Hewitson)	(チョウ目、シジミチョウ科) 東洋区
angled pierrot		*Caleta decidia* (Hewitson)	(チョウ目、シジミチョウ科) 東洋区
angled red forester	メスチャヒカゲ	*Lethe chandica* (Moore)	(チョウ目、タテハチョウ科) 旧北区、東洋区
angled sulphur			yellow brimstone を見よ
angled sunbeam		*Curetis acuta* Moore	(チョウ目、シジミチョウ科) 東洋区
angled wave moth		*Scopula ancellata* (Hulst)	(チョウ目、シャクガ科) 新北区
angler's curse		*Caenis*	(カゲロウ目、ヒメカゲロウ科) の昆虫の総称
angles		*Caprona*	(チョウ目、セセリチョウ科) の昆虫の総称
anglewings	ヒオドシチョウ属	*Nymphalis*	(チョウ目、タテハチョウ科) の昆虫の総称

英名	和名	学名	所属、分布、ほか
Angochlora sweat bee		*Angochlora pura* (Say)	(ハチ目、コハナバチ科) 新北区
Angola white-lady swordtail	アカネタイマイ	*Graphium angolanus* (Goeze)	(チョウ目、アゲハチョウ科) エチオピア区。Angola white-lady ともいう
Angora goat biting louse		*Bovicola crassipes* (Rudow)	(ハジラミ目、ケモノハジラミ科) 新北区。Angora-goat biting louse とも記す
Angora goat biting louse (1)		*Bovicola limbatus* (Gervais)	(ハジラミ目、ケモノハジラミ科) 全北区
Angora goat chewing louse			Angora goat biting louse (1) を見よ
Angora goat louse			Angora goat biting louse (1) を見よ
Angoumois grain moth	バクガ	*Sitotroga cerealella* (Olivier)	(チョウ目、キバガ科) 日本、汎世界。ムギ害虫。Angoumois はフランスの地名
Angoumois grain neb			Angoumois grain moth を見よ
angraecum scale		*Conchaspis angraeci* Cockerell	(カメムシ目、Conchaspididae) 大洋区
anguina moth		*Dasylophia anguina* J. E. Smith	(チョウ目、シャチホコガ科) 新北区
angular metalmark		*Amphiselenis chama* (Staudinger)	(チョウ目、シジミタテハ科) 新熱帯区
angular owlet moth		*Carea subtilis* Walker	(チョウ目、ヤガ科) 東洋区
angular winged grasshoppers			long-horned grasshoppers (1) を見よ
angular-winged katydid (1)		*Microcentrum rhombifolium* (Saussure)	(バッタ目、キリギリス科) 新北区。カンキツ害虫
angular-winged katydid (2)		*Microcentrum retinerve* (Burmeister)	(バッタ目、キリギリス科) 新北区。カンキツ害虫
angular-winged katydids		*Microcentrum*	(バッタ目、キリギリス科) の昆虫の総称
angulate fan-foot			brown-lined owlet moth を見よ
angulate leafhopper		*Acinopterus angulatus* Lawson	(カメムシ目、オオヨコバイ科) 新北区
angulate spreadwing			falcate skipper を見よ
angulate-winged katydid			angular-winged katydid (2) を見よ
angulated bark beetle	マツノツノキクイムシ	*Orthotomicus anguatus* (Eichhoff)	(コウチュウ目、キクイムシ科) 日本、旧北区、東洋区
angulated cutworm			salt-and-pepper looper moth を見よ
angulose prominent moth		*Peridea angulosa* (Smith)	(チョウ目、シャチホコガ科) 新北区
Angus' underwing moth		*Catocala angusi* Grote	(チョウ目、ヤガ科) 新北区。Angus' underwing ともいう
Angus's Datana moth		*Datana angusii* Grote et Robinson	(チョウ目、シャチホコガ科) 新北区
Angusta crescent		*Castilia angusta* (Hewitson)	(チョウ目、タテハチョウ科) 新熱帯区
Anicia checkerspot		*Euphydryas anicia* (Doubleday)	(チョウ目、タテハチョウ科) 新北区

英名	和名	学名	所属、分布、ほか
anise swallowtail	アメリカキアゲハ	*Papilio zelicaon* Lucas	(チョウ目、アゲハチョウ科) 新北区
Anna carpenterworm moth		*Givira anna* (Dyar)	(チョウ目、ボクトウガ科) 新北区
Anna tiger moth			little virgin moth を見よ
Annam bushbrown		*Mycalesis annamitica* Fruhstorfer	(チョウ目、タテハチョウ科) 東洋区
Annaphilas		*Annaphila*	(チョウ目、ヤガ科) の昆虫の総称
Anna's blue			Idas blue を見よ
Anna's eighty-eight			eighty-eight butterfly (1) を見よ
Anne's Epitola		*Stempfferia annae* Libert	(チョウ目、シジミチョウ科) エチオピア区
annona seed wasp		*Bephratelloides cubensis* (Ashmead)	(ハチ目、カタビロコバチ科) 大洋区
annual white grub		*Ochrosidia villosa* Burmeister	(コウチュウ目、コガネムシ科) 新北区。芝害虫
annulate beauty			annulate mosaic を見よ
annulate-legged earwig			ring-legged earwig を見よ
annulate mosaic		*Colobura annulata* Willmott, Constantino et Hall	(チョウ目、タテハチョウ科) 新熱帯区
annulate spittle bug		*Aphrophora annulata* Ball	(カメムシ目、アワフキムシ科) 新北区
annulated smudge		*Rhigognostis annulatella* (Curtis)	(チョウ目、スガ科) 旧北区
annulet		*Charissa obscurata* (Denis et Schiffermüller)	(チョウ目、シャクガ科) 旧北区。annulet moth ともいう
anomalous		*Stilbia anomala* (Haworth)	(チョウ目、ヤガ科) 旧北区。anomalous moth ともいう
anomalous blue		*Polyommatus admetus* (Esper)	(チョウ目、シジミチョウ科) 旧北区
anomalous crickets		Pentacentrinae	(バッタ目、コオロギ科) の昆虫の総称
anomalous wainscot			anomalous を見よ
anomalus checkerspot		*Texola anomalus* (Godman et Salvin)	(チョウ目、タテハチョウ科) 新熱帯区
anomis fruit moth			hibiscus looper を見よ
anopheline mosquitoes	ハマダラカ亜科	Anophelinae	(ハエ目、カ科) の昆虫の総称
anopheline mosquitoes			malaria mosquitoes を見よ　南アフリカでの英名
Ansorge's Charaxes		*Charaxes ansorgei* Rothschild	(チョウ目、タテハチョウ科) エチオピア区
Ansorge's danaid		*Amauris ellioti* Butler	(チョウ目、タテハチョウ科) エチオピア区
Ansorge's hawklet		*Sphingonaepiopsis ansorgei* Rothschild	(チョウ目、スズメガ科) エチオピア区
Ansorge's leaf butterfly		*Junonia ansorgei* (Rothschild)	(チョウ目、タテハチョウ科) エチオピア区
ant beetle		*Thanasimus formicarius* (Linnaeus)	(コウチュウ目、カッコウムシ科) 旧北区

英名	和名	学名	所属、分布、ほか
ant-beetles	ヒゲブトオサムシ科	Paussidae	（コウチュウ目）の昆虫の総称　中南米での英名
ant-bird skipper		*Phanus vitreus* (Stoll)	（チョウ目、セセリチョウ科）新熱帯区
ant-blues		*Acrodipsas*	（チョウ目、シジミチョウ科）の昆虫の総称
ant bug		*Dacerla mediospinosa* Signoret	（カメムシ目、カスミカメムシ科）新北区。アリに似る
ant-bugs		*Dacerla*	（カメムシ目、カスミカメムシ科）の昆虫の総称
ant butterflies	トンボマダラ亜科	Ithomininae	（チョウ目、タテハチョウ科）の昆虫の総称 army ants の分泌物を吸うことから
ant capsid		*Myrmecoris gracilis* (Sahlberg)	（カメムシ目、カスミカメムシ科）旧北区
ant cow	アブラムシ		aphids (1) を見よ
ant-cricket (1)		*Myrmecophilus acervorum* (Panzer)	（バッタ目、アリヅカコオロギ科）旧北区
ant cricket (2)		*Myrmecophilus oregonensis* Bruner	（バッタ目、アリヅカコオロギ科）新北区
ant crickets	アリヅカコオロギ科	Myrmecophilidae	（バッタ目）の昆虫の総称
ant crickets			crickets (1) を見よ
ant damsel bug		*Himacerus myrmicoides* (Costa)	（カメムシ目、マキバサシガメ科）旧北区
ant flies		*Microdon*	（ハエ目、ハナアブ科）の昆虫の総称
ant flies			black scavenger flies を見よ　豪州での英名
ant-heap white		*Dixeia pigea* (Boisduval)	（チョウ目、シロチョウ科）エチオピア区
ant-like flower beetles	アリモドキ科	Anthicidae	（コウチュウ目）の昆虫の総称
ant-like leaf beetles		Euglenidae	（コウチュウ目）の昆虫の総称
ant-like leaf beetles	ニセクビボソムシ科	Aderidae	（コウチュウ目）の昆虫の総称
ant-like stone beetles	コケムシ科	Scydmaenidae	（コウチュウ目）の昆虫の総称
ant-like weevils		Myrmecinae	（コウチュウ目、ゾウムシ科）の昆虫の総称　アリモドキゾウムシ亜科 Cyladinae とは別
antlion	コウスバカゲロウ	*Myrmeleon formicarius* (Linnaeus)	（アミメカゲロウ目、ウスバカゲロウ科）日本、旧北区
ant-lion flies			antlions (1) を見よ
antlion lacewing		*Glenoleon falsus* (Walker)	（アミメカゲロウ目、ウスバカゲロウ科）豪州区
ant-lion lacewing		*Weeleus acutus* (Walker)	（アミメカゲロウ目、ウスバカゲロウ科）豪州区
antlion lacewings			antlions (1) を見よ
antlions (1)	ウスバカゲロウ科	Myrmeleontidae	（アミメカゲロウ目）の昆虫の総称
ant-lions (2)		*Brachynemurus*	（アミメカゲロウ目、ウスバカゲロウ科）の昆虫の総称
antloving beetles (1)	アリヅカムシ科	Pselaphidae	（コウチュウ目）の昆虫の総称　朽木、こけ、アリの巣でみつかる
ant-loving beetles	ヒゲナガアリヅカムシ亜科	Pselaphinae	（コウチュウ目、アリヅカムシ科）の昆虫の総称
ant-loving beetles (2)		Clavigeridae	（コウチュウ目）の昆虫の総称

英名	和名	学名	所属、分布、ほか
ant-loving crickets		Myrmecophilinae	（バッタ目、コオロギ科）の昆虫の総称　4 mm 以下で無翅、アリの巣に生息
ant-mantis		*Mantillica*	（カマキリ目、Thespidae）の昆虫の総称　アリに似た南米のカマキリ
ant-mimicking treehopper		*Cyphonia clavata* Fabricius	（カメムシ目、ツノゼミ科）新熱帯区
ant mimic bugs			red bugs （1）を見よ
ant nest beetle		*Paussus favieri* Fairmaire	（コウチュウ目、ヒゲブトオサムシ科）旧北区、エチオピア区
ant scarabs		*Cremastocheilus*	（コウチュウ目、コガネムシ科）の昆虫の総称　新北区。時にアリの巣でアリの幼虫を捕食する
Anteas Actinote			common Actinote を見よ
antelope beetle		*Dorcus parallelus* (Say)	（コウチュウ目、クワガタムシ科）新北区
Antestia bugs		*Antestiopsis*	（カメムシ目、カメムシ科）の昆虫の総称
Anthaxias		*Anthaxia*	（コウチュウ目、タマムシ科）の昆虫の総称
anthelids			Australian lappet moths を見よ
anthocorid bugs			flower bugs を見よ
anthocorids			flower bugs を見よ
anthomyiid flies	ハナバエ科	Anthomyiidae	（ハエ目）の昆虫の総称　一部の種は root maggots, seed maggots ともいわれる。世界に 1,000 種
anthomyiine flies		Anthomyiinae	（ハエ目、ハナバエ科）の昆虫の総称
anthomyzid flies		Anthomyzidae	（ハエ目）の昆虫の総称
anthomyzids			anthomyzid flies を見よ
anthophorid bees			long-tongued bees を見よ
anthophorids			long-tongued bees を見よ
anthracinus skipper		*Cynea anthracinus* (Mabille)	（チョウ目、セセリチョウ科）新熱帯区
anthribid beetles			fungus weevils を見よ
anthurium thrips			orchid thrips を見よ
anthurium whitefly		*Aleurotubulus anthuricola* Nakahara	（カメムシ目、コナジラミ科）大洋区
Antillean blue	チビウラナミシジミ	*Hemiargus ceraunus* (Fabricius)	（チョウ目、シジミチョウ科）新北区、新熱帯区
Antillean checkered-skipper		*Pyrgus crisia* Herrich-Schäffer	（チョウ目、セセリチョウ科）新北区
Antillean daggerwing		*Marpesia eleuchea* Hübner	（チョウ目、タテハチョウ科）新北区
Antillean mapwing		*Hypanartia paullus* (Fabricius)	（チョウ目、タテハチョウ科）新熱帯区
Antillean sister		*Adelpha gelania* (Godart)	（チョウ目、タテハチョウ科）新熱帯区
Antilles blue			Antillean blue を見よ
Antiochus longwing		*Heliconius antiochus* (Linnaeus)	（チョウ目、タテハチョウ科）新熱帯区
antique sap beetle		*Carpophilus antiquus* (Melsheimer)	（コウチュウ目、ケシキスイ科）新北区
antique tussock moth			common vapourer を見よ

英名	和名	学名	所属、分布、ほか
antirrhinum beetle		*Brachypterolus pulicarius* (Linnaeus)	（コウチュウ目、ケシキスイ科）旧北区
antirrhinum beetle		*Brachypterolus vestitus* (Kiesenwetter)	（コウチュウ目、ケシキスイ科）旧北区
antirrhinum brocade		*Calophasia platyptera* (Esper)	（チョウ目、ヤガ科）旧北区
antler moth		*Cerapteryx graminis* (Linnaeus)	（チョウ目、ヤガ科）旧北区。antler ともいう
antler sawfly			bristly rose slug sawfly を見よ
antler sawfly			rose sawfly (1) を見よ
antlered treehoppers		*Sphongophorus*	（カメムシ目、ツノゼミ科）の昆虫の総称
Antra skipper		*Euphyes antra* Evans	（チョウ目、セセリチョウ科）新熱帯区
ants (1)	アリ上科	Formicoidea	（ハチ目）の昆虫の総称
ants	アリ科	Formicidae	（ハチ目）の昆虫の総称　世界に 14,000 種
ants			black ants を見よ
ants-nest cricket			ant-cricket (1) を見よ　ant's-nest cricket とも記す
Aon moth		*Aon noctuiformis* Neumoegen	（チョウ目、ヤガ科）新北区
apache bush katydid		*Insara apache* (Rehn)	（バッタ目、キリギリス科）新北区
apache cicada		*Diceroprocta apache* Davis	（カメムシ目、セミ科）新北区、新熱帯区
apache dancer		*Argia munda* Calvert	（トンボ目、イトトンボ科）新北区
apache paper wasp		*Polistes apachus* Saussure	（ハチ目、スズメバチ科）新北区
Apache skipper		*Hesperia woodgatei* Williams	（チョウ目、セセリチョウ科）新北区、新熱帯区
apache spiketail		*Cordulegaster diadema* Selys	（トンボ目、オニヤンマ科）新北区、新熱帯区
Apama hairstreak			canyon green hairstreak を見よ
apate borer		*Apate monachus* (Fabricius)	（コウチュウ目、ナガシンクイムシ科）旧北区
apatelodid moths		Apateloididae	（チョウ目）の昆虫の総称
Apaturina dartwhite		*Catasticta apaturina* Butler	（チョウ目、シロチョウ科）新熱帯区
apeflies		*Spalgis*	（チョウ目、シジミチョウ科）の昆虫の総称
apefly		*Spalgis epius* (Westwood)	（チョウ目、シジミチョウ科）東洋区
apefly moth			apefly を見よ
aphananthe prominent	ムクツマキシャチホコ	*Phalera angustipennis* Matsumura	（チョウ目、シャチホコガ科）日本、東洋区
aphelinids			fun flies を見よ
aphid			aphids (1) を見よ
aphid destroyer		*Deraecoris ruber* (Linnaeus)	（カメムシ目、カスミカメムシ科）旧北区
aphid destructor			aphid destroyer を見よ
aphid flies		*Leucopis*	（ハエ目、アブラコバエ科）の昆虫の総称
aphid flies (1)	アブラコバエ科	Chamaemyiidae	（ハエ目）の昆虫の総称
aphid lion			common green lacewing を見よ　幼虫の英名

英名	和名	学名	所属、分布、ほか
aphid lion			golden-eye lacewing (1)(2) を見よ　幼虫の英名
aphid midge		*Aphidoletes aphidimyza* (Rondani)	(ハエ目、タマバエ科) 新北区。幼虫はアブラムシを捕食し、温室で生物的防除に使用
aphid parasite		*Aphelinus asychis* (Walker)	(ハチ目、ツヤコバチ科) 豪州区
aphid parasites (1)	アブラバチ科	Aphidiidae	(ハチ目) の昆虫の総称
aphid parasites	アブラバチ亜科	Aphidiinae	(ハチ目、コマユバチ科) の昆虫の総称
aphid parasites		*Aphidius*	(ハチ目、コマユバチ科) の昆虫の総称
aphid predatory midge			aphid midge を見よ
aphid wasps (1)	コシボソアナバチ科	Pemphredonidae	(ハチ目) の昆虫の総称
aphid wasps (2)	ヒメコシボソバチ亜科	Pemphredoninae	(ハチ目、コシボソアナバチ科) の昆虫の総称
aphides			aphids (1) を見よ
aphidiids			aphid parasites (1) を見よ
aphid-lions			green lacewings (2) を見よ
aphids (1)	アブラムシ上科	Aphidoidea	(カメムシ目、アブラムシ科) の昆虫の総称
aphids (2)	アブラムシ科	Aphididae	(カメムシ目) の昆虫の総称　世界に 3,500 種
aphids (3)		Aphidinae	(カメムシ目、アブラムシ科) の昆虫の総称
aphids		Aphidina	(カメムシ目、アブラムシ科) の昆虫の総称
aphids			bark aphids を見よ　英国での英名
aphis			aphids (1) を見よ　複数型は aphides
aphis lions			brown lacewings (1) を見よ　幼虫の英名
aphis-lions			green lacewings (1) を見よ
aphis wolves			brown lacewings (1) を見よ　幼虫の英名
Aphodian dung beetles	マグソコガネ亜科	Aphodiinae	(コウチュウ目、コガネムシ科) の昆虫の総称
aphodine dung beetles			Aphodian dung beetles を見よ
Aphrodite	ツヤアカギンボシヒョウモン	*Speyeria aphrodite* Fabricius	(チョウ目、タテハチョウ科) 新北区
Aphrodite fritillary			Aphrodite を見よ
apical prominent moth		*Clostera apicalis* (Walker)	(チョウ目、シャチホコガ科) 新北区。apical prominent ともいう
Apollo	アポロウスバアゲハ	*Parnassius apollo* (Linnaeus)	(チョウ目、アゲハチョウ科) 旧北区。Apollo butterfly ともいう
Apollo butterflies			Apollos を見よ
Apollo butterflies			Parnassians を見よ
Apollo butterflies			Apollos を見よ　米国での英名
Apollo jewel	アポロニシキシジミ	*Hypochrysops apollo* Miskin	(チョウ目、シジミチョウ科) 豪州区
Apollo shieldback		*Idiostatus apollo* Rentz	(バッタ目、キリギリス科) 新北区
Apollos	ウスバアゲハ属	*Parnassius*	(チョウ目、アゲハチョウ科) の昆虫の総称
Apollos			Parnassians を見よ

英名	和名	学名	所属、分布、ほか
Apollot			Parnassians を見よ
Appalachian azure		*Celastrina neglecta major* (Edwards)	(チョウ目、シジミチョウ科) 新北区
Appalachian brown		*Satyrodes appalachia* Chermock	(チョウ目、タテハチョウ科) 新北区
Appalachian eyed brown		*Satyrodes appalachia leeuwi* (Gatrelle et Arbogast)	(チョウ目、タテハチョウ科) 新北区。Appalachian brown を参照
Appalachian jewelwing		*Calopteryx angustipennis* (Selys)	(トンボ目、イトトンボ科) 新北区
Appalachian skipper			northern grizzled skipper を見よ
Appalachian snaketail		*Ophiogomphus incurvatus* Carle	(トンボ目、サナエトンボ科) 新北区
Appalachian tiger swallowtail		*Papilio appalachiensis* (Pavulaan et D. Wright)	(チョウ目、アゲハチョウ科) 新北区
apple ambrosia beetle	サクキクイムシ	*Xylosandrus crassiusculus* (Motschulsky)	(コウチュウ目、キクイムシ科) 日本、旧北区。クリを加害
apple ambrosia beetle (1)	ニレザイノキクイムシ	*Xyleborus apicalis* Blandford	(コウチュウ目、キクイムシ科) 日本、旧北区、東洋区。リンゴ、カキなどを加害
apple and pear bryobia		*Bryobia rubrioculus* (Scheuten)	(アザミウマ目、アザミウマ科) 旧北区
apple and plum casebearer		*Coleophora nigricella* (Stephens)	(チョウ目、ツツミノガ科) 旧北区
apple and plum casebearer			apple casebearer を見よ
apple and plum case-bearer			cigar casebearer (2) を見よ
apple and plum casebearer moth		*Coleophora spinella* (Schrank)	(チョウ目、ツツミノガ科) 全北区。apple and plum casebearer ともいう
apple and potato leafhopper			potato leafhopper (1) を見よ
apple and thorn fruit weevil			apple fruit rhynchites を見よ
apple and thorn skeletonizer		*Prochoreutis pariana* (Clerck)	(チョウ目、ハマキモドキガ科) 新北区
apple and thorn skeletonizer			apple leaf skeletonizer (1) を見よ
apple-anthriscus aphid			foxglove aphid (1) を見よ
apple aphid (1)	ヨーロッパリンゴアブラムシ	*Aphis pomi* (De Geer)	(カメムシ目、アブラムシ科) 旧北区、新北区。リンゴ害虫
apple aphid (2)		*Rhopalosiphum crataegellum* Theobald	(カメムシ目、アブラムシ科) 旧北区
apple aphid			spirea aphid を見よ
apple aphid			woolly apple aphid を見よ
apple argid sawfly	リンゴハバチ	*Arge mali* (Takahashi)	(ハチ目、ミフシハバチ科) 日本、旧北区。リンゴ害虫。apple saw fly ともいう
apple bark beetle	シイノホソキクイムシ	*Xyleborus defensus* Blandford	(コウチュウ目、キクイムシ科) 日本
apple bark beetle			large shot-hole borer を見よ

英名	和名	学名	所属、分布、ほか
apple bark borer		*Synanthedon pyri* (Harris)	（チョウ目、スカシバガ科）新北区。apple bark boxer moth ともいう
apple barkminer			apple leafminer (3) を見よ　apple barkminer moth ともいう
apple black cosmet		*Blastodacna atra* (Haworth)	（チョウ目、カザリバガ科）旧北区
apple black-margined longicorn	ヘリグロリンゴカミキリ	*Nupserha marginella* (Bates)	（コウチュウ目、カミキリムシ科）、日本、旧北区
apple blossom midge	リンゴツボミタマバエ	*Contarinia mali* Barnes	（ハエ目、タマバエ科）日本、旧北区。リンゴ害虫
apple blossom tineid			brindled argent を見よ
apple blossom weevil	リンゴハナゾウムシ	*Anthonomus pomorum* (Linnaeus)	（コウチュウ目、ゾウムシ科）日本、旧北区。リンゴ害虫。apple weevil ともいう
apple blossom weevil parasite		*Ephialtes pomorum* (Ratzeburg)	（ハチ目、ヒメバチ科）旧北区
apple blotch leafminer		*Phyllonorycter crataegella* (Clemens)	（チョウ目、ホソガ科）新北区。apple blotch leafminer moth ともいう
apple blunt-tipped moth	リンゴツマキリアツバ	*Pangrapta obscurata* (Butler)	（チョウ目、ヤガ科）日本、旧北区
apple borer			apple clearwing moth を見よ
apple borer			quince borer を見よ
apple brown tortrix			brown tortrix を見よ
apple bucculatrix		*Bucculatrix pomifoliella* Clemens	（チョウ目、チビガ科）新北区。リンゴの枝を加害
apple bud aphid			grain aphid (1) を見よ
apple bud moth			eyespotted bud moth を見よ
apple bud weevil		*Anthonomus pyri* Kollar	（コウチュウ目、ゾウムシ科）旧北区
apple budworm			apple black cosmet を見よ　幼虫の英名
apple bumble bee		*Bombus pomorum* (Panzer)	（ハチ目、ミツバチ科）旧北区
apple capsid		*Plesiocoris rugicollis* (Fallén)	（カメムシ目、カスミカメムシ科）旧北区
apple capsid bug			apple capsid を見よ
apple casebearer	リンゴツツミノガ	*Coleophora serratella* (Linnaeus)	（チョウ目、ツツミノガ科）日本、旧北区。リンゴ害虫
apple caterpillar	リンゴカレハ	*Odonestis pruni japonensis* Tams	（チョウ目、カレハガ科）日本。リンゴ害虫
apple-chervil aphid		*Dysaphis chaerophylli* (Börner)	（カメムシ目、アブラムシ科）旧北区
apple chrysomelid		*Merista sexmaculata* Hope	（コウチュウ目、ハムシ科）東洋区
apple clearwing moth		*Synanthedon myopaeformis* (Borkhausen)	（チョウ目、スカシバガ科）全北区。apple clearwing ともいう
apple codling moth			codling moth を見よ
apple curculio		*Tachypterellus quadrigibbus* (Say)	（コウチュウ目、ゾウムシ科）新北区。リンゴ他害虫
apple curculio			apple weevil (1) を見よ

英名	和名	学名	所属、分布、ほか
apple dagger moth	リンゴケンモン	*Triaena intermedia* (Warren)	（チョウ目、ヤガ科）日本、旧北区。リンゴ害虫
apple dock aphid		*Dysaphis radicola* (Mordvilko)	（カメムシ目、アブラムシ科）豪州区
apple dumpling bug		*Campylomma liebknechti* (Girault)	（カメムシ目、カスミカメムシ科）豪州区
apple ermine moth	リンゴスガ	*Yponomeuta mallinellus* Zeller	（チョウ目、スガ科）日本、全北区。リンゴ害虫。apple ermine ともいう
apple fall cankerworm	クロテンフユシャク	*Inurois fletcheri* Inoue	（チョウ目、シャクガ科）日本
apple flea weevil		*Rhynchaenus pallicornis* (Say)	（コウチュウ目、ゾウムシ科）新北区。リンゴ害虫
apple frosted green weevil	リンゴコフキゾウムシ	*Phyllobium armatus* Roelofs	（コウチュウ目、ゾウムシ科）日本
apple frosted leaf beetle	コフキサルハムシ	*Lypesthes ater* (Motschulsky)	（コウチュウ目、ハムシ科）日本、旧北区
apple fruit borer		*Idiophantis chiridota* Meyrick	（チョウ目、キバガ科）旧北区、東洋区
apple fruit fly	リンゴミバエ	*Rhagoletis pomonella* (Walsh)	（ハエ目、ミバエ科）新北区。リンゴ害虫
apple fruit licker	リンゴハイイロハマキ	*Spilonota lechriaspis* Meyrick	（チョウ目、ハマキガ科）日本、旧北区
apple fruitminer		*Marmara pomonella* Busck	（チョウ目、ホソガ科）新北区。apple fruitminer moth ともいう
apple fruit miner			apple fruit moth を見よ　幼虫の英名
apple fruit moth	リンゴヒメシンクイ	*Argyresthia conjugella* Zeller	（チョウ目、メムシガ科）日本、全北区。リンゴ害虫
apple fruit rhynchites		*Rhynchites aequatus* (Linnaeus)	（コウチュウ目、オトシブミ科）旧北区
apple fruit sawfly		*Hoplocampa testudinea* (Klug)	（ハチ目、ハバチ科）新北区。apple sawfly ともいう
apple geometrid	ウスジロエダシャク	*Ectropis obliqua* (Prout)	（チョウ目、シャクガ科）日本
apple geometrid	タテスジナミシャク	*Pareulype consanguinea* (Butler)	（チョウ目、シャクガ科）日本、旧北区
apple grain aphid		*Rhopalosiphum fitchii* (Sanderson)	（カメムシ目、アブラムシ科）新北区。リンゴ他害虫
apple grain aphid		*Rhopalosiphum prunifoliae* (Fitch)	（カメムシ目、アブラムシ科）新北区
apple-grain aphid			oat bird-cherry aphid を見よ
apple-grain aphis			oat bird-cherry aphid を見よ
apple-green swallowtail		*Papilio phorcas* Cramer	（チョウ目、アゲハチョウ科）エチオピア区
apple greenish geometrid			green pug moth を見よ
apple hairy caterpillar		*Lymantria obfuscata* Walker	（チョウ目、ドクガ科）東洋区
apple hawk moth			giant hawk moth を見よ
apple heliodinid	キイロマイコガ	*Stathmopoda auriferella* (Walker)	（チョウ目、ニセマイコガ科）日本、東洋区、エチオピア区、豪州区

英名	和名	学名	所属、分布、ほか
apple horned looper	リンゴツノエダシャク	*Phthonosema tendinosaria* (Bremer)	（チョウ目、シャクガ科）日本、旧北区
apple humble bee			apple bumble bee を見よ
apple-leaf aphid			apple aphid (1) を見よ
apple-leaf blister moth		*Parornix petiolella* (Frey)	（チョウ目、ホソガ科）旧北区
apple leaf bug	リンゴクロカスミカメムシ	*Heterocordylus flavipes* Nitobe	（カメムシ目、カスミカメムシ科）日本
apple leaf casebearer	ツツマダラメイガ	*Acrobasis tokiella* (Ragonot)	（チョウ目、メイガ科）日本、旧北区
apple leaf crumpler		*Acrobasis indigenella* (Zeller)	（チョウ目、メイガ科）新北区
apple leaf-curling aphid	リンゴコブアブラムシ	*Ovatus malisuctus* Matsumura	（カメムシ目、アブラムシ科）日本、東洋区、旧北区。リンゴ害虫
apple leaf-curling midge			apple leaf midge を見よ
apple leaf-curling moth		*Adoxophyes privatana* (Walker)	（チョウ目、ハマキガ科）旧北区、東洋区
apple leaf jassid			apple leafhopper (4) を見よ
apple leaf midge		*Dasineura mali* (Kieffer)	（ハエ目、タマバエ科）旧北区、豪州区
apple-leaf midget		*Phyllonorycter cydoniella* (Fabricius)	（チョウ目、ホソガ科）旧北区
apple leaf miner (1)		*Phyllonorycter blancardella* Fabricius	（チョウ目、ホソガ科）全北区
apple leaf miner (2)		*Phyllonorycter corylifoliella* Hübner	（チョウ目、ホソガ科）旧北区
apple leafminer (3)		*Tischeria malifoliella* Clemens	（チョウ目、ムモンハモグリガ科）新北区
apple leafminer (4)		*Marmara elotella* (Busck)	（チョウ目、ホソガ科）新北区
apple leafminer	キンモンホソガ	*Phyllonorycter ringoniella* (Matsumura)	（チョウ目、ホソガ科）日本、旧北区
apple leafminer			peach leafminer を見よ
apple leaf miner			apple blotch leafminer を見よ
apple leaf moth			apple fruit moth を見よ
apple leaf roller			pine tube moth (1) を見よ
apple leaf roller			light brown apple moth を見よ
apple leaf roller			appe leaf sewer を見よ
apple leaf roller moth			light brown apple moth を見よ
apple leaf sewer		*Ancylis nubeculana* Clemens	（チョウ目、ハマキガ科）新北区
apple leaf skeletonizer		*Psorosina hammondi* (Riley)	（チョウ目、メイガ科）新北区
apple leaf skeletonizer	リンゴハマキモドキ	*Choreutis vinosa* (Diakonoff)	（チョウ目、ハマキモドキガ科）日本。リンゴ害虫
apple leaf skeletonizer (1)	ニセリンゴハマキモドキ	*Choreutis pariana* (Clerck)	（チョウ目、ハマキモドキガ科）日本、全北区。apple leaf skeletonizer moth ともいう
apple leaf sucker			apple sucker を見よ

英名	和名	学名	所属、分布、ほか
apple leaf trumpet miner			apple leafminer (3) を見よ　apple leaf trumpet miner moth ともいう
apple leaf trumpet miner			apple leafminer (4) を見よ　米国での英名
apple leaf weevil			green leaf weevil (1) を見よ
apple leafhopper (1)		*Edwardsiana australis* (Froggatt)	(カメムシ目、オオヨコバイ科) 豪州区
apple leafhopper (2)	リンゴマダラヨコバイ	*Orientus ishidae* (Matsumura)	(カメムシ目、オオヨコバイ科) 日本、旧北区。リンゴ害虫
apple leafhopper (3)		*Empoasca maligna* (Walsh)	(カメムシ目、オオヨコバイ科) 新北区。リンゴ他害虫
apple leafhopper (4)		*Typhlocyba froggatti* Baker	(カメムシ目、オオヨコバイ科) 旧北区、豪州区
apple leafhopper (5)		*Edwardsiana crataegi* (Douglas)	(カメムシ目、オオヨコバイ科) 豪州区
apple leafhopper			potato leafhopper (1) を見よ
apple leafmining casebearer			cherry casebearer を見よ
apple leafroller	アミメキイロハマキ	*Ptycholoma imitator* (Walsingham)	(チョウ目、ハマキガ科) 日本、旧北区。イチゴ、果樹類の害虫
apple leafroller (1)	カクモンハマキ	*Archips xylosteanus* (Linnaeus)	(チョウ目、ハマキガ科) 日本、旧北区。果樹類の害虫
apple longicorn beetle			apple stem borer を見よ
apple looper			feathered moth を見よ
apple lyonetid			Northants bentwing を見よ
apple maggot			apple fruit fly を見よ
apple maggot fly			apple fruit fly を見よ
apple marmorated leafhopper			apple leafhopper (2) を見よ
apple mealybug		*Phenacoccus aceris* (Signoret)	(カメムシ目、コナカイガラムシ科) 全北区。北米での英名
apple mealybug			pear mealybug を見よ
apple midge			apple leaf midge を見よ
apple minute bark beetle	リンゴノコキクイムシ	*Cryphalus malus* Niijima	(コウチュウ目、キクイムシ科) 日本、旧北区
apple minute weevil		*Rhamphus pullus* Hustache	(コウチュウ目、ゾウムシ科) 旧北区
apple moth			apple fruit moth を見よ
apple moth			apple ermine moth を見よ
apple mussel scale			oystershell scale (1) を見よ
apple narrow longicorn	ホソツツリンゴカミキリ	*Oberea nigriventris* Bates	(コウチュウ目、カミキリムシ科) 日本
apple oystershell scale			oystershell scale (1) を見よ
apple peel tortricid			summer fruit tortrix (2) を見よ
apple pigmy			apple pygmy を見よ

英名	和名	学名	所属、分布、ほか
apple pistol casebearer	リンゴピストルミノガ	*Coleophora ringoniella* Oku	（チョウ目、ツツミノガ科）日本。リンゴ害虫
apple pistol casebearer (1)		*Coleophora malivorella* Riley	（チョウ目、ツツミノガ科）新北区
apple pith moth			hawthorn cosmet を見よ
apple pith moth			apple black cosmet を見よ
apple psylla			apple sucker を見よ
apple psyllid			apple sucker を見よ
apple pygmy		*Nepticula malella* (Stainton)	（チョウ目、モグリチビガ科）旧北区。apple pygmy moth ともいう
apple red bug		*Lygidea mendax* Reuter	（カメムシ目、カスミカメムシ科）新北区。リンゴ他害虫
apple root aphid			woolly apple aphid を見よ
apple root borer		*Dorysthenes hugelii* Redtenbach	（コウチュウ目、カミキリムシ科）東洋区
apple root weevils		*Perperus*	（コウチュウ目、ゾウムシ科）の昆虫の総称
apple roselia	リンゴコブガ	*Mimerastria mandschuriana* (Oberthür)	（チョウ目、コブガ科）日本、旧北区
apple round bark beetle			apple ambrosia beetle (1) を見よ
apple seed chalcid (1)		*Torymus varians* (Walker)	（ハチ目、オナガコバチ科）新北区
apple seed chalcid (2)		*Syntomaspis druparum* Boheman	（ハチ目、オナガコバチ科）旧北区
apple-seed chalcis			apple seed chalcid (1)(2) を見よ
apple shoot borer		*Apriona cinerea* Chevrolet	（コウチュウ目、カミキリムシ科）東洋区
apple shot hole borer		*Scolytoplatypus raja* Blandford	（コウチュウ目、キクイムシ科）東洋区
apple skeletonizer			apple leaf skeletonizer (1) を見よ
apple-skin miner			apple fruitminer を見よ
apple-skin worm		*Tortrix franciscana* Walsingham	（チョウ目、ハマキガ科）新北区
apple skinworm			orange tortrix (1) を見よ
apple sphinx		*Sphinx gordius* (Cramer)	（チョウ目、スズメガ科）新北区。apple sphinx moth ともいう
apple stem borer	リンゴカミキリ	*Oberea japonica* (Thunberg)	（コウチュウ目、カミキリムシ科）日本、旧北区。リンゴ害虫
apple-stem piercer			pear weevil を見よ
apple sucker	リンゴキジラミ	*Psylla mali* (Schmidberger)	（カメムシ目、キジラミ科）日本、全北区。リンゴ害虫
apple surface-eating tortricid			great brown twist を見よ
apple tent caterpilar			American tent caterpillar を見よ
apple thrips			blossom thrips (1) を見よ
apple tortoise beetle	セモンジンガサハムシ	*Cassida versicolor* (Boheman)	（コウチュウ目、ハムシ科）日本、旧北区、東洋区
apple tortoise beetle		*Metriona thais* Boheman	（コウチュウ目、ハムシ科）東洋区

英名	和名	学名	所属、分布、ほか
apple tortrix	ミダレカクモンハマキ	*Archips fuscocupreanus* Walsingham	(チョウ目、ハマキガ科) 日本、旧北区。果樹類の害虫
apple-tree aphid			apple aphid (1) を見よ
apple tree aphis			apple aphid (1) を見よ
apple tree borer			flatheaded apple tree borer を見よ
apple tree web-moth		*Hyponomeuta malinellus* Zeller	(チョウ目、スガ科) 新北区
apple-trunk borer			quince borer を見よ
apple twig beetle			apple twig borer (1) を見よ
apple twig borer (1)		*Hypothenemus obscurus* (Fabricius)	(コウチュウ目、キクイムシ科) 新北区
apple twig borer (2)		*Amphicerus bicaudatus* (Say)	(コウチュウ目、ナガシンクイムシ科) 新北区。リンゴ他害虫
apple twig cutter		*Rhynchites coeruleus* (De Geer)	(コウチュウ目、オトシブミ科) 旧北区
apple-valerian aphid		*Dysaphis brancoi* (Börner)	(カメムシ目、アブラムシ科) 新北区
apple web-spinning sawfly		*Neurotoma nemoralis* Linnaeus	(ハチ目、ヒラタハバチ科) 旧北区
apple weevil	リンゴアナアキゾウ	*Dyscerus shikokuensis* (Kono)	(コウチュウ目、ゾウムシ科) 日本
apple weevil		*Hylobius freyi* Zumpt	(コウチュウ目、ゾウムシ科) 日本
apple weevil (1)		*Otiorhynchus cribricollis* Gyllenhal	(コウチュウ目、ゾウムシ科) 豪州区
apple worm			codling moth を見よ 米国での幼虫の英名
apricot and vine moth			narrow red-barred twist を見よ
apricot borer			cherry tree borer (2) を見よ
apricot crescent		*Tegosa claudina* Escholtz	(チョウ目、タテハチョウ科) 新熱帯区
apricot leaf caterpillar		*Parasa repunda* Hampson	(チョウ目、イラガ科) 旧北区
apricot moth			narrow red-barred twist を見よ
apricot noctuid		*Cosmia subtilis* Staudinger	(チョウ目、ヤガ科) 旧北区
apricot playboy		*Deudorix dinochares* Grose-Smith	(チョウ目、シジミチョウ科) エチオピア区
apricot scale			European fruit lecanium を見よ
apricot scale			Italian pear scale を見よ
apricot sulphur		*Phoebis argante* (Fabricius)	(チョウ目、シロチョウ科) 新北区、新熱帯区
apricot weevil		*Rhynchites aurautus* Scopoli	(コウチュウ目、オトシブミ科) 旧北区
April fritillary			pearl-bordered fritillary を見よ
aproned Cenopis moth		*Cenopis niveana* (Walsingham)	(チョウ目、ハマキガ科) 新北区
apterous earwig			short-winged earwig (1) (2) を見よ
apterous gall wasp	ハネナシタマバチ	*Callirhytis tobiiro* Ashmead	(ハチ目、タマバチ科) 日本
apterous insects			apterygote insects を見よ

英名	和名	学名	所属、分布、ほか
apterygote insects	無翅亜綱	Apterygota	の昆虫の総称
Apuleja mountain satyr		Eretris apuleja Felder	(チョウ目、タテハチョウ科) 新熱帯区
Apulia mountain satyr		Lymanopoda apulia Hopffer	(チョウ目、タテハチョウ科) 新熱帯区
aquamarine hairstreak		Oenomaus ortygnus (Cramer)	(チョウ目、シジミチョウ科) 新熱帯区
aquatic leaf beetles		Donacia	(コウチュウ目、ハムシ科) の昆虫の総称
Arab leopard			tawny silverline を見よ
Arab mantis			devil's flower mantis (1) を見よ
Arabian grizzled skipper		Spialia mangana (Rebel)	(チョウ目、セセリチョウ科) エチオピア区
Arabian jasmine pyralid		Hendecasis duplifascialis Hampson	(チョウ目、メイガ科) 東洋区
Arabian mantis		Eremiaphila arabica Saussure	(カマキリ目、Eremiaphilidae) 旧北区、エチオピア区
arabis midge		Dasineura alpestris (Kieffer)	(ハエ目、タマバエ科) 旧北区
Arabs			orange tips を見よ
Arachne checkerspot		Poladryas arachne (Edwards)	(チョウ目、タテハチョウ科) 新北区
araeopine planthoppers			(カメムシ目、ウンカ科) ウンカ科の異名 Araeopidae に由来
Arakawa oak aphid	クロトゲマダラアブラムシ	Tuberculatus stigmatis (Matsumura)	(カメムシ目、アブラムシ科) 日本、旧北区、東洋区
aralia leafroller	ウドノメイガ	Udonomeiga vicinalis (South)	(チョウ目、メイガ科) 日本、東洋区。ウド、タラノキを加害
aralia shoot borer moth		Papaipema araliae Bird et Jones	(チョウ目、ヤガ科) 新北区
Aranda copper		Aloeides aranda (Wallengren)	(チョウ目、シジミチョウ科) エチオピア区
araucaria aphid		Neophyllaphis araucariae Takahashi	(カメムシ目、アブラムシ科) 新北区
araucaria mealybug		Eriococcus araucariae Maskell	(カメムシ目、フクロカイガラムシ科) 大洋区
araucaria scale		Rhizococcus araucariae Maskell	(カメムシ目、コナカイガラムシ科) 旧北区
Araxes skipper		Apyrrothrix araxes (Hewitson)	(チョウ目、セセリチョウ科) 新北区
arboreal ant		Azteca alfari Emery	(ハチ目、アリ科) 新熱帯区
arboreal ant			territoreal ant を見よ
arborvitae aphid		Cinara thujaphilina (del Guercio)	(カメムシ目、アブラムシ科) 全北区、新熱帯区、豪州区
arborvitae bark beetle		Phloeosinus thujae (Perris)	(コウチュウ目、キクイムシ科) 旧北区
arborvitae leaf miner		Argyresthia thuiella (Packard)	(チョウ目、メムシガ科) 全北区。arborvitae leafminer moth ともいう
arborvitae weevil	ヒラズネヒゲボソゾウムシ	Phyllobius intrusus Kono	(コウチュウ目、ゾウムシ科) 日本、新北区

英名	和名	学名	所属、分布、ほか
archaic bell moths		Neopseustidae	(チョウ目) の昆虫の総称
archaic moths			mandibulate moths を見よ
archaic sun moths		Acanthopteroctetidae	(チョウ目) の昆虫の総称
archduke	ヤマオオイナズマ	*Lexias dirtea* (Fabricius)	(チョウ目、タテハチョウ科) 東洋区
archduke		*Lexias pardalis dirteana* (Corbet)	(チョウ目、タテハチョウ科) 東洋区
archdukes	オオイナズマ属	*Lexias*	(チョウ目、タテハチョウ科) の昆虫の総称
arched hooktip		*Drepana arcuata* Walker	(チョウ目、カギバガ科) 新北区。arched hooktip moth ともいう
arched marble		*Olethreutes arcuellus* (Clerck)	(チョウ目、ハマキガ科) 旧北区
Archer's dart		*Agrotis vestigialis* (Hufnagel)	(チョウ目、ヤガ科) 旧北区。Archer dart ともいう
Archer's dart moth			Archer's dart を見よ
arches moth		*Lacinipolia quadrilineata* (Grote)	(チョウ目、ヤガ科) 新北区
arctic alpine	カンボウベニヒカゲ	*Erebia rossi* Curtis	(チョウ目、タテハチョウ科) 全北区
arctic Argus		*Paroeneis sikkimensis* (Staudinger)	(チョウ目、タテハチョウ科) 東洋区
arctic blue		*Agriades glandon* (de Prunner)	(チョウ目、シジミチョウ科) 新北区。A. g. aquilo (Boisduval) も同英名
arctic blue			high mountain blue を見よ
arctic bluet		*Coenagrion johanssoni* (Wallengren)	(トンボ目、イトトンボ科) 旧北区
arctic fritillary	ホッキョクヒョウモン	*Clossiana chariclea* (Schneider)	(チョウ目、ドクチョウ科) 全北区。新北区の *C. c. arctica* (Zetterstedt) も同英名
arctic grayling	ホッキョクタカネヒカゲ	*Oeneis bore* (Schneider)	(チョウ目、タテハチョウ科) 全北区
arctic green sulphur	ブースモンキチョウ	*Colias boothi* Curtis	(チョウ目、シロチョウ科) 新北区
arctic ridge fritillary			Astarte fritillary を見よ
arctic ringlet			spruce bog alpine を見よ
arctic skipper		*Carterocephalus mandan* (Edwards)	(チョウ目、セセリチョウ科) 新北区
arctic skipper			chequered skipper を見よ
arctic skipperling			chequered skipper を見よ
arctic sulphur			Moorland clouded yellow を見よ
arctic sulphur			Labrador sulphur を見よ
arctic white		*Pieris angelika* Eitschberger	(チョウ目、シロチョウ科) 新北区
arctic woodland ringlet	ホクオウベニヒカゲ	*Erebia polaris* Staudinger	(チョウ目、タテハチョウ科) 旧北区
arctics	タカネヒカゲ属	*Oeneis*	(チョウ目、タテハチョウ科) の昆虫の総称
arctics			satyrs を見よ
arctiids			tiger moths (1) を見よ

英名	和名	学名	所属、分布、ほか
Ardent crescent		*Phyciodes ardys* (Hewitson)	（チョウ目、タテハチョウ科）新北区、新熱帯区
ardisia ensign scale			bladhia scale を見よ
Ardys crescent		*Anthanassa ardys ardys* (Hewitson)	（チョウ目、タテハチョウ科）新熱帯区。*A. a. subota* Godman et Salvin も同英名
areca nut weevil			coffee weevil を見よ
Arela dart		*Semalea arela* (Mabille)	（チョウ目、セセリチョウ科）エチオピア区
Arela skipper			Arela dart を見よ
arenate syrphid		*Syrphus arenatus* (Fallén)	（ハエ目、ハナアブ科）新北区
Arene sailor		*Dynamine arene* Hübner	（チョウ目、タテハチョウ科）新熱帯区
Ares Emesis			Ares metalmark を見よ
Ares metalmark		*Emesis ares* (Edwards)	（チョウ目、シジミタテハ科）新北区
Argante giant sulphur			apricot sulphur を見よ
arge moth		*Apantesis arge* Hampson	（チョウ目、ヒトリガ科）新北区
argent and sable		*Rheumaptera hastata* (Linnaeus)	（チョウ目、シャクガ科）旧北区。日本亜種は *R. h. rikovskensis* (Matsumura) オオシロオビクロナミシャク
argent and sable dwarf		*Elachista canapennella* (Hübner)	（チョウ目、クサモグリガ科）旧北区
Argentine ant	アルゼンチンアリ	*Iridomyrmex humilis* (Mayr)	（ハチ目、アリ科）日本、新北区、新熱帯区、豪州区
Argentine bagworm		*Oiketicus platensis* Berg	（チョウ目、ミノガ科）新熱帯区
Argentine bagworm (1)		*Oiketicus kirbyi* Guilding	（チョウ目、ミノガ科）新熱帯区。カンキツ害虫
Argentine moth borer			cactus moth を見よ
Argentine rice water weevil		*Lissorhoptrus bosqi* Kuschel	（コウチュウ目、ゾウムシ科）新熱帯区
Argentine stem weevil		*Listronotus bonariensis* (Kuschel)	（コウチュウ目、ゾウムシ科）豪州区
Argentine tree cricket			prairie tree cricket を見よ　Argentina tree cricket とも記す
Argentine white-crested mantis		*Stagmatoptera hyaloptera* (Perty)	（カマキリ目、カマキリ科）新熱帯区
Argentinian wood cockroach			South American cockroach を見よ
argents			shiny head-standing moths を見よ
argid flies			argid sawflies を見よ
argid sawflies	ミフシハバチ科	Argidae	（ハチ目）の昆虫の総称
argids			argid sawflies を見よ
Argon skipper		*Argon lota* (Hewtitson)	（チョウ目、セセリチョウ科）新熱帯区
Argus		*Albulina*	（チョウ目、シジミチョウ科）の昆虫の総称
Argus		*Aricia*	（チョウ目、シジミチョウ科）の昆虫の総称
Argus		*Pseudaricia*	（チョウ目、シジミチョウ科）の昆虫の総称
argus tortoise beetle		*Chelymorpha cassidea* (Fabricius)	（コウチュウ目、ハムシ科）新北区。サツマイモ害虫

英名	和名	学名	所属、分布、ほか
argyresthiid moths			shiny head-standing moths を見よ
argyresthiids			shiny head-standing moths を見よ
Argyrodines Calephelis		*Calephelis argyrodines* (H. Bates)	（チョウ目、シジミタテハ科）新熱帯区
arid bluet		*Africallagma sinuatum* (Ris)	（トンボ目、イトトンボ科）エチオピア区
arid bronze azure		*Ogyris subterrestris* Field	（チョウ目、シジミチョウ科）豪州区
arid copper		*Aloeides arida* Tite et Dickson	（チョウ目、シジミチョウ科）エチオピア区
arid Eudesmia moth		*Eudesmia arida* (Skinner)	（チョウ目、ヒトリガ科）新北区、新熱帯区
arid-land katydids		*Neobarrettia*	（バッタ目、キリギリス科）の昆虫の総称
arid-land subterranean termite		*Reticulitermes tibialis* Banks	（シロアリ目、ミゾガシラシロアリ科）新北区
arid lands honey ants			honey ants を見よ
Aristotele's skipper		*Sophista aristoteles* (Westwood)	（チョウ目、セセリチョウ科）新熱帯区
Aristotle's duke		*Siseme aristoteles* (Latreille)	（チョウ目、シジミタテハ科）新熱帯区
Arita skipper		*Arita arita* (Schaus)	（チョウ目、セセリチョウ科）新熱帯区
arixenids			earwigs を見よ　ハサミムシ目の Arixeniidae に由来
Arizona Araxes skipper		*Apyrrothrix araxes arizonae* (Godman et Salvin)	（チョウ目、セセリチョウ科）新北区、新熱帯区
Arizona bird-dropping moth		*Ponometia elegantula* (Harvey)	（チョウ目、ヤガ科）新北区
Arizona blister beetle			desert blister beetle (1) を見よ
Arizona Calephelis			Arizona metalmark を見よ
Arizona checkerspot		*Texola perse* (Edwards)	（チョウ目、タテハチョウ科）新熱帯区
Arizona clearwing moth		*Carmenta auritincta* (Engelhardt)	（チョウ目、スカシバガ科）新北区
Arizona cotton stainer		*Dysdercus mimulus* Hussey	（カメムシ目、ホシカメムシ科）新北区
Arizona cricket		*Gryllita arizonae* Hebard	（バッタ目、コオロギ科）新北区
Arizona desert miller		*Conochares arizonae* H. Edwards	（チョウ目、ヤガ科）新北区
Arizona duskywing		*Erynnis juvenalis clitus* (W. H. Edwards)	（チョウ目、セセリチョウ科）新北区、新熱帯区
Arizona eyed click beetle		*Alaus lusciosus* Hope	（コウチュウ目、コメツキムシ科）新北区
Arizona giant-skipper		*Agathymus aryxna* (Dyar)	（チョウ目、セセリチョウ科）新北区、新熱帯区
Arizona gold beetle		*Physonota arizonae* Schaeffer	（コウチュウ目、ハムシ科）新北区
Arizona golden beetle			Arizona gold beetle を見よ
Arizona hairstreak		*Erora quaderna* (Hewitson)	（チョウ目、シジミチョウ科）新北区

英名	和名	学名	所属、分布、ほか
Arizona longwing		*Capnobotes arizonensis* Rehn	（バッタ目、キリギリス科）新北区
Arizona mantis			Arizona mantis, Arizona unicorn mantis, Arizona tan mantis の総称
Arizona mantis		*Stagmomantis limbata* Hahn	（カマキリ目、カマキリ科）新北区。bordered mantis ともいう
Arizona meadow katydid		*Orchelimum unispina* (Saussure et Pictet)	（バッタ目、キリギリス科）新北区
Arizona metalmark		*Calephelis arizonensis* McAlpine	（チョウ目、シジミタテハ科）新北区
Arizona mottled-skipper			Arizona skipper を見よ
Arizona navel orangeworm			navel orangeworm (1) を見よ
Arizona net-winged beetle		*Lycus arizonensis* Green	（コウチュウ目、ベニボタル科）新北区。多くの擬態者のモデル
Arizona pine beetle			southern pine beetle (1) を見よ
Arizona pine-satyr			pine satyr を見よ
Arizona powdered skipper		*Systasea zampa* (Edwards)	（チョウ目、セセリチョウ科）新北区、新熱帯区
Arizona red-spotted purple		*Limenitis arthemis arizonensis* Edwards	（チョウ目、タテハチョウ科）新熱帯区
Arizona silver-spotted skipper		*Epargyreus clarus huachuca* Dixon	（チョウ目、セセリチョウ科）新熱帯区
Arizona sister	アリゾナイチモンジ	*Adelpha eulalia* (Doubleday)	（チョウ目、タテハチョウ科）新熱帯区
Arizona skipper		*Codatractus arizonensis* (Skinner)	（チョウ目、セセリチョウ科）新北区
Arizona snaketail		*Ophiogomphus arizonicus* Kennedy	（トンボ目、サナエトンボ科）新北区
Arizona tan mantis		*Stagmomantis gracilipes* Rehn	（カマキリ目、カマキリ科）新北区
Arizona unicorn mantis		*Pseudovates arizonae* Hebard	（カマキリ目、カマキリ科）新北区、新熱帯区
Arizona walkingstick		*Diapheromera arizonensis* Caudell	（ナナフシ目、Bacunculidae）新北区
Arkansas roadside skipper		*Amblyscirtes linda* Freeman	（チョウ目、セセリチョウ科）新北区
Armand's sailer	ヒメアサクラミスジ	*Neptis armandia* Oberthür	（チョウ目、タテハチョウ科）東洋区
armed horse bot fly			horse bot fly を見よ
armed springtail		*Xenyllodes armatus* (Axelson)	（トビムシ目、イボトビムシ科）日本、全北区
armed springtail			mushroom springtail (1) を見よ
armored lice			small-mammal sucking lice を見よ
armored mayflies		Baetiscidae	（カゲロウ目）の昆虫の総称
armored scale		*Aspidiotus acutiformis* Cockerell	（カメムシ目、マルカイガラムシ科）旧北区。カンキツ害虫
armored scale		*Lepidosaphes lasianthi* (Green)	（カメムシ目、マルカイガラムシ科）新熱帯区。カンキツ害虫

英名	和名	学名	所属、分布、ほか
armored scale			Ross's black scale (1) を見よ
armored scales	マルカイガラムシ科	Diaspididae	(カメムシ目) の昆虫の総称　カイガラムシ類で最大の科。世界に1500種
armored stink beetle		*Eleodes armata* LeConte	(コウチュウ目、ゴミムシダマシ科) 新北区
armoured crickets		Bradyporidae	(バッタ目) の昆虫の総称
armoured scales			armored scales を見よ
armourtail		*Armagomphus armiger* (Tillyard)	(トンボ目、サナエトンボ科) 豪州区
army ant		*Anomma nigricans* Illiger	(ハチ目、アリ科) エチオピア区
army ants	ヒメサスライアリ亜科	Aenictinae	(ハチ目、アリ科) の昆虫の総称
army ants		*Onychomyrmex*	(ハチ目、アリ科) の昆虫の総称
army ants		*Aomma*	(ハチ目、アリ科) の昆虫の総称
army ants	グンタイアリ亜科	Ectioninae	(ハチ目、アリ科) の昆虫の総称　サスライアリともいう。*Ection* 属はグンタイアリ属といわれる
army ants (1)	サスライアリ亜科	Dorylinae	(ハチ目、アリ科) の昆虫の総称
army ants			safari ants を見よ
army cutworm		*Euxoa auxiliaris* (Grote)	(チョウ目、ヤガ科) 新北区。トウモロコシ、アルファルファ害虫。成虫は army cutworm moth
army cutworm			western armyworm を見よ
army leaf beetle		*Dactylispa manteroi* Gestro	(コウチュウ目、ハムシ科) 旧北区
army weevil			rice hispa を見よ
armyworm (1)		*Mythimna unipuncta* (Haworth)	(チョウ目、ヤガ科) 全北区、エチオピア区。成虫は armyworm moth
armyworm (2)	アワヨトウ	*Mythimna separata* (Walker)	(チョウ目、ヤガ科) 日本、旧北区、東洋区、豪州区。イネ、アワなどの著名害虫
army worm (3)		*Lycoria militaris* Nowicki	(ハエ目、クロバネキノコバエ科) 旧北区
armyworm			beet armyworm を見よ
armyworm egg parasite		*Telenomus nawai* Ashmead	(ハチ目、タマゴクロバチ科) 大洋区
armyworm moth			armyworm (1) を見よ
armyworm moths	カラスヨトウ亜科	Amphipyrinae	(チョウ目、ヤガ科) の昆虫の総称
armyworms			owlet moths を見よ　幼虫の英名
Arnaca blue ringlet		*Chloreuptychia arnaca* Fabricius	(チョウ目、タテハチョウ科) 新熱帯区
arnly			dobsonflies (1) を見よ　幼虫の英名
Arogos skipper			beard-grass skipper を見よ
Arola ministreak		*Ministrymon arola* (Hewitson)	(チョウ目、シジミチョウ科) 新熱帯区
Arota		*Tharsalea arota* (Boisduval)	(チョウ目、シジミチョウ科) 新北区
Arran brown	クマモベニヒカゲ	*Erebia ligea takanonis* Matsumura	(チョウ目、タテハチョウ科) 日本。E. ligea (Linnaeus) 旧北区の英名
arran carpet		*Chloroclysta concinnata* (Stephens)	(チョウ目、シャクガ科) 旧北区

英名	和名	学名	所属、分布、ほか
arrindy silkworm		*Philosamia ricini* (Boisduval)	(チョウ目、ヤママユガ科) 旧北区、東洋区
arrow clubtail		*Stylurus spiniceps* (Walsh)	(トンボ目、サナエトンボ科) 新北区
arrow sphinx		*Lophostethus dumolinii* (Angas)	(チョウ目、スズメガ科) エチオピア区
arrowhead blue	ヤジリシジミ	*Glaucopsyche piasus* (Boisduval)	(チョウ目、シジミチョウ科) 新北区
arrowhead moth		*Gretchena deludana* (Clemens)	(チョウ目、ハマキガ科) 新北区
arrowhead scale	ヤノネカイガラムシ	*Unaspis yanonensis* (Kuwana)	(カメムシ目、マルカイガラムシ科) 日本、旧北区、東洋区。カンキツ害虫
arrow-head-shaped round scale	クロホシマルカイガラムシ	*Lindingaspis setiger* (Maskell)	(カメムシ目、マルカイガラムシ科) 日本
arrowhead skipper		*Teniorhinus harona* (Westwood)	(チョウ目、セセリチョウ科) エチオピア区
arrowhead spiketail		*Cordulegaster obliqua* (Say)	(トンボ目、オニヤンマ科) 新北区
arrowwood clearwing moth			lesser viburnum clearwing moth を見よ
Arroyo bluet		*Enallagma praevarum* (Hagen)	(トンボ目、イトトンボ科) 新北区
Arroyo darner		*Rhionaeschna dugesi* (Calvert)	(トンボ目、ヤンマ科) 新北区
Arsalte skipper		*Heliopetes arsalte* (Linnaeus)	(チョウ目、セセリチョウ科) 新北区、新熱帯区
artemisia gall midge	オトコヨモギヒメツボミタマバエ	*Rhopalomyia caterva* Monzen	(ハエ目、タマバエ科) 日本、旧北区
artemisia sailor		*Dynamine artemisia* (Fabricius)	(チョウ目、タテハチョウ科) 新熱帯区
artemisia swallowtail			common yellow swallowtail を見よ
Artena clearwing		*Pteronymia artena artena* (Hewitson)	(チョウ目、タテハチョウ科) 新熱帯区
artichoke aphid	ゴボウクギケアブラムシ	*Capitophorus elaeagni* (Del Guercio)	(カメムシ目、アブラムシ科) 日本、旧北区、豪州区。アザミ害虫
artichoke gall		*Andricus fecundator* (Hartig)	(ハチ目、タマバチ科) がカシ類につくる虫えいで、アザミのつぼみのような形。hop gall ともいわれる。旧北区
artichoke moth		*Gortyna xanthenes* Germar	(チョウ目、ヤガ科) 旧北区
artichoke moth			burdock leaf roller を見よ
artichoke pear-shaped weevil		*Apion carduorum* Kirby	(コウチュウ目、ホソクチゾウムシ科) 旧北区
artichoke plume moth		*Platyptilia carduidactyla* (Riley)	(チョウ目、トリバガ科) 新北区
artichoke root louse			artichoke tuber aphid を見よ
artichoke tortoise beetle		*Cassida deflorata* Suffrian	(コウチュウ目、ハムシ科) 旧北区
artichoke tuber aphid		*Protrama radicis* (Kaltenbach)	(カメムシ目、アブラムシ科) 旧北区

英名	和名	学名	所属、分布、ほか
artichoke tuber aphid			Jerusalem artichoke tuber aphid を見よ
artichoke weevil			sluggish weevil を見よ
artillery beetle			bombardier beetle (1) を見よ
artillery beetle			lesser bombardier beetle を見よ
artocarpus long-horned beetle		*Pterolophia bigibbera* (Newman)	（コウチュウ目、カミキリムシ科）大洋区
artocarpus scale			Florida red scale を見よ
arum aphid			mottled arum aphid を見よ
aruncus sawfly			spiraea sawfly を見よ
Aryxna giant skipper			Arizona giant-skipper を見よ
Asa hairstreak		*Hypostrymon asa* (Hewitson)	（チョウ目、シジミチョウ科）新熱帯区
ascalaphid		*Libelloides longicornis* (Linnaeus)	（アミメカゲロウ目、ツノトンボ科）旧北区
ascalaphid owlfly		*Libelloides macaronius* (Scopoli)	（アミメカゲロウ目、ツノトンボ科）旧北区
ascalaphids			owlflies を見よ
ascalaphus flies			owlflies を見よ
ascalaphus fly			ascalaphid owlfly を見よ
asclepias reddish longicorn	トウワタベニカミキリ	*Tetraopes tetrophthalmus* Forster	（コウチュウ目、カミキリムシ科）新北区
Asella sphinx		*Sphinx asella* (Rothschild et Jordan)	（チョウ目、スズメガ科）新北区
ash bark beetle		*Hylastinus fraxini* (Panzer)	（コウチュウ目、キクイムシ科）旧北区
ash bark beetle (1)		*Leperisinus varius* Fabricius	（コウチュウ目、キクイムシ科）旧北区
ash borer		*Podosesia syringae fraxini* (Lugger)	（チョウ目、スカシバガ科）新北区。ash トネリコ、リラ につく。ash borer moth ともいう
ash bud moth		*Prays fraxinella* (Bjerkander)	（チョウ目、スガ科）旧北区
ash bullet gall midge		*Dasineura pellex* (Osten Sacken)	（ハエ目、タマバエ科）新北区
ash-coloured sober		*Acompsia cinerella* (Clerck)	（チョウ目、キバガ科）旧北区
ash-gray blister beetle		*Epicauta fabricii* (LeConte)	（コウチュウ目、ツチハンミョウ科）新北区。ジャガイモ害虫
ash gray ladybird			ashy grey lady beetle を見よ
ash-gray ladybird beetle		*Olla sayi* (Crotch)	（コウチュウ目、テントウムシ科）新北区
ash-gray ladybird beetle			ashy grey lady beetle を見よ
ash-gray leaf bugs	チビカメムシ科	Piesmatidae	（カメムシ目）の昆虫の総称
ash leaf cone roller moth			privet leaf miner を見よ
ash leaf gall sucker			ash sucker を見よ
ash leaf sucker			ash sucker を見よ

英名	和名	学名	所属、分布、ほか
ash leaf weevil		*Stereonychus fraxini* (De Geer)	（コウチュウ目、ゾウムシ科）旧北区
ash meadows bug		*Ambrysus amargosus* La Rivers	（カメムシ目、コバンムシ科）新北区。絶滅危惧種
ash midrib gall midge		*Contarinia canadensis* Felt	（ハエ目、タマバエ科）新北区
ash mid-rib pouch-gall midge		*Dasineura fraxini* (Bremi)	（ハエ目、タマバエ科）旧北区
ash plant bug		*Tropidosteptes amoenus* Reuter	（カメムシ目、カスミカメムシ科）新北区。ash トネリコ 害虫
ash psyllid			ash sucker を見よ　豪州での英名
ash pug		*Eupithecia fraxinata* Crewe	（チョウ目、シャクガ科）旧北区
ash scale		*Pseudochermes fraxini* (Kaltenbach)	（カメムシ目、フクロカイガラムシ科）旧北区
ash shoulder-knot		*Scotochrosta pulla* Denis et Schiffermüller	（チョウ目、ヤガ科）旧北区
ash sphinx		*Manduca jasminearum* (Guérin)	（チョウ目、スズメガ科）新北区。ash sphinx moth ともいう
ash sucker		*Psyllopsis fraxini* (Linnaeus)	（カメムシ目、キジラミ科）全北区
ash tip borer moth		*Papaipema furcata* (Smith)	（チョウ目、ヤガ科）新北区
ash-tree aphid		*Prociphilus fraxini* Fabricius	（カメムシ目、アブラムシ科）新北区
ash underwing			clifden nonpareil を見よ
ash whitefly	ザクロシロトゲコナジラミ	*Siphoninus phillyreae* (Haliday)	（カメムシ目、コナジラミ科）日本
ashen pinion moth			green fruitworm (1) を見よ
ashen smoky blue		*Euchrysops subpallida* Bethune-Baker	（チョウ目、シジミチョウ科）エチオピア区
Ashworth's rustic		*Xestia ashworthii* (Doubleday)	（チョウ目、ヤガ科）旧北区
ashy button		*Acleris sparsana* (Denis et Schiffermüller)	（チョウ目、ハマキガ科）旧北区
ashy clubtail		*Gomphus lividus* Selys	（トンボ目、サナエトンボ科）新北区
ashy gray lady beetle		*Olla v-nigrum* (Mulsant)	（コウチュウ目、テントウムシ科）新北区、新熱帯区
ashy gray ladybird			ashy grey lady beetle を見よ
ashy gray ladybird beetle			ashy grey lady beetle を見よ
ashy grey lady beetle		*Olla abdominalis* (Say)	（コウチュウ目、テントウムシ科）新北区
ashy Meganola moth		*Meganola spodia* Franclemont	（チョウ目、コブガ科）新北区
ashy Pleromelloida moth		*Pleromelloida cinerea* (Smith)	（チョウ目、ヤガ科）新北区
Asian admiral			Indian painted lady を見よ　米国での英名
Asian cabbage white			Indian cabbage white を見よ

英名	和名	学名	所属、分布、ほか
Asian citrus psyllid			citrus psylla (1) を見よ
Asian cockroach	オキナワチャバネゴキブリ	*Blattella asahinai* Mizukubo	(ゴキブリ目、チャバネゴキブリ科) 日本、新北区
Asian comma	キタテハ	*Polygonia c-aureum* (Linnaeus)	(チョウ目、タテハチョウ科) 日本、旧北区
Asian corn borer			Oriental corn borer を見よ
Asian cotton leafworm			common cutworm を見よ
Asian cycad scale		*Aulacaspis yasumatsui* Takagi	(カメムシ目、マルカイガラムシ科) 東洋区、新北区
Asian dwarf honey bee			lesser honey bee を見よ
Asian fritillary	カラフトヒョウモンモドキ	*Hypodryas intermedia* (Ménétriès)	(チョウ目、タテハチョウ科) 旧北区
Asian fruit fly			cherry drosophila を見よ
Asian giant hornet			giant hornet を見よ
Asian groundling			ditch jewel を見よ
Asian gypsy moth			gypsy moth を見よ
Asian honey bee			Indian honey bee を見よ
Asian horntail	タイワンヒラアシキバチ	*Eriotremex formosanus* (Matsumura)	(ハチ目、キバチ科) 日本、東洋区、新北区
Asian long-horned beetle		*Anoplophora glabripennis* (Motschulsky)	(コウチュウ目、カミキリムシ科) 日本、全北区
Asian multicolored lady beetle		*Harmonia axyridis* (Pallas)	(コウチュウ目、テントウムシ科) 全北区、大洋区。multicolored Asian lady beetle、Asian multicolored ladybug ともいう
Asian palm weevil			Asiatic palm weevil を見よ
Asian pintail			pintail を見よ
Asian predatory wasp		*Vespa velutina* Lepeletier	(ハチ目、スズメバチ科) 東洋区。Asian hornet ともいう
Asian rice bug			rice bug (2) を見よ
Asian rice gall midge			rice stem gall midge を見よ
Asian rock pool mosquito		*Ochlerotatus japonicus* (Theobald)	(ハエ目、カ科) 日本、旧北区
Asian sapphires			coppers (2) を見よ
Asian skimmer		*Orchetrum glaucum* (Brauer)	(トンボ目、トンボ科) 東洋区
Asian slim		*Aciagrion occidentale* Laidlaw	(トンボ目、イトトンボ科) 東洋区
Asian swallowtail			citrus swallowtail (1) を見よ　Asia swallowtail も使われる
Asian tiger mosquito			tiger mosquito を見よ
Asian widow		*Palpopleura sexmaculata sexmaculata* (Fabricius)	(トンボ目、トンボ科) 東洋区
Asiatic beetle	セマダラコガネムシ	*Anomala orientalis* Waterhouse	(コウチュウ目、コガネムシ科) 日本、東洋区、新北区
Asiatic cockroach			common cockroach を見よ

英名	和名	学名	所属、分布、ほか
Asiatic garden beetle	アカビロウドコガネ	*Maladera castanea* (Arrow)	（コウチュウ目、コガネムシ科）日本、新北区。野菜害虫
Asiatic ladybird		*Chilocorus similis* (Rossi)	（コウチュウ目、テントウムシ科）旧北区
Asiatic leafroller	ホソアトキハマキ	*Archips breviplicanus* Walsingham	（チョウ目、ハマキガ科）日本、旧北区
Asiatic locust			Asiatic migratory locust を見よ
Asiatic migratory locust	トノサマバッタ	*Locusta migratoria migratoria* (Linnaeus)	（バッタ目、バッタ科）日本、旧北区、東洋区。大群で移動する飛蝗は本種の群生相
Asiatic oak weevil	クリイロクチブトゾウムシ	*Cyrtepistomus castaneus* (Roelofs)	（コウチュウ目、ゾウムシ科）日本、新北区。米国での英名
Asiatic palm weevil	ヤシオオオサゾウムシ	*Rhynchophorus ferrugineus* (Olivier)	（コウチュウ目、オサゾウムシ科）日本、全北区、東洋区、エチオピア区。ヤシ害虫
Asiatic red palm weevil			Asiatic palm weevil を見よ
Asiatic red scale	マキアカマルカイガラムシ	*Aonidiella taxus* Leonardi	（カメムシ目、マルカイガラムシ科）日本、全北区
Asiatic rhinoceros beetle			rhinoceros beetle (1) を見よ
Asiatic rice borer			rice stem borer を見よ
Asiatic rose scale			rose scale を見よ
Asimina webworm moth		*Omphalocera munroei* Martin	（チョウ目、メイガ科）新北区
Asine longtail		*Polythrix asine* (Hewitson)	（チョウ目、セセリチョウ科）新北区、新熱帯区
asparagus aphid		*Brachycorynella asparagi* (Mordvilko)	（カメムシ目、アブラムシ科）全北区
asparagus beetle		*Crioceris asparagi* (Linnaeus)	（コウチュウ目、ハムシ科）新北区
asparagus beetle			spotted asparagus beetle を見よ
asparagus beetles	クビボソハムシ亜科	Criocerinae	（コウチュウ目、ハムシ科）の昆虫の総称
asparagus borer		*Parahypopta caestrum* Reutti	（チョウ目、ボクトウガ科）旧北区
asparagus caterpillar			beet armyworm を見よ
asparagus fern caterpillar			beet armyworm を見よ
asparagus fly		*Platypara poeciloptera* (Schrank)	（ハエ目、ミバエ科）旧北区。幼虫は asparagus maggot
asparagus miner		*Ophiomyia simplex* (Loew)	（ハエ目、ハモグリバエ科）全北区
asparagus stem miner	アスパラガスハモグリバエ	*Ophiomyia asparagi* Spencer	（ハエ目、ハモグリバエ科）日本。アスパラガス害虫
aspen blotch miner		*Phyllonorycter tremuloidiella* (Braun)	（チョウ目、ホソガ科）新北区
aspen blotch miner		*Phyllonorycter salicifoliella* Chambers	（チョウ目、ホソガ科）新北区
aspen burncow		*Poecilonota variolosa* (Paykull)	（コウチュウ目、タマムシ科）旧北区
aspen carpenterworm moth		*Acossus populi* (Walker)	（チョウ目、ボクトウガ科）新北区

英名	和名	学名	所属、分布、ほか
aspen dusky wing			dreamy duskywing を見よ
aspen gall aphid		*Pachypappa grandis* Tullgren	（カメムシ目、アブラムシ科）旧北区
aspen leaf beetle		*Chrysomela crotchi* Brown	（コウチュウ目、ハムシ科）新北区
aspen leaf-beetle			unspotted aspen leaf-beetle を見よ
aspen leaf blotch miner moth		*Phyllonorycter apparella* (Herrich-Schäffer)	（チョウ目、ホソガ科）全北区
aspen leaf-curler		*Pseudexentera aregonana* Walsingham	（チョウ目、ハマキガ科）新北区
aspen leaf-eating beetle			unspotted aspen leaf-beetle を見よ
aspen leaf gall midge		*Harmandia tremulae* (Winnertz)	（ハエ目、タマバエ科）旧北区
aspen leaf roller weevil			poplar leaf roller を見よ
aspen leafminer			poplar leafminer moth を見よ
aspen leaftier			poplar leafroller moth を見よ
aspen petiole gall midge			poplar gall midge を見よ
aspen petiole gall moth		*Ectoedemia populella* Busck	（チョウ目、モグリチビガ科）新北区
aspen serpentine leafminer moth		*Phyllocnistis populiella* Chambers	（チョウ目、コハモグリガ科）新北区
aspen twoleaf tier			pale Enargia moth を見よ
aspen webworm moth		*Pococera aplastella* (Hulst)	（チョウ目、メイガ科）新北区
Aspernatus skipper		*Thespieus aspernatus* Draudt	（チョウ目、セセリチョウ科）新熱帯区
aspidistra scale			fern scale (1) を見よ　南アフリカでの英名
Assam Cornelian		*Deudorix gaetulia* de Nicéville	（チョウ目、シジミチョウ科）東洋区
Assam darter		*Ochlodes siva* Moore	（チョウ目、セセリチョウ科）東洋区
Assam Faun		*Faunis assama* (Westwood)	（チョウ目、タテハチョウ科）東洋区
Assam flash		*Rapala tara* de Nicéville	（チョウ目、シジミチョウ科）東洋区
Assam lascar	アッサムキンミスジ	*Pantoporia assamica* (Moore)	（チョウ目、タテハチョウ科）東洋区
Assam nettle grub		*Thosea cervina* Moore	（チョウ目、イラガ科）東洋区
Assam pierrot		*Tarucus waterstradti dharta* Bethune-Baker	（チョウ目、シジミチョウ科）東洋区
assassin bug (1)		*Melanolestes picipes* (Herrich-Schäffer)	（カメムシ目、サシガメ科）新北区
assassin bug (2)		*Pristhesancus plagipennis* Walker	（カメムシ目、サシガメ科）豪州区
assassin bug			masked hunter bug を見よ　英名は捕食性に由来
assassin bugs		*Pselliopus*	（カメムシ目、サシガメ科）の昆虫の総称

英名	和名	学名	所属、分布、ほか
assassin bugs (1)	サシガメ科	Reduviidae	(カメムシ目) の昆虫の総称　捕食性。世界に5,000種以上
assassin flies			robber flies (1) を見よ
Assegai sprite		*Pseudagrion assegaii* Pinhey	(トンボ目、イトトンボ科) エチオピア区
assembly moth		*Samea ecclesialis* Guénee	(チョウ目、メイガ科) 新北区
Assmann's fritillary	コヒョウモンモドキ	*Melitaea britomartus niphona* (Butler)	(チョウ目、タテハチョウ科) 日本。*M. britomartus* Assmann の英名
Assyrian		*Terinos clarissa* Boisduval	(チョウ目、タテハチョウ科) 東洋区
Astala eighty-eight	ルリモンウラスジタテハ	*Diaethria astala astala* Guérin-Méneville	(チョウ目、タテハチョウ科) 新熱帯区。*D. a. asteroide* Maza et Maza も同英名。
astaline wasps		Astalinae	(ハチ目、アナバチ科) の昆虫の総称
Astarte eighty-eight	アオウズマキタテハ	*Callicore astarte* (Cramer)	(チョウ目、タテハチョウ科) 新熱帯区
Astarte fritillary		*Clossiana astarte* (Doubleday)	(チョウ目、タテハチョウ科) 新北区
asteiid flies		Asteiidae	(ハエ目) の昆虫の総称
aster borer moth		*Papaipema impecuniosa* (Grote)	(チョウ目、ヤガ科) 新北区
aster checkerspot		*Chlosyne hoffmanni* (Behr)	(チョウ目、タテハチョウ科) 新北区
aster-head Phaneta moth		*Phaneta tomonana* (Kearfott)	(チョウ目、ハマキガ科) 新北区
aster leafhopper		*Macrosteles quadrilineatus* Forbes	(カメムシ目、オオヨコバイ科) 新北区
aster leafhopper (1)	フタテンヨコバイ	*Macrosteles fascifrons* (Stål)	(カメムシ目、オオヨコバイ科) 日本、新北区
aster leafminer	ヨメナシロハモグリバエ	*Calycomyza humeralis* (von Roser)	(ハエ目、ハモグリバエ科) 日本、新北区
aster leafminer	ヨメナスジハモグリバエ	*Liriomyza asterivora* Sasakawa	(ハエ目、ハモグリバエ科) 日本
Asterodia sandman			star sandman を見よ
Asthene wave moth		*Pleuroprucha asthenaria* (Walker)	(チョウ目、シャクガ科) 新北区、新熱帯区
astronomer moth		*Olethreutes astrologana* (Zeller)	(チョウ目、ハマキガ科) 新北区
astur moth		*Carales astur* (Cramer)	(チョウ目、ヒトリガ科) 新熱帯区
Astyalus swallowtail			broad-banded swallowtail を見よ
Atala	クロウラホシシジミ	*Eumaeus atala* (Poey)	(チョウ目、シジミチョウ科) 新北区、新熱帯区。Atala butterfly、Atala moth ともいう
Atesa hairstreak		*Oenomaus atesa* (Hewitson)	(チョウ目、シジミチョウ科) 新熱帯区
atherbell			dragonflies (1) を見よ
atherbill			dragonflies (1) を見よ
ather-cap			dragonflies (1) を見よ
athericid flies		Athericidae	(ハエ目) の昆虫の総称　シギアブ科に入れることあり
Athis longwing		*Heliconius athis* Doubleday	(チョウ目、タテハチョウ科) 新熱帯区

英名	和名	学名	所属、分布、ほか
atis borer			bullocks heart fruit borer を見よ
atis moth borer			bullocks heart fruit borer を見よ
Atkinson's bob		*Arnetta atkinsoni* (Moore)	(チョウ目、セセリチョウ科) 東洋区
Atlantic azure		*Celastrina ladon idella* D. Wright et Pavulaan	(チョウ目、シジミチョウ科) 新北区。spring azure を参照
Atlantic bluet		*Enallagma doubledayi* (Selys)	(トンボ目、イトトンボ科) 新北区
Atlantis fritillary	アトグロギンボシヒョウモン	*Speyeria atlantis atlantis* (Edwards)	(チョウ目、シジミチョウ科) 新北区
Atlas beetle	オオカブトムシ	*Chalcosoma atlas* (Linnaeus)	(コウチュウ目、コガネムシ科) 東洋区。ギリシャ神話の巨人神アトラスに由来
atlas blue		*Polyommatus atlanticus* (Elwes)	(チョウ目、シジミチョウ科) 旧北区
Atlas moth	ヨナクニサン	*Attacus atlas* (Linnaeus)	(チョウ目、ヤママユガ科) 日本、東洋区。オオアヤニシキとの別名あり。最大級の蛾
Atmore's twist	ウスキカクモンハマキ	*Cornicacoecia lafauryana* (Ragonot)	(チョウ目、ハマキガ科) 日本、旧北区
Atnius groundstreak		*Calycopis atnius* (Herrich-Schäffer)	(チョウ目、シジミチョウ科) 新熱帯区
atocala underwing			Brou's underwing moth を見よ
Atossa nymph	ヒゲナガタテハ	*Euriphene atossa* (Hewitson)	(チョウ目、タテハチョウ科) エチオピア区
Atta ant cockroach		*Attaphila fungicola* Wheeler	(ゴキブリ目、ブラベルスゴキブリ科) 新北区。体長 2.7mm。ハキリアリ (*Atta* 属) の巣にすむ
attelabid beetles			attelabid weevils を見よ
attelabid weevils	オトシブミ科	Attelabidae	(コウチュウ目) の昆虫の総称
attenuated bluet		*Enallagma daeckii* (Calvert)	(トンボ目、イトトンボ科) 新北区
attin ant			fungus growing ant (1) を見よ
Atys hairstreak		*Atlides atys* (Cramer)	(チョウ目、シジミチョウ科) 新熱帯区
Aubyn Rogers' Acraea		*Acraea aubyni* Eltringham	(チョウ目、タテハチョウ科) エチオピア区
Aubyn's Charaxes		*Charaxes aubyni* van Someren et Jackson	(チョウ目、タテハチョウ科) エチオピア区
Auckland tree weta		*Hemideina thoracica* (White)	(バッタ目、Stenopelmatidae) 豪州区
aucuba fruit midge	アオキミタマバエ	*Asphondylia aucubae* Yukawa et Ohsaki	(ハエ目、タマバエ科) 日本。英名は aucuba アオキ に由来
aucuba sawfly	ハラナガハバチ	*Tenthredo hilaris* Smith	(ハチ目、ハバチ科) 日本、旧北区
aucuba scale			oleander scale を見よ
aucuba whitefly	アオキコナジラミ	*Aleurotuberculatus aucubae* (Kuwana)	(カメムシ目、コナジラミ科) 日本
Auda hairstreak		*Johnsonita auda* (Hewitson)	(チョウ目、シジミチョウ科) 新熱帯区
Audeoud's bush blue		*Cacyreus audeoudi* Stempffer	(チョウ目、シジミチョウ科) エチオピア区
auger beetle		*Xylodeleis obsipa* Germar	(コウチュウ目、ナガシンクイムシ科) 豪州区。穿孔性から螺旋形錐という英名

英名	和名	学名	所属、分布、ほか
auger beetles			false powderpost beetles を見よ　豪州での英名
Auginus longtail		*Polythrix auginus* (Hewitson)	(チョウ目、セセリチョウ科) 新熱帯区
August day birds			dog-day cicadas を見よ
August skipper			silver-spotted skipper (1) を見よ
August thorn		*Ennomos quercinaria* (Hufnagel)	(チョウ目、シャクガ科) 旧北区
Augustinula hairstreak		*Theritas augustinula* (Goodson)	(チョウ目、シジミチョウ科) 新熱帯区
Aulacean tiger-moth			black woolly-bear を見よ
aulacid wasps	セダカヤセバチ科	Aulacidae	(ハチ目) の昆虫の総称
aulacids			aulacid wasps を見よ
aulacigastrid flies	ナガショウジョウバエ科	Aulacigastridae	(ハエ目) の昆虫の総称
Aulestes doctor	ニジイロシジミタテハ	*Ancyluris aulestes* (Cramer)	(チョウ目、シジミタテハ科) 新熱帯区
Aulestis flasher		*Astraptes aulestis* (Cramer)	(チョウ目、セセリチョウ科) 新熱帯区
Aura hairstreak		*Erora aura* (Godman et Salvin)	(チョウ目、シジミチョウ科) 新熱帯区
aurea bagworm moth	ニトベミノガ	*Mahasena aurea* (Butler)	(チョウ目、ミノガ科) 日本。aureous bagworm moth ともいう
aureate hornbeam midget		*Phyllonorycter quinnata* (Geoffroy)	(チョウ目、ホソガ科) 旧北区
aurelia			butterflies and moths を見よ　蛹の英名
Aureola leafwing		*Memphis aureola* (Bates)	(チョウ目、タテハチョウ科) 新熱帯区
auricled leafhopper	ミミズク	*Ledra auditura* Walker	(カメムシ目、ミミズク科) 日本、旧北区、東洋区
auricula root aphid		*Thecabius auriculae* (Murray)	(カメムシ目、アブラムシ科) 旧北区
auriferous pearl		*Mecyna flavalis* (Denis et Schiffermüller)	(チョウ目、メイガ科) 旧北区
Aurivillius' Acraea		*Acraea aurivillii* Staudinger	(チョウ目、タテハチョウ科) エチオピア区
Aurivillius recluse		*Caenides hidaroides* Aurivillius	(チョウ目、セセリチョウ科) エチオピア区
aurora bluetail	キバライトトンボ	*Ischnura aurora* Brauer	(トンボ目、イトトンボ科) 日本、東洋区、豪州区
aurora damsel		*Chromagrion conditum* (Selys)	(トンボ目、イトトンボ科) 新北区
Austaut's Algerian heath		*Coenonympha austauti* Oberthür	(チョウ目、タテハチョウ科) 旧北区
Austaut's grayring		*Hipparchia hansii* (Austaut)	(チョウ目、タテハチョウ科) 旧北区
Austen's swift			colon swift を見よ
Austin's Aguna		*Aguna coeloides* Austin et Mielke	(チョウ目、セセリチョウ科) 新熱帯区
Austin's Anastrus		*Anastrus virens albopannus* Austin	(チョウ目、セセリチョウ科) 新熱帯区

英名	和名	学名	所属、分布、ほか
Austin's Mylon		*Mylon cristata* Austin	（チョウ目、セセリチョウ科）新熱帯区
Austin's shadowdamsel		*Drepanosticta austeni* Lieftinck	（トンボ目、Platystictidae）東洋区
Australian admiral	モンキアカタテハ	*Vanessa itea* (Fabricius)	（チョウ目、タテハチョウ科）豪州区
Australian alfalfa weevil			clover head weevil を見よ
Australian ant		*Nothomyrmecia macrops* Clark	（ハチ目、アリ科）豪州区。絶滅危惧種
Australian archaic sun moths		Lophocoronidae	（チョウ目）の昆虫の総称
Australian atlas moth		*Attacus wardi* (Rothschild)	（チョウ目、ヤママユガ科）豪州区
Australian atlas moth			Hercules moth を見よ
Australian beak	ムラサキテングチョウ	*Libythea geoffroy philippina* Staudinger	（チョウ目、タテハチョウ科）日本、東洋区、豪州区。*L. geoffroy* (Godart) の英名
Australian Bogong			Bogong moth を見よ　Australian Bugong とも記す
Australian brown lacewing		*Micromus timidus* (Hagen)	（アミメカゲロウ目、ヒメカゲロウ科）大洋区、豪州区
Australian bug			cottony cushion scale を見よ
Australian bug dorthesia			cottony cushion scale を見よ
Australian bull ant			bulldog ant (1) を見よ
Australian cabbage looper			looper caterpillar (1) を見よ
Australian carpet beetle		*Anthrenocerus australis* (Hope)	（コウチュウ目、カツオブシムシ科）旧北区、豪州区
Australian citrus whitefly		*Orchamoplatus citri* (Takahashi)	（カメムシ目、コナジラミ科）豪州区
Australian cockroach	コワモンゴキブリ	*Periplaneta australasiae* (Fabricius)	（ゴキブリ目、ゴキブリ科）日本、全北区、豪州区。屋内外に生息
Australian common armyworm		*Mythimna convecta* (Walker)	（チョウ目、ヤガ科）全北区、豪州区、新熱帯区。イネ害虫
Australian crow			common Indian crow を見よ
Australian desert ant		*Melophorus bagoti* Lubbock	（ハチ目、アリ科）豪州区
Australian devil's coachhorse			devil's coach horse (1) を見よ
Australian dog louse			dog louse を見よ
Australian duskhawker		*Anaciaeschna jaspidea* (Burmeister)	（トンボ目、ヤンマ科）豪州区
Australian emperor dragonfly		*Hemianax papuensis* Burmeister	（トンボ目、ヤンマ科）豪州区。Australian emperor ともいう
Australian fern weevil			fern weevil (1) を見よ
Australian flower wasps		Thynnidae	（ハチ目）の昆虫の総称
Australian fritillary			Indian fritillary を見よ　豪州での英名

英名	和名	学名	所属、分布、ほか
Australian goat moth		*Culama caliginosa* (Walker)	（チョウ目、ボクトウガ科）豪州区
Australian grain borer			Australian spider beetle (1) を見よ
Australian grass leafhopper		*Nesoclutha pallida* (Evans)	（カメムシ目、オオヨコバイ科）豪州区
Australian gull	フトヘリシロチョウ	*Cepora perimale* (Donovan)	（チョウ目、シロチョウ科）豪州区
Australian harlequin			harlequin metalmark を見よ
Australian hedge blue		*Udara tenella* (Miskin)	（チョウ目、シジミチョウ科）豪州区
Australian hornet		*Abispa ephippium* (Fabricius)	（ハチ目、スズメバチ科）豪州区
Australian housefly		*Musca retustissima* Walker	（ハエ目、イエバエ科）豪州区
Australian jumper		*Myrmecia nigrocincta* Smith	（ハチ目、アリ科）豪州区
Australian jumper ant			jack jumper ant を見よ
Australian king cricket		*Anastostoma australasiae* Gray	（バッタ目、Stenopelmatidae）豪州区
Australian king-crickets		*Australostoma*	（バッタ目、Stenopelmatidae）の昆虫の総称
Australian lady beetle			vedalia を見よ
Australian lady beetles		Noviini	（コウチュウ目、テントウムシ科）の昆虫の総称
Australian ladybird			vedalia を見よ
Australian ladybird beetles			Australian lady beetles を見よ
Australian lappet moths		Anthelidae	（チョウ目）の昆虫の総称
Australian leafhopper egg parasite		*Anagrus optabilis* (Perkins)	（ハチ目、ホソハネコバチ科）大洋区
Australian leafroller tachinid		*Trigonospila brevifacies* (Hardy)	（ハエ目、ヤドリバエ科）豪州区
Australian leafwing	イワサキコノハチョウ	*Doleschallia bisaltide* (Cramer)	（チョウ目、タテハチョウ科）東洋区、豪州区。leafwing ともいう
Australian leafwing (1)		*Doleschallia polibete* (Cramer)	（チョウ目、タテハチョウ科）東洋区、豪州区
Australian lurcher	キオビコノハ	*Yoma sabina podium* Tsukada	（チョウ目、タテハチョウ科）日本、東洋。豪州の *Y. sabina parva* (Butler) の英名
Australian lurcher			lurcher を見よ　台湾亜種 *Y. s. podium* Tsukada も同英名
Australian magpie moth			senecio moth を見よ
Australian malaria mosquito		*Anopheles farauti* Laveran	（ハエ目、カ科）豪州区
Australian mantid			purplewinged mantid を見よ
Australian mantis			stick mantis を見よ

英名	和名	学名	所属、分布、ほか
Australian minute egg parasite		*Trichogramma australicum* Girault	(ハチ目、タマゴコバチ科) 豪州区
Australian painted lady	ゴウシュウヒメアカタテハ	*Vanessa kershawi* (McCoy)	(チョウ目、タテハチョウ科) 豪州区
Australian paper wasp		*Polistes humilis* (Fabricius)	(ハチ目、スズメバチ科) 豪州区
Australian parasite moths			parasite moths を見よ
Australian pine borer		*Chrysobothris tranquebarica* (Gmelin)	(コウチュウ目、タマムシ科) 新北区
Australian plague locust			wandering grasshopper を見よ
Australian plane	パラオオナガシジミ	*Bindahara phocides* (Fabricius)	(チョウ目、シジミチョウ科) 豪州区
Australian primitive ghost moths		Anomosetidae	(チョウ目) の昆虫の総称
Australian privet hawk moth			privet hawk moth を見よ
Australian rat flea			rat flea を見よ
Australian regent skipper			regent skipper をみよ
Australian rice beetle		*Dicranolaius bellulus* (Guérin)	(コウチュウ目、ジョウカイモドキ科) 豪州区。イネ害虫
Australian roach			Australian cockroach を見よ
Australian rustic		*Cupha prosope* (Fabricius)	(チョウ目、タテハチョウ科) 豪州区
Australian sheep blowfly			bronze bottle fly を見よ
Australian shield bug			harlequin bug (2) を見よ
Australian silkworm moths		Carthaeidae	(チョウ目) の昆虫の総称
Australian soldier-fly			sugarcane soldier fly を見よ
Australian spider beetle		*Ptinus tectus* Boieldieu	(コウチュウ目、ヒョウホンムシ科) 全北区、豪州区
Australian spider beetle (1)		*Ptinus ocellus* Brown	(コウチュウ目、ヒョウホンムシ科) 新北区
Australian termite			giant northern termite を見よ
Australian termites	ムカシシロアリ科	Mastotermitidae	(シロアリ目) の昆虫の総称
Australian tortoise beetle		*Trachymela sloanei* (Blackburn)	(コウチュウ目、ハムシ科) 新北区、豪州区
Australian upside down fly		*Neurochaeta inversa* McAlpine	(ハエ目、Neurochaetidae) 豪州区
Australian vagrant	オナガタテハ	*Vagrans egista* (Cramer)	(チョウ目、タテハチョウ科) 東洋区、豪州区。*V. e. macromalayana* Fruhstorfer も同英名
Australian walking stick			giant prickly stick insect を見よ
Australian wheat weevil			lesser grain borer を見よ　米国での英名
autostichid moth		*Symmoca signatella* Herrich-Schäffer	(チョウ目、Symmocidae) 全北区

英名	和名	学名	所属、分布、ほか
autumn dun		*Ecdyonurus dispar* (Curtis)	(カゲロウ目、ヒラタカゲロウ科) 旧北区
autumn fly			face fly を見よ
autumn green carpet			red-green carpet (1) を見よ
autumn gum moth		*Mnesampela privata* (Guenée)	(チョウ目、シャクガ科) 豪州区
autumn hawker			migrant hawker を見よ
autumn leaf			Australian leafwing (1) を見よ
autumn leaf roller	ハイトビスジハマキ	*Syndemis musculana* Hübner	(チョウ目、ハマキガ科) 日本。マメ科牧草の害虫
autumn leaf vagrant			orange-and-lemon を見よ
autumn meadowhawk			yellow-legged meadowhawk を見よ
autumn ringlet	アキベニヒカゲ	*Erebia neoridas* (Boisduval)	(チョウ目、タテハチョウ科) 旧北区
autumn silkworm moths		Lemoniidae	(チョウ目) の昆虫の総称
Autumna skipper		*Cobalopsis autumna* (Plötz)	(チョウ目、セセリチョウ科) 新熱帯区
autumnal breeze fly		*Tabanus autumnalis* Linnaeus	(ハエ目、アブ科) 旧北区
autumnal moth		*Epirrita autumnata* (Borkhausen)	(チョウ目、シャクガ科) 全北区。日本には *E. a. autumna* (Bryk) アキナミシャク が分布
autumnal rustic		*Paradiarsia glareosa* (Esper)	(チョウ目、ヤガ科) 旧北区
autumnal shade		*Exapate congelatella* Bradley et Martin	(チョウ目、ハマキガ科) 旧北区
autumnal snout		*Schrankia intermedialis* Reid	(チョウ目、ヤガ科) 旧北区
Auxo bent-skipper		*Camptopleura auxo* (Möschler)	(チョウ目、セセリチョウ科) 新熱帯区
Avalon hairstreak		*Strymon avalona* (Wright)	(チョウ目、シジミチョウ科) 新北区
Avalon scrub-hairstreak			Avalon hairstreak を見よ
avare pod borer		*Adisura atkinsoni* Moore	(チョウ目、ヤガ科) 東洋区
avocado bark beetle		*Palaticus* sp.	(コウチュウ目、ゾウムシ科) 豪州区
avocado blossom thrips		*Frankliniella cephalica* (Crawford)	(アザミウマ目、アザミウマ科) 新北区。アボカド害虫
avocado lacebug		*Acysta perseae* Heidemann	(カメムシ目、グンバイムシ科) 新北区。アボカド害虫.
avocado leaf roller		*Gracilaria perseae* Busck	(チョウ目、ホソガ科) 新北区。アボカド害虫
avocado leafroller		*Homona spargotis* Meyrick	(チョウ目、ハマキガ科) 豪州区
avocado looper		*Sabulodes aegrotata* Guenée	(チョウ目、シャクガ科) 新北区。多くの植物を加害
avocado mealybug			coconut mealybug を見よ 米国での英名
avocado scale	コノハカイガラムシ	*Fiorinia fioriniae* (Targioni)	(カメムシ目、マルカイガラムシ科) 日本、大洋区、新北区、汎熱帯
avocado seed moth		*Stenoma catenifer* Walsingham	(チョウ目、マルハキバガ科) 新熱帯区

英名	和名	学名	所属、分布、ほか
avocado spanworm moth		*Epimecis detexta* (Walker)	（チョウ目、シャクガ科）新北区
avocado thrips		*Scirtothrips persicae* Nakahara	（アザミウマ目、アザミウマ科）新北区。アボカド害虫
avocado tree girdler		*Heilips squamosus* LeConte	（コウチュウ目、ゾウムシ科）新北区。アボカド害虫
avocado treehopper		*Metcalfiella monogromma* Germar	（コウチュウ目、ゾウムシ科）新北区。アボカド害虫
avocado whitefly			Florida whitefly を見よ
awame borer moth		*Papaipema aweme* (Lyman)	（チョウ目、ヤガ科）新北区
awlking			green velvet skipper を見よ
awlkings		*Choaspes*	（チョウ目、セセリチョウ科）の昆虫の総称
awls	タイワンアオバセセリ属	*Badamia*	（チョウ目、セセリチョウ科）の昆虫の総称
awls	ビロウドセセリ属	*Hasora*	（チョウ目、セセリチョウ科）の昆虫の総称
awls	クジャクセセリ属	*Allora*	（チョウ目、セセリチョウ科）の昆虫の総称
awls		Coeliadinae	（チョウ目、セセリチョウ科）の昆虫の総称
axehead orange			axehead skipper を見よ
axehead skipper		*Acada biseriata* (Mabille)	（チョウ目、セセリチョウ科）エチオピア区
axures	ルリシジミ属	*Celastrina*	（チョウ目、シジミチョウ科）の昆虫の総称　米国で使用
axymyiid flies	クチキカ科	Axymyiidae	（ハエ目）の昆虫の総称
Ayaya ringlet		*Magneuptychia iris* Hewitson	（チョウ目、タテハチョウ科）新熱帯区
Aymaran metalmark		*Calephelis aymaran* McAlpine	（チョウ目、シジミタテハ科）新熱帯区
azalea aphid	ツツジアブラムシ	*Vesiculaphis caricis* (Fullaway)	（カメムシ目、アブラムシ科）日本、旧北区、東洋区。大洋区
azalea argid sawfly	ルリチュウレンジ	*Arge similis* (Vollenhoven)	（ハチ目、ミフシハバチ科）日本、旧北区、東洋区。ツツジ害虫
azalea bark scale		*Eriococcus azaleae* Comstock	（カメムシ目、フクロカイガラムシ科）新北区
azalea caterpillar		*Datana major* Grote et Robinson	（チョウ目、シャチホコガ科）新北区
azalea cottony scale			azalea mealybug を見よ
azalea flower looper		*Cithecia excisa* Butler	（チョウ目、シャクガ科）旧北区
azalea globular treehopper	ミヤママルツノゼミ	*Gargara rhodendrona* Kato	（カメムシ目、ツノゼミ科）日本
azalea gracilarid			azalea leafminer (1) を見よ
azalea granulated leaf beetle	ムシクソハムシ	*Chlamisus spilotus* (Baly)	（コウチュウ目、ハムシ科）日本、旧北区
azalea lace bug	ツツジグンバイ	*Stephanitis pyrioides* (Scott)	（カメムシ目、グンバイムシ科）日本、全北区、東洋区、豪州区。ツツジ害虫
azalea leafminer (1)	リンゴハマキホソガ	*Caloptilia zachrysa* Meyrick	（チョウ目、ホソガ科）日本、旧北区、東洋区
azalea leaf miner (2)	ツツジハマキホソガ	*Caloptilia azaleella* (Brants)	（チョウ目、ホソガ科）日本、全北区、豪州区。azalea leafminer moth ともいう

英名	和名	学名	所属、分布、ほか
azalea mealybug	ツツジコナカイガラムシ	*Phenacoccus azaleae* Kuwana	(カメムシ目、コナカイガラムシ科) 日本。ツツジ害虫
azalea plant bug		*Rhinocapsus vanduzeei* Uhler	(カメムシ目、カスミカメムシ科) 新北区
azalea rough bollworm	ベニモンアオリンガ	*Earias roseifera* Butler	(チョウ目、ヤガ科) 日本、旧北区。ツツジ害虫
azalea sawfly		*Arge simillima* (Smith)	(ハチ目、ミフシハバチ科) 日本、旧北区
azalea sphinx		*Darapsa pholus* (Cramer)	(チョウ目、スズメガ科) 新北区
azalea sphinx moth		*Darapsa choerilus* (Cramer)	(チョウ目、スズメガ科) 新北区。azalea sphinx ともいう
azalea whitefly	ツツジコナジラミ	*Pealius azaleae* (Baker et Moles)	(カメムシ目、コナジラミ科) 日本、旧北区。ツツジ害虫
azalea whitefly (1)		*Pealius hibisci* (Kotinsky)	(カメムシ目、コナジラミ科) 新北区
azeca banner	トンボマダラモドキ	*Vila azeca* Doubleday et Hewitson	(チョウ目、タテハチョウ科) 新熱帯区
Azia hairstreak		*Tmolus azia* (Hewitson)	(チョウ目、シジミチョウ科) 新熱帯区
Aziza firetip		*Pyrrhopyge aziza* Hewitson	(チョウ目、セセリチョウ科) 新熱帯区
Azores grayling		*Hipparchia azorina* (Strecker)	(チョウ目、タテハチョウ科) 旧北区
Aztec Calephelis		*Calephelis azteca* McAlpine	(チョウ目、シジミタテハ科) 新熱帯区
Aztec dancer		*Argia nahuana* (Calvert)	(トンボ目、イトトンボ科) 新北区
Aztec grouse locust		*Telmatettix aztecus* (Saussure)	(バッタ目、ヒシバッタ科) 新北区
Aztec scallopwing		*Staphylus azteca* (Scudder)	(チョウ目、セセリチョウ科) 新北区、新熱帯区
Aztec thread-waisted wasp		*Ammophila azteca azteca* Cameron	(ハチ目、アナバチ科) 新北区
Azuma bamboo scale	アズマフクロカイガラムシ	*Eriococcus azumae* Kanda	(カメムシ目、フクロカイガラムシ科) 日本
azure blue			holly blue を見よ
azure bluet		*Enallagma aspersum* (Hagen)	(トンボ目、イトトンボ科) 新北区
azure bluet			azure damselfly を見よ
azure chalk-hill blue		*Polyommatus caelestissima* (Verity)	(チョウ目、シジミチョウ科) 旧北区
azure damselfly		*Coenagrion puella* (Linnaeus)	(トンボ目、イトトンボ科) 旧北区
azure dartlet		*Enallagma parvum* Selys	(トンボ目、イトトンボ科) 東洋区
azure gem		*Chloroselas azurea* Butler	(チョウ目、シジミチョウ科) エチオピア区
azure hairstreak		*Hemiolaus caeculus* (Hopffer)	(チョウ目、シジミチョウ科) エチオピア区
azure hawker		*Aeschna caerulea* (Strom)	(トンボ目、ヤンマ科) 旧北区
azure moonbeam		*Philiris azure* (Wind et Clench)	(チョウ目、シジミチョウ科) 豪州区
azure mountain blue		*Albulina asiatica* (Elwes)	(チョウ目、シジミチョウ科) 東洋区。
azure opal		*Chrysoritis azurius* (Swanepoel)	(チョウ目、シジミチョウ科) エチオピア区

英名	和名	学名	所属、分布、ほか
azure sapphire	サファイアウラフチベニシジミ	*Heliophorus moorei* (Hewitson)	（チョウ目、シジミチョウ科）東洋区
azure Theope		*Theope pedias* Herrich-Schäffer	（チョウ目、シジミチョウ科）新熱帯区
azure-winged Eurybia			Unxia metalmark を見よ

英名	和名	学名	所属、分布、ほか
B			
Babault's zebra blue		*Leptotes bebaulti* (Stempfer)	（チョウ目、シジミチョウ科）エチオピア区
Babul blues		*Azanus*	（チョウ目、シジミチョウ科）の昆虫の総称
baby fourring		*Ypthima philomela* (Linnaeus)	（チョウ目、タテハチョウ科）東洋区
baby lascar	ヒメキンミスジ	*Pantoporia aurelia* (Staudinger)	（チョウ目、タテハチョウ科）東洋区
baby royal		*Bullis buto* (de Nicéville)	（チョウ目、シジミチョウ科）東洋区
Baccharis borer plume moth		*Hellinsia balanotes* (Meyrick)	（チョウ目、トリバガ科）新北区。Baccharis borer ともいう
Baccharis carpenterworm moth		*Prionoxystus piger* (Grote)	（チョウ目、ボクトウガ科）新北区
backbugs		Aneurinae	（カメムシ目、ヒラタカメムシ科）の昆虫の総称
back-striped katydid	セスジツユムシ	*Ducetia japonica* (Thunberg)	（バッタ目、キリギリス科）日本、旧北区、東洋区、豪州区
back swimmer		*Notonecta viridis* Delcourt	（カメムシ目、マツモムシ科）旧北区
back swimmer			water boatman (1) を見よ　backswimmer とも記す
backswimmer			common backswimmer を見よ
backswimmer bugs			backswimmers を見よ
backswimmers	マツモムシ科	Notonectidae	（カメムシ目）の昆虫の総称　世界に200種以上
backswimmers			large backswimmers を見よ
backthorn aphid			buckthorn-potato aphid を見よ
bacon beetle			larder beetle を見よ
bacon beetles			larder beetles を見よ
bacon flies			skipper flies を見よ
bacon fly			cheese skipper をみよ
bad man's needle			dragonflies (1) を見よ
bad-wing		*Dyspteris abortivaria* Herrich-Schäffer	（チョウ目、シャクガ科）新北区。badwing moth ともいう
badger biting louse		*Trichodectes melis* (Fabricius)	（ハジラミ目、ケモノハジラミ科）旧北区
Badham's blue		*Lepidochrysops badhami* van Son	（チョウ目、シジミチョウ科）エチオピア区
badlands cricket		*Gryllus personatus* Uhler	（バッタ目、コオロギ科）新北区
Badplaas sprite		*Pseudagrion inopinatum* Balinsky	（トンボ目、イトトンボ科）エチオピア区
Baeotus beauty		*Baeotus baeotus* (Doubleday)	（チョウ目、タテハチョウ科）新熱帯区
bag			clothes moth (1) を見よ
bag moth			ribbed case moth を見よ
bag moths			bagworm moths を見よ
bag-shelter moth			processionary caterpillar (1) を見よ

英名	和名	学名	所属、分布、ほか
Baggett's spanworm moth		*Nematocampa baggettaria* Ferguson	（チョウ目、シャクガ科）新北区
bagpipe cicada		*Lembeja paradoxa* (Karsch)	（カメムシ目、セミ科）豪州区
bagrada bug			painted bug を見よ
bagworm		*Cryptothelea variegata* Snellen	（チョウ目、ミノガ科）旧北区。カンキツ害虫
bagworm (1)		*Thyridopteryx ephemeraeformis* (Haworth)	（チョウ目、ミノガ科）新北区
bagworm			tea bagworm を見よ
bagworm			bagworm moth (1) を見よ
bagworm moth		*Eumeta variegata* Snellen	（チョウ目、ミノガ科）東洋区
bagworm moth		*Oeceticus omnivorus* Fereday	（チョウ目、ミノガ科）豪州区
bagworm moth (1)	オオキタクロミノガ	*Canephora unicolor* (Hufnagel)	（チョウ目、ミノガ科）日本、旧北区
bagworm moths	ミノガ科	Psychidae	（チョウ目）の昆虫の総称
bagworms		*Oiketicus*	（チョウ目、ミノガ科）の昆虫の総称
bagworms			bagworm moths を見よ
Bahaman swallowtail		*Papilio andraemon* (Hübner)	（チョウ目、アゲハチョウ科）新熱帯区
Bahamian palm owlet moth		*Litoprosopus bahamensis* Hampson	（チョウ目、ヤガ科）新北区
Bailey's Charaxes		*Charaxes baileyi* van Someren	（チョウ目、タテハチョウ科）エチオピア区
Baird's swallowtail		*Papilio machaon bairdii* E. H. Edwards	（チョウ目、アゲハチョウ科）新北区、新熱帯区。old world swallowtail (1) を参照
Baja acacia skipper		*Cogia hippalus peninsularis* Miller et MacNeill	（チョウ目、セセリチョウ科）新熱帯区。acacia skipper を参照
Baja azure		*Celastrina echo* ssp.	（チョウ目、シジミチョウ科）新熱帯区。Pacific azure を参照
Baja Calephelis		*Calephelis bajaensis* McAlpine	（チョウ目、シジミタテハ科）新熱帯区
Baja California bluet		*Enallagma eiseni* Calvert	（トンボ目、イトトンボ科）新北区、新熱帯区
Baja hairstreak		*Habrodais poodiae* Brown et Faulkner	（チョウ目、シジミチョウ科）新熱帯区
Baja metalmark		*Apodemia virgulti dialeuca* Opler et Powell	（チョウ目、シジミタテハ科）新熱帯区
Baker's mealybug			grape mealybug を見よ
balance fly			dragonflies (1) を見よ
bald cypress coneworm moth		*Dioryctria pygmaeella* Ragonot	（チョウ目、メイガ科）新北区。bald cypress coneworm ともいう
bald cypress sphinx moth		*Isoparce cupressi* (Boisduval)	（チョウ目、スズメガ科）新北区、新熱帯区。baldcypress sphinx ともいう
baldcypress webworm			green hemlock needleminer moth を見よ

英名	和名	学名	所属、分布、ほか
bald duskywing			afranius duskywing を見よ
bald hornet			baldfaced hornet を見よ
baldfaced hornet		*Vespula (Dolichovespula) maculata* (Linnaeus)	(ハチ目、スズメバチ科) 新北区
Balinsky's sprite			Badplaas sprite を見よ
Balkan clouded yellow	バルカンベニモンキチョウ	*Colias balcanica* Rebel	(チョウ目、シロチョウ科) 旧北区
Balkan clouded yellow	コーカサスベニモンキチョウ	*Colias caucasica* Staudinger	(チョウ目、シロチョウ科) 旧北区
Balkan copper		*Palaeochrysophanus candens* Herrich-Schäffer	(チョウ目、シジミチョウ科) 旧北区
Balkan fritillary	バルカンヒメヒョウモン	*Boloria graeca* (Staudinger)	(チョウ目、タテハチョウ科) 旧北区
Balkan goldenring		*Cordulegaster heros* Theischinger	(トンボ目、オニヤンマ科) 旧北区
Balkan green-veined white		*Artogeia balcana* Lorkovic	(チョウ目、シロチョウ科) 旧北区
Balkan heath		*Coenonympha orientalis* Rebel	(チョウ目、タテハチョウ科) 旧北区
Balkan marbled white		*Melanargia larissa* Geyer	(チョウ目、タテハチョウ科) 旧北区
Balkan pierrot			little tiger blue を見よ
ball-bearer leafhopper			globe-bearing treehopper を見よ
ball-bearing treehopper			globe-bearing treehopper を見よ
ball gall-fly			goldenrod gall fly を見よ
ball gallmaker			goldenrod gall fly を見よ
ball-rollers			chafer beetles を見よ
ballast bell		*Epiblema grandaevana* (Liebig et Zeller)	(チョウ目、ハマキガ科) 旧北区
balloon fly		*Hilara sartor* Becker	(ハエ目、オドリバエ科) 旧北区
Ball's blue		*Lepidochrysops balli* Dickson	(チョウ目、シジミチョウ科) エチオピア区
Ball's little shieldback		*Decticita balli* Hebard	(バッタ目、キリギリス科) 新北区
Ball's shieldback		*Eremopedes balli* Caudell	(バッタ目、キリギリス科) 新北区
balm-crickets			free beaks を見よ
balsam carpet		*Xanthorhoe biriviata* (Borkhausen)	(チョウ目、シャクガ科) 旧北区。日本亜種は *X. b. angularia* (Leech) ナカシロスジナミシャク
balsam fir sawfly		*Neodiprion abietis* (Harris)	(ハチ目、マツハバチ科) 新北区
balsam fir sawyer		*Monochamus marmorator* Kirby	(コウチュウ目、カミキリムシ科) 新北区
balsam gall midge		*Paradiplosis tumifex* Gagneé	(ハエ目、タマバエ科) 新北区
balsam poplar leaf blotch miner moth		*Phyllonorycter nipigon* (Freeman)	(チョウ目、ホソガ科) 新北区

英名	和名	学名	所属、分布、ほか
balsam shootboring sawfly		*Pleroneura brunneicornis* Rohwer	（ハチ目、ナギナタハバチ科）新北区
balsam striped hawk		*Hippotion balsaminae* (Walker)	（チョウ目、スズメガ科）エチオピア区
balsam twig aphid	トドハトジワタアブラムシ	*Mindarus abietinus* Koch	（カメムシ目、アブラムシ科）日本、全北区
balsam woolly adelgid		*Adelges piceae* (Ratzeburg)	（カメムシ目、カサアブラムシ科）全北区
balsam woolly aphid			balsam woolly adelgid を見よ
Baltic grayling	カラフトタカネヒカゲ	*Oeneis jutta* (Hübner)	（チョウ目、タテハチョウ科）全北区。新北区の亜種 *O. j. ascerta* Masters et Sorenson も同英名
Baltic hawker		*Aeschna serrata* Hagen	（トンボ目、ヤンマ科）旧北区
Baltimore	ヒガシヒョウモン	*Euphydryas phaeton* (Drury)	（チョウ目、タテハチョウ科）新北区。Baltimore butterfly ともいう
Baltimore Bomolocha moth		*Hypena baltimoralis* Guenée	（チョウ目、ヤガ科）新北区。Baltimore Bomolocha, Baltimore Hypena ともいう
Baltimore butterfly		*Euryphura achlys* Hopffer	（チョウ目、タテハチョウ科）エチオピア区
Baltimore checkerspot			Baltimore を見よ
Baluchi fritillary		*Melitaea robertsi* Butler	（チョウ目、タテハチョウ科）東洋区
Baluchi hairstreak		*Neolycaena connae* Evans	（チョウ目、シジミチョウ科）東洋区
Baluchi heath		*Lyela myops* (Staudinger)	（チョウ目、タテハチョウ科）旧北区
Baluchi yellow swallowtail			southern swallowtail を見よ
Baly's earth boring beetle			dorbeetle (2) を見よ
Bamba spreadwing		*Carrhenes bamba* Evans	（チョウ目、セセリチョウ科）新熱帯区
bamboo aphid	コウシュンツノアブラムシ	*Pseudoregma koshunensis* (Takahashi)	（カメムシ目、アブラムシ科）日本
bamboo aphid (1)	タケアブラムシ	*Melanaphis bambusae* (Fullaway)	（カメムシ目、アブラムシ科）日本、東洋区、豪州区、エチオピア区
bamboo aphid (2)	タケツノアブラムシ	*Pseudoregma bambucicola* (Takahashi)	（カメムシ目、アブラムシ科）日本。タケ害虫
bamboo aphid (3)	ササッパヒゲマダラアブラムシ	*Takecallis arundinariae* (Essig)	（カメムシ目、アブラムシ科）日本、全北区、東洋区
bamboo aphid			bamboo long-horned aphid を見よ
bamboo asterolecanium	セダカフサカイガラムシ	*Asterolecanium bambusicola* Kuwana	（カメムシ目、フサカイガラムシ科）日本、旧北区、東洋区
bamboo beetle			bamboo powderpost beetle を見よ
bamboo borer			bamboo longhorn beetle を見よ
bamboo borer			bamboo powderpost beetle を見よ　豪州、南アフリカでの英名
bamboo borer beetle			bamboo powderpost beetle を見よ
bamboo bubok			bamboo powderpost beetle を見よ
bamboo bug	ヒゲナガヘリカメムシ	*Notobitus meleagris* (Fabricius)	（カメムシ目、ヘリカメムシ科）日本、東洋区

英名	和名	学名	所属、分布、ほか
bamboo bug	ヒメクモヘリカメムシ	*Paraplesius unicolor* Scott	(カメムシ目、ホソヘリカメムシ科) 日本
bamboo chinch bug	ホソコバネナガカメムシ	*Macropes obnubilus* (Distant)	(カメムシ目、ナガカメムシ科) 日本
bamboo diaspidid scale	タケシロナガカイガラムシ	*Kuwanaspis pseudoleucaspis* (Kuwana)	(カメムシ目、マルカイガラムシ科) 日本、全北区、東洋区。タケ害虫
bamboo elongate leafhopper			bamboo leafhopper を見よ
bamboo eurytomid			bamboo jointworm を見よ
bamboo false cottony scale	タケワタカイガラモドキ	*Heliococcus takae* (Kuwana)	(カメムシ目、コナカイガラムシ科) 日本
bamboo felt scale			bamboo scale (1) を見よ
bamboo florinia scale	ササヒメシロカイガラムシ	*Unachionaspis tenuis* (Maskell)	(カメムシ目、マルカイガラムシ科) 日本
bamboo forester		*Lethe kansa* (Moore)	(チョウ目、タテハチョウ科) 東洋区
bamboo fringed scale	タケフサカイガラムシ	*Asterolecanius bambusae* (Boisduval)	(カメムシ目、フサカイガラムシ科) 日本、汎熱帯。タケ害虫
bamboo gall chalcid	モウソウタマゴバチ	*Aiolomorphus rhopalvides* Walker	(ハチ目、カタビロコバチ科) 日本、旧北区
bamboo jointworm	マタケコバチ	*Gahaniola phyllostachitis* Gahan	(ハチ目、カタビロコバチ科) 日本、新北区。タケ害虫
bamboo lasiocampid	タケカレハ	*Euthrix albomaculata japonica* Lajonquier	(チョウ目、カレハガ科) 日本。タケ害虫
bamboo leaf roller		*Microstega pandalis* (Hübner)	(チョウ目、メイガ科) 旧北区
bamboo leafhopper	タケナガヨコバイ	*Bambusana banbusae* (Matsumura)	(カメムシ目、オオヨコバイ科) 日本。タケ害虫
bamboo longhorn beetle	タケトラカミキリ	*Chlorophorus annularis* (Fabricius)	(コウチュウ目、カミキリムシ科) 日本、新北区、大洋区。タケ害虫。bamboo longicorn, bamboo longicorn beetle ともいう
bamboo long-horned aphid	タケヒゲマダラアブラムシ	*Takecallis arundicolens* (Clarke)	(カメムシ目、アブラムシ科) 日本、全北区、東洋区
bamboo marmorated aphid			bamboo woolly aphid を見よ
bamboo mealybug		*Chaetococcus bambusae* (Maskell)	(カメムシ目、コナカイガラムシ科) 新北区
bamboo myzocallis			bamboo aphid (3) を見よ
bamboo page	アサギドクチョウ	*Philaethria dido* (Linnaeus)	(チョウ目、ドク科) 新熱帯区
bamboo planthopper	タケウンカ	*Epeurysa nawaii* Matsumura	(カメムシ目、ウンカ科) 日本、東洋区
bamboo planthopper	タイワンヒゲブトウンカ	*Purohita taiwanensis* Muir	(カメムシ目、ウンカ科) 日本、東洋区、大洋区
bamboo planthopper (1)	タケヒゲブトウンカ	*Purohita cervina* Distant	(カメムシ目、ウンカ科) 日本、東洋区、大洋区
bamboo powderpost beetle	チビタケナガシンクイムシ	*Dinoderus minutus* (Fabricius)	(コウチュウ目、ナガシンクイムシ科) 日本、汎世界
bamboo pyralid	キムジノメイガ	*Prodasycnemis inornata* (Butler)	(チョウ目、メイガ科) 日本、旧北区

英名	和名	学名	所属、分布、ほか
bamboo pyrausta	タケノメイガ	*Coclebotys coclesalis* (Walker)	(チョウ目、メイガ科) 日本、東洋区
bamboo rivula	タケアツバ	*Rivula leucanioides* (Walker)	(チョウ目、ヤガ科) 日本、東洋区
bamboo root aphid	タケネハマキワタムシ	*Prociphilus take* (Shinji)	(カメムシ目、アブラムシ科) 日本
bamboo round scale			penicillate scale を見よ
bamboo scale	タケカタカイガラモドキ	*Aclerda tokionis* (Cockerell)	(カメムシ目、カタカイガラモドキ科) 日本、新北区
bamboo scale (1)	タケフクロカイガラムシ	*Eriococcus onukii* Kuwana	(カメムシ目、フクロカイガラムシ科) 日本。タケ害虫
bamboo scale (2)	タケシロオカイガラムシ	*Antonina crawii* Cockerell	(カメムシ目、フクロカイガラムシ科) 日本、旧北区、大洋区、タケ害虫
bamboo scale			bamboo fringed scale を見よ
bamboo scurly scale	ササシロナガカイガラムシ	*Unachionaspis bambusae* (Cockerell)	(カメムシ目、マルカイガラムシ科) 日本、旧北区
bamboo shoot borer	ハジマヨトウ	*Bambusiphila vulgaris* (Butler)	(チョウ目、ヤガ科) 日本、旧北区。タケ害虫
bamboo-shoot cutworm			bamboo shoot borer を見よ
bamboo sylph		*Metisella syrinx* (Trimen)	(チョウ目、セセリチョウ科) エチオピア区
bamboo tiger longicorn			bamboo longhorn beetle を見よ
bamboo tree-brown	シロオビヒカゲ	*Lethe europa* (Fabricius)	(チョウ目、タテハチョウ科) 日本、東洋区。*L. e. malaya* Corbet, *L. e. pavida* Fruhstorfer も同英名
bamboo weevil		*Cyrtotrachelus longimanus* Fabricius	(コウチュウ目、ゾウムシ科) 東洋区
bamboo white scale	ハワードシロナガカイガラムシ	*Kuwanaspis howardi* (Cooley)	(カメムシ目、マルカイガラムシ科) 日本、旧北区。タケ害虫
bamboo woolly aphid	タケヒメツノアブラムシ	*Astegopteryx bambusifoliae* (Takahashi)	(カメムシ目、アブラムシ科) 日本、東洋区
bamboo woolly aphid			bamboo aphid (2) を見よ
bamboo zygaenid	タケノホソクロバ	*Balataea funeralis* (Butler)	(チョウ目、マダラガ科) 日本、旧北区。タケ害虫
bambootails		*Disparoneura*	(トンボ目、Protoneuridae) の昆虫の総称
Bampton's copper		*Aloeides bamptoni* Tite et Dickson	(チョウ目、シジミチョウ科) エチオピア区
banana aphid	クロスジコバネアブラムシ	*Pentalonia nigronervosa* Coquerel	(カメムシ目、アブラムシ科) 日本、汎世界。バナナ害虫
banana borer			banana weevil を見よ
banana corn weevil			banana weevil を見よ
banana flies			vinegar flies (1) を見よ
banana flower thrips		*Frankliniella parvula* Hood	(アザミウマ目、アザミウマ科) 新熱帯区
banana flower thrips			flower thrips (1) を見よ　豪州での英名
banana fruit caterpillar		*Tiracola plagiata* (Walker)	(チョウ目、ヤガ科) 豪州区

英名	和名	学名	所属、分布、ほか
banana fruit fly		*Dacus musae* (Tryon)	（ハエ目、ミバエ科）豪州区
banana fruit-scarring beetle		*Colaspis hypochlora* Lefebvre	（コウチュウ目、ハムシ科）新熱帯区
banana lace bug		*Stephanitis typica* (Distant)	（カメムシ目、グンバイムシ科）日本、旧北区、東洋区
banana leaf roller			banana skipper（2）を見よ
banana mealybug			Comstock mealybug を見よ
banana moth		*Antichloris viridis* Druce	（チョウ目、ヒトリガ科）新熱帯区
banana moth		*Hieroxestis subcervinella* (Walker)	（チョウ目　ハモグリガ科）旧北区
banana moth (1)		*Opogona sacchari* (Bojer)	（チョウ目、ヒロズコガ科）全北区、大洋区、エチオピア区、新熱帯区
banana moth			banana scab moth (1) を見よ
banana nightfighter		*Moltena fiara* (Butler)	（チョウ目、セセリチョウ科）エチオピア区
banana pseudostem weevil	バナナツヤオサゾウムシ	*Odoiporus longicollis* (Olivier)	（コウチュウ目、オサゾウムシ科）日本、東洋区
banana rind thrips			rind thrips を見よ
banana root borer			banana weevil を見よ
banana rust thrips		*Scirtothrips signipennis* Bagnall	（アザミウマ目、アザミウマ科）豪州区。カンキツ害虫
banana rust thrips			orchid thrips を見よ
banana scab moth		*Lamprosema charesalis* Walker	（チョウ目、メイガ科）東洋区
banana scab moth (1)		*Nacoleia octasema* (Meyrick)	（チョウ目、メイガ科）東洋区、豪州区、大洋区
banana silvering thrips			banana thrips を見よ　豪州での英名。banana-silvering thrips とも記す
banana skipper (1)		*Pelopides thrax* (Hübner)	（チョウ目、セセリチョウ科）新北区
banana skipper (2)	バナナセセリ	*Erionota torus* Evans	（チョウ目、セセリチョウ科）日本、東洋区
banana skipper			palm redeye (1) を見よ
banana skipper egg parasite		*Ooencyrtus erionotae* Ferriere	（ハチ目、トビコバチ科}。大洋区
banana skipper larval parasite		*Apanteles erionotae* McKenzie	（ハチ目、コマユバチ科）大洋区
banana-spotting bug		*Amblypelta lutescens lutescens* (Distant)	（カメムシ目、ヘリカメムシ科）豪州区
banana stalk borer		*Castniomera humboldti* Boisduval	（チョウ目、カストニアガ科）新熱帯区
banana stalk fly		*Derocephalus anusticollis* Enderlein	（ハエ目、ナガズヤセバエ科）豪州区
banana stem borer		*Telchin licus* (Drury)	（チョウ目、カストニアガ科）新熱帯区、大洋区
banana stem borer			giant sugarcane borer を見よ
banana stem weevil			banana pseudostem weevil を見よ
banana stowaway		*Antichloris eriphia* (Fabricius)	（チョウ目、カノコガ科）旧北区、新熱帯区
banana thrips		*Hercinothrips bicinctus* (Bagnall)	（アザミウマ目、アザミウマ科）エチオピア区、豪州区、全北区、新熱帯区。バナナ害虫

英名	和名	学名	所属、分布、ほか
banana-tree night-fighter			banana nightfighter を見よ
banana weevil	バショウゾウムシ	*Cosmopolites sordidus* (Germar)	（コウチュウ目、ゾウムシ科）東洋区、新北区、エチオピア区。大洋区、豪州区、新熱帯区。バナナ害虫
banana weevil borer			banana weevil を見よ
band-celled sister			Mexican sister (1) をみよ
band-eyed brown horsefly			small horse-fly を見よ
band-eyed drone fly		*Eristalinus taeniops* (Wiedemann)	（ハエ目、ハナアブ科）全北区、エチオピア区、東洋区
band-spotted skipper		*Xeniades chalestra pteras* Godman	（チョウ目、セセリチョウ科）新熱帯区
band-winged dragonlet		*Erythrodiplax umbrata* (Linnaeus)	（トンボ目、トンボ科）新北区、新熱帯区
band-winged grasshoppers		Oedipodinae	（バッタ目、バッタ科）の昆虫の総称　多くの種の後翅は明るい色
band-winged meadowhawk		*Sympetrum semicinctum* (Say)	（トンボ目、トンボ科）新北区
banded ace	ウラオビコチャバネセセリ	*Halpe zema* (Hewitson)	（チョウ目、セセリチョウ科）東洋区
banded Achaea		*Achaea catella* Guenée	（チョウ目、ヤガ科）エチオピア区
banded Agrion		*Calopteryx splendens* (Harris)	（トンボ目、イトトンボ科）旧北区
banded Agrion damselflies			banded Agrions を見よ
banded Agrions		*Calopteryx*	（トンボ目、イトトンボ科）の昆虫の総称
banded alder borer		*Rosalia funebris* Motschulsky	（コウチュウ目、カミキリムシ科）新北区。banded alder borer beetle ともいう
banded alpine	ホッキョクベニヒカゲ	*Erebia fasciata* Butler	（チョウ目、タテハチョウ科）新北区
banded angle	ツバメアツバセセリ	*Odontoptilum pygela* (Hewitson)	（チョウ目、セセリチョウ科）東洋区
banded angle			chestnut angle を見よ
banded antlion		*Glenoleon pulchellus* Rambur	（アミメカゲロウ目、ウスバカゲロウ科）豪州区
banded apple pigmy			apple pygmy を見よ
banded ash borer		*Neoclytus caprea* (Say)	（コウチュウ目、カミキリムシ科）新北区
banded ash clearwing		*Podosesia aureocincta* Purrington et Nielsen	（チョウ目、スカシバガ科）新北区。banded ash clearwing moth ともいう
banded awl			common banded awl を見よ
banded bagnest			Boisduval silkworm を見よ
banded banner		*Pyrrhogyra neaerea* (Linnaeus)	（チョウ目、タテハチョウ科）新北区。新熱帯区の *P. n. hypsenor* Godman et Salvin も同英名
banded blowfly		*Chrysomya albiceps* (Wiedemann)	（ハエ目、クロバエ科）旧北区

英名	和名	学名	所属、分布、ほか
banded blue pierrot		*Discolampa ethion* (Westwood)	(チョウ目、シジミチョウ科) 東洋区。*D. e. thalimar* Fruhstorfer も同英名
banded blue swallowtail		*Papilio oribazus* Boisduval	(チョウ目、アゲハチョウ科) エチオピア区
banded caterpillar parasite		*Ichneumon promissorius* Erichson	(ハチ目、ヒメバチ科) 豪州区。banded caterpillar parasite wasp ともいう
banded cereal aphid		*Sitobion jeelamaniae* (David)	(カメムシ目、アブラムシ科) 東洋区
banded checkerspot			banded patch を見よ
banded clover seedbearer		*Coleophora spissicornis* Meyrick	(チョウ目、ツツミノガ科) 全北区、豪州区
banded cucumber beetle		*Diabrotica balteata* LeConte	(コウチュウ目、ハムシ科) 新北区
banded daggerwing			road page を見よ　米国での英名
banded dandy		*Laringa horsfieldii* (Boisduval)	(チョウ目、タテハチョウ科) 東洋区
banded darter	ミヤマアカネ	*Sympetrum pelemontanum elatum* (Selys)	(トンボ目、トンボ科) 日本。原名亜種も同英名
banded demoiselle			banded Agrion を見よ
banded demoiselle damselflies			banded Agrions を見よ
banded demoiselles			banded Agrions を見よ
banded demon		*Notocrypta waigensis proserpina* (Butler)	(チョウ目、セセリチョウ科) 豪州区
banded demon		*Trina geometrina* (Felder et Felder)	(チョウ目、セセリチョウ科) 新熱帯区
banded demon			common banded demon を見よ　*N. p. varians* Plotz も同英名
banded drepanid			dusty hooktip を見よ
banded duffer		*Discophora deo* de Nicéville	(チョウ目、タテハチョウ科) 東洋区
banded dusk-darter		*Parazyxomma flavicans* (Martin)	(トンボ目、トンボ科) エチオピア区
banded dusk-flat			banded red-eye (1) を見よ
banded elfin			pine elfin を見よ
banded elm bark beetle		*Scolytus schevyrewi* Semenov	(コウチュウ目、キクイムシ科) 旧北区
banded Euselasia		*Euselasia clithra* (Bates)	(チョウ目、シジミタテハ科) 新熱帯区
banded evening brown		*Melanitis amabilis* (Boisduval)	(チョウ目、タテハチョウ科) 豪州区
banded evening brown			yellow-banded evening brown を見よ
banded false-brome dwarf		*Elachista gangabella* Zeller	(チョウ目、クサモグリガ科) 旧北区
banded flesh fly		*Eumacronychia scitula* Reinhard	(ハエ目、ニクバエ科) 新北区。本種はハチの巣に産卵する
banded flower mantis		*Theopropus elegans* Westwood	(カマキリ目、Hymenopodidae) 東洋区

英名	和名	学名	所属、分布、ほか
banded flutterer			graphic flutterer を見よ
banded forester		*Bebearia oxione* (Hewitson)	(チョウ目、タテハチョウ科) エチオピア区
banded general		*Stratiomys potamida* Meigen	(ハエ目、ミズアブ科) 旧北区
banded glasshouse thrips	クリオバネアザミウマ	*Hercinothrips femoralis* (Reiter)	(アザミウマ目、アザミウマ科) 日本、豪州区、全北区、新熱帯区、大洋区
banded gold-tip	ナカグロツマアカシロチョウ	*Colotis eris* (Klug)	(チョウ目、シロチョウ科) エチオピア区
banded grass-veneer		*Pediasia fascelinella* (Hübner)	(チョウ目、メイガ科) 旧北区
banded green moth		*Jaspidea celsia* Linnaeus	(チョウ目、ヤガ科) 旧北区
banded greenhouse thrips			banded glasshouse thrips を見よ　北米での英名
banded groundling		*Brachythemis leucosticta* (Burmeister)	(トンボ目、トンボ科) 旧北区、エチオピア区
banded hairstreak	オビカラスシジミ	*Satyrium calanus falacer* (Godart)	(チョウ目、シジミチョウ科) 新北区。*S. calanus* (Hübner) も同英名
banded harlequin		*Paralaxita orphna laocoon* (de Nicéville)	(チョウ目、シジミタテハ科) 東洋区
banded Helen			yellow Helen (1) を見よ
banded hickory borer		*Knulliana cincta* (Drury)	(コウチュウ目、カミキリムシ科) 新北区
banded hopper		*Platylesches picanini* (Holland)	(チョウ目、セセリチョウ科) エチオピア区
banded hornet			greater banded hornet を見よ
banded horntail			horntail (1) を見よ
banded house mosquito		*Theobaldia annulata* Schrank	(ハエ目、カ科) 旧北区
banded Jack-pine needleminer moth		*Coleotechnites canusella* (Freeman)	(チョウ目、キバガ科) 新北区
banded jewelwing			banded Agrion を見よ
banded King shoemaker			King shoemaker (1) を見よ
banded leafhopper	オビヒメヨコバイ	*Naratettix zonatus* (Matsumura)	(カメムシ目、オオヨコバイ科) 日本、旧北区
banded lichen moth		*Eutane terminalis* Walker	(チョウ目、ヒトリガ科) 豪州区
banded lineblue	アカシジミ	*Japonica lutea lutea* (Hewitson)	(チョウ目、シジミチョウ科) 日本、旧北区
banded lineblue		*Prosotas aluta* (Druce)	(チョウ目、シジミチョウ科) 東洋区
banded long-horned beetle		*Xestoleptura crassipes* LeConte	(コウチュウ目、カミキリムシ科) 新北区。banded longhorn beetle ともいう
banded longwing			banded orange heliconian を見よ
banded mapwing		*Hypanartia dione disjuncta* Willmott, Hall et Lamas	(チョウ目、タテハチョウ科) 新熱帯区
banded Marquis		*Bassarona teuta* (Doubleday)	(チョウ目、タテハチョウ科) 東洋区。*B. t. goodrichi* (Distant) も同英名
banded mosquito			banded house mosquito を見よ

英名	和名	学名	所属、分布、ほか
banded net-wing beetle			reticulated netwinged beetle を見よ
banded net-winged beetle		*Calopteron discrepans* (Newman)	（コウチュウ目、ベニボタル科）新北区
banded orange heliconian	オビモンドクチョウ	*Dryadula phaetusa* (Linnaeus)	（チョウ目、タテハチョウ科）新北区、新熱帯区。orange barred tiger, banded orange, orange tiger ともいう
banded patch		*Chlosyne endeis* (Godman et Salvin)	（チョウ目、タテハチョウ科）新北区、新熱帯区
banded peacock			Fatima を見よ
banded peacock			moss peacock を見よ
banded peacock fatima			Fatima を見よ
banded pennant		*Celithemis fasciata* Kirby	（トンボ目、トンボ科）新北区
banded pine moth			barred red を見よ
banded pine weevil			minor pine weevil を見よ
banded pine weevil			larger pine weevil を見よ
banded puffin	オオクモガタシロチョウ	*Appias pandione lagela* (Moore)	（チョウ目、シロチョウ科）東洋区
banded pug		*Eupithecia placidata* Taylor	（チョウ目、シャクガ科）新北区
banded purple			white admiral (1) を見よ
banded purple wing			whitened bluewing を見よ
banded quaker moth		*Protorthodes incincta* (Morrison)	（チョウ目、ヤガ科）新北区
banded redeye		*Gangara lebadea* (Hewitson)	（チョウ目、セセリチョウ科）東洋区
banded red-eye (1)	キシタマエオビボカシセセリ	*Chaetocneme critomedia* (Guérin-Méneville)	（チョウ目、セセリチョウ科）豪州区
banded roach			brown-banded cockroach (1) を見よ
banded rose sawfly		*Allantus cinctus* (Linnaeus)	（ハチ目、ハバチ科）旧北区
banded royal		*Rachana jalindra* (Horsfield)	（チョウ目、シジミチョウ科）東洋区。*R. j. burdona* (Hewitson) も同英名
banded royal		*Charana jalindra* Moore	（チョウ目、シジミチョウ科）東洋区
banded scaly cricket		*Cycloptilum albocircum* Love et Walker	（バッタ目、コオロギ科）新北区
banded Scythris moth		*Scythris trivinctella* (Zeller)	（チョウ目、キヌバコガ科）新北区
banded similar-wing			pine false looper moth を見よ
banded sphinx moth		*Eumorpha fasciatus* (Sulzer)	（チョウ目、スズメガ科）新北区、新熱帯区
banded straw ace		*Pithauria marsena* (Hewitson)	（チョウ目、セセリチョウ科）東洋区
banded sugar ant			sugar ant (1) を見よ
banded sunflower moth		*Aethes hospes* (Walsingham)	（チョウ目、ホソハマキガ科）新北区
banded swallowtail	オビモンアゲハ	*Papilio demolion* Cramer	（チョウ目、アゲハチョウ科）東洋区

英名	和名	学名	所属、分布、ほか
banded Theope		*Theope eurygonina* Bates	（チョウ目、シジミチョウ科）新熱帯区
banded thrips	シマアザミウマ科	Aeolothripidae	（アザミウマ目）の昆虫の総称　多くの種が捕食性
banded thrips			striped thrips を見よ
banded tiger moth		*Apantesis vittata* (Fabricius)	（チョウ目、ヒトリガ科）新北区
banded tigerwing		*Aeria eurimedia* (Cramer)	（チョウ目、タテハチョウ科）新熱帯区
banded treebrown	シナシロオビヒカゲ	*Lethe confusa* Aurivillius	（チョウ目、タテハチョウ科）東洋区。*L. c. enima* Fruhstorfer も同英名
banded tussock moth			checkered tussock moth を見よ
banded tussock moth			pale tussock を見よ
banded vapourer		*Aroa discalis* Walker	（チョウ目、ドクガ科）東洋区
banded white ringlet			white satyr を見よ
banded wing grasshopper		*Trimerotropis pallidipennis* (Burmeister)	（バッタ目、バッタ科）大洋区、新北区
banded-wing whitefly			banded-winged whitefly を見よ
banded-winged palm thrips		*Parthenothrips dracaenae* (Heeger)	（アザミウマ目、アザミウマ科）全北区、大洋区
banded-winged thrips			banded thrips を見よ
banded-winged whitefly		*Trialeurodes abutilonea* (Haldeman)	（カメムシ目、コナジラミ科）新北区
banded woollybear		*Pyrrharctia isabella* (Smith et Abbott)	（チョウ目、ヒトリガ科）新北区、大洋区。成虫は banded woollybear moth といい、開帳　5 cm
banded Yeoman		*Cirrochroa orissa orissa* Felder	（チョウ目、タテハチョウ科）東洋区
Bang-Haas' white spots		*Osmodes banghaasii* Holland	（チョウ目、セセリチョウ科）エチオピア区
bank click beetle		*Hypnoidus riparius* (Fabricius)	（コウチュウ目、コメツキムシ科）旧北区
bank note blue	オオクジャクシジミ	*Evenus regalis* (Cramer)	（チョウ目、シジミチョウ科）新熱帯区
bank wireworm			bank click beetle を見よ　幼虫の英名
Bankes's knot-horn		*Epischnia banksiella* (Richardson)	（チョウ目、メイガ科）旧北区
Bank's brown		*Heteronympha banksii* (Leach)	（チョウ目、タテハチョウ科）豪州区
Banks Peninsula tree weta		*Hemideina ricta* Hutton	（バッタ目、Stenopelmatidae）豪州区
banksia gall midge		*Dasineura banksiae* Kolesik	（ハエ目、タマバエ科）豪州区
banksia jewel beetle		*Cyria imperialis* (Fabricius)	（コウチュウ目、タマムシ科）豪州区
banksia longicorn		*Paroplites australis* (Erichson)	（コウチュウ目、カミキリムシ科）豪州区
banksia moth		*Danima banksiae* (Lewin)	（チョウ目、シャチホコガ科）豪州区
banner		*Eublemma anachoresis* (Wallengren)	（チョウ目、ヤガ科）エチオピア区、大洋区

英名	和名	学名	所属、分布、ほか
banner clubtail		*Gomphus apomyius* Donnelly	（トンボ目、サナエトンボ科）新北区
banner metalmark		*Thisbe lycorias lycorias* (Hewitson)	（チョウ目、シジミタテハ科）新熱帯区
banyan aphid		*Thoracaphis fici* (Takahashi)	（カメムシ目、アブラムシ科）新北区
bar-celled Oleria			zea clearwing を見よ
bar-faced ground crickets		*Pictonemobius*	（バッタ目、コオロギ科）の昆虫の総称
bar-sided darner		*Gynacantha mexicana* Selys	（トンボ目、ヤンマ科）新北区、新熱帯区
Baracoa skipper		*Polites baracoa* (Lucas)	（チョウ目、セセリチョウ科）新北区
Barajo hairstreak		*Laothus barajo* (Reakirt)	（チョウ目、シジミチョウ科）新熱帯区
Barbara's copper		*Aloeides barbarae* Henning et Henning	（チョウ目、シジミチョウ科）エチオピア区
Barbary skipper		*Muschampia mohammed* (Oberthür)	（チョウ目、セセリチョウ科）旧北区
barber		*Pseudenargia ulicis* (Staudinger)	（チョウ目、ヤガ科）旧北区
barber bug			kissing bug を見よ
barber moth			barber を見よ
barberpole caterpillar moth		*Mimoschinia rufofascialis* (Stephens)	（チョウ目、メイガ科）新北区、新熱帯区。幼虫は barberpole caterpillar
barberry aphid	ヘビノボラズツトアブラムシ	*Liosomaphis berberidis* (Kaltenbach)	（カメムシ目、アブラムシ科）日本、旧北区、豪州区
barberry carpet		*Pareulype berberata* (Denis et Schiffermüller)	（チョウ目、シャクガ科）旧北区
barberry fly			berberis seed fly を見よ
barberry geometer moth		*Coryphista meadi* (Packard)	（チョウ目、シャクガ科）新北区。barberry looper ともいう
barberry seed fly			berberis seed fly を見よ
barberry skipper		*Syrichtus mohammed* Oberthür	（チョウ目、セセリチョウ科）旧北区
Barber's Acraea		*Acraea barberi* Trimen	（チョウ目、タテハチョウ科）エチオピア区
Barber's brown lacewing		*Sympherobius barberi* (Banks)	（アミメカゲロウ目、ヒメカゲロウ科）新北区、大洋区。Barber brown lacewing ともいう
Barber's ranger		*Kedestes barberae* (Trimen)	（チョウ目、セセリチョウ科）エチオピア区
barbet			luminous percher を見よ　barbet percher ともいう
Barbut's cuckoo bee		*Bombus barbutellus* (Kirby)	（ハチ目、ミツバチ科）旧北区
Barce metalmark		*Baeotis barce barce* Hewitson	（チョウ目、シジミタテハ科）新熱帯区
bardee grub		*Bardistus cibarius* Newman	（コウチュウ目、カミキリムシ科）豪州区
bardee grub		*Abantiades marcidus* Tindale	（チョウ目、コウモリガ科）豪州区

英名	和名	学名	所属、分布、ほか
bardee grub (1)		*Trictena atripalpis* (Walker)	(チョウ目、コウモリガ科) 豪州区。アボリジニの人たちが幼虫を食べる
bardi grub			bardee grub (1) を見よ
bare-patched leafroller moth		*Pseudexentera spoliana* (Clemens)	(チョウ目、ハマキガ科) 新北区
bark ambrosia beetle	ハンノスジキクイムシ	*Xyleborus seriatus* Blandford	(コウチュウ目、キクイムシ科) 日本、旧北区、東洋区。木材害虫
bark anobiid			pine bark anobiid を見よ　南アフリカでの英名
bark aphids	ネアブラムシ科	Phylloxeridae	(カメムシ目) の昆虫の総称
bark beetle		*Hypothenemus aspericollis* (Wollaston)	(コウチュウ目、キクイムシ科) 豪州区。カンキツ害虫
bark beetle		*Xyleborus testaceus* Walker	(コウチュウ目、キクイムシ科) 東洋区。カンキツ害虫
bark beetle			shot-hole borer (2) を見よ
bark beetles		*Scolytus*	(コウチュウ目、キクイムシ科) の昆虫の総称
bark beetles (1)	キクイムシ科	Scolytidae	(コウチュウ目) の昆虫の総称
bark beetles (2)		*Hylesinus*	(コウチュウ目、キクイムシ科) の昆虫の総称
bark beetles			true bark beetles を見よ
bark beetles			weevils (2) を見よ
bark borers			bark beetles (1) を見よ
bark borers			true bark beetles を見よ
bark bugs			flat bugs を見よ　豪州での英名
bark cicada		*Burbunga gilmorei* (Distant)	(カメムシ目、セミ科) 豪州区
bark cockroach		*Laxta granicollis* (Saussure)	(ゴキブリ目、ブラベルスゴキブリ科) 豪州区
bark crevice scale insects		Cryptococcidae	(カメムシ目) の昆虫の総称
bark eating borer		*Comocritis pieria* Meyrick	(チョウ目、スガ科) 東洋区
bark-eating borer		*Indarbella (Arbela) quadrinotata* Walker	(チョウ目、Metarbelidae) 東洋区。カンキツ害虫
bark-eating borer		*Indarbella (Arbela) tetraonis* Moore	(チョウ目、Metarbelidae) 東洋区。カンキツ害虫
bark eating caterpillar			bark eating borer を見よ
bark-eating termites		*Macrotermes*	(シロアリ目、シロアリ科) の昆虫の総称
bark-eating termites		*Odontotermes*	(シロアリ目、シロアリ科) の昆虫の総称
bark-gnawing beetles		Ostomidae	(コウチュウ目) の昆虫の総称
bark-gnawing beetles			gnawing beetles (1) を見よ
bark louse		*Caecilius pilipennis* (Lienhard)	(チャタテムシ目、ケチャタテ科) 旧北区
bark-mimicking grasshopper		*Coryphistes ruricola* (Burmeister)	(バッタ目、バッタ科) 豪州区
bark tortrix			cherry-bark moth を見よ　米国での英名
bark weevils	キクイゾウムシ亜科	Cossoninae	(コウチュウ目、ゾウムシ科) の昆虫の総称

英名	和名	学名	所属、分布、ほか
Barker's cupid		*Euchrysops barkeri* (Trimen)	(チョウ目、シジミチョウ科) エチオピア区
barklice			booklice (2) を見よ
barklike longhorn		*Enaphalodes niveitectus* Schaeffer	(コウチュウ目、カミキリムシ科) 新北区
bark-like moth	キノカワガ	*Blenina senex* (Butler)	(チョウ目、ヤガ科) 日本、東洋区
Barkly's copper		*Aloeides barklyi* (Trimen)	(チョウ目、シジミチョウ科) エチオピア区
barley aphid		*Sipha maydis* Passerini	(カメムシ目、アブラムシ科) 旧北区、東洋区、エチオピア区
barley corn aphid			corn leaf aphid を見よ
barley crub			southern armyworm (2) を見よ
barley flea beetle	ムギヒサゴトビハムシ	*Chaetocnema cylindrica* (Baly)	(コウチュウ目、ハムシ科) 日本、旧北区。ムギ害虫
barley flea beetle		*Phyllotreta vittula* (Redtenbacher)	(コウチュウ目、ハムシ科) 旧北区
barley fly		*Delia arambourgi* (Séguy)	(ハエ目、ハナバエ科) エチオピア区
barley gout fly			chloropid gout fly を見よ
barley grub			southern armyworm (2) を見よ
barley jointworm		*Harmolita hordei* (Harris)	(ハチ目、カタビロコバチ科) 全北区
barley leaf beetle			cereal leaf beetle (1)(2) を見よ
barley leafminer	ムギクロハモグリバエ	*Agromyza albipennis* Meigen	(ハエ目、ハモグリバエ科) 日本、旧北区。ムギ害虫
barley leafminer	オカザキハモグリバエ	*Cerodontha incisa* (Meigen)	(ハエ目、ハモグリバエ科) 日本。ムギ害虫
barley leafminer	ムギキベリハモグリバエ	*Cerodontha lateralis* (Macquart)	(ハエ目、ハモグリバエ科) 日本、全北区。ムギ害虫
barley leafminer (1)	ヤノハモグリバエ	*Agromyza yanonis* (Matsumura)	(ハエ目、ハモグリバエ科) 日本。ムギ害虫
barley leafminer (2)	ムギキイロハモグリバエ	*Cerodontha denticornis* (Panzer)	(ハエ目、ハモグリバエ科) 日本、旧北区。ムギ害虫
barley-spotted silverdrop		*Epargyreus brodkorbi* Freeman	(チョウ目、セセリチョウ科) 新熱帯区
barley stem gall midge		*Mayetiola hordei* Kieffer	(ハエ目、タマバエ科) 旧北区、エチオピア区
barley straw worm			barley jointworm を見よ
barley wireworm	トビイロムナボソコメツキ	*Agriotes ograe fuscicollis* Miwa	(コウチュウ目、コメツキムシ科) 日本、旧北区。ムギ害虫
barley yellow leaf-miner fly			barley leafminer (2) を見よ
barnacle scale	フジツボロウムシ	*Ceroplastes cirripediformis* Comstock	(カメムシ目、カタカイガラムシ科) 日本、新北区。カンキツ害虫
Barnard's azure		*Ogyris barnardi* Miskin	(チョウ目、シジミチョウ科) 豪州区
Barnes' sister		*Adelpha barnesia* Schaus	(チョウ目、タテハチョウ科) 新熱帯区
Barne's tiger moth		*Cisthene barnesii* (Dyar)	(チョウ目、ヒトリガ科) 新北区
barnyard grass stem borer	ヒエホソメイガ	*Enesima leucotaeniella* (Ragonot)	(チョウ目、メイガ科) 日本。ヒエ害虫

英名	和名	学名	所属、分布、ほか
baron	トガリバイナズマ	*Euthalia aconthea* (Cramer)	（チョウ目、タテハチョウ科）東洋区。*E. a. gurda* Fruhstorfer も同英名。baron butterfly ともいう
Baron Terloo's sphinx moth		*Proserpinus terlooii* Edwards	（チョウ目、スズメガ科）新北区。Terloo sphinx moth ともいう
baronet		*Euthalia nais* (Forster)	（チョウ目、タテハチョウ科）東洋区
barons	イナズマチョウ属	*Euthalia*	（チョウ目、タテハチョウ科）の昆虫の総称
barred angle moth		*Speranza subcessaria* (Walker)	（チョウ目、シャクガ科）新北区
barred apple pigmy		*Ectoedemia atricollis* (Stainton)	（チョウ目、モグリチビガ科）旧北区
barred blues			leopard butterflies を見よ
barred carpet	ウスカバスジエダシャク	*Perizoma taeniata* (Stephens)	（チョウ目、シャクガ科）日本、旧北区
barred chestnut	エゾオオバコヤガ	*Diarsia dahlii* (Hübner)	（チョウ目、ヤガ科）日本、旧北区
barred chestnut clay			barred chestnut を見よ
barred conch		*Cochylimorpha alternana* (Stephens)	（チョウ目、ハマキガ科）旧北区
barred dwarf conch		*Cochylis nana* (Haworth)	（チョウ目、ホソハマキガ科）旧北区
barred false sailer			Malagasy sailer を見よ
barred fruit tree tortrix moth			currant twist を見よ　barred fruit tree tortrix ともいう
barred gold conch	ミヤマコホソハマキ	*Aethes rutilana* (Hübner)	（チョウ目、ホソハマキガ科）日本、旧北区
barred grass-skipper			barred skipper を見よ
barred grass-veneer		*Agriphila inquinatellus* (Denis et Schiffermüller)	（チョウ目、メイガ科）旧北区
barred green colonel		*Odontomyia hydroleon* (Linnaeus)	（ハエ目、ミズアブ科）旧北区
barred groundling			crepuscular rock-rose moth を見よ
barred hook-tip		*Drepana cultraria* (Fabricius)	（チョウ目、カギバガ科）旧北区
barred light-green			light emerald を見よ
barred marble	ツマオビキホソハマキ	*Eupoecillia angustana* (Hübner)	（チョウ目、ホソハマキガ科）日本、旧北区
barred marble conch			barred marble を見よ
barred pink looper moth			barred marble を見よ
barred red		*Hylaea fasciaria* (Linnaeus)	（チョウ目、シャクガ科）旧北区
barred rivulet		*Perizoma bifaciata* (Haworth)	（チョウ目、シャクガ科）旧北区
barred sailer		*Neptis trigonophora* Butler	（チョウ目、タテハチョウ科）エチオピア区
barred sallow		*Xanthia aurago* (Denis et Schiffermüller)	（チョウ目、ヤガ科）旧北区
barred short-tail skipper		*Polythrix gyges* Evans	（チョウ目、セセリチョウ科）新熱帯区

英名	和名	学名	所属、分布、ほか
barred skipper		*Dispar compacta* (Butler)	（チョウ目、セセリチョウ科）豪州区
barred snout		*Nemotelus uliginosus* (Linnaeus)	（ハエ目、ミズアブ科）旧北区
barred straw chevron			barred straw moth を見よ
barred straw moth		*Eulithis pyraliata* (Denis et Schiffermüller)	（チョウ目、シャクガ科）旧北区。barred straw ともいう
barred sulphur			fairy yelow を見よ
barred tooth-striped		*Trichopteryx polycommata* (Denis et Schiffermüller)	（チョウ目、シャクガ科）旧北区。日本亜種は *T. p. anna* Inoue ハネナガコバネナミシャク
barred umber		*Plagodis pulveraria* (Linnaeus)	（チョウ目、シャクガ科）旧北区。日本に2亜種 *P. p. japonica* Butler コナフキエダシャク（本、四、九）、*P. p. jezoensis* (Inoue)（北）
barred umber thorn			barred umber を見よ
barred upland looper moth		*Xanthorhoe orophyla* Meyrick	（チョウ目、シャクガ科）豪州区
barred woodbrown	マイトリアヤマヒカゲ	*Lethe maitrya* de Nicéville	（チョウ目、タテハチョウ科）東洋区
barred yellow		*Cidaria fulvata* (Forster)	（チョウ目、シャクガ科）旧北区
barred yellow			fairy yelow を見よ
barred yellow winged Temnora		*Temnora pylas* (Cramer)	（チョウ目、スズメガ科）エチオピア区
Barrens moth		*Properigea costa* (Barnes et Benjamin)	（チョウ目、ヤガ科）新北区
Barrens underwing			Jair underwing moth を見よ
barretti drywood termite		*Incisitermes barretti* (Gay)	（シロアリ目、レイビシロアリ科）豪州区
Barrett's marbled coronet		*Hadena luteago barrettii* (Doubleday)	（チョウ目、ヤガ科）旧北区
Barry's hairstreak		*Callophrys gryneus plicataria* Johnson	（チョウ目、シジミチョウ科）新北区。juniper hairstreak (1) を参照
Barry's hairstreak		*Mitoura barryi* Johnson	（チョウ目、シジミチョウ科）新北区
Bartram's hairstreak		*Strymon acis bartrami* (Comstock et Huntington)	（チョウ目、シジミチョウ科）新北区
Bartram's round winged katydid		*Amblycorypha bartrami* Walker	（バッタ目、キリギリス科）新北区
Bartram's scrub-hairstreak			Bartram's hairstreak を見よ
basal-dash Glaphyria moth		*Glaphyria basiflavalis* Barnes et McDunnough	（チョウ目、メイガ科）新北区
base-spotted pigmy		*Stigmella basiguttella* (Heinemann)	（チョウ目、モグリチビガ科）旧北区
basker		*Euchromia lethe* (Fabricius)	（チョウ目、ヒトリガ科）エチオピア区
basker moth			basker を見よ
baskers		*Urothemis*	（トンボ目、トンボ科）の昆虫の総称
basket longhorn beetle		*Gracilia minuta* (Fabricius)	（コウチュウ目、カミキリムシ科）旧北区、豪州区。ヨーロッパの最小カミキリムシの1種
basket moths			bagworm moths を見よ

英名	和名	学名	所属、分布、ほか
basket worm		*Bambalina consorta* Templeton	(チョウ目、ミノガ科) 旧北区
basketworm		*Kophene snelleni* Heylaerts	(チョウ目、ミノガ科) 東洋区
basket worms			bagworm moths を見よ　幼虫の英名
Basoches ruby-eye		*Carystoides basoches* (Latreille)	(チョウ目、セセリチョウ科) 新熱帯区
basswood aphid			lime leaf aphid を見よ　北米での英名
basswood lace bug		*Gargaphia tiliae* (Walsh)	(カメムシ目、グンバイムシ科) 新北区体長 4～5 mm。basswood アメリカシナノキ他につく
basswood leaf miner		*Baliosus nervosus* Panzer	(コウチュウ目、ハムシ科) 新北区
basswood leaf miner		*Baliosus ruber* (Weber)	(コウチュウ目、ハムシ科) 新北区
basswood leaf roller		*Pantographa limata* Grote et Robinson	(チョウ目、メイガ科) 新北区。幼虫はシナノキにつく。basswood leafroller moth ともいう
basswood miner moth		*Phyllonorycter lucetiella* (Clemens)	(チョウ目、コハモグリガ科) 新北区
basswood Olethreutes moth		*Olethreutes tilianum* (Heinrich)	(チョウ目、ハマキガ科) 新北区
basswood round-blotch miner moth		*Phyllocnistis tiliacella* (Chambers)	(チョウ目、コハモグリガ科) 新北区
Basuto magpie		*Thestor basutus* Wallengren	(チョウ目、シジミチョウ科) エチオピア区。斑紋がカササギに似ることから。Basuto は南アフリカの地名
bat bug		*Cimex pipistrelli* Jenyns	(カメムシ目、トコジラミ科) 旧北区
bat bugs	コウモリヤドリカメムシ科	Polyctenidae	(カメムシ目) の昆虫の総称　豪州区。コウモリに外部寄生
bat bugs			bed bugs を見よ
bat fleas	コウモリノミ科	Ischnopsyllidae	(ノミ目) の昆虫の総称
bat flies (1)	クモバエ科	Nycteribiidae	(ハエ目) の昆虫の総称
bat flies (2)	コウモリバエ科	Streblidae	(ハエ目) の昆虫の総称
bat flies			pupiparous flies を見よ
bat lice			bat flies (1) を見よ
bat louse flies			bat flies (1) を見よ　南アフリカでの英名
bat tick flies			bat flies (1) を見よ
bat willow gall midge			willow terminal leaf midge を見よ
Bate's crescentspot			tawny crescent を見よ
Bates' crow		*Euploea batesii* (C. et R. Felder)	(チョウ目、タテハチョウ科) 豪州区
Bates' hairstreak		*Evenus batesii* (Hewitson)	(チョウ目、シジミチョウ科) 新熱帯区
Bates' sister			Massilia sister を見よ
Bates swordtail		*Protesilaus glaucolaus* (Bates)	(チョウ目、アゲハチョウ科) 新熱帯区
bath white	チョウセンモンシロチョウ	*Pontia daplidice* (Linnaeus)	(チョウ目、シロチョウ科) 日本、旧北区、東洋区、エチオピア区。bath white butterfly ともいう
bath whites	チョウセンシロチョウ属	*Pontia*	(チョウ目、シロチョウ科) の昆虫の総称
bathroom fly		*Telmatoscopus albipunctatus* (Williston)	(ハエ目、チョウバエ科) 全北区

英名	和名	学名	所属、分布、ほか
Bathurst burr seed fly		*Euaresta bullans* (Wiedemann)	(ハエ目、ミバエ科) 旧北区、豪州区
Bathurst copper			purple copper を見よ
Batia moth			lesser tawny tubic を見よ
batman moth		*Coelostathma discopunctana* Clemens	(チョウ目、ハマキガ科) 新北区
Baton blue		*Pseudophilotes baton* (Bergstrasser)	(チョウ目、シジミチョウ科) 旧北区
battle twig			common earwig を見よ
battling glider		*Cymothoe alcimeda* (Godart)	(チョウ目、タテハチョウ科) エチオピア区
Baumann's Charaxes		*Charaxes baumanni* Rogenhofer	(チョウ目、タテハチョウ科) エチオピア区
Baumann's conch		*Aethes hartmanniana* (Clerck)	(チョウ目、ホソハマキガ科) 旧北区
Baviannskloof blue		*Lepidochrysops poseidon* Pringle	(チョウ目、シジミチョウ科) エチオピア区
Bavius blue		*Pseudophilotes bavius* (Eversmann)	(チョウ目、シジミチョウ科) 旧北区
Baxter's polyptychus		*Polyptychus baxteri* Rothschild et Jordan	(チョウ目、スズメガ科) エチオピア区
Bay checkerspot		*Euphydryas editha bayensis* Sternitz	(チョウ目、タテハチョウ科) 新北区。Bay checkerspot butterfly ともいう
bay-feather case		*Coleophora badiipennella* (Duponchel)	(チョウ目、ツツミノガ科) 旧北区
bay flea louse			bay sucker を見よ
bay psyllid			bay sucker を見よ
Bay skipper		*Euphyes bayensis* Shuey	(チョウ目、セセリチョウ科) 新北区
bay sucker		*Trioza alacris* Flor	(カメムシ目、トガリキジラミ科) 旧北区。bay 月桂樹 に由来
bay-tree scale		*Dynaspidiotus britannicus* (Newstead)	(カメムシ目、マルカイガラムシ科) 全北区
bay underwing moth		*Catocala badia* Grote et Robinson	(チョウ目、ヤガ科) 新北区。bay underwing, bayberry underwing ともいう
bay whitefly		*Paraleyrodes perseae* (Quaintance)	(カメムシ目、コナジラミ科) 新北区、新熱帯区。カンキツ害虫
bayberry leaftier moth		*Strepsicrates smithiana* Walsingham	(チョウ目、ハマキガ科) 新北区、大洋区
bayou clubtail		*Arigomphus maxwelli* (Ferguson)	(トンボ目、サナエトンボ科) 新北区
bazar fly	フタスジイエバエ	*Musca sorbens* Wiedemann	(ハエ目、イエバエ科) 日本、旧北区、東洋区、豪州区、大洋区、エチオピア区
BD butterfly	ベニオビムラサキタテハ	*Callicore cynosura* (Doubleday et Hewitson)	(チョウ目、タテハチョウ科) 新熱帯区。後翅裏面の斑紋が BD に似る
beach flies		*Fucellia*	(ハエ目、イエバエ科) の昆虫の総称
beach flies (1)	ニセミギワバエ科	Canaceidae	(ハエ目) の昆虫の総称
beach ground cricket	ハマスズ	*Pteronemobius csikii* (Bolivar)	(バッタ目、ケラ科) 日本、旧北区、東洋区

英名	和名	学名	所属、分布、ほか
beach pea borer			white-edged Pima moth を見よ
beach skipper			obscure skipper を見よ
beach trip		*Anaxipha litarena* Fulton	（バッタ目、クサヒバリ科）新北区
beachgrass scale		*Acanthococcus carolinae* (Williams)	（カメムシ目、フクロカイガラムシ科）新北区
beaded chestnut		*Agrochola lychnidis* (Denis et Schiffermüller)	（チョウ目、ヤガ科）旧北区
beaded lacewings	ケカゲロウ科	Berothidae	（アミメカゲロウ目）の昆虫の総称
beak			nettle-tree butterfly を見よ
beaked insects			hemipteroids を見よ
beaks	テングチョウ属	*Libythea*	（チョウ目、タテハチョウ科）の昆虫の総称
bean and poppy aphid			bean aphid を見よ
bean aphid	マメクロアブラムシ	*Aphis fabae* Scopoli	（カメムシ目、アブラムシ科）日本、全北区。エンドウ、ソラマメなどの害虫
bean aphid			cowpea aphid を見よ
bean beetle			broadbean weevil を見よ
bean beetle			bean weevil を見よ
bean blister beetle	マメハンミョウ	*Epicauta gorhami* Marseul	（コウチュウ目、ツチハンミョウ科）日本。マメ、ナスなどの害虫
bean blossom thrips		*Megalurothrips usitatus* (Bagnall)	（アザミウマ目、アザミウマ科）日本、東洋区、豪州区
bean blue			common grass-blue (1) を見よ
bean bruchid			bean weevil を見よ
bean bug	ホソヘリカメムシ	*Riptortus clavatus* (Thunberg)	（カメムシ目、ホソヘリカメムシ科）日本、旧北区、東洋区。イネ他各種作物、果樹の害虫
bean bug	キスジホソヘリカメムシ	*Riptortus linearis* (Fabricius)	（カメムシ目、ホソヘリカメムシ科）日本、旧北区、東洋区。ダイズなどマメ類害虫
bean bug	キボシホソヘリカメムシ	*Riptortus pedestris* (Fabricius)	（カメムシ目、ホソヘリカメムシ科）日本、東洋区
bean butterfly			pea blue を見よ
bean capsid		*Pycnoderes quadrimaculatus* Guérin-Méneville	（カメムシ目、カスミカメムシ科）新北区
bean flower caterpillar			dark Cerulean（2）を見よ
bean flower thrips		*Taeniothrips sjostedti* (Trybom)	（アザミウマ目、アザミウマ科）エチオピア区
bean flower thrips			bean blossom thrips を見よ
bean fly			French bean fly を見よ　豪州での英名。bean-fly　とも記される。
bean frosted weevil	コフキゾウムシ	*Eugnathus distinctus* Roelofs	（コウチュウ目、ゾウムシ科）日本、旧北区、東洋区
bean gall			willow bean-gall sawfly （ハチ目、ハバチ科）がヤナギの葉につくる虫えいで 12 mm の大きさ。旧北区
bean lace bug			coffee lace bug をみよ

英名	和名	学名	所属、分布、ほか
bean lady beetle			Mexican bean beetle (1) を見よ
bean leaf beetle		*Cerotoma trifurcata* (Forster)	(コウチュウ目、ハムシ科) 新北区。ダイズ害虫
bean leaf beetle (1)	ヒメキバネサルハムシ	*Pagria signata* (Motschulsky)	(コウチュウ目、ハムシ科) 日本、旧北区、東洋区、大洋区
bean leaf roller			long-tailed skipper を見よ
bean leaf skeletonizer moth		*Autoplusia egena* (Guenée)	(チョウ目、ヤガ科) 新北区、新熱帯区。幼虫は bean-leaf skeletonizer
bean leaf webber			bean pyralid (1) を見よ
bean-leaf webworm moth		*Omiodes indicata* (Fabricius)	(チョウ目、メイガ科) 全北区、新熱帯区、東洋区、豪州区、エチオピア区
bean leafhopper		*Empoasca kraemeri* Ross et Moore	(カメムシ目、オオヨコバイ科) 新熱帯区
bean leafhopper			potato leafhopper (1) を見よ
bean leafminer	ツメクサハモグリバエ	*Liriomyza congesta* (Becker)	(ハエ目、ハモグリバエ科) 日本。ダイズ他マメ類害虫
bean leafminer beetle	マメチビタマムシ	*Trachys reitteri* Obenberger	(コウチュウ目、タマムシ科) 日本、旧北区。ダイズ害虫
bean leafroller		*Omiodes diemenalis* (Guenée)	(チョウ目、メイガ科) 東洋区、豪州区
bean lycaenid			gray hairstreak (1) を見よ
bean miner	ニセインゲンハモグリバエ	*Ophiomyia centrosematis* (de Meijere)	(ハエ目、ハモグリバエ科) 日本、東洋区、豪州区。インゲンなどマメ類害虫
bean narrow-mouth weevil	マメホソクチゾウムシ	*Apion collare* Schilsky	(コウチュウ目、ホソクチゾウムシ科) 日本、東洋区
bean pod borer (1)	マメノメイガ	*Maruca testulalis* (Hübner)	(チョウ目、メイガ科) 日本、旧北区、東洋区、エチオピア区、豪州区、新熱帯区。ダイズ他マメ類害虫
bean pod borer (2)		*Maruca vitrata* (Fabricius)	(チョウ目、メイガ科) 東洋区、豪州区。bean pod borer moth ともいう
bean pod borer			lima-bean pod borer を見よ
bean pod moth			bean pod borer (2) を見よ
bean pyralid	ミスジノメイガ	*Protonoceros capitalis* (Fabricius)	(チョウ目、メイガ科) 日本、東洋区。ダイズ他マメ類害虫
bean pyralid (1)	マエウスキノメイガ	*Hedylepta indicata* (Fabricius)	(チョウ目、メイガ科) 日本、東洋区、新北区、エチオピア区。ダイズ他マメ類害虫
bean root aphid		*Smynthurodes betae* Westwood	(カメムシ目、アブラムシ科) 汎世界
bean seed beetle			broadbean weevil を見よ
bean seed fly	タネバエ	*Delia platura* (Meigen)	(ハエ目、ハナバエ科) 日本、汎世界。マメ類ほか多くの作物害虫
bean seed maggot			seed potato maggot を見よ
bean shoot moth		*Epinotia aporema* (Walsingham)	(チョウ目、ハマキガ科) 新熱帯区
bean stalk weevil		*Sternechus paludatus* (Casey)	(コウチュウ目、ゾウムシ科) 新北区
bean stem miner			soybean stem miner を見よ
bean thrips	マメクダアザミウマ	*Liothrips glycinicola* Okamoto	(アザミウマ目、クダアザミウマ科) 日本

英名	和名	学名	所属、分布、ほか
bean thrips		*Caliothrips fasciatus* (Pergande)	(アザミウマ目、アザミウマ科) 全北区。カンキツ害虫
bean thrips			pea thrips (1) を見よ
bean tussock moth	マメドクガ	*Cifuna locuples confusa* (Bremer)	(チョウ目、ドクガ科) 日本、旧北区。マメ類、果樹などの害虫
bean webworm	ウコンノメイガ	*Pleuroptya ruralis* (Scopoli)	(チョウ目、メイガ科) 日本、旧北区、東洋区。マメ類などの害虫
bean weevil	インゲンマメゾウムシ	*Acanthoscelides obtectus* (Say)	(コウチュウ目、ハムシ科) 日本、汎世界。貯穀害虫
bean weevil			adzuki weevil を見よ
bean weevil			cowpea weevil を見よ
bean weevils			pea weevils を見よ
bean weevils			seed weevils (1) を見よ
Bean's tiger-moth		*Neoarctia beanii* (Neumoegen)	(チョウ目、ヒトリガ科) 新北区
bear			stout spanworm moth を見よ
bear beetle		*Paracotalpa ursina* Horn	(コウチュウ目、コガネムシ科) 新北区。バラ科植物に多い
bear moths		Brachodidae	(チョウ目) の昆虫の総称
bear sphinx		*Aretonotus lucidus* Boisduval	(チョウ目、スズメガ科) 新北区
bearberry aphid		*Tamalia (Phillaphis) coweni* (Cockerell)	(カメムシ目、アブラムシ科) 新北区
beard-grass skipper		*Atrytone arogos* (Boisduval et LeConte)	(チョウ目、セセリチョウ科) 新北区
bearded robber fly		*Efferia pogonias* (Wiedemann)	(ハエ目、ムシヒキアブ科) 新北区
bearded skolly		*Thestor barbatus* Henning et Henning	(チョウ目、シジミチョウ科) エチオピア区
bearded weevil		*Rhinostomus barbirostris* (Fabricius)	(コウチュウ目、ゾウムシ科) 新熱帯区
Beardsley leafhopper		*Balclutha beardsleyi* Namba	(カメムシ目、オオヨコバイ科) 新北区、大洋区。ハワイの昆虫研究者に由来
Beardsley leafhopper		*Balclutha saltuella* (Kirschbaum)	(カメムシ目、オオヨコバイ科) 新北区、大洋区
Beaufort opal		*Chrysoritis beaufortius* (Dickson)	(チョウ目、シジミチョウ科) エチオピア区
beaulieu dung beetle		*Aphodius niger* Illiger	(コウチュウ目、コガネムシ科) 旧北区
beauties		Boarmiinae	(チョウ目、シャクガ科) の昆虫の総称
beautiful Antinephele		*Antinephele maculifera* Holland	(チョウ目、スズメガ科) エチオピア区
beautiful arches		*Blepharita satura* (Denis et Schiffermüller)	(チョウ目、ヤガ科) 旧北区
beautiful banner		*Epiphile hermosa* Moza et Frances	(チョウ目、タテハチョウ科) 新熱帯区
beautiful-bordered pearl	ウスムラサキノメイガ	*Agrotera nemoralis* (Scopoli)	(チョウ目、メイガ科) 日本、旧北区

英名	和名	学名	所属、分布、ほか
beautiful brindled groundling		*Recurvaria nanella* (Hübner)	(チョウ目、キバガ科) 全北区。リンゴ害虫
beautiful brocade	ムラサキヨトウ	*Lacanobia contigua* (Denis et Schiffermüller)	(チョウ目、ヤガ科) 日本、旧北区
beautiful brocade			light brocade を見よ
beautiful carpet	イチゴナミシャク	*Mesoleuca albicillata* (Linnaeus)	(チョウ目、シャクガ科) 日本、旧北区
beautiful china mark		*Nymphula stagnata* (Donovan)	(チョウ目、メイガ科) 旧北区
beautiful Cosmopterix moth		*Cosmopterix pulchrimella* Chambers	(チョウ目、カザリバガ科) 新北区
beautiful crescent piercer		*Cydia jungiella* (Clerck)	(チョウ目、ハマキガ科) 旧北区
beautiful demoiselle			demoiselle Agrion を見よ
beautiful Epicallima moth		*Epicallima formosella* (Denis et Schiffermüller)	(チョウ目、マルハキバガ科) 新北区
beautiful ermel		*Ethmia pusiella* (Linnaeus)	(チョウ目、スエヒロキバガ科) 旧北区
beautiful Eutelia moth		*Eutelia pulcherrimus* Grote	(チョウ目、ヤガ科) 新北区
beautiful glasswing		*Ithomia terra* Hewitson	(チョウ目、タテハチョウ科) 新熱帯区
beautiful Glyphodes moth			mulberry pyralid を見よ
beautiful golden Y		*Autographa pulchrina* (Haworth)	(チョウ目、ヤガ科) 旧北区
beautiful golden Y		*Autographa v-aureum* Guenée	(チョウ目、ヤガ科) 旧北区
beautiful gothic		*Leucochlaena oditis* (Hübner)	(チョウ目、ヤガ科) 旧北区
beautiful Hawaiian damselfly		*Megalagrion calliphya* (McLachlan)	(トンボ目、イトトンボ科) 大洋区
beautiful hawk		*Leptoclanis pulchra* Rothschild et Jordan	(チョウ目、スズメガ科) エチオピア区
beautiful hook-tip	カギアツバ	*Laspeyria flexula* (Denis et Schiffermüller)	(チョウ目、ヤガ科) 日本、旧北区
beautiful hook-wing			beautiful hook-tip を見よ
beautiful jewelwing			demoiselle Agrion を見よ
beautiful Junea	カギバシラホシジャノメ	*Junea dorinda* (C. et R. Felder)	(チョウ目、タテハチョウ科) 新熱帯区
beautiful knot-horn		*Pempelia formosa* (Haworth)	(チョウ目、メイガ科) 旧北区
beautiful leopard		*Timelaea albescens* Oberthür	(チョウ目、タテハチョウ科) 旧北区
beautiful marbled bell	ニセモンシロスソモンヒメハマキ	*Eucosma campoliliana* (Denis et Schiffermüller	(チョウ目、ハマキガ科) 日本、旧北区
beautiful marbled groundling		*Calyocolum marmoreum* (Haworth)	(チョウ目、キバガ科) 旧北区
beautiful Methiola		*Methiola* sp.	(バッタ目、バッタ科) 豪州区

英名	和名	学名	所属、分布、ほか
beautiful mining moths			ermine moths (1) を見よ
beautiful monarch			forest monarch を見よ
beautiful Mylon		*Mylon zephus* (Butler)	（チョウ目、セセリチョウ科）新熱帯区
beautiful parasites		Callimomidae	（ハチ目）の昆虫の総称　Torymidae の異名。米国での英名
beautiful parasites			seed chalcids (1) を見よ
beautiful petaltail		*Petalura pulcherrima* Tillyard	（トンボ目、ムカシヤンマ科）豪州区
beautiful Phaneta moth		*Phaneta formosana* (Clemens)	（チョウ目、ハマキガ科）新北区
beautiful plume		*Amblyptilia acanthodactyla* (Hübner)	（チョウ目、トリバガ科）旧北区
beautiful snout		*Hypena crassalis* (Fabricius)	（チョウ目、ヤガ科）旧北区
beautiful Sparganothis moth		*Sparganothis pulcherrimana* (Walsingham)	（チョウ目、ハマキガ科）新北区
beautiful sylph		*Metisella formosas* (Butler)	（チョウ目、セセリチョウ科）エチオピア区
beautiful tiger		*Amphicallia bellatrix* (Dalman)	（チョウ目、ヒトリガ科）旧北区
beautiful tiger			forest monarch を見よ
beautiful tiger beetle		*Cicindela formosa* Say	（コウチュウ目、ハンミョウ科）新北区
beautiful underwing			beautiful yellow underwing を見よ
beautiful Utetheisa			bella moth を見よ
beautiful wood-nymph moth		*Eudryas grata* Fabricius	（チョウ目、ヤガ科）新北区。beautiful wood-nymph ともいう
beautiful yellow underwing		*Anarta myrtilli* (Linnaeus)	（チョウ目、ヤガ科）旧北区
beautiful zebra blue		*Leptotes pulcher* (Murray)	（チョウ目、シジミチョウ科）エチオピア区
beauty (beauties)			anglewings を見よ
beaver beetle			beaver parasite を見よ
beaver beetles			mammal nest beetles を見よ
beaver-nest beetle		*Leptinus validus* Horn	（コウチュウ目、Leptinidae）新北区
beaver parasite		*Platypsyllus castoris* Ritsema	（コウチュウ目、Platypsyllidae）新北区。体長2～3 mm。ビーバーに外部寄生
beaver parasites		Platypsillidae	（コウチュウ目）の昆虫の総称
beaverpond baskettail		*Epitheca canis* (McLachlan)	（トンボ目、エゾトンボ科）新北区
beaverpond clubtail		*Gomphus borealis* Needham	（トンボ目、サナエトンボ科）新北区
Bechina purplewing		*Eunica bechina* Hewitson	（チョウ目、タテハチョウ科）新熱帯区
Bechtel's shieldback		*Idiostatus bechteli* Rentz	（バッタ目、キリギリス科）新北区
Becker's creamy yellow glider	ベッケリーキイロタテハ	*Cymothoe beckeri* (Herrich-Schäffer)	（チョウ、タテハチョウ科）エチオピア区
Becker's white		*Pontia beckerii* (Edwards)	（チョウ目、シロチョウ科）新北区

英名	和名	学名	所属、分布、ほか
bed bug		*Cimex rotundatus* Signoret	（カメムシ目、トコジラミ科）東洋区、エチオピア区。人に寄生
bed bug (1)	トコジラミ	*Cimex lectularius* Linnaeus	（カメムシ目、トコジラミ科）日本、汎世界。人に寄生
bed bug (2)		*Cimex hemipterus* Fabricius	（カメムシ目、トコジラミ科）東洋区
bedbug			bed bug (1) を見よ
bed bugs	トコジラミ科	Cimicidae	（カメムシ目）の昆虫の総称　人、鳥から吸血
bedeguar			mossy rose gall wasp を見よ
bedeguar gall			mossy rose gall wasp を見よ
bedeguar gall wasp			mossy rose gall wasp を見よ
bedford blue			small blue butterfly を見よ
Bedford Russell's Idea		*Idea tambusisiana* Bedford-Russell	（チョウ目、タテハチョウ科）東洋区
Bedford Russell's tree-nymph			Bedford Russell's Idea を見よ
bedrule brocade		*Blepharita solieri* Boisduval	（チョウ目、ヤガ科）旧北区
bedstraw aphid		*Dysaphis pyri* (Boyer de Fonscolombe)	（カメムシ目、アブラムシ科）旧北区
bedstraw hawk	イブキスズメ	*Hyles gallii* (Rottenburg)	（チョウ目、スズメガ科）日本、全北区、東洋区
bedstraw hawk moth			bedstraw hawk を見よ　米国での英名
bedstraw hawklet		*Microsphinx pumilum* (Boisduval)	（チョウ目、スズメガ科）エチオピア区
bee		*Trigona amalthea* Fabricius	（ハチ目、ミツバチ科）新熱帯区
bee		*Chorinea faunus* Fabricius	（チョウ目、シジミタテハ科）新熱帯区
bee			honey bee を見よ　一般にミツバチやハチを指す
bee assassin		*Apiomerus crassipes* (Fabricius)	（カメムシ目、サシガメ科）新北区
bee beetle		*Trichodes apiarius* (Linnaeus)	（コウチュウ目、カッコウムシ科）旧北区。ミツバチの巣につく欧州の小甲虫
bee beetle			bee chafer を見よ
bee chafer		*Trichius fasciatus* (Linnaeus)	（コウチュウ目、コガネムシ科）旧北区。bee beetle ともいう
bee flies	ツリアブ科	Bombyliidae	（ハエ目）の昆虫の総称
bee flies			woolly bee flies を見よ
bee fly		*Anthrax analis* Say	（ハエ目、ツリアブ科）新北区
bee fly (1)	ビロウドツリアブ	*Bombylius major* Linnaeus	（ハエ目、ツリアブ科）日本、全北区
bee hawk moth		*Cephanodes kingii* (Macleay)	（チョウ目、スズメガ科）豪州区
bee hawk moth			larger pellucid hawk moth を見よ
bee hawk moths		*Hemaris*	（チョウ目、スズメガ科）の昆虫の総称
bee hawks		*Cephonodes*	（チョウ目、スズメガ科）の昆虫の総称
bee-hunting digger wasp			bee-killer wasp を見よ

英名	和名	学名	所属、分布、ほか
bee killer			robber flies (1) を見よ
bee-killer			assassin bug (2) を見よ
bee-killer			bee-killer wasp を見よ
bee-killer wasp		*Philanthus triangulus* (Fabricius)	（ハチ目、アナバチ科）旧北区
bee-killer wasps (1)	フシダカバチ亜科	Philanthinae	（ハチ目、アナバチ科）の昆虫の総称
bee-killer wasps (2)		Philanthidae	（ハチ目）の昆虫の総称　学名はアナバチ科の異名
bee-like tachinid fly		*Hystricia abrupta* (Wiedemann)	（ハエ目、ヤドリバエ科）新北区
bee lice	ミツバチシラミバエ科	Braulidae	（ハエ目）の昆虫の総称
bee louse	ミツバチシラミバエ	*Braula coeca* Nitzsch	（ハエ目、ミツバチシラミバエ科）日本、全北区。ミツバチの巣に生息
bee louse flies			bee lice を見よ
bee louse flies			bee lice を見よ
bee moth		*Aphomia sociella* Linnaeus	（チョウ目、メイガ科）全北区
bee moth			greater wax moth を見よ
bee moths			wax moths を見よ
bee parasite beetle		*Stylops vandykei* Bohart	（ネジレバネ目、ハチネジレバネ科）新北区
bee pirate			bee-killer wasp を見よ　米国、南アフリカでの英名
bee pirates			bee-killer wasps (2) を見よ　米国、南アフリカでの英名
bee wolf			bee-killer wasp を見よ
bee wolf			bee beetle の幼虫
bee-wolf wasps			bee-killer wasps (1) を見よ
beech aphid			woolly beech aphid を見よ
beech blight aphid		*Fagiphagus imbricator* (Fitch)	（カメムシ目、アブラムシ科）新北区
beech borer			beech splendour beetle を見よ
beech bud gall midge		*Contarinia fagi* Rubsaamen	（ハエ目、タマバエ科）旧北区
beech caterpillar	ブナアオシャチホコ	*Quadricalcarifera punctatella* (Motschulsky)	（チョウ目、シャチホコガ科）日本。幼虫はブナを食う
beech coccus			beech scale（1）を見よ
beech flea weevil			beech leaf miner beetle を見よ
beech-furniture borer		*Ptilinus pectinicornis* (Linnaeus)	（コウチュウ目、シバンムシ科）旧北区
beech gall-midge		*Oligotrophus fagineus* Kieffer	（ハエ目、タマバエ科）がブナの葉の上面につくる球形の虫えいで高さ 10 mm。旧北区
beech green carpet		*Colostygia olivata* (Denis et Schiffermüller)	（チョウ目、シャクガ科）旧北区。beech-green carpet とも記す
beech hairy-pouch-gall midge		*Hartigola annulipes* (Hartig)	（ハエ目、タマバエ科）旧北区
beech leaf gall midge			Japanese beech gall midge を見よ　beech-leaf gall midge とも記す

英名	和名	学名	所属、分布、ほか
beech leaf miner			beech leaf miner beetle を見よ
beech leaf miner beetle		*Rhynchaenus fagi* (Linnaeus)	(コウチュウ目、ゾウムシ科) 旧北区
beech leafhopper		*Fagocyba cruenta* (Herrich-Schäffer)	(カメムシ目、オオヨコバイ科) 旧北区
beech leaf-mining weevil			beech leaf miner beetle を見よ
beech leaftier			black-fringed leaftier moth を見よ
beech longicorn			beech longicorn beetle を見よ
beech longicorn beetle	ヨコヤマヒゲナガカミキリ	*Dolichoprosopus yokoyamai* (Gressitt)	(コウチュウ目、カミキリムシ科) 日本
beech louse		*Lachnus exsicator* Altum	(カメムシ目、アブラムシ科) 旧北区
beech scale		*Cryptococcus fagi* (Hartig)	(カメムシ目、Cryptococcidae) 旧北区
beech scale (1)		*Cryptococcus fagisuga* Lindinger	(カメムシ目、Cryptococcidae) 全北区
beech smooth-pouch gall midge			Japanese beech gall midge を見よ
beech splendour beetle	ヤナギナガタマムシ	*Agrilus viridis* (Linnaeus)	(コウチュウ目、タマムシ科) 日本、全北区
beech tortrix			smoky marbled piercer を見よ　米国での英名。beech tortricid ともいう
beech-tree beetles		Perothopidae	(コウチュウ目) の昆虫の総称
beech winter moth			northern winter moth (1) を見よ
beedle miner moths			shiny head-standing moths を見よ
beef chewing louse			cattle chewing louse を見よ
beehive honey moth			dried-fruit moth を見よ
beehole borer		*Xyleutes ceramicus* Walker	(チョウ目、ボクトウガ科) 東洋区
bees	ミツバチ上科	Apoidea	(ハチ目) の昆虫の総称
bees (1)	ミツバチ科	Apidae	(ハチ目) の昆虫の総称
beet and mangold fly			beet leaf miner (1) を見よ
beet armyworm	シロイチモジヨトウ	*Spodoptera exigua* (Hübner)	(チョウ目、ヤガ科) 日本、全北区、東洋区、豪州区。サトイモなど各種作物の害虫。成虫は beet armyworm moth
beet beetle			pygmy mangold beetle を見よ
beet bug	テンサイヒメヒラタカメムシ	*Piesma quadratum* (Fieber)	(カメムシ目、チビカメムシ科) 旧北区
beet bugs			ash-gray leaf bugs を見よ
beet carrion beetle		*Blitophaga opaca* (Linnaeus)	(コウチュウ目、シデムシ科) 旧北区
beet caterpillar			white dot を見よ
beet flea beetle		*Chaetocnema tibialis* (Illiger)	(コウチュウ目、ハムシ科) 旧北区
beet fly			beet leaf miner (1) を見よ
beet lace bug			beet bug を見よ
beet leaf beetle		*Erynephala puncticollis* (Say)	(コウチュウ目、ハムシ科) 新北区

英名	和名	学名	所属、分布、ほか
beet leaf bug			beet bug を見よ
beet leaf miner (1)		*Pegomya betae* (Curtis)	（ハエ目、ハナバエ科）全北区。beet leafminer とも記す
beet leafminer		*Liriomyza chenopodii* (Watt)	（ハエ目、ハモグリバエ科）豪州区
beet leafminer	テンサイモグリハナバエ	*Pegomya cunicularia* (Rondani)	（ハエ目、ハナバエ科）日本、全北区。テンサイ、ホウレンソウなどの害虫
beet leaf weevil		*Tanymecus palliatus* (Fabricius)	（コウチュウ目、ゾウムシ科）旧北区
beet leafhopper	テンサイヨコバイ	*Circulifer tenellus* (Baker)	（カメムシ目、オオヨコバイ科）全北区
beet lixus		*Lixus junci* Boheman	（コウチュウ目、ゾウムシ科）旧北区
beet moth			obscure-barred coast groundling を見よ
beet moth			sugarbeet moth を見よ
beet root weevil		*Cleonus punctiventris* (Germar)	（コウチュウ目、ゾウムシ科）旧北区
beet semi-looper	タマナギンウワバ	*Autographa nigrisigna* (Walker)	（チョウ目、ヤガ科）日本、東洋区
beet tortoise beetle (1)	カメノコハムシ	*Cassida nebulosa* Linnaeus	（コウチュウ目、ハムシ科）日本、旧北区
beet tortoise beetle (2)	スジミドリカメノコハムシ	*Cassida vittata* (de Villers)	（コウチュウ目、ハムシ科）旧北区
beet webworm	シロオビノメイガ	*Hymenia recurvalis* (Fabricius)	（チョウ目、メイガ科）日本、東洋区、豪州区、全北区、大洋区、新熱帯区、エチオピア区。テンサイ、ホウレンソウなどの害虫。beet webworm moth ともいう
beet weevil		*Cleonus mendicus* (Gyllenhal)	（コウチュウ目、ゾウムシ科）旧北区
beet worm			silver Y moth (1) を見よ　幼虫の英名
beetle cricket			smaller yellow-legged grass cricket を見よ
beetle crickets			bush crickets (3) を見よ
beetle flies	ヨロイバエ科	Celyphidae	（ハエ目）の昆虫の総称
beetle roach		*Diploptera dytiscoides* (Serville)	（ゴキブリ目、ブラベルスゴキブリ科）豪州区
beetle wasp		*Eucerceris canaliculata* (Say)	（ハチ目、アナバチ科）新北区
beetle wasps			cicada killers を見よ
beetles	コウチュウ目	Coleoptera	の昆虫の総称　世界に約 370,000 種
beggar moth		*Eubaphe mendica* (Walker)	（チョウ目、シャクガ科）新北区。beggar ともいう
begonia thrips		*Scirtothrips parvus* Moulton	（アザミウマ目、アザミウマ科）旧北区
Behren's rain beetle		*Pleocoma behrensi* LeConte	（コウチュウ目、コガネムシ科）新北区
Behr's blue		*Glaucopsyche lygdamus incognita* Tilden	（チョウ目、シジミチョウ科）新北区。silvery blue (1) を参照
Behr's hairstreak	バーカラスシジミ	*Satyrium behrii* (Edwards)	（チョウ目、シジミチョウ科）新北区
Behr's metalmark		*Apodemia virgulti virgulti* (Behr)	（チョウ目、シジミタテハ科）新熱帯区
Behr's sulphur	ミドリモンキチョウ	*Colias behrii* Edwards	（チョウ目、シロチョウ科）新北区

英名	和名	学名	所属、分布、ほか
Belfrage's cricket		*Trigonidomimus belfragei* Caudell	（バッタ目、コオロギ科）新北区
Belfrage's plume moth		*Pselnophorus belfragei* (Fish)	（チョウ目、トリバガ科）新北区。
Belkin's dune tabanid fly		*Brennania belkini* (Philip)	（ハエ目、アブ科）新北区。絶滅危惧種
bell bird psyllid		*Glycaspis baileyi* Moore	（カメムシ目、キジラミ科）豪州区
bell-flower pug			campanula pug を見よ
bell moths			leafroller moths を見よ
bell moths			archaic bell moths を見よ
Bella mapwing		*Hypanartia bella* (Fabricius)	（チョウ目、タテハチョウ科）新熱帯区
Bella moth		*Utetheisa bella* (Linnaeus)	（チョウ目、ヒトリガ科）新北区。rattle-box moth ともいわれる。
belladonna flea beetle		*Epitrix atropae* Foudras	（コウチュウ目、ハムシ科）旧北区
belle point			salix noctuid を見よ
Beller's ground beetle		*Agonum belleri* (Hatch)	（コウチュウ目、オサムシ科）新北区
Belle's sanddragon		*Progomphus bellei* Knopf et Tennessen	（トンボ目、サナエトンボ科）新北区
Bellona metalmark		*Necyria bellona* Westwood	（チョウ目、シジミチョウ科）新熱帯区
Bell's cloudy wing			King page を見よ
Bell's giant-skipper		*Agathymus belli* (Freeman)	（チョウ目、セセリチョウ科）新熱帯区
Bell's longtail		*Urbanus belli* (Hayward)	（チョウ目、セセリチョウ科）新北区、新熱帯区
Bell's paradise skipper		*Phocides metrodorus* Bell	（チョウ目、セセリチョウ科）新熱帯区
Bell's roadside skipper		*Amblyscirtes belii* Freeman	（チョウ目、セセリチョウ科）新北区
Bell's skipper		*Pachyneuria duidae* (Bell)	（チョウ目、セセリチョウ科）新熱帯区
Bell's skipper		*Saturnus reticulata obscurus* (Bell)	（チョウ目、セセリチョウ科）新熱帯区
belted beauty		*Lycia zonaria britannica* Harrison	（チョウ目、シャクガ科）旧北区。原名亜種 *L. zonaria* (Denis et Schiffermuller) も同英名
belted beauty			swallow-tailed moth を見よ
belted brindled-beauty			belted beauty を見よ
belted burnet			variable burnet moth (1) を見よ
belted skimmer			stream cruiser を見よ
belted skimmers		*Didymops*	（トンボ目、ヤマトンボ科）
belted skimmers			river skimmers (1) を見よ
belted skipper		*Atrytonopsis cestus* (Edwards)	（チョウ目、セセリチョウ科）新北区、新熱帯区
Belti skipper		*Zera belti* (Godman et Salvin)	（チョウ目、セセリチョウ科）新熱帯区
Belt's Myscelus		*Myscelus belti* Godman et Salvin	（チョウ目、セセリチョウ科）新熱帯区
Belus skipper		*Phocides belus* (Godman et Salvin)	（チョウ目、セセリチョウ科）新北区、新熱帯区

英名	和名	学名	所属、分布、ほか
Belus swallowtail	ベルスアオジャコウ	*Battus belus* (Cramer)	（チョウ目、アゲハチョウ科）新熱帯区
bembicid wasps	ハナダカバチ亜科	Bembicinae	（ハチ目、ハナダカバチ科）の昆虫の総称
Bengal albatross			eastern striped albatross を見よ
Bengal swift	トガリチャバネセセリ	*Pelopidas agna* (Moore)	（チョウ目、セセリチョウ科）日本、東洋区
Benito's skipper		*Poanes benito* Freeman	（チョウ目、セセリチョウ科）新熱帯区
bent-line carpet moth		*Costaconvexa centrostrigaria* (Wollaston)	（チョウ目、シャクガ科）全北区。bent-line carpet ともいう
bent-line gray moth		*Iridopsis larvaria* (Guenée)	（チョウ目、シャクガ科）新北区
bent-lined tan moth		*Oxycilla malaca* (Grote)	（チョウ目、ヤガ科）新北区
bentwing ghost moth		*Zelotypia stacyi* Scott	（チョウ目、コウモリガ科）豪州区
bentwing swift moth			bentwing ghost moth を見よ
bent-winged owlet moth		*Blaptina caradrinalis* Guenée	（チョウ目、ヤガ科）新北区。bent-winged owlet ともいう
bentwings		Leucopterinae	（チョウ目、ハモグリガ科）の昆虫の総称
bentwings			leaf skeletonizer moths (1) を見よ
benzoin aphid	アオキコブアブラムシ	*Acyrthosiphon linderae* (Shinji)	（カメムシ目、アブラムシ科）日本
ber defoliator			ber moth を見よ
ber hairy caterpillar			ber moth を見よ
ber leaf thrips		*Florithrips fraegardhi* Trybom	（アザミウマ目、アザミウマ科）東洋区
ber moth		*Thiacidas postica* Walker	（チョウ目、ヤガ科）東洋区
beraeids			small eastern caddisflies を見よ
berberis aphid			barberry aphid を見よ
berberis sawfly	ハサミチュウレンジ	*Arge berberides* (Klug)	（ハチ目、ミフシハバチ科）日本、旧北区
berberis seed fly	ワモンハマダラミバエ	*Phagocarpus permundus* (Harris)	（ハエ目、ミバエ科）旧北区
Berger's clouded yellow		*Colias sareptensis* Staudinger	（チョウ目、シロチョウ科）旧北区
Berger's clouded yellow	ニセモトモンキチョウ	*Colias alfacariensis* (Berger)	（チョウ目、シロチョウ科）旧北区
Berger's clouded yellow			new-clouded yellow を見よ
Berger's dusky dart		*Paracleros substrigata* (Holland)	（チョウ目、セセリチョウ科）エチオピア区
Bergmann's button		*Croesia bergmanniana* Linnaeus	（チョウ目、ハマキガ科）旧北区
Berg's Alatuncusia moth		*Alatuncusia bergii* (Möschler)	（チョウ目、メイガ科）新北区
Berkeley's Charaxes		*Charaxes berkeleyi* van Someren et Jackson	（チョウ目、タテハチョウ科）エチオピア区
Bermuda fruit fly			Mediterranean fruit fly を見よ
Bermudagrass leafhopper		*Carneocephala sagittifera* (Uhler)	（カメムシ目、オオヨコバイ科）大洋区

英名	和名	学名	所属、分布、ほか
Bermudagrass scale			Ruth grass scale を見よ
Bermudagrass shootborer			Bermudagrass stem maggot を見よ
Bermudagrass stem maggot	ギョウギシバクキイエバエ	*Atherigona reversura* Villeneure	（ハエ目、イエバエ科）日本、東洋区、大洋区
Bernardino blue		*Euphilotes bernardino bernardino* (Barnes et McDunnough)	（チョウ目、シジミチョウ科）新熱帯区
berothids			beaded lacewings を見よ
berry-gall			lenticular gall wasp の虫えいの英名
berry plume moth			Himmelman's plume moth を見よ
Berry's skipper			Florida swamp skipper を見よ
bertha armyworm			bertha armyworm moth を見よ
bertha armyworm moth		*Mamestra configurata* Walker	（チョウ目、ヤガ科）新北区
bespeckled leafhopper			brown speckled leafhopper を見よ
besprinkled cricket		*Gryllus conspersus* Schaum	（バッタ目、コオロギ科）旧北区。中国で鳴声鑑賞のため飼育
bess-beetle		*Odontotaenius disjunctus* (Illiger)	（コウチュウ目、クロツヤムシ科）新北区
bess beetles			peg beetles (1) を見よ
bessbug			bess-beetle を見よ
bessbugs			peg beetles (1) を見よ
Besta palm dart		*Telicota besta* Evans	（チョウ目、セセリチョウ科）東洋区
beth blue		*Hemiargus thomasi* Clench	（チョウ目、シジミチョウ科）新北区
Bethune's Zale moth		*Zale bethunei* (Smith)	（チョウ目、ヤガ科）新北区
bethylid wasps	アリガタバチ科	Bethylidae	（ハチ目）の昆虫の総称
bethylids			bethylid wasps を見よ
bethyloid wasps	アリガタバチ上科	Bethyloidea	（ハチ目）の昆虫の総称
betrothed underwing moth		*Catocala innubens* Guenée	（チョウ目、ヤガ科）新北区
Betsy beetle			bess-beetle を見よ
Betsy beetles			peg beetles (1) を見よ
betula aphid	カバヒラタアブラムシ	*Hormaphis betulae* (Mordvilko)	（カメムシ目、アブラムシ科）日本、旧北区
betula argid sawfly	カンバチュウレンジ	*Arge pullata* (Zaddach)	（ハチ目、ミフシハバチ科）日本、旧北区。カンバ類の害虫
betula leafminer	カンバハモグリバエ	*Agromyza betulae* Sasakawa	（ハエ目、ハモグリバエ科）日本。カンバ類の害虫
betula pyralid	キササゲノメイガ	*Sinomphisa plagialis* (Wileman)	（チョウ目、メイガ科）日本、旧北区
Beulah's opal		*Chrysoritis beulah* (Quickelberge)	（チョウ目、シジミチョウ科）エチオピア区
Bevan's swift		*Borbo bevani* Moore	（チョウ目、セセリチョウ科）豪州区

英名	和名	学名	所属、分布、ほか
Bevan's swift		*Pseudoborbo bevani* (Moore)	（チョウ目、セセリチョウ科）東洋区、豪州区
Beyer's scarab	ベイエリーウグイスコガネ	*Plusiotis beyeri* Skinner	（コウチュウ目、コガネムシ科）新北区
Bhang aphid			hemp aphid を見よ
Bhatula beetle of India		*Heteronychus* sp.	（コウチュウ目、コガネムシ科）東洋区。イネ害虫
bherwa		*Schizodactylus monstrosus* Drury	（バッタ目、Schizodactylidae）東洋区。インドでの英名
Bhutan blackvein		*Aporia harrietae* (de Nicéville)	（チョウ目、シロチョウ科）東洋区
Bhutan flat		*Celaenorrhinus flavocincta* (de Nicéville)	（チョウ目、セセリチョウ科）東洋区
Bhutan glory	シボリアゲハ	*Bhutanitis lidderdalii* Atkinson	（チョウ目、アゲハチョウ科）東洋区
Bhutan sergeant	ニトベミスジ	*Athyma jina* Moore	（チョウ目、タテハチョウ科）東洋区
Bhutan treebrown		*Lethe margaritae* (Elwes)	（チョウ目、タテハチョウ科）東洋区
Bhutya lineblue		*Prosotas bhutea* (de Nicéville)	（チョウ目、シジミチョウ科）東洋区
bi-spot royal		*Ancema ctesia* (Hewitson)	（チョウ目、シジミチョウ科）東洋区
Bia owl			confused owlet を見よ
Biak dark crow		*Euploea albicosta* Joicey et Noakes	（チョウ目、タテハチョウ科）東洋区
Biak threespot crow		*Euploea tripunctata* Joicey et Noakes	（チョウ目、タテハチョウ科）東洋区
Biak tiger		*Parantica marcia* (Joicey et Talbot)	（チョウ目、タテハチョウ科）東洋区
bianca-rossa			dictyospermum scale を見よ
bibindandy		*Borocera madagascariensis* Boisduval	（チョウ目、カレハガ科）エチオピア区
bibionid fly			march fly を見よ
bibit fly			rice seedling fly を見よ
bicolor bark beetle	フタイロキクイムシ	*Xyleborus bicolor* Blandford	（コウチュウ目、キクイムシ科）日本、東洋区
bicolor geometrid			blue-bordered carpet を見よ
bicolored angle moth			dingy angle を見よ　bicolored angle ともいう
bicolored bog damsel			eastern red damsel を見よ
bicolored Chloraspilates moth		*Chloraspilates bicoloraria* Packard	（チョウ目、シャクガ科）新北区
bicolored moth		*Eilema bicolor* (Grote)	（チョウ目、ヒトリガ科）新北区
bicolored Pyrausta moth		*Pyrausta bicoloralis* (Guenée)	（チョウ目、メイガ科）新北区
bicolour ace		*Sovia hyrtacus* (de Nicéville)	（チョウ目、セセリチョウ科）東洋区
bicolour commodore	オオチャイロイチモンジ（ザイライチモンジ）	*Limenitis zayla* (Doubleday)	（チョウ目、タテハチョウ科）東洋区
bicolour moonbeam			purple moonbeam を見よ

英名	和名	学名	所属、分布、ほか
bicoloured Cedarinia		*Cedarinia* sp.	（バッタ目、バッタ科）豪州区
biddie			yellow-backed biddie を見よ
biddies	オニヤンマ科	Cordulegasteridae	（トンボ目）の昆虫の総称
Bidens borer		*Epiblema obtiosana* (Clemens)	（チョウ目、ハマキガ科）新北区。*Bidens* センダングサの属 を食べることに由来。bidens borer moth ともいう
Biden's grass-veneer moth		*Crambus bidens* Zeller	（チョウ目、メイガ科）新北区
bidentate scarab		*Euetheola bidentata* (Burmeister)	（コウチュウ目、コガネムシ科）新熱帯区。カンキツ害虫
Biedouw Mantophasma	ビドーカカトアルキ	*Karoophasma biedouwensis* Klass, Picker, Damgaard, van Noort et Tojo	（カカトアルキ目、Austrophasmatidae）エチオピア区
bifasciculate scale	トビイロマルカイガラムシ	*Chrysomphalus bifasciculatus* Ferris	（カメムシ目、マルカイガラムシ科）日本、全北区、東洋区、大洋区。カンキツ、チャなど多くの植物を加害
bifid plushblue		*Arhopala diardi* (Hewitson)	（チョウ目、シジミチョウ科）東洋区
bifid plushblue			shining plushblue (1) を見よ
bifid-tongued bees			obtuse-tongued bees を見よ
bifid-tongued bees			yellow-faced bees (1) を見よ
big bedbug			bloodsucking conenose を見よ
big bend false katydid		*Amblycorypha insolita* Rehn et Hebard	（バッタ目、キリギリス科）新北区
big black horse fly			western horsefly を見よ
big bluet		*Enallagma durum* (Hagen)	（トンボ目、イトトンボ科）新北区
big dipper firefly			common eastern firefly を見よ
big-eye brown		*Pseudonympha loxophthalma* (Väri)	（チョウ目、タテハチョウ科）エチオピア区
bigeyed bug		*Geocoris lubra* Kirkaldy	（カメムシ目、ナガカメムシ科）豪州区
bigeyed bug (1)		*Geocoris bullatus* Say	（カメムシ目、ナガカメムシ科）新北区
big-eyed bugs		Geocorinae	（カメムシ目、ナガカメムシ科）の昆虫の総称
big-eyed bugs (1)		*Geocoris*	（カメムシ目、ナガカメムシ科）新北区
big-eyed click beetle			eastern eyed click beetle を見よ
big-eyed flies			big-headed flies を見よ
big-eyed gemmed-satyr		*Cyllopsis suivalenoides* Miller	（チョウ目、タテハチョウ科）新熱帯区
big greasy	ウスバジャコウアゲハ	*Cressida cressida* (Fabricius)	（チョウ目、アゲハチョウ科）豪州区。big greasy butterfly ともいう
big green pine-tree katydid			pine katydid を見よ
big-headed ant (1)	ツヤオオズアリ	*Pheidole megacephala* (Fabricius)	（ハチ目、アリ科）日本、全北区、豪州区。働きアリで大きな頭部をもつものあり
big-headed ant	オオズアリ	*Pheidole nodus* F. Smith	（ハチ目、アリ科）日本、旧北区、東洋区
big-headed ants	オオズアリ属	*Pheidole*	（ハチ目、アリ科）の昆虫の総称

英名	和名	学名	所属、分布、ほか
big-headed flies	アタマアブ科	Pipunculidae	（ハエ目）の昆虫の総称　英名は大きな頭部に由来
bigheaded grasshopper		*Aulocara elliotti* (Thomas)	（バッタ目、バッタ科）新北区
big-headed stick insect		*Megacrania alpheus* (Westwood)	（ナナフシ目、ナナフシ科）豪州区
big red skimmer			red skimmer を見よ
big rhombic planthopper	オオヒシウンカ	*Oliarus subnubilus* Matsumura	（カメムシ目、ヒシウンカ科）日本
big-spiked gemmed-satyr		*Cyllopsis pephredo* (Godman)	（チョウ目、タテハチョウ科）新熱帯区
big-spotted cleg		*Haematopota bigoti* Gobert	（ハエ目、アブ科）旧北区
big-spotted scarlet-eye		*Dyscophellus phraxanor lama* Evans	（チョウ目、セセリチョウ科）新熱帯区
big-tooth anabrus		*Anabrus cerciata* Caudell	（バッタ目、キリギリス科）新北区
Bigg's brownie		*Miletus biggsii* (Distant)	（チョウ目、シジミチョウ科）東洋区
bilberry brind			golden-rod brindle を見よ
bilberry bumblebee			mountain bumblebee を見よ
bilberry pigmy		*Stigmella myrtillella* (Stainton)	（チョウ目、モグリチビガ科）旧北区
bilberry pug		*Chloroclystis debiliata* (Hübner)	（チョウ目、シャクガ科）旧北区
bilberry tortrix		*Aphelia viburniana* (Denis et Schiffermüller)	（チョウ目、ハマキガ科）旧北区。bilberry tortrix moth ともいう
billbug			La Plata weevil を見よ
bill bugs		*Calendra*	（コウチュウ目、ゾウムシ科）の昆虫の総称
billbugs		*Sphenophorus*	（コウチュウ目、オサゾウムシ科）の昆虫の総称
billbugs		*Stenophorus*	（コウチュウ目、ゾウムシ科）の昆虫の総称
billbugs (1)		Rhynchophorinae	（コウチュウ目、ゾウムシ科）の昆虫の総称
billbugs (2)		Calandrinae	（コウチュウ目、ゾウムシ科）の昆虫の総称
billbugs			weevils (1) を見よ　米国での英名
Billy witch			June beetle (1) を見よ
bilobed Dichomeris moth		*Dichomeris bilobella* (Zeller)	（チョウ目、キバガ科）新北区
bilobed looper			bilobed looper moth を見よ
bilobed looper moth		*Autographa biloba* (Stephens)	（チョウ目、ヤガ科）全北区、新熱帯区、大洋区
bilobed semilooper			bilobed looper moth を見よ
bindweed carl			peach leafminer を見よ
bindweed noctuid			yellow-spotted small noctuid を見よ
bindweed plume moth	ヒルガオトリバ	*Emmelina jezonica* (Matsumura)	（チョウ目、トリバガ科）日本。幼虫はヒルガオ、サツマイモなどを摂食
bioculate predacious bug			two-spotted stink bug を見よ
biphyllid beetles			false skin beetles を見よ

英名	和名	学名	所属、分布、ほか
birch angle moth		*Macaria notata* (Linnaeus)	(チョウ目、シャクガ科) 全北区
birch aphid	カバワタフキアブラムシ	*Euceraphis punctipennis* (Zetterstedt)	(カメムシ目、アブラムシ科) 日本、全北区、東洋区。カンバ類の害虫
birch aphid		*Clethrobius comes* (Walker)	(カメムシ目、アブラムシ科) 日本、旧北区
birch aphid		*Calaphis flava* Mordvilko	(カメムシ目、アブラムシ科) 旧北区、豪州区
birch-aspen leafroller moth			Solander's bell を見よ
birch bark beetle		*Dryocoetes betulae* Hopkins	(コウチュウ目、キクイムシ科) 新北区
birch bark beetle			birch sapwood borer を見よ
birch bug			birch shield bug を見よ
birch casebearer			apple casebearer を見よ 米国での英名
birch casebearer moth		*Coleophora comptoniella* (McDunnough)	(チョウ目、ツツミノガ科) 新北区
birch clearwing moth			large red-belted clearwing を見よ
birch dagger			unmarked dagger moth を見よ
birch dagger moth		*Acronicta betulae* Riley	(チョウ目、ヤガ科) 新北区
birch Epinotia moth	フタシロモンヒメハマキ	*Epinotia trigonella* (Linnaeus)	(チョウ目、ハマキガ科) 日本、全北区
birch ermel	ニセウスグロヒメスガ	*Swammerdamia caesiella* (Hübner)	(チョウ目、スガ科) 日本、全北区
birch flea beetle			elm leaf-mining weevil を見よ
birch knot-horn	イイジマクロマダラメイガ	*Metriostola betulae* (Goeze)	(チョウ目、メイガ科) 旧北区
birch lasiocampid		*Eriogaster lanestris* Linnaeus	(チョウ目、カレハガ科) 旧北区
birch leaf blotchminer moth		*Cameraria betulivora* (Walsingham)	(チョウ目、ホソガ科) 新北区
birch leaf gall midge		*Anisostephus betulinum* (Kieffer)	(ハエ目、タマバエ科) 旧北区
birch leaf-miner			birch leaf-mining sawfly (1) を見よ birch leaf miner, birch leafminer とも記す
birch leaf-mining sawfly (1)	シラカバノクロボシハムグリハバチ	*Fenusa pusilla* (Lepeletier)	(ハチ目、ハバチ科) 日本、旧北区
birch leaf-mining sawfly (2)		*Heterarthrus nemoratus* (Fallén)	(ハチ目、ハバチ科) 全北区
birch leaf-mining weevil		*Rhynchaenus rusci* (Hübner)	(コウチュウ目、ゾウムシ科) 旧北区
birch leaf roller weevil			palm weevil (1) を見よ
birch leaffolder			yellow birch leafroller moth を見よ
birch little midget		*Phyllonorycter anderidae* (Fletcher)	(チョウ目、ホソガ科) 旧北区
birch looper			argent and sable を見よ カナダでの英名
birch mocha		*Cyclophora albipunctata* (Hufnagel)	(チョウ目、シャクガ科) 日本亜種は *C. a. griseolata* (Staudinger) ヨツメヒメシャク
birch psylla	カバキジラミ	*Psylla betulae* (Linnaeus)	(カメムシ目、キジラミ科) 日本、旧北区

英名	和名	学名	所属、分布、ほか
birch red midget	ツマスジキンモンホソガ	*Phyllonorycter ulmifoliella* (Hübner)	(チョウ目、ホソガ科) 日本、旧北区
birch sapwood borer		*Scolytus ratzeburgi* Janson	(コウチュウ目、キクイムシ科) 旧北区
birch sawfly		*Arge pectoralis* (Leach)	(ハチ目、ミフシハバチ科) 新北区
birch sawfly (1)	カラフトモモブトハバチ	*Cimbex femoratus femoratus* (Linnaeus)	(ハチ目、コンボウハバチ科) 日本、旧北区
birch sawfly			hazel sawfly を見よ
birch seed midge		*Semudobia betulae* (Winnertz)	(ハエ目、タマバエ科) 旧北区
birch shield bug	セグロベニモンツノカメムシ	*Elasmostethus interstinctus* (Linnaeus)	(カメムシ目、ツノカメムシ科) 日本、旧北区
birch shoot borer moth		*Epinotia solicitana* (Walker)	(チョウ目、ハマキガ科) 新北区
birch skeletonizer		*Bucculatrix canadensisella* Chambers	(チョウ目、チビガ科) 新北区。birch skeletonizer moth ともいう
birch sweep		*Proutia betulina* Zeller	(チョウ目、ミノガ科) 旧北区
birch tube maker		*Acrobasis betulella* Hulst	(チョウ目、メイガ科) 新北区。birch tubemaker moth ともいう
birch twig weevil		*Magdalis carbonaria* (Linnaeus)	(コウチュウ目、ゾウムシ科) 旧北区
birch weevil		*Curculio betulae* (Stephens)	(コウチュウ目、ゾウムシ科) 旧北区
Birchim's shieldback		*Idiostatus birchimi* Rentz	(バッタ目、キリギリス科) 新北区
bird bugs			bed bugs を見よ
bird-cherry aphid			oat bird-cherry aphid を見よ
bird-cherry ermine	サクラスガ	*Yponomeuta evonymellus* (Linnaeus)	(チョウ目、スガ科) 日本、旧北区。bird-cherry ermine moth ともいう
bird cherry-oat aphid			oat bird-cherry aphid を見よ
bird fleas	ナガノミ科	Ceratophyllidae	(ノミ目) の昆虫の総称
bird hoverfly		*Eupeodes volucris* Osten Sacken	(ハエ目、ハナアブ科) 新北区
bird lice (1)	チョウカクハジラミ科	Philopteridae	(ハジラミ目) の昆虫の総称
bird lice (2)	タネハジラミ科	Ricinidae	(ハジラミ目) の昆虫の総称　ハチドリと燕雀目のトリに寄生
bird lice			chewing lice を見よ　英国での英名
bird lice			hawk lice を見よ
bird lice			poultry lice を見よ
bird lice			bird lice (1) を見よ
bird nest barklice		Psoquilidae	(チャタテムシ目) の昆虫の総称　トリの巣、樹幹、住宅などに生息
bird nest moth		*Monopis crocicapitella* (Clemens)	(チョウ目、ヒロズコガ科) 新北区
bird nest parasites		*Protocalliphora*	(ハエ目、クロバエ科) の昆虫の総称
bird of paradise flies		*Callipappus*	(カメムシ目、ワタフキカイガラムシ科) の昆虫の総称
bird parasite flies			louse flies を見よ

英名	和名	学名	所属、分布、ほか
bird's-blood fly		*Protocalliphora sordida* Zetterstedt	（ハエ目、クロバエ科）旧北区
birds-foot trefoil thrips		*Odontothrips loti* (Haliday)	（アザミウマ目、アザミウマ科）旧北区
bird's-wing		*Dypterygia scabriuscula* (Linnaeus)	（チョウ目、ヤガ科）旧北区
birdwings	トリバネアゲハ属	*Ornithoptera*	（チョウ目、アゲハチョウ科）の昆虫の総称
birdwings	アカエリトリバネアゲハ属	*Trogonoptera*	（チョウ目、アゲハチョウ科）の昆虫の総称
birdwings	キシタアゲハ属	*Troides*	（チョウ目、アゲハチョウ科）の昆虫の総称
biscuit beetle	ジンサンシバンムシ	*Stegobium paniceum* (Linnaeus)	（コウチュウ目、シバンムシ科）日本、汎世界。貯蔵食品、動植物標本の害虫
biscuit beetle			rusty grain beetle を見よ
biscuit beetle			flat grain beetle (1) を見よ
bisected honey locust moth		*Sphingicampa bisecta* (Lintner)	（チョウ目、ヤママユガ科）新北区
bishey-barneybee			ladybird beetles を見よ
bishop Barnaby			seven-spot ladybird を見よ
Bishop bug			tarnished plant bug (1) を見よ
Bishop's mitre		*Aelia acuminata* (Linnaeus)	（カメムシ目、カメムシ科）旧北区
Bishop's mitre shield-bug			Bishop's mitre を見よ
Bismarck's paradise skipper		*Abantis bismarcki* Karsch	（チョウ目、セセリチョウ科）エチオピア区
bison snaketail		*Ophiogomphus bison* Selys	（トンボ目、サナエトンボ科）新北区
bison snout beetles		Thecesterminae	（コウチュウ目、ゾウムシ科）の昆虫の総称
Bitias hairstreak		*Panthiades bitias* (Cramer)	（チョウ目、シジミチョウ科）新北区、新熱帯区
biting ant		*Macromischoides aculeatus* (Mayr)	（ハチ目、アリ科）エチオピア区
biting bird lice			chewing lice を見よ
biting bird lice			poultry lice を見よ
biting cat louse			cat biting louse を見よ　米国での英名
biting dog louse			dog chewing louse を見よ　米国での英名
biting goat louse			goat louse を見よ　米国での英名
biting Guinea pig lice			rodent chewing lice を見よ
biting horse louse			horse biting louse を見よ
biting house fly			stable fly を見よ
biting kangaroo lice			marsupial chewing lice を見よ
biting lice			chewing lice を見よ
biting lice			mammal chewing lice を見よ
biting louse of cattle			cattle chewing louse を見よ
biting louse of goat			goat louse を見よ
biting louse of horse			pilose biting horse louse を見よ

英名	和名	学名	所属、分布、ほか
biting midge		*Culicoides pulicaris* (Linnaeus)	（ハエ目、ヌカカ科）日本
biting midges		*Austroconops, Lasiohelea, Leptoconops, Styloconops*	（ハエ目、ヌカカ科）の昆虫の総称
biting midges (1)	ヌカカ科	Ceratopogonidae	（ハエ目）の昆虫の総称　昆虫に寄生したり、温血動物から吸血する微小昆虫
biting midges			punkies を見よ
biting sheep louse			sheep biting louse を見よ
biting snipe flies			common biting flies を見よ
biting stable fly			stable fly を見よ　豪州での英名
bitou seed fly		*Mesoclanis polana* Munro	（ハエ目、ミバエ科）豪州区、エチオピア区
bittacids			hanging flies を見よ
bitter-bush blue		*Theclinesthes albocincta* (Waterhouse)	（チョウ目、シジミチョウ科）豪州区
bitter-cress smudge		*Eidophasia messingiella* (Fischer von Röslerstamm)	（チョウ目、スガ科）旧北区
bitter gourd fruit fly			melon fly を見よ
bituberculate scale		*Eulecanium bituberculatum* Targioni	（カメムシ目、カタカイガラムシ科）旧北区
bizarre caddisflies	カタツツトビケラ科	Lepidostomatidae	（トビケラ目）の昆虫の総称
bizarre looper		*Anisozyga pieroides* (Walker)	（チョウ目、シャクガ科）豪州区
Bjad's ciliate blue		*Annthene bjoernstadi* Collins et Larsen	（チョウ目、シジミチョウ科）エチオピア区
black			crimson-banded black を見よ
black alder sphinx		*Dolba hyloeus* (Drury)	（チョウ目、スズメガ科）新北区
black alfalfa leaf-beetle		*Colaspidema atrum* (Olivier)	（コウチュウ目、ハムシ科）旧北区
black and orange		*Venessula milca* (Hewitson)	（チョウ目、タテハチョウ科）エチオピア区
black-and-orange playboy		*Virachola dariaves* (Hewitson)	（チョウ目、シジミチョウ科）エチオピア区
black and red froghopper		*Cercopis vulnerata* (Rossi)	（カメムシ目、コガシラアワフキ科）旧北区
black and red harlequin		*Eresiomera bicolor* (Grose Smith et Kirby)	（チョウ目、シジミチョウ科）エチオピア区
black and red leaf miner		*Throscoryssa citri* Maulik	（コウチュウ目、ハムシ科）旧北区。カンキツ害虫
black and red longhorn beetle		*Pedostrangalia revestita* (Linnaeus)	（コウチュウ目、カミキリムシ科）旧北区
black-and-red rose aphid		*Eomacrosiphum nigromaculosum* (McDougall)	（カメムシ目、アブラムシ科）新北区
black-and-white Acraea		*Acraea oreas* Sharpe	（チョウ目、タテハチョウ科）エチオピア区
black-and-white aeroplane		*Neptis praslini staudingereana* de Nicéville	（チョウ目、タテハチョウ科）豪州区

英名	和名	学名	所属、分布、ほか
black and white-barred		*Scoparia aspidota* Meyrick	（チョウ目、メイガ科）豪州区
black and white-barred scoparia moth			black and white-barred を見よ
black and white citrus borer		*Melanauster chinensis* Forster	（コウチュウ目、カミキリムシ科）旧北区。カンキツ害虫
black and white click beetle		*Chalcolepidius webbi* LeConte	（コウチュウ目、コメツキムシ科）新北区
black-and-white damsel		*Apanisagrion lais* (Selys)	（トンボ目、イトトンボ科）新北区
black and white flat		*Tagiades japetus janetta* Butler	（チョウ目、セセリチョウ科）豪州区。common snow flat を参照
black and white flat		*Gerosis limax dirae* (de Nicéville)	（チョウ目、セセリチョウ科）東洋区
black and white Helen			yellow Helen（1）を見よ
black and white Helen		*Papilio nephelus sunatus* Corbet	（チョウ目、アゲハチョウ科）東洋区
black-and-white ringlet		*Hypocysta angustata* Waterhouse et Lyell	（チョウ目、タテハチョウ科）豪州区
black and white skipper		*Antipodia atralba* (Tepper)	（チョウ目、セセリチョウ科）豪州区
black-and-white swift		*Sabera caesina* (Hewitson)	（チョウ目、セセリチョウ科）豪州区
black-and-white tit	タスキネッタイシジミ	*Hypolycaena danis* (C. et R. Felder)	（チョウ目、シジミチョウ科）豪州区
black and white tiger			Malay tiger を見よ
black-and-yellow crane fly		*Nephrotoma maculosa* (Meigen)	（ハエ目、ガガンボ科）新北区
black-and-yellow lichen moth			lichen moth (2) をみよ
black-and-yellow longhorn			spotted longhorn を見よ
black-and-yellow mud dauber			mud dauber を見よ
black and yellow mud wasp		*Sceliphron spiriflex* Linnaeus	（ハチ目、アナバチ科）旧北区
black and yellow spotted longhorn beetle			spotted longhorn を見よ
black angle		*Tapena thwaitesi* Moore	（チョウ目、セセリチョウ科）東洋区。*T. t. bornea* Evans も同英名
black ant	クロヤマアリ	*Formica* (*Serviformica*) *japonica* Motschulsky	（ハチ目、アリ科）日本、旧北区、東洋区
black ant (1)	トビイロケアリ	*Lasius niger* (Linnaeus)	（ハチ目、アリ科）日本、全北区。最も普通のアリ
black ant-blue		*Acrodipsas hirtipes* (Sands)	（チョウ目、シジミチョウ科）豪州区
black anthonomus			strawberry blossom weevil (2) を見よ
black ants		*Lasius*	（ハチ目、アリ科）の昆虫の総称
black aphid			cherry aphid (1) を見よ

英名	和名	学名	所属、分布、ほか
black Apollo			clouded Apollo を見よ
black apple capsid		*Atractotomus mali* (Meyer)	(カメムシ目、ハナカメムシ科) 旧北区
black apple sucker	クロリンゴキジラミ	*Psylla malivorella* (Sasaki)	(カメムシ目、キジラミ科) 日本。リンゴ害虫
black araucaria scale			Ross's black scale (1) を見よ 豪州での英名
black archduke		*Lexias dirtea merguia* Tytler	(チョウ目、タテハチョウ科) 東洋区。archduke を参照
black-arched tussock			nun moth を見よ
black-arched tussock moth			nun moth を見よ
black arches			nun moth を見よ
black arches			nolid moths を見よ
black arches moth			nun moth を見よ
black army cutworm moth	アトウスヤガ	*Actebia fennica* (Tauscher)	(チョウ目、ヤガ科) 日本、新北区。幼虫は black army curtworm
black armyworm	ツマジロクサヨトウ	*Spodoptera frugiperda* (J. E. Smith)	(チョウ目、ヤガ科) 新北区、新熱帯区。イネ害虫
black army worm			black army cutworm moth を見よ
black Australian anopheline		*Anopheles bancroftii* Giles	(ハエ目、カ科) 豪州区
black-back green sawfly	セグロアオハバチ	*Tenthredo mesomelas* Linnaeus	(ハチ目、ハバチ科) 日本、旧北区
black-back leafhopper	セグロアオズキンヨコバイ	*Jassus dorsalis* (Matsumura)	(カメムシ目、オオヨコバイ科) 日本
black-back prominent			poplar prominent (2) を見よ
black-back vegetable sawfly			cabbage sawfly (1) を見よ
black-backed meadow ant		*Formica pratensis* Retzius	(ハチ目、アリ科) 旧北区
black bamboo borer		*Bastrychopsis parallela* (Lesne)	(コウチュウ目、ナガシンクイムシ科) 東洋区
black bambootail		*Prodasineura verticalis* (Selys)	(トンボ目、モノサシトンボ科) 東洋区
black-band-headed leafhopper	オサヨコバイ	*Tartessus ferrugineus ferrugineus* (Walker)	(カメムシ目、オオヨコバイ科) 日本、東洋区
black-banded		*Polymixis xanthomista* (Hübner)	(チョウ目、ヤガ科) 旧北区。亜種 *P. x. statices* (Gregson) も同英名。black-banded ともいう
black-banded brocade moth		*Oligia modica* (Guenée)	(チョウ目、ヤガ科) 新北区
black-banded carpet moth		*Eustroma semiatrata* Hulst	(チョウ目、シャクガ科) 新北区
black-banded hadena			black-banded brocade moth を見よ
black-banded Holomelina			ruddy Holomelina を見よ
black-banded orange moth		*Epelis truncataria* (Walker)	(チョウ目、シャクガ科) 新北区
black-banded owlet moth		*Phalaenostola larentioides* Grote	(チョウ目、ヤガ科) 新北区。black-banded owlet ともいう

英名	和名	学名	所属、分布、ほか
black-banded wasp moth		*Syntomeida melanthus* (Cramer)	（チョウ目、ヒトリガ科）新北区、新熱帯区
black-banded white		*Itaballia demophile* (Linnaeus)	（チョウ目、シロチョウ科）新熱帯区。*I. d. centralis* Joicey et Talbot も同英名
black-banded white geometrid	クロオビシロナミシャク	*Trichopteryx ustata* (Christoph)	（チョウ目、シャクガ科）日本
black-barred brown moth		*Plagiomimicus pityochromus* Grote	（チョウ目、ヤガ科）新北区
black barred gold tip			banded gold-tip を見よ
black-barred mantid		*Mantis octospilota* Westwood	（カマキリ目、カマキリ科）豪州区
black bean aphid		*Aphis robiniae* Macchiat	（カメムシ目、アブラムシ科）旧北区
black bean aphid			cowpea aphid を見よ
black bean aphid			bean aphid を見よ
black bean aphid			dock aphid (1) を見よ
black bean blister beetle	チョウセンマメハンミョウ	*Epicauta chinensis taishoensis* (Lewis)	（コウチュウ目、ツチハンミョウ科）日本、旧北区
black bees			andrenas を見よ
black beetle		*Metanastes vulgivagus* (Olliff)	（コウチュウ目、コガネムシ科）豪州区
black beetle		*Heteronychus sanctae-helanae* Blanchard	（コウチュウ目、コガネムシ科）豪州区。カンキツ害虫
black beetle		*Omaseus vulgaris* Linnaeus	（コウチュウ目、ゴミムシダマシ科）旧北区
black beetle			common cockroach を見よ
black beetle			African black beetle を見よ
black beetles			Hercules beetles を見よ
black beetles			cockroaches (1)(2) を見よ　black beetle はゴキブリの俗称
black-bellied Anteros		*Anteros formosus micon* Druce	（チョウ目、シジミタテハ科）新熱帯区
black-bellied clerid		*Enoclerus lecontei* (Wolcott)	（コオウチュウ目、カッコウムシ科）新北区
black-belted moth		*Melanchra prionistis* Meyrick	（チョウ目、ヤガ科）豪州区
black bit moth		*Celiptera frustulum* Guenée	（チョウ目、ヤガ科）新北区
black blister beetle		*Epicauta pennsylvanica* (De Geer)	（コウチュウ目、ツチハンミョウ科）新北区
black-blotched Bactra moth			dusty marble を見よ
black-blotched Schizura moth		*Schizura leptinoides* (Grote)	（チョウ目、シャチホコガ科）新北区
black blow fly	クロキンバエ	*Phormia regina* (Meigen)	（ハエ目、クロバエ科）日本、全北区、大洋区。幼虫は動物死体、便池などに生息
black-bodied sawfly			small gooseberry sawfly (1) を見よ
black bog ant		*Formica picea* Melander	（ハチ目、アリ科）旧北区

英名	和名	学名	所属、分布、ほか
black-bordered Babul blue		*Azanus moriqua* (Wallengren)	（チョウ目、シジミチョウ科）エチオピア区
black-bordered Charaxes		*Charaxes pollux* (Cramer)	（チョウ目、タテハチョウ科）エチオピア区
black-bordered crescent		*Tegosa anieta cluvia* (Godman et Salvin)	（チョウ目、タテハチョウ科）新熱帯区。*T. a. luka* Higgins も同英名
black-bordered lemon moth		*Marimatha nigrofimbria* (Guenée)	（チョウ目、ヤガ科）新北区
black-bordered piercer		*Pammene argyrana* (Hübner)	（チョウ目、ハマキガ科）旧北区
black-bordered Tegosa		*Tegosa anieta* (Hewitson)	（チョウ目、タテハチョウ科）新熱帯区
black-bordered Tegosa			black-bordered crescent を見よ
black-bordered zebra blue		*Leptotes marginalis* (Stempffer)	（チョウ目、シジミチョウ科）エチオピア区
black borer wasp		*Trypoxylon figulus* (Linnaeus)	（ハチ目、アナバチ科）旧北区
black borers		*Apate*	（コウチュウ目、ナガシンクイムシ科）の昆虫の総称
black borers			false powderpost beetles を見よ
black bramble pigmy		*Ectoedemia agrimoniae* (Frey)	（チョウ目、モグリチビガ科）旧北区
black bramble pigmy		*Ectoedemia rubivora* (Wocke)	（チョウ目、モグリチビガ科）旧北区
black-branded royal		*Tajuria culta* (de Nicéville)	（チョウ目、シジミチョウ科）東洋区
black-branded swift			small branded swift を見よ
black-branded swift			Chinese swift を見よ
black branded swift			small branded swift を見よ
black buckeye	ジャマイカタテハモドキ（カリブタテハモドキ）	*Junonia genoveva* (Cramer)	（チョウ目、タテハチョウ科）新北区、新熱帯区。亜種 *J. g. nigrosuffusa* Barnes et McDunnough も同英名
black bug			bean capsid を見よ
black bugs		*Scotinophara*	（カメムシ目、カメムシ科）イネ害虫
black bulldog ant			bulldog ant (2) を見よ
black burying beetle		*Necrophorus nigritus* Mannerheim	（コウチュウ目、シデムシ科）新北区
black burying beetle		*Necrodes littoralis* (Linnaeus)	（コウチュウ目、シデムシ科）旧北区
black burying beetle		*Necrophorus humator* Fabricius	（コウチュウ目、シデムシ科）旧北区
black cactus longhorn		*Moneilema armatam* LeConte	（コウチュウ目、カミキリムシ科）新北区。成虫幼虫ともにサボテンを食う
black calosoma		*Calosoma semilaeve* LeConte	（コウチュウ目、オサムシ科）新北区
black camphor thrips			camphor thrips を見よ
black-caped groundling		*Teleiodes decorella* (Haworth)	（チョウ目、キバガ科）旧北区

英名	和名	学名	所属、分布、ほか
black carnation thrips			western flower thrips (2) を見よ
black carpenter ant		*Camponotus pennsylvanicus* (De Geer)	(ハチ目、アリ科) 新北区
black carpenter ant	クロオオアリ	*Camponotus japonicus* Mayr	(ハチ目、アリ科) 日本、旧北区
black carpenter ant			red carpenter ant (1) を見よ
black carpet beetle		*Attagenus megatoma* (Fabricius)	(コウチュウ目、カツオブシムシ科) 新北区。カーペット、貯蔵穀類などの害虫
black carpet beetle		*Attagenus unicolor* (Brahm)	(コウチュウ目、カツオブシムシ科) 全北区、豪州区
black carpet beetle	ヒメカツオブシムシ	*Attagenus japonicus* Reitter	(コウチュウ目、カツオブシムシ科) 日本、全北区。貯蔵食品、動植物標本の害虫
black carpet beetle (1)		*Attagenus piceus* (Olivier)	(コウチュウ目、カツオブシムシ科) 旧北区
black carpet beetle			fur beetle を見よ
black carrion fly		*Hydrotaea rostrata* (Robineau-Desvoidy)	(ハエ目、イエバエ科) 豪州区
black chafer	クロコガネ	*Holotrichia kiotonensis* Brenske	(コウチュウ目、コガネムシ科) 日本、旧北区
black chafer	リュウキュウクロコガネ	*Holotrichia loochooana loochooana* (Sawada)	(コウチュウ目、コガネムシ科) 日本
black checkerspot		*Chlosyne melanarge* (Bates)	(チョウ目、タテハチョウ科) 新熱帯区。black patch ともいう。
black checkerspot			Cyneas checkerspot を見よ
black cherry aphid			cherry aphid (1) を見よ 米国での英名
black cherry fruit fly	クロオウトウミバエ	*Rhagoletis fausta* (Osten Sacken)	(ハエ目、ミバエ科) 新北区
black chrysanthemum aphid			chrysanthemum aphid (1) を見よ 米国での英名
black citrus aphid	コミカンアブラムシ	*Toxoptera aurantii* (Boyer de Fonscolombe)	(カメムシ目、アブラムシ科) 日本、全北区、東洋区、汎熱帯。ミカン、チャなどを加害
black citrus aphid			citrus brown aphid を見よ
black citrus fly		*Drosophila punctipennis* van der Wulp	(ハエ目、ショウジョウバエ科) 東洋区。カンキツ害虫
black citrus leaf beetle	クロウスバハムシ	*Luperus moorii* Baly	(コウチュウ目、ハムシ科) 日本
black-clothed gliding ant		*Cephalotes atratus* (Linnaeus)	(ハチ目、アリ科) 新熱帯区
black-clouded longhorn beetle		*Leiopus nebulosus* (Linnaeus)	(コウチュウ目、カミキリムシ科) 旧北区
black clubtail		*Hagenius brevistylus* Selys	(トンボ目、サナエトンボ科) 新北区。北米のサナエトンボ科で最大種。開帳 11.5 cm
black-coated clothes moth			carpet moth を見よ
black cockroach		*Platyzosteria novaeseelandia* (Brunner von Wattenwyl)	(ゴキブリ目、ゴキブリ科) 豪州区

英名	和名	学名	所属、分布、ほか
black cockroach wasp		*Dolichurus stantoni* (Ashmead)	（ハチ目、アナバチ科）新北区
black cocoon ant		*Dolichoderus bituberculatus* May	（ハチ目、アリ科）東洋区
black collar		*Ochropleura flammatra* (Denis et Schiffermüller)	（チョウ目、ヤガ科）旧北区、東洋区
black colonel		*Odontomyia tigrina* (Fabricius)	（ハエ目、ミズアブ科）旧北区
black cotton bug		*Oxycarenus lugubris* Motschulsky	（カメムシ目、ナガカメムシ科）東洋区
black crazy ant			hairy ant を見よ
black crescent		*Phyciodes ptolyca* (Bates)	（チョウ目、タテハチョウ科）新北区
black crescent			Ptolyca crescentspot を見よ
black cucurbit leaf beetle	クロウリハムシ	*Aulacophora nigripennis* Motschulsky	（コウチュウ目、ハムシ科）日本、旧北区、東洋区
black cupid		*Elkalyce kala* (de Nicéville)	（チョウ目、シジミチョウ科）東洋区
black currant aphid (1)		*Rhopalosiphoninus ribesinus* van der Goot	（カメムシ目、アブラムシ科）旧北区
black currant aphid (2)		*Cryptomyzus galeopsidis* (Kaltenbach)	（カメムシ目、アブラムシ科）旧北区
black currant flower midge		*Dasineura sampaina* Tavares	（ハエ目、タマバエ科）旧北区
black currant leaf-curling midge			black currant leaf midge を見よ
black currant leaf midge		*Dasineura tetensi* (Rübsaamen)	（ハエ目、タマバエ科）旧北区
black currant sawfly		*Nematus alfaciens* Benson	（ハチ目、ハバチ科）旧北区
black currant-sow thistle aphid			currant-lettuce aphid を見よ
black currant tiger longicorn	フタオビミドリトラカミキリ	*Chlorophorus muscosus* (Bates)	（コウチュウ目、カミキリムシ科）日本、旧北区
black cutworm	タマナヤガ	*Agrotis ipsilon* (Hufnagel)	（チョウ目、ヤガ科）日本、全北区、東洋区、豪州区。作物類の著名害虫。成虫は black cutworm moth
black dadap bug		*Cyclopelta obscura* Le Peletier et Serville	（カメムシ目、カメムシ科）東洋区。dadap はデイゴのインドネシア語
black darter	ムツアカネ	*Sympetrum danae* (Sulzer)	（トンボ目、トンボ科）日本、全北区。
black dash Epinotia moth		*Epinotia sotipena* Brown	（チョウ目、ハマキガ科）新北区
black-dashed Apamea moth		*Apamea nigrior* Smith	（チョウ目、ヤガ科）新北区。black-dashed Apamea ともいう
black-dashed Hydriomena moth		*Hydriomena divisaria* (Walker)	（チョウ目、シャクガ科）新北区
black dasher		*Micrathyria atra* (Martin)	（トンボ目、トンボ科）大洋区
black deer fly		*Chrysops niger* Macquart	（ハエ目、アブ科）新北区
black deerfly		*Chrysops sepulchralis* (Fabricius)	（ハエ目、アブ科）旧北区

英名	和名	学名	所属、分布、ほか
black desert grasshopper			horse lubber を見よ
black devil		*Viola violella* (Mabille)	（チョウ目、セセリチョウ科）新熱帯区
black diamond moth		*Nyctemera annulata* Boisduval	（チョウ目、ヒトリガ科）豪州区
black digger wasp	クロアナバチ	*Sphex argentatus* Fabricius	（ハチ目、アナバチ科）日本、東洋区、豪州区
black diving beetle	クロゲンゴロウ	*Cybister brevis* Aube	（コウチュウ目、ゲンゴロウ科）日本、旧北区
black dolphin			cowpea aphid を見よ
black dolphin			bean aphid を見よ
black-dotted birch leaftier moth		*Nites betulella* (Busck)	（チョウ目、マルハキバガ科）新北区
black-dotted brown moth		*Cissusa spadix* (Cramer)	（チョウ目、ヤガ科）新北区
black-dotted buprestid	クロボシタマムシ	*Ovalisia virgata* (Motschulsky)	（コウチュウ目、タマムシ科）日本、旧北区
black-dotted caddice fly			rice caddisfly (1)(2) を見よ
black-dotted groundling		*Stenolechia gemmella* (Linnaeus)	（チョウ目、キバガ科）旧北区
black-dotted Hemeroplanis moth		*Hemeroplanis habitalis* (Walker)	（チョウ目、ヤガ科）新北区
black-dotted Lithacodia moth		*Maliattha synochitis* (Grote et Robinson)	（チョウ目、ヤガ科）新北区
black-dotted ruddy moth		*Ilecta intractata* (Walker)	（チョウ目、シャクガ科）新北区
black-dotted shieldback		*Idiostatus fuscopunctatus* (Scudder)	（バッタ目、キリギリス科）新北区
black-dotted skipper		*Mucia zygia* (Plotz)	（チョウ目、セセリチョウ科）新熱帯区
black-dotted Spragueia moth		*Spragueia onagrus* (Guenée)	（チョウ目、ヤガ科）新北区
black double-blotched bell	オオナガバヒメハマキ	*Epinotia maculana* (Caradja)	（チョウ目、ハマキガ科）日本、旧北区
black dung beetle		*Copris incertus prociduus* (Say)	（コウチュウ目、コガネムシ科）新北区
black dush		*Euphyes conspicuus* (Edwards)	（チョウ目、セセリチョウ科）新北区
black dwarf honey bee		*Apis andreniformis* South	（ハチ目、ミツバチ科）東洋区
black dynastor		*Dynastor macrosiris* Westwood	（チョウ目、タテハチョウ科）新熱帯区
black earwig		*Chelisoches morio* (Fabricius)	（ハサミムシ目、ネッタイハサミムシ科）汎世界
black earwigs	ネッタイハサミムシ科	Chelisochidae	（ハサミムシ目）の昆虫の総称
blackedged Apollo	チビタカネウスバアゲハ	*Parnassius simo* Gray	（チョウ目、アゲハチョウ科）東洋区
black-edged Carbatina moth		*Dichomeris picrocarpa* (Meyrick)	（チョウ目、キバガ科）日本、全北区、東洋区。black-edged Carbatina ともいう

英名	和名	学名	所属、分布、ほか
black-edged Euselasia		*Euselasia cataleuca* (R. Felder)	（チョウ目、シジミタテハ科）新熱帯区
black-edged fairy playboy		*Paradeudorix marginata* (Stempffer)	（チョウ目、シジミチョウ科）エチオピア区
black-edged marble	クロマダラシンムシガ	*Endothenia nigricostana* (Haworth)	（チョウ目、ハマキガ科）日本、旧北区
black-edged spirit		*Leptosia marginea* (Mabille)	（チョウ目、シロチョウ科）エチオピア区
black elm bark weevil		*Magdalis barbita* (Say)	（コウチュウ目、ゾウムシ科）新北区
black elongate ant			harvester ant (4) を見よ
black emperor		*Anax tristis* Hagen	（トンボ目、ヤンマ科）エチオピア区
black-ended bear			banded woollybear を見よ
black-etched prominent moth		*Cerura scitiscripta* Walker	（チョウ目、シャチホコガ科）新北区。black-etched prominent ともいう
black evening moth	クロホウジャク	*Macroglossum sega* Butler	（チョウ目、スズメガ科）日本、旧北区、東洋区
black eye		*Leptomyrina gorgias* (Stoll)	（チョウ目、シジミチョウ科）エチオピア区
black-eyed blue	アフリカヤマトシジミ	*Glaucopsyche melanops* (Boisduval)	（チョウ目、シジミチョウ科）旧北区
black-eyed plane	シロモンアトグロミスジ	*Pantoporia venilia* (Linnaeus)	（チョウ目、タテハチョウ科）豪州区
black-eyed Zale moth		*Zale curema* (Smith)	（チョウ目、ヤガ科）新北区。black-eyed Zale ともいう
blackeyes pea weevil			cowpea weevil を見よ
black-faced bumble bee		*Bombus californicus* Smith	（ハチ目、ミツバチ科）新北区
black-faced bush cricket		*Orocharis nigrifrons* Walker	（バッタ目、コオロギ科）新北区
blackfaced leafhopper		*Graminella nigrifrons* (Forbes)	（カメムシ目、オオヨコバイ科）新北区
black-faced scaly cricket		*Cycloptilum kelainopum* Love et Walker	（バッタ目、コオロギ科）新北区
black fairy hairstreak		*Hypolycaena nigra* Bethune-Baker	（チョウ目、シジミチョウ科）エチオピア区
black falcon		*Corades pannonia* Hewitson	（チョウ目、タテハチョウ科）新熱帯区
black fantasy		*Pseudaletis batesi* Druce	（チョウ目、シジミチョウ科）エチオピア区
black-feather clay case			apple casebearer を見よ
black fern aphid			fern aphid (1) を見よ
black field cricket		*Teleogryllus commodus* (Walker)	（バッタ目、コオロギ科）豪州区
black field earwig	ヒメハサミムシ	*Nala lividipes* (Dufour)	（ハサミムシ目、オオハサミムシ科）日本、豪州区
black flat		*Celaenorrhinus spilothyrus* (Felder)	（チョウ目、セセリチョウ科）東洋区
black flat		*Eagris tetrastigma* (Mabille)	（チョウ目、セセリチョウ科）エチオピア区

英名	和名	学名	所属、分布、ほか
black flattened leafhopper	クロヒラタヨコバイ	*Penthimia nitida* Lethierry	(カメムシ目、オオヨコバイ科) 日本、旧北区、東洋区
black flattened rhombic planthopper	クロヒラタヒシウンカ	*Oliarus iguchii* Matsumura	(カメムシ目、ヒシウンカ科) 日本
black flea beetle			cabbage flea beetle (1) を見よ
black flies (1)		*Simulium*	(ハエ目、ブユ科) の昆虫の総称
black flies (2)	ブユ科	Simuliidae	(ハエ目) の昆虫の総称　温血動物から吸血する小型昆虫
black flies			thrips (2) を見よ
black flour beetle		*Tribolium madens* (Charpentier)	(コウチュウ目、ゴミムシダマシ科) 旧北区
black flour beetle (1)		*Tribolium audax* Halstead	(コウチュウ目、ゴミムシダマシ科) 新北区
black flower chafer	クロハナムグリ	*Glycyphana fulvistemma* Motschulsky	(コウチュウ目、コガネムシ科) 日本
black flower thrips		*Haplothrips gowdeyi* (Franklin)	(アザミウマ目、クダアザミウマ科) 新北区
black fly		*Simulium equinum* (Linnaeus)	(ハエ目、ブユ科) 旧北区
black fly			ブユ科の昆虫を指すが、ミカンバエにも使われる
black fly			bean aphid を見よ
black fly			pea thrips (1) を見よ
black forester		*Lethe vindhya* (C. et R. Felder)	(チョウ目、タテハチョウ科) 東洋区
black friar	ヘカテシロモンマダラ	*Amauris hecate* Butler	(チョウ目、タテハチョウ科) エチオピア区
black Friday		*Psaltoda pictibasis* (Walker)	(カメムシ目、セミ科) 豪州区
black-fringed leaftier moth		*Psilocorsis cryptolechiella* (Chambers)	(チョウ目、マルハキバガ科) 新北区
black-fringed moss-snipefly		*Ptiolina obscura* (Fallén)	(ハエ目、シギアブ科) 旧北区
black-fringed Psilocorsis moth			black-fringed leaftier moth を見よ
black froghopper	クロフアワフキ	*Sinophora submacula* Metcalf et Horton	(カメムシ目、アワフキムシ科) 日本、旧北区
black-fronted forktail		*Ischnura denticollis* (Burmeister)	(トンボ目、イトトンボ科) 新北区
black-fronted threadtail		*Protoneura capillaris* (Rambur)	(トンボ目、ミナミイトトンボ科) 新熱帯区
black fungus beetle	ヒメゴミムシダマシ	*Alphitobius laevigatus* (Fabricius)	(コウチュウ目、ゴミムシダマシ科) 日本、汎世界
black fungus gnat	チビクロバネキノコバエ	*Bradysia agrestis* Sasakawa	(ハエ目、クロバネキノコバエ科) 日本。ウリ類、シイタケの害虫
black fungus gnats			dark-winged fungus flies を見よ　豪州での英名
black fungus moth		*Metalectra tantillus* (Grote)	(チョウ目、ヤガ科) 新北区
black garbage flies		*Ophyra*	(ハエ目、ハナバエ科) の昆虫の総称

英名	和名	学名	所属、分布、ほか
black garbage fly	ヒメクロバエ	*Hydrotaea ignava* (Harris)	（ハエ目、イエバエ科）日本、全北区、東洋区。幼虫は動物死体、便池などに生息
black garden ant			black ant (1) を見よ
black gem		*Microchrysa cyaneiventris* (Zetterstedt)	（ハエ目、ミズアブ科）旧北区
black giant-skipper		*Agathymus rethon* (Dyar)	（チョウ目、セセリチョウ科）新熱帯区
black gibbous weevil	クロコブゾウムシ	*Niphades variegatus* (Roelofs)	（コウチュウ目、ゾウムシ科）日本、旧北区
black gnats			biting midges (1) を見よ
black grain stem sawfly		*Trachelus tabidus* (Fabricius)	（ハチ目、クキバチ科）全北区
black grass aphid		*Schizaphis nigerrima* (Hille Ris Lambers)	（カメムシ目、アブラムシ科）旧北区
black grass bugs		*Irbisia*	（カメムシ目、カスミカメムシ科）の昆虫の総称
black grass skipper		*Monza alberti* (Holland)	（チョウ目、セセリチョウ科）エチオピア区
black grass veneer			tobacco webworm を見よ
black ground stink bug	ツチカメムシ	*Macroscytus japonensis* (Scott)	（カメムシ目、ツチカメムシ科）日本、東洋区
black hairstreak	リンゴシジミ	*Strymonidia pruni jezoensis* (Matsumura)	（チョウ目、シジミチョウ科）日本、旧北区。*S. pruni* Linnaeus の英名
black hairstreak (1)		*Ocaria ocrisia* (Hewitson)	（チョウ目、シジミチョウ科）新北区。Hewitson's blackstreak、black hairstreak butterfly ともいう
black hairy aphid	クロニタイケアブラムシ	*Periphyllus testudinaceus* (Fernie)	（カメムシ目、アブラムシ科）日本、全北区
black hawk moth	クロスズメ	*Hyloicus caligineus* Butler	（チョウ目、スズメガ科）日本、旧北区
black hawk moth			pine hawk moth を見よ
black-headed ant		*Neivamyrmex melanocephalus* (Emery)	（ハチ目、アリ科）新北区
black-headed ash sawfly		*Tethida barda* (Say)	（ハチ目、ハバチ科）新北区
black-headed birch leaffolder moth			Logan's button (1) を見よ
black-headed budworm		*Acleris variana* (Fernald)	（チョウ目、ハマキガ科）新北区。開帳 17 〜 20 mm
black-headed caterpillar			coconut leaf-eating caterpillar を見よ
black-headed conch		*Cochylis atricapitana* (Stephens)	（チョウ目、ホソハマキガ科）旧北区
black-headed cranberry worm			black-headed fireworm を見よ　幼虫の英名
black-headed dwarf		*Elachista atricomella* Stainton	（チョウ目、クサモグリガ科）旧北区
black-headed fireworm	クロネハイイロヒメハマキ	*Rhopobota naevana* (Hübner)	（チョウ目、ハマキガ科）日本、全北区。ツルコケモモ類害虫
black-headed leaf-cutting bee	ズグロハキリバチ	*Chalicodoma monticola* (Smith)	（ハチ目、ハキリバチ科）日本、東洋区
blackheaded pasture cockchafer		*Aphodius pseudotasmaniae* Given	（コウチュウ目、コガネムシ科）豪州区

英名	和名	学名	所属、分布、ほか
blackheaded pasture cockchafer		*Aphodius tasmaniae* Hope	(コウチュウ目、コガネムシ科) 豪州区
black-headed pigmy		*Stigmella atricapitella* (Haworth)	(チョウ目、モグリチビガ科) 旧北区
blackheaded pine sawfly		*Neodiprion excitans* Rohwer	(ハチ目、マツハバチ科) 新北区
black headed sod webworm			vagabond Crambus を見よ
black-headed soft-winged flower beetle		*Collops nigriceps* (Say)	(コウチュウ目、ジョウカイモドキ科) 新北区
black-headed webworm			striped webworm を見よ
black heart		*Uranothauma nubifer* (Trimen)	(チョウ目、シジミチョウ科) エチオピア区
black heart playboy		*Deudorix otraeda* Hewitson	(チョウ目、シジミチョウ科) エチオピア区
black hills beetle			mountain pine beetle (1) を見よ
black honey ant			honey ant (1) を見よ
black-hook Hawaiian damselfly			nigrohamatum damselfly を見よ
black hopper			white-tail hopper を見よ
black-horned cleg		*Haematopota crassicornis* Wahlberg	(ハエ目、アブ科) 新北区
black-horned gem		*Microchrysa polita* (Linnaeus)	(ハエ目、ミズアブ科) 旧北区
black-horned katydid	ツユムシ	*Phaneroptera falcata* (Poda)	(バッタ目、キリギリス科) 日本、旧北区
black-horned pine borer		*Callidium antennatum hesperum* Casey	(コウチュウ目、カミキリムシ科) 新北区
black-horned tree cricket		*Oecanthus nigricornis* Walker	(バッタ目、カンタン科) 新北区。果樹害虫
black hornet			baldfaced hornet を見よ
black horse fly		*Tabanus atratus* Fabricius	(ハエ目、アブ科) 新北区
black house ant		*Ochetellus glaber* (Mayr)	(ハチ目、アリ科) 東洋区、豪州区。black ant ともいう
black house ants		*Ochetellus*	(ハチ目、アリ科) の昆虫の総称
black household ant			white-footed house ant を見よ
black hunter		*Leptothrips mali* (Fitch)	(アザミウマ目、クダアザミウマ科) 新北区。捕食性
black hunter thrips			black hunter を見よ
black hyponomeutid			euonymus ermine moth を見よ
black imported fire ant		*Solenopsis richteri* Forel	(ハチ目、アリ科) 新北区、新熱帯区
black-jack			mustard beetle (2) を見よ
black jacket		*Vespula consobrina* (Saussure)	(ハチ目、スズメバチ科) 新北区。blackjacket とも記される
black Jezebel			common Jezebel (1) を見よ

英名	和名	学名	所属、分布、ほか
black-kneed capsid		*Blepharidopterus angulatus* (Fallén)	（カメムシ目、カスミカメムシ科）旧北区
black-kneed gum leaf grasshopper		*Goniaea carinata* (Stål)	（バッタ目、バッタ科）豪州区
black lady beetle	ハラアカクロテントウ	*Rhyzobius ventralis* (Erichson)	（コウチュウ目、テントウムシ科）新北区、豪州区
black ladybird		*Rhyzobius forestieri* (Mulsant)	（コウチュウ目、テントウムシ科）日本、豪州区
black ladybird beetle			black lady beetle を見よ
black Langessa moth		*Langessa nomophilalis* (Dyar)	（チョウ目、メイガ科）新北区
black larder beetle	トビカツオブシムシ	*Dermestes ater* De Geer	（コウチュウ目、カツオブシムシ科）日本、汎世界。貯蔵食品害虫
black larder beetle	カドマルカツオブシムシ	*Dermestes haemorrhoidalis* (Kuster)	（コウチュウ目、カツオブシムシ科）日本、旧北区
black lawn beetle			African black beetle を見よ
black leaf beetle		*Rhyparida nitida* Clark	（コウチュウ目、ハムシ科）豪州区
black leaf beetle		*Rhyparida morosa* Jacoby	（コウチュウ目、ハムシ科）豪州区
black leaf-cutting bee	シロオビツツハナバチ	*Osmia excavata* Alfken	（ハチ目、ハキリバチ科）日本、旧北区
black leaf folder			sweetpotato webworm moth を見よ
black leafminer		*Phytobia maculosa* (Malloch)	（ハエ目、ハモグリバエ科）大洋区。
blackleg tortoiseshell			large tortoiseshell を見よ
blacklegged ham beetle			common bone beetle を見よ
black-legged horsefly		*Hybomitra micans* (Meigen)	（ハエ目、アブ科）旧北区
black-legged meadow katydid		*Orchelimum nigripes* Scudder	（バッタ目、キリギリス科）新北区
black-legged tortoise beetle		*Jonthonota nigripes* (Olivier)	（コウチュウ目、ハムシ科）新北区。サツマイモ害虫
black-legged water-snipefly		*Ibisia marginata* (Fabricius)	（ハエ目、ナガレアブ科）旧北区
black legume aphid			cowpea aphid を見よ
black-letter Pococera moth		*Pococera melanogrammos* (Zeller)	（チョウ目、メイガ科）新北区
blackline Hawaiian damselfly			nigrohamatum damselfly を見よ
black line scale			black thread scale を見よ
black-lined carpenterworm moth		*Inguromorpha basalis* Walker	（チョウ目、ボクトウガ科）新北区
black-lined cutworm			black army cutworm moth を見よ
black-lined eggar		*Grammodora nigrolineata* Aurivillius	（チョウ目、シャクガ科）豪州区
black-lined greenish geometrid	クロスジアオシャク	*Geometra valida* Felder et Rogenhofer	（チョウ目、シャクガ科）日本、旧北区

英名	和名	学名	所属、分布、ほか
black-lined noctuid			small fanfoot (1) を見よ
black-lined orthetrum			black-tailed skimmer を見よ
black locust borer			carpenter moth (1) を見よ
black locust gall midge			locust gall midge を見よ
black longicorn			black pine longhorn beetle を見よ
black longicorn beetle			black pine longhorn beetle を見よ
black maize beetle			African black beetle を見よ
black maize beetles		*Heteronychus*	（コウチュウ目、コガネムシ科）の昆虫の総称
black margined aphid		*Monellia costalis* (Fitch)	（カメムシ目、アブラムシ科）新北区
black margined aphid			little hickory aphid を見よ
black-margined red longicorn	ヘリグロベニカミキリ	*Purpuricenus spectabilis* Motschulsky	（コウチュウ目、カミキリムシ科）日本、旧北区
black-margined shieldback		*Pediodectes nigromarginatus* (Caudell)	（バッタ目、キリギリス科）新北区
black-margined yellow pecan aphid			little hickory aphid を見よ
black-marked Ancylis moth		*Ancylis metamelana* (Walker)	（チョウ目、ハマキガ科）新北区
black-marked birch leafminer			birch leaf-mining sawfly (2) を見よ
black-marked brown scoparia moth		*Scoparia ustimacula* Felder	（チョウ目、メイガ科）豪州区
black-marked Inga moth		*Inga sparsiciliella* (Clemens)	（チョウ目、マルハキバガ科）新北区
black-marked orange tip		*Colotis daira* (Klug)	（チョウ目、シロチョウ科）エチオピア区
black-marked prominent			cherry caterpillar を見よ
black-marked tussock moth	クロモンドクガ	*Pida niphonis* (Butler)	（チョウ目、ドクガ科）日本、旧北区
black-marked yellow hindwinged noctuid	カバフキシタバ	*Catocala mirifica* Butler	（チョウ目、ヤガ科）日本
black-mottled delicate geometrid	クロフヒメエダシャク	*Peratophyga hyalinata aerata* (Moore)	（チョウ目、シャクガ科）日本
black marmorate long-winged planthopper	クロフハネナガウンカ	*Mysidioides sapporensis* (Matsumura)	（カメムシ目、ハネナガウンカ科）日本、旧北区、東洋区
black marsh trotter			ferruginous glider を見よ
black meadowhawk			black darter を見よ
black metallic beetle	クロタマムシ	*Buprestis haemorrhoidalis japonensis* E. Saunders	（コウチュウ目、タマムシ科）日本
black minute weevil	タバゲササラゾウムシ	*Demimaea fascicularis* (Roelofs)	（コウチュウ目、ゾウムシ科）日本、東洋区

英名	和名	学名	所属、分布、ほか
black moth	カラスヨトウ	*Amphipyra livida corvina* Motschulsky	（チョウ目、ヤガ科）日本
black mounntain moth		*Psodos coracina* (Esper)	（チョウ目、シャクガ科）旧北区。日本亜種は *P. c. daisetsuzana* Matsumura ダイセツタカネエダシャク。black mountain ともいう
black muslin sweep		*Pachythelia villosella* (Ochsenheimer)	（チョウ目、ミノガ科）旧北区
blackneck	エゾクビグロクチバ	*Lygephila pastinum* (Treitschke)	（チョウ目、ヤガ科）日本、旧北区。blackneck moth ともいう
black-nosed conehead		*Neoconocephalus melanorhinus* (Rehn et Hebard)	（バッタ目、キリギリス科）新北区
black odoreous ant			field ant を見よ
black-olive caterpillar moth		*Garella nilotica* (Regenhofer)	（チョウ目、ヤガ科）豪州区、大洋区、新北区、新熱帯区。幼虫は black-olive caterpillar
black olive scale			black scale を見よ
black orange aphid			black citrus aphid を見よ　black orange aphis ともいう
black paddy bug		*Scotinophara coarctata* (Thunberg)	（カメムシ目、カメムシ科）東洋区
black paintbrush swift		*Baoris unicolor* Moore	（チョウ目、セセリチョウ科）東洋区
black palmer			cabbage sawfly (3) を見よ　幼虫の英名
black pansy			blue pansy (1) を見よ
black parlatoria			citrus scale を見よ
black parlatoria scale			chaff scale (1) を見よ
black parlatoria scale			citrus scale を見よ　豪州での英名
black-patch piercer	コトドマツヒメハマキ	*Pammene ochsenheimeriana* (Lienig et Zeller)	（チョウ目、ハマキガ科）日本、旧北区
black-patched Clepsis moth		*Clepsis melaleucanus* (Walker)	（チョウ目、ハマキガ科）新北区
black-patched cracker		*Hamadryas atlantis* (Bates)	（チョウ目、タテハチョウ科）新北区、新熱帯区
black-patched Glaphyria moth		*Glaphyria fulminalis* (Grote)	（チョウ目、メイガ科）新北区。black-patched Glaphyria ともいう
black-patched graylet moth		*Hyperstrotia secta* (Grote)	（チョウ目、ヤガ科）新北区
black pavement ant			pavement ant を見よ　米国での英名
black peach aphid		*Brachycaudus persicae* (Passerini)	（カメムシ目、アブラムシ科）新北区
black peach aphid (1)		*Brachycaudus persicaecola* (Boisduval)	（カメムシ目、アブラムシ科）旧北区
black peach aphid			waterlily aphid を見よ
black pecan aphid		*Melanocallis caryaefoliae* (Davis)	（カメムシ目、アブラムシ科）新北区。ペカン害虫
black Penestola moth		*Penestola bufalis* (Guenée)	（チョウ目、メイガ科）新北区
black percher		*Diplacodes lefebvrii* (Rambur)	（トンボ目、トンボ科）エチオピア区

英名	和名	学名	所属、分布、ほか
black petaltail		*Tanypteryx hageni* (Selys)	(トンボ目、ムカシヤンマ科) 新北区
black pie		*Tuxentius melaena* (Trimen)	(チョウ目、シジミチョウ科) エチオピア区
black piercer			sector-spotted piercer を見よ
black pierrot		*Tarucus grammicus* (Grose-Smith et Kirby)	(チョウ目、シジミチョウ科) エチオピア区
black pine bark beetle	マツノクロキクイムシ	*Hylestes ater* (Paykull)	(コウチュウ目、キクイムシ科) 日本、旧北区、豪州区
black pine leaf scale		*Nuculaspis californica* (Coleman)	(カメムシ目、マルカイガラムシ科) 新北区
black pine longhorn beetle	クロカミキリ	*Spondylis buprestoides* Linnaeus	(コウチュウ目、カミキリムシ科) 日本、旧北区、東洋区。black pine longhorn ともいう
black pine sawyer		*Monochamus scutellatus* (Say)	(コウチュウ目、カミキリムシ科) 新北区
black plague thrips		*Haplothrips froggatti* Hood	(アザミウマ目、クダアザミウマ科) 豪州区
black-polished bark beetle	クロツヤキクイムシ	*Trypodendron proximum* (Niijima)	(コウチュウ目、キクイムシ科) 日本、旧北区
black pondhawk		*Erythemis attala* (Selys)	(トンボ目、トンボ科) 新熱帯区
black potter wasp		*Delta pyriformis philippinensis* (Bequaert)	(ハチ目、スズメバチ科) 新北区
black prince		*Psaltoda plaga* (Walker)	(カメムシ目、セミ科) 豪州区
black prince		*Rohana parisatis* (Westwood)	(チョウ目、タテハチョウ科) 東洋区
black rain beetle		*Pleocoma puncticollis* Rivers	(コウチュウ目、コガネムシ科) 新北区
black Rajah		*Charaxes setan* Detani	(チョウ目、タテハチョウ科) 新熱帯区
black Rajah			wise Rajah を見よ
black rhombic planthopper	クロヒシウンカ	*Oliarus angusticeps* Horvath	(カメムシ目、ヒシウンカ科) 日本、旧北区
black rice bug	イネクロカメムシ	*Scotinophara lurida* (Burmeister)	(カメムシ目、カメムシ科) 日本、東洋区。イネ害虫
black rice planthopper	シロオビウンカ	*Unkanodes albifascia* (Matsumura)	(カメムシ目、ウンカ科) 日本、東洋区。イネ害虫
black rice stink bug			black rice bug を見よ
black rice worm	コメノシマメイガ	*Aglossa dimidiata* (Haworth)	(チョウ目、メイガ科) 日本、旧北区、東洋区。幼虫は貯蔵食品、乾燥動植物質を食う
black rice-plant weevil	クロイネゾウモドキ	*Notaris oryzae* (Ishida)	(コウチュウ目、ゾウムシ科) 日本
black-rimmed hunchback		*Ogcodes pallipes* Latreille	(ハエ目、コガシラアブ科) 旧北区
black-rimmed prominent moth			false sphinx を見よ
black-ringed ochre			common white-spot skipper を見よ
black ringlet	バルカンベニヒカゲ	*Erebia melas* (Herbst)	(チョウ目、タテハチョウ科) 旧北区
black root-bug		*Cyrtomenus mirabilis* (Perty)	(カメムシ目、ツチカメムシ科) 新熱帯区。ピーナッツ害虫

英名	和名	学名	所属、分布、ほか
black rustic		*Aporophyla nigra* (Haworth)	（チョウ目、ヤガ科）旧北区
black rustic		*Aporophyla lunula* Strom	（チョウ目、ヤガ科）旧北区
black saddlebags skimmer		*Tramea lacerata* (Hagen)	（トンボ目、トンボ科）新北区。black saddlebags ともいう
black salt marsh mosquito		*Aedes taeniorhynchus* (Wiedemann)	（ハエ目、カ科）新北区
black satyr		*Satyrus actaea* Esper	（チョウ目、タテハチョウ科）旧北区
black sawfly		*Tenthredo atra* Linnaeus	（ハチ目、ハバチ科）旧北区
black sawfly	ヤチダモハバチ	*Tomostethus nigritus* (Fabricius)	（ハチ目、ハバチ科）日本、旧北区
black scale	オリーブカタカイガラムシ	*Saissetia oleae* (Bernard)	（カメムシ目、カタカイガラムシ科）日本、新北区、エチオピア区。オリーブ害虫
black scale			Florida red scale を見よ
black scale			nigra scale を見よ
black scale of coffee			nigra scale を見よ
black scavenger flies	ツヤホソバエ科	Sepsidae	（ハエ目）の昆虫の総称
black-shaped Platynota moth		*Platynota flavedana* Clemens	（チョウ目、ハマキガ科）新北区
black shield leafroller			gray Archips moth を見よ
black shining ant			field ant を見よ
black-shoulder chafer	カタモンコガネ	*Blitopertha conspurcata* (Harold)	（コウチュウ目、コガネムシ科）日本
black-shouldered spinyleg		*Dromogomphus spinosus* Selys	（トンボ目、サナエトンボ科）新北区
black shore bugs		*Saldula*	（カメムシ目、ミズギワカメムシ科）の昆虫の総称
black-sided grouse locust			black-sided pygmy grasshopper を見よ
black-sided meadow katydid		*Conocephalus nigropleurum* (Brunner)	（バッタ目、キリギリス科）新北区
black-sided pygmy grasshopper		*Tettigidea lateralis* (Say)	（バッタ目、ヒシバッタ科）豪州区
black-sided shieldback		*Clinopleura melanopleura* (Scudder)	（バッタ目、キリギリス科）新北区
black silkworm uji-fly			silkworm tachina fly (1) を見よ
black skipper		*Toxidia melania* (Waterhouse)	（チョウ目、セセリチョウ科）豪州区
black slug cup moth		*Doratifera casta* Scott	（チョウ目、イラガ科）豪州区
black-smudged Chionodes moth		*Chionodes mediofuscella* (Clemens)	（チョウ目、キバガ科）新北区
black snipefly		*Chrysopilus cristatus* (Fabricius)	（ハエ目、シギアブ科）旧北区
black snout			green cloverworm moth を見よ
black snow mosquito		*Aedes ventrovittis* Dyar	（ハエ目、カ科）新北区
black sober		*Anacampsis temerella* (Lienig et Zeller)	（チョウ目、キバガ科）旧北区

英名	和名	学名	所属、分布、ほか
blacksoil scarab		*Othnonius batesii* Olliff	(コウチュウ目、コガネムシ科) 豪州区
black soldier fly	アメリカミズアブ	*Hermetia illucens* (Linnaeus)	(ハエ目、ミズアブ科) 日本、汎世界。塵芥などに発生
black-speckled flunkey		*Coscinia cribraria* (Linnaeus)	(チョウ目、ヒトリガ科) 旧北区
black-speckled grey groundling		*Adrasteia proximella* (Hübner)	(チョウ目、キバガ科) 旧北区
black spider beetle		*Mezium americanum* (Laporte)	(コウチュウ目、シバンムシ科) 汎世界
black spined ant			weaver ant (1) を見よ
black-spined weevil		*Scolopterus penicillatus* White	(コウチュウ目、ゾウムシ科) 豪州区
black splash		*Tetrathemis polleni* (Selys)	(トンボ目、トンボ科) エチオピア区
black-spot Bematistes			Alcione Bematistes を見よ
black-spot hairstreak		*Parrhasius polibetes* (Stoll)	(チョウ目、シジミチョウ科) 新熱帯区
black-spot Remella			two-tone skipper を見よ
black-spotted and blue-winged longicorn			black-spotted blue longicorn beetle を見よ
black-spotted beech longicorn beetle	ヒゲナガゴマフカミキリ	*Palimua liturata* (Bates)	(コウチュウ目、カミキリムシ科) 日本。木材害虫。black-spotted beech longicorn ともいう
black-spotted blue longicorn beetle	ルリボシカミキリ	*Rosalia batesi* Harold	(コウチュウ目、カミキリムシ科) 日本。black-spotted blue longicorn beetle, black-spotted longicorn ともいう
black-spotted bug	ヒサゴナガカメムシ	*Mizaldus lewisii* Distant	(カメムシ目、ナガカメムシ科) 日本、東洋区
black-spotted buprestid	マスダクロホシタマムシ	*Ovalisia vivata* (Lewis)	(コウチュウ目、タマムシ科) 日本。幼虫がスギ、ヒノキに食入
black-spotted cantharid	セボシジョウカイ	*Athemus vitellinus* (Kiesenwetter)	(コウチュウ目、ジョウカイボン科) 日本
black-spotted dagger moth	ゴマケンモン	*Moma alpium* (Osbeck)	(チョウ目、ヤガ科) 日本、旧北区。クリ害虫
black-spotted flash			common tit (1) を見よ
black-spotted grass-blue	クロホシヒメシジミ	*Famegana alsulus* (Herrich-Schäffer)	(チョウ目、シジミチョウ科) 日本、東洋区、豪州区
black-spotted grasshopper		*Stenobothrus nigromaculatus* (Herrich-Schäffer)	(バッタ目、バッタ科) 旧北区
black-spotted hairstreak		*Satyrium oenone* Leech	(チョウ目、シジミチョウ科) 旧北区
black-spotted lady beetles		Coccinellini	(コウチュウ目、テントウムシ科) の昆虫の総称
black-spotted ladybird beetles			black-spotted lady beetles を見よ
black-spotted leaf beetle	クロボシツツハムシ	*Cryptocephalus signaticeps* Baly	(コウチュウ目、ハムシ科) 日本、旧北区
black-spotted leafhopper	ホシハトムネヨコバイ	*Macropsis scutellata* (Boheman)	(カメムシ目、オオヨコバイ科) 日本、旧北区

英名	和名	学名	所属、分布、ほか
black-spotted leafroller moth		*Sciota virgatella* (Clemens)	(チョウ目、メイガ科) 新北区
black-spotted pierrot		*Tarucus balkanicus nigra* Bethune-Baker	(チョウ目、シジミチョウ科) 東洋区。little tiger blue を参照
black-spotted pine sawfly			pine sawfly (1) を見よ
blackspotted pliers support beetle		*Rhagium mordax* (De Geer)	(コウチュウ目、カミキリムシ科) 旧北区
black-spotted prominent moth			anguina moth を見よ　black-spotted prominent ともいう
black-spotted red longicorn	モンクロベニカミキリ	*Purpuricenus lituratus* Ganglebauer	(コウチュウ目、カミキリムシ科) 日本　旧北区
black-spotted Schrankia moth		*Schrankia macula* (Druce)	(チョウ目、ヤガ科) 新北区、新熱帯区
black-spotted silvery prominent	モンクロギンシャチホコ	*Wilemanus bidentatus* Matsumura	(チョウ目、シャチホコガ科) 日本
black-spotted white			psyche butterfly を見よ
black-spotted white looper moth			magpie moth (1) を見よ
black sprite		*Pseudagrion commoniae* Förster	(トンボ目、イトトンボ科) エチオピア区
black spruce beetle	トドマツカミキリ	*Tetropium castaneum* (Linnaeus)	(コウチュウ目、カミキリムシ科) 日本、旧北区。European spruce longhorn beetle ともいう
black squeaker		*Paulopsalta encaustica* (Germar)	(カメムシ目、セミ科) 豪州区
black stem-borer of coffee			apate borer を見よ
black stink bug		*Coptosoma xanthogramma* (White)	(カメムシ目、マルカメムシ科) 新北区
black stink roach			black cockroach を見よ
black strawberry beetle		*Clivina taemaniensis* Sloane	(コウチュウ目、オサムシ科) 豪州区
black-streaked blue		*Uranothauma poggei poggei* (Dewitz)	(チョウ目、シジミチョウ科) エチオピア区
black-striped arctiid	クロスジヒトリ	*Crematonotos gangis* (Linnaeus)	(チョウ目、ヒトリガ科) 日本、東洋区、豪州区
black striped bark beetle			striped ambrosia beetle を見よ
black-striped bark-like moth	クロスジキノカワガ	*Nycteola asiatica* (Krulikovsky)	(チョウ目、ヤガ科) 日本、旧北区
black striped-edge piercer			pea moth を見よ
black-striped elongate froghopper	クロスジホソアワフキ	*Aphilaenus nigripectus* (Matsumura)	(カメムシ目、アワフキムシ科) 日本、旧北区
black-striped flower longicorn	カエデノヘリクロハナカミキリ	*Eustrangalis distenioides* Bates	(コウチュウ目、カミキリムシ科) 日本、旧北区
black-striped froghopper	クロスジアワフキ	*Aphrophora vittatus* Matsumura	(カメムシ目、アワフキムシ科) 日本
black-striped hairtail		*Anthene amarah* (Guérin-Méneville)	(チョウ目、シジミチョウ科) エチオピア区

英名	和名	学名	所属、分布、ほか
black striped leaf bug	ナカグロカスミカメムシ	*Adelphocoris suturalis* (Jakovlev)	(カメムシ目、カスミカメムシ科)、日本、旧北区
black striped longhorn beetle		*Stenurella melanura* (Linnaeus)	(コウチュウ目、カミキリムシ科) 旧北区
black-striped long-necked leaf beetle	ルイスクビナガハムシ	*Lilioceris lewisi* (Jacoby)	(コウチュウ目、ハムシ科) 日本
black-striped mosquito			southern saltmarsh mosquito を見よ
black-striped prominent	クロスジシャチホコ	*Lophocosma sarantuja* Schintlmeister et Kinoshita	(チョウ目、シャチホコガ科) 日本、旧北区
black-striped roselia	クロスジコブガ	*Meganola fumosa* (Butler)	(チョウ目、コブガ科) 日本、旧北区
black sunflower scarab		*Pseudoheteronyx basicollis* Lea	(コウチュウ目、コガネムシ科) 豪州区
black sunflower stem weevil		*Apion (Fallapion) occidentale* Fall	(コウチュウ目、ホソクチゾウムシ科) 新北区
black swallowtail	クロキアゲハ (メスクロキアゲハ)	*Papilio polyxenes* Fabricius	(チョウ目、アゲハチョウ科) 新北区
black swallowtails	アゲハ属	*Papilio*	(チョウ目、アゲハチョウ科) の昆虫の総称
black swamp ash bark beetle	ヤチダモノクロキクイムシ	*Hylesinus tristis* Blandford	(コウチュウ目、キクイムシ科) 日本、旧北区、東洋区
black swarming leaf beetle		*Rhyparida discopunctulata* Blackburn	(コウチュウ目、ハムシ科) 豪州区
black swordtail			mamba を見よ
black Sympetrum		*Sympetrum scoticum* Donovan	(トンボ目、トンボ科) 旧北区
black Sympetrum			black darter を見よ
black-tailed bee fly			bee fly (1) を見よ
black-tailed bluet		*Azuragrion nigridorsum* (Selys)	(トンボ目、イトトンボ科) エチオピア区
black-tailed click beetle		*Ampedus apicatus* (Say)	(コウチュウ目、コメツキムシ科) 新北区
black-tailed false-skimmer			eastern blacktail を見よ　black-tailed skimmer ともいう
black-tailed mosquito		*Culiseta melanura* (Coquillett)	(ハエ目、カ科) 新北区
black-tailed skimmer		*Orthetrum cancellatum* (Linnaeus)	(トンボ目、トンボ科) 旧北区、東洋区
black-tailed skimmer			eastern blacktail を見よ
black tea thrips			greenhouse thrips を見よ
black termite		*Hospitalitermes monoceros* (Koenig)	(シロアリ目、シロアリ科) 東洋区
black-thorax leafhopper	ムナグロズキンヨコバイ	*Idiocerus nigripectus* Matsumura	(カメムシ目、オオヨコバイ科) 日本、旧北区
black thread scale	クロイトカイガラムシ	*Ischnaspis longirostris* (Signoret)	(カメムシ目、マルカイガラムシ科)日本 (小笠原) 全北区, エチオピア区。温室害虫
black threadtail		*Prodasineura autumnalis* (Fraser)	(トンボ目、モノサシトンボ科) 東洋区

英名	和名	学名	所属、分布、ほか
black thrips		Melanthripidae	（アザミウマ目）の昆虫の総称　多くの種が捕食性
black thrips			red clover thrips を見よ
black tiger longicorn			kokeshi longicorn beetle を見よ
black tinder fungus beetle	ホソカブトゴミムシダマシ	*Bolitophagus reticulatus* (Linnaeus)	（コウチュウ目、ゴミムシダマシ科）日本、旧北区
black tip Acraea			black-tipped Acraea を見よ
blacktip sulphur			mimosa yellow を見よ
black-tipped Acraea		*Acraea caldarena* Hewitson	（チョウ目、タテハチョウ科）エチオピア区
black tipped archduke			black archduke を見よ
black-tipped cedemerid			wharf borer を見よ
black-tipped darner		*Aeschna tuberculifera* Walker	（トンボ目、ヤンマ科）新北区
black-tipped diadem	シロモンサザナミムラサキ	*Hypolimnas monteironis* (Druce)	（チョウ目、タテハチョウ科）エチオピア区
black-tipped flashwing		*Vestalis apicalis nigrescens* Fraser	（トンボ目、イトトンボ科）東洋区
black-tipped forest glory		*Vestalis apicalis* Selys	（トンボ目、イトトンボ科）東洋区
black-tipped hangingfly		*Hylobittacus apicallis* (Hagen)	（シリアゲムシ目、ガガンボモドキ科）新北区
black-tipped leafhopper	ツマグロオオヨコバイ	*Bothrogonia ferruginea* (Fabricius)	（カメムシ目、オオヨコバイ科）日本、東洋区、エチオピア区。多くの作物、果樹の害虫
black-tipped pellucid planthopper	ツマグロスケバ	*Orithopagus lunifer* Uhler	（カメムシ目、テングスケバ科）日本、旧北区、東洋区
black-tipped percher		*Diplacodes nebulosa* (Fabricius)	（トンボ目、トンボ科）東洋区
black-tipped Ptichodis moth		*Ptichodis vinculum* (Guenée)	（チョウ目、ヤガ科）新北区
black-tipped Rudenia moth		*Rudenia leguminana* (Busck)	（チョウ目、ハマキガ科）新北区、新熱帯区
black-tipped sawfly	アトグロヒラタハバチ	*Acantholyda nemoralis* (Thomson)	（ハチ目、ヒラタハバチ科）日本、旧北区
black-tipped scarlet		*Axiocerses coalescens* Henning et Henning	（チョウ目、シジミチョウ科）エチオピア区
black-tipped soldier beetle			blood sucker を見よ
black-topped Euselasia		*Euselasia bettina* (Hewitson)	（チョウ目、シジミタテハ科）新熱帯区
black tree cricket		*Homaeogryllus japonicus* Haan	（バッタ目、コオロギ科）日本、旧北区
black treeticker		*Birrima varians* (Germar)	（カメムシ目、セミ科）豪州区
black-tufted white-spots		*Osmodes lindseyi* Miller	（チョウ目、セセリチョウ科）エチオピア区
black turfgrass Ataenius		*Ataenius spretulus* (Haldeman)	（コウチュウ目、コガネムシ科）新北区

英名	和名	学名	所属、分布、ほか
black turnip moth		*Syntheta nigerrima* (Guenée)	（チョウ目、ヤガ科）豪州区
black turpentine beetle		*Dendroctonus terebrans* (Olivier)	（コウチュウ目、キクイムシ科）新北区
black twig borer			castanopsis ambrosia beetle (2) を見よ　米国での英名
black-unispotted leafhopper	ヒトツメヨコバイ	*Phlogotettix cyclops* (Mulsant et Rey)	（カメムシ目、オオヨコバイ科）日本、旧北区
black urban bibionid		*Dilophus orbatus* (Say)	（ハエ目、ケバエ科）新北区
black V moth		*Arctornis l-nigrum* (Müller)	（チョウ目、ドクガ科）旧北区。black V ともいう
blackvein fritillary		*Melitaea arcesia* Bremer	（チョウ目、タテハチョウ科）旧北区、東洋区
blackvein sergeant	ランガオオミスジ	*Athyma ranga* Moore	（チョウ目、タテハチョウ科）東洋区
black-veined ant-blue		*Acrodispas arcana* (Miller et Edwards)	（チョウ目、シジミチョウ科）豪州区
black-veined Brangas		*Brangas coccineifrons* (Godman et Salvin)	（チョウ目、シジミチョウ科）新熱帯区
black-veined brown			monarch butterfly を見よ
black-veined Epitola		*Stempfferia congoana* (Aurivillius)	（チョウ目、シジミチョウ科）エチオピア区
black-veined hairstreak			large striped blue を見よ
black-veined leafwing		*Consul excellens excellens* (Bates)	（チョウ目、タテハチョウ科）新熱帯区。*C. e. genini* (Le Cerf) も同英名
black-veined looper			black-veined moth を見よ
black veined moth		*Cyclosia papillionaris* Drury	（チョウ目、マダラガ科）東洋区
black-veined moth		*Siona lineata* (Scopoli)	（チョウ目、シャクガ科）旧北区。black-veined ともいう
black-veined redeye		*Matapa sasivarna* (Moore)	（チョウ目、セセリチョウ科）東洋区
black-veined ruby-eye		*Synale cynaxa* (Hewitson)	（チョウ目、セセリチョウ科）新熱帯区
black-veined skipper		*Mnasinous patage* Godman	（チョウ目、セセリチョウ科）新熱帯区
black veined tiger			common tiger (1) を見よ　*D. m. hegesippus* Cramer の英名
black-veined white		*Dixeia doxo parva* Talbot	（チョウ目、シロチョウ科）エチオピア区。African small white (1) を参照
black-veined white (1)	エゾシロチョウ	*Aporia crataegi adherbal* Fruhstorfer	（チョウ目、シロチョウ科）日本、旧北区。*A. crataegi* (Linnaeus) の英名。black-veined white butterfly ともいう
black-veined white spots		*Osmodes costatus* Aurivillius	（チョウ目、セセリチョウ科）エチオピア区
blackveins	エゾシロチョウ属	*Aporia*	（チョウ目、シロチョウ科）の昆虫の総称
black velvet			green-banded swallowtail を見よ
black vicetail		*Hemigomphus atratus* Watson	（トンボ目、サナエトンボ科）豪州区
black vine thrips			common black vine thrips を見よ

英名	和名	学名	所属、分布、ほか
black vine weevil	キンケクチブトゾウムシ	*Otiorhynchus sulcatus* (Fabricius)	（コウチュウ目、ゾウムシ科）日本、全北区、豪州。キクなど多くの植物につく
black walnut curculio		*Conotrachelus retentus* Say	（コウチュウ目、ゾウムシ科）新北区
black wasp moth		*Ceryx imaon* (Cramer)	（チョウ目、カノコガ科）東洋区
blackwater bluet		*Enallagma weewa* Byers	（トンボ目、イトトンボ科）新北区
blackwater clubtail		*Gomphus dilatatus* Rambur	（トンボ目、サナエトンボ科）新北区
black-waved brown moth		*Selidosema suavis* Butler	（チョウ目、シャクガ科）豪州区
black-waved flannel moth			yellow flannel moth (1) を見よ
black webspinner		*Oligotoma nigra* (Hagen)	（シロアリモドキ目、シロアリモドキ科）全北区、豪州区
black wedge-spot moth		*Homophoberia apicosa* (Haworth)	（チョウ目、ヤガ科）新北区
black weevil		*Liparus coronatus* Goeze	（コウチュウ目、ゾウムシ科）旧北区
black weevil (1)		*Otiorhynchus niger* Fabricius	（コウチュウ目、ゾウムシ科）旧北区
black weevil			black vine weevil を見よ
black wheat gall midge	ムギクロタマバエ	*Contarinia* sp.	（ハエ目、タマバエ科）日本。ムギ害虫
black wheat thrips			grain thrips を見よ
black willow aphid		*Pterocomma salicis* (Linnaeus)	（カメムシ目、アブラムシ科）全北区
black willow scale			willow scale (2) を見よ
black windmill		*Byasa crassipes* (Oberthür)	（チョウ目、アゲハチョウ科）東洋区
black wing			ebony jewelwing を見よ　black-wing とも記す。
black-wing Epitola		*Geritola virginea* (Bethune-Baker)	（チョウ目、シジミチョウ科）エチオピア区
blackwing subterranean termite		*Odontotermes formosanus* (Shiraki)	（シロアリ目、シロアリ科）東洋区。blackwinged subterranean termite ともいう
black-winged Acraea	アトキホソチョウ	*Acraea asboloplintha* Karsch	（チョウ目、タテハチョウ科）エチオピア区
black-winged Dahana		*Dahana atripennis* Grote	（チョウ目、ヒトリガ科）新北区。black-winged Dahana moth ともいう
black-winged damselfly			ebony jewelwing を見よ
black-winged dragonlet		*Erythrodiplax funerea* (Hagen)	（トンボ目、トンボ科）新北区、新熱帯区
black-winged shieldback		*Zacycloptera atripennis* Caudell	（バッタ目、キリギリス科）新北区
black winter weevil		*Ceutorhynchus picitarsis* Gyllenhal	（コウチュウ目、ゾウムシ科）旧北区
black witch		*Ascalapha odorata* (Linnaeus)	（チョウ目、ヤガ科）新北区、新熱帯区、東洋区、エチオピア区
black witch moth			black witch を見よ
black woolly-bear		*Arctia caja* (Linnaeus)	（チョウ目、ヒトリガ科）日本、旧北区。日本亜種は *A. c. phaeosoma* (Butler) ヒトリガ

英名	和名	学名	所属、分布、ほか
black xylocopid	クマバチ	*Xylocopa appendiculata circumvolans* Smith	(ハチ目、ミツバチ科) 日本
black Zale moth		*Zale undularis* (Drury)	(チョウ目、ヤガ科) 新北区
black zigzag moth		*Panthea acronyctoides* (Walker)	(チョウ目、ヤガ科) 新北区。black zigzag ともいう
blackberry aphid		*Sitobion fragariae* (Walker)	(カメムシ目、アブラムシ科) 全北区、豪州区、エチオピア区
blackberry aphid			scarce blackberry aphid を見よ
blackberry blossom midge			blackberry flower midge を見よ
blackberry bud weevil			strawberry weevil (1) を見よ
blackberry bug		*Elasmucha ferrugata* (Fabricius)	(カメムシ目、ツノカメムシ科) 旧北区
blackberry-cereal aphid			blackberry aphid を見よ
blackberry clearwing moth			raspberry crown borer を見よ
blackberry flower midge		*Contarinia rubicola* Rubsaamen	(ハエ目、タマバエ科) 旧北区
blackberry leaf midge		*Dasineura plicatrix* (Loew)	(ハエ目、タマバエ科) 旧北区
blackberry leaf skeletonizer			narrow-winged false feather を見よ
blackberry leafhopper		*Dikrella californica* (Lawson)	(カメムシ目、オオヨコバイ科) 新北区
blackberry leafminer moth		*Coptotriche aenea* (Frey et Boll)	(チョウ目、ムモンハムグリガ科) 新北区
blackberry looper moth		*Chlorochlamys chloroleucaria* (Guenée)	(チョウ目、シャクガ科) 全北区。blackberry looper ともいう
blackberry scale			scurfy scale を見よ
blackberry skeletonizer			narrow-winged false feather を見よ　blackberry skeletonizer moth ともいう
blackberry spanworm		*Hame argillacearia* Packard	(チョウ目、シャクガ科) 新北区
blackberry stem gall midge	イチゴウロコタマバエ	*Lasioptera rubi* (Schrank)	(ハエ目、タマバエ科) 日本、旧北区
blackbottle fly		*Phormia terraenovae* Robineau-Desvoidy	(ハエ目、クロバエ科) 全北区。blackbottle ともいう
Blackburn butterfly			Hawaiian blue を見よ
Blackburn damsel bug		*Nabis blackburni* White	(カメムシ目、マキバサシガメ科) 新北区
Blackburn dragonfly		*Nesogonia blackburni* (McLachlan)	(トンボ目、トンボ科) 新北区
Blackburn's blue			Hawaiian blue を見よ　Blackburn's bluet ともいう
Blackburn's Hawaiian damselfly		*Megalagrion blackburni* McLachlan	(トンボ目、イトトンボ科) 大洋区

英名	和名	学名	所属、分布、ほか
Blackburn's sphinx moth		*Manduca blackburni* Butler	(チョウ目、スズメガ科) 大洋区。絶滅危惧種
blackbutt leafminer		*Acrocercops laciniella* (Meyrick)	(チョウ目、ホソガ科) 豪州区
blackened bluewing			dark wave を見よ
blackish hornet	クロスズメバチ	*Vespula flaviceps lewisii* (Cameron)	(ハチ目、スズメバチ科) 日本
blackish meadow katydid		*Conocephalus semivittatus* Walker	(バッタ目、キリギリス科) 豪州区
bladder cicada		*Cystosoma saundersii* Westwood	(カメムシ目、セミ科) 豪州区
bladder-feet			thrips (1) を見よ
bladder flies			small-headed flies を見よ　豪州での英名
bladder-footed insects			thrips (2) を見よ
bladder grasshoppers		Pneumoridae	(バッタ目) の昆虫の総称
bladder pod midge			cabbage gall midge を見よ
bladhia scale	ヤブコウジハマカイガラムシ	*Nipponorthezia ardisiae* Kuwana	(カメムシ目、ハマカイガラムシ科) 日本
Blair's mocha		*Cyclophora puppillaria* (Hübner)	(チョウ目、シャクガ科) 旧北区
Blair's shoulder knot moth		*Lithophane leautieri hesperica* Boursin	(チョウ目、ヤガ科) 旧北区。Blair's shoulder knot ともいう
Blair's wainscott	テンモントガリヨトウ	*Sedina buettneri* (Hering)	(チョウ目、ヤガ科) 日本、旧北区
Blake's tiger-moth		*Grammia blakei* (Grote)	(チョウ目、ヒトリガ科) 新北区
Blanchard's Ideopsis		*Ideopsis vitrea* (Blanchard)	(チョウ目、タテハチョウ科) 東洋区
Blanchard's sphinx moth		*Adhemarius blanchardorum* (Hodges)	(チョウ目、スズメガ科) 新北区、新熱帯区
Blanchard's wood nymph			Blanchard's Ideopsis を見よ
blanched hornbeam midget		*Phyllonorycter tenerella* (Joannis)	(チョウ目、ホソガ科) 旧北区
Blandford fly		*Simulium posticatum* Meigen	(ハエ目、ブユ科) 旧北区
blank swift		*Caltoris kumara* (Moore)	(チョウ目、セセリチョウ科) 東洋区
blastobasid moths			scavenger moths を見よ
blastophaga			fig wasp を見よ
blazing star borer moth		*Papaipema beeriana* Bird	(チョウ目、ヤガ科) 新北区
blazing star clearwing moth		*Carmenta anthracipennis* (Boisduval)	(チョウ目、スカシバガ科) 新北区
bleached pug		*Eupithecia expallidata* Doubleday	(チョウ目、シャクガ科) 旧北区
blight			胴枯病をおこす菌や害虫
blight			bean aphid を見よ
blind beetle			June beetle (1) を見よ

英名	和名	学名	所属、分布、ほか
blind buzzart			June beetle (1) を見よ
blind click beetle		*Alaus myops* (Fabricius)	(コウチュウ目、コメツキムシ科) 新北区
blind Eurybia		*Eurybia elvina elvina* Stichel	(チョウ目、シジミタテハ科) 新熱帯区
blind-eye bushbrown		*Mycalesis mamerta* (Stoll)	(チョウ目、タテハチョウ科) 旧北区、東洋区
blind-eyed sphinx moth		*Paonias excaecatus* (J. E. Smith)	(チョウ目、スズメガ科) 新北区。blinded sphinx、blind sphinx ともいう
blind flies			clegs を見よ
blind leafhoppers		*Typhlocyba*	(カメムシ目、オオヨコバイ科) の昆虫の総称
blind purplewing	ミグドニアアイイロタテハ	*Eunica mygdonia omoa* Hall	(チョウ目、タテハチョウ科) 新熱帯区
blind ringlet		*Erebia pharte* (Hübner)	(チョウ目、タテハチョウ科) 旧北区
blind springtails	シロトビムシ科	Onychiuridae	(トビムシ目) の昆虫の総称　日本から約25種
blind underleaf		*Eurybia albiseriata* Weymer	(チョウ目、シジミタテハ科) 新熱帯区
blinding breeze fly			horse-fly (1) を見よ
blinding gadfly			horse-fly (1) を見よ
blinding storm-fly			horse-fly (1) を見よ
blister beetle		*Lytta vesicatoria* (Linnaeus)	(コウチュウ目、ツチハンミョウ科) 全北区。カンタリジンの材料
blister beetles		*Mylabris*	(コウチュウ目、ツチハンミョウ科) の昆虫の総称
blister beetles (1)	ツチハンミョウ科	Meloidae	(コウチュウ目) の昆虫の総称　幼虫は他昆虫に寄生し、成虫からカンタリジンを抽出する
blister beetles (2)		*Epicauta, Macrobasis*	(コウチュウ目、ツチハンミョウ科) の昆虫の総称
blister beetles (3)		*Parisopalpus*	(コウチュウ目、カミキリモドキ科) の昆虫の総称
blister coneworm		*Dioryctria clarioralis* (Walker)	(チョウ目、メイガ科) 新北区。blister coneworm moth ともいう
blister fly			blister beetle を見よ
blister leaf-miner of wheat and barley			blotch leaf-miner of wheat and barley を見よ
Blomer's rivulet	キモンハイイロナミシャク	*Venusia blomeri* (Curtis)	(チョウ目、シャクガ科) 日本、旧北区
Blomfield's bark wing			Blomfield's beauty を見よ
Blomfield's beauty	オオカバタテハ	*Smyrna blomfildia* (Fabricius)	(チョウ目、タテハチョウ科) 新北区、新熱帯区
blond glider		*Cymothoe coranus* Grose-Smith	(チョウ目、タテハチョウ科) エチオピア区
blood Acraea			blood-red Acraea をみよ
blood-red Acraea		*Acraea petraea* Boisduval	(チョウ目、タテハチョウ科) エチオピア区
blood-red longhorn beetle		*Anastrangalia sanguinolenta* (Linnaeus)	(コウチュウ目、カミキリムシ科) 旧北区
blood-red slave-maker			slave-maker ant を見よ　blood-red slave-maker ant ともいう
blood sucker		*Rhagonycha fulva* Scopoli	(コウチュウ目、ジョウカイボン科) 旧北区

英名	和名	学名	所属、分布、ほか
blood-sucking bedbugs			bed bugs を見よ
blood-sucking body louse			sheep face louse を見よ
blood-sucking body louse of sheep			sheep face louse を見よ
blood-sucking buffalo fly		*Lyperosia exigua* (de Meijere)	（ハエ目、イエバエ科）東洋区
bloodsucking conenose		*Triatoma sanguisuga* (LeConte)	（カメムシ目、サシガメ科）新北区。ナンキンムシを捕食。シナガス病を媒介
bloodsucking cone-nose			masked hunter bug を見よ
blood-sucking foot louse			sheep foot louse を見よ
bloodtails		*Lathrecista*	（トンボ目、トンボ科）の昆虫の総称
blood-vein		*Timandra comae* Schmidt	（チョウ目、シャクガ科）旧北区
blood-vein moth		*Timandra griseata* Petersen	（チョウ目、シャクガ科）旧北区。blood-vein ともいう
bloodworm		*Chironomus cavazzai* Kieffer	（ハエ目、ユスリカ科）旧北区。幼虫の英名。赤ミミズと俗称される
bloodworm			plumed gnat を見よ
bloodworms	ユスリカ属	*Chironomus*	（ハエ目、ユスリカ科）の昆虫の総称　幼虫はアカムシ
bloody-nosed beetle		*Timarcha tenebricosa* (Fabricius)	（コウチュウ目、ハムシ科）旧北区
bloody-winged grasshopper		*Arphia pseudonietana* (Thomas)	（バッタ目、バッタ科）新北区
blossom Anomala		*Anomala undulata* Melsheimer	（コウチュウ目、コガネムシ科）新北区。カンキツやアボカド害虫
blossom beetle			pollen beetle を見よ
blossom beetles			pollen beetles (1) を見よ
blossom feeder		*Epicometis (Tropinota) hirtella* Poda	（コウチュウ目、コガネムシ科）旧北区。カンキツ害虫
blossom feeding scarab			citrus flower chafer を見よ
blossom fly			fever fly を見よ
blossom midge	ランツボミタマバエ	*Contarinia maculipennis* Felt	（ハエ目、タマバエ科）日本、大洋区。ラン害虫
blossom thrips		*Megalurothrips distalis* (Karny)	（アザミウマ目、アザミウマ科）東洋区
blossom thrips (1)		*Thrips imaginis* Bagnall	（アザミウマ目、アザミウマ科）豪州区
blossom underwing moth		*Orthosia miniosa* (Denis et Schiffermüller)	（チョウ目、ヤガ科）旧北区。blossom underwing ともいう
blossom weevil			apple blossom weevil を見よ
blossom weevils		*Anthonomus*	（コウチュウ目、ゾウムシ科）の昆虫の総称
blotch leaf miner		*Amauromyza maculosa* (Malloch)	（ハエ目、ハモグリバエ科）大洋区

英名	和名	学名	所属、分布、ほか
blotch leaf-miner of wheat and barley		*Agromyza nigripes* Meigen	（ハエ目、ハモグリバエ科）旧北区
blotch miners			leafblotch miners (1) を見よ
blotched argent			cherry fruit moth を見よ
blotched blue	ニブイロミナミシジミ	*Candalides acastus* (Cox)	（チョウ目、シジミチョウ科）豪州区
blotched dusky-blue			blotched blue を見よ
blotched emerald		*Comibaena bajularia* (Denis et Schiffermüller)	（チョウ目、シャクガ科）旧北区
blotched emerald	ヨツモンマエシロアオシャク	*Comibaena pustulata* Hufnagel	（チョウ目、シャクガ科）旧北区
blotched leaf sitter		*Gorgyra aretina* (Hewitson)	（チョウ目、セセリチョウ科）エチオピア区
blotched leopard		*Lachnoptera ayresii* Trimen	（チョウ目、タテハチョウ科）エチオピア区
blotched long-horned owlfly		*Tmesibasis lacerata* (Hagen)	（アミメカゲロウ目、ツノトンボ科）エチオピア区
blotched marble		*Endothenia quadrimaculana* (Haworth)	（チョウ目、ハマキガ科）旧北区
blotched Nephele		*Nephele bipartita* Butler	（チョウ目、スズメガ科）エチオピア区
blotched obscure piercer	ウスグロヒメハマキ	*Pammene obscurana* (Stephens)	（チョウ目、ハマキガ科）日本、旧北区
blotched obscure piercer			obscure blotched piercer を見よ
blow flies	クロバエ科	Calliphoridae	（ハエ目）の昆虫の総称　とくに *Calliphora* の種
blowflies			blow flies を見よ
blowflies			flesh flies を見よ
blow fly		*Cynomyopsis cadaverina* (Robineau-Desvoidy)	（ハエ目、クロバエ科）新北区
blow fly			green botfly を見よ
blow fly			green blow fly (1) を見よ
blowfly			blue blowfly を見よ
blowfly			bluebottle fly を見よ
blows			blow flies を見よ
bloxworth blue			short-tailed blue を見よ
bloxworth snout		*Hypena obsitalis* (Hübner)	（チョウ目、ヤガ科）旧北区
blue admiral	ルリタテハ	*Kaniska canace nojaponicum* (von Siebold)	（チョウ目、タテハチョウ科）日本。旧北区、東洋区の *K. canace* (Linnaeus) の英名。*K. c. drilon* (Fruhstorfer)、*K. c. perakana* (Distant) も同英名
blue aeschna			azure hawker を見よ
blue albatross			common migrant (1) を見よ
blue alfalfa aphid	コンドウヒゲナガアブラムシ	*Acyrthosiphon kondoi* Shinji	（カメムシ目、アブラムシ科）日本、全北区、大洋区、豪州区。マメ類害虫

英名	和名	学名	所属、分布、ほか
blue-and-orange eighty-eight			Tolima numberwing を見よ　C. t. guatemalena (H. Bates), C. t. pacifica (H. Bates), C. t. tehuana Maza et Maza も同英名
blue ant		Diamma bicolor Westwood	（ハチ目、コツチバチ科）豪州区
blue ant-slug		Aslauga vininga Hewitson	（チョウ目、シジミチョウ科）エチオピア区
blue Argus		Aricia anteros (Freyer)	（チョウ目、シジミチョウ科）旧北区
blue Argus			eyed pansy を見よ　豪州での英名
blue baboon beetle		Saxinis omogera Lacordaire	（コウチュウ目、ハムシ科）新北区
blue-banded bee		Amegilla cingulata (Fabricius)	（ハチ目、ミツバチ科）東洋区、豪州区
blue-banded bee		Amegilla pulchra (Smith)	（ハチ目、ミツバチ科）豪州区
bluebanded bees		Amegilla	（ハチ目、ミツバチ科）の昆虫の総称　豪州での英名
blue-banded damsel		Ischnura evansi Morton	（トンボ目、イトトンボ科）旧北区
blue-banded egg-fly		Hypolimnas alimena (Linnaeus)	（チョウ目、タテハチョウ科）豪州区
blue banded forest brown		Bicyclus hewitsoni Doumet	（チョウ目、タテハチョウ科）エチオピア区
blue-banded purplewing			whitened bluewing を見よ
blue-banded skipper		Gesta gesta (Herrich-Schaffer)	（チョウ目、セセリチョウ科）新北区
blue-banded skipper			false duskywing を見よ
blue barley flea beetle	ヒサゴトビハムシ	Chaetocnema ingenua (Baly)	（コウチュウ目、ハムシ科）日本、旧北区
blue baron		Euthalia telchinia (Ménétriès)	（チョウ目、タテハチョウ科）東洋区
blue-barred forest brown		Bicyclus sebetus (Hewitson)	（チョウ目、タテハチョウ科）エチオピア区
blue-barred leafwing		Anaea ryphea Cramer	（チョウ目、タテハチョウ科）新熱帯区
blue-based flutterer		Rhyothemis triangularis Kirby	（トンボ目、トンボ科）東洋区
blue-based skipper		Neoxeniades molion (Godman)	（チョウ目、セセリチョウ科）新熱帯区
blue-based Theope		Theope virgilius (Fabricius)	（チョウ目、シジミタテハ科）新熱帯区
blue basker		Urothemis edwardsii (Selys)	（トンボ目、トンボ科）エチオピア区
blue Begum	ルリオビヤイロタテハ	Prothoe franck (Godart)	（チョウ目、タテハチョウ科）東洋区
blue blowflies			blow flies を見よ
blue blowfly	ホホアカクロバエ	Calliphora vicina Robeneau-Desvoidy	（ハエ目、クロバエ科）日本、旧北区、豪州区
blue blowfly			bluebottle fly を見よ
blue blowfly			common bluebottle fly を見よ
blue-bordered carpet		Plemyria rubiginata (Denis et Schiffermüller)	（チョウ目、シャクガ科）旧北区。日本亜種は P. r. japonica Inoue トビモンシロナミシャク

英名	和名	学名	所属、分布、ほか
bluebottle		*Neomyia cornicina* (Fabricius)	(ハエ目、イエバエ科) 旧北区
bluebottle			common bluebottle fly を見よ
blue bottle flies			blow flies を見よ
bluebottle fly	ミヤマクロバエ	*Calliphora vomitoria* (Linnaeus)	(ハエ目、クロバエ科) 日本、全北区、東洋区、大洋区。アオバエの俗称あり
blue bottle fly			green botfly を見よ
blue bottle fly			common bluebottle fly を見よ
bluebottle fly			blow fly を見よ
bluebottle fly			blue blowfly を見よ
bluebottles			blow flies を見よ
blue branded king crow		*Euploea eunice leucogonis* (Butler)	(チョウ目、タテハチョウ科) 東洋区
blue brilliant		*Simiskina deolina* (Fruhstorfer)	(チョウ目、シジミチョウ科) 東洋区
blue bug	ルリクチブトカメムシ	*Zicrona caerulea* (Linnaeus)	(カメムシ目、カメムシ科) 日本、旧北区、東洋区、新熱帯区
blue bug			rosy apple aphid (1) を見よ　bluebug とも記す
blue bug			large chicken louse を見よ
blue bugs		*Calidea*	(カメムシ目、カメムシ科) の昆虫の総称
blue cactus borer			North American cactus moth を見よ
blue caliph		*Enispe cycnus* Westwood	(チョウ目、タテハチョウ科) 東洋区
blue cascader		*Zygonyx natalensis* (Martin)	(トンボ目、トンボ科) エチオピア区
blue-celled purplewing		*Eunica volumna* Godart	(チョウ目、タテハチョウ科) 新熱帯区
blue cereal leaf beetle		*Lema gallaeciana* Heyden	(コウチュウ目、ハムシ科) 旧北区
blue Charaxes	ミズアオフタオチョウ	*Charaxes bohemani* Felder	(チョウ目、タテハチョウ科) エチオピア区
blue cimbex	ルリモモブトハバチ	*Cimbex japonica* Kirby	(ハチ目、コンボウハバチ科) 日本、旧北区
blue coconut leaf beetle		*Brontispa chalybeipennis* (Zacher)	(コウチュウ目、ハムシ科) 大洋区
blue-collared firetip		*Mysoria amra* (Hewitson)	(チョウ目、セセリチョウ科) 新熱帯区
blue copper	メスベニルリシジミ	*Lycaena heteronea* Boisduval	(チョウ目、シジミチョウ科) 新北区、新熱帯区
blue Cornelian		*Deudorix epirus* (C. et R. Felder)	(チョウ目、シジミチョウ科) 豪州区
blue crow	クルギイルリマダラ	*Euploea klugii* Moore	(チョウ目、タテハチョウ科) 東洋区
blue damselflies			narrow-winged damselflies を見よ
blue damselfly		*Enallagma civile* (Hagen)	(トンボ目、イトトンボ科) 大洋区、新北区
blue dandy		*Laringa castelnaui* (C. et R. Felder)	(チョウ目、タテハチョウ科) 東洋区
blue darkie		*Allotinus subviolaceus* Felder	(チョウ目、シジミチョウ科) 東洋区。

英名	和名	学名	所属、分布、ほか
blue dasher		*Pachydiplax longipennis* (Burmeister)	(トンボ目、トンボ科) 新北区
blue demon Charaxes		*Charaxes virilis* van Someren et Jackson	(チョウ目、タテハチョウ科) エチオピア区
blue diadem	サザナミムラサキ	*Hypolimnas salmacis* (Drury)	(チョウ目、タテハチョウ科) エチオピア区
blue doctor	ペリアンデルツバメシジミタテハ	*Rhetus periander* Cramer	(チョウ目、シジミタテハ科) 新熱帯区
blue dog louse			dog sucking louse を見よ　南アフリカでの英名
blue Duchess	ニジオビイナズマ	*Euthalia duda* Staudinger	(チョウ目、タテハチョウ科) 東洋区
blue duke		*Bassarona durga* (Moore)	(チョウ目、タテハチョウ科) 東洋区
blue duskywing		*Anastrus neaeris* Möschler	(チョウ目、セセリチョウ科) 新熱帯区
blue-dusted roadside skipper		*Amblyscirtes alternata* (Grote et Robinson)	(チョウ目、セセリチョウ科) 新北区
blue-edged playboy		*Pilodeudorix zela* (Hewitson)	(チョウ目、シジミチョウ科) エチオピア区
blue emperor			emperor dragonfly を見よ
blue emperor			Ulysses butterfly を見よ
blue evening brown		*Melanitis ansorgei* Rothschild	(チョウ目、タテハチョウ科) エチオピア区
blue-eyed darner		*Rhionaeschna multicolor* (Hagen)	(トンボ目、ヤンマ科) 新北区、新熱帯区
blue-eyed darners		*Rhionaeschna*	(トンボ目、ヤンマ科) の昆虫の総称
blue-eyed grayling			large wood nymph を見よ
blue-eyed grayling			wood nymph を見よ
blue-eyed pondcruise		*Epophthalmia vittata cyanocephala* Hagen	(トンボ目、エゾトンボ科) 東洋区
blue-eyed sailor			Dyonis checkerspot を見よ
blue eyes lacewing		*Nymphes myrmeleonides* Leach	(アミメカゲロウ目、Nymphidae) 豪州区
blue-faced darner		*Coryphaeschna adnexa* (Hagen)	(トンボ目、ヤンマ科) 新北区、新熱帯区
blue-faced ringtail		*Erpetogomphus eutainia* Calvert	(トンボ目、サナエトンボ科) 新北区
blue false flasher		*Neodeniades bajula* (Schaus)	(チョウ目、セセリチョウ科) 新熱帯区
blue-flash skipper		*Rachelia extrusus* C. et R. Felder	(チョウ目、セセリチョウ科) 豪州区
blue flasher			flashing Astraptes を見よ
blue flattened longicorn			violet tanbark beetle を見よ
blue forester		*Lethe scanda* (Moore)	(チョウ目、タテハチョウ科) 東洋区
blue fresh-fly			blue blowfly を見よ
blue fresh-fly			bluebottle fly を見よ
blue-fronted dancer		*Argia apicalis* (Say)	(トンボ目、イトトンボ科) 新北区

英名	和名	学名	所属、分布、ほか
blue-frosted banner		*Catonephele numilia esite* (R. Felder)	（チョウ目、タテハチョウ科）新熱帯区。*C. n. immaculata* Jenkins も同英名。blue-spotted firewing を参照
blue gem	キララシジミ	*Poritia erycinoides* (C. et R. Felder)	（チョウ目、シジミチョウ科）東洋区
blue glassy tiger		*Ideopsis vulgaris* (Butler)	（チョウ目、タテハチョウ科）東洋区。*I. v. macrina* (Fruhstorfer) も同英名
blue glassy tiger	リュウキュウアサギマダラ	*Radena similis similis* (Linnaeus)	（チョウ目、タテハチョウ科）日本、東洋区
blue glassy tiger (1)	ウスコモンマダラ	*Tirumala limniace* (Cramer)	（チョウ目、タテハチョウ科）日本、東洋区
blue-glossed skipper		*Onophas columbaria columbaria* (Herrich-Schäffer)	（チョウ目、セセリチョウ科）新熱帯区
blue goat louse			sucking goat louse を見よ　南アフリカでの英名
bluegrass billbug		*Calendra parvulus* (Gyllenhal)	（コウチュウ目、ゾウムシ科）新北区。トウモロコシ、ブルーグラス害虫
bluegrass plant bug			meadow plant bug を見よ　米国での英名
bluegrass webworm	シバツトガ	*Parapediasia teterrella* (Zeller)	（チョウ目、メイガ科）日本。新北区。トウモロコシ害虫。bluegrass webworm moth, bluegrass sod webworm, bluegrass sod webworm moth ともいう
blue-gray Lasaia		*Lasaia maria maria* Clench	（チョウ目、シジミタテハ科）新熱帯区。*L. m. anna* Clench　も同英名
blue-gray satyr			Libye ringlet を見よ
blue-green aphid			blue alfalfa aphid を見よ　豪州での英名
blue-green citrus nibbler		*Colasposoma fulgidum* Lefevre	（コウチュウ目、ハムシ科）豪州区。カンキツ害虫
blue-green corn aphid			corn leaf aphid を見よ
blue-green orange weevil		*Pachneus citri* Marshall	（コウチュウ目、ゾウムシ科）新熱帯区。カンキツ害虫。亜種 *P. c. litus* Germar も同英名
blue-green reflector		*Doxocopa cherubina* (C. et R. Felder)	（チョウ目、タテハチョウ科）新熱帯区
blue-green sharpshooter		*Graphocephala atropunctata* (Signoret)	（カメムシ目、オオヨコバイ科）新北区
blue-gum borer			common eucalypt longicorn を見よ
bluegum eulophid		*Ophelimus eucalypti* (Gahan)	（ハチ目、ヒメコバチ科）豪州区
bluegum psyllid			eucalyptus sucker を見よ　blue gum psyllid とも記す
blue gum scale			gumtree scale を見よ
blue ground beetle		*Carabus intricatus* Linnaeus	（コウチュウ目、オサムシ科）旧北区
blue ham beetle			common bone beetle を見よ
blue hawker			southern hawker を見よ
blue head			figure of eight を見よ
blue heart playboy		*Pilodeudorix caerulea* (Druce)	（チョウ目、シジミチョウ科）エチオピア区。亜種 *P. c. obscurata* (Trimen) も同英名

英名	和名	学名	所属、分布、ほか
blue Helen		*Papilio prexaspes prexaspes* C. et R. Felder	（チョウ目、アゲハチョウ科）東洋区
blue Helen			yellow Helen (1) を見よ
blue horntail		*Sirex cyaneus* Fabricius	（ハチ目、キバチ科）全北区
blue horntail			polished horntail を見よ　カナダでの英名
blue imperial		*Ticherra acte* (Moore)	（チョウ目、シジミチョウ科）東洋区
blue Jay		*Graphium evemon eventus* (Fruhstorfer)	（チョウ目、アゲハチョウ科）東洋区。lesser Jay を参照
blue jewel		*Hypochrysops delicia* Hewitson	（チョウ目、シジミチョウ科）豪州区
blue Judy		*Abisara talantus* Aurivillius	（チョウ目、シジミタテハ科）エチオピア区
blue Kaiser	ダルリサマダラジャノメ	*Penthema darlisa* Moore	（チョウ目、タテハチョウ科）東洋区
blue king crow			blue crow を見よ
blue Lasaia			blue metalmark を見よ
blue leaf beetle	ルリハムシ	*Linaeidea aenea* (Linnaeus)	（コウチュウ目、ハムシ科）日本、旧北区
blue leaf-cut weevil	ルリオトシブミ	*Euops punctatostriatus* (Motschulsky)	（コウチュウ目、オトシブミ科）日本
blue leafwing		*Anaea oenomaiis* (Boisduval)	（チョウ目、タテハチョウ科）新熱帯区
blue leafwing (1)		*Anaea pithyusa* (R. Felder)	（チョウ目、タテハチョウ科）新北区、新熱帯区
blue-legged grasshopper		*Metator pardalinus* (Saussure)	（バッタ目、バッタ科）新北区
blue-lit argent		*Argyresthia glaucinella* Zeller	（チョウ目、メムシガ科）旧北区
blue-lit pear pigmy		*Stigmella pyri* (Glitz)	（チョウ目、モグリチビガ科）旧北区
blue longicorn beetle			pear borer (1) を見よ
blue louse			sucking goat louse を見よ
blue louse			short nosed cattle louse を見よ　米国での英名
blue marsh hawk		*Orthetrum glaucum* (Brauer)	（トンボ目、トンボ科）東洋区
blue metalmark		*Lasaia sula* Staudinger	（チョウ目、シジミタテハ科）新北区
blue milkweed beetle			cobalt milkweed beetle を見よ
blue mimic		*Euriphene iris* (Aurivillius)	（チョウ目、タテハチョウ科）エチオピア区
blue minute flea beetle	イヌノフグリトビハムシ	*Longitarsus hosaticus* (Linnaeus)	（コウチュウ目、ハムシ科）日本、旧北区
blue monarch			chestnut tiger を見よ
blue monarch			blue glassy tiger (1) を見よ
blue monarch			African blue tiger を見よ
blue monkey beetle		*Scelophysa trimeni* Peringuey	（コウチュウ目、コガネムシ科）エチオピア区
blue moon			greengrocer を見よ
blue moon butterfly			common eggfly を見よ　豪州での英名

英名	和名	学名	所属、分布、ほか
blue moonbeam		*Philiris nitens* (Grose-Smith)	（チョウ目、シジミチョウ科）豪州区
blue mormon	テンジクアゲハ	*Menelaides polymnestor* (Cramer)	（チョウ目、アゲハチョウ科）東洋区
blue Morpho (1)	レテノールモルフォ	*Morpho rhetenor* Cramer	（チョウ目、タテハチョウ科）新熱帯区
blue Morpho (2)	メネラウスモルフォ	*Morpho menelaus* (Linnaeus)	（チョウ目、タテハチョウ科）新熱帯区
blue mother-of-pearl		*Protogoniomorpha temora* (Felder et Felder)	（チョウ目、タテハチョウ科）エチオピア区
Blue Mountain firefly		*Luciola lychnus* (Olliff)	（コウチュウ目、ホタル科）豪州区
Blue Mountain swallowtail			Ulysses butterfly を見よ　Blue Mountain butterfly ともいう
blue mud dauber		*Chlorion aerarium* Patton	（ハチ目、アナバチ科）新北区
blue mud dauber			blue mud wasp (1) を見よ
blue mud wasp		*Chalybion caeruleum* (Linnaeus)	（ハチ目、アナバチ科）新北区
blue mud wasp (1)		*Chalybion californicum* (Saussure)	（ハチ目、アナバチ科）新北区
blue nawab	イチモンジフタオチョウ	*Polyura schreiberi* (Godart)	（チョウ目、タテハチョウ科）東洋区。*P. s. tisamenus* (Fruhstorfer) も同英名
blue needle			dragonflies (1) を見よ
blue night butterfly		*Cepheuptychia cephus* (Fabricius)	（チョウ目、タテハチョウ科）新熱帯区
bluenose		*Acharia horrida* Dyar	（チョウ目、イラガ科）新熱帯区
blue oakleaf (1)	アイイロコノハチョウ	*Kallima philarchus* (Westwood)	（チョウ目、タテハチョウ科）東洋区
blue oakleaf (2)	アオコノハチョウ	*Kallima horsfieldi* Kollar	（チョウ目、タテハチョウ科）東洋区
blue pansy (1)	アフリカタテハモドキ	*Junonia oenone* (Linnaeus)	（チョウ目、タテハチョウ科）エチオピア区
blue pansy (2)	アオタテハモドキ	*Junonia orithya* (Linnaeus)	（チョウ目、タテハチョウ科）日本、東洋区、豪州区、エチオピア区。*J. o. wallacei* Distant も同英名
blue pansy			eyed pansy を見よ
blue patch Charaxes		*Charaxes lactetinctus* Karsch	（チョウ目、タテハチョウ科）エチオピア区
blue-patched eyed-metalmark		*Mesosemia carissima* Bates	（チョウ目、シジミタテハ科）新熱帯区
blue peacock	オオクジャクアゲハ	*Papilio arcturus* Westwood	（チョウ目、アゲハチョウ科）東洋区
blue percher			chalky percher を見よ
blue Perisama		*Perisama philinus* Doubleday	（チョウ目、タテハチョウ科）新熱帯区
blue pied pierrot		*Zintha hintza* (Trimen)	（チョウ目、シジミチョウ科）エチオピア区
blue pirate			blue dasher を見よ
blue policeman			policeman を見よ
blue polished chafer			purplish chafer を見よ
blue posy		*Biduanda melisa* (Hewitson)	（チョウ目、シジミチョウ科）東洋区

英名	和名	学名	所属、分布、ほか
blue pursuer		*Potamarcha congener* (Rambur)	（トンボ目、トンボ科）東洋区
blue quaker		*Pithecops fulgens* Doherty	（チョウ目、シジミチョウ科）東洋区
blue-rayed metalmark		*Lyropteryx apollonia* Westwood	（チョウ目、シジミタテハ科）新熱帯区
blue-rayed playboy		*Deudorix camerona* Plötz	（チョウ目、シジミチョウ科）エチオピア区
blue-ringed dancer		*Argia sedula* (Hagen)	（トンボ目、イトトンボ科）新北区
blue riverdamsel			blue sprite を見よ
blue rose leaf beetle			rose leaf beetle (1) を見よ
blue royal	ネグロフタオシジミ	*Pratapa icetoides* (Elwes)	（チョウ目、シジミチョウ科）東洋区
blue royal Assam			blue royal を見よ
blue sailer		*Pseudoneptis bugandensis* Stoneham	（チョウ目、タテハチョウ科）エチオピア区
blue sharpshooter		*Hordnia circellata* (Signoret)	（カメムシ目、オオヨコバイ科）新北区
blue shieldbug			blue bug を見よ
blue slim		*Aciagrion fragilis* (Tillyard)	（トンボ目、イトトンボ科）豪州区
blue soldier fly		*Exaireta spinigera* (Wiedemann)	（ハエ目、ミズアブ科）新北区、豪州区
blue spangled Charaxes	シロモンクロフタオチョウ	*Charaxes guderiana* (Dewitz)	（チョウ目、タテハチョウ科）エチオピア区
blue speckled butterfly			holly blue を見よ
blue spider wasp		*Psammochares luctuosus* (Cresson)	（ハチ目、Psammocharidae）新北区
blue spot pansy	ベニモンタテハモドキ	*Junonia westermanni* (Westwood)	（チョウ目、タテハチョウ科）エチオピア区
blue spotted crow	ミダムスルリマダラ	*Euploea midamus* (Linnaeus)	（チョウ目、タテハチョウ科）東洋区。*E. m. singapura* Moore も同英名。blue-spotted crow とも記す
blue sprite		*Pseudagrion microcephalum* (Rambur)	（トンボ目、イトトンボ科）東洋区
blue spruce gall aphid			Douglas fir adelges を見よ
blue stag beetle		*Platycerus caraboides* (Linnaeus)	（コウチュウ目、クワガタムシ科）旧北区
blue stemborer		*Patagoniodes farinaria* (Turner)	（チョウ目、メイガ科）豪州区
blue stem-borer		*Homoeosoma vagellum* Zeller	（チョウ目、メイガ科）豪州区
blue studded skipper		*Sostrata nordica* Evans	（チョウ目、セセリチョウ科）新北区、新熱帯区
blue swallowtail			pipevine swallowtail を見よ
blue swallowtail			giant blue swallowtail を見よ
blue tiger (1)		*Tarucus theophrastus* (Fabricius)	（チョウ目、シジミチョウ科）豪州区、エチオピア区

英名	和名	学名	所属、分布、ほか
blue tiger		*Danaus hamatus* (Macleay)	（チョウ目、タテハチョウ科）豪州区
blue tiger			blue tiger butterfly を見よ
blue tiger			dark blue tiger (1) を見よ
blue tiger			blue glassy tiger (1) を見よ
blue tiger butterfly	ミナミコモンマダラ	*Tirumala hamata orientalis* (Semper)	（チョウ目、タテハチョウ科）日本、東洋区。*T. hamata* (Macleay) dark blue tiger の英名でもある
blue tigers	コモンマダラ属	*Tirumala*	（チョウ目、タテハチョウ科）の昆虫の総称
blue tit	フタオルリシジミ	*Chliaria kina* (Hewitson)	（チョウ目、シジミチョウ科）東洋区
blue transparent		*Ithomia pellucida* Weymer	（チョウ目、タテハチョウ科）新熱帯区
blue triangle			common bluebottle を見よ
blue underwing moth			clifden nonpareil を見よ 米国での英名
blue vagrant			Cambridge vagrant を見よ
blue wanderer			dark blue tiger (1) を見よ
blue wave			whitened bluewing を見よ
blue willow beetle		*Phyllodecta vulgatissima* (Linnaeus)	（コウチュウ目、ハムシ科）旧北区
blue willow leaf beetle			blue willow beetle を見よ
blue willow weevil		*Phyllodecta tibialis* (Suffrian)	（コウチュウ目、ハムシ科）旧北区
blue wing		*Myscelia ethusa* (Doyere)	（チョウ目、タテハチョウ科）エチオピア区
blue-winged dry-wooden longicorn			tanbark beetle を見よ
blue-winged dun			sherry spinner を見よ
blue-winged duns			blue-winged olives を見よ
blue-winged Eurybia		*Eurybia lycisca* Westwood	（チョウ目、シジミタテハ科）新熱帯区
blue-winged grasshopper		*Oedipoda caerulescens* (Linnaeus)	（バッタ目、バッタ科）旧北区
blue-winged grasshopper			Sierran blue-winged grasshopper を見よ
blue-winged locust		*Sphingonotus caerulans* (Linnaeus)	（バッタ目、バッタ科）旧北区
blue-winged olives		*Ephemerella*	（カゲロウ目、マダラカゲロウ科）の昆虫の総称
blue-winged olive-sherry spinner			sherry spinner を見よ
blue-winged sheenmark			blue-winged Eurybia を見よ
blue-winged wasp		*Scolia dubia* Say	（ハチ目、ツチバチ科）新北区
blue winged olive		*Ephemerella ignita* (Poda)	（カゲロウ目、マダラカゲロウ科）旧北区
blue winged olive		*Baetis tricaudatus* Dodds	（カゲロウ目、コカゲロウ科）新北区
blue wood wasp			polished horntail を見よ
blueberry budworm moth		*Abagrotis anchocelioides* (Guenée)	（チョウ目、ヤガ科）新北区

英名	和名	学名	所属、分布、ほか
blueberry case beetle		*Neochlamisus cribripennis* (LeConte)	(コウチュウ目、ハムシ科) 新北区
blueberry dart moth		*Coenophila opacifrons* Grote	(チョウ目、ヤガ科) 新北区。blueberry dart ともいう
blueberry flagleaf webworm moth		*Dasytoma salicella* (Hübner)	(チョウ目、Chimabachidae) 新北区
blueberry flea beetle		*Altica sylvia* Malloch	(コウチュウ目、ハムシ科) 新北区
blueberry gall midge		*Dasineura oxycoccana* (Johnson)	(ハエ目、タマバエ科) 新北区
blueberry gray moth		*Glena cognataria* Hübner	(チョウ目、シャクガ科) 新北区
blueberry leafroller			blueberry flagleaf webworm moth を見よ
blueberry leafroller			Sparganothis fruitworm moth を見よ
blueberry leaftier		*Croesia curvalana* (Kearfoot)	(チョウ目、ハマキガ科) 新北区。成虫は blueberry leaftier moth
blueberry maggot		*Rhagoletis mendax* Curran	(ハエ目、ミバエ科) 新北区
blueberry maggot			apple fruit fly を見よ
blueberry sulphur		*Colias pelidne pelidne* Boisduval et LeConte	(チョウ目、シロチョウ科) 新北区
blueberry thrips		*Frankliniella vaccinii* Morgan	(アザミウマ目、アザミウマ科) 新北区
blueberry tip borer		*Hendecaneura shawiana* Kearfott	(チョウ目、ハマキガ科) 新北区。成虫は blueberry tip borer moth
blueberry tip midge		*Prodiplosis vaccinii* (Felt)	(ハエ目、タマバエ科) 新北区
blue-black syrphid fly		*Pseudodoros clavatus* (Fabricius)	(ハエ目、ハナアブ科) 新北区
bluebodied blowfly			lesser brown blowfly (1)(2) を見よ
blues (1)	ヒメシジミ亜科	Polyommatinae	(チョウ目、シジミチョウ科) の昆虫の総称
blues		Plebeiinae	(チョウ目、シジミチョウ科) の昆虫の総称
blues	ミヤマシジミ属	*Lycaeides*	(チョウ目、シジミチョウ科) の昆虫の総称
blues	ヒメシジミ属	*Plebejus*	(チョウ目、シジミチョウ科) の昆虫の総称
blues			bluets を見よ
blues			axures を見よ　英国で使用
blues			narrow-winged damselflies を見よ
blues			coppers (1) を見よ
blues and Arguses			blues (1) を見よ
blues and coppers			coppers (1) を見よ
blueshouldered cornuted shadowdamsel		*Drepanosticta subtropica* (Fraser)	(トンボ目、Platystictidae) 東洋区
blue-spot hairstreak	ヒメツマアカカラスシジミ	*Satyrium spini* (Schiffermüller)	(チョウ目、シジミチョウ科) 旧北区
blue-spotted Arab	モモイロツマアカシロチョウ	*Colotis phisadia* (Godart)	(チョウ目、シロチョウ科) 東洋区、エチオピア区
blue-spotted Charaxes	メスアカフタオチョウ	*Charaxes cithaeron* C. et R. Felder	(チョウ目、タテハチョウ科) エチオピア区
blue-spotted comet darner		*Anax concolor* Brauer	(トンボ目、ヤンマ科) 新北区、新熱帯区

英名	和名	学名	所属、分布、ほか
blue-spotted crow	ミダムスルリマダラ	*Euploea midamus* (Linnaeus)	（チョウ目、タテハチョウ科）日本、東洋区
blue-spotted emperor			blue-spotted Charaxes を見よ
blue-spotted firewing	ミツボシタテハ	*Catonephele numilia* (Cramer)	（チョウ目、タテハチョウ科）新北区、新熱帯区
blue-spotted hairstreak		*Ocaria arpoxais* (Godman et Salvin)	（チョウ目、シジミチョウ科）新熱帯区
blue-spotted hawker		*Adversaeschna brevistyla* (Rambur)	（トンボ目、ヤンマ科）豪州区、大洋区
blue-spotted pink lady			Australian painted lady を見よ
blue-spotted satyr		*Drucina championi championi* Godman et Salvin	（チョウ目、タテハチョウ科）新熱帯区
blue-spotted scrub hopper		*Aeromachus kali* (de Nicéville)	（チョウ目、セセリチョウ科）東洋区
blue-stained satyr		*Lymanopoda cinna* Godman et Salvin	（チョウ目、タテハチョウ科）新熱帯区
blue-stiched eighty-eight		*Callicore astarte casta* (Salvin)	（チョウ目、タテハチョウ科）新熱帯区。*C. a. patelina* (Hewitson) も同英名。Astarte eighty-eight を参照
blue-striped mime	ビロードマネシアゲハ	*Chilasa slateri* Hewitson	（チョウ目、アゲハチョウ科）東洋区
blue-striped nettle-grub moth		*Parasa lepida* (Cramer)	（チョウ目、イラガ科）日本、東洋区、エチオピア区。幼虫は blue-striped nettle-grub
blue-striped palmfly	オオルリモンジャノメ	*Elymnias patna* (Westwood)	（チョウ目、タテハチョウ科）東洋区
blue-striped spreadwing		*Lestes tenuatus* Rambur	（トンボ目、アオイトトンボ科）新北区、新熱帯区
blue-studded skipper		*Sostrata bifasciata* (Ménétriès)	（チョウ目、セセリチョウ科）新熱帯区
blue-tail jester		*Symbrenthia niphanda* Moore	（チョウ目、タテハチョウ科）東洋区
blue-tailed damselfly	マンシュウイトトンボ	*Ischnura elegans* (van der Linden)	（トンボ目、イトトンボ科）日本、旧北区、東洋区。common bluetip ともいう
blue-tailed forest hawk			triangle skimmer を見よ
bluetails		*Ischnura*	（トンボ目、イトトンボ科）の昆虫の総称
blue-tipped dancer		*Argia tibialis* (Rambur)	（トンボ目、イトトンボ科）新北区
blue-topped satyr		*Chloreuptychia sericeella* (Bates)	（チョウ目、タテハチョウ科）新熱帯区
blue-washed metalmark		*Napaea heteroea* (H. Bates)	（チョウ目、シジミタテハ科）新熱帯区
bluets		*Enallagma*	（トンボ目、イトトンボ科）の昆虫の総称
bluets			narrow-winged damselflies を見よ
bluff weta		*Deinacrida elegans* Gibbs	（バッタ目、Stenopelmatidae）豪州区
bluish Eurybia			Druce's underleaf を見よ
bluish green hairstreak		*Callophrys viridis* (Edwards)	（チョウ目、シジミチョウ科）新北区

英名	和名	学名	所属、分布、ほか
bluish longtail		*Urbanus viterboana* (Ehrmann)	(チョウ目、セセリチョウ科) 新熱帯区
bluish-spotted geometrid	ルリモンエダシャク	*Cleora insolita* (Butler)	(チョウ目、シャクガ科) 日本、旧北区
bluish spring moth		*Lomographa semiclarata* (Walker)	(チョウ目、シャクガ科) 新北区
bluish wanderer		*Leptophobia caesia* (Lucas)	(チョウ目、シロチョウ科) 新熱帯区
bluish zygaenid	ルリハダホソクロバ	*Rhagades pruni esmeralda* (Butler)	(チョウ目、マダラガ科) 日本、旧北区
blunt-nosed cranberry leafhopper		*Scleroracus vaccinii* (Van Duzee)	(カメムシ目、オオヨコバイ科) 新北区
blunt-tipped purplish noctuid			latin を見よ
blunt-winged knot-horn			carob moth (2) を見よ
Blunt's flat-body		*Depressaria depressana* (Fabricius)	(チョウ目、マルハキバガ科) 旧北区
blurred bent-skipper		*Ebrietas evanidus* Mabille	(チョウ目、セセリチョウ科) 新熱帯区
blurry Anastrus		*Anastrus tolimus tolimus* (Plötz)	(チョウ目、セセリチョウ科) 新熱帯区
blurry chocolate angle moth		*Macaria transitaria* Walker	(チョウ目、シャクガ科) 新北区。blurry chocolate angle ともいう
blurry forestwraith		*Platysticta maculata* Hagen	(トンボ目、Platystictidae) 東洋区
blurry-patched Nola moth		*Nola cilicoides* (Grote)	(チョウ目、コブガ科) 新北区。blurry-patched Nola ともいう
blurry skipper		*Vettius tertianus* (Herrich-Schäffer)	(チョウ目、セセリチョウ科) 新熱帯区
blush Conchylodes moth		*Conchylodes salamisalis* Druce	(チョウ目、メイガ科) 新北区
blushing phantom		*Cithaerias pireta* (Stoll)	(チョウ目、タテハチョウ科) 新熱帯区
boar louse		*Haematopinus apri* Goureau	(シラミ目、ケモノジラミ科) 旧北区
boat bug			water boatman (1) を見よ
boat bug			back swimmer を見よ
boat bugs			backswimmers を見よ
boatflies			backswimmers を見よ
boatfly			water boatman (1) を見よ
boatfly			back swimmer を見よ
boatman flies			broadmouthed flies を見よ　豪州での英名
boatman fly		*Pogonortalis doclea* (Walker)	(ハエ目、ヒロクチバエ科) 豪州区
boatman fly			island fruit fly (1) を見よ
bobs		*Arnetta*	(チョウ目、セセリチョウ科) の昆虫の総称
Bob's hairstreak		*Aubergina* sp.	(チョウ目、シジミチョウ科) 新熱帯区
bock beetles			long-horned beetles を見よ

英名	和名	学名	所属、分布、ほか
Bocquet's demon Charaxes		*Charaxes bocqueti* Minig	（チョウ目、タテハチョウ科）エチオピア区
body and head louse			body louse を見よ 、
body lice			human lice を見よ
body louse	ヒトジラミ	*Pediculus humanus* Linnaeus	（シラミ目、ヒトジラミ科）日本、汎世界。ヒトに寄生
body louse			crab louse を見よ
boehmeria noctuid	クロキシタアツバ	*Hypena amica* (Butler)	（チョウ目、ヤガ科）日本、旧北区
Boeotia sister		*Adelpha boeotia* (Felder et Felder)	（チョウ目、タテハチョウ科）新熱帯区
bog Bibarrambla moth		*Bibarrambla allenella* (Walsingham)	（チョウ目、マルハキバガ科）新北区
bog bush-cricket		*Metrioptera brachyptera* (Linnaeus)	（バッタ目、キリギリス科）旧北区
bog copper	ヒメベニシジミ	*Lycaena epixanthe* (Boisduval et LeConte)	（チョウ目、シジミチョウ科）新北区
bog Deltote			bog Lithacodia moth を見よ
bog elfin	メインコツバメ	*Callophrys lanoraieensis* (Sheppart)	（チョウ目、シジミチョウ科）新北区
bog fritillary		*Proclossiana eunomia* (Esper)	（チョウ目、タテハチョウ科）全北区
bog ground beetles		Elaphrini	（コウチュウ目、オサムシ科）の昆虫の総称
bog hawker		*Aeschna subarctica* (Walker)	（トンボ目、ヤンマ科）旧北区
bog Holomelina		*Holomelina lamae* (Freeman)	（チョウ目、ヒトリガ科）新北区。bog Holomelina moth ともいう
bog katydid		*Metrioptera sphagnorum* (Walker)	（バッタ目、キリギリス科）新北区
bog Lithacodia moth		*Deltote bellicula* (Hübner)	（チョウ目、ヤガ科）新北区
bog Lygropia moth		*Lygropia rivulalis* Hampson	（チョウ目、メイガ科）新北区
bog-myrtle twist			bilberry tortrix を見よ
Bogong moth		*Agrotis infusa* (Boisduval)	（チョウ目、ヤガ科）豪州区。冬は高地に移動し越冬する。Bogon moth ともいう
bogus yucca moth		*Prodoxus decipiens* Riley	（チョウ目、マガリガ科）新北区
bogus yucca moths			yucca moths (2) を見よ
Boisduval scale			orchid scale を見よ　Boisduval's scale とも記す
Boisduval silkmoth			Boisduval silkworm を見よ
Boisduval silkworm		*Anaphe panda* Boisduval	（チョウ目、シャチホコガ科）エチオピア区
Boisduval's autumnal moth		*Oenosandra boisduvali* Newman	（チョウ目、シャチホコガ科）豪州区
Boisduval's blue			common blue (1) を見よ
Boisduval's brownie			Boisduval's Miletus を見よ
Boisduval's carpenter moth			witjuti grub を見よ
Boisduval's cracker			pale cracker を見よ

英名	和名	学名	所属、分布、ほか
Boisduval's evening brown	オビコノマチョウ	*Melanitis boisduvalia* (C. et R. Felder)	（チョウ目、タテハチョウ科）東洋区
Boisduval's false Acraea	オオベニイロホソチョウモドキ	*Pseudacraea boisduvali* (Doubleday)	（チョウ目、タテハチョウ科）エチオピア区
Boisduval's leafwing		*Memphis moruus boisduvali* (Comstock)	（チョウ目、タテハチョウ科）新熱帯区。Moruus leafwing を参照
Boisduval's Miletus	ヒメウラシモフリシジミ	*Miletus boisduvali* Moore	（チョウ目、シジミチョウ科）東洋区
Boisduval's Raven			darkened white を見よ
Boisduval's sulphur		*Aphrissa boisduvalii* (C. et R. Felder)	（チョウ目、シロチョウ科）新熱帯区
Boisduval's tree nymph		*Sevenia boisduvali* (Wallengren)	（チョウ目、タテハチョウ科）エチオピア区
Boisduval's white-lady	アダマストールタイマイ	*Graphium adamastor* (Boisduval)	（チョウ目、アゲハチョウ科）エチオピア区
Boisduval's yellow		*Eurema arbela boisduvaliana* (C. et R. Felder)	（チョウ目、シロチョウ科）新北区、新熱帯区
Boland brown		*Melampias huebneri* (van Son)	（チョウ目、タテハチョウ科）エチオピア区
Boland rocksitter		*Durbaniopsis saga* (Trimen)	（チョウ目、シジミチョウ科）エチオピア区
Boland sandman		*Spialia sataspes* (Trimen)	（チョウ目、セセリチョウ科）エチオピア区
Boland Thestor		*Thestor protumnus* (Linnaeus)	（チョウ目、シジミチョウ科）エチオピア区
bold-based Zale moth		*Zale lunifera* (Hübner)	（チョウ目、ヤガ科）新北区。bold-based Zale ともいう
bold faceted-skipper			faceted skipper を見よ
bold-feathered grass moth		*Herpetogramma pertextalis* (Lederer)	（チョウ目、メイガ科）新北区、エチオピア区
bold leaftail		*Phyllogomphus brunneus* Pinhey	（トンボ目、サナエトンボ科）エチオピア区
bold line-blue			large 4-line blue を見よ
bold medicine moth		*Chrysendeton medicinalis* (Grote)	（チョウ目、メイガ科）新北区
bold Mylon		*Mylon lassia* (Hewitson)	（チョウ目、セセリチョウ科）新熱帯区
bold pug		*Eupithecia rotundopuncta* Packard	（チョウ目、シャクガ科）新北区
bold-spotted hairstreak		*Siderus philinna* (Hewitson)	（チョウ目、シジミチョウ科）新熱帯区
bold-veined skipperling		*Oarisma era* Dyar	（チョウ目、セセリチョウ科）新熱帯区
boldly-marked Archips moth		*Archips dissitana* (Grote)	（チョウ目、ハマキガ科）新北区
Bolivian grass mantis		*Thesprotia macilenta* Saussure et Zehntner	（カマキリ目、カマキリ科）新熱帯区
Bolivian Perisama		*Perisama canoma* Druce	（チョウ目、タテハチョウ科）新熱帯区
Bolivian sanddragon		*Progomphus bolivensis* Belle	（トンボ目、サナエトンボ科）新熱帯区

英名	和名	学名	所属、分布、ほか
boll weevil	ワタミハナゾウムシ	*Anthonomus grandis* Boheman	（コウチュウ目、ゾウムシ科）新北区。ワタ害虫
Boll's sand cockroach		*Arenivaga bolliana* (Saussure)	（ゴキブリ目、Polyphagidae）新北区。ネズミの巣に生息
bollworm			corn earworm（1）（2）を見よ　豪州での英名
bolting-cloth beetle			cadelle を見よ
Bolvian sapphire		*Lasaia moeros* Staudinger	（チョウ目、シジミタテハ科）新熱帯区
bomb flies			warble flies（1）（3）を見よ
bomb fly			cattle warble fly (1) を見よ
bombardier beetle		*Pheropsophus versicalis* (Dejean)	（コウチュウ目、ホソクビゴミムシ科）豪州区
bombardier beetle		*Brachinus tschernikhi* (Mannerheim)	（コウチュウ目、ホソクビゴミムシ科）新北区
bombardier beetle (1)		*Brachinus crepitans* (Linnaeus)	（コウチュウ目、ホソクビゴミムシ科）旧北区
bombardier beetle			lesser bombardier beetle を見よ
bombardier beetles		*Brachinus*	（コウチュウ目、ホソクビゴミムシ科）の昆虫の総称　へっぴり虫の俗称あり
bombardier beetles (1)	ホソクビゴミムシ科	Brachinidae	（コウチュウ目）の昆虫の総称
Bombay canary			American cockroach を見よ
Bombay fly			cattle warble fly (1) を見よ
Bombay locust	タイワンツチイナゴ	*Patanga succincta* (Linnaeus)	（バッタ目、バッタ科）日本、東洋区。イネ、トウモロコシなどの害虫
bombycid			カイコガ科の蛾。silkworm moths を見よ
bombylids			bee flies を見よ
Bonasia Acraea	カザリヒメホソチョウ	*Acraea bonasia* Fabricius	（チョウ目、タテハチョウ科）エチオピア区
bonavist plume moth	フジマメトリバ	*Sphenarches anisodactylus* (Walker)	（チョウ目、トリバガ科）日本、東洋区、豪州区、エチオピア区、新熱帯区。フジマメ、インゲンマメなどの害虫
Bond's shade		*Cnephasia communana* Herrich-Schäffer	（チョウ目、ハマキガ科）旧北区
Bond's sweep		*Masonia crassiorella* Bruand	（チョウ目、ミノガ科）旧北区
Bond's wainscot			Morris's wainscot を見よ
bone beetles			checkered beetles を見よ
bone flies			skipper flies を見よ　豪州での英名
bone-house earwig		*Marava arachidis* (Versin)	（ハサミムシ目、チビハサミムシ科）汎世界
boneset borer moth		*Carmenta pyralidiformis* (Walker)	（チョウ目、スカシバガ科）新北区。幼虫は boneset borer
Bonthain tiger	スレワッタンアサギマダラ	*Parantica sulewattan* (Fruhstorfer)	（チョウ目、タテハチョウ科）東洋区
booklice (1)	コナチャタテ科	Liposcelidae	（チャタテムシ目）の昆虫の総称　コナムシ、コナチャタテムシといわれる
booklice (2)	チャタテムシ目	Psocoptera	の昆虫の総称　世界に約3,000種

英名	和名	学名	所属、分布、ほか
booklice (3)		*Liposcelis*	(チャタテムシ目、コナチャタテ科) の昆虫の総称
booklouse	ヒラタチャタテ	*Liposcelis bostrychophilus* Badonell	(チャタテムシ目、コナチャタテ科) 日本、汎世界。無翅で貯蔵食品や本の害虫
booklouse	ウスグロチャタテ	*Liposcelis subfuscus* Broadhead	(チャタテムシ目、コナチャタテ科) 日本、旧北区、新熱帯区。貯蔵食品や本の害虫
booklouse		*Liposcelis corrodens* Heymons	(チャタテムシ目、コナチャタテ科) 新北区
booklouse (1)		*Liposcelis terricolis* (Badonnel)	(チャタテムシ目、コナチャタテ科) 旧北区
booklouse (2)	コチャタテ	*Trogium pulsatorium* (Linnaeus)	(チャタテムシ目、コチャタテ科) 日本、汎世界。貯蔵食品や標本類につく
booklouse (3)	コナチャタテ	*Liposcelis divinatorius* (Müller)	(チャタテムシ目、コナチャタテ科) 日本、汎世界
booklouse			granary booklouse を見よ
bookworm			biscuit beetle を見よ
bookworms			silverfish (4) を見よ
boomerang owl			Oedipus owl butterfly を見よ
Booth's sulphur		*Colias tyche boothii* J. Curtis	(チョウ目、シロチョウ科) 新北区。pale arctic clouded yellow を参照
Borbo skipper		*Borbo borbonica* (Boisduval)	(チョウ目、セセリチョウ科) エチオピア区
Bordeaux cricket		*Tartarogryllus bordigalensis* (Latreille)	(バッタ目、コオロギ科) 旧北区
border copper		*Aloeides caffrariae* Henning	(チョウ目、シジミチョウ科) エチオピア区
bordered Apamea moth			wheat earworm を見よ
bordered beauty		*Epione repandaria* (Hufnagel)	(チョウ目、シャクガ科) 旧北区
bordered buff		*Ptelina carnuta* (Hewitson)	(チョウ目、シジミチョウ科) エチオピア区
bordered chequer		*Nematocampa limbata* (Haworth)	(チョウ目、シャクガ科) 旧北区
bordered conch		*Tischeria marginea* (Haworth)	(チョウ目、ムモンハモグリガ科) 旧北区
bordered echium ermel		*Ethmia bipunctella* (Fabricius)	(チョウ目、スエヒロキバガ科) 旧北区
bordered fawn moth		*Sericosema juturnaria* (Guenée)	(チョウ目、シャクガ科) 新北区
bordered gothic		*Heliophobus reticulata* (Goeze)	(チョウ目、ヤガ科) 旧北区
bordered grey		*Selidosema brunnearia* (Villers)	(チョウ目、シャクガ科) 旧北区
bordered grey beauty			bordered grey を見よ
bordered knob-tiped shadowdamsel		*Drepanosticta submontana* (Fraser)	(トンボ目、Platystictidae) 東洋区
bordered orange			bordered sallow を見よ
bordered patch		*Chlosyne lacinia* (Geyer)	(チョウ目、タテハチョウ科) 新北区、新熱帯区

英名	和名	学名	所属、分布、ほか
bordered pearl		*Paratalanta nubialis* (Hübner)	（チョウ目、メイガ科）旧北区
bordered pearl			bamboo leaf roller を見よ
bordered plant bug		*Euryophthalmus cinctus* (Herrich-Schäffer)	（カメムシ目、ホシカメムシ科）新北区
bordered plant bugs			red bugs (2) を見よ
bordered pug		*Eupithecia succenturiata* (Linnaeus)	（チョウ目、シャクガ科）日本、旧北区
bordered rustic			Australian rustic を見よ
bordered sallow	キタバコガ	*Pyrrhia umbra* (Hufnagel)	（チョウ目、ヤガ科）日本、旧北区
bordered sallow moth		*Pyrrhia cilisca* Guenée	（チョウ目、ヤガ科）新北区。bordered sallow ともいう
bordered straw		*Heliothis peltigera* (Denis et Schiffermüller)	（チョウ目、ヤガ科）旧北区
bordered thorn			filament borer を見よ
bordered white			bordered white beauty を見よ
bordered white beauty		*Bupalus piniarius* (Linnaeus)	（チョウ目、シャクガ科）旧北区
bordered white moth			bordered white beauty を見よ
bordered white pine geometer			bordered white beauty を見よ
boreal bluet		*Enallagma boreale* Selys	（トンボ目、イトトンボ科）新北区
boreal shieldback		*Steiroxys borealis* Scudder	（バッタ目、キリギリス科）新北区
boreal snaketail		*Ophiogomphus colubrinus* Selys	（トンボ目、サナエトンボ科）新北区
borer			植物に穿孔する昆虫を指す
boring cutworm			common cutworm を見よ　南アフリカでの英名
boronia psyllid		*Ctenarytaina thysanura* Ferris et Klyver	（カメムシ目、キジラミ科）豪州区
Bosnian blue		*Plebejus dardanus* (Freyer)	（チョウ目、シジミチョウ科）旧北区
Boston ivy tiger moth	トビイロトラガ	*Sarbanissa subflava* (Moore)	（チョウ目、トラガ科）日本、旧北区
bostrichid powder-post beetles			false powderpost beetles を見よ
bostrichids			false powderpost beetles を見よ
bostrychid			lesser grain borer を見よ
bot			horse bot fly を見よ　幼虫の英名
bot			cattle warble fly (1) を見よ　幼虫の英名
botflies (1)		Oestrinae	（ハエ目、ヒツジバエ科）の昆虫の総称　bot flies とも記す
bot flies (2)	ヒツジバエ科	Oestridae	（ハエ目）の昆虫の総称
botflies			blow flies を見よ
botflies			horse flies (1) を見よ
bot-flies			horse bot flies を見よ　英国での英名
bot-flies			warble flies (3) を見よ

英名	和名	学名	所属、分布、ほか
bot-fly			horse bot fly を見よ
bot grub			cattle warble fly (1) を見よ　幼虫の英名
bots			blow flies を見よ
bots			horse bot flies を見よ　幼虫の英名
bots			warble flies (3) を見よ　幼虫の英名
Botany Bay weevil			diamond beetle を見よ
Bottimer's Mompha moth		*Mompha bottimeri* Busck	(チョウ目、カザリバガ科) 新北区
bottle brush weevil			bearded weevil を見よ
bottlebush weevil			bearded weevil を見よ
bottle cicada		*Glaucopsaltria viridis* Goding et Froggatt	(カメムシ目、セミ科) 豪州区
bottle flies			blow flies を見よ
bottom-spotted skipper		*Papias dictys* Godman	(チョウ目、セセリチョウ科) 新熱帯区
bougainvillea caterpillar moth		*Asciodes gordialis* Guenée	(チョウ目、メイガ科) 新北区
bougainvillea looper			somber carpet moth を見よ　somber carpet ともいう
boulder copper		*Lycaena boldenarum* White	(チョウ目、シジミチョウ科) 東洋区、豪州区
boulder jewel			Fitzsimon's jewel を見よ
bourbon scale			coconut scale (1) を見よ　米国での英名
bow-winged Actinote		*Actinote melampeplos melampeplos* Godman et Salvin	(チョウ目、タテハチョウ科) 新熱帯区
bow-winged grasshopper			field grasshopper を見よ
Bowker's blue		*Tarucus bowkeri* (Trimen)	(チョウ目、シジミチョウ科) エチオピア区
Bowker's hairstreak		*Stugeta bowkeri* (Trimen)	(チョウ目、シジミチョウ科) エチオピア区
Bowker's marbled sapphire			Bowker's hairstreak を見よ
Bowker's sapphire			Bowker's hairstreak を見よ
Bowker's tailed blue			Bowker's hairstreak を見よ
Bowker's widow		*Dingana bowkeri* (Trimen)	(チョウ目、タテハチョウ科) エチオピア区
boxelder aphid		*Periphyllus negundinis* (Thomas)	(カメムシ目、アブラムシ科) 新北区
boxelder bug		*Leptocoris trivittatus* (Say)	(カメムシ目、ヒメヘリカメムシ科) 新北区。boxelder ツルアリドオシ の種子を食べる
boxelder gall midge		*Contarinia negundinis* (Gillette)	(ハエ目、タマバエ科) 新北区
box elder leaf roller		*Caloptilia negundella* (Chambers)	(チョウ目、ホソガ科) 新北区。boxelder leafroller moth ともいう
boxelder leafworm moth		*Chionodes obscurusella* (Chambers)	(チョウ目、キバガ科) 新北区
boxelder psyllid		*Psylla negundinis* Mally	(カメムシ目、キジラミ科) 新北区

英名	和名	学名	所属、分布、ほか
box-elder tussock moth		*Orgyia leuschneri* Riotte	（チョウ目、ドクガ科）新北区
boxelder twig borer		*Proteoteras willingana* (Kearfott)	（チョウ目、ハマキガ科）新北区
box leaf midge			box leaf-mining midge を見よ
box leaf miner			box leaf-mining midge を見よ
box leaf-mining midge		*Monarthropalpus buxi* (Labourlbene)	（ハエ目、タマバエ科）旧北区
box midge			box leaf-mining midge を見よ
box scale		*Eriococcus buxi* (Fonscolombe)	（カメムシ目、フクロカイガラムシ科）旧北区
box sucker			boxwood psyllid を見よ
box-thorn aphid		*Myzus persicae dyslycialis* Müller	（カメムシ目、アブラムシ科）旧北区
boxtree midge			box leaf-mining midge を見よ
boxtree moth		*Entometa guttularis* (Walker)	（チョウ目、カレハガ科）豪州区
box-tree pyralid	ツゲノメイガ	*Glyphodes perspectalis* (Walker)	（チョウ目、メイガ科）日本、旧北区、東洋区。ツゲ害虫
boxweed psyllid			boxwood psyllid を見よ
boxwood leafminer		*Monarthropalpus flavus* (Schrank)	（ハエ目、タマバエ科）新北区
boxwood leaf miner			box leaf-mining midge を見よ　米国での英名
boxwood leaftier moth		*Galasa nigrinodis* (Zeller)	（チョウ目、メイガ科）新北区
boxwood psylla			boxwood psyllid を見よ　米国での英名
boxwood psyllid		*Cacopsylla buxi* (Linnaeus)	（カメムシ目、キジラミ科）全北区
boxwood webworm			boxwood leaftier moth を見よ
boxer mantis			South American dead leaf mantis (1) を見よ
brachycentrids			humpless casemake caddisflies を見よ
brachymerous beetles		*Brachymera*	（コウチュウ目）の昆虫の総称
bracken aphid	ワラビツメナシアブラムシ	*Shinjia orientalis* (Mordvilko)	（カメムシ目、アブラムシ科）日本、旧北区、豪州区
bracken borer moth		*Papaipema pterisii* Bird	（チョウ目、ヤガ科）新北区
bracken bug	ズアカシダカスミカメムシ	*Monalocoris filicis* (Linnaeus)	（カメムシ目、カスミカメムシ科）日本、旧北区
bracken clock			garden chafer を見よ
bracken neb		*Paltodora cytisella* (Curtis)	（チョウ目、キバガ科）旧北区
bracon flies			braconid wasps を見よ
braconid flies			braconid wasps を見よ
braconid wasps	コマユバチ科	Braconidae	（ハチ目）の昆虫の総称
braconids			braconid wasps を見よ
Bradley's meadow katydid		*Orchelimum bradleyi* Rehn et Hebard	（バッタ目、キリギリス科）新北区

英名	和名	学名	所属、分布、ほか
brahmaeids			Brahmin moths を見よ
Brahmin moths	イボタガ科	Brahmaeidae	(チョウ目)の昆虫の総称
bramble fly			green hairstreak を見よ
bramble green hairstreak	シロモンウラアオシジミ	*Callophrys dumetorum* (Boisduval)	(チョウ目、シジミチョウ科)新北区
bramble leaf midge			blackberry leaf midge を見よ
bramble leafhopper		*Ribautiana tenerrima* (Herrich-Schäffer)	(カメムシ目、オオヨコバイ科)全北区
bramble leaf-miner			golden pigmy を見よ
bramble sawfly		*Philomastix macleaii* (Westwood)	(ハチ目、Pergidae)豪州区
bramble shoot moth		*Epiblema udmanniana* (Linnaeus)	(チョウ目、ハマキガ科)旧北区。幼虫は bramble shoot borer
bramble shoot webber			bramble shoot moth を見よ　幼虫の英名
bran bug			confused flour beetle を見よ　北米での英名
branch-and-leaf buprestids	ナガタマムシ亜科	Agrilinae	(コウチュウ目、タマムシ科)の昆虫の総称
branch and limb beetles			false powderpost beetles を見よ
branch and limb borers			false powderpost beetles を見よ
branch and trunk borer		*Citriphaga mixta* Lea	(コウチュウ目、カミキリムシ科)豪州区。カンキツ害虫
branch and trunk borer			citrus branch borer を見よ
branch and twig borer			California branch borer を見よ
branch and twig borers			false powderpost beetles を見よ
branch-barred groundling		*Calyocolum vicinella* (Douglas)	(チョウ目、キバガ科)旧北区
branded awlking		*Choaspes plateni* (Staudinger)	(チョウ目、セセリチョウ科)東洋区
branded dart		*Potanthus rectifasiata* (Elwes et Edwards)	(チョウ目、セセリチョウ科)東洋区
branded evening brown		*Cyllogenes suradeva* (Moore)	(チョウ目、タテハチョウ科)東洋区
branded imperial		*Eooxylides tharis distanti* Riley	(チョウ目、シジミチョウ科)東洋区
branded large fox		*Rhabdomantis galatia* (Hewitson)	(チョウ目、セセリチョウ科)エチオピア区
branded meadowbrown			Oriental meadow brown を見よ
branded orange awlet		*Burana oedipodea* (Swainson)	(チョウ目、セセリチョウ科)東洋区
branded royal		*Tajuria melastigma* de Nicéville	(チョウ目、シジミチョウ科)東洋区

英名	和名	学名	所属、分布、ほか
branded yamfly		*Yasoda tripunctata* (Hewitson)	（チョウ目、シジミチョウ科）東洋区
branded Yeomen		*Algia fasciata* (C. et R. Felder)	（チョウ目、タテハチョウ科）東洋区
Brander's great marble	ドロヒメハマキ	*Pseudosciaphila branderiana* (Linnaeus)	（チョウ目、ハマキガ科）日本、旧北区
brass beetle			rose chafer (1) を見よ
brass beetle			golden spider beetle を見よ
brassica bug			cabbage bug (2) を見よ
brassica flea beetle			striped flea beetle (2) を見よ
brassica flower midge		*Gephyraulus raphanistri* Kieffer	（ハエ目、タマバエ科）旧北区
brassica leaf beetle	ダイコンハムシ	*Phaedon brassicae* Baly	（コウチュウ目、ハムシ科）日本、東洋区。ダイコン害虫
brassica pod midge			cabbage gall midge を見よ
brassy			brassy flea beetle (1) を見よ
brassy cutworm		*Eriopyga rufula* (Grote)	（チョウ目、ヤガ科）新北区。ブドウ害虫
brassy flea beetle (1)	テンサイトビハムシ	*Chaetocnema concinna* (Marshall)	（コウチュウ目、ハムシ科）日本、旧北区。テンサイ害虫
brassy flea beetle (2)		*Chaetocnema pulicaria* Melsheimer	（コウチュウ目、ハムシ科）新北区。トウモロコシの病害を伝搬
brassy roadside-skipper		*Amblyscirtes fluonia* Godman	（チョウ目、セセリチョウ科）新熱帯区
brassy willow beetle			brassy willow leaf beetle を見よ
brassy willow leaf beetle		*Phyllodecta vitellinae* (Linnaeus)	（コウチュウ目、ハムシ科）旧北区
brassy-toothed flea beetle			brassy flea beetle (1) を見よ
Brasus skipper		*Lindra brasus* (Mielke)	（チョウ目、セセリチョウ科）新熱帯区
Brauer's arrowhead			Brauer's blue を見よ
Brauer's blue		*Lepidochrysops braueri* Dickson	（チョウ目、シジミチョウ科）エチオピア区
Brauer's copper		*Aloeides braueri* Tite et Dickson	（チョウ目、シジミチョウ科）エチオピア区
Brauer's opal		*Chrysoritis braueri* (Pennington)	（チョウ目、シジミチョウ科）エチオピア区
Braun's skolly		*Thestor braunsi* van Son	（チョウ目、シジミチョウ科）エチオピア区
Brazil bean weevil	ブラジルマメゾウムシ	*Zabrotes subfasciatus* (Boheman)	（コウチュウ目、ハムシ科）新北区、新熱帯区。マメ類害虫
Brazilian Dynastor	ナポレオンフクロウチョウ	*Dynastor napoleon* Doubleday	（チョウ目、タテハチョウ科）新熱帯区
Brazilian grass mantis		*Thesprotia brasiliensis* Scudder	（カマキリ目、カマキリ科）新熱帯区
Brazilian lacewing		*Plesiochrysa brasiliensis* (Schneider)	（アミメカゲロウ目、クサカゲロウ科）新熱帯区
Brazilian leafhopper		*Protalebrella brasiliensis* (Baker)	（カメムシ目、オオヨコバイ科）新北区、新熱帯区
Brazilian Morpho	エガモルフォ	*Morpho aega* (Hübner)	（チョウ目、タテハチョウ科）新熱帯区

英名	和名	学名	所属、分布、ほか
Brazilian orchid thrips		*Anaphothrips orchidaceus* Bognall	（アザミウマ目、アザミウマ科）新熱帯区
Brazilian owl		*Caligo brasiliensis* (C. Felder)	（チョウ目、タテハチョウ科）新熱帯区
Brazilian painted lady	ナンベイヒメアカタテハ	*Vanessa braziliensis* (Moore)	（チョウ目、タテハチョウ科）新熱帯区。Brazilian lady ともいう
Brazilian poplar moth			Alamo moth を見よ
Brazilian skipper		*Calpodes ethlius* (Stoll)	（チョウ目、セセリチョウ科）新北区、新熱帯区
bread beetle			cadelle を見よ
bread beetle			biscuit beetle を見よ
bread eater			common cockroach を見よ
breadfruit mealybug			Egyptian cottony-cussion scale を見よ
Breck robberfly		*Machimus arthriticus* (Zeller)	（ハエ目、ムシヒキアブ科）旧北区。英国の地名に由来
Bredow's sister			California sister (1) を見よ
breeze			cattle warble fly (1)(2) を見よ
breeze flies			horse bot flies を見よ
breeze flies			clegs を見よ
breeze-flies			horse flies (1) を見よ
breezes			horse flies (1) を見よ
Brenda's Hypagyrtis moth		*Hypagyrtis brendae* Heitzman	（チョウ目、シャクガ科）新北区
brenthid beetle		*Brenthus anchorago* Linnaeus	（コウチュウ目、ゾウムシ科）新北区
brentid beetles			straight snouted weevils を見よ
Brenton blue butterfly		*Orachrysops niobe* (Trimen)	（チョウ目、シジミチョウ科）エチオピア区
brick		*Agrochola circellaris* (Hufnagel)	（チョウ目、ヤガ科）旧北区。成虫は brick moth
brick-red borer moth		*Papaipema marginidens* (Guenée)	（チョウ目、ヤガ科）新北区
brick silver spot		*Aphnaeus asterius* Plötz	（チョウ目、シジミチョウ科）エチオピア区
bride underwing moth		*Catocala neogama* (Smith)	（チョウ目、ヤガ科）新北区。bride ともいう
bridge roller	ウスベニカギバヒメハマキ	*Ancylis uncella* (Denis et Schiffermüller)	（チョウ目、ハマキガ科）日本、全北区
Bridgham's brocade moth		*Oligia bridghamii* (Grote et Robinson)	（チョウ目、ヤガ科）新北区
brigadier		*Agathymus evansi* (Freeman)	（チョウ目、セセリチョウ科）新北区、新熱帯区。種名を奉献された W. H. Evans は陸軍准将 (brigadier) であったことに由来
brigadier moth			brigadier を見よ
Briggs' outer barklouse		*Ectopsocus briggsi* (McLachlan)	（チャタテムシ目、マドチャタテ科）旧北区
bright Babul blue			desert Babul blue を見よ

英名	和名	学名	所属、分布、ほか
bright blue hister beetle	ルリエンマムシ	*Saprinus speciosus* Erichson	(コウチュウ目、エンマムシ科) 日本、東洋区、豪州区、エチオピア区
bright Brangas		*Brangas getus* (Fabricius)	(チョウ目、シジミチョウ科) 新熱帯区
bright Cerulean		*Jamides aleuas* (C. et R. Felder)	(チョウ目、シジミチョウ科) 豪州区
bright chalk blue		*Thermoniphas togara* (Plötz)	(チョウ目、シジミチョウ科) エチオピア区
bright copper		*Paralucia aurifera* (Blanchard)	(チョウ目、シジミチョウ科) 豪州区
bright Cornelian		*Deudorix diovis* Hewitson	(チョウ目、シジミチョウ科) 豪州区
bright-eye bushbrown		*Mycalesis nicotia* Westwood	(チョウ目、タテハチョウ科) 東洋区
bright-eyed brown		*Heteronympha cordace* (Geyer)	(チョウ目、タテハチョウ科) 豪州区
bright-eyed ringlet	モリシロジャノメ	*Melanargia lugens epimede* (Staudinger)	(チョウ目、タテハメチョウ科) 日本、旧北区
bright-eyed ringlet	チロルベニヒカゲ	*Erebia oeme* (Hübner)	(チョウ目、タテハチョウ科) 旧北区
bright forest-blue			Cephenes blue を見よ
bright four-spined legionnaire		*Chorisops nagatomii* Rozkosny	(ハエ目、ミズアブ科) 旧北区
bright golden-rod plume		*Leioptilus osteodactylus* (Zeller)	(チョウ目、トリバガ科) 日本、旧北区
bright green forest moth		*Tatosoma tipulata* Walker	(チョウ目、シャクガ科) 豪州区
bright horsefly		*Hybomitra distinguenda* (Verrall)	(ハエ目、アブ科) 日本、旧北区
bright-line brown-eye	シロスジヨトウ	*Lacanobia oleracea* (Linnaeus)	(チョウ目、ヤガ科) 日本、全北区、エチオピア区。成虫は bright-line brown-eye moth
bright marble tortrix			splended piercer を見よ
bright Mellana		*Quasimellana aurora* (Bell)	(チョウ目、セセリチョウ科) 新熱帯区
bright metallic tiger beetle		*Tetracha fulgida* (Klug)	(コウチュウ目、オサムシ科) 新熱帯区
bright moths			yucca moths (1) を見よ
bright oak blue		*Arhopala madytus* Fruhstorfer	(チョウ目、シジミチョウ科) 豪州区
bright purple azure			Barnard's azure を見よ
bright red velvet bob		*Koruthaialos sindu* (C. et R. Felder)	(チョウ目、セセリチョウ科) 東洋区
bright shield skipper		*Signeta flammeata* (Butler)	(チョウ目、セセリチョウ科) 豪州区
bright silver-barred bell		*Epinotia nemorivaga* (Tengstrom)	(チョウ目、ハマキガ科) 旧北区
bright sunbeam		*Curetis bulis* (Westwood)	(チョウ目、シジミチョウ科) 東洋区。*C. b. stigmata* (Moore) も同英名
bright wave			pale ochraceous wave を見よ
Brighton sober		*Syncopacma vinella* (Bankes)	(チョウ目、キバガ科) 旧北区
Brighton wainscot		*Oria musculosa* Hübner	(チョウ目、ヤガ科) 旧北区。成虫は Brighton wainscot moth

英名	和名	学名	所属、分布、ほか
Brighton's Epiblema moth		*Epiblema brightonana* (Kearfott)	(チョウ目、ハマキガ科) 新北区
brights			yucca moths (1) を見よ
Brigid's elfin		*Sarangesa brigida* (Plötz)	(チョウ目、セセリチョウ科) エチオピア区
brilliant Anastrus			blue duskywing を見よ
brilliant blue		*Junonia rhadama* (Boisduval)	(チョウ目、タテハチョウ科) エチオピア区
brilliant blue			star blue を見よ
brilliant emerald		*Somatochlora metallica* (van der Linden)	(トンボ目、エゾトンボ科) 旧北区
brilliant flash		*Rapala sphinx* Fabricius	(チョウ目、シジミチョウ科) 東洋区
brilliant gem			gem (2) を見よ
brilliant greenmark			green mantle を見よ
brilliant groundstreak		*Calycopis xeneta* (Hewitson)	(チョウ目、シジミチョウ科) 新熱帯区
brilliant meadow blue		*Polyommatus saria* (Alpheraky)	(チョウ目、シジミチョウ科) 東洋区
brilliant nymph			shining blue nymph を見よ
brilliant silverpatch		*Aides brilla* Freeman	(チョウ目、セセリチョウ科) 新熱帯区
brimps			horse flies (1) を見よ
brimstone butterflies			cabbage whites を見よ
brimstone clubtail		*Stylurus intricatus* (Selys)	(トンボ目、サナエトンボ科) 新北区
brimstone moth		*Opisthograptis luteolata* (Linnaeus)	(チョウ目、シャクガ科) 全北区。brimstone ともいう
brimstone	ヤマキチョウ	*Gonepteryx rhamni* Latreille	(チョウ目、シロチョウ科) 東洋区。日本亜種 *G. rhamni maxima* Butler も英名。brimstone butterfly ともいう
brimstones			old world brimstones を見よ
Brinck's shadowdamsel		*Drepanosticta brincki* Lieftinck	(トンボ目、Platystictidae) 東洋区
brindesmaid		*Catocala paranympha* (Linnaeus)	(チョウ目、ヤガ科) 旧北区
brindle-barred yellow		*Trichopteryx viretata* (Hübner)	(チョウ目、シャクガ科) 旧北区
brindled argent		*Argyresthia cornella* Fabricius	(チョウ目、メムシガ科) 旧北区
brindled-beauties		Bistoninae	(チョウ目、シャクガ科) の昆虫の総称
brindled beauty		*Lycia hirtaria* (Clerck)	(チョウ目、シャクガ科) 旧北区。成虫は brindled beauty moth。日本亜種は *L. h. parallelaria* Inoue ムクゲエダシャク
brindled bell moth		*Epalxiphora axenana* Meyrick	(チョウ目、ハマキガ科) 豪州区
brindled green		*Dryobotodes protea* Denis et Schiffermüller	(チョウ目、ヤガ科) 旧北区
brindled green (1)		*Dryobotodes eremita* (Fabricius)	(チョウ目、ヤガ科) 旧北区
brindled green mottle			brindled green (1) を見よ

英名	和名	学名	所属、分布、ほか
brindled marbled bell	ポプラヒメハマキ	*Gypsonoma minutana* (Hübner)	(チョウ目、ハマキガ科) 日本、旧北区
brindled marbled conch		*Phtheochroa sodaliana* (Haworth)	(チョウ目、ハマキガ科) 旧北区
brindled ochre		*Dasypolia templi* (Thunberg)	(チョウ目、ヤガ科) 旧北区
brindled pug		*Eupithecia abbreviata* (Stephens)	(チョウ目、シャクガ科) 旧北区
brindled straw flat-body		*Agonopterix arenella* (Denis et Schiffermüller)	(チョウ目、マルハキバガ科) 旧北区
brindled white-spot		*Paradarisa extersaria* (Hübner)	(チョウ目、シャクガ科) 旧北区
brindled white-spot (1)		*Parectropis similaria* (Hufnagel)	(チョウ目、シャクガ科) 旧北区。日本亜種は *P. s. japonica* Sato　シロモンキエダシャク
brindles		Apameinae	(チョウ目、ヤガ科) の昆虫の総称
brine fly		*Ephydra riparia* Fallén	(ハエ目、ミギワバエ科) 全北区、東洋区、エチオピア区
brine fly (1)		*Ephydra cinerea* Jones	(ハエ目、ミギワバエ科) 新北区
brineflies			shore flies を見よ　米国での英名
brinjal leaf-webber			southern beet webworm を見よ
Brinkman's blue		*Orachrysops brinkmani* Heath	(チョウ目、シジミチョウ科) エチオピア区
briseis underwing moth		*Catocala briseis* Edwards	(チョウ目、ヤガ科) 新北区。briseis underwing ともいう
bristle flies			tachina flies を見よ　豪州での英名
bristle-legged moths	ホソマイコガ科	Schreckensteiniidae	(チョウ目) の昆虫の総称
bristletail			silverfish (2) を見よ
bristletails			silverfish (1)(3) を見よ
bristle-tails			horse flies (1) を見よ
bristletails			long-tailed silverfish を見よ
bristle thrips		Panchaetothripidae	(アザミウマ目) の昆虫の総称
bristle-tipped sanddragon		*Progomphus tennesseni* Daigle	(トンボ目、サナエトンボ科) 新熱帯区
bristly cutworm			kidney-spotted miner を見よ
bristly cutworm moth			kidney-spotted miner を見よ
bristly rose slug			rose sawfly (1) を見よ　米国での幼虫の英名
bristly rose slug sawfly		*Cladius difformis* (Panzer)	(ハチ目、ハバチ科) 全北区
bristly roseslug			lesser antler sawfly を見よ
Brittania moth		*Acleris britannia* (Kearfott)	(チョウ目、ハマキガ科) 新北区
Brixton beauty		*Acontia nitidula* (Fabricius)	(チョウ目、ヤガ科) 旧北区
brize			cattle warble fly (1) (2) を見よ
brizes			horse flies (1) を見よ

英名	和名	学名	所属、分布、ほか
broadband swallowtail			broad-banded swallowtail を見よ
broad-banded awl		*Hasora hurama* Butler	（チョウ目、セセリチョウ科）豪州区
broad-banded brilliant		*Simiskina phalena* (Hewitson)	（チョウ目、シジミチョウ科）東洋区
broad-banded crow	ヒロオビルリマダラ	*Euploea latifasciata* Weymer	（チョウ目、タテハチョウ科）東洋区
broad-banded demon		*Chamaelimnas briola* H. Bates	（チョウ目、シジミタテハ科）新熱帯区
broad-banded Eulogia moth		*Eulogia ochrifrontella* (Zeller)	（チョウ目、メイガ科）新北区
broad-banded flasher		*Astraptes apastus apastus* (Cramer)	（チョウ目、セセリチョウ科）新熱帯区
broad-banded giant-skipper		*Megathymus beulahae beulahae* Stallings et Turner	（チョウ目、セセリチョウ科）新熱帯区。*M. b. gayleae* Stallings, Turner et Stallings も同英名
broad-banded page		*Siproeta superba superba* (Bates)	（チョウ目、タテハチョウ科）新熱帯区
broad-banded sailer		*Neptis sankara* (Kollar)	（チョウ目、タテハチョウ科）東洋区
broad banded-skipper			common banded skipper を見よ
broad-banded swallowtail		*Papilio astyalus* Godart	（チョウ目、アゲハチョウ科）新北区
broad-barred button moth			broad-barred marble (2) を見よ
broad-barred knot-horn		*Acrobasis consociella* Hübner	（チョウ目、メイガ科）旧北区
broad-barred marble (1)	ヤナギハマキ	*Acleris latifasciana* (Haworth)	（チョウ目、ハマキガ科）日本、旧北区
broad-barred marble (2)		*Acleris laterana* (Fabricius)	（チョウ目、ハマキガ科）旧北区
broad-barred white		*Hecatera bicolorata* (Hufnagel)	（チョウ目、ヤガ科）旧北区
broad-barred white gothic			broad-barred white を見よ
broadbean beetle			broadbean weevil を見よ
broadbean weevil	ソラマメゾウムシ	*Bruchus rufimanus* Boheman	（コウチュウ目、ハムシ科）日本、汎世界。ソラマメ害虫
broad blue-banded swallowtail	ブロミウスルリアゲハ	*Papilio bromius* Doubleday	（チョウ目、アゲハチョウ科）エチオピア区
broad-bodied chaser		*Libellula depressa* Linnaeus	（トンボ目、トンボ科）旧北区
broad-bodied leaf beetles			leaf beetles (3) を見よ
broad-bodied libellula			broad-bodied chaser を見よ
broad-bordered Acraea	ヒイロホソチョウ	*Acraea anemosa* Hewitson	（チョウ目、タテハチョウ科）エチオピア区
broad-bordered bee hawk			broad-bordered bee hawkmoth (1) を見よ

英名	和名	学名	所属、分布、ほか
broad-bordered bee hawk-moth		*Hemaris bombyliformis* Linnaeus	（チョウ目、スズメガ科）旧北区
broad-bordered bee hawkmoth (1)		*Hemaris fuciformis* (Linnaeus)	（チョウ目、スズメガ科）旧北区。broad bordered bee hawkmoth とも記す
broad-bordered five-spot burnet			five-spot burnet を見よ
broad-bordered grass yellow			small grass yellow (1) を見よ
broad-bordered white			broad-bordered white underwing を見よ
broad-bordered white underwing	コイズミヨトウ	*Anarta melanopa* (Thunberg)	（チョウ目、ヤガ科）日本、全北区
broad-bordered yellow underwing		*Noctua fimbriata* Schreber	（チョウ目、ヤガ科）旧北区
broad-breasted Adhemarius		*Adhemarius eurysthenes* (R. Felder)	（チョウ目、スズメガ科）新熱帯区
broad bug	ハラビロヘリカメムシ	*Homoeocerus dilatatus* Horvath	（カメムシ目、ヘリカメムシ科）日本、旧北区
broad centurion		*Chloromyia formosa* (Scopoli)	（ハエ目、ミズアブ科）全北区
broad damsel bug		*Nabis flavomarginatus* Scholz	（カメムシ目、マキバサシガメ科）旧北区
broad-faced cricket		*Loxoblemmus taicoun* Saussure	（バッタ目、コオロギ科）旧北区。中国で鳴声鑑賞のため飼育
broad green-banded swallowtail		*Papilio chrapkowskii* Suffert	（チョウ目、アゲハチョウ科）エチオピア区
broad-headed bugs	ホソヘリカメムシ科	Alydidae	（カメムシ目）の昆虫の総称
broad-headed horsefly		*Hybomitra lurida* (Fallén)	（ハエ目、アブ科）日本、全北区
broad-horned flour beetle	オオツノコクヌストモドキ	*Gnathocerus cornutus* (Fabricius)	（コウチュウ目、ゴミムシダマシ科）日本、汎世界
broadland horsefly		*Hybomitra muehlfeldi* (Brauer)	（ハエ目、アブ科）旧北区
broad-lined Erastria moth		*Erastria coloraria* (Fabricius)	（チョウ目、シャクガ科）新北区
broad-lined sallow moth		*Sympistis infixa* (Walker)	（チョウ目、ヤガ科）新北区
broad-margined azure			Olane azure を見よ
broad-margined grass yellow		*Eurema candida* (Stoll)	（チョウ目、シロチョウ科）豪州区
broad margined grass yellow		*Eurema puella* (Boisduval)	（チョウ目、シロチョウ科）豪州区
broad-mouth click beetle	クチブトコメツキ	*Silesis musculus* Candèze	（コウチュウ目、コメツキムシ科）日本、旧北区
broadmouthed flies	ヒロクチバエ科	Platystomatidae	（ハエ目）の昆虫の総称
broad-necked root borer		*Prionus (Prionellus) laticollis* (Drury)	（コウチュウ目、カミキリムシ科）新北区

英名	和名	学名	所属、分布、ほか
broad-nosed bark beetles			bark weevils を見よ
broad-nosed grain weevil		*Caulophilus oryzae* (Gyllenhal)	（コウチュウ目、ゾウムシ科）全北区。アボカド害虫
broad-nosed snout beetles		*Cossonus*	（コウチュウ目、ゾウムシ科）の昆虫の総称
broad-nosed weevils		Leptopiinae	（コウチュウ目、ゾウムシ科）の昆虫の総称
broad-nosed weevils		Brachyrhininae	（コウチュウ目、Brachyrhinidae）の昆虫の総称
broad-nosed weevils		Thylacitinae	（コウチュウ目、ゾウムシ科）の昆虫の総称
broad-nosed weevils		Entiminae	（コウチュウ目、ゾウムシ科）の昆虫の総称
broad-nosed weevils		*Otiorhynchus*	（コウチュウ目、ゾウムシ科）の昆虫の総称
broad-patch Carolella moth		*Carolella sartana* (Hübner)	（チョウ目、ハマキガ科）新北区
broad pigeon louse	ハトマルハジラミ	*Campanulotes compar* (Burmeister)	（ハジラミ目、チョウカクハジラミ科）日本、汎世界
broad pigeon louse			small pigeon louse (1) を見よ
broad sallow moth		*Xylotype capax* (Grote)	（チョウ目、ヤガ科）新北区
broad scarlet			scarlet darter を見よ
broad-shouldered water-striders			ripple bugs を見よ
broad spark		*Sinthusa chandrana* (Moore)	（チョウ目、シジミチョウ科）東洋区
broadstick sailer		*Neptis narayana* Moore	（チョウ目、タテハチョウ科）東洋区
broad-streaked sallow midget	フトオビキンモンホソガ	*Phyllonorycter spinotella* (Duponchel)	（チョウ目、ホソガ科）日本、旧北区
broad-streaked sallow midget		*Phyllonorycter hilarella* (Zetterstedt)	（チョウ目、ホソガ科）旧北区
broad-streaked smudge		*Ypsolopha ustella* (Clerck)	（チョウ目、スガ科）旧北区
broad-striped forceptail		*Aphylla angustifolia* Garrison	（トンボ目、サナエトンボ科）新北区、新熱帯区
broad striped hawk		*Rhodafra opheltes* (Cramer)	（チョウ目、スズメガ科）エチオピア区
broadtail royal		*Creon cleobis* (Godart)	（チョウ目、シジミチョウ科）東洋区。*C. c. queda* Corbet も同英名
broad-tailed shadowdragon		*Neurocordulia michaeli* Brunelle	（トンボ目、エゾトンボ科）新北区
broad-thorax click beetle	クロクシコメツキ	*Melanotus senilis* Candèze	（コウチュウ目、コメツキムシ科）日本
broad-tipped clearwing		*Pteronymia cotytto* (Guérin-Méneville)	（チョウ目、タテハチョウ科）新北区、新熱帯区
broad-tipped conehead		*Neoconocephalus triops* (Linnaeus)	（バッタ目、キリギリス科）新北区
broad-tipped weevil	シリブトチョッキリ	*Chokkirius truncatus* (Sharp)	（コウチュウ目、ゾウムシ科）日本、旧北区、東洋区
broad wedge-shaped planthopper	カタビロクサビウンカ	*Issus harimensis* Matsumura	（カメムシ目、マルウンカ科）日本
broad weevils			straight snouted weevils を見よ

英名	和名	学名	所属、分布、ほか
broadwing			angler's curse を見よ
broad-winged bush katydid		*Scudderia pistillata* Brunner	（バッタ目、キリギリス科）新北区
broad-winged damselflies	カワトンボ科	Calopterygidae	（トンボ目）の昆虫の総称
broad-winged damselflies			narrow-winged damselflies を見よ
broad-winged damselfly			ebony jewelwing を見よ
broad-winged honey		*Lamoria anella* (Denis et Schiffermüller)	（チョウ目、メイガ科）旧北区
broad-winged katydid			angular-winged katydid (1) を見よ
broad-winged skipper		*Poanes viator* (Edwards)	（チョウ目、セセリチョウ科）新北区
broadwinged thrips			banded thrips を見よ　豪州での英名
broad-winged tree cricket		*Oecanthus latipennis* Riley	（バッタ目、カンタン科）新北区
broadly green-banded swallowtail		*Papilio chrapkowskoides* Storace	（チョウ目、アゲハチョウ科）エチオピア区。*P. c. nurettini* Kocak も同英名
Brock's roadside-skipper		*Amblyscirtes brocki* Freeman	（チョウ目、セセリチョウ科）新熱帯区
brokenbacked bug		*Taylorilygus pallidulus* (Blanchard)	（カメムシ目、カスミカメムシ科）豪州区
brokenbacked bug	ウスモンミドリカスミカメ	*Taylorilygus apicalis* (Fieber)	（カメムシ目、カスミカメムシ科）日本、汎世界
broken-banded flat		*Celaenorrhinus eligius* (Stoll)	（チョウ目、セセリチョウ科）新熱帯区
broken-banded leafroller moth		*Choristoneura fractivittana* (Clemens)	（チョウ目、ハマキガ科）新北区
broken-banded Y			large looper moth を見よ
broken-barred bell		*Rhopobota stagnana* (Denis et Schiffermüller)	（チョウ目、ハマキガ科）旧北区
broken-barred blue		*Spindasis homeyeri* Dewitz	（チョウ目、シジミチョウ科）エチオピア区
broken-barred carpet		*Electrophaes corylata* Thunberg	（チョウ目、シャクガ科）旧北区。日本亜種は *E. c. granitalis* (Butler) キンオビナミシャク
broken-barred roller		*Ancylis unguicella* (Linnaeus)	（チョウ目、ハマキガ科）旧北区
broken-belted bumblebee		*Bombus soroensis* Fabricius	（ハチ目、ミツバチ科）旧北区
broken dash		*Wallengrenia otho* Abbot et Smith	（チョウ目、セセリチョウ科）新北区
broken-line Hypenodes moth		*Hypenodes fractilinea* (Smith)	（チョウ目、ヤガ科）新北区。broken-line Hypenodes ともいう
broken-line Zomaria moth		*Zomaria interruptolineana* (Fernald)	（チョウ目、ハマキガ科）新北区。broken-line Zomaria ともいう
broken-lined brocade moth		*Mesapamea fractilinea* (Grote)	（チョウ目、ヤガ科）新北区

英名	和名	学名	所属、分布、ほか
broken-lined giant chafer		*Polyphylla diffracta* Casey	（コウチュウ目、コガネムシ科）新北区
broken silverdrop			broken silverspot を見よ
broken silverspot		*Epargyreus exadeus* (Cramer)	（チョウ目、セセリチョウ科）新北区。broken silverdrop ともいう
brome aphid		*Holcaphis bromicola* (Hille Ris Lambers)	（カメムシ目、アブラムシ科）旧北区
brome grass cutworm		*Apamea finitima* Guenée	（チョウ目、ヤガ科）新北区
bromegrass seed midge		*Contarinia bromicola* (Marikovskij et Agafonova)	（ハエ目、タマバエ科）新北区
bromeliad dragonlet		*Erythrodiplax bromelicola* Westfall	（トンボ目、トンボ科）新熱帯区
bromeliad mosquito		*Wyeomyia mitchelli* (Theobald)	（ハエ目、カ科）大洋区
bromeliad pod borer moth		*Epimorius testaceellus* Ragonot	（チョウ目、メイガ科）新北区、新熱帯区
Bromeliad scrub-hairstreak		*Strymon serapio* (Godman et Salvin)	（チョウ目、シジミチョウ科）新北区
bronze alder moth			Greek-lettered argent を見よ
bronze ant-blue			large ant-blue を見よ
bronze apple pigmy		*Stigmella desperatella* (Frey)	（チョウ目、モグリチビガ科）旧北区
bronze apple tree weevil		*Magdalis aenescens* LeConte	（コウチュウ目、ゾウムシ科）新北区
bronze beetle		*Eucolaspis brunnea* (Fabricius)	（コウチュウ目、ハムシ科）豪州区
bronze birch borer		*Agrilus anixus* Gory	（コウチュウ目、タマムシ科）新北区
bronze black-horn moth		*Protosynaema steropucha* Meyrick	（チョウ目、クチブサガ科）豪州区
bronze bottle fly	ヒツジキンバエ	*Phaenicia cuprina* (Wiedemann)	（ハエ目、クロバエ科）日本、全北区、東洋区、大洋区、豪州区、新熱帯区
bronze copper		*Lycaena hyllus* (Cramer)	（チョウ目、シジミチョウ科）新北区
bronze cutworm			brown cutworm (1) を見よ
bronze duke	ナライナズマ	*Euthalia nara* (Moore)	（チョウ目、タテハチョウ科）東洋区
bronze dump fly			dump fly を見よ
bronze dump fly			black garbage fly を見よ
bronze flat			eastern flat を見よ
bronze juniper argent		*Blastotere arceuthina* Zeller	（チョウ目、メムシガ科）旧北区
bronze leaf beetle		*Diachus auratus* (Fabricius)	（コウチュウ目、ハムシ科）新北区
bronze line-blue			pointed lineblue を見よ
bronze ochre		*Trapezites macqueeni* Kerr et Sands	（チョウ目、セセリチョウ科）豪州区
bronze orange bug		*Rhoecocoris sulciventris* (Stål)	（カメムシ目、カメムシ科）豪州区。カンキツ害虫

英名	和名	学名	所属、分布、ほか
bronze poplar borer		*Agrilus liragus* Barter et Brown	（コウチュウ目、タマムシ科）新北区
bronze roadside skipper		*Amblyscirtes aenus* Edwards	（チョウ目、セセリチョウ科）新北区
bronze rowan pigmy		*Stigmella nylandriella* (Tengstrom)	（チョウ目、モグリチビガ科）旧北区
bronze small case		*Coleophora potentillae* (Elisha)	（チョウ目、ツツミノガ科）旧北区
bronze tiger beetle		*Cicindela (Cicindela) repanda* Dejean	（コウチュウ目、ハンミョウ科）新北区。blonzed tiger beetle ともいう
bronze underwing		*Catocala cara* Guenée	（チョウ目、ヤガ科）新北区。幼虫はヤナギを食う
bronzed copper		*Hyllolycaena hyllus* (Cramer)	（チョウ目、シジミチョウ科）新北区
bronzed cutworm		*Nephelodes minians* Guenée	（チョウ目、ヤガ科）新北区
bronzed cutworm moth		*Prorella emmedonia* (Grossbeck)	（チョウ目、ヤガ科）新北区。トウモロコシ害虫。幼虫は bronzed cutworm
bronzed field beetle		*Adelium brevicorne* Blessig	（コウチュウ目、ゴミムシダマシ科）豪州区
bronzed river cruiser		*Macromia annulata* Hagen	（トンボ目、ヤマトンボ科）新北区
bronzy Macrochilo moth		*Macrochilo orciferalis* (Walker)	（チョウ目、ヤガ科）新北区
brook hooktail		*Paragomphus henryi* (Laidlaw)	（トンボ目、サナエトンボ科）東洋区
brook snaketail		*Ophiogomphus apersus* Morse	（トンボ目、サナエトンボ科）新北区
Brook's opal		*Chrysoritis brooksi* (Riley)	（チョウ目、シジミチョウ科）エチオピア区
broom brocade	マメヨトウ	*Mamestra pisi* Linnaeus	（チョウ目、ヤガ科）旧北区。ダイズなどマメ類害虫
broom caterpillar	オオケンモン	*Acronicta major* (Bremer)	（チョウ目、ヤガ科）日本、旧北区
broom flower bug		*Anthocoris sarothamni* Douglas et Scott	（カメムシ目、ハナカメムシ科）旧北区
broom moth			broom brocade を見よ
broom seed beetle		*Bruchidius villosus* (Fabricius)	（コウチュウ目、ハムシ科）旧北区
broom tip		*Chesias rufata* (Fabricius)	（チョウ目、シャクガ科）旧北区
broomcorn-stem fly		*Atherigona varia* (Meigen)	（ハエ目、イエバエ科）旧北区
broomcorn stem maggot			broomcorn stem-fly を見よ　幼虫の英名
broom-tip chevron			broom tip を見よ
broomtip moth			broom tip を見よ
Broteas scarlet-eye		*Nascus broteas* (Cramer)	（チョウ目、セセリチョウ科）新熱帯区
brother		*Raphia frater* Grote	（チョウ目、ヤガ科）全北区。brother moth ともいう
Brou's sphinx moth		*Lapara phaeobrachycerous* Brou	（チョウ目、スズメガ科）新北区
Brou's underwing moth		*Catocala atocala* Brou	（チョウ目、ヤガ科）新北区。Brou's underwing ともいう

英名	和名	学名	所属、分布、ほか
brown accented butterfly			violet lacewing を見よ
brown aeschna			brown hawker を見よ
brown African swallowtail			mocker swallowtail を見よ
brown ambrosia aphid		*Uroleucon ambrosiae* (Thomas)	（カメムシ目、アブラムシ科）新北区、新熱帯区
brown ambrosia aphid		*Uroleucon rudbeckiae* (Fitch)	（カメムシ目、アブラムシ科）新北区
brown and black beetle	ブドウサルハムシ	*Bromius obscurus* (Linnaeus)	（コウチュウ目、ハムシ科）日本、旧北区。ブドウ害虫
brown and white cuckoo bee		*Epeolus compactus* Cresson	（ハチ目、コシブトハナバチ科）新北区
brown-and-white metalmark		*Calociasma nycteus* (Godman et Salvin)	（チョウ目、シジミタテハ科）新熱帯区
brown-and-yellow fruit chafer			garden fruit chafer を見よ
brown angle shades moth		*Phlogophora periculosa* Guenée	（チョウ目、ヤガ科）新北区。brown angle shades ともいう
brown ant			fire ant (2) を見よ
brown ant-blue		*Acrodipsas mortoni* (Sands, Miller et Kerr)	（チョウ目、シジミチョウ科）豪州区
brown aphid		*Acyrthosiphon spartii* (Koch)	（カメムシ目、アブラムシ科）旧北区
brown apple budworm			eyespotted bud moth を見よ
brown apricot scale			European fruit lecanium を見よ　米国での英名
brown arborvitae leafminer			brown cedar leafminer moth を見よ
brown arches			gypsy moth を見よ
brown arches moth		*Lacinipolia stricta* (Walker)	（チョウ目、ヤガ科）新北区
brown Argus		*Ypthima hyagriva* Moore	（チョウ目、タテハチョウ科）東洋区
brown Argus (1)	ハマベシジミ	*Aricia agestis* Denis et Schiffermüller	（チョウ目、シジミチョウ科）旧北区。brown Argus butterfly ともいう
brown Argus blue			brown Argus (1) を見よ
brown awl	タイワンアオバセセリ	*Badamia exclamationis* (Fabricius)	（チョウ目、セセリチョウ科）日本、旧北区、東洋区、豪州区
brown-backed rice planthopper			brown rice planthopper を見よ
brown bamboo scale	タケトビイロマルカイガラムシ	*Odonaspis bambusarum* (Cockerell)	（カメムシ目、マルカイガラムシ科）日本
brown-banded carder bee		*Bombus humilis* Illiger	（ハチ目、ミツバチ科）旧北区
brown-banded carder bee		*Bombus helferanus* Seidl	（ハチ目、ミツバチ科）旧北区
brown-banded carpet moth		*Hydriomena rixata* Felder	（チョウ目、シャクガ科）豪州区

英名	和名	学名	所属、分布、ほか
brown-banded cockroach (1)		*Supella supellectillum* Serville	(ゴキブリ目、チャバネゴキブリ科) 新北区
brown-banded cockroach (2)	チャオビゴキブリ	*Supella longipalpa* (Fabricius)	(ゴキブリ目、チャバネゴキブリ科) 日本、全北区、豪州区。体長 14 mm
brown-banded roach			brown-banded cockroach (1) を見よ　カナダでの英名
brown banded skipper		*Timochares ruptifasciata* (Plötz)	(チョウ目、セセリチョウ科) 新北区、新熱帯区
brown bark borer			orange stem borer (1) を見よ
brown bark carpet moth		*Horisme intestinata* (Guenée)	(チョウ目、シャクガ科) 新北区
brown-barred hairstreak		*Thaeides theia* (Hewitson)	(チョウ目、シジミチョウ科) 新熱帯区
brown basket lerp		*Cardiaspina fiscella* Taylor	(カメムシ目、カイガラキジラミ科) 豪州区
brown bean bug (1)		*Riptortus serripes* (Fabricius)	(カメムシ目、ホソヘリカメムシ科) 豪州区
brown bean bug (2)		*Melanacanthus scutellaris* (Dallas)	(カメムシ目、ホソヘリカメムシ科) 豪州区
brown beetle			June beetle (1) を見よ
brown black-square Epitola		*Stempfferia gordoni* (Druce)	(チョウ目、シジミチョウ科) エチオピア区
brown blowfly		*Calliphora stygia* (Fabricius)	(ハエ目、クロバエ科) 豪州区
brown blue			brown Argus (1) を見よ
brown-bordered geometer moth		*Eumacaria madopata* (Guenée)	(チョウ目、シャクガ科) 新北区
brown-bordered piercer		*Pammene inquilina* Fletcher	(チョウ目、ハマキガ科) 旧北区
brown-bordered white		*Itaballia pandosia* (Hewitson)	(チョウ目、シロチョウ科) 新熱帯区。*I. p. kicada* (Reakirt) も同英名
brown brassy owlet		*Scythris fuscoaenea* (Haworth)	(チョウ目、キヌバコガ科) 旧北区
brown brindled flat-body		*Depressaria badiella* (Hübner)	(チョウ目、マルハキバガ科) 旧北区
brown broad-winged planthopper	トビイロハゴロモ	*Mimophantia maritima* Matsumura	(カメムシ目、アオバハゴロモ科) 日本、東洋区
brown bug			hemispherical scale (1) を見よ
brown bullet		*Agathymus estelleae* (Stallings et Turner)	(チョウ目、セセリチョウ科) 新北区、新熱帯区
brown bunyip		*Tamasa tristigma* (Germar)	(カメムシ目、セミ科) 豪州区
brown butterflies			satyrs を見よ
brown carpenter ant	ツヤシリアゲアリ	*Crematogaster laboriosa* F. Smith	(ハチ目、アリ科) 日本、旧北区
brown carpet beetle		*Attagenus smirnovi* Zhantiev	(コウチュウ目、カツオブシムシ科) エチオピア区
brown cedar leafminer moth		*Coleotechnites thujaella* (Kearfott)	(チョウ目、キバガ科) 新北区
brown chafer		*Hoplia dispar* LeConte	(コウチュウ目、コガネムシ科) 新北区

英名	和名	学名	所属、分布、ほか
brown chafer		*Serica brunnea* (Linnaeus)	(コウチュウ目、コガネムシ科) 旧北区
brown chafer (1)	コイチャコガネ	*Adoretus tenuimaculatus* Waterhouse	(コウチュウ目、コガネムシ科) 日本
brown chicken louse	カクアゴハジラミ	*Goniodes dissimilis* Denny	(ハジラミ目、チョウカクハジラミ科) 日本、全北区
brown China mark			China mark moth (1) を見よ
brown China mark moth			China mark moth (1) を見よ
brown citrus aphid			citrus brown aphid を見よ 米国での英名
brown click beetle	サビキコリ	*Agrypnus binodulus* (Motschulsky)	(コウチュウ目、コメツキムシ科) 日本
brown climbing cutworm		*Rhynchagrotis cupida* Grote	(チョウ目、ヤガ科) 新北区
brown climbing cutworm			cupid dart moth を見よ
brown cockchafer		*Rhopaea magnicornis* Blackburn	(コウチュウ目、コガネムシ科) 豪州区
brown cockroach	トビイロゴキブリ	*Periplaneta brunnea* Burmeister	(ゴキブリ目、ゴキブリ科) 日本、新北区。屋内、野外に生息
brown coffee scale			hemispherical scale (1) を見よ
brown commodore	ナタリカタテハモドキ	*Junonia natalica* (C. et R. Felder)	(チョウ目、タテハチョウ科) エチオピア区
brown-copper case		*Coleophora fuscocuprella* Herrich-Schäffer	(チョウ目、ツツミノガ科) 旧北区
brown-copper ermel		*Roeslerstammia erxlebella* (Fabricius)	(チョウ目、ヒカリバコガ科) 旧北区
brown cotton leafworm		*Acontia dacia* Druce	(チョウ目、ヤガ科) 新北区
brown crescent		*Anthanassa atronia* (Bates)	(チョウ目、タテハチョウ科) 新熱帯区
brown crescent			crescent を見よ
brown ctenucha moth		*Ctenucha brunnea* Stretch	(チョウ目、カノコガ科) 新北区。brown ctenucha ともいう。
brown ctenuchid moth			brown ctenucha moth を見よ
brown cupid		*Euchrysops brunneus* Bethune-Baker	(チョウ目、シジミチョウ科) エチオピア区
brown cutworm		*Agrotis munda* Walker	(チョウ目、ヤガ科) 豪州区
brown cutworm		*Agrotis radians* Guenée	(チョウ目、ヤガ科) 豪州区
brown cutworm		*Agrotis longidentifera* (Hampson)	(チョウ目、ヤガ科) エチオピア区、東洋区
brown cutworm (1)		*Agrochola emmedonia* Cramer	(チョウ目、ヤガ科) 新北区
brown darkling beetle		*Ecnolagria grandis* (Gyllenhal)	(コウチュウ目、ゴミムシダマシ科) 豪州区
brown dart moth			flame shoulder を見よ 米国での英名
brown day-moth			common sheep moth を見よ
brown Donacaula moth		*Donacaula roscidellus* (Dyar)	(チョウ目、メイガ科) 新北区

英名	和名	学名	所属、分布、ほか
brown-dotted clothes moth		*Niditinea fuscella* (Linnaeus)	（チョウ目、ヒロズコガ科）全北区、豪州区
brown-dotted clothes moth			poultry house moth (1) を見よ
brown-dotted grey dwarf		*Biselachista cinereopunctella* (Haworth)	（チョウ目、クサモグリガ科）旧北区
brown-dotted leafhopper	ホシヨコバイ	*Xestocephalus japonicus* Ishihara	（カメムシ目、オオヨコバイ科）日本、旧北区
brown-dotted yellow tussock moth	ゴマフリキドクガ	*Euproctis pulverea* (Leech)	（チョウ目、ドクガ科）日本
brown double streak			brown hairstreak を見よ
brown dung beetle		*Onthophagus gazella* Fabricius	（コウチュウ目、コガネムシ科）新北区、豪州区、エチオピア区
brown dung fly		*Scatophaga furcata* (Say)	（ハエ目、フンバエ科）旧北区
brown dungball roller		*Sisyphus rubrus* Paschalidis	（コウチュウ目、コガネムシ科）豪州区
brown elfin	チャイロコツバメ	*Callophrys augustinus* (Westwood)	（チョウ目、シジミチョウ科）新北区。亜種 *C. a. annettae* (dos Passos) , *C. a. iroides* (Boisduval) も同英名
brown emerald damselfly		*Sympecma fusca* (van der Linden)	（トンボ目、アオイトトンボ科）旧北区
brown eucalypt beetle		*Lepidiota rothei* Blackburn	（コウチュウ目、コガネムシ科）豪州区
brown evening moth		*Selidosema dejectaria* Walker	（チョウ目、シャクガ科）豪州区
brown fanfoot		*Zanclognatha tarsipennalis* (Treitschke)	（チョウ目、ヤガ科）旧北区
brown feathered			brown rustic を見よ
brown fern moth		*Selidosema pelurgata* Walker	（チョウ目、シャクガ科）豪州区
brown fiddle-shaped planthopper	トビイログンバイウンカ	*Ommatissus lofuensis* Muir	（カメムシ目、グンバイウンカ科）日本、旧北区、東洋区
brown field crickets		*Lepidogryllus*	（バッタ目、コオロギ科）の昆虫の総称
brown flower beetle		*Glycyphana stolata* (Fabricius)	（コウチュウ目、コガネムシ科）豪州区
brown forest moth		*Selidosema productata* Walker	（チョウ目、シャクガ科）豪州区
brown forest swift		*Melphina noctula* (Druce)	（チョウ目、セセリチョウ科）エチオピア区
brown forest sylph		*Ceratrichia brunnea* Bethune-Baker	（チョウ目、セセリチョウ科）エチオピア区
brown forester		*Lethe serbonis* (Hewitson)	（チョウ目、タテハチョウ科）東洋区
brown fruit scale			European fruit lecanium を見よ
brown goose-berry scale			nut scale を見よ
brown Gorgon		*Meandrusa gyas* (Westwood)	（チョウ目、アゲハチョウ科）東洋区
brown Gorgon	スキロンアゲハ	*Meandrusa sciron* Leech	（チョウ目、アゲハチョウ科）東洋区

英名	和名	学名	所属、分布、ほか
brown gourd-shaped weevil	トビイロヒョウタンゾウムシ	*Scepticus uniformis* Kono	(コウチュウ目、ゾウムシ科) 日本。イモ、マメ類など多くの作物害虫
brown grain beetle			red flour beetle を見よ
brown granulated planthopper	コブウンカ	*Tropidocephala brunneipennis* Signoret	(カメムシ目、ウンカ科) 日本、東洋区、豪州区、エチオピア区
brown-green fanner		*Glyphipterix fuscoviridella* (Haworth)	(チョウ目、ホソハマキモドキガ科) 旧北区
brown greenstreak		*Cyanophrys fusius* (Godman et Salvin)	(チョウ目、シジミチョウ科) 新熱帯区
brown grey		*Scoparia ambigualis* Treitschke	(チョウ目、メイガ科) 旧北区
brown-grey case		*Coleophora siccifolia* Stainton	(チョウ目、ツツミノガ科) 旧北区
brown grizzled skipper			large grizzled skipper (2) を見よ
brown hairstreak	チョウセンメスアカシジミ	*Thecla betulae* (Linnaeus)	(チョウ目、シジミチョウ科) 旧北区。brown hairstreak butterfly ともいう
brown hawker		*Aeschna grandis* (Linnaeus)	(トンボ目、ヤンマ科) 旧北区
brownheaded ash sawfly		*Tomostethus multicinctus* (Rohwer)	(ハチ目、ハバチ科) 新北区
brownheaded Jack pine sawfly		*Neodiprion dubiosus* Schedl	(ハチ目、マツハバチ科) 新北区
brown-headed rice planthopper			brown rice planthopper を見よ
brown hemlock needleminer moth		*Coleotechnites macleodi* (Freeman)	(チョウ目、キバガ科) 新北区。browm hemlock needleminer ともいう
brown-hooded cockroach		*Cryptocerus punctulatus* Scudder	(ゴキブリ目、ヒメゴキブリ科) 新北区。最も原始的な種で、湿った朽木に生息
brown-hooded cockroaches	ヒメゴキブリ科	Cryptocercidae	(ゴキブリ目) の昆虫の総称
brown hooded owlet		*Cucullia convexipennis* Grote et Robinson	(チョウ目、ヤガ科) 新北区、新熱帯区。brown hooded owlet moth ともいう
brown hornet			hornet (1) を見よ
brown house ant		*Crematogaster darwiniana* (Forel)	(ハチ目、アリ科) 豪州区
brown house ant			big-headed ant (2) を見よ
brown house mosquito			southern house mosquito を見よ　豪州での英名
brown house moth			large common tubic を見よ
brown king crow		*Euploea klugii erichsonii* C. et R. Felder	(チョウ目、タテハチョウ科) 東洋区。blue crow を参照
brown knot-horn			brown muslin sweep を見よ
brown lacewing		*Kimminsia subnebulosa* (Stephens)	(アミメカゲロウ目、ヒメカゲロウ科) 旧北区
brown lacewing		*Hemerobius stigma* Stephens	(アミメカゲロウ目、ヒメカゲロウ科) 新北区
brown lacewing		*Sympherobius angustatus* (Banks)	(アミメカゲロウ目、ヒメカゲロウ科) 新北区

英名	和名	学名	所属、分布、ほか
brown lacewing		*Megalomus hirtus* (Linnaeus)	(アミメカゲロウ目、ヒメカゲロウ科) 旧北区
brown lacewing		*Micromus tasmaniae* Walker	(アミメカゲロウ目、ヒメカゲロウ科) 豪州区
brown lacewing		*Micromus variolosus* Hagen	(アミメカゲロウ目、ヒメカゲロウ科) 新北区
brown lacewing			Pacific brown lacewing を見よ
brown lacewings		*Micromus*	(アミメカゲロウ目、ヒメカゲロウ科) の昆虫の総称
brown lacewings		*Hemerobius*	(アミメカゲロウ目、ヒメカゲロウ科) の昆虫の総称
brown lacewings (1)	ヒメカゲロウ科	Hemerobiidae	(アミメカゲロウ目) の昆虫の総称
brown lacewings			spongilla-flies を見よ　英国での英名
brown lacewings			lesser brown lacewings を見よ
brown leaf beetle		*Ootheca mutabilis* (Sahlberg)	(コウチュウ目、ハムシ科) エチオピア区
brown leaf cicada		*Lembeja vitticollis* (Ashton)	(カメムシ目、セミ科) 豪州区
brown leaf weevil		*Phyllobius oblongus* (Linnaeus)	(コウチュウ目、ゾウムシ科) 旧北区
brown leatherwing		*Cantharis consors* LeConte	(コウチュウ目、ジョウカイボン科) 新北区
brown-line bright-eye		*Mythimna conigera* (Denis et Schiffermüller)	(チョウ目、ヤガ科) 旧北区
brown-line sapphire		*Iolaus alienus* Trimen	(チョウ目、シジミチョウ科) エチオピア区
brown-line wainscot			brown-line bright-eye を見よ
brown-lined looper moth		*Neoalcis californiaria* (Packard)	(チョウ目、シャクガ科) 新北区。brown-lined looper ともいう
brown-lined owlet moth		*Macrochilo litophora* (Grote)	(チョウ目、ヤガ科) 新北区
brown-lined sallow moth		*Sympistis badistriga* (Grote)	(チョウ目、ヤガ科) 新北区
brown locust	チャイロトビバッタ	*Locusta pardalina* (Walker)	(バッタ目、バッタ科) エチオピア区
brown longhorn beetle		*Obrium brunneum* (Fabricius)	(コウチュウ目、カミキリムシ科) 旧北区
brown longtail		*Urbanus procne* (Plötz)	(チョウ目、セセリチョウ科) 新北区
brown looper			sinister moth を見よ
brown looper			larger irregular-marked noctuid を見よ
brown louse			brown chicken louse を見よ　豪州での英名
brown lyctid beetle			powdered beetle (1) を見よ　brown lyctus beetle とも記す。
brown mantisfly		*Climaciella brunnea* (Say)	(アミメカゲロウ目、カマキリモドキ科) 新北区、新熱帯区
brown mantispid		*Plega signata* (Hagen)	(アミメカゲロウ目、カマキリモドキ科) 新北区
brown marmorated longicorn	チャゴマフカミキリ	*Mesosa perplexa* Pascoe	(コウチュウ目、カミキリムシ科) 日本、旧北区、東洋区
brown marmorated prominent			large dark prominent を見よ

英名	和名	学名	所属、分布、ほか
brown marmorated stink bug	クサギカメムシ	*Halyomorpha halys* (Stål)	（カメムシ目、カメムシ科）日本、全北区、東洋区。クサギや果樹類につく
brown mayfly		*Ephemera vulgata* Linnaeus	（カゲロウ目、モンカゲロウ科）旧北区
brown mirid		*Creontiades pacificus* (Stål)	（カメムシ目、カスミカメムシ科）豪州区
brown mort bleu		*Catoblepia berecynthia* (Cramer)	（チョウ目、タテハチョウ科）新熱帯区
brown mountain grasshopper		*Podisma pedestris* (Linnaeus)	（バッタ目、Catantopidae）旧北区
brown muslin sweep	ウスグロマダラメイガ	*Pyla fusca* (Haworth)	（チョウ目、メイガ科）日本、全北区
brown nettle moth		*Mecyna marmariana* Meyrick	（チョウ目、メイガ科）豪州区
brown oak tortrix		*Archips crataegana* (Hübner)	（チョウ目、ハマキガ科）旧北区
brown oak tortrix moth			apple leafroller (1) を見よ
brown ochre		*Trapezites iacchus* (Fabricius)	（チョウ目、セセリチョウ科）豪州区
brown olive scale			black scale を見よ
brown onyx		*Horaga albimacula albistigmata* Moulton	（チョウ目、シジミチョウ科）東洋区
brown onyx		*Horaga viola* Moore	（チョウ目、シジミチョウ科）東洋区。violet onyx を参照
brown owl-butterfly		*Narope testacea* Godman et Salvin	（チョウ目、タテハチョウ科）新熱帯区
brown owlet moth			dark arches を見よ
brown Panopoda moth		*Panopoda carneicosta* Guenée	（チョウ目、ヤガ科）新北区
brown pansy	シロモンタテハモドキ	*Junonia stygia* (Aurivillius)	（チョウ目、タテハチョウ科）エチオピア区
brown pansy (1)		*Junonia hedonia* Linnaeus	（チョウ目、タテハチョウ科）東洋区
brown pasture looper		*Cimpla aridaria* (Guenée)	（チョウ目、シャクガ科）豪州区
brown-patch Ceres forester		*Euphaedra viridicaerulea* Bartel	（チョウ目、タテハチョウ科）エチオピア区
brown-patched looper			large looper moth を見よ
brown peach aphid			black peach aphid (1) を見よ
brown peacock			scarlet peacock を見よ
brown pectinate-horned click beetle	クロツヤクシコメツキ	*Melanotus annosus* Candèze	（コウチュウ目、コメツキムシ科）日本、旧北区
brown pied flat		*Coladenia agni* (de Nicéville)	（チョウ目、セセリチョウ科）東洋区
brown pine bark beetle	マツノカバイロキクイムシ	*Hylurgops glabratus* (Zetterstedt)	（コウチュウ目、キクイムシ科）日本、旧北区、東洋区
brown pine looper moth		*Caripeta angustiorata* Walker	（チョウ目、シャクガ科）新北区
brown pine weevil	マツトビゾウムシ	*Scythropus scutellaris* Roelofs	（コウチュウ目、ゾウムシ科）日本

英名	和名	学名	所属、分布、ほか
brown pineapple scale	パイナップルクロマルカイガラムシ	*Melanaspis bromeliae* (Leonardi)	(カメムシ目、マルカイガラムシ科) 日本、新北区。パイナップ害虫
brown planthopper			brown rice planthopper を見よ
brown playboy		*Deudorix antalus* (Hopffer)	(チョウ目、シジミチョウ科) エチオピア区
brown powder post beetle			powderpost beetle (1) を見よ
brown prince		*Rohana parvata* (Moore)	(チョウ目、タテハチョウ科) 東洋区
brown prionid		*Orthosoma brunneum* (Forster)	(コウチュウ目、カミキリムシ科) 新北区
brown rice planthopper	トビイロウンカ	*Nilaparvata lugens* (Stål)	(カメムシ目、ウンカ科) 日本、旧北区、東洋区、豪州区。著名なイネ害虫
brown rice stink bug	イワサキカメムシ	*Starioides iwasakii* Matsumura	(カメムシ目、カメムシ科) 日本
brown roach			brown cockroach を見よ
brown rose grub			greater brown cloaked bell を見よ 幼虫の英名
brown rustic		*Rusina ferruginea* (Esper)	(チョウ目、ヤガ科) 旧北区
brown salt marsh mosquito		*Ochlerotatus cantator* (Coquillett)	(ハエ目、カ科) 新北区
brown saltern groundling		*Scrobipalpa instabilella* (Douglas)	(チョウ目、キバガ科) 旧北区
brown saltmarsh mosquito (1)	セスジヤブカ	*Aedes dorsalis* Meigen	(ハエ目、カ科) 日本、旧北区
brown saltmarsh mosquito		*Ochlerotatus cantator* Coquillett	(ハエ目、カ科) 新北区
brown scale			European fruit lecanium を見よ
brown scale of coffee			hemispherical scale (2) を見よ
brown scallop		*Philereme vetulata* (Denis et Schiffermüller)	(チョウ目、シャクガ科) 旧北区
brown scavenger beetles			plaster beetles を見よ
brown scoopwing moth		*Calledapteryx dryopterata* Grote	(チョウ目、ツバメガ科) 新北区
brown seed chafer	ナエドロチャイロコガネ	*Sericania mimica* Lewis	(コウチュウ目、コガネムシ科) 日本
brown semilooper			larger irregular-marked noctuid を見よ
brown-shaded carpet moth		*Venusia compataria* (Walker)	(チョウ目、シャクガ科) 新北区
brown-shaded gray moth		*Iridopsis defectaria* Guenée	(チョウ目、シャクガ科) 新北区
brown shield bug		*Dictyotus caenosus* (Westwood)	(カメムシ目、カメムシ科) 豪州区
brown shieldback		*Idionotus brunneus* Scudder	(バッタ目、キリギリス科) 新北区
brown-shouldered bell moth		*Cnephasia incessana* Walker	(チョウ目、ハマキガ科) 豪州区
brown-shouldered Scoparia moth		*Scoparia minuscularia* Walker	(チョウ目、メイガ科) 豪州区

英名	和名	学名	所属、分布、ほか
brown silky skipper			Arela dart を見よ
brown silver-line	シダエダシャク	*Petrophora chlorosata* (Scopoli)	(チョウ目、シャクガ科) 日本、旧北区。brown silver-lined ともいう
brown Siproeta	ツマアカシロビタテハ	*Siproeta epaphus* (Latreille)	(チョウ目、タテハチョウ科) 新北区、新熱帯区
brown skipper			umber skipper を見よ
brown small longicorn	ヒゲナガヒメカミキリ	*Ceresium longicorne* Pic	(コウチュウ目、カミキリムシ科) 日本、東洋区
brown smudge bug		*Deraeocoris signatus* (Distant)	(カメムシ目、カスミカメムシ科) 豪州区
brown soft scale	ヒラタカタカイガラムシ	*Coccus hesperidum* Linnaeus	(カメムシ目、カタカイガラムシ科) 日本、新北区、豪州区。多くの果樹、植物の害虫
brown soft scale			nut scale を見よ
brown soldier	イワサキタテハモドキ	*Junonia hedonia iwasakii* (Matsumura)	(チョウ目、タテハチョウ科) 日本、東洋区。豪州の *J. hedonia zelima* (Fabricius) の英名。brown pansy (1) を参照
brown speckled leafhopper		*Paraphlepsius irroratus* (Say)	(カメムシ目、オオヨコバイ科) 新北区
brown spider beetle	ヒメヒョウホンムシ	*Ptinus clavipes* Panzer	(コウチュウ目、ヒョウホンムシ科) 日本、汎世界。貯蔵食品の害虫
brown spider beetle			white-marked spider beetle を見よ　南アフリカでの英名
brown spiketail		*Cordulegaster bilineata* (Carle)	(トンボ目、オニヤンマ科) 新北区
brown-spot chestnut			brown-spot pinion を見よ
brown-spot pinion		*Agrochola litura* (Linnaeus)	(チョウ目、ヤガ科) 旧北区
brown spot Temnora		*Temnora plagiata* Walker	(チョウ目、スズメガ科) エチオピア区
brown-spotted bush cricket		*Platycleis tessellata* (Charpentier)	(バッタ目、キリギリス科) 旧北区
brown-spotted grasshopper		*Psoloessa delicatula* (Scudder)	(バッタ目、バッタ科) 新北区
brown-spotted prominent	トビモンシャチホコ	*Drymonia dodonides* (Staudinger)	(チョウ目、シャチホコガ科) 日本、旧北区
brown-spotted Zale moth		*Zale helata* McDunnough	(チョウ目、ヤガ科) 新北区。brown-spotted Zale ともいう
brown spruce aphid			spruce shoot aphid を見よ
brown spruce-bast borer			brown pine bark beetle を見よ
brown stag beetle		*Rhyssonotus nebulosus* Kirby	(コウチュウ目、クワガタムシ科) 豪州区
brown stem bug		*Thyanta perditor* (Fabricius)	(カメムシ目、カメムシ科) 新熱帯区。イネ害虫
brown stink bug		*Euschistus servus* (Say)	(カメムシ目、カメムシ科) 新北区
brown stream mayflies		*Rhithrogena*	(カゲロウ目、ヒラタカゲロウ科) の昆虫の総称
brown stripe-edged piercer		*Cydia tenebrosana* Duponchel	(チョウ目、ハマキガ科) 旧北区

英名	和名	学名	所属、分布、ほか
brown striped hawk		*Basiothia schenki* (Möschler)	(チョウ目、スズメガ科) エチオピア区
brown-striped matchstick		*Moritala hmta* (Nieminski)	(バッタ目、Eumastacidae) 豪州区
brown sugarcane cicada		*Cicadetta crucifera* (Ashton)	(カメムシ目、セミ科) 豪州区
brown swarming leaf beetle		*Rhyparida limbatipennis* Jacoby	(コウチュウ目、ハムシ科) 豪州区
brown sword-tail crickets		*Anaxipha*	(バッタ目、クサヒバリ科) の昆虫の総称
brown tachinid		*Jurinella lutzi* Curran	(ハエ目、ヤドリバエ科) 新北区
brown tail			brown-tail moth (1) を見よ
brown-tail moth (1)		*Euproctis chrysorrhoea* (Linnaeus)	(チョウ目、ドクガ科) 全北区。brown-tail ともいう
browntail moth (2)	モンシロドクガ	*Euproctis similis* (Fuessly)	(チョウ目、ドクガ科) 日本、全北区。多くの作物、果樹を加害
brown-tailed moth			brown-tail moth (1) を見よ
brown tiger		*Obeidia tigrata* Thierry-Mieg	(チョウ目、シャクガ科) 東洋区
brown tiger	ウスグロカバマダラ	*Danaus philene* Stoll	(チョウ目、タテハチョウ科) 豪州区
brown tiger moth		*Spilosoma pteridis* H. Edwards	(チョウ目、ヒトリガ科) 新北区
brown-tipped brownie		*Allotinus taras* (Doherty)	(チョウ目、シジミチョウ科) 東洋区
brown-tipped metalmark		*Lamphiotes velazquezi* (Beutelspacher)	(チョウ目、シジミタテハ科) 新熱帯区
brown-tipped pear pigmy		*Stigmella minusculella* (Herrich-Schäffer)	(チョウ目、モグリチビガ科) 旧北区
brown-tipped skipper		*Netrobalane canopus* (Trimen)	(チョウ目、セセリチョウ科) エチオピア区
brown tit		*Hypolycaena thecloides* (Felder)	(チョウ目、シジミチョウ科) 東洋区
brown tortrix	トビハマキ	*Pandemis heparana* Denis et Schiffermüller	(チョウ目、ハマキガ科) 日本、旧北区。ダイズ、リンゴなど果樹、バラ、サクラなどの害虫
brown tree ant		*Lasius brunneus* (Latreille)	(ハチ目、アリ科) 旧北区
brown tree nymph		*Sallya boisduvali* (Wallengren)	(チョウ目、タテハチョウ科) エチオピア区
brown treehopper	トビイロツノゼミ	*Machaerotypus sibiricus* (Lethierry)	(カメムシ目、ツノゼミ科) 日本、旧北区。ウメなどにつく
brown twig borer		*Xyleborus morigerus* Blandford	(コウチュウ目、キクイムシ科) 東洋区、エチオピア区。コーヒー害虫
brown-veined neb		*Metzneria neuropterella* Zeller	(チョウ目、キバガ科) 旧北区
brown-veined wainscot		*Archanara dissoluta* (Treitschke)	(チョウ目、ヤガ科) 旧北区
brown-veined white		*Belenois aurota* (Fabricius)	(チョウ目、シロチョウ科) 旧北区、東洋区、エチオピア区
brown velvety chafer	カバイロビロウドコガネ	*Nipponoserica similis* (Lewis)	(コウチュウ目、コガネムシ科) 日本

英名	和名	学名	所属、分布、ほか
brown veronica carpet moth		*Chloroclystis lunata* Philpott	(チョウ目、シャクガ科) 豪州区
brown wavy line Argyrostrotis			woodland chocolate moth を見よ
brown white-spot			mountain Argus (1) を見よ
brown willow beetle			yellow willow leaf beetle を見よ
brown willow leaf beetle			yellow willow leaf beetle を見よ
brown-winged bush cricket		*Mastophyllum scabricolle* (Serville)	(バッタ目、キリギリス科) 旧北区
brown-winged dry-wooden longicorn			blue-winged dry-wooden longicorn を見よ
brown-winged green bug	チャバネアオカメムシ	*Plautia crossota stali* Scott	(カメムシ目、カメムシ科) 日本、旧北区。多くの作物、果樹を加害
brown-winged shield-back			sooty longwing を見よ
brown wingless cockroach			brown-hooded cockroach を見よ
brown wood plume		*Stenoptilia pterodactyla* (Linnaeus)	(チョウ目、トリバガ科) 旧北区
brown woodling moth		*Egira perlubens* (Grote)	(チョウ目、ヤガ科) 新北区
brown yamfly		*Drina donina* (Hewitson)	(チョウ目、シジミチョウ科) 東洋区
brownies	アシナガシジミ属	*Miletus*	(チョウ目、シジミチョウ科) の昆虫の総称
brownish cracker		*Hamadryas iphthime* (Bates)	(チョウ目、タテハチョウ科) 新北区、新熱帯区。*H. i. joannae* Jenkins も同英名
brownish gelechid	カバイロキバガ	*Carbatina picrocarpa* Meyrick	(チョウ目、キバガ科) 日本、東洋区
brownish leaf beetle	トビサルハムシ	*Trichochrysea japana japana* (Motschulsky)	(コウチュウ目、ハムシ科) 日本、旧北区、東洋区
brownish long-horned weevil	ツチイロヒゲボソゾウムシ	*Phyllobius incomptus* Sharp	(コウチュウ目、ゾウムシ科) 日本
brownish long-necked leaf beetle	アカクビナガハムシ	*Lilioceris subpolita* (Motschulsky)	(コウチュウ目、ハムシ科) 日本
brownish stink bug	チャイロカメムシ	*Eurygaster testudinaria* (Geoffroy)	(カメムシ目、キンカメムシ科) 日本、旧北区
brownish yellow cidaria			broken-barred carpet を見よ
browns (1)	クロヒカゲ属	*Lethe*	(チョウ目、タテハチョウ科) の昆虫の総称
browns (2)	クロコノマチョウ属	*Melanitis*	(チョウ目、タテハチョウ科) の昆虫の総称
browns			satyrs を見よ
browns			ringlets (1) を見よ
Brown's Calephelis		*Calephelis browni* McAlpine	(チョウ目、シジミタテハ科) 新熱帯区
Brown's grayling		*Pseudochazara amymone* Brown	(チョウ目、タテハチョウ科) 旧北区
Brownsville meadow katydid		*Conocephalus resacensis* Rehn et Hebard	(バッタ目、キリギリス科) 新北区

英名	和名	学名	所属、分布、ほか
Brownsville tree cricket		*Neoxabea formosa* (Walker)	(バッタ目、コオロギ科) 新北区
Bruce's measuring worm			Bruce's spanworm を見よ
Bruce's spanworm		*Operophtera bruceata* (Hulst)	(チョウ目、シャクガ科) 新北区。雌の翅は退化。Bruce spanworm, Bruce spanworm moth ともいう
Bruce's tiger		*Neoarctia brucei* (H. Edwards)	(チョウ目、ヒトリガ科) 新北区
bruchid egg parasite		*Uscana semifumipennis* Girault	(ハチ目、タマゴコバチ科) 大洋区
bruchids		Bruchinae	(コウチュウ目、ハムシ科) の昆虫の総称
bruised skipper		*Zera hyacinthinius hyacinthinus* (Mabille)	(チョウ目、セセリチョウ科) 新熱帯区
bruner slantfaced grasshopper		*Bruneria brunnea* (Thomas)	(バッタ目、バッタ科) 新北区
bruner spurthroated grasshopper		*Melanoplus bruneri* Scudder	(バッタ目、バッタ科) 新北区
Bruner's longwing		*Capnobotes bruneri* Scudder	(バッタ目、キリギリス科) 新北区
Bruner's shieldback		*Pediodectes bruneri* (Caudell)	(バッタ目、キリギリス科) 新北区
brush flitter		*Hyarotis microsticta* (Wood-Mason et de Nicéville)	(チョウ目、セセリチョウ科) 東洋区
brush-footed butterflies	タテハチョウ科	Nymphalidae	(チョウ目) の昆虫の総称　世界に 8,000 種以上
brushfoots		Nymphalinae	(チョウ目、タテハチョウ科) の昆虫の総称
brushfoots			brush-footed butterflies を見よ
brushlegged mayflies	ヒトリガカゲロウ科	Oligoneuriidae	(カゲロウ目) の昆虫の総称
brush-winged honey			stored nut moth を見よ
Brussels lace		*Cleorodes lichenaria* (Hufnagel)	(チョウ目、シャクガ科) 旧北区
bryony ladybird		*Henosepilachna argus* (Geoffory)	(コウチュウ目、テントウムシ科) 旧北区
bryony leafminer	ナスハモグリバエ	*Liriomyza bryoniae* (Kaltenbach)	(ハエ目、ハモグリバエ科) 日本．旧北区。ナスなど多くの作物を加害
Bubaris skipper		*Ouleus bubaris* (Godman et Salvin)	(チョウ目、セセリチョウ科) 新熱帯区
Bubobon skipperling		*Dalla bubobon* (Dyar)	(チョウ目、セセリチョウ科) 新熱帯区
Buchanan frosted green weevil			arborvitae weevil を見よ
buck moth		*Hemileuca maia* (Drury)	(チョウ目、ヤママユガ科) 新北区
buck moths		Hemileucinae	(チョウ目、ヤママユガ科) の昆虫の総称
Buckell's grig			Buckell's hump-winged cricket を見よ
Buckell's hump-winged cricket		*Cyphoderris buckelli* Hebard	(バッタ目、Prophalangopsidae) 新北区
buckeye	アメリカタテハモドキ	*Junonia coenia* (Hübner)	(チョウ目、タテハチョウ科) 新北区。buckeyes butterfly ともいう

英名	和名	学名	所属、分布、ほか
buckeye butterflies			buckeyes を見よ
buckeye butterflies			pansies を見よ
buckeye petiole borer moth		*Zeiraphera claypoleana* (Riley)	(チョウ目、ハマキガ科) 新北区
buckeyes	タテハモドキ属	*Junonia*	(チョウ目、タテハチョウ科) の昆虫の総称
Buck's-horn groundling		*Scrobipalpa samadensis* (Paffenzeller)	(チョウ目、キバガ科) 旧北区
buckthorn aphid		*Aphis mammulata* Ghimingham et Hille Ris Lambers	(カメムシ目、アブラムシ科) 旧北区
buckthorn aphid			buckthorn-potato aphid を見よ
buckthorn dusky wing			pacuvius duskywing を見よ
buckthorn pigmy		*Stigmella catharticella* (Stainton)	(チョウ目、モグリチビガ科) 旧北区
buckthorn sucker		*Trichochermes walkeri* (Foerster)	(カメムシ目、トガリキジラミ科) 旧北区
buckthorn-potato aphid		*Aphis nasturtii* Kaltenbach	(カメムシ目、アブラムシ科) 全北区、東洋区
buckwheat blue			square spotted blue を見よ
buckwheat borer moth		*Synanthedon polygoni* (Edwards)	(チョウ目、スカシバガ科) 新北区
buckwheat cutworm	シロスジアオヨトウ	*Trachea atriplicis gnoma* (Butler)	(チョウ目、ヤガ科) 日本
bud curculio			apple twig cutter を見よ
bud deformer			gooseberry bud midge を見よ
bud leaf-roller		*Rhopobota ustomaculana* Curtis	(チョウ目、ハマキガ科) 旧北区
bud midge			citrus gall midge を見よ
bud moth			arrowhead moth を見よ
bud moth			eyespotted bud moth を見よ
bud weevil			apple bud weevil を見よ
bud weevils			blossom weevils を見よ
bud worm			corn earworm (2) を見よ　幼虫の英名
budded lappet		*Eucraera gemmata* Distant	(チョウ目、カレハガ科) エチオピア区
buddleia budworm moth		*Pyramidobela angelarum* Keifer	(チョウ目、マルハキバガ科) 新北区
budwing mantis		*Parasphendale agrionina* Wermer	(カマキリ目、カマキリ科) エチオピア区
budwing mantis		*Parasphendale affinis* Wermer	(カマキリ目、カマキリ科) エチオピア区
budworm			eye-spotted bud moth を見よ　幼虫の英名
budworm moths	タバコガ亜科	Heliothinae	(チョウ目、ヤガ科) の昆虫の総称
buff arches		*Habrosyne pyritoides* (Hufnagel)	(チョウ目、トガリバガ科) 旧北区。日本亜種は *H. p. derasoides* (Butler) アヤトガリバ
buff ermine		*Spilosoma lutea* (Hufnagel)	(チョウ目、ヒトリガ科) 旧北区
buff ermine			white ermine moth を見よ

英名	和名	学名	所属、分布、ほか
buff ermine moth			buff ermine を見よ
buff footman		*Eilema depressa* (Esper)	(チョウ目、ヒトリガ科) 旧北区。日本亜種は *E. d. pavescens* (Butler) ムジホソバ
buff rough-winged button		*Acleris permutana* (Duponchel)	(チョウ目、ハマキガ科) 旧北区
buff-tailed bumble bee		*Bombus terrestris* (Linnaeus)	(ハチ目、ミツバチ科) 旧北区
buff tip		*Phalera bucephala* (Linnaeus)	(チョウ目、シャチホコガ科) 旧北区。buff-tip とも記す
buff-tip moth		*Sceliodes cordalis* Doubleday	(チョウ目、メイガ科) 豪州区
buff-tip moth			buff tip を見よ
buff-tipped Phaneta moth		*Phaneta ochroterminana* (Kearfott)	(チョウ目、ハマキガ科) 新北区
buff-tipped skipper			brown-tipped skipper を見よ
buffalo beetle			buffalo carpet beetle を見よ
buffalo beetles			carpet beetles を見よ
buffalo bug			buffalo carpet beetle を見よ
buffalo carpet beetle		*Anthrenus scrophulariae* (Linnaeus)	(コウチュウ目、カツオブシムシ科) 全北区
buffalo carpet beetle			varied carpet beetle を見よ
buffalo flower beetle		*Orphilus subnitidus* LeConte	(コウチュウ目、カツオブシムシ科) 新北区
buffalo fly		*Haematobia exigua* (de Meijere)	(ハエ目、イエバエ科) 豪州区
buffalo fly			horn fly (1) を見よ
buffalo gnat (1)		*Cnephia pecuarum* (Riley)	(ハエ目、ブユ科) 新北区
buffalo gnat (2)		*Simulium venustum* Say	(ハエ目、ブユ科) 新北区
buffalo gnat (3)		*Prosimulium fulvum* (Coquillett)	(ハエ目、ブユ科) 新北区
buffalo gnats			black flies (1) (2) を見よ
buffalo louse	スイギュウジラミ	*Haematopinus tuberculatus* (Burmeister)	(シラミ目、ケモノジラミ科) 日本、汎世界。スイギュウに寄生
buffalo louse		*Haematopinus bufalieuropaei* (Latreille)	(シラミ目、ケモノジラミ科) 旧北区
buffalo moth		*Parapamea buffaloensis* (Grote)	(チョウ目、ヤガ科) 新北区
buffalo treehopper		*Stictocephala bisonia* Kopp et Yonke	(カメムシ目、ツノゼミ科) 全北区。体長 7〜9 mm
buffalo treehopper		*Stictocephala bubalus* (Fabricius)	(カメムシ目、ツノゼミ科) 新北区
buffalo treehoppers		*Stictocephala*	(カメムシ目、ツノゼミ科) の昆虫の総称
buffalograss seed caterpillar		*Mampava rhodoneura* (Turner)	(チョウ目、メイガ科) 豪州区
buffalograss webworm		*Surattha indentella* Kearfott	(チョウ目、メイガ科) 新北区
buffs		*Baliochila*	(チョウ目、シジミチョウ科) の昆虫の総称

英名	和名	学名	所属、分布、ほか
buffs		*Cnodontes*	(チョウ目、シジミチョウ科) の昆虫の総称
buffs		*Teriomima*	(チョウ目、シジミチョウ科) の昆虫の総称
buffy Dichomeris moth		*Dichomeris aleatrix* Hodges	(チョウ目、キバガ科) 新北区
bug			カメムシ目の昆虫。カブトムシ、ナンキンムシにも使われる (英国)。bugs を見よ
bug		*Leptoglossus membranaceus* Fabricius	(カメムシ目、ヘリカメムシ科) エチオピア区。カンキツ害虫
bug		*Leptoglossus stigma* (Herbst)	(カメムシ目、ヘリカメムシ科) 新北区、新熱帯区。カンキツ害虫
bug			bed bug (1) を見よ
bug			citron bug を見よ
bugloss spear-wing			Viper's bugloss spearwing を見よ
bugs			昆虫を一般に bug という。homopterous insects と true bugs を見よ
bugs			true bugs を見よ
bugs			bed bugs を見よ
bugs			psyllines を見よ
bugus yucca moths			yucca moths (3) を見よ
bulb and potato aphid		*Rhopalosiphoninus latysiphon* (Davidson)	(カメムシ目、アブラムシ科) 旧北区、豪州区
bulb aphid			bulb and potato aphid を見よ
bulb fly			narcissus bulb fly を見よ
bulb fly			onion bulb fly を見よ
bulb snout beetles		*Brachycerus*	(コウチュウ目、ゾウムシ科) の昆虫の総称
Bulgarian ringlet		*Erebia orientalis* Elwes	(チョウ目、タテハチョウ科) 旧北区
bull adder			dragonflies (1) を見よ
bull ant			bulldog ant (1) を見よ
bull ants			bulldog ants を見よ
bulldog ant		*Myrmecia brevinoda* Forel	(ハチ目、アリ科) 豪州区。女王は 2.5 cm の大型種
bulldog ant		*Myrmecia nigriceps* Mayr	(ハチ目、アリ科) 豪州区
bulldog ant (1)	ブルドッグアリ	*Myrmecia gulosa* (Fabricius)	(ハチ目、アリ科) 豪州区
bulldog ant (2)		*Myrmecia forficata* Smith	(ハチ目、アリ科) 旧北区、豪州区
bulldog ants		*Myrmecia*	(ハチ目、アリ科) の昆虫の総称　inchmen jumper ants, sergeant ants ともいわれる。強力な大腮と刺針をもつ
bulldog ants			keleps を見よ
bulldog flies			horse-flies (2) を見よ
bulldogs			keleps を見よ　豪州での英名
bullet ant	サシハリアリ	*Paraponera clavata* (Fabricius)	(ハチ目、アリ科) 新熱帯区
bulloak jewel		*Hypochrysops piceata* (Kerr, Macqueen et Sands)	(チョウ目、シジミチョウ科) 豪州区

英名	和名	学名	所属、分布、ほか
bullocks heart fruit borer		*Anonaepestis bengalella* Ragonot	(チョウ目、メイガ科) 東洋区
Bull's eye moth			Io moth を見よ
bullseye			Io moth を見よ
bullseye borer		*Tryphocaria acanthocera* (Macleay)	(コウチュウ目、カミキリムシ科) 豪州区
bullstang			dragonflies (1) を見よ
bull-ting			dragonflies (1) を見よ
bulrush wainscot		*Nonagria arundinis* (Fabricius)	(チョウ目、ヤガ科) 旧北区
bulrush wainscot		*Nonagria typhae* (Thunberg)	(チョウ目、ヤガ科) 旧北区
bumblebee metalmark			Felder's pigmy を見よ
bumble bee moth		*Hemaris diffinis* (Boisduval)	(チョウ目、スズメガ科) 新北区
bumblebee robberfly		*Laphria flava* (Linnaeus)	(ハエ目、ムシヒキアブ科) 旧北区
bumble bee robber fly		*Mallophora fautrix* Osten Sacken	(ハエ目、ムシヒキアブ科) 新北区
bumble bee wax moth			bee moth を見よ
bumble bees (1)	マルハナバチ属	*Bombus*	(ハチ目、ミツバチ科) の昆虫の総称　Apinae, Bombinae の種にも使用される
bumble bees (2)		Bombini	(ハチ目、ミツバチ科) の昆虫の総称
bumble bees (3)		Bombinae	(ハチ目、ミツバチ科) の昆虫の総称
bumble flower beetle		*Euphoria inda* (Linnaeus)	(コウチュウ目、コガネムシ科)、新北区、新熱帯区
Bumelia fruit fly		*Anastrepha pallens* (Coquillett)	(ハエ目、ミバエ科) 新北区
Bumelia leafworm moth		*Lactura popula* (Hübner)	(チョウ目、マダラガ科) 新北区
Bumelia webworm moth		*Urodus parvula* (Edwards)	(チョウ目、スガ科) 新北区
bunchberry leaffolder moth		*Olethreutes connectum* (McDunnough)	(チョウ目、ハマキガ科) 新北区
bunch-grass skipper		*Problema byssus* Edwards	(チョウ目、セセリチョウ科) 新北区
Bundaberg canegrub		*Lepidiota crinita* Brenske	(コウチュウ目、コガネムシ科) 豪州区
bungee caterpillar			black-olive caterpillar moth を見よ
buprestid beetles			flat-headed beetles を見よ
Buquet's vagrant	ウラキウスアオシロチョウ	*Nepheronia buquetii* (Boisduval)	(チョウ目、シロチョウ科) エチオピア区
Burchell's brown		*Pseudonympha hippia* (Cramer)	(チョウ目、タテハチョウ科) エチオピア区
Burchell's crescent		*Telenassa teletusa burchelli* Moulton	(チョウ目、タテハチョウ科) 新熱帯区
burdock borer		*Papaipema cataphracta* (Grote)	(チョウ目、ヤガ科) 新北区。burdock borer moth ともいう

英名	和名	学名	所属、分布、ほか
burdock bug	ヘリカメムシ	*Coreus marginatus orientalis* Kiritschenko	（カメムシ目、ヘリカメムシ科）日本、旧北区
burdock fruit fly	キイロホシミバエ	*Sitarea vibrissata* (Coquillett)	（ハエ目、ミバエ科）日本
burdock Japanese weevil	ハスジゾウムシ	*Cleonus japonicus* Faust	（コウチュウ目、ゾウムシ科）日本
burdock leaf roller	ギンボシハマキモドキ	*Tebenna bjerkandrella* (Thunberg)	（チョウ目、ハマキモドキガ科）日本、旧北区
burdock leaf skeletonizer	ゴボウハマキモドキ	*Tebenna micalis* (Mann)	（チョウ目、ハマキモドキガ科）日本。ゴボウ害虫
burdock leafminer	ゴボウハモグリバエ	*Phytomyza lappae* Goureau	（ハエ目、ハモグリバエ科）日本、旧北区。ゴボウ害虫
burdock long-horned aphid	ゴボウヒゲナガアブラムシ	*Uroleucon gobonis* (Matsumura)	（カメムシ目、アブラムシ科）日本、旧北区、東洋区
burdock pyralid	ゴマダラメイガ	*Myelois cribrella* (Hübner)	（チョウ目、メイガ科）日本、旧北区
burdock root miner	ゴボウネモグリバエ	*Ophiomyia lappivora* (Koizumi)	（ハエ目、ハモグリバエ科）日本。ゴボウ害虫
burdock seedhead moth			dingy straw neb を見よ
burdock silver-striped phalonid	エダオビホソハマキ	*Aethes rubigana* (Treitschke)	（チョウ目、ホソハマキガ科）日本、旧北区
burdock weevil	ゴボウゾウムシ	*Larinus latissimus* Roelofs	（コウチュウ目、ゾウムシ科）日本
burdock worm	ゴボウトガリヨトウ	*Gortyna fortis* (Butler)	（チョウ目、ヤガ科）日本、旧北区。ゴボウ害虫
Burgundy bluet		*Enallagma dubium* Root	（トンボ目、イトトンボ科）新北区
Burmeister mantid		*Orthodera burmeisteri* Wood-Mason	（カマキリ目、カマキリ科）東洋区、新北区
Burmeister's glider		*Tramea basilaris burmeisteri* Kirby	（トンボ目、トンボ科）東洋区
Burmese banded peacock			moss peacock を見よ
Burmese bushblue		*Arhopala birmana* (Moore)	（チョウ目、シジミチョウ科）東洋区
Burmese Cerulean		*Jamides philatus* (Snellen)	（チョウ目、シジミチョウ科）東洋区
Burmese dart		*Potanthus juno* (Evans)	（チョウ目、セセリチョウ科）東洋区
Burmese lascar		*Lasippa heliodore* (Fabricius)	（チョウ目、タテハチョウ科）東洋区。*L. h. dorelia* Butler も同英名
Burmese lascar (1)		*Lasippa tiga* (Moore)	（チョウ目、タテハチョウ科）東洋区
Burmese Raven		*Papilio mahadeva* Moore	（チョウ目、アゲハチョウ科）東洋区
Burmese sunbeam		*Curetis saronis* Moore	（チョウ目、シジミチョウ科）東洋区
burnet companion		*Euclidia glyphica* (Linnaeus)	（チョウ目、ヤガ科）旧北区
burnet moth	ベニモンマダラガ	*Zygaena niphona* Butler	（チョウ目、マダラガ科）日本、旧北区
burnet moths			leaf skeletonizer moths (2) を見よ
burnet pigmy		*Stigmella poterii* (Stainton)	（チョウ目、モグリチビガ科）旧北区
burnets			leaf skeletonizer moths (2) を見よ
burnets and foresters			leaf skeletonizer moths (2) を見よ

英名	和名	学名	所属、分布、ほか
Burney's longwing		*Heliconius burneyi* (Hübner)	(チョウ目、タテハチョウ科) 新熱帯区
burnished brass	エゾヒサゴキンウワバ	*Diachrysia chrysitis* (Linnaeus)	(チョウ目、ヤガ科) 日本、旧北区。成虫は burnished brass moth
burnished pigmy		*Stigmella aeneofasciella* (Herrich-Schäffer)	(チョウ目、モグリチビガ科) 旧北区
Burns' scarlet-eye		*Cephise nuspesez* Burns	(チョウ目、セセリチョウ科) 新熱帯区
burnt pine longhorn		*Arhopalus ferus* (Mulsant)	(コウチュウ目、カミキリムシ科) 旧北区、豪州区。burnt pine longhorn beetle ともいう
burnt tip Grammoptera		*Grammoptera ustulata* (Schaller)	(コウチュウ目、カミキリムシ科) 旧北区
burrel flies			horse flies (2) を見よ
burrel-flies			clegs を見よ
burrelfly			cattle warble fly (1)(2) を見よ
burren green		*Calamia virens* Linnaeus	(チョウ目、ヤガ科) 旧北区
burren green (1)		*Calamia tridens* (Hufnagel)	(チョウ目、ヤガ科) 旧北区
Burro-weed mealybug		*Spirococcus ventralis* McKenzie	(カメムシ目、コナカイガラムシ科) 新北区
burrow bugs			ground bugs を見よ　南アフリカでの英名
burrowbush borer moth		*Hymenoclea palmii* (Beutenmüller)	(チョウ目、スカシバガ科) 新北区
burrower bugs			ground bugs を見よ
burrowing and tusked mayflies	モンカゲロウ上科	Ephemeroidea	(カゲロウ目) の昆虫の総称
burrowing bees			andrenid bees を見よ
burrowing bees			andrenas を見よ
burrowing bugs			ground bugs を見よ
burrowing cockroach		*Pycnoscelus indicus* (Fabricius)	(ゴキブリ目、ブラベルスゴキブリ科) 大洋区
burrowing cockroach			Surinam cockroach を見よ
burrowing flea			chigoe flea を見よ
burrowing ground beetle		*Geopinus incrassatus* Dejean	(コウチュウ目、オサムシ科) 新北区。砂中にもぐる
burrowing mayflies	モンカゲロウ科	Ephemeridae	(カゲロウ目) の昆虫の総称
burrowing mayflies			burrowing and tusked mayflies を見よ
burrowing mayfly		*Ephoron virgo* (Olivier)	(カゲロウ目、オオシロカゲロウ科) 旧北区
burrowing mayfly		*Hexagenia bilineata* (Say)	(カゲロウ目、モンカゲロウ科) 新北区
burrowing roach			Surinam cockroach を見よ
burrowing wasp		*Philanthus ventilabris* Fabricius	(ハチ目、アナバチ科) 新北区
burrowing wasps			sandloving wasps を見よ
burrowing wasps			mason wasps を見よ
burrowing water beetles	コツブゲンゴロウ科	Noteridae	(コウチュウ目) の昆虫の総称

英名	和名	学名	所属、分布、ほか
burrowing water beetles		*Noterus*	(コウチュウ目、コツブゲンゴロウ科) の昆虫の総称
burrowing water beetles		*Hydrocanthus*	(コウチュウ目、コツブゲンゴロウ科) の昆虫の総称
burrowing water beetles		*Suphicellus*	(コウチュウ目、コツブゲンゴロウ科) の昆虫の総称
burrowing webworm moths		Acrolophidae	(チョウ目) の昆虫の総称
burrowing webworms			burrowing webworm moths を見よ
Buru opalescent birdwing	ブルーキシタアゲハ	*Troides chimaera prattorum* Joicey et Talbot	(チョウ目、アゲハチョウ科) 東洋区
burying beetle			common burying beetle を見よ
burying beetles (1)	シデムシ科	Silphidae	(コウチュウ目) の昆虫の総称
burying beetles (2)		*Necrophorus*	(コウチュウ目、シデムシ科) の昆虫の総称
burying manntis		*Sphodropoda tristis* (Saussure)	(カマキリ目、カマキリ科) 豪州区
bush and round headed katydids			bush katydids (1) を見よ
bush beauty		*Paralethe dendrophilus* (Trimen)	(チョウ目、タテハチョウ科) エチオピア区
bush blue		*Cacyreus lingeus* (Stoll)	(チョウ目、シジミチョウ科) エチオピア区
bush browns		*Bicyclus*	(チョウ目、タテハチョウ科) の昆虫の総称
bushbrowns	コジャノメ属	*Mycalesis*	(チョウ目、タテハチョウ科) の昆虫の総称
bush Charaxes	カバイロフタオチョウ	*Charaxes achaemenes* C. et R. Felder	(チョウ目、タテハチョウ科) エチオピア区
bush cicadas			dog-day cicadas を見よ
bush clover aphid			potato aphid (2) を見よ
bush-clover globular treehopper	ワキジロマルツノゼミ	*Gargara desmodiuma* Kato	(カメムシ目、ツノゼミ科) 日本
bush-clover leaf-cut weevil	ハギリオトシブミ	*Euops lespedezae* Sharp	(コウチュウ目、オトシブミ科) 日本
bush-clover noctuid	カクモンキシタバ	*Chrysorithrum amatum* (Bremer et Grey)	(チョウ目、ヤガ科) 日本、旧北区
bush clover pyralid	オオウスベニトガリメイガ	*Endotricha icelusalis* (Walker)	(チョウ目、メイガ科) 日本、旧北区
bush cricket		*Pholidoptera griseoaptera* (De Geer)	(バッタ目、キリギリス科) 旧北区
bush cricket			saddle-backed bush cricket を見よ
bush cricket			common garden katydid を見よ
bush-crickets		Phaneropteridae	(バッタ目) の昆虫の総称
bush crickets (1)	キリギリス亜目	Ensifera	(バッタ目) の昆虫の総称
bush crickets (2)		Eneopterinae	(バッタ目、コオロギ科) の昆虫の総称
bush crickets (3)	クサヒバリ科	Trigonidiidae	(バッタ目) の昆虫の総称
bush crickets (4)		*Barbitistes*	(バッタ目、キリギリス科) の昆虫の総称

英名	和名	学名	所属、分布、ほか
bush crickets and long-horned grasshoppers			long-horned grasshoppers (1) を見よ
bush fly		*Musca vetustissima* Walker	(ハエ目、イエバエ科) 豪州区
bush gad-fly		*Hybomitra sexfasciata* (Hine)	(ハエ目、アブ科) 豪州区
bush gad-fly		*Scaptia adrel* (Walker)	(ハエ目、アブ科) 豪州区
bush hopper	チビキマダラセセリ	*Ampittia dioscorides* (Fabricius)	(チョウ目、セセリチョウ科) 東洋区。*A. d. camertes* (Hewitson) も同英名
bush hoppers		*Ampittia*	(チョウ目、セセリチョウ科) の昆虫の総称
bush katydid		*Scudderia curvicauda* (De Geer)	(バッタ目、キリギリス科) 新北区
bush katydids (1)	ツユムシ亜科	Phaneropterinae	(バッタ目、キリギリス科) の昆虫の総称
bush katydids		*Scudderia, Microcentrum*	(バッタ目、キリギリス科) の昆虫の総称
bush kite			forest swallowtail を見よ
bush locust		*Phymateus leprosus* Fabricius	(バッタ目、オンブバッタ科) エチオピア区。カンキツ害虫
bush mantis			giant African mantis を見よ
bush mosquitoes			field mosquitoes を見よ　南アフリカでの英名
bush nightfighter		*Artitropa erinnys erinnys* (Trimen)	(チョウ目、セセリチョウ科)
bush night-fighter			white dart を見よ
bush scarlet		*Axiocerses amanga* (Westwood)	(チョウ目、シジミチョウ科) エチオピア区
bushtailed caddisflies	ケトビケラ科	Sericostomatidae	(トビケラ目) の昆虫の総称
bushveld Charaxes			bush Charaxes を見よ
bushveld emperor			bush Charaxes を見よ
bushveld orange tip	フチナシツマアカシロチョウ	*Colotis pallene* (Hopffer)	(チョウ目、シロチョウ科) エチオピア区
bushveld ringlet		*Ypthima impura paupera* Ungemach	(チョウ目、タテハチョウ科) エチオピア区。impure ringlet を参照
bushveld sandman		*Spialia colotes transvaaliae* (Trimen)	(チョウ目、セセリチョウ科) エチオピア区
bushy knot-horn		*Acrobasis tumidana* (Denis et Schiffermüller)	(チョウ目、メイガ科) 旧北区
Butler's alpine			common alpine を見よ
Butler's brahmin	アフリカイボタガ	*Dactyloceras swanzii* (Butler)	(チョウ目、イボタガ科) エチオピア区
Butler's ciliate blue			pale hairtail を見よ
Butler's dwarf	チベットシロチョウ	*Baltia butleri* Alpheraky	(チョウ目、シロチョウ科) 東洋区
Butler's metalmark		*Lemonias caliginea* (Butler)	(チョウ目、シジミタテハ科) 新熱帯区
Butler's Perisama		*Perisama tristrigosa* Butler	(チョウ目、タテハチョウ科) 新熱帯区
Butler's ringlet		*Cissia terrestris* Butler	(チョウ目、タテハチョウ科) 新熱帯区
Butler's satyr		*Cissia palladia* (Butler)	(チョウ目、タテハチョウ科) 新熱帯区
Butler's satyr		*Pseudodebis zimri* (Butler)	(チョウ目、タテハチョウ科) 新熱帯区

英名	和名	学名	所属、分布、ほか
Butler's short-tail skipper		*Polythrix hirtius* (Butler)	(チョウ目、セセリチョウ科) 新熱帯区
Butler's skipper		*Mnasilus allubita* (Butler)	(チョウ目、セセリチョウ科) 新熱帯区
butterbur		*Hydraecia petasitis* Doubleday	(チョウ目、ヤガ科) 旧北区
butterbur ear			butterbur を見よ
butterbur flower miner	フキハモグリバエ	*Phytomyza hasegawai* Sasakawa	(ハエ目、ハモグリバエ科) 日本。幼虫がフキの葉にもぐる
butterbur moth			butterbur を見よ
butterbur plume moth	フキトリバ	*Pselnophorus vilis* (Butler)	(チョウ目、トリバガ科) 日本、旧北区。フキ害虫
butterbur sawfly	フキシマハバチ	*Pachyprotasis fukii* Okutani	(ハチ目、ハバチ科) 日本
butterbur sawfly	フキクロハバチ	*Tenthredo bipunctula malaisei* (Takeuchi)	(ハチ目、ハバチ科) 日本
butterflies	蝶類	Rhopalocera	(チョウ目) セセリチョウ上科とアゲハチョウ上科の昆虫
butterflies and moths	チョウ目	Lepidoptera	蝶蛾類で世界に約 120,000 種
butterfly chrysalis chalcid		*Brachymeria ovata* (Say)	(ハチ目、アシブトコバチ科) 新北区
butterfly moths			sun moths (1) を見よ
butternut curculio		*Conotrachelus juglandis* LeConte	(コウチュウ目、ゾウムシ科) 新北区
button beetle			button weevil を見よ
buttonbush leafminer moth		*Mompha cephalonthiella* (Chambers)	(チョウ目、カザリバガ科) 新北区
buttonbush owlet			lost owlet moth を見よ
button-gall			spangle gall wasp の虫えい
button-top midge			willow button-top midge を見よ
button weevil		*Coccotrypes dactyliperda* (Fabricius)	(コウチュウ目、キクイムシ科) 旧北区
buttoned moth		*Hypena rostralis* (Linnaeus)	(チョウ目、ヤガ科) 旧北区
buttoned snout			buttoned moth を見よ
Butt's hawk		*Hoplistopus butti* Rothschild et Jordan	(チョウ目、スズメガ科) エチオピア区
Buxton's hairstreak		*Hypolycaena buxtoni* Hewitson	(チョウ目、シジミチョウ科) エチオピア区
buzzard louse			hawk louse を見よ
buzzers			midges を見よ
Byssus skipper			bunch-grass skipper を見よ

英名	和名	学名	所属、分布、ほか
C			
cabbage aphid	ダイコンアブラムシ	*Brevicoryne brassicae* (Linnaeus)	（カメムシ目、アブラムシ科）日本、汎世界。ダイコン、アブラナ科野菜の害虫。cabbage aphis ともいう
cabbage aphid			green peach aphid を見よ
cabbage aphid			turnip aphid (1) を見よ
cabbage aphis			green peach aphid を見よ
cabbage armyworm	ヨトウガ	*Mamestra brassicae* (Linnaeus)	（チョウ目、ヤガ科）日本、旧北区。多くの作物の著名害虫
cabbage borer			cabbage webworm (1) を見よ
cabbage budworm		*Hellula phidilealis* (Walker)	（チョウ目、メイガ科）新北区、新熱帯区、大洋区。成虫は cabbage budworm morth
cabbage bug (1)	ナガメ	*Eurydema rugosa* Motschulsky	（カメムシ目、カメムシ科）日本、旧北区。アブラナ科野菜害虫
cabbage bug (2)		*Eurydema oleraceum* (Linnaeus)	（カメムシ目、カメムシ科）旧北区。アブラナ科野菜害虫
cabbage butterfly			common white を見よ　*P. rapae* Linnaeus の米国での英名
cabbage-centre grub		*Hellula hydralis* Guenée	（チョウ目、メイガ科）豪州区（ニューギニア）
cabbage-centre grub			cabbage webworm (1) を見よ
cabbage cluster caterpillar			cabbage head caterpillar を見よ
cabbage crowngall fly			swede midge を見よ
cabbage curculio		*Ceutorhynchus rapae* Gyllenhal	（コウチュウ目、ゾウムシ科）新北区
cabbage dot			cabbage armyworm を見よ
cabbage flea beetle (1)	ムモンキスジノミハムシ	*Phyllotreta atra* (Fabricius)	（コウチュウ目、ハムシ科）日本、旧北区
cabbage flea beetle (2)	ナトビハムシ	*Psylliodes punctifrons* Baly	（コウチュウ目、ハムシ科）日本、東洋区。アブラナ科野菜害虫
cabbage flea beetles		*Phyllotreta*	（コウチュウ目、ハムシ科）の昆虫の総称
cabbage gall midge	ダイコンタマバエ	*Dasineura brassicae* (Winnertz)	（ハエ目、タマバエ科）日本、旧北区
cabbage gall midge			swede midge を見よ
cabbage gall weevil		*Ceutorhynchus sulcicollis* Paykull	（コウチュウ目、ゾウムシ科）旧北区
cabbage gall weevil			turnip gall weevil を見よ
cabbage head caterpillar	ケブカノメイガ	*Crocidolomia pavonana* (Fabricius)	（チョウ目、メイガ科）日本、東洋区、大洋区、豪州区。アブラナ科野菜害虫
cabbage leaf miner		*Phytomyza rufipes* Meigen	（ハエ目、ハモグリバエ科）旧北区
cabbage leaf miner			diamondback moth を見よ
cabbage leafminer			crucifer leafminer を見よ　豪州での英名
cabbage leafminer			vegetable leafminer を見よ　米国での英名
cabbage leaf-tier			cabbage webworm (1) を見よ

英名	和名	学名	所属、分布、ほか
cabbage looper (1)	イラクサギンウワバ	*Trichoplusia ni* (Hübner)	(チョウ目、ヤガ科) 日本、東洋区、全北区、エチオピア区、豪州区。亜種 *T. n. brassicae* Riley も同英名。ナス、アブラナ科野菜害虫。成虫は cabbage looper moth
cabbage looper (2)		*Plusia brassicae* (Linnaeus)	(チョウ目、ヤガ科) 全北区
cabbage looper			silver Y moth (1) を見よ　北米での幼虫の英名
cabbage looper moth			cabbage looper (1) を見よ
cabbage louse			cabbage aphid を見よ
cabbage maggot			cabbage root fly (2)(3) を見よ　米国での幼虫の英名
cabbage midge		*Contarinia torquens* Meijere	(ハエ目、タマバエ科) 旧北区
cabbage midge			swede midge を見よ
cabbage moth			cabbage armyworm を見よ
cabbage moth			diamondback moth を見よ
cabbage palm caterpillar			palmetto borer moth を見よ
cabbage palm moth		*Epiphryne verriculata* Felder	(チョウ目、シャクガ科) 豪州区
cabbage plant-louse			cabbage aphid を見よ
cabbage Plutella			diamondback moth を見よ
cabbage root fly (1)	ダイコンバエ	*Delia floralis* (Fallén)	(ハエ目、ハナバエ科) 日本、全北区。ダイコンなどアブラナ科野菜害虫
cabbage root fly (2)		*Hylemya brassicae* (Wiedemann)	(ハエ目、ハナバエ科) 旧北区
cabbage root fly (3)		*Delia radicum* (Linnaeus)	(ハエ目、ハナバエ科) 全北区
cabbage root maggot			cabbage root fly (2) を見よ
cabbage sawfly (1)	セグロカブラハバチ	*Athalia infumata* (Marlatt)	(ハチ目、ハバチ科) 日本、旧北区。アブラナ科野菜害虫
cabbage sawfly (2)	ニホンカブラハバチ	*Athalia japonica* (Klug)	(ハチ目、ハバチ科) 日本、東洋区。アブラナ科野菜害虫
cabbage sawfly (3)	カブラハバチ	*Athalia rosae ruficornis* Jakovlev	(ハチ目、ハバチ科) 日本、東洋区。アブラナ科野菜害虫
cabbage seed-pod curculio			cabbage seed weevil を見よ　米国での英名
cabbage seedpod weevil			cabbage seed weevil を見よ　カナダでの英名
cabbage seedstalk curculio			cabbage stem weevil を見よ
cabbage seed-stalk weevil			cabbage stem weevil を見よ
cabbage seed-stalked curculio			cabbage stem weevil を見よ　米国での英名
cabbage seed weevil		*Ceutorhynchus assimilis* (Paykull)	(コウチュウ目、ゾウムシ科) 全北区
cabbage semi-looper			cabbage looper (1) を見よ
cabbage shoot weevil			cabbage seed weevil を見よ

英名	和名	学名	所属、分布、ほか
cabbage snow-fly			cabbage whitefly (1) を見よ
cabbage stem flea beetle		*Psylliodes chrysocephala* (Linnaeus)	（コウチュウ目、ハムシ科）旧北区
cabbage stem weevil		*Ceutorhynchus quadridens* (Panzer)	（コウチュウ目、ゾウムシ科）全北区
cabbage stink bug			cabbage bug (1) を見よ
cabbage thrips		*Thrips angusticeps* Uzel	（アザミウマ目、アザミウマ科）旧北区
cabbage tree emperor moth			common emperor を見よ
cabbage web moth			diamondback moth を見よ
cabbage webworm		*Hellula rogatalis* (Hulst)	（チョウ目、メイガ科）新北区。アブラナ科野菜害虫
cabbage webworm (1)	ハイマダラノメイガ	*Hellula undalis* (Fabricius)	（チョウ目、メイガ科）日本、全北区。米国での英名。アブラナ科野菜害虫。成虫は cabbage webworm moth
cabbage webworm			cabbage budworm を見よ
cabbage white			large white を見よ
cabbage white			common white を見よ　*P. rapae* Linnaeus の豪州での英名
cabbage white			Indian cabbage white を見よ
cabbage white butterfly			large white を見よ
cabbage white butterfly			common white を見よ　*P. rapae* Linnaeus の米国での英名
cabbage white butterfly parasite	アオムシサムライコマユバチ	*Apanteles glomeratus* (Linnaeus)	（ハチ目、コマユバチ科）日本、豪州区。モンシロチョウ幼虫に寄生
cabbage white butterfly pupal parasite			cabbage white pupal parasite を見よ
cabbage white pupal parasite	アオムシコバチ	*Pteromalus puparum* (Linnaeus)	（ハチ目、コガネコバチ科）日本、東洋区、豪州区、全北区。モンシロチョウ蛹に寄生
cabbage whitefly		*Aleyrodes brassicae* Walker	（カメムシ目、コナジラミ科）豪州区
cabbage whitefly (1)		*Aleurothrixus proletella* (Linnaeus)	（カメムシ目、コナジラミ科）旧北区。カンキツ害虫
cabbage whites	シロチョウ科	Pieridae	（チョウ目）の昆虫の総称　世界に 2,000 種
cabbage worm			シロチョウなどの幼虫
cabinet beetle			furniture beetle (1) を見よ
cabinet beetle			buffalo carpet beetle を見よ
cabinet beetle			varied carpet beetle を見よ
cacao bean moth			tobacco moth (1) を見よ
cacao beetle		*Steirastoma breve* (Sulzer)	（コウチュウ目、カミキリムシ科）新熱帯区
cacao capsid		*Distantiella theobroma* Distant	（カメムシ目、カスミカメムシ科）エチオピア区
cacao capsid (1)		*Sahlbergella singularis* Haglund	（カメムシ目、カスミカメムシ科）エチオピア区
cacao moth			tobacco moth (1) を見よ
cacao thrips			red-banded thrips を見よ

英名	和名	学名	所属、分布、ほか
cacao weevil			coffee weevil を見よ
Cachar mandarine blue		*Charana cepheis* de Nicéville	（チョウ目、シジミチョウ科）東洋区
cacticans mealybug	サボテンネコナカイガラムシ	*Rhizoecus cacticans* (Hambleton)	（カメムシ目、コナカイガラムシ科）日本、新北区、新熱帯区。サボテン害虫
cactoblastis			cactus moth を見よ
cactus borers			cactus longhorn beetles を見よ
cactus buprestid		*Acmaeodera scalaris* Mannerheim	（コウチュウ目、タマムシ科）新北区。サボテン害虫
cactus felt scale	サボテンフクロカイガラムシ	*Eriococcus coccineus* Cockerell	（カメムシ目、フクロカイガラムシ科）日本、汎世界。サボテン害虫
cactus flies	ナガズヤセバエ科	Neriidae	（ハエ目）の昆虫の総称
cactus flower beetle		*Carpophilus pallipennis* (Say)	（コウチュウ目、ケシキスイ科）新北区
cactus fly		*Volucella mexicana* Macquart	（ハエ目、ハナアブ科）新北区
cactus fruit midge		*Asphondylia opuntiae* Felt	（ハエ目、タマバエ科）新北区、大洋区
cactus longhorn		*Coenopoeus palmeri* LeConte	（コウチュウ目、カミキリムシ科）新北区
cactus longhorn beetles		*Moneilema*	（コウチュウ目、カミキリムシ科）の昆虫の総称
cactus mealybug		*Dactylopius confusus* (Cockerell)	（カメムシ目、コチニールカイガラムシ科）大洋区、新北区、豪州区
cactus mealybug			cactus felt scale を見よ
cactus mealybug			prickly pear cochineal を見よ
cactus moth		*Cactoblastis cactorum* (Bergroth)	（チョウ目、メイガ科）新北区、新熱帯区、豪州区。ウチワサボテンの防除に各地に導入
cactus scale	サボテンシロカイガラムシ	*Diaspis echinocacti* (Bouché)	（カメムシ目、マルカイガラムシ科）日本、全北区、汎熱帯。サボテン害虫
Cadbury's lichen moth		*Comachara cadburyi* Franclemont	（チョウ目、ヒトリガ科）新北区
caddice			caddisflies (1) を見よ
caddiceflies			caddisflies (1) を見よ
caddis worm	イサゴムシ		トビケラの幼虫
caddis worms			caddisflies (1) を見よ
caddises			caddisflies (1) を見よ
caddisflies		*Psychoglypha*	（トビケラ目、エグリトビケラ科）の昆虫の総称
caddisflies (1)	トビケラ目	Trichoptera	の昆虫の総称　単数は caddisfly。世界に約 7,000 種
caddisflies			northern caddisflies を見よ
caddisfly			great red sedge を見よ
cadelle	コクヌスト	*Tenebroides mauritanicus* (Linnaeus)	（コウチュウ目、ヒラタキクイムシ科）日本、汎世界。貯穀害虫
cadelle beetle			cadelle を見よ
cadelle bread beetle			cadelle を見よ
cadmus	ヒメオリオンタテハ	*Historis acheronta* (Fabricius)	（チョウ目、タテハチョウ科）新北区。cadmus butterfly ともいう。

英名	和名	学名	所属、分布、ほか
caeciliid dustlice			lizard barklice を見よ
Cahela moth		*Cahela ponderosella* (Barnes et McDunnough)	（チョウ目、メイガ科）新北区
caicus skipper		*Cogia caicus* (Herrich-Schäffer)	（チョウ目、セセリチョウ科）新北区
caicus sphinx moth		*Phryxus caicus* (Cramer)	（チョウ目、スズメガ科）新北区
Cairns birdwing		*Ornithoptera euphorion* (Gray)	（チョウ目、アゲハチョウ科）豪州区。豪州最大の固有チョウ
Cairns birdwing			Priam's birdwing を見よ
Cairns Hamadryad	ゴウシュウトンボマダラ	*Tellervo zoilus* (Fabricius)	（チョウ目、タテハチョウ科）豪州区
Cairns Hamadryas			Cairns Hamadryad を見よ
Cajun virtuoso katydid		*Amblycorypha cajuni* Walker	（バッタ目、キリギリス科）新北区
cake knot-horn			almond moth を見よ
Calachinia hairstreak		*Nesiostrymon calchinia* (Hewitson)	（チョウ目、シジミチョウ科）新熱帯区
caladium root mealybug		*Rhizoecus caladii* Green	（カメムシ目、コナカイガラムシ科）大洋区
calamoceratids			comblipped casemaker caddisflies を見よ
Caledon copper		*Aloeides caledoni* Tite et Dickson	（チョウ目、シジミチョウ科）エチオピア区
Caledonia seed bug		*Nysius caledoniae* Distant	（カメムシ目、ナガカメムシ科）新北区
Caledonian button		*Acleris caledoniana* (Stephens)	（チョウ目、ハマキガ科）旧北区
calendula fly		*Napomyza lateralis* (Fallén)	（ハエ目、ハモグリバエ科）日本、全北区
Calephelis metalmark		*Calephelis* sp.	（チョウ目、シジミタテハ科）新熱帯区
calicoback			harlequin bug (1) を見よ
calico bug			harlequin bug (1) を見よ　北米での英名
calico moth			crimson speckled を見よ　米国での英名
calico pennant			Elisa skimmer を見よ
calico scale	サラサカタカイガラムシ	*Eulecanium cerasorum* (Cockerell)	（カメムシ目、カタカイガラムシ科）日本、新北区。果樹など多くの植物の害虫
calicoes	カスリタテハ属	*Hamadryas*	（チョウ目、タテハチョウ科）の昆虫の総称
California acorn weevil		*Curculio uniformis* (LeConte)	（コウチュウ目、ゾウムシ科）新北区
California alderfly		*Sialis californica* Banks	（アミメカゲロウ目、センブリ科）新北区
California angle-wing		*Microcentrum californicum* Hebard	（バッタ目、キリギリス科）新北区
California arctic		*Oeneis ivallda* (Mead)	（チョウ目、タテハチョウ科）新北区
California black tiger beetle		*Omus californicus* Eschscholtz	（コウチュウ目、ハンミョウ科）新北区
California branch borer		*Polycaon confertus* LeConte	（コウチュウ目、ナガシンクイムシ科）新北区

英名	和名	学名	所属、分布、ほか
California brown lacewing		*Sympherobius californicus* Banks	（アミメカゲロウ目、ヒメカゲロウ科）新北区
California camel cricket		*Ceuthophilus californianus* Scudder	（バッタ目、カマドウマ科）新北区
California cankerworm		*Paleacrita longiciliata* Hulst	（チョウ目、シャクガ科）新北区
California carpenter bee		*Xylocopa californica* Cresson	（ハチ目、ミツバチ科）新北区
California cloudy wing			King page を見よ
California creeping water bug		*Ambrysus californicus* Montandon	（カメムシ目、コバンムシ科）新北区
California crescent			Orseis crescentspot を見よ
California dancer		*Argia agrioides* Calvert	（トンボ目、イトトンボ科）新北区
California darner		*Rhionaeschna californica* (Calvert)	（トンボ目、ヤンマ科）新北区
California dobson		*Neohermes californicus* (Walker)	（アミメカゲロウ目、ヘビトンボ科）新北区
California dobsonflies		*Neohermes*	（アミメカゲロウ目、ヘビトンボ科）の昆虫の総称
California dogface	アメリカモンキチョウ	*Colias eurydice* Boisduval	（チョウ目、シロチョウ科）新北区、新熱帯区。Californian dogface ともいう。カリフォルニアイヌモンキチョウの和名もあり
California five-spined Ips		*Ips paraconfusus* Lanier	（コウチュウ目、キクイムシ科）新北区
California five-spined Ips			pinyon pine engraver を見よ
California flat-headed borer		*Melanophila californica* Van Dyke	（コウチュウ目、タマムシ科）新北区
California giant skipper		*Agathymus stephensi* (Skinner)	（チョウ目、セセリチョウ科）新北区、新熱帯区
California giant skipper (1)		*Agathymus neumoegeni* (Edwards)	（チョウ目、セセリチョウ科）新北区、新熱帯区。Neumoegen's giant skipper を参照
California giant stonefly			giant stonefly (1) を見よ
California glowworm			western firefly を見よ
California granite moth		*Digrammia californiaria* (Packard)	（チョウ目、シャクガ科）新北区
California grape rootworm			brown and black beetle を見よ
California grapeberry rootworm			brown and black beetle を見よ
California grass-veneer moth		*Euchromius californicalis* (Packard)	（チョウ目、メイガ科）新北区。California grass veneer ともいう
California green lacewing			green lacewing (1) を見よ
California ground cricket		*Neonemobius eurynotus* (Rehn et Hebard)	（バッタ目、コオロギ科）新北区

英名	和名	学名	所属、分布、ほか
California ground-squirrel flea			ground squirrel flea を見よ
California hairstreak	カリフォルニアカラスシジミ	*Satyrium californica* (Edwards)	(チョウ目、シジミチョウ科) 新北区
California harvester ant			seed harvester ant を見よ
California horntail		*Urocerus californicus* Norton	(ハチ目、キバチ科) 新北区
California lady beetle		*Coccinella californica* Mannerheim	(コウチュウ目、テントウムシ科) 新北区
California ladybird beetle			California lady beetle を見よ
California mantid		*Stagmomantis californica* (Rehn et Hebard)	(カマキリ目、カマキリ科) 新北区。California mantis ともいう
California maple aphid	モミジニタイケアブラムシ	*Periphyllus californiensis* (Shinji)	(カメムシ目、アブラムシ科) 日本、全北区、東洋区、豪州区。カエデ類につく
California marble			pearly marblewing を見よ
California oak gall wasp		*Andricus quercus-californicus* Bassett	(ハチ目、タマバチ科) 新北区
California oak moth		*Phryganidia californica* Packard	(チョウ目、シャチホコガ科) 新北区
California oakworm			California oak moth を見よ
California oakworm moth			California oak moth を見よ
California Pacuvius duskywing		*Erynnis pacuvius callidus* (Grinnell)	(チョウ目、セセリチョウ科) 新熱帯区
California palm borer		*Dinapate wrightii* Horn	(コウチュウ目、ナガシンクイムシ科) 新北区
California patch		*Chlosyne californica* (Wright)	(チョウ目、タテハチョウ科) 新北区
California pear sawfly		*Pristiphora abbreviata* (Hartig)	(ハチ目、ハバチ科) 新北区
California pear-slug		*Pristiphora californica* Marlatt	(ハチ目、ハバチ科) 旧北区
California phengodid			glowworm (3) を見よ
California pistol case bearer		*Coleophora sacramenta* Heinrich	(チョウ目、ツツミノガ科) 新北区
California pleasing fungus beetle		*Cypherotylus californica* (Lacordaire)	(コウチュウ目、オオキノコムシ科) 新北区
California praying mantis			California mantid を見よ
California Prionus		*Prionus (Prionellus) californicus* Motschulsky	(コウチュウ目、カミキリムシ科) 新北区
California Pyrausta moth		*Pyrausta californicalis* (Packard)	(チョウ目、メイガ科) 新北区
California red scale	アカマルカイガラムシ	*Aonidiella aurantii* (Maskell)	(カメムシ目、マルカイガラムシ科) 日本、新北区、汎熱帯
California ringlet	カリフォルニアヒメヒカゲ	*Coenonympha tullia california* Westwood	(チョウ目、タテハチョウ科) 新北区。large heath を参照

英名	和名	学名	所属、分布、ほか
California salmonfly			giant stonefly (1) を見よ
California salt marsh mosquito		*Aedes (Ochlerotatus) squamiger* (Coquillett)	(ハエ目、カ科) 新北区
California scale			San Jose scale を見よ
California shieldback		*Eremopedes californica* Rentz	(バッタ目、キリギリス科) 新北区
California silver-spotted skipper		*Epargyreus clarus californicus* MacNeil	(チョウ目、セセリチョウ科) 新熱帯区
California sister (1)	ツマキイチモンジ (アメリカイチモンジ)	*Adelpha bredowii* Geyer	(チョウ目、タテハチョウ科) 新北区、新熱帯区
California sister	カリフォルニアイチモンジ	*Adelpha californica* (Butler)	(チョウ目、タテハチョウ科) 新北区、新熱帯区
California skipper			rural skipper を見よ
California snow scorpionfly		*Boreus californicus* Packard	(シリアゲムシ目、ユキシリアゲムシ科) 新北区
California spongilla-fly		*Climacia californica* Chandler	(アミメカゲロウ目、ミズカゲロウ科) 新北区
California spread-wing		*Archilestes californica* McLachlan	(トンボ目、アオイトトンボ科) 新北区
California sycamore borer		*Synanthedon resplendens* (H. Edwards)	(チョウ目、スカシバガ科) 新北区
California tent caterpillar		*Malacosoma californicum* (Packard)	(チョウ目、カレハガ科) 新北区
California Timema		*Timema californica* Scudder	(ナナフシ目、Timematidae) 新北区
California tortoiseshell	カリフォルニアヒオドシ	*Nymphalis californica* (Boisduval)	(チョウ目、タテハチョウ科) 新北区
California tussock moth		*Hemerocampa vetusta* (Boisduval)	(チョウ目、ドクガ科) 新北区
California white			spring white を見よ
California yucca moth		*Tegeticula maculata* (Riley)	(チョウ目、マガリガ科) 新北区
Californian Acroneuria		*Acroneuria californica* Banks	(カワゲラ目、カワゲラ科) 新北区
Californian mantis			California manntid を見よ
Californian orchid thrips			orchid thrips を見よ
Californian scale			San Jose scale を見よ
caligo butterflies	フクロチョウ亜科	Brassolinae	(チョウ目、タテハチョウ科) の昆虫の総称
Calima skipperling		*Dalla calima* Steinhauser	(チョウ目、セセリチョウ科) 新熱帯区
Callanga blue		*Leptotes callanga* Dyar	(チョウ目、シジミチョウ科) 新熱帯区
Calleta silkmoth		*Eupackardia calleta* (Westwood)	(チョウ目、ヤママユガ科) 新北区
callidus deer fly		*Chrysops callidus* Osten Sacken	(ハエ目、アブ科) 新北区
calligrapha		*Calligrapha*	(コウチュウ目、ハムシ科) の昆虫の総称　キャリグラファ甲虫といわれる

英名	和名	学名	所属、分布、ほか
callimomids			beautiful parasites を見よ
Calline false sergeant		*Pseudathyma callina* (Grose-Smith)	（チョウ目、タテハチョウ科）エチオピア区
Calliomma satyr		*Amphidecta calliomma* Felder et Felder	（チョウ目、タテハチョウ科）新熱帯区
Callipe fritillary	カリッペギンボシヒョウモン	*Speyeria callipe* (Boisduval)	（チョウ目、タテハチョウ科）新北区。S. c. comstocki (Gunder) ウスイロギンボシヒョウモンも同英名
Calosomas			caterpillar hunters を見よ
caltrop moth		*Ephysteris subdiminutella* (Stainton)	（チョウ目、キバガ科）豪州区
calus groundstreak			Godart's groundstreak を見よ
Calusa trig		*Anaxipha* sp.	（バッタ目、クサヒバリ科）新北区
Calypso caper white	カリプソシロチョウ	*Belenois calypso* (Drury)	（チョウ目、シロチョウ科）エチオピア区
calyptrate flies	有弁翅類	Calyptrata	（ハエ目）の昆虫の総称
calyptrate muscoid flies		Calyptratae	（ハエ目）の昆虫の総称
calyptrate muscoids			calypterate muscoid flies を見よ
camarade knot-horn		*Acrobasis sodalella* Zeller	（チョウ目、メイガ科）旧北区
Camberwell beauty	キベリタテハ	*Nymphalis antiopa asopos* (Fruhstorfer)	（チョウ目、タテハチョウ科）日本。全北区の *N. antiopa* (Linnaeus) の英名。Camberwell beauty butterfly ともいう。ロンドンの地区名 Camberwell に由来
Cambrian carpet		*Horisme aquata* (Hübner)	（チョウ目、シャクガ科）旧北区。イングランド北西部の Cambria に由来
Cambrian umber			Cambrian carpet を見よ
Cambridge blue			Marsyas hairstreak を見よ
Cambridge small fritillary			Duke of Burgundy fritillary を見よ
Cambridge vagrant		*Nepheronia thalassina* (Boisduval)	（チョウ目、シロチョウ科）エチオピア区
Camdeboo brown		*Cassionympha camdeboo* (Dickson)	（チョウ目、タテハチョウ科）エチオピア区
Camdeboo skolly		*Thestor camdeboo* Dickson et Wykeham	（チョウ目、シジミチョウ科）エチオピア区
camel crickets (1)	カマドウマ科	Rhaphidophoridae	（バッタ目）の昆虫の総称　ラクダのように背が丸く突きでていることに由来
camel crickets (2)		*Ceuthophilus*	（バッタ目、カマドウマ科）の昆虫の総称
camel crickets			cave crickets (1)（2）を見よ
camel crickets			long-horned grasshoppers (3) を見よ
camel flies			horse flies (1) を見よ
camel head-bot			North African camel oestrid を見よ
camellia cottony scale			camellia scale (1) を見よ
camellia curculio	ツバキシギゾウムシ	*Curculio camelliae* (Roelofs)	（コウチュウ目、ゾウムシ科）日本
camellia false cottony scale	ツバキカタカイガラモドキ	*Metaceroneme japonica* (Maskell)	（カメムシ目、カタカイガラムシ科）日本、旧北区

英名	和名	学名	所属、分布、ほか
camellia flower moth	スギタニモンキリガ	*Sugitania lepida* (Butler)	(チョウ目、ヤガ科) 日本
camellia fruit fly	ツバキハマダラミバエ	*Staurella camelliae* Ito	(ハエ目、ミバエ科) 日本、旧北区。ツバキ害虫
camellia leaf miner			camellia webworm を見よ
camellia leaf vein midge	ヤブツバキウロコタマバエ	*Lasioptera camelliae* Ohno et Yukawa	(ハエ目、タマバエ科) 日本。ツバキ害虫
camellia mining scale	クロカタマルカイガラムシ	*Duplaspidiotus claviger* (Cockerell)	(カメムシ目、マルカイガラムシ科) 日本、新北区、汎熱帯
camellia oystershell scale			camellia scale (2) を見よ
camellia parlatoria scale			camellia scale (2) を見よ
camellia phenacaspis scale			false oleander scale を見よ
camellia pulvinaria			camellia scale (1) を見よ　米国での英名
camellia scale	ツバキカキカイガラムシ	*Lepidosaphes camelliae* Hoke	(カメムシ目、マルカイガラムシ科) 日本、新北区、豪州区。ツバキ害虫
camellia scale (1)	ツバキワタカイガラムシ	*Pulvinaria floccifera* (Westwood)	(カメムシ目、マルカイガラムシ科) 日本、汎熱帯。ツバキ害虫
camellia scale (2)	ツバキクロホシカイガラムシ	*Parlatoria camelliae* Comstock	(カメムシ目、マルカイガラムシ科) 日本、汎世界。ツバキ、ヒサカキなどの害虫
camellia scale			avocado scale を見よ　米国での英名
camellia scale			greedy scale (1) を見よ
camellia spiny whitefly	チャトゲコナジラミ	*Aleurocanthus camelliae* Kanmiya et Kasai	(カメムシ目、コナジラミ科) 日本、東洋区。チャ害虫
camellia tortrix		*Homona menciana* (Walker)	(チョウ目、ハマキガ科) 旧北区
camellia webworm	トビマダラメイガ	*Samaria ardentella* Ragonot	(チョウ目、メイガ科) 日本。カキ、チャなどの害虫
camellia whitefly	ツバキコナジラミ	*Aleurotrachelus camelliae* (Kuwana)	(カメムシ目、コナジラミ科) 日本。ツバキ害虫
camelneck flies			snake-flies (1) を見よ
cameo		*Crypsedra gemmea* Treitschke	(チョウ目、ヤガ科) 旧北区
Cameroon bug		*Antestiopsis faceta* Germar	(カメムシ目、カメムシ科) エチオピア区。コーヒー害虫
Cameroon mantis		*Alalomantis muta* Wood-Mason	(カマキリ目、カマキリ科) エチオピア区
camillid flies		Camillidae	(ハエ目) の昆虫の総称
camouflaged looper			wavy-lined emerald moth を見よ
camouflaged skipper		*Diaeus varna* Evans	(チョウ目、セセリチョウ科) 新熱帯区
campanula pug		*Eupithecia denotata* (Hübner)	(チョウ目、シャクガ科) 旧北区
camphor ambrosia beetle	クスノキオオキクイムシ	*Xyleborus mutilatus* Blandford	(コウチュウ目、キクイムシ科) 日本、東洋区、新北区。木材害虫
camphor aphid	クスアブラムシ	*Sinomegoura citricola* (Van der Goot)	(カメムシ目、アブラムシ科) 日本、旧北区

英名	和名	学名	所属、分布、ほか
camphor aulacaspis	ヤブニッケイシロカイガラムシ	*Aulacaspis yabunikkei* Kuwana	(カメムシ目、マルカイガラムシ科) 日本、旧北区
camphor bark beetle	クスノキクイムシ	*Phloeosinus seriatus* Blandford	(コウチュウ目、キクイムシ科) 日本、旧北区
camphor green geometrid	クスアオシャク	*Thalasodes subquardraria* Inoue	(チョウ目、シャクガ科) 日本、東洋区
camphor lace bug	クスグンバイ	*Stephanitis fasciicarina* Takeya	(カメムシ目、グンバイムシ科) 日本。クスノキ害虫
camphor lace bug			Andromeda lace bug を見よ
camphor leaf miner	クスノハムグリガ		(チョウ目、ホソガ科) 日本
camphor leafhopper	クスサジヨコバイ	*Favintiga camphorae* (Matsumura)	(カメムシ目、オオヨコバイ科) 日本。クスノキ害虫
camphor longicorn beetle	クスベニカミキリ	*Pyrestes nipponicus* Hayashi	(コウチュウ目、カミキリムシ科) 日本。クスノキ害虫
camphor reddish longicorn			camphor longicorn beetle を見よ
camphor scale	ミカンマルカイガラムシ	*Pseudaonidia duplex* (Cockerell)	(カメムシ目、マルカイガラムシ科) 日本、汎世界。果樹、樹木の害虫
camphor shoot beetle		*Xylosandrus mutilatus* (Blandford)	(コウチュウ目、ゾウムシ科) 全北区
camphor shot borer			camphor ambrosia beetle を見よ
camphor spiny whitefly	クストゲコナジラミ	*Aleurocanthus cinnamomi* Takahashi	(カメムシ目、コナジラミ科) 日本、東洋区。クスノキ害虫
camphor spoon-shaped leafhopper			camphor leafhopper を見よ
camphor sucker	クストガリキジラミ	*Trioza camphorae* Sasaki	(カメムシ目、トガリキジラミ科) 日本、東洋区。クスノキ害虫
camphor thrips	クスクダアザミウマ	*Liothrips floridensis* (Watson)	(アザミウマ目、クダアザミウマ科) 日本、東洋区、新北区、豪州区
camphor tree borer			swift moth を見よ
camphor tree weevil	クスアナアキゾウムシ	*Pimelocerus hylobioides* (Desbrochers)	(コウチュウ目、ゾウムシ科) 日本、東洋区
camphorweed owlet moth		*Cucullia alfarata* Strecker	(チョウ目、ヤガ科) 新北区
campion	フサクビヨトウ	*Sideridis honeyi* (Yoshimoto)	(チョウ目、ヤガ科) 日本。カーネーション害虫。ナデシコ科の総称 campion に由来
campion			campion moth を見よ
campion coronet		*Hadena cucubali* Schiffermüller	(チョウ目、ヤガ科) 旧北区
campion moth		*Hadena rivularis* (Fabricius)	(チョウ目、ヤガ科) 旧北区。campion ともいう
campion moth			campion を見よ
campodeids			slender entotrophs を見よ
campodeids			entotrophs を見よ
Canada arctic			Macoun's arctic を見よ
Canada darner		*Aeschna canadensis* Walker	(トンボ目、ヤンマ科) 新北区

英名	和名	学名	所属、分布、ほか
Canadian Agonopterix moth		*Agonopterix canadensis* (Busck)	（チョウ目、マルハキバガ科）新北区。Canadian Agonopterix ともいう
Canadian apple mealybug			apple mealybug を見よ　米国での英名
Canadian chicken flea		*Ceratophyllus gibsoni* Fox	（ノミ目、トゲノミ科）新北区
Canadian giant moth		*Andropolia contacta* Walker	（チョウ目、ヤガ科）新北区。Canadian giant ともいう
Canadian Melanolophia moth		*Melanolophia canadaria* (Guenée)	（チョウ目、シャクガ科）新北区。Canadian Melanolophia ともいう
Canadian owlet moth		*Calyptra canadensis* (Bethune)	（チョウ目、ヤガ科）新北区。Canadian owlet ともいう
Canadian Petrophila moth		*Petrophila canadensis* (Munroe)	（チョウ目、メイガ科）新北区。black-patched Glaphyria ともいう
Canadian Sonia moth		*Sonia canadana* McDunnough	（チョウ目、ハマキガ科）新北区
Canadian sphinx		*Sphinx canadensis* Boisduval	（チョウ目、スズメガ科）新北区。Canadian sphinx moth ともいう
Canadian tiger swallowtail		*Papilio canadensis* Rothschild et Jordan	（チョウ目、アゲハチョウ科）新北区
canarium moth			Cricula silkmoth を見よ
Canary blue		*Cyclyrius webbianus* (Brull)	（チョウ目、シジミチョウ科）旧北区
Canary brimstone		*Gonepteryx cleobule* (Hübner)	（チョウ目、シロチョウ科）旧北区
canary fly			apple leafhopper (1)(4)(5) を見よ　豪州での英名
Canary grayling	カナリージャノメ	*Hipparchia wyssii* (Christ)	（チョウ目、タテハチョウ科）旧北区
Canary Island brimstone		*Gonepteryx cleopatra cleobule* (Hübner)	（チョウ目、シロチョウ科）旧北区。Cleopatra を参照
Canary Islands large white		*Pieris cheiranthi* (Hübner)	（チョウ目、シロチョウ科）旧北区、エチオピア区
Canary red admiral		*Vanessa vulcania* (Godart)	（チョウ目、タテハチョウ科）旧北区
canary-shouldered thorn		*Ennomos alniaria* (Linnaeus)	（チョウ目、シャクガ科）全北区。canary-shouldered thorn moth ともいう
Canary skipper		*Thymelicus christi* Rebel	（チョウ目、セセリチョウ科）旧北区
Canary speckled wood	カナリーウラジャノメ	*Pararge xiphioides* Staudinger	（チョウ目、タテハチョウ科）旧北区
Canary Ypsolopha moth		*Ypsolopha canariella* (Walsingham)	（チョウ目、スガ科）新北区
Candelaria skipper		*Euphyes canda* Steinhauser et Warren	（チョウ目、セセリチョウ科）新熱帯区
candy-striped leafhopper		*Graphocephala coccinea* (Forster)	（カメムシ目、オオヨコバイ科）新北区。canary-striped leafhopper, candy-stripe leafhopper ともいう
cane aphid	ヒエアブラムシ	*Melanaphis sacchari* (Zehntner)	（カメムシ目、アブラムシ科）日本、汎世界。ヒエ、サトウキビ、トウモロコシ害虫。cane aphis の使用例もあり
cane leaf scale		*Chionaspis saccharifolii* Zehntner	（カメムシ目、マルカイガラムシ科）東洋区

英名	和名	学名	所属、分布、ほか
cane maggot			loganberry cane fly を見よ　幼虫の英名
cane-moth borer		*Bathytricha truncata* (Walker)	（チョウ目、ヤガ科）東洋区、豪州区
cane weevil borer			sugarcane weevil borer を見よ
cane weevil tachinid		*Lixophaga sphenophori* (Villeneuve)	（ハエ目、ヤドリバエ科）大洋区
Caninus hairstreak		*Gargina caninius* (Druce)	（チョウ目、シジミチョウ科）新熱帯区
canker-worm moths			geometer moths を見よ
canker worms		*Paleacrita*	（チョウ目、シャクガ科）の幼虫の総称
cankerworms		Aenochromatidae	（チョウ目）の昆虫の総称
canker-worms			butterflies and moths を見よ　幼虫の英名
canker worms			geometer moths を見よ
Canna skipper		*Quinta cannae* (Herrich-Schäffer)	（チョウ目、セセリチョウ科）新熱帯区
Canna skipper			Brazilian skipper を見よ
cannabis aphid			hemp aphid を見よ
canopus butterfly		*Papilio canopus* Westwood	（チョウ目、アゲハチョウ科）豪州区
canopy skipper		*Nothodiplax dendrophila* Belle	（トンボ目、トンボ科）新熱帯区
Cantab bell			fir budmoth (1) を見よ
Canterbury tree weta		*Hemideina femorata* Hutton	（バッタ目、Stenopelmatidae）豪州区
cantharid beetles	ホタル上科	Cantharoidea	（コウチュウ目）の昆虫の総称
cantharides			blister beetle を見よ
cantharis			blister beetle を見よ
Cantobrica tiger		*Rhodussa cantobrica* (Hewitson)	（チョウ目、タテハチョウ科）新熱帯区
Cantra ruby-eye		*Talides cantra* Evans	（チョウ目、セセリチョウ科）新熱帯区
canyon green hairstreak	アリゾナウラアオシジミ	*Callophrys apama* (Edwards)	（チョウ目、シジミチョウ科）新北区
canyon rubyspot		*Hetaerina vulnerata* Hagen	（トンボ目、イトトンボ科）新北区、新熱帯区
Canyonland gemmed-satyr		*Cyllopsis pertepida pertepida* (Dyar)	（チョウ目、タテハチョウ科）新熱帯区。*C. p. avicula* (Nabokov), *C. p. intermedia* Miller, *C. p. maniola* (Nabokov) も同英名。Canyonland satyr ともいう
Capaneus butterfly		*Papilio fuscus capaneus* Westwood	（チョウ目、アゲハチョウ科）豪州区。yellow Helen (1) を参照
cape ascalaphid		*Proctarrelabis capensis* (Thunberg)	（アミメカゲロウ目、ツノトンボ科）エチオピア区
cape autumn widow		*Dira clytus* (Linnaeus)	（チョウ目、タテハチョウ科）エチオピア区
cape black-eye		*Leptomyrina lara* (Linnaeus)	（チョウ目、シジミチョウ科）エチオピア区
cape blue			cupreous blue を見よ
cape bluet		*Proischnura polychromatica* (Barnard)	（トンボ目、イトトンボ科）エチオピア区
cape brown		*Cassionympha detecta* (Trimen)	（チョウ目、タテハチョウ科）エチオピア区

英名	和名	学名	所属、分布、ほか
cape gooseberry budworm			Oriental tobacco budworm を見よ
cape grey		*Crudaria capensis* van Son	（チョウ目、シジミチョウ科）エチオピア区
cape hawk		*Theretra capensis* (Linnaeus)	（チョウ目、スズメガ科）エチオピア区
cape sprite			Palmiet sprite を見よ
cape thorntail		*Ceratogomphus triceraticus* Balinsky	（トンボ目、サナエトンボ科）エチオピア区
Cape York aeroplane		*Pantoporia venilia moorei* (Macleay)	（チョウ目、タテハチョウ科）豪州区
Cape York aeroplane			black-eyed plane を見よ
Cape York birdwing			Priam's birdwing を見よ　豪州での英名
Cape York dragon		*Antipodogomphus edentulus* Watson	（トンボ目、サナエトンボ科）豪州区
Cape York Hamadryad		*Tellervo zoilus gelo* Waterhouse et Lyell	（チョウ目、タテハチョウ科）豪州区。Cairns Hamadryad を参照
Cape York pearl-white		*Elodina claudia* de Baar et Hancock	（チョウ目、シロチョウ科）豪州区
Capenas bluewing	シロモンルリツヤタテハ	*Myscelia capenas* (Hewitson)	（チョウ目、タテハチョウ科）新熱帯区
caper gull			Australian gull を見よ
caper-leaf webworm moth		*Dichogama redtenbacheri* Ledeter	（チョウ目、メイガ科）新北区
caper white	ジャワシロチョウ	*Belenois java* (Linnaeus)	（チョウ目、シロチョウ科）東洋区、豪州区
caper white			brown-veined white を見よ
caper whites	ヘリグロシロチョウ属	*Belenois*	（チョウ目、シロチョウ科）の昆虫の総称
capillate cattle louse			small blue cattle louse を見よ　米国での英名
capri fig wasp			fig wasp を見よ　豪州での英名
capricorn beetle		*Cerambyx scopolii* Fuessly	（コウチュウ目、カミキリムシ科）旧北区
capricorn beetles			long-horned leaf beetles を見よ
caprifiers			fig wasps (1) を見よ
caprine flat-body		*Agonopterix capreolella* (Zeller)	（チョウ目、マルハキバガ科）旧北区
Capronnier's red-eye		*Pteroteinon capronnieri* (Plötz)	（チョウ目、セセリチョウ科）エチオピア区
capsid bug			apple capsid を見よ
capsid bugs			plant bugs (1) を見よ　英国での英名
capsids			plant bugs (1) を見よ
capsule borer			gingelly borer moth を見よ
capsule moth		*Hadena capsularis* (Guenée)	（チョウ目、ヤガ科）新北区
capuchin beetle		*Bostrichus capucinus* (Linnaeus)	（コウチュウ目、ナガシンクイムシ科）旧北区
Capucinus sister		*Adelpha capucinus* Walch	（チョウ目、タテハチョウ科）新熱帯区
carabid beetles			ground beetles を見よ

英名	和名	学名	所属、分布、ほか
caragana aphid		*Acyrthosiphon caraganae* (Cholodkovsky)	（カメムシ目、アブラムシ科）全北区。caragana ムレスズメ に由来
caragana blister beetle		*Epicauta subglabra* (Fall)	（コウチュウ目、ツチハンミョウ科）新北区
caragana plant bug		*Lopidea dakota* Knight	（カメムシ目、カスミカメムシ科）新北区
carambola fruit-borer		*Eucosma notanthes* Meyrick	（チョウ目、ハマキガ科）旧北区
caraway webworm			carrot moth (1) を見よ　南アフリカでの英名
carbuncle clothes moth		*Euplocamus anthracinalis* (Scopoli)	（チョウ目、ヒロズコガ科）旧北区
carcass beetle		*Omorgus amictus* (Haaf)	（コウチュウ目、コブスジコガネムシ科）豪州区
carcass beetle		*Omorgus alternans* (Macleay)	（コウチュウ目、コブスジコガネムシ科）豪州区
carcass beetle		*Omorgus costatus* (Wiedemann)	（コウチュウ目、コブスジコガネムシ科）豪州区
carcass beetles			hide beetles (1) を見よ　豪州での英名
Carcasson's streaked sailer		*Neptis carcassoni* van Son	（チョウ目、タテハチョウ科）エチオピア区
cardamine aphid	ミズタカラシアブラムシ	*Aphis mizutakarashi* Shinji	（カメムシ目、アブラムシ科）日本
cardamon root borer		*Hilarographa caminodes* Meyrick	（チョウ目、ハマキガ科）東洋区
cardamon thrips		*Sciothrips cardamoni* (Ramakrishna)	（アザミウマ目、アザミウマ科）大洋区、東洋区、新熱帯区
cardenolide sequestering moth			polka dot moth を見よ
carder bee		*Bombus pascuorum* (Scopoli)	（ハチ目、ミツバチ科）旧北区
carder bees			bumble bees (2)(3) を見よ
carder bees			leaf-cutting bees (1) を見よ
cardinal	パンドラヒョウモン	*Argynnis pandora* (Denis et Schiffermüller)	（チョウ目、タテハチョウ科）旧北区
cardinal			Niobe fritillary を見よ　cardinal butterfly ともいう
cardinal Apollo		*Parnassius cardinal* Grum-Grshimailo	（チョウ目、アゲハチョウ科）旧北区
cardinal beetle		*Pyrochroa serraticornis* (Scopoli)	（コウチュウ目、アカハネムシ科）旧北区
cardinal beetle			fire beetle (1) を見よ
cardinal beetle			vedalia を見よ
cardinal beetles	アカハネムシ科	Pyrochroidae	（コウチュウ目）の昆虫の総称
cardinal click beetle		*Ampedus cardinalis* (Schiödte)	（コウチュウ目、コメツキムシ科）旧北区
cardinal ladybird			vedalia を見よ
cardinal meadowhawk		*Sympetrum illotum* (Hagen)	（トンボ目、トンボ科）新北区、新熱帯区
Carex mealybug		*Trionymus caricis* McConnell	（カメムシ目、コナカイガラムシ科）新北区

英名	和名	学名	所属、分布、ほか
Caria hairstreak		*Erora caria* (Schaus)	（チョウ目、シジミチョウ科）新熱帯区
Caribbean black scale		*Saissetia neglecta* De Lotto	（カメムシ目、カタカイガラムシ科）新北区、大洋区
Caribbean conehead		*Neoconocephalus maxilliosus* (Fabricius)	（バッタ目、キリギリス科）新北区、新熱帯区
Caribbean dainty white			shy yellow を見よ
Caribbean darner		*Triacanthagyna caribbea* Williamson	（トンボ目、ヤンマ科）新北区
Caribbean dasher		*Micrathyria dissocians* (Calvert)	（トンボ目、トンボ科）新熱帯区
Caribbean dried fruit moth		*Ectomyelois decolor* (Zeller)	（チョウ目、メイガ科）新北区、新熱帯区
Caribbean fruit fly	カリブミバエ	*Anastrepha suspensa* (Loew)	（ハエ目、ミバエ科）大洋区、新熱帯区
Caribbean meadow katydid		*Conocephalus cinereus* (Thunberg)	（バッタ目、キリギリス科）新北区。新熱帯区
Caribbean mole cricket			changa を見よ
Caribbean peacock			Cuban peacokck を見よ
Caribbean pod borer		*Fundella pellucens* Zeller	（チョウ目、メイガ科）新北区、新熱帯区。Caribbean pod borer moth ともいう
Caribbean ruby-eye		*Perichares philetes* (Gmelin)	（チョウ目、セセリチョウ科）新北区、新熱帯区
Caribbean scavenger moth		*Erechthias minuscula* (Walsingham)	（チョウ目、ヒロズコガ科）新北区、新熱帯区
Caribbean skipper		*Pyrrhocalles antiqua* (Herrich-Schäffer)	（チョウ目、セセリチョウ科）新北区、新熱帯区
Caribbean swallowtail		*Papilio pelaus* Fabricius	（チョウ目、アゲハチョウ科）新北区
Caribbean swampdamsel		*Leptobasis candelaria* Alayo	（トンボ目、イトトンボ科）新熱帯区
Caribbean yellowface		*Neoerythromma cultellatum* (Hagen)	（トンボ目、イトトンボ科）新北区、新熱帯区
caribou nostril fly		*Cephenomyia trompe* Linnaeus	（ハエ目、ヒツジバエ科）旧北区
caribou warble fly		*Oedemagena tarandi* Linnaeus	（ハエ目、ヒツジバエ科）旧北区
carinate katydid		*Amblycorypha carinata* Rehn et Hebard	（バッタ目、キリギリス科）新北区
carinate meadow katydid		*Orchelimum carinatum* Walker	（バッタ目、キリギリス科）新北区
carinated shield-back katydid			keeled shieldback を見よ
Carissima underwing moth		*Catocala carissima* Hulst	（チョウ目、ヤガ科）新北区。Carissima underwing ともいう
Caristanius moth		*Caristanius decoloralis* (Walker)	（チョウ目、メイガ科）新北区、大洋区
carita thread-legged katydid		*Arethaea carita* Scudder	（バッタ目、キリギリス科）新北区

英名	和名	学名	所属、分布、ほか
carl moths			trumpet leafminer moths を見よ
carline flat-body		*Agonopterix nanatella* (Stainton)	（チョウ目、マルハキバガ科）旧北区
carline neb			chalk knot-horn を見よ
Carline skipper		*Pyrgus carlinae* (Rambur)	（チョウ目、セセリチョウ科）旧北区
Carlos' mottled-skipper		*Codatractus carlos carlos* Evans	（チョウ目、セセリチョウ科）新熱帯区
Carlotta's tiger moth		*Apantesis carlotta* Ferguson	（チョウ目、ヒトリガ科）新北区
carls			trumpet leafminer moths を見よ
Carme crescent		*Eresia carme* (Doubleday)	（チョウ目、タテハチョウ科）新熱帯区
carmine falcon		*Corades chelonis* Hewitson	（チョウ目、タテハチョウ科）新熱帯区
carmine skimmer		*Orthemis discolor* (Burmeister)	（トンボ目、トンボ科）新北区、新熱帯区
carmine snout moth		*Peoria approximella* (Walker)	（チョウ目、メイガ科）新北区
Carnarvon darner		*Austroaeschna muelleri* Theischinger	（トンボ目、ヤンマ科）豪州区
carnation fly		*Delia cardui* (Meigen)	（ハエ目、ハナバエ科）旧北区
carnation fly (1)		*Delia brunnescens* (Zetterstedt)	（ハエ目、ハナバエ科）旧北区、新北区
carnation leafminer	カーネーションハモグリバエ	*Liriomyza dianthicola* (Venturi)	（ハエ目、ハモグリバエ科）日本。カーネーション害虫
carnation leafroller			Mediterranean carnation leaf roller を見よ
carnation maggot			carnation fly (1) を見よ　米国での幼虫の英名
carnation thrips		*Thrips atratus* Haliday	（アザミウマ目、アザミウマ科）旧北区
carnation tip maggot	ハコベハナバエ	*Delia echinata* (Séguy)	（ハエ目、ハナバエ科）日本、全北区。ホウレンソウ、カーネーションなどの害虫
carnation tortrix			Mediterranean carnation leaf roller を見よ　carnation tortrix moth ともいう
carnation twist			Mediterranean carnation leaf roller を見よ
carnation worm			campion を見よ
Carnica hairstreak		*Dicya carnica* (Hewitson)	（チョウ目、シジミチョウ科）新熱帯区
carnicera wasp		*Agelaia panamaensis* (Cameron)	（ハチ目、スズメバチ科）新熱帯区
carnid flies		Carnidae	（ハエ目）の昆虫の総称
Carniolian honey bee		*Apis mellifica carnica* Linnaeus	（ハチ目、ミツバチ科）旧北区
carnivore fleas	ケナガノミ科	Vermipsyllidae	（ノミ目）の昆虫の総称
carnivorous beetles	食肉亜目	Adephaga	（コウチュウ目）の昆虫の総称
carob bean moth			locust bean moth を見よ　豪州での英名
carob gall midge			carob midge を見よ
carob midge		*Eumarchalia gennadii* (Marchal)	（ハエ目、タマバエ科）旧北区。carob イナゴマメに由来
carob moth (1)		*Ephestia calidella* (Guenée)	（チョウ目、メイガ科）旧北区

英名	和名	学名	所属、分布、ほか
carob moth (2)		*Spectrobates ceratoniae* (Zeller)	（チョウ目、メイガ科）全北区、豪州区、エチオピア区
carob moth			locust bean moth を見よ　豪州での英名
Carole's silverspot			spring mountain fritillary を見よ
Carolina chocolate moth		*Argyrostrotis carolina* (Smith)	（チョウ目、ヤガ科）新北区
Carolina cloak grasshopper			Carolina grasshopper を見よ
Carolina conifer aphid		*Cinara atlantica* (Wilson)	（カメムシ目、アブラムシ科）新北区
Carolina coniferous aphid		*Cinara carolina* Tissot	（カメムシ目、アブラムシ科）大洋区
Carolina grasshopper		*Dissosteira carolina* (Linnaeus)	（バッタ目、バッタ科）新北区、大洋区。トウモロコシ害虫。後翅は黒色で縁は淡色。米国で普通
Carolina ground cricket		*Eunemobius carolinus* (Scudder)	（バッタ目、コオロギ科）新北区
Carolina leafwing		*Memphis xenocles carolina* (Comstock)	（チョウ目、タテハチョウ科）新熱帯区。Westwood's leafwing を参照
Carolina locust			Carolina grasshopper を見よ
Carolina mantid		*Stagmomantis carolina* (Johannson)	（カマキリ目、カマキリ科）新北区、新熱帯区。米国の普通種
Carolina mantis			Carolina mantid を見よ
Carolina roadside skipper		*Amblyscirtes carolina* (Skinner)	（チョウ目、セセリチョウ科）新北区
Carolina satyr		*Hermeuptychia sosybius* (Fabricius)	（チョウ目、タテハチョウ科）新熱帯区
Carolina satyr			Hermes ringlet を見よ
Carolina sawyer		*Monochamus carolinensis* Olivier	（コウチュウ目、カミキリムシ科）新北区
Carolina sphinx			tobacco worm を見よ　Carolina sphinx moth ともいう
Carolina tiger beetle		*Megacephala carolina* (Linnaeus)	（コウチュウ目、ハンミョウ科）新北区
Caroline grasshopper			Carolina grasshopper を見よ
Carol's fritillary		*Speyeria carolae* (dos Passos et Grey)	（チョウ目、タテハチョウ科）新北区
Carolynn's copper		*Aloeides carolynnae* Dickson	（チョウ目、シジミチョウ科）エチオピア区
carousing Anteros		*Anteros carausius* Westwood	（チョウ目、シジミタテハ科）新熱帯区
Carpasia hairstreak			jeweled hairstreak を見よ
carpenter ant	ミカドオオアリ	*Camponotus kiusiuensis* Santschi	（ハチ目、アリ科）日本。材木、竹切株に営巣
carpenter ant			large carpenter ant を見よ
carpenter ant			herculean ant (1)(2) を見よ
carpenter ant			red carpenter ant (1)(2) を見よ
carpenter ants	オオアリ属	*Camponotus*	（ハチ目、アリ科）の昆虫の総称

英名	和名	学名	所属、分布、ほか
carpenter bee			violet carpenter bee を見よ
carpenter bee			eastern carpenter bee を見よ
carpenter bees	クマバチ属	*Xylocopa*	（ハチ目、ミツバチ科）の昆虫の総称　コシブトハナバチ科に入れられることもある。Xylocopinae, Xylocopini, Xylocopidae も carpenter bees という
carpenter bees			cuckoo bees (1) を見よ
carpenter moth			goat moth (1) を見よ
carpenter moth		*Arbela dea* Swinhoe	（チョウ目、Metarbelidae）東洋区
carpenter moth (1)		*Prionoxystus robiniae* (Peck)	（チョウ目、ボクトウガ科）全北区
carpenter moths (1)	ボクトウガ科	Cossidae	（チョウ目）の昆虫の総称
carpenter moths (2)		*Prionoxystus*	（チョウ目、ボクトウガ科）の昆虫の総称　北米での英名
carpenter worm			carpenter moth (1) を見よ
carpenter-worm			goat moth (1) を見よ
carpenter worm moth			goat moth (1) を見よ
carpenterworm moths			carpenter moths (1) を見よ
carpenters			carpenter moths (1) を見よ
carpenter's sapphire		*Stugeta carpenteri* Stempffer	（チョウ目、シジミチョウ科）エチオピア区
carpenterworms			carpenter moths (1)(2) を見よ
carpet beetle		*Anthrenus sophonisba* Beal	（コウチュウ目、カツオブシムシ科）新北区
carpet beetle			varied carpet beetle を見よ
carpet beetle			fur beetle を見よ
carpet beetles		*Anthrenus*	（コウチュウ目、カツオブシムシ科）の昆虫の総称
carpet beetles			larder beetles を見よ
carpet clothes moth			carpet moth を見よ
carpet fly			house windowfly を見よ
carpet-grass webworm moth		*Fissicrambus hyatiellus* (Zincken)	（チョウ目、メイガ科）新北区、新熱帯区
carpet moth	ジュウタンガ	*Trichophaga tapetzella* (Linnaeus)	（チョウ目、ヒロズコガ科）日本、汎世界
carpet moths	ナミシャク亜科	Larentiinae	（チョウ目、シャクガ科）の昆虫の総称
carpet thorn		*Bapta distinctata* Herrich-Schäffer	（チョウ目、シャクガ科）旧北区
carpets			geometer moths を見よ
carpets			carpet moths を見よ
carpinus aphid	クマシデアブラムシ	*Greenidia carpini* Takahashi	（カメムシ目、アブラムシ科）日本
carpinus bark beetle	サワシバノキクイムシ	*Scolytus claviger* Blandford	（コウチュウ目、キクイムシ科）日本、旧北区
carpinus marmorate-horned aphid			filbert aphid を見よ

英名	和名	学名	所属、分布、ほか
carpinus round bark beetle	シデノマルキクイムシ	*Sphaerotrypes carpini* Eggers	（コウチュウ目、キクイムシ科）日本
carposinid moths			fruitworm moths を見よ
carrion beetle			Japanese carrion beetle を見よ
carrion beetles			burying beetles (1)(2) を見よ
carrion flies			blow flies を見よ
carrionflower moth		*Acrolepiopsis incertella* (Chambers)	（チョウ目、アトヒゲコガ科）新北区
carrot aphid		*Semiaphis dauci* (Fabricius)	（カメムシ目、アブラムシ科）旧北区
carrot aphid (1)	ニンジンフタオアブラムシ	*Cavariella aegopodii* (Scopoli)	（カメムシ目、アブラムシ科）日本、全北区、豪州区。ニンジン、ミツバ害虫
carrot beetle			muck beetle を見よ
carrot budworm	クロモンシロハマキ	*Epinotia majorana* (Caradja)	（チョウ目、ハマキガ科）日本、旧北区。ニンジン害虫
carrot bug	アカヒメヘリカメムシ	*Aeschynteles maculatus* (Fieber)	（カメムシ目、ヒメヘリカメムシ科）日本、旧北区。イネ、ムギ、ニンジンなどを加害
carrot flies			rust flies を見よ
carrot fly			carrot rust fly を見よ　carrot-fly とも記す
carrot gall midge		*Kiefferia pimpinellae* Loew	（ハエ目、タマバエ科）旧北区
carrot mealy aphid			celery aphid (2) を見よ
carrot moth (1)		*Depressaria nervosa* Haworth	（チョウ目、マルハキバガ科）旧北区、エチオピア区
carrot moth (2)		*Depressaria daucella* (Denis et Schiffermüller)	（チョウ目、マルハキバガ科）旧北区
carrot plant bug		*Orthops basalis* (Costa)	（カメムシ目、カスミカメムシ科）旧北区
carrot psyllid		*Trioza apicalis* Foerster	（カメムシ目、トガリキジラミ科）旧北区
carrot root aphid		*Pemphigus phenax* Borner et Blunck	（カメムシ目、アブラムシ科）旧北区
carrot rust fly	ニンジンサビバエ	*Psila rosae* (Fabricius)	（ハエ目、ハネオレバエ科）新北区
carrot sawfly	オオツマグロハバチ	*Tenthredo providens* Smith	（ハチ目、ハバチ科）日本。ニンジン、ミツバ害虫
carrot weevil	ニンジンゾウムシ	*Listronotus oregonensis* (LeConte)	（コウチュウ目、ゾウムシ科）新北区。ニンジン害虫
carrot weevil			spotted vegetable weevil を見よ
carrot weevil			vegetable weevil (1) を見よ
Carswell's little blue		*Cupido carswelli* Stempffer	（チョウ目、シジミチョウ科）旧北区
Carus skipper		*Polites carus* (Edwards)	（チョウ目、セセリチョウ科）新熱帯区
cascaders		*Zygonyx*	（トンボ目、トンボ科）の昆虫の総称
cascades skipper			Mardon skipper を見よ
case bearers			casebearer moths を見よ　casebearers とも記す
case bearers			leefblotch miners (1) を見よ　米国での英名
casebearer moths	ツツミノガ科	Coleophoridae	（チョウ目）の昆虫の総称

英名	和名	学名	所属、分布、ほか
casebearing clothes moth			casemaking clothes moth (2) を見よ
case-bearing clothes moth			casemaking clothes moth (1)(3) を見よ
case-bearing leaf beetle			locust leaf miner を見よ
case-bearing leaf beetles		*Pachybrachys*	（コウチュウ目、ハムシ科）の昆虫の総称
case-bearing leaf beetles (1)	ナガツツハムシ亜科	Clytrinae	（コウチュウ目、ハムシ科）の昆虫の総称
case-bearing leaf beetles (2)	ツツハムシ亜科	Cryptocephalinae	（コウチュウ目、ハムシ科）の昆虫の総称
case moths			casebearer moths を見よ
case moths			bagworm moths を見よ
caseflies			caddisflies (1) を見よ
casemaking clothes moth		*Tinea flavescentella* Haworth	（チョウ目、ヒロズコガ科）豪州区
casemaking clothes moth		*Tinea murairella* Staudinger	（チョウ目、ヒロズコガ科）豪州区
casemaking clothes moth (1)		*Tinea pellionella* (Linnaeus)	（チョウ目、ヒロズコガ科）汎世界
casemaking clothes moth (2)	イガ	*Tinea translucens* Meyrick	（チョウ目、ヒロズコガ科）日本、東洋区。衣類害虫。コイガの英名 clothes moth を使うことあり
casemaking clothes moth (3)		*Tinea dubiella* Stainton	（チョウ目、ヒロズコガ科）豪州区
casemaking clothes moth			clothes moth (1) を見よ
casemaking clothes moth			sparrow-nest clothes moth を見よ
cases			casebearer moths を見よ
Casey's June beetle		*Dinacoma caseyi* Blaisdell	（コウチュウ目、コガネムシ科）新北区
Casey's ladybird beetle		*Cycloneda polita* Casey	（コウチュウ目、テントウムシ科）新北区
cashew canarium moth			Cricula silkmoth を見よ
cashew Helopeltis		*Helopeltis anacardii* Miller	（カメムシ目、カスミカメムシ科）エチオピア区
cashew stem girdler		*Paranaleptes reticulata* (Thomson)	（コウチュウ目、カミキリムシ科）エチオピア区
cashew weevil		*Mecocorynus loripes* Chevrolat	（コウチュウ目、ゾウムシ科）エチオピア区
casks			yellow V carl を見よ
cassava hawk moth			ello sphinx を見よ
cassava hornworm		*Erinnyis ello* (Linnaeus)	（チョウ目、スズメガ科）新熱帯区
cassava mealybug		*Phenacoccus manihoti* Matile-Ferrero	（カメムシ目、コナカイガラムシ科）エチオピア区
cassava scale		*Aonidomytilus albus* (Cockerell)	（カメムシ目、マルカイガラムシ科）エチオピア区、東洋区、新北区、新熱帯区
cassava whitefly			tobacco whitefly を見よ

英名	和名	学名	所属、分布、ほか
cassia bark scale			herculean scale を見よ
cassia butterfly			lemon migrant を見よ
cassia webworm moth		*Anabasis ochrodesma* (Zeller)	（チョウ目、メイガ科）新北区。cassia webworm ともいう
Cassia's owl-butterfly		*Opsiphanes cassiae mexicana* Bristow	（チョウ目、タテハチョウ科）新熱帯区
Cassini			（カメムシ目、セミ科）アメリカの周期セミのうち、*Magicicada cassini* (Fisher) と *M. tredecassini* Alexander et Moore (13年セミ) の2種の俗称
Cassin's 17-year cicada		*Magicicada cassini* (Fisher)	（カメムシ目、セミ科）新北区。Cassini を参照
Cassin's 13-year cicada		*Magicicada tredecassini* Alexander et Moore	（カメムシ目、セミ科）新北区。Cassini を参照
Cassiope blue		*Plebejus cassiope* (Emmel et Emmel)	（チョウ目、シジミチョウ科）新北区
Cassiope blue			heather blue を見よ
cassius blue	ニシインドシジミ	*Leptotes cassius* (Cramer)	（チョウ目、シジミチョウ科）新北区、新熱帯区
Cassotis clearwing		*Hypoleria lavinia cassotis* (Bates)	（チョウ目、タテハチョウ科）新熱帯区。Lavinia glasswing を参照
Cassus roadside skipper		*Amblyscirtes cassus* Edwards	（チョウ目、セセリチョウ科）新北区、新熱帯区
Castalia green mantle		*Caria castalia* (Ménétriès)	（チョウ目、シジミタテハ科）新熱帯区
castanea bark beetle	ウラジロカシノキクイムシ	*Xyleborus pelliculosus* Eichhoff	（コウチュウ目、キクイムシ科）日本
castaneus garden beetle			velvety chafer を見よ
castanopsis ambrosia beetle (1)	シイノキクイムシ	*Xyleborus exesus* Blandford	（コウチュウ目、キクイムシ科）日本、東洋区。クリ害虫
castanopsis ambrosia beetle (2)	シイノコキクイムシ	*Xylosandrus compactus* (Eichhoff)	（コウチュウ目、キクイムシ科）日本、東洋区、大洋区、新北区、エチオピア区。カキ、チャなどの害虫
castanopsis gall thrips	シイオナガクダアザミウマ	*Varshneyia pasanii* (Mukaigawa)	（アザミウマ目、クダアザミウマ科）日本。シイノキ類につく
castanopsis thrips	シイマルクダアザミウマ	*Litotetothrips pasaniae* Kurosawa	（アザミウマ目、クダアザミウマ科）日本、東洋区。シイノキ類につく
Castitas hairstreak		*Iaspis castitas* (Druce)	（チョウ目、シジミチョウ科）新熱帯区
castniids			sun moths (1) を見よ
Castolus tufted-skipper		*Nisoniades castolus* (Hewitson)	（チョウ目、セセリチョウ科）新熱帯区
castor hairly caterpillar		*Euproctis lunata* Walker	（チョウ目、ドクガ科）東洋区
castor oil looper			croton caterpillar を見よ
castor semilooper			croton caterpillar を見よ　castor oil semi-looper ともいう
castor semi-looper moth			croton caterpillar を見よ
castor stem borer		*Xyleutes capensis* (Walker)	（チョウ目、ボクトウガ科）旧北区、エチオピア区
castors	カバタテハ属	*Ariadne*	（チョウ目、タテハチョウ科）の昆虫の総称

英名	和名	学名	所属、分布、ほか
casuarina mealybug		*Pseudoripersia turgipes* (Maskell)	（カメムシ目、コナカイガラムシ科）豪州区
cat biting louse	ネコハジラミ	*Felicola subrostratus* (Burmeister)	（ハジラミ目、ケモノハジラミ科）日本、汎世界。ネコに寄生
cat chewing louse			cat biting louse を見よ
cat flea	ネコノミ	*Ctenocephalides felis felis* (Bouché)	（ノミ目、ヒトノミ科）日本、汎世界。ヒトや多くの獣類に寄生
cat louse			cat biting louse を見よ
catabena moth		*Neogalea esula* (Druce)	（チョウ目、ヤガ科）豪州区、新熱帯区
catabena moth			lantana stick moth を見よ
Catalina cherry moth		*Cydia latiferreana* (Walsingham)	（チョウ目、ハマキガ科）新北区。クルミ害虫
Catalina orangetip			desert orange-tip (1) を見よ
Catalina shield-back cricket			Propst's shieldback を見よ
Catalonian rosy pearl			gingelly borer moth を見よ
catalpa mealybug			Comstock mealybug を見よ
catalpa midge		*Contarinia catalpae* (Comstock)	（ハエ目、タマバエ科）新北区
catalpa moth			catalpa sphinx を見よ
catalpa scale			white peach scale を見よ
catalpa sphinx		*Ceratomia catalpae* (Boisduval)	（チョウ目、スズメガ科）新北区。catalpa キササゲの類 の害虫。catalpa sphinx moth ともいう
catantopine grasshoppers			spine-breasted grasshoppers を見よ
Catasticta mimic		*Dismorphia lygdamis doris* Baumann et Reissinger	（チョウ目、シロチョウ科）新熱帯区
caterpillar			イモムシ、アオムシ、毛虫、カイコを指す
caterpillar destroyer			lesser armyworm parasite fly を見よ
caterpillar hunter		*Calosoma scrutator* Fabricius	（コウチュウ目、オサムシ科）新北区。灯火によく来る
caterpillar hunters		*Calosoma*	（コウチュウ目、オサムシ科）の昆虫の総称　捕食性でチョウ目などの幼虫を捕食
caterpillar wasp		*Podalonia communis* (Cresson)	（ハチ目、アナバチ科）新北区
caterpillars			butterflies and moths を見よ　幼虫の英名
cathayeia cottony scale	イイギリワタカイガラムシ	*Pulvinaria idesiae* Kuwana	（カメムシ目、カタカイガラムシ科）日本
cathedral termite		*Nasutitermes triodiae* (Froggatt)	（シロアリ目、シロアリ科）豪州区
catopid beetles		Catopidae	（コウチュウ目）の昆虫の総称
Cator's sapphire		*Etesiolaus catori* (Bethune-Baker)	（チョウ目、シジミチョウ科）エチオピア区
cats under cat			brown Siproeta を見よ
cat's-eye sapphire		*Lasaia arsis* Staudinger	（チョウ目、シジミタテハ科）新熱帯区
catseyes		*Zipaetis*	（チョウ目、タテハチョウ科）の昆虫の総称

英名	和名	学名	所属、分布、ほか
catshead sprite		*Pseudagrion coeleste* Longfield	（トンボ目、イトトンボ科）エチオピア区
cattail billbug		*Calendra pertinax* (Olivier)	（コウチュウ目、ゾウムシ科）新北区
cattail borer moth		*Bellura obliqua* (Walker)	（チョウ目、ヤガ科）新北区
cattail conehead		*Bucrates malivolans* (Scudder)	（バッタ目、キリギリス科）新北区
cat-tail moth			shy cosmet を見よ
cat-tail stem fly			rice stem fly (1)(3) を見よ
cattle biting fly		*Siphona stimulans* (Meigen)	（ハエ目、ヤドリバエ科）旧北区
cattle biting louse			cattle chewing louse を見よ
cattle-biting louse			cattle warble fly (1) を見よ
cattle chewing louse	ウシハジラミ	*Bovicola bovis* (Linnaeus)	（ハジラミ目、ケモノハジラミ科）日本、汎世界。ウシに寄生
cattle flat-fly			cattle fly を見よ
cattle fly		*Hippobosca variegata* Megerle	（ハエ目、シラミバエ科）旧北区
cattle grub			cattle warble fly (2) を見よ
cattle grubs			warble flies (2) を見よ
cattle heart		*Parides eurimedes* (Stoll)	（チョウ目、アゲハチョウ科）新北区、新熱帯区
cattleheart mimic		*Mimoides xeniades* (Hewitson)	（チョウ目、アゲハチョウ科）新熱帯区
cattleheart white		*Archonias brassolis approximata* (Butler)	（チョウ目、シロチョウ科）新熱帯区
cattle heel fly			cattle warble fly (1) を見よ
cattle ked			cattle fly を見よ
cattle louse			cattle chewing louse を見よ
cattle-poisoning sawfly		*Lophyrotoma interrupta* (Klug)	（ハチ目、Pergidae）豪州区
cattle red louse		*Trichodectes bovis* (Linnaeus)	（ハジラミ目、ケモノハジラミ科）新北区
cattle red louse			cattle chewing louse を見よ
cattle-tail louse	ペッカリジラミ	*Haematopinus quadripertusus* Fahrenholz	（シラミ目、ケモノジラミ科）新北区、新熱帯区、豪州区
cattle warble			cattle warble fly (1) を見よ
cattle warble fly (1)	ウシバエ	*Hypoderma bovis* (Linnaeus)	（ハエ目、ウシバエ科）日本、全北区、豪州区。ウシ、ウマの皮下に寄生
cattle warble fly (2)	キスジウシバエ	*Hypoderma lineata* (De Villiers)	（ハエ目、ウシバエ科）日本、汎世界。cattle warble-fly とも記す
cattleya aphid			orchid aphid (2) を見よ
cattleya bug		*Tenthecoris bicolor* Scott	（カメムシ目、カスミカメムシ科）旧北区
cattleya fly			orchid fly を見よ
cattleya gall midge		*Parallelodiplosis cattleyae* (Molliard)	（ハエ目、タマバエ科）旧北区
cattleya midge			cattleya gall midge を見よ

英名	和名	学名	所属、分布、ほか
cattleya scale	ナガクロホシカイガラムシ	*Parlatoria proteus* (Curtis)	(カメムシ目、マルカイガラムシ科) 日本、旧北区、大洋区、豪州区、新熱帯区。カンキツ、ランの害虫
Caucana metalmark		*Juditha caucana* (Stichel)	(チョウ目、シジミタテハ科) 新熱帯区
Caucasus beetle	コーカサスオオカブトムシ	*Chalcosoma caucasus* Fabricius	(コウチュウ目、コガネムシ科) 東洋区。アジア最大のカブトムシ
caudata canegrub		*Lepidiota caudata* Blackburn	(コウチュウ目、コガネムシ科) 豪州区
Caudell's conehead		*Neoconocephalus caudellianus* (Davis)	(バッタ目、キリギリス科) 新北区
cave barklice	セマガリチャタテ科	Psyllipsocidae	(チャタテムシ目) の昆虫の総称 洞穴、住宅、シロアリの巣に生息
cave cricket		*Troglophilus cavicola* (Kollar)	(バッタ目、カマドウマ科) 旧北区
cave crickets (1)	コロギス科	Gryllacrididae	(バッタ目) の昆虫の総称 翅は退化
cave crickets (2)		Stenopelmatidae	(バッタ目) の昆虫の総称
cave crickets			camel crickets (1)(2) を見よ
cave-dwelling crickets			cave crickets (2) を見よ
cave weta		*Gymnoplectron uncata* (Richards)	(バッタ目、カマドウマ科) 豪州区
cave weta			camel crickets (1) を見よ
cavity-dwelling ant		*Leptothorax curvispinosus* Mayr	(ハチ目、アリ科) 新北区
Cayenne tiger beetle		*Odontocheila cayennensis* (Fabricius)	(コウチュウ目、ハンミョウ科) 新熱帯区
ceanothus borer moth		*Synanthedon mellinipennis* (Boisduval)	(チョウ目、スカシバガ科) 新北区
ceanothus Nola moth		*Nola minna* Butler	(チョウ目、コブガ科) 新北区
ceanothus silk moth		*Hyalophora euryalus* (Boisduval)	(チョウ目、ヤママユガ科) 新北区
ceanothus silkworm			ceanothus silk moth を見よ
cebrionid beetles			robust click beetles を見よ
cecidomyiid fly		*Harmandiola globuli* (Rubsaamen)	(ハエ目、タマバエ科) 旧北区
cecids			gall midges を見よ
cecropia moth	セクロピアサン (アカスジシンジュサン)	*Hyalophora cecropia* (Linnaeus)	(チョウ目、ヤママユガ科) 新北区
cecropia sister			Phylaca sister を見よ
cedar aphid		*Cupressobium maui* Bradley	(カメムシ目、アブラムシ科) 大洋区
cedar aphid			cypress aphid (2) を見よ
cedar bark beetle		*Phloeosinus punctatus* LeConte	(コウチュウ目、キクイムシ科) 新北区
cedar bark beetle			peach bark beetle を見よ
cedar beetle		*Zenoa picea* (Beauvois)	(コウチュウ目、Rhipiceridae) 新北区

英名	和名	学名	所属、分布、ほか
cedar beetles	ホソクシヒゲムシ科	Callithipidae	（コウチュウ目）の昆虫の総称
cedar beetles (1)	クシヒゲムシ科	Rhipiceridae	（コウチュウ目）の昆虫の総称　幼虫がセミに寄生する種がある
cedar bush brown		*Mycalesis sirius* (Fabricius)	（チョウ目、タテハチョウ科）豪州区
cedar flat-headed borer		*Chrysobothris nixa* (Horn)	（コウチュウ目、タマムシ科）新北区
Cedar Island Bernardino blue		*Euphilotes bernardino garthi* Mattoni	（チョウ目、シジミチョウ科）新熱帯区
Cedar Island juniper hairstreak		*Callophrys gryneus cedrosensis* (Brown et Faulkner)	（チョウ目、シジミチョウ科）新熱帯区。juniper hairstreak (1) を参照
cedar leaf-miner		*Blastotere thujella* Packard	（チョウ目、メムシガ科）旧北区
cedar leaf moth		*Acleris undulana* Walsingham	（チョウ目、ハマキガ科）旧北区
cedar pinion moth		*Lithophane thujae* Webster et Thomas	（チョウ目、ヤガ科）新北区
cedar shoot caterpillar			mahogany shoot borer (1) を見よ
cedar tree borer		*Semanotus ligneus* (Fabricius)	（コウチュウ目、カミキリムシ科）新北区
cedar tussock		*Leptocneria reducta* (Walker)	（チョウ目、ドクガ科）豪州区
cedar tussock moth	スギドクガ	*Calliteara argentata* (Butler)	（チョウ目、ドクガ科）日本、旧北区。スギ、イブキなどの害虫
cedar twig moth			mahogany shoot borer (1) を見よ
cedar wood wasp			incense-cedar wasp を見よ
cedar wood wasps		Anaxyelidae	（ハチ目）の昆虫の総称
Cederberg copper		*Aloeides monticola* Pringe	（チョウ目、シジミチョウ科）エチオピア区
Celadon sister			Celerio sister を見よ
Celebes castor		*Ariadne celebensis* Holland	（チョウ目、タテハチョウ科）東洋区
Celebes map	シロオビイシガケチョウ	*Cyrestis strigata* C. et R. Felder	（チョウ目、タテハチョウ科）東洋区
Celebes sailer		*Neptis celebica* (Moore)	（チョウ目、タテハチョウ科）東洋区
Celerio sister		*Adelpha serpa* (Boisduval)	（チョウ目、タテハチョウ科）新熱帯区。*A. s. celerio* (Bates) も同英名
celery aphid	コウノフタオアブラムシ	*Cavariella konoi* Takahashi	（カメムシ目、アブラムシ科）日本、全北区。カナダでの英名。ポプラ害虫
celery aphid (1)		*Cavariella archangelicae* (Scopoli)	（カメムシ目、アブラムシ科）旧北区
celery aphid (2)	ニンジンアブラムシ	*Brachycaudus heraclei* Takahashi	（カメムシ目、アブラムシ科）日本、新北区。ニンジン、ミツバなどの害虫
celery aphid			bean aphid を見よ
celery fly		*Melanagromyza apii* Hering	（ハエ目、ハモグリバエ科）豪州区
celery-fly (1)		*Euleia heraclei* (Linnaeus)	（ハエ目、ミバエ科）旧北区。celery fly とも記す
celery leafminer			legume leafminer を見よ　米国での英名

英名	和名	学名	所属、分布、ほか
celery leaftier (1)	クロモンキノメイガ	*Udea testacea* (Butler)	（チョウ目、メイガ科）日本。ダイズ、アブラナ科野菜、花類の害虫
celery leaf tier (2)		*Udea rubigallis* (Guenée)	（チョウ目、メイガ科）新北区。celery leaftier moth ともいう
celery leaftier		*Phlyctaenia ferrugalis* Hübner	（チョウ目、メイガ科）新北区
celery leaftier			rusty-dot pearl (1) を見よ
celery looper			celery looper moth を見よ
celery looper moth		*Anagrapha falcifera* (Kirby)	（チョウ目、ヤガ科）新北区
celery pine moth		*Selidosema monacha* Hudson	（チョウ目、シャクガ科）豪州区
celery stalkworm			spruce pyralid を見よ
celery stalkworm			lucerne moth を見よ
celery stem maggot			celery fly (1) を見よ　幼虫の英名
celeryworm			black swallowtail を見よ
Celia's roadside skipper			roadside rambler を見よ
cell-barred metalmark			Cramer's Mesene を見よ
cellar beetle		*Blaps mucronata* Latreille	（コウチュウ目、ゴミムシダマシ科）旧北区
cellar beetle (1)		*Blaps mortisaga* Linnaeus	（コウチュウ目、ゴミムシダマシ科）旧北区
cellar beetles		*Blaps*	（コウチュウ目、ゴミムシダマシ科）の昆虫の総称
cellar Melipotis moth		*Melipotis cellaris* (Guenée)	（チョウ目、ヤガ科）新北区
Celmus hairstreak		*Celmia celmus* (Cramer)	（チョウ目、シジミチョウ科）新熱帯区
Celona hairstreak		*Nesiostrymon celona* (Hewitson)	（チョウ目、シジミチョウ科）新熱帯区
celtis aphid	エノキアブラムシ	*Toxoptera celtis* (Shinji)	（カメムシ目、アブラムシ科）日本
celtis cutworm	モクメクチバ	*Perinaenea accipiter* (Felder et Rogenhofer)	（チョウ目、ヤガ科）日本、東洋区
celtis elongate scale	エノキシロカイガラムシ	*Pseudaulacaspis celtis* (Kuwana)	（カメムシ目、マルカイガラムシ科）日本
Celypha moth	ウスクリイロヒメハマキ	*Celypha cespitana* (Hübner)	（チョウ目、ハマキガ科）日本、全北区
cembra pine bark beetle			larch ips を見よ
ceneraria leafminer		*Chromatomyia syngenesiae* Hardy	（ハエ目、ハモグリバエ科）豪州区
Centaur oakblue			dull oak blue を見よ
Centaurus beetle		*Augosoma centaurus* (Fabricius)	（コウチュウ目、コガネムシ科）エチオピア区
Central African mantis			African mantis (3) を見よ
Central American banded-skipper		*Autochton vectilucis* (Butler)	（チョウ目、セセリチョウ科）新熱帯区
Central American checkered-skipper			Colombian chequered skipper を見よ

英名	和名	学名	所属、分布、ほか
Central American giant cave cockroach			giant drummer を見よ
Central American paper wasp			nocturnal paper wasp を見よ
Central American potato tuber moth			potato tuber moth (2) を見よ
Central American sootywing			Mauve scallopwing を見よ
Central American umber skipper		*Poanes melane poa* (Evans)	（チョウ目、セセリチョウ科）新熱帯区。umber skipper を参照
central caper white	ネキヘリグロシロチョウ	*Belenois theuszi* (Dewitz)	（チョウ目、シロチョウ科）エチオピア区
central emperor swallowtail	オオサカハチアゲハ	*Papilio lormieri* Distant	（チョウ目、アゲハチョウ科）エチオピア区
central mountain blue		*Harpendyreus marungensis* (Joicey et Talbot)	（チョウ目、シジミチョウ科）エチオピア区
central sprite		*Celaenorrhinus chrysoglossa* (Mabille)	（チョウ目、セセリチョウ科）エチオピア区
centre-barred sallow		*Atethmia centrago* (Haworth)	（チョウ目、ヤガ科）旧北区
centre-barred sallow		*Atethmia xerampelina* Hübner	（チョウ目、ヤガ科）旧北区
Cepero's groundhopper		*Tetrix ceperoi* (Bolivar)	（バッタ目、ヒシバッタ科）旧北区。Cepero's grasshopper ともいう
Cephena Epitola		*Cephetola cephena* (Hewitson)	（チョウ目、シジミチョウ科）エチオピア区
Cephenes blue		*Pseudodipsas cephenes* Hewitson	（チョウ目、シジミチョウ科）豪州区
Cephoides skipper		*Mnasitheus cephoides* Hayward	（チョウ目、セセリチョウ科）新熱帯区
Cephus blue ringlet			blue night butterfly を見よ
cerambycids			long-horned beetles を見よ
cerambycine longhorns			long-horned beetles を見よ
ceraphronid wasps	ヒゲナガクロバチ科	Ceraphronidae	（ハチ目）の昆虫の総称
ceraphronids			ceraphronid wasps を見よ
Cerata groundstreak		*Calycopis cerata* (Hewitson)	（チョウ目、シジミチョウ科）新熱帯区
ceratinid bees			small carpenter bees (1) を見よ
Ceraunus blue			Antillean blue を見よ
ceravitreous coccids			pit scales を見よ
Cercene Epitola		*Stempfferia cercene* (Hewitson)	（チョウ目、シジミチョウ科）エチオピア区
cercerine wasps		Cercerinae	（ハチ目、アナバチ科）の昆虫の総称
Cerea metalmark butterfly			spangled golden Emesis を見よ

英名	和名	学名	所属、分布、ほか
cereal and dried fruit moths			cereal moths を見よ
cereal aphid			Oriental grassroot aphid を見よ
cereal armyworm			antler moth を見よ
cereal bug		*Eurygaster austriaca* (Schrank)	(カメムシ目、キンカメムシ科) 旧北区
cereal bug			Hottentot bug (2) を見よ
cereal chafer		*Anisoplia segetum* Herbst	(コウチュウ目、コガネムシ科) 旧北区
cereal chafers		*Anisoplia*	(コウチュウ目、コガネムシ科) の昆虫の総称
cereal curculio			banana weevil を見よ
cereal curculio			spinetailed weevil を見よ
cereal flea beetle		*Chaetocnema aridula* (Gyllenhal)	(コウチュウ目、ハムシ科) 旧北区
cereal leaf aphid			corn leaf aphid を見よ
cereal leaf aphid			oat bird-cherry aphid を見よ
cereal leaf beetle (1)		*Oulema melanopus* (Linnaeus)	(コウチュウ目、ハムシ科) 全北区
cereal leaf beetle (2)		*Lema cyanella* (Linnaeus)	(コウチュウ目、ハムシ科) 旧北区
cereal leaf-beetles			asparagus beetles を見よ
cereal leaf-miner		*Scythris temperatella* Lederer	(チョウ目、キヌバコガ科) 旧北区
cereal leaf miner			rice leafminer (2) を見よ
cereal mealybug		*Trionymus aseripticius* Williams	(カメムシ目、コナカイガラムシ科) 豪州区
cereal moths	マダラメイガ亜科	Phycitinae	(チョウ目、メイガ科) の昆虫の総称
cereal psocid			booklouse (3) を見よ　米国での英名
cereal stem moth		*Ochsenheimeria vacculella* Fischer von Röslerstamm	(チョウ目、Ochsenheimeriidae) 新北区
cereal stem moths		Ochsenheimeriidae	(チョウ目) の昆虫の総称
cereal thrips		*Haplothrips ganglbaueri* Schmutz	(アザミウマ目、クダアザミウマ科) 東洋区
cereal thrips			grain thrips を見よ
cereal whitefly		*Vasdavidius indicus* (David et Subramaniam)	(カメムシ目、コナジラミ科) 東洋区、エチオピア区
Ceres streamjack		*Metacnemis angusta* (Selys)	(トンボ目、モノサシトンボ科) エチオピア区
ceresium long-horned beetle		*Ceresium unicolor* (Fabricius)	(コウチュウ目、カミキリムシ科) 大洋区
Cerisy's sphinx			willow hawk moth (2) を見よ　北米での英名
Ceromia groundstreak		*Ziegleria ceromia* (Hewitson)	(チョウ目、シジミチョウ科) 新熱帯区
cerophytid beetles		Cerophytidae	(コウチュウ目) の昆虫の総称
cerophytids			cerophytid beetles を見よ
Cerulean hairstreak		*Chrysozephyrus suroia* (Tytler)	(チョウ目、シジミチョウ科) 東洋区

英名	和名	学名	所属、分布、ほか
Ceruleans	ルリウラナミシジミ属	*Jamides*	(チョウ目、シジミチョウ科) の昆虫の総称
Cervara skipper		*Vacerra cervara* Steinhauser	(チョウ目、セセリチョウ科) 新熱帯区
cerylonid beetles			minute bark beetles を見よ
Cestri hairstreak		*Strymon cestri* (Reakirt)	(チョウ目、シジミチョウ科) 新北区、新熱帯区
Cestus skipper			belted skipper を見よ
Ceylon ace		*Halpe egena* (Felder)	(チョウ目、セセリチョウ科) 東洋区
Ceylon blue glassy tiger	リュウキュウアサギマダラ	*Ideopsis similis* (Linnaeus)	(チョウ目、タテハチョウ科) 東洋区
Ceylon blue oakleaf			blue oakleaf (1) を見よ
Ceylon Cerulean		*Jamides coruscans* (Moore)	(チョウ目、シジミチョウ科) 東洋区
Ceylon forester		*Lethe dynsate* (Hewitson)	(チョウ目、タテハチョウ科) 東洋区
Ceylon hedge blue		*Udara lanka* (Moore)	(チョウ目、シジミチョウ科) 東洋区
Ceylon lesser albatross	ナミエシロチョウ	*Appias paulina minato* (Fruhstorfer)	(チョウ目、シロチョウ科) 日本、東洋区、豪州区。A. paulina (Cramer) の英名
Ceylon palmfly		*Elymnias singhala* Moore	(チョウ目、タテハチョウ科) 東洋区
Ceylon rose	ジョホンベニモンアゲハ	*Atrophaneura jophon* (Gray)	(チョウ目、アゲハチョウ科) 東洋区
Ceylon snow flat		*Tagiades distans* Moore	(チョウ目、セセリチョウ科) 東洋区
Ceylon swift	ヒメイチモンジセセリ	*Parnara naso bada* (Moore)	(チョウ目、セセリチョウ科) 日本、東洋区。straight swift (1) も参照
Ceylon treebrown		*Lethe daretis* (Hewitson)	(チョウ目、タテハチョウ科) 東洋区
chaenomedes sawfly	ボケヒメハバチ	*Priophorus cydoniae* (Takeuchi)	(ハチ目、ハバチ科) 日本
chafer			horse chestnut cockchafer を見よ
chafer beetle		*Adoretus bicaudatus* Arrow	(コウチュウ目、コガネムシ科) 東洋区
chafer beetles	コガネムシ科	Scarabaeidae	(コウチュウ目) の昆虫の総称 世界に 12,000 種。一般に chafer コガネムシといわれる
chafer beetles			May beetles (3) を見よ
chafer grubs			chafer beetles を見よ
chafer grubs		*Schizanycha*	(コウチュウ目、コガネムシ科) の昆虫の総称
chafers		*Macrodactylus*	(コウチュウ目、コガネムシ科) の昆虫の総称
chafers			chafer beetles を見よ 英国での英名
chafers			May beetles (3) を見よ
chaff scale (1)	マルクロホシカイガラムシ	*Parlatoria pergandii* Comstock	(カメムシ目、マルカイガラムシ科) 日本、汎世界。カンキツ類、サンゴジュ害虫
chaff scale (2)	マツリクロホシカイガラムシ	*Parlatoria cinerea* Doane et Hadden	(カメムシ目、マルカイガラムシ科) 日本、汎熱帯
chain-dot moth		*Cingilia catenaria* (Drury)	(チョウ目、シャクガ科) 新北区
chain fern borer moth		*Papaipema stenocelis* Dyar	(チョウ目、ヤガ科) 新北区
chain-spotted geometer			chain-dot moth を見よ

英名	和名	学名	所属、分布、ほか
chain-spotted geometer moth			chain-dot moth を見よ
chain swordtail			five-bar swordtail (1) を見よ
chaitophorine plantlice	ケアブラムシ亜科	Chaitophorinae	(カメムシ目、アブラムシ科) の昆虫の総称
chalcedectids		Chalcedectidae	(ハチ目) の昆虫の総称
Chalcedon	カリフォルニアヒョウモンモドキ	*Euphydryas chalcedona* (Doubleday)	(チョウ目、タテハチョウ科) 新北区
Chalcedon checkerspot			Chalcedon を見よ
chalcid flies			chalcid wasps (1) (2) を見よ
chalcid fly			clover seed chalcid (1) を見よ
chalcid wasps (1)	アシブトコバチ科	Chalcididae	(ハチ目) の昆虫の総称　copper のギリシャ語は khalk で、その金属色に因む。単に chalcid や chalcid fly といわれる
chalcid wasps (2)	コバチ上科	Chalcidoidea	(ハチ目) の昆虫の総称
chalcidids			chalcid wasps (1) を見よ
chalcidoid flies			chalcid wasps (2) を見よ
chalcids			chalcid wasps (1)(2) を見よ
chalcis-flies			chalcid wasps (2) を見よ　米国での英名
chalk carpet		*Ortholitha bipunctaria* (Denis et Schiffermüller)	(チョウ目、シャクガ科) 旧北区
chalk-fliff conch		*Cochylidia rupicola* (Curtis)	(チョウ目、ホソハマキガ科) 旧北区
chalk-hill blue	コリドンヒメシジミ	*Lysandra coridon* (Poda)	(チョウ目、シジミチョウ科) 旧北区。chalk-hill blue butterfly ともいう
chalk-hill lancewing		*Epermenia insecurella* (Stainton)	(チョウ目、ササベリガ科) 旧北区
chalk knot-horn		*Phycitodes maritima* Tengstrom	(チョウ目、メイガ科) 旧北区
chalk-marked skipper		*Thespieus dalman* (Latreille)	(チョウ目、セセリチョウ科) 新熱帯区
chalk white		*Elodina parthia* (Hewitson)	(チョウ目、シロチョウ科) 豪州区
chalky bird-dropping moth		*Acontia cretata* (Grote et Robinson)	(チョウ目、ヤガ科) 新北区
chalky Inga moth		*Inga cretacea* (Zeller)	(チョウ目、マルハキバガ科) 新北区
chalky percher		*Diplacodes trivialis* (Rambur)	(トンボ目、トンボ科) 東洋区
chalky spreadwing		*Lestes sigma* Calvert	(トンボ目、アオイトトンボ科) 新北区
chalky wave moth		*Scopula purata* (Guenée)	(チョウ目、シャクガ科) 新北区。chalky wave ともいう
chamaecyparia torymid	ヒノキモンオナガコバチ	*Megastigmus chamaecyparidis* Kamijo	(ハチ目、オナガコバチ科) 日本
chamaecyparis scale	チャボヒバフクロカイガラムシ	*Eriococcus chabohiba* Kuwana et Nitobe	(カメムシ目、フクロカイガラムシ科) 日本
Chamaeleon hopper		*Platylesches chamaeleon* (Mabille)	(チョウ目、セセリチョウ科) エチオピア区

英名	和名	学名	所属、分布、ほか
chamaemyiid flies			aphid flies (1) を見よ
Chamberlain's yellow		*Eurema chamberlaini* Butler	(チョウ目、シロチョウ科) 新北区
chamois biting louse		*Bovicola alpinus* Keler	(ハジラミ目、ケモノハジラミ科) 旧北区
chamois fly			chamois ked を見よ　カモシカの1種 chamois シャモア に由来
chamois ked		*Melophagus rupicaprinus* Rondani	(ハエ目、シラミバエ科) 旧北区
chamomile conch		*Cochylidia implicitana* (Wocke)	(チョウ目、ホソハマキガ科) 旧北区
chamomile shark		*Cucullia chamomillae* (Denis et Schiffermüller)	(チョウ目、ヤガ科) 旧北区
Chamul skipper		*Euphyes chamuli* Freeman	(チョウ目、セセリチョウ科) 新熱帯区
Chanchamayo sabre-wing		*Parelbella peruana* Mielke	(チョウ目、セセリチョウ科) 新熱帯区
Chanchamayo sabre-wing			Ahira sabre-wing を見よ
changa		*Scapteriscus vicinus* Scudder	(バッタ目、ケラ科) 新北区
changa			mole crickets (1) を見よ
changa mole cricket		*Scapteriscus didactylus* (Latreille)	(バッタ目、ケラ科) 新熱帯区
changeable grass veneer moth		*Fissicrambus mutabilis* (Clemens)	(チョウ目、メイガ科) 新北区。changeable grass-veneer ともいう
Chanler's Charaxes		*Charaxes chanleri* Holland	(チョウ目、タテハチョウ科) エチオピア区
Channel Islands pug		*Eupithecia ultimaria* Boisduval	(チョウ目、シャクガ科) 旧北区
Chan's megastick		*Phobaeticus chani* Bragg	(ナナフシ目、ナナフシ科) 東洋区。56 cm に達する最長の昆虫。ボルネオには本種のほか 32 cm 長の *P. kirbyi* Brunner von Wattenwyl も産する
Chanstola skipper		*Hesperilla chaostola* Meyrick	(チョウ目、セセリチョウ科) 豪州区
Chaonitis metalmark		*Chalodeta chaonitis* (Hewitson)	(チョウ目、シジミタテハ科) 新熱帯区
chaparral camel crickets		*Gammarotettix*	(バッタ目、コロギス科) 新北区
chaparral false katydid		*Platylyra californica* Scudder	(バッタ目、キリギリス科) 新北区
chaparral mealybug		*Spirococcus quercinus* McKenzie	(カメムシ目、コナカイガラムシ科) 新北区
chaparral shieldback		*Cyrtophyllicus chlorum* Hebard	(バッタ目、キリギリス科) 新北区
Chapman's blue		*Polyommatus thersites* (Cantener)	(チョウ目、シジミチョウ科) 旧北区
Chapman's cupid		*Cupido argiades diporides* (Chapman)	(チョウ目、シジミチョウ科) 東洋区
Chapman's cupid			short-tailed blue を見よ

英名	和名	学名	所属、分布、ほか
Chapman's green hairstreak		*Callophrys avis* Chapman	（チョウ目、シジミチョウ科）旧北区
Chapman's hedge blue		*Notarthrinus binghami* Chapman	（チョウ目、シジミチョウ科）東洋区
Chapman's ringlet	オオベニヒカゲ	*Erebia palarica* Chapman	（チョウ目、タテハチョウ科）旧北区
Chara checkerspot		*Dymasia chara* (Edwards)	（チョウ目、タテハチョウ科）新北区
charaxes			leafwings を見よ
charcoal beetle			charcoal borer を見よ
charcoal borer		*Melanophila consputa* LeConte	（コウチュウ目、タマムシ科）新北区
charcoal-winged percher			black-tipped percher を見よ
charipid wasps		Charipidae	（ハチ目）の昆虫の総称
charlock weevil		*Ceutorhynchus contractus* (Marsham)	（コウチュウ目、ゾウムシ科）旧北区
charming underwing moth		*Catocala blandula* Hulst	（チョウ目、ヤガ科）新北区
charred dagger moth		*Acronicta brumosa* Guenée	（チョウ目、ヤガ科）新北区
chasers			common skimmers を見よ　英名は追跡するような飛翔に由来
checkered Apogeshna moth		*Apogeshna stenialis* (Guenée)	（チョウ目、メイガ科）新北区
checkered beetle		*Enoclerus rosmarus* (Say)	（コウチュウ目、カッコウムシ科）新北区
checkered beetles	カッコウムシ科	Cleridae	（コウチュウ目）の昆虫の総称
checkered flower beetles			checkered beetles を見よ
checkered gem		*Zeritis neriene* Boisduval	（チョウ目、シジミチョウ科）エチオピア区
checkered sootywing		*Bolla cylindus* (Godman et Salvin)	（チョウ目、セセリチョウ科）新熱帯区
checkered swallowtail			chequered swallowtail を見よ
checkered tussock moth		*Halysidota tessellaris* J. E. Smith	（チョウ目、ヒトリガ科）新北区
checkered tussock moth			walnut caterpillar moth を見よ
checkered white			southern cabbage worm を見よ
cheese flies			skipper flies を見よ
cheese fly			cheese skipper を見よ
cheese hopper			cheese skipper を見よ　幼虫の英名
cheese maggot			cheese skipper を見よ　幼虫の英名
cheese maggots			skipper flies を見よ　幼虫の英名
cheese skipper	チーズバエ	*Piophila (Piophila) casei* (Linnaeus)	（ハエ目、チーズバエ科）日本、汎世界。動物性食品に発生
cheese skipper flies			skipper flies を見よ　米国での英名
cheese skippers			skipper flies を見よ
cheetah		*Argina amanda* (Boisduval)	（チョウ目、ヒトリガ科）エチオピア区

英名	和名	学名	所属、分布、ほか
Chelidonis dartwhite		*Catasticta chelidonis* (Hopffer)	(チョウ目、シロチョウ科) 新熱帯区
Chelmos blue		*Polyommatus iphigenia* (Herrich-Schäffer)	(チョウ目、シジミチョウ科) 旧北区
chelonariid beetle		*Chelonarium lecontei* Thomson	(コウチュウ目、ダエンマルトゲムシ科) 新北区
chelonariid beetles			turtle beetles を見よ
chenopod aphid			purple Lamb's quarters mealy aphid を見よ
chenopodium aphid			chenopod aphid を見よ
chenopodium long-horned beetle			Hawaiian aweoweo long-horned beetle を見よ
chenopodium Scythris moth		*Scythris limbella* (Fabricius)	(チョウ目、キヌバコガ科) 全北区
Chen's whitefly		*Aleurocanthus cheni* B. Young	(カメムシ目、コナジラミ科) 旧北区。カンキツ害虫
chequered ace		*Sovia separata* (Moore)	(チョウ目、セセリチョウ科) 東洋区
chequered beetles			checkered beetles を見よ
chequered blue	ジョウザンシジミ	*Scolitandites orion jezoensis* (Matsumura)	(チョウ目、シジミチョウ科) 日本。*S. orion* (Pallas) 旧北区の英名
chequered blue (1)		*Theclinesthes serpentata* (Herrich-Schäffer)	(チョウ目、シジミチョウ科) 豪州区
chequered blue			eastern baron blue を見よ
chequered copper			small copper (2) を見よ
chequered cuckoo bee		*Thyreus caeruleopunctatus* Blanchard	(ハチ目、ミツバチ科) 豪州区
chequered fruit-tree tortrix	ウスアミメトビハマキ	*Pandemis corylana* (Fabricius)	(チョウ目、ハマキガ科) 日本、旧北区
chequered grass veneer		*Catoptria falsella* (Denis et Schiffermüller)	(チョウ目、メイガ科) 旧北区
chequered grass-skipper			Tillyard's skipper を見よ
chequered hooked smudge		*Ypsolophus asperellu* Linnaeus	(チョウ目、クチブサガ科) 旧北区
chequered lancer			silver spotted lancer を見よ
chequered ranger		*Kedestes lepenula* (Wallengren)	(チョウ目、セセリチョウ科) エチオピア区
chequered skipper	タカネキマダラセセリ	*Carterocephalus phalaemon akaishianus* Fujioka	(チョウ目、セセリチョウ科) 日本。*C. phalaemon* (Pallas) 旧北区の英名
chequered skipper			chequered ranger を見よ
chequered skipper butterfly			chequered skipper を見よ
chequered straw pearl		*Evergestis pallidata* Hufnagel	(チョウ目、メイガ科) 全北区。キャベツ害虫
chequered swallowtail	オナシアゲハ	*Papilio demoleus* (Linnaeus)	(チョウ目、アゲハチョウ科) 日本、東洋区、豪州区。*P. d. libanius* (Fruhstorfer) 日本、東洋区、*P. d. sthenelus* (Macleay) 豪州区も同英名
chermes bugs			psyllines を見よ

英名	和名	学名	所属、分布、ほか
Chermock's gemmed-satyr		*Cyllopsis pseudopephredo* Chermock	（チョウ目、タテハチョウ科）新熱帯区
cherokee clubtail		*Gomphus consanguis* Selys	（トンボ目、サナエトンボ科）新北区
cheroot beetle			cigarette beetle を見よ
cherry and hawthorn sawfly		*Profenusa canadensis* (Marlatt)	（ハチ目、ハバチ科）新北区
cherry aphid	サクラコブアブラムシ	*Tuberocephalus sakurae* (Matsumura)	（カメムシ目、アブラムシ科）日本
cherry aphid (1)	ニワウメクロコブアブラムシ	*Myzus cerasi* (Fabricius)	（カメムシ目、アブラムシ科）日本、汎世界。ウメ、サクラ害虫
cherry bark beetle	サクラノキクイムシ	*Polygraphus ssiori* Niijima	（コウチュウ目、キクイムシ科）日本、旧北区
cherry-bark moth		*Enarmonia formosana* (Scopoli)	（チョウ目、ハマキガ科）旧北区
cherry bark scale	サクラアカカイガラムシ	*Kuwanina parva* (Maskell)	（カメムシ目、フクロカイガラムシ科）日本。サクラ害虫
cherry bark tortrix			cherry-bark moth を見よ
cherry bark tortrix moth			cherry-bark moth を見よ
cherry black aphid			cherry aphid (1) を見よ
cherry blackfly			cherry aphid (1) を見よ
cherry-blossom moth			cherry fruit moth を見よ
cherry blotch miner moth		*Phyllonorycter propinquinella* (Braun)	（チョウ目、ホソガ科）新北区
cherry bluet		*Enallagma concisum* Williamson	（トンボ目、イトトンボ科）新北区
cherry borer moth			cherry tree borer (2) を見よ
cherry brown tortrix	ウストビハマキ	*Pandemis chlorograpta* (Meyrick)	（チョウ目、ハマキガ科）日本、旧北区
cherry casebearer		*Coleophora pruniella* Clemens	（チョウ目、ツツミノガ科）新北区。cherry casebearer moth ともいう
cherry caterpillar	モンクロシャチホコ	*Phalera flavescens* (Bremer et Grey)	（チョウ目、シャチホコガ科）日本、旧北区、東洋区。リンゴなど果樹、サクラなどの害虫
cherry chafer	サクラコガネ	*Anomala daimiana* Harold	（コウチュウ目、コガネムシ科）日本
cherry curculio		*Tachypterellus consors cerasi* List	（コウチュウ目、ゾウムシ科）新北区。リンゴ他害虫
cherry dagger moth	サクラケンモン	*Hyboma adaucta* (Warren)	（チョウ目、ヤガ科）日本、旧北区。リンゴ、モモ、サクラの害虫
cherry dagger moth			speared dagger moth を見よ
cherry drosophila	オウトウショウジョウバエ	*Drosophila suzukii* (Matsumura)	（ハエ目、ショウジョウバエ科）日本、旧北区、東洋区。オウトウ、モモなど果樹害虫
cherry ermine			orchard ermine を見よ
cherry ermine moth			apple ermine moth を見よ
cherry-eye sprite		*Pseudagrion sublacteum* (Karsch)	（トンボ目、イトトンボ科）エチオピア区
cherry fruit fly (1)	シロオビオウトウミバエ	*Rhagoletis cingulata* (Loew)	（ハエ目、ミバエ科）新北区。オウトウ害虫

英名	和名	学名	所属、分布、ほか
cherry fruit fly (2)	ヨーロッパオウトウミバエ	*Rhagoletis cerasi* (Linnaeus)	(ハエ目、ミバエ科) 旧北区。オウトウ害虫
cherry fruit moth		*Argyresthia pruniella* (Clerck)	(チョウ目、メムシガ科) 全北区
cherry fruit moth			apple fruit moth を見よ
cherry fruit sawfly		*Hoplocampa cookei* (Clarke)	(ハチ目、ハバチ科) 新北区
cherry fruit weevil			apricot weevil を見よ
cherry fruit worm		*Grapholita packardi* Zeller	(チョウ目、ハマキガ科) 新北区。cherry fruitworm moth ともいう
cherry gall wasp		*Cynips quercusfolii* Linnaeus	(ハチ目、タマバチ科) 旧北区。本種がカシの葉の下面につくる虫えいを cherry gall といい、径 20 mm のサクランボ状
cherry gelechiid	サクラキバガ	*Anacampsis anisogramma* Meyrick	(チョウ目、キバガ科) 日本、旧北区。モモ、サクラなどの害虫
cherry horn worm	ウチスズメ	*Smerinthus planus* Walker	(チョウ目、スズメガ科) 日本、旧北区。リンゴ、サクラ、ポプラなどの害虫
cherry kuwania scale			cherry bark scale を見よ
cherry leaf beetle		*Tricholochmaea cavicollis* (LeConte)	(コウチュウ目、ハムシ科) 新北区
cherry leaf-cone caterpillar moth		*Caloptilia invariabilis* (Braun)	(チョウ目、ホソガ科) 新北区
cherry leafhopper			silver leafhopper を見よ
cherry leaf-miner beetle	アカガネチビタマムシ	*Trachys tsushimae* (Obenberger)	(コウチュウ目、タマムシ科) 日本、旧北区
cherry leaf roller		*Archips cerasivoranus* (Fitch)	(チョウ目、ハマキガ科) 新北区
cherry leaf worm	アカバキリガ	*Orthosia carnipennis* (Butler)	(チョウ目、ヤガ科) 日本、旧北区、東洋区。リンゴなど果樹、サクラの害虫
cherry looper		*Chloroclysta approximata* (Walker)	(チョウ目、シャクガ科) 豪州区
cherry maggot			cherry fruit fly (1)(2) を見よ　幼虫の英名
cherry narrow bark beetle	サクラノホソキクイムシ	*Xyleborus attenuatus* Blandford	(コウチュウ目、キクイムシ科) 日本、旧北区、東洋区
cherry nepticulid	オオイシチビガ	*Trifurcula oishiella* Matsumura	(チョウ目、モグリチビガ科) 日本
cherrynose		*Macrotristria angularis* (Germar)	(カメムシ目、セミ科) 豪州区
cherry plant-louse			cherry aphid (1) を見よ
cherry rhopalosiphum			rice root aphid を見よ
cherry sawfly	サクラセグロハバチ	*Allantus nakabusensis* Takeuchi	(ハチ目、ハバチ科) 日本。サクラ害虫
cherry sawfly	サクラホソハバチ	*Pareophora gracilis* Takeuchi	(ハチ目、ハバチ科) 日本。サクラ害虫
cherry sawfly	サクラヒメハバチ	*Trichiocampus pruni* Takeuchi	(ハチ目、ハバチ科) 日本、旧北区。サクラ害虫
cherry sawfly			pear sawfly (1) を見よ　豪州での英名
cherry scale			Forbes scale を見よ

英名	和名	学名	所属、分布、ほか
cherry scallop-shell moth			scallop shell を見よ　北米での英名
cherry shoot borer moth		*Argyresthia oreasella* Chambers	（チョウ目、メムシガ科）新北区
cherry slug			pear sawfly (1) を見よ　豪州での英名
cherry slugworm			pear sawfly (1) を見よ
cherry spinner			brindled beauty を見よ
cherry-spotted metalmark		*Adelotypa annuliera* (Godman)	（チョウ目、シジミタテハ科）新熱帯区
cherry stink bug			forest bug を見よ
cherry tent caterpillar			cherry leaf roller を見よ
cherry tiger moth	カクモンヒトリ	*Thanatarctia inaequalis* (Butler)	（チョウ目、ヒトリガ科）日本。カンキツなど果樹やサクラ害虫
cherry-tree bark beetle	ヨツメキクイムシ	*Polygraphus polygraphus* (Linnaeus)	（コウチュウ目、キクイムシ科）日本、旧北区
cherry tree borer (1)	コスカシバ	*Synanthedon hector* (Butler)	（チョウ目、スカシバガ科）日本、旧北区。モモ、ウメ、サクラなどの害虫
cherry tree borer (2)		*Maroga unipuncta* Donovan	（チョウ目、マルハキバガ科）旧北区。カンキツ害虫
cherry tree borer (3)		*Maroga melanostigma* Wallengren	（チョウ目、マルハキバガ科）豪州区
cherry tree tortricid			cherry leaf roller を見よ
cherry vinegar fly			cherry drosophila を見よ
cherry web-spinning sawfly	サクラヒラタハバチ	*Neurotoma iridescens* (Andre)	（ハチ目、ヒラタハバチ科）日本、旧北区
cherry weevil		*Anthonomus rectirostris* (Linnaeus)	（コウチュウ目、ゾウムシ科）旧北区
cherry weevil		*Curculio cerasorum* Paykull	（コウチュウ目、ゾウムシ科）旧北区
cherry wood borer			cherry tree borer (1) を見よ
cherry worm			cherry leaf roller を見よ
chersia sphinx			great ash sphinx を見よ
Cherubina emperor		*Doxocopa laurentia cherubina* (C. et R. Felder)	（チョウ目、タテハチョウ科）新熱帯区
Cheshire horsefly		*Atylotus plebejus* (Fallén)	（ハエ目、アブ科）旧北区。日本には *A. p. sibiricus* (Olsouflev) シベリアキイロアブが分布する
chestnut			chestnut moth を見よ
chestnut and black royal		*Tajuria yajna* (Doherty)	（チョウ目、シジミチョウ科）東洋区
chestnut angle	アツバセセリ	*Odontoptilum angulata* (Felder)	（チョウ目、セセリチョウ科）東洋区
chestnut anomalous blue		*Polyommatus pelopi* (Brown)	（チョウ目、シジミチョウ科）旧北区
chestnut aphid	クリヒゲマダラアブラムシ	*Myzocallis kuricola* (Matsumura)	（カメムシ目、アブラムシ科）日本、旧北区、東洋区
chestnut bark borer		*Anoplodera mites* (Forster)	（コウチュウ目、カミキリムシ科）新北区

英名	和名	学名	所属、分布、ほか
chestnut bob		*Iambrix salsala* (Moore)	（チョウ目、セセリチョウ科）東洋区
chestnut borer moth		*Synanthedon castaneae* (Busck)	（チョウ目、スカシバガ科）新北区
chestnut brown chafer			brown chafer (1) を見よ
chestnut button		*Acleris hyemana* (Haworth)	（チョウ目、ハマキガ科）旧北区
chestnut clearwing moth			chestnut borer moth を見よ
chestnut-coloured carpet		*Thera cognata* (Thunberg)	（チョウ目、シャクガ科）旧北区
chestnut crescent		*Anthanassa argentea* (Godman et Salvin)	（チョウ目、タテハチョウ科）新北区、新熱帯区
chestnut curuculio	コナラシギゾウムシ	*Curculio dentipes* (Roelofs)	（コウチュウ目、ゾウムシ科）日本、旧北区
chestnut fruit noctuid	ネスジキノカワガ	*Garella ruficirra* (Hampson)	（チョウ目、ヤガ科）日本、東洋区。クリ害虫
chestnut gall wasp	クリタマバチ	*Dryocosmus kuriphilus* Yasumatsu	（ハチ目、タマバチ科）日本、全北区。クリ害虫
chestnut heath	ユーラシアヒメヒカゲ	*Coenonympha glycerion* Borkhausen	（チョウ目、タテハチョウ科）旧北区
chestnut lace bug	ヒメグンバイ	*Uhlerites debilis* (Uhler)	（カメムシ目、グンバイムシ科）日本、旧北区、東洋区。クリ、カシ類の害虫
chestnut leaf-cut weevil	オトシブミ	*Apoderus jekelii* Roelofs	（コウチュウ目、オトシブミ科）日本、旧北区
chestnut leafwing		*Anaea echemus* (Doubleday et Hewitson)	（チョウ目、タテハチョウ科）新北区
chestnut-leaved oak aphid	クヌギエダアブラムシ	*Diphyllaphis quercus* Takahashi	（カメムシ目、アブラムシ科）日本
chestnut-leaved oak bark beetle	クヌギノキクイムシ	*Eidophelus imitans* Eichhoff	（コウチュウ目、キクイムシ科）日本、旧北区
chestnut-leaved oak broad-mouth weevil	カシワクチブトゾウムシ	*Myllocerus griseus* Roelofs	（コウチュウ目、ゾウムシ科）日本
chestnut-leaved oak bug			quercus stink bug を見よ
chestnut-leaved oak gall midge			quercus gall midge を見よ
chestnut leaved oak gall wasp			quercus gall wasp (1) を見よ
chestnut-leaved oak heliozelid	クヌギチビガモドキ	*Heliozela subpurpurea* Meyrick	（チョウ目、ツヤコガ科）日本
chestnut-leaved oak leaf worm			hibiscus looper を実よ
chestnut-leaved oak leafminer	クヌギキハモグリガ	*Tischeria quercifolia* Kuroko	（チョウ目、ムモンハモグリガ科）日本
chestnut-leaved oak scale	ヒメタマカイガラムシ	*Kermes miyasakii* Kuwana	（カメムシ目、タマカイガラムシ科）日本

英名	和名	学名	所属、分布、ほか
chestnut-marked pondweed moth		*Parapoynx badiusalis* (Walker)	(チョウ目、メイガ科) 新北区
chestnut-marked skipper		*Thespieus macareus* (Herrich-Schäffer)	(チョウ目、セセリチョウ科) 新北区、新熱帯区
chestnut moth		*Conistra vaccinii* (Linnaeus)	(チョウ目、ヤガ科) 旧北区
chestnut moth			splended piercer を見よ
chestnut Presba		*Syncordulia venator* (Barnard)	(トンボ目、エゾトンボ科) エチオピア区
chestnut pyralid	ナカトビフトメイガ	*Orthaga achatina* (Butler)	(チョウ目、メイガ科) 日本、旧北区
chestnut Rajah	ヘリボシフタオチョウ	*Charaxes durnfordi* Distant	(チョウ目、タテハチョウ科) 東洋区
chestnut sawfly		*Apethymus kuri* Takeuchi	(ハチ目、ハバチ科) 日本。クリ害虫
chestnut Schizura moth		*Schizura badia* (Packard)	(チョウ目、シャチホコガ科) 新北区
chestnut shieldback		*Neduba castanea* (Scudder)	(バッタ目、キリギリス科) 新北区
chestnut short-winged katydid		*Dichopetala castanea* Rehn et Hebard	(バッタ目、キリギリス科) 新北区
chestnut-streaked sailer		*Neptis jumbah* Moore	(チョウ目、タテハチョウ科) 東洋区
chestnut Takachiho scale	タカチホカタカイガラムシ	*Lecanium takachihoi* Kuwana	(カメムシ目、カタカイガラムシ科) 日本、旧北区
chestnut thyridid	アカジママドガ	*Striglina cancellata* (Christoph)	(チョウ目、マドガ科) 日本、旧北区。クリ、クルミなどの害虫
chestnut thyridid		*Striglina scitaria* Walker	(チョウ目、マドガ科) 東洋区、大洋区
chestnut tiger	アサギマダラ	*Parantica sita niphonica* (Moore)	(チョウ目、タテハチョウ科) 日本、東洋区。chestnut tiger butterfly ともいう。mountain tiger を参照
chestnut timberworm		*Melittomma sericeum* (Harris)	(コウチュウ目、ツツシンクイ科) 新北区
chestnut tortrix		*Cydia splendana* Hübner	(チョウ目、ハマキガ科) 旧北区
chestnut-tree louse		*Lachnus longipes* Buckton	(カメムシ目、アブラムシ科) 旧北区
chestnut-tree louse			large daimyo oak aphid を見よ
chestnut trunk borer	ミヤマカミキリ	*Massicus raddei* (Blessig)	(コウチュウ目、カミキリムシ科)日本、旧北区。クリ、カシ類など多くの樹木害虫
chestnut tussock moth	アカヒゲドクガ	*Calliteara lunulata* (Butler)	(チョウ目、ドクガ科) 日本、旧北区。クリ害虫
chestnut twig borer	クリタマムシ	*Taxoscelus auriceps* (E. Saunders)	(コウチュウ目、タマムシ科) 日本、旧北区。カシ類害虫
chestnut weevil		*Curculio elephas* (Gyllenhal)	(コウチュウ目、ゾウムシ科) 旧北区
chestnuts		Orthosiinae	(チョウ目、ヤガ科) の昆虫の総称
chevron	キマダラナミシャク	*Eulithis testata* (Linnaeus)	(チョウ目、シャクガ科) 日本、旧北区
chevron cutworm		*Diarsia intermixta* (Guenée)	(チョウ目、ヤガ科) 豪州区
chevron moth			chevron を見よ

英名	和名	学名	所属、分布、ほか
Chewaucan shieldback		*Idiostatus chewaucan* Rentz	（バッタ目、キリギリス科）新北区
chewing lice	ハジラミ目	Mallophaga	の昆虫の総称　世界に約 5000 種
chi		*Antitype chi* (Linnaeus)	（チョウ目、ヤガ科）旧北区。ギリシャ文字の x に由来
chi moth			chi を見よ
Chiapan blue-skipper		*Quadrus francesius* Freeman	（チョウ目、セセリチョウ科）新熱帯区
Chiapan giant-skipper		*Stallingsia jacki* Stallings, Turner et Stallings	（チョウ目、セセリチョウ科）新熱帯区
Chiapan silver-plated skipper		*Vettius coryna argentus* Freeman	（チョウ目、セセリチョウ科）新熱帯区。Coryna skipper を参照
Chiapan skipper		*Mimia chiapaensis* Freeman	（チョウ目、セセリチョウ科）新熱帯区
Chiapan skipperling		*Dalla nubes* Steinhauser	（チョウ目、セセリチョウ科）新熱帯区
Chiapan white		*Perrhybris pamela chajulensis* Maza et Maza	（チョウ目、シロチョウ科）新熱帯区。*P. p. mapa* Maza et Maza も同英名。Pamela を参照
Chiapas cattleheart		*Parides anchises marthilia* (Maza)	（チョウ目、アゲハチョウ科）新熱帯区
Chiapas crescent		*Castilia chiapaensis* (Beutelspacher)	（チョウ目、タテハチョウ科）新熱帯区
Chiapas hairstreak		*Callophrys scaphia* Clench	（チョウ目、シジミチョウ科）新熱帯区
chicken body louse	ニワトリオオハジラミ	*Menacanthus stramineus* (Nitzsch)	（ハジラミ目、タンカクハジラミ科）日本、汎世界。ニワトリに寄生
chicken dung fly		*Fannia pusio* (Wiedemann)	（ハエ目、ヒメイエバエ科）新北区
chicken flea	ニワトリノミ	*Ceratophyllus gallinae* (Schrank)	（ノミ目、ナガノミ科）日本、全北区。日本亜種は *C. g. dilatus* Dudolkina
chicken flea			sticktight flea を見よ
chicken fluff louse			fluff louse (1) を見よ
chicken head louse	ハバビロナガハジラミ	*Cuclotogaster heterographis* (Nitzsch)	（ハジラミ目、チョウカクハジラミ科）日本、汎世界
chicken louse	ニワトリハジラミ	*Menopon gallinae* (Linnaeus)	（ハジラミ目、タンカクトリハジラミ科）日本、汎世界。ニワトリ、ハトなどに寄生
chicken louse			chicken head louse を見よ
chicken shaft louse			chicken louse を見よ
chicken wing louse	ニワトリナガハジラミ	*Lipeurus caponis* (Linnaeus)	（ハジラミ目、チョウカクハジラミ科）日本、汎世界。ニワトリに寄生
chickweed geometer		*Haematopis grataria* (Fabricius)	（チョウ目、シャクガ科）新北区
chickweed geometer moth			chickweed geometer を見よ
chief butterfly		*Anayrus echeria* (Stoll)	（チョウ目、タテハチョウ科）エチオピア区
chief earwig		*Prolabia arachidis* (Yersin)	（ハサミムシ目、クロハサミムシ科）汎世界
chief earwig			bone-house earwig を見よ　米国での英名
chigger			chigoe flea を見よ
chigoe			chigoe flea を見よ

英名	和名	学名	所属、分布、ほか
chigoe flea	スナノミ	*Tunga penetrans* (Linnaeus)	(ノミ目、スナノミ科) 新北区、新熱帯区。家畜、ヒトの足に寄生
chigoe fleas	スナノミ科	Tungidae	(ノミ目) の昆虫の総称
chigoes			common fleas を見よ
chigoes			fleas を見よ
Chihuahuan juniper hairstreak		*Callophrys gryneus turkingtoni* Johnson	(チョウ目、シジミチョウ科) 新熱帯区。juniper hairstreak (1) を参照
Chilcott's Coleotechnites moth		*Coleotechnites chilcotti* (Freeman)	(チョウ目、キバガ科) 新北区
Childers canegrub		*Antitrogus parvulus* Britton	(コウチュウ目、コガネムシ科) 豪州区
children's stick insect		*Tropidoderus childrenii* Gray	(ナナフシ目、ナナフシ科) 豪州区
Chilean lady		*Vanessa terpsichore* Philippi	(チョウ目、タテハチョウ科) 新熱帯区
Chilean stag beetle			Darwin's beetle を見よ
chilli gall midge			carob midge を見よ
chillie thrips			yellow tea thrips (1) を見よ
Chimaera birdwing	キマエラトリバネアゲハ	*Ornithoptera chimaera* (Rothschild)	(チョウ目、アゲハチョウ科) 東洋区
chimney-sweep			chimney sweeper を見よ
chimney sweeper		*Odezia atrata* (Linnaeus)	(チョウ目、シャクガ科) 旧北区
chimney sweeper moth			chimney sweeper を見よ
chin bot			throat bot fly を見よ　南アフリカでの幼虫の英名
chin fly			throat bot fly を見よ
China aster plume moth	エゾギクトリバ	*Platyptilia farfarella* (Zeller)	(チョウ目、トリバガ科) 日本、旧北区、東洋区。キンセンカなどキク科花類の害虫
China grass banded caterpillar			ramie moth を見よ
China mark moth		*Nymphuliella daeckealis* (Haimbach)	(チョウ目、メイガ科) 新北区
China mark moth (1)		*Nymphula nymphaeata* (Linnaeus)	(チョウ目、メイガ科) 旧北区。イネ害虫
China mark moths	ミズメイガ亜科	Nymphulinae	(チョウ目、メイガ科) の昆虫の総称
China mark moths		Nymphula	(チョウ目、メイガ科) の昆虫の総称
China nawab	ヒメフタオチョウ	*Polyura narcaeus* (Hewitson)	(チョウ目、タテハチョウ科) 旧北区、東洋区
Chinantlan crescent		*Castilia chinantlensis* (de la Maza)	(チョウ目、タテハチョウ科) 新熱帯区
Chinati checkerspot		*Chlosyne chinatiensis* (Tinkham)	(チョウ目、タテハチョウ科) 新北区
chinch			bed bug (1) を見よ
chinch bug	チンチナガカメムシ	*Blissus leucopterus* (Say)	(カメムシ目、ナガカメムシ科) 新北区。体長 4 mm ほど。ワタ他の大害虫
chinch bug			German cockroach を見よ

英名	和名	学名	所属、分布、ほか
chinch bug			bed bug (1) を見よ　米国での英名
chinch bugs		*Blissus*	（カメムシ目、ナガカメムシ科）の昆虫の総称
chinch bugs (1)	ナガカメムシ科	Lygaeidae	（カメムシ目）の昆虫の総称　世界に3000種以上
chincha		*Leptoglossus zonatus* (Dallas)	（カメムシ目、ヘリカメムシ科）新北区
chinche			bed bug (1) を見よ
chinchilla rat louse		*Abrocomophaga chilensis* Emerson et Price	（ハジラミ目、ケモノタンカクハジラミ科）新熱帯区
Chinese angelica aphid	タラフタオアブラムシ	*Cavariella araliae* Takahashi	（カメムシ目、アブラムシ科）日本、旧北区、東洋区
Chinese arrowed stemborer	クワイホソハマキ	*Phalonidia mesotypa* Razowsky	（チョウ目、ホソハマキガ科）日本、旧北区。クワイ害虫
Chinese ash woolly aphid			ash-tree aphid を見よ
Chinese bean weevil			adzuki weevil を見よ
Chinese blistering cicada		*Huechys sanguinea* (De Geer)	（カメムシ目、セミ科）東洋区。中国ではChu-kiという
Chinese bushbrown	ヒメジャノメ	*Mycalesis gotama* Moore	（チョウ目、タテハチョウ科）旧北区、東洋区
Chinese cabbage cutworm			rice armyworm を見よ
Chinese character		*Cilix glaucata* (Scopoli)	（チョウ目、カギバガ科）旧北区
Chinese cochlid	クロシタアオイラガ	*Parasa sinica* (Moore)	（チョウ目、イラガ科）日本、旧北区。リンゴ、ウメなど果樹、サクラやカシ類など樹木害虫
Chinese cricket		*Gryllus chinensis* Weber	（バッタ目、コオロギ科）旧北区
Chinese dart			confucian dart を見よ
Chinese dryinid		*Pseudogonatopus hospes* Perkins	（ハチ目、カマバチ科）新北区
Chinese grasshopper			rice grasshopper (3) を見よ
Chinese hairstreak		*Amblopala avidiena* (Hewitson)	（チョウ目、シジミチョウ科）東洋区
Chinese hairy aphid	シナタケアブラムシ	*Chaitophorus saliniger* Shinji	（カメムシ目、アブラムシ科）日本、旧北区、東洋区
Chinese juniper round scale	ヒノキマルカイガラムシ	*Nuculaspis pseudomyeri* (Kuwana)	（カメムシ目、マルカイガラムシ科）日本、旧北区
Chinese-junk caterpillars			Chinese junks を見よ
Chinese junks		*Doratifera*	（チョウ目、イラガ科）の昆虫の総称
Chinese labyrinth (1)		*Neope armandii* (Oberthür)	（チョウ目、タテハチョウ科）東洋区
Chinese labyrinth		*Lethe armandina* (Oberthür)	（チョウ目、タテハチョウ科）東洋区
Chinese lantern		*Symmachia rubina* H. Bates	（チョウ目、シジミタテハ科）新熱帯区
Chinese luna moth		*Actias dubernardi* (Oberthür)	（チョウ目、ヤママユガ科）旧北区
Chinese mantid	オオカマキリ	*Tenodera aridifolia sinensis* Saussure	（カマキリ目、カマキリ科）日本、新北区

英名	和名	学名	所属、分布、ほか
Chinese mantis			Chinese mantid を見よ
Chinese map butterfly		*Araschnia davidis* Poujade	（チョウ目、タテハチョウ科）旧北区
Chinese moon moth			Indian moon moth を見よ
Chinese moon moth			Chinese luna moth を見よ
Chinese nettle caterpillar			slug caterpillar を見よ
Chinese oak silkmoth	サクサン	*Anthraea pernyi* (Guérin-Méneville)	（チョウ目、ヤママユガ科）日本、旧北区、東洋区、豪州区。ヤママユ同様飼育され、マユから天蚕糸をとる
Chinese oak silkworm			Chinese oak silkmoth を見よ
Chinese obscure scale	シナクロホシカイガラムシ	*Parlatoreopsis chinensis* (Marlatt)	（カメムシ目、マルカイガラムシ科）日本、東洋区。ナシ、サクラなど多くの果樹、樹木害虫
Chinese parlatoria		*Parlatoria sinensis* Maskell	（カメムシ目、マルカイガラムシ科）全北区。カンキツ害虫
Chinese peacock		*Papilio bianor* Cramer	（チョウ目、アゲハチョウ科）旧北区。日本亜種は *P. b. dehaanii* C. et R. Felder カラスアゲハ
Chinese peacock black swallowtail emerald			Chinese peacock を見よ
Chinese planthopper	ウシウンカ	*Perkinsiella sinensis* Kirkaldy	（カメムシ目、ウンカ科）日本、東洋区
Chinese plushblue		*Arhopala chinensis* Felder	（チョウ目、シジミチョウ科）東洋区
Chinese red scale			Asiatic red scale を見よ　米国での英名
Chinese rose beetle	シナコイチャコガネ	*Adoretus sinicus* Burmeister	（コウチュウ目、コガネムシ科）日本、東洋区、新北区、大洋区、豪州区
Chinese royal		*Tajuria luculentus* (Leech)	（チョウ目、シジミチョウ科）東洋区
Chinese scale			San Jose scale を見よ
Chinese silkworm			silkworm を見よ
Chinese sumac gall aphid	ヌルデイボフシ	*Nurudea ibofushi* Matsumura	（カメムシ目、アブラムシ科）日本
Chinese sumac pyralid	ツマグロシマメイガ	*Arippara indicator* Walker	（チョウ目、メイガ科）日本、旧北区、東洋区
Chinese sumac rosy gall aphid	ヌルデベニフシアブラムシ	*Nurudea yanoniella* Matsumura	（カメムシ目、アブラムシ科）日本
Chinese swallowtail			citrus swallowtail (1) を見よ
Chinese swift	タイワンチャバネセセリ	*Pelopidas sinensis* (Mabille)	（チョウ目、セセリチョウ科）東洋区
Chinese tasar moth			Chinese oak silkmoth を見よ　Chinese tussah moth ともいう
Chinese thrips	シナクダアザミウマ	*Haplothrips chinensis* Priesner	（アザミウマ目、クダアザミウマ科）日本、旧北区、東洋区
Chinese water scorpion	ミズカマキリ	*Ramatra chinensis* Mayer	（カメムシ目、タイコウチ科）日本、旧北区、東洋区
Chinese wax scale (1)	イボタロウムシ	*Ericerus pela* (Chavanner)	（カメムシ目、カタカイガラムシ科）日本、全北区
Chinese wax scale (2)		*Ceroplastes sinensis* Del Guercio	（カメムシ目、カタカイガラムシ科）新北区

英名	和名	学名	所属、分布、ほか
Chinese weevil			adzuki weevil を見よ
Chinese white pine beetle		*Dendroctonus armandi* Tsai et Li	(コウチュウ目、キクイムシ科) 旧北区
Chinese windmill	ジャコウアゲハ	*Byasa alcinous alcinous* (Klug)	(チョウ目、アゲハチョウ科) 日本、旧北区、東洋区
Chinese yellow-breast flat		*Gerosis sinica* (C. et R. Felder)	(チョウ目、セセリチョウ科) 東洋区
Chinese yellow sailer		*Neptis cydippe* Leech	(チョウ目、タテハチョウ科) 旧北区、東洋区
Chinese yellow swallowtail			citrus swallowtail (1) を見よ
chink bug			bed bug (1) を見よ
chintz			bed bug (1) を見よ
Chiricahua pine white		*Neophasia terlooii* Behr	(チョウ目、シロチョウ科) 新北区
Chiriqui flasher		*Astraptes chiriquensis chiriquensis* (Staudinger)	(チョウ目、セセリチョウ科) 新熱帯区
chiron sphinx		*Xylophanes chiron* (Drury)	(チョウ目、スズメガ科) 新熱帯区
chironomid midge			plumed gnat を見よ
chirping trig		*Anaxipha delicatula* (Scudder)	(バッタ目、クサヒバリ科) 新北区
Chisos banded-skipper		*Autochton cincta* (Plötz)	(チョウ目、セセリチョウ科) 新北区
Chisos katydid		*Paracyrtophyllus excelsus* (Rehn et Hebard)	(バッタ目、キリギリス科) 新北区
Chisos metalmark		*Apodemia chisoensis* Freeman	(チョウ目、シジミタテハ科) 新北区
Chisos skipperling		*Piruna haferniki* Freeman	(チョウ目、セセリチョウ科) 新北区、新熱帯区
Chisoya sphinx		*Sphinx chisoya* (Schaus)	(チョウ目、スズメガ科) 新北区
Chitral flash		*Rapala extensa* Evans	(チョウ目、シジミチョウ科) 東洋区
Chitral fritillary		*Melitaea chitralensis* Moore	(チョウ目、タテハチョウ科) 東洋区
chloropid flies			frit flies を見よ
chloropid gout fly		*Chlorops pumilionis* (Bjerkander)	(ハエ目、キモグリバエ科) 全北区、エチオピア区
cho blue			holly blue を見よ　米国での英名
chocolate albatross		*Appias lyncida* (Cramer)	(チョウ目、シロチョウ科) 東洋区。*A. l. formosana* (Wallace) タイワンシロチョウ, *A. l. vasava* (Cramer) も同英名
chocolate albatross		*Appias placidia* (Stoll)	(チョウ目、シロチョウ科) 東洋区
chocolate Argus			brown soldier を見よ
chocolate demon		*Ancistroides nigrita* (Latreille)	(チョウ目、セセリチョウ科) 東洋区。*A. n. maura* Snellen も同英名
chocolate fan-foot			bronzy Macrochilo moth を見よ
chocolate flat		*Eagris subalbida* (Holland)	(チョウ目、セセリチョウ科) エチオピア区
chocolate grass yellow	サリキチョウ	*Eurema sari sodalis* (Moore)	(チョウ目、シロチョウ科) 東洋区
chocolate Idia			rotund Idia moth を見よ

英名	和名	学名	所属、分布、ほか
chocolate jungle queen		*Stichophthalma nourmahal* (Westwood)	（チョウ目、タテハチョウ科）東洋区
chocolate moth		*Acherdoa ferraria* Walker	（チョウ目、ヤガ科）新北区
chocolate moth			tobacco moth (1) を見よ
chocolate pansy		*Junonia hedonia ida* Cramer	（チョウ目、タテハチョウ科）東洋区。brown pansy を参照
chocolate pansy	クロタテハモドキ	*Junonia iphita* (Cramer)	（チョウ目、タテハチョウ科）日本、東洋区。*J. i. horsfieldi* Moore も同英名
chocolate prominent moth		*Peridea ferruginea* (Packard)	（チョウ目、シャチホコガ科）新北区
chocolate Renia moth		*Renia nemoralis* Barnes et McDunnough	（チョウ目、ヤガ科）新北区
chocolate royal	フタオムラサキシジミ	*Remelana jangala* (Horsfield)	（チョウ目、シジミチョウ科）東洋区。*R. j. travana* (Hewitson) も同英名
chocolate sailor		*Neptis harita harita* (Moore)	（チョウ目、タテハチョウ科）東洋区
chocolate skipperling		*Piruna brunnea* (Scudder)	（チョウ目、セセリチョウ科）新熱帯区
chocolate soldier			greengrocer を見よ
chocolate tiger	タイワンアサギマダラ	*Parantica melaneus swinhoei* (Moore)	（チョウ目、タテハチョウ科）日本、東洋区。*P. melaneus* (Cramer) の英名。*P. m. sinopion* (Fruhstorfer) も同英名
chocolate-tip		*Clostera curtula* (Linnaeus)	（チョウ目、シャチホコガ科）旧北区
chocolate-tip moth			chocolate-tip を見よ
Choctaw ground cricket		*Allonemobius shalontaki* Braswell, Birge et Howard	（バッタ目、コオロギ科）新北区
chokeberry underwing			hawthorn underwing moth を見よ
chokecherry leafroller		*Sparganothis directana* (Walker)	（チョウ目、ハマキガ科）新北区。成虫は chokecherry leafroller moth
chokecherry midge		*Contarinia virginiae* (Felt)	（ハエ目、タマバエ科）新北区
chondrylla gall midge			skeleton gall midge を見よ
choreutid moths			metalmark moths を見よ
chorus cicada		*Amphipsalta zealandica* (Boisduval)	（カメムシ目、セミ科）豪州区
Chrapkowski's green-banded swallowtail			broad blue-banded swallowtail を見よ
Chrinda Acraea	サチスメスジロホソチョウ	*Acraea satis* Ward	（チョウ目、タテハチョウ科）エチオピア区
Christiernin's flat-body		*Hypercallia citrinalis* (Scopoli)	（チョウ目、マルハキバガ科）旧北区
Christmas bees			cicadas を見よ
Christmas beetle		*Anoplognathus pallidicollis* Blanchard	（コウチュウ目、コガネムシ科）豪州区
Christmas beetles		*Anoplognathus*	（コウチュウ目、コガネムシ科）の昆虫の総称
Christmas berry webworm			honeydew moth を見よ

英名	和名	学名	所属、分布、ほか
Christmas berry webworm moth		*Loxostege floridalis* Barnes et McDunnough	（チョウ目、メイガ科）新北区
Christmas butterfly			citrus swallowtail (2) を見よ
Christmas cicadas			cicadas を見よ
Christmas emperor		*Polyura andrewsi* (Butler)	（チョウ目、タテハチョウ科）豪州区。オーストラリアの Christmas 島に由来
Christmas forester		*Celaenorrhinus mokeezi* (Wallengren)	（チョウ目、セセリチョウ科）エチオピア区
Christmas Island cicada		*Oxypleura calypso* (Kirby)	（カメムシ目、セミ科）豪州区
Christmas Island white			Ceylon lesser albatross を見よ
Christmas singers			cicadas を見よ
Christmas swallowtail			great Mormon を見よ　オーストラリアの Christmas 島に由来
Christmasberry moth			honeydew moth を見よ
Christmas-berry seed wasp		*Megastigmus transvaalensis* (Hussey)	（ハチ目、オナガコバチ科）大洋区、新北区、エチオピア区
Christmas-berry thrips		*Rhynchothrips ilex* (Moulton)	（アザミウマ目、クダアザミウマ科）新北区
Christmas-tree Acraea			broad-bordered Acraea を見よ
Christoricha skipper		*Hesperilla chrysotricha* Meyrick	（チョウ目、セセリチョウ科）豪州区
Christy's carpet		*Oporinia chrystyi* Prout	（チョウ目、シャクガ科）旧北区
Chromis firewing	クロミスミツボシタテハ	*Catonephele chromis* Doubleday	（チョウ目、タテハチョウ科）新熱帯区
chrysalis			butterflies and moths を見よ　蛹の英名でもあり、chrysalid（複数形 chrysalides）ともいう
chrysanthemum aphid		*Macrosiphoniella oblonga* (Mordvilko)	（カメムシ目、アブラムシ科）旧北区
chrysanthemum aphid (1)	キクヒメヒゲナガアブラムシ	*Macrosiphoniella sanborni* (Gillette)	（カメムシ目、アブラムシ科）日本、汎世界。キク害虫
chrysanthemum arctiid	ゴマベニシタヒトリ	*Rhyparia purpurata gerda* Warnecke	（チョウ目、ヒトリガ科）日本、旧北区
chrysanthemum blotch miner		*Euribia zoe* (Meigen)	（ハエ目、ミバエ科）旧北区
chrysanthemum casebearer	キクツツミノガ	*Coleophora kurokoi* Oku	（チョウ目、ツツミノガ科）日本。キク害虫
chrysanthemum flea beetle		*Longitarsus succineus* (Foudras)	（コウチュウ目、ハムシ科）旧北区
chrysanthemum flower borer		*Lorita abornana* Busck	（チョウ目、ハマキガ科）新北区
chrysanthemum flower borer (1)		*Lorita scarificata* Meyrick	（チョウ目、ハマキガ科）新北区、大洋区。chrysanthemum flower borer moth ともいう
chrysanthemum fly			drone fly を見よ
chrysanthemum fruit fly	ミスジハマダラミバエ	*Trypeta artemisicola* Hendel	（ハエ目、ミバエ科）日本、旧北区。キク害虫

英名	和名	学名	所属、分布、ほか
chrysanthemum gall fly		*Diarthronomyia hypogaea* (Loew)	(ハエ目、タマバエ科) 新北区。温室のキク害虫
chrysanthemum gall fly			chrysanthemum gall midge を見よ
chrysanthemum gall midge		*Diarthronomyia chrysanthemi* Ahlberg	(ハエ目、タマバエ科) 全北区、豪州区。キク害虫
chrysanthemum gall midge			Japanese chrysanthemum gall midge を見よ
chrysanthemum gallnut fly			Japanese chrysanthemum gall midge を見よ
chrysanthemum golden plusia	キクキンウワバ	*Thysanoplusia intermixta* (Warren)	(チョウ目、ヤガ科) 日本。キクの他野菜、カンキツ類の害虫
chrysanthemum grained moth	キクセダカモクメ	*Cucullia elongata* Butler	(チョウ目、ヤガ科) 日本、東洋区
chrysanthemum greenish geometrid	ヨツメアオシャク	*Thetidia albocostaria* (Bremer)	(チョウ目、シャクガ科) 日本、旧北区。キク害虫
chrysanthemum lace bug		*Corythucha marmorata* (Uhler)	(カメムシ目、グンバイムシ科) 新北区
chrysanthemum lace bug	キクグンバイ	*Galeatus spinifrons* (Fallén)	(カメムシ目、グンバイムシ科) 日本、旧北区、東洋区。キク害虫
chrysanthemum leafhopper		*Eupteryx melissae* Curtis	(カメムシ目、オオヨコバイ科) 旧北区
chrysanthemum leaf miner (1)		*Phytomyza atricornis* Meigen	(ハエ目、ハモグリバエ科) 旧北区、豪州区
chrysanthemum leafminer	キクハモグリバエ	*Phytomyza albiceps* Meigen	(ハエ目、ハモグリバエ科) 日本、全北区
chrysanthemum leafminer		*Phytomyza syngenesiae* (Hardy)	(ハエ目、ハモグリバエ科) 新北区
chrysanthemum leafminer	ニッポンキクハモグリバエ	*Phytomyza japonica* Sasakawa	(ハエ目、ハモグリバエ科) 日本。キク害虫
chrysanthemum leaf miner			garden pea leafminer を見よ
chrysanthemum longicorn			chrysanthemum longicorn beetle を見よ
chrysanthemum longicorn beetle	キクスイカミキリ	*Phytoecia rufiventris* Gautier	(コウチュウ目、カミキリムシ科) 日本、東洋区。キク害虫
chrysanthemum longihorn	ネジロカミキリ	*Pogonocherus seminiveus* Bates	(コウチュウ目、カミキリムシ科) 日本、旧北区。ヤツデ害虫
chrysanthemum looper	シロヒメシャク	*Scopula nivearia* (Leech)	(チョウ目、シャクガ科) 日本、旧北区
chrysanthemum midge			chrysanthemum gall midge を見よ
chrysanthemum silver phytometra	キクギンウワバ	*Macdunnoughia confusa* (Stephens)	(チョウ目、ヤガ科) 日本、旧北区
chrysanthemum stem fly		*Paroxyna misella* (Loew)	(ハエ目、ミバエ科) 旧北区
chrysanthemum stool fly		*Psila nigricornis* Meigen	(ハエ目、ハネオレバエ科) 旧北区
chrysanthemum stool miner			chrysanthemum stool fly を見よ

英名	和名	学名	所属、分布、ほか
chrysanthemum thrips	クロゲハナアザミウマ	*Thrips nigropilosus* Uzel	(アザミウマ目、アザミウマ科) 日本、全北区、大洋区、豪州区。キク、バラなど花類害虫
chrysidoid wasps	セイボウ上科	Chrysidoidea	(ハチ目) の昆虫の総称
Chrysites banner		*Epiphile chrysites* Latreille	(チョウ目、タテハチョウ科) 新熱帯区
Chrysogone wood nymph		*Taygetis chrysogone* Doubleday	(チョウ目、タテハチョウ科) 新熱帯区
chrysomelid beetles			leaf beetles (1) を見よ
Chrysophrys skipper		*Mnasitheus chrysophrys* (Mabille)	(チョウ目、セセリチョウ科) 新熱帯区
chrysotricha skipper			Plebeia skipper を見よ
Chryxus arctic	カナダタカネヒカゲ	*Oeneis chryxus* (Doubleday)	(チョウ目、タテハチョウ科) 新北区
Chumbi Argus		*Lycaena semiargus* Stauder	(チョウ目、シジミチョウ科) 東洋区
Chumbi green underwing		*Lycaena younghusbandi* Elwes	(チョウ目、シジミチョウ科) 東洋区
Chumbi wall		*Chonala masoni* (Elwes)	(チョウ目、タテハチョウ科) 東洋区
Chumbi white		*Pieris chumbiensis* de Nicéville	(チョウ目、シロチョウ科) 東洋区
churchyard beetle		*Blaps mucronata* Latreille	(コウチュウ目、ゴミムシダマシ科) 旧北区
churchyard beetle			cellar beetle (1) を見よ
churchyard beetles			cellar beetles を見よ
Churkin's apollo		*Parnassius davydovi* Churkin	(チョウ目、アゲハチョウ科) 旧北区
churr worm			mole cricket (3) を見よ
churrworm			field cricket (1) を見よ
churr-worms			mole crickets (1) を見よ
chyromyid flies		Chyromyidae	(ハエ目) の昆虫の総称 幼虫はトリの巣、朽木などに生息
cibrate weevil			apple weevil (1) を見よ
cicada killer		*Exeirus lateritius* Shukard	(ハチ目、アナバチ科) 豪州区
cicada killer			cicada-killer wasp を見よ
cicada-killer wasp		*Sphecius speciosus* (Drury)	(ハチ目、アナバチ科) 新北区
cicada killers	アナバチ科	Sphecidae	(ハチ目) の昆虫の総称 digger wasp ジガバチといわれる
cicada parasite beetles			cedar beetles (1) を見よ
cicadas	セミ科	Cicadidae	(カメムシ目) の昆虫の総称 世界に 1,500 種
cicadas			free beaks を見よ
cicadines		Cicadina	(カメムシ目) の昆虫の総称
cicadoid leafhoppers	セミ上科	Cicadoidea	(カメムシ目) の昆虫の総称
cigar beetle			cigarette beetle を見よ
cigar case bearer (1)		*Coleophora fletcherella* Fernald	(チョウ目、ツツミノガ科) 新北区
cigar casebearer (2)		*Coleophora cerasivorella* Packard	(チョウ目、ツツミノガ科) 全北区。北米での英名。リンゴ他害虫

英名	和名	学名	所属、分布、ほか
cigar case-bearer		*Coleophora occidentis* Zeller	（チョウ目、ツツミノガ科）新北区
cigar casebearer			apple and plum casebearer moth を見よ
cigar casebearer moth			apple casebearer を見よ　cigar casebearer ともいう
cigarette beetle	タバコシバンムシ	*Lasioderma serricorne* (Fabricius)	（コウチュウ目、シバンムシ科）日本、汎世界。貯蔵食品、動植物標本の害虫
cigarette beetle			tobacco beetle を見よ
ciliate blue		*Anthene emolus* (Godart)	（チョウ目、シジミチョウ科）東洋区。*A. e. goberus* (Fruhstorfer) も同英名
cimbicid sawfles	コンボウハバチ科	Cimbicidae	（ハチ目）の昆虫の総称
cimbicida sawflies			cimbicid sawflies を見よ
cimbicids			cimbicid sawflies を見よ
cinarine plantlice		Cinarinae	（カメムシ目、アブラムシ科）の昆虫の総称
Cinderella admiral		*Hypanartia cinderella* Lamas, Willmott et Hall	（チョウ目、タテハチョウ科）新熱帯区
cineraria leaf-miner			chrysanthemum leaf miner (1) を見よ　豪州での英名
cineraria moth			senecio moth を見よ
cinereous cockroach	ハイイロゴキブリ	*Nauphoeta cinerea* (Olivier)	（ゴキブリ目、ハイイロゴキブリ科）日本、新北区。穀類貯蔵場所などに生息
cinereous knot-horn			tobacco moth (1) を見よ
cinereous pearl		*Pyla fuscalis* Schiffermüller	（チョウ目、メイガ科）旧北区
cinereous silver-barred bell		*Epinotia fraternana* (Haworth)	（チョウ目、ハマキガ科）旧北区
cingalese bushbrown		*Mycalesis rama* (Moore)	（チョウ目、タテハチョウ科）東洋区
cinnabar		*Tyria jacobaeae* (Linnaeus)	（チョウ目、ヒトリガ科）旧北区
cinnabar moth			cinnabar を見よ　米国での英名
cinnamon longicorn			camphor longicorn beetle を見よ
cinnamon prionid		*Callipogon cinnamomeus* (Linnaeus)	（コウチュウ目、カミキリムシ科）新熱帯区
cinnamon shadowdragon		*Neurocordulia virginiensis* Davis	（トンボ目、エゾトンボ科）新北区
cinnamon tussock moth		*Dasychira cinnamomea* (Grote et Robinson)	（チョウ目、ドクガ科）新北区
cinnamonum scale	ヤブニッケイマルカイガラムシ	*Diaonidia yabunikkei* (Kuwana)	（カメムシ目、マルカイガラムシ科）日本
cinquefoil copper			Dorcas copper を見よ
cinquefoil skipper		*Pyrgus cirsii* (Rambur)	（チョウ目、セセリチョウ科）旧北区
Circe	カバシタゴマダラ	*Hestina nama* Doubleday	（チョウ目、タテハチョウ科）東洋区。ギリシャ神話の魔女キルケーに由来
Circe (1)	アカホシゴマダラ	*Hestina assimilis shirakii* Shirozu	（チョウ目、タテハチョウ科）日本。*H. assimilis* (Linnaeus) の英名
Circe			Sirens を見よ
circular black scale parasite		*Aphytis holoxanthus* DeBach	（ハチ目、ツヤコバチ科）豪州区

英名	和名	学名	所属、分布、ほか
circular black scale			Florida red scale を見よ
circular purple scale			Florida red scale を見よ　南アフリカでの英名
circular scale			Florida red scale を見よ
circular-seamed flies	環縫群	Cyclorrhapha	(ハエ目) の昆虫の総称
circular white scale			oleander scale を見よ
circumducta satyr		*Pedaliodes circumducta* Thieme	(チョウ目、タテハチョウ科) 新熱帯区
circumscript Mompha moth		*Mompha circumscriptella* (Zeller)	(チョウ目、カザリバガ科) 新北区
Cisseis Morpho		*Morpho cisseis* C. et R. Felder	(チョウ目、タテハチョウ科) 新熱帯区
cistus forester		*Adscita geryon* (Hübner)	(チョウ目、マダラガ科) 旧北区
citricola scale	カンキツカタカイガラムシ	*Coccus pseudomagnoliarum* (Kuwana)	(カメムシ目、カタカイガラムシ科) 日本、全北区。カンキツ害虫
citricola scale			gray citrus scale を見よ
citrine forktail		*Ischnura hastata* (Say)	(トンボ目、イトトンボ科) 新北区
citriphilous mealybug			citrophilus mealybug (1) を見よ
citron bug		*Leptoglossus gonagra* (Fabricius)	(カメムシ目、ヘリカメムシ科) 新北区、新熱帯区。カンキツ害虫
citron Perisama		*Perisama oppelii xanthica* (Hewitson)	(チョウ目、タテハチョウ科) 新熱帯区
citrophilous mealybug			citrophilus mealybug (2) を見よ
citrophilus mealybug (1)	カハニコナカイガラムシ	*Pseudococcus fragilis* Brain	(カメムシ目、コナカイガラムシ科) 汎世界
citrophilus mealybug (2)		*Pseudococcus calceolariae* (Maskell)	(カメムシ目、コナカイガラムシ科) 汎世界
citrus agrilus			flatheaded citrus borer を見よ
citrus ambrosia beetle			Saxena ambrosia beetle を見よ
citrus ant			red ant (1) を見よ
citrus aphid			citrus brown aphid を見よ　米国での英名
citrus aphid			black citrus aphid を見よ
citrus aphis			black citrus aphid を見よ
citrus bark beetle		*Hypothenemus citri* Ebeling	(コウチュウ目、キクイムシ科) 新熱帯区。カンキツ害虫
citrus bark borer		*Agrilus occipitalis* Eschscholtz	(コウチュウ目、タマムシ科) 東洋区。カンキツ害虫
citrus bark borers			twig girdlers を見よ
citrus black scale			black scale を見よ
citrus blackfly	ミカンクロトゲコナジラミ	*Aleurocanthus woglumi* Ashby	(カメムシ目、コナジラミ科) 新北区、エチオピア区。カンキツ害虫
citrus-blossom bug		*Austropeplus* sp.	(カメムシ目、カスミカメムシ科) 豪州区
citrus blossom fly		*Lonchaea gibbosa* de Meijere	(ハエ目、クロツヤバエ科) 東洋区。カンキツ害虫

英名	和名	学名	所属、分布、ほか
citrus blossom midge		*Cecidomya* sp.	(ハエ目、タマバエ科) 豪州区
citrus bollworm			corn earworm (2) を見よ　幼虫の英名
citrus branch borer		*Uracanthus cryptophagus* Olliff	(コウチュウ目、カミキリムシ科) 豪州区
citrus branch girdler		*Trigona trinidadensis* Provancher	(ハチ目、ミツバチ科) 新熱帯区
citrus broad-banded geometrid	クロフオオシロエダシャク	*Pogonopygia nigralbata* Warren	(チョウ目、シャクガ科) 日本
citrus brown aphid	ミカンクロアブラムシ	*Toxoptera citricidus* (Kirkaldy)	(カメムシ目、アブラムシ科) 日本、新北区、豪州区。カンキツ害虫
citrus butterfly			citrus swallowtail (2) を見よ
citrus coccus			citricola scale を見よ
citrus cottony scale			cottony citrus scale (1) を見よ
citrus crane fly	カンキツヒメガガンボ	*Limonia amatrix* (Alexander)	(ハエ目、ガガンボ科) 日本。カンキツ害虫
citrus cutworm			cutworm (1) を見よ
citrus dog			spangle を見よ
citrus flatheaded borer			flatheaded citrus borer を見よ
citrus flea beetle		*Prodagricomela nigricollis* Chen	(コウチュウ目、ハムシ科) 東洋区
citrus flower chafer	コアオハナムグリ	*Oxycetonia jucunda* (Faldermann)	(コウチュウ目、コガネムシ科) 日本、全北区。果樹、バラなどの害虫
citrus flower gall midge			Japanese citrus flower-bud midge を見よ
citrus flower moth		*Prays nephelomima* Meyrick	(チョウ目、スガ科) 豪州区
citrus flower moth (1)		*Prays citri* (Milliere)	(チョウ目、スガ科) 旧北区、東洋区、豪州区
citrus flower thrips	ハナクダアザミウマ	*Haplothrips kurdjumovi* Karny	(アザミウマ目、クダアザミウマ科) 日本、旧北区
citrus fruit borer		*Citripestis sagittiferella* (Moore)	(チョウ目、メイガ科) 東洋区。カンキツ害虫
citrus fruit borer			orange moth (1) を見よ
citrus fruit fly	ミカンバエ	*Tetradacus tsuneonis* Miyake	(ハエ目、ミバエ科) 日本、東洋区。black fly ともいう。カンキツ害虫
citrus fruit-piercer moth		*Gonodonta nutrix* Cramer	(チョウ目、ヤガ科) 新北区。citrus fruit-piercer ともいう
citrus fruit weevil		*Neomerimnetes sobrinus* Lea	(コウチュウ目、ゾウムシ科) 豪州区
citrus gall midge		*Prodiplosis longifila* Gagné	(ハエ目、タマバエ科) 新北区、新熱帯区
citrus gall wasp		*Bruchophagus fellis* (Girault)	(ハチ目、カタビロコバチ科) 豪州区。カンキツ害虫
citrus green stinkbug			citrus stink bug を見よ
citrus green tree hopper			citrus katydid (1) を見よ
citrus ground mealybug	ミカンネコナカイガラムシ	*Rhizoecus kondonis* Kuwana	(カメムシ目、コナカイガラムシ科) 日本、新北区。カンキツ害虫

英名	和名	学名	所属、分布、ほか
citrus jassid			citrus leafhopper (1) を見よ
citrus jewel beetles			twig girdlers を見よ
citrus katydid	ヤマクダマキモドキ	*Halochlora longifissa* Matsumura et Shiraki	（バッタ目、キリギリス科）日本
citrus katydid (1)		*Caedicia strenua* (Walker)	（バッタ目、キリギリス科）豪州区。カンキツ害虫
citrus leaf-eating cricket		*Tamborina australis* (Walker)	（バッタ目、コオロギ科）豪州区
citrus leafeating weevil		*Eutinophaea bicristata* Lea	（コウチュウ目、ゾウムシ科）豪州区。カンキツ害虫
citrus leafhopper	カンキツヒメヨコバイ	*Apheliona ferruginea* (Matsumura)	（カメムシ目、オオヨコバイ科）日本。カンキツ害虫
citrus leafhopper (1)		*Empoasca smithi* (Fletcher et Donaldson)	（カメムシ目、オオヨコバイ科）豪州区
citrus leafminer	ミカンハモグリガ	*Phyllocnistis citrella* (Stainton)	（チョウ目、コハモグリガ科）日本、豪州区。カンキツ害虫
citrus leaf-miner			citrus flea beetle を見よ
citrus leafroller	ミカンマルハキバガ	*Psorosticha melanocrepida* Clarke	（チョウ目、マルハキバガ科）日本。カンキツ害虫
citrus leafroller (1)		*Psorosticha zizyphi* (Stainton)	（チョウ目、マルハキバガ科）東洋区、豪州区。カンキツ害虫
citrus locust		*Chondacris rosae* De Geer	（バッタ目、バッタ科）旧北区。カンキツ害虫
citrus locust		*Schistocera flavo-fasciata* (De Geer)	（バッタ目、バッタ科）新熱帯区。カンキツ害虫
citrus longhorn		*Monochamus versteegi* Ritsema	（コウチュウ目、カミキリムシ科）東洋区。カンキツ害虫
citrus longhorn beetle		*Anoplophora chinensis* (Forster)	（コウチュウ目、カミキリムシ科）日本、東洋区、旧北区。citrus long-horned beetle ともいう
citrus longicorn		*Skeletodes tetrops* Newman	（コウチュウ目、カミキリムシ科）豪州区
citrus looper		*Iridopsis fragilaria* (Grossbeck)	（チョウ目、シャクガ科）新北区
citrus looper			fragile gray を見よ
citrus mealybug (1)	ミカンコナカイガラムシ	*Planococcus citri* (Risso)	（カメムシ目、コナカイガラムシ科）日本、東洋区、新北区、新熱帯区、エチオピア区。カンキツ、コーヒーなどの害虫
citrus mealybug (2)	ミカンヒメコナカイガラムシ	*Pseudococcus citriculus* Green	（カメムシ目、コナカイガラムシ科）日本、東洋区、大洋区、新北区。カンキツ害虫
citrus mealybug predator		*Oligochrysa lutea* (Walker)	（アミメカゲロウ目、クサカゲロウ科）豪州区
citrus moth		*Cryptoblabes plagioleuca* Turner	（チョウ目、メイガ科）豪州区、大洋区
citrus mussel scale			purple scale を見よ　南アフリカでの英名
citrus orthezia		*Orthezia praelonga* Douglass	（カメムシ目、ハマカイガラムシ科）新熱帯区。カンキツ害虫
citrus parlatoria			citrus scale を見よ
citrus parlatoria scale			citrus scale を見よ
citrus peel miner		*Marmara salictella* Clemens	（チョウ目、ホソガ科）新北区。カンキツ害虫

英名	和名	学名	所属、分布、ほか
citrus peelminer moth		*Marmara gulosa* Guillen et Davis	（チョウ目、ホソガ科）新北区、新熱帯区。citrus peelminer ともいう
citrus planthopper		*Metcalfa pruinosa* (Say)	（カメムシ目、アオバハゴロモ科）新北区。カンキツ害虫
citrus planthopper		*Colgar peracutum* (Walker)	（カメムシ目、アオバハゴロモ科）豪州区
citrus psylla (1)	ミカンキジラミ	*Diaphorina citri* Kuwayama	（カメムシ目、キジラミ科）日本、東洋区。カンキツ害虫
citrus psylla (2)		*Trioza (Spanioza) erythreae* (Del Guercio)	（カメムシ目、トガリキジラミ科）エチオピア区。カンキツ害虫
citrus psyllid			citrus psylla (2) を見よ
citrus purple scale			purple scale を見よ
citrus red scale			California red scale を見よ
citrus rind borer		*Adoxophyes templana* (Pag)	（チョウ目、ハマキガ科）豪州区
citrus rind borer		*Prays endocarpa* Meyrick	（チョウ目、スガ科）旧北区
citrus root-bark channeller		*Pseudomydaus citriperda* (Tryon)	（コウチュウ目、ゾウムシ科）豪州区
citrus root mealybug	ジモグリコナカイガラムシ	*Geococcus citrinus* Kuwana	（カメムシメ、コナカイガラムシ科）日本。カンキツ、チャの害虫
citrus root mealybug			citrus ground mealybug を見よ
citrus root scale			citrus root mealybug を見よ
citrus root weevil		*Pachnaeus litus* (Germar)	（コウチュウ目、ゾウムシ科）新北区
citrus root weevil			blue-green orange weevil を見よ
citrus rust thrips			orchid thrips を見よ
citrus sawyer	ヒメヒゲナガカミキリ	*Monochamus subfasciatus subfasciatus* (Bates)	（コウチュウ目、カミキリムシ科）日本
citrus scale	ヒメクロカイガラムシ	*Parlatoria ziziphi* (Lucas)	（カメムシ目、マルカイガラムシ科）日本、旧北区、大洋区、豪州区。カンキツ害虫
citrus scale			citrus mealybug (1) を見よ
citrus shield bugs		*Rhynchocoris*	（カメムシ目、カメムシ科）の昆虫の総称
citrus snout beetle		*Sciobius granosus* Fahrer	（コウチュウ目、ゾウムシ科）エチオピア区。カンキツ害虫
citrus snout beetle		*Maleuterpes dentipes* Heller	（コウチュウ目、ゾウムシ科）豪州区。カンキツ害虫
citrus snow scale	ミカンナガカイガラムシ	*Unaspis citri* (Comstock)	（カメムシ目、マルカイガラムシ科）日本、汎世界。カンキツ害虫
citrus spiny whitefly	ミカントゲコナジラミ	*Aleurocanthus spiniferus* (Quaintance)	（カメムシ目、コナジラミ科）日本、東洋区、大洋区。カンキツ害虫
citrus stink bug	ミカントゲカメムシ	*Rhynchocoris humeralis* (Thunberg)	（カメムシ目、カメムシ科）日本、東洋区。カンキツ害虫
citrus string cottony scale			cottony citrus scale (2) を見よ
citrus swallowtail (1)	ナミアゲハ	*Papilio xuthus* Linnaeus	（チョウ目、アゲハチョウ科）日本、旧北区
citrus swallowtail (2)	アフリカオナシアゲハ	*Papilio demodocus* Esper	（チョウ目、アゲハチョウ科）エチオピア区
citrus swallowtail			King page を見よ
citrus swallowtail			chequered swallowtail を見よ

英名	和名	学名	所属、分布、ほか
citrus thrips		*Scirtothrips citri* (Moulton)	(アザミウマ目、アザミウマ科) 新北区
citrus thrips			South African citrus thrips を見よ
citrus trunkborer		*Platyomopsis pulverulens* (Boisduval)	(コウチュウ目、カミキリムシ科) 豪州区
citrus whitefly		*Dialeurodes citricola* B. Young	(カメムシ目、コナジラミ科) 旧北区。カンキツ害虫
citrus whitefly (1)	ミカンコナジラミ	*Dialeurodes citri* (Ashmead)	(カメムシ目、コナジラミ科) 日本、全北区。カンキツ害虫
citrus whitefly (2)		*Dialeurodes citrifolii* (Morgan)	(カメムシ目、コナジラミ科) 日本、東洋区、新北区、新熱帯区。カンキツ害虫
citrus white fly			woolly whitefly を見よ
citrus white-striped longicorn	アトモンマルケシカミキリ	*Exocentrus lineatus* Bates	(コウチュウ目、カミキリムシ科) 日本
citrus yellow scale			yellow scale を見よ
city longicorn beetle		*Aeolesthes sarta* (Solsky)	(コウチュウ目、カミキリムシ科) 旧北区、東洋区
civil blue			blue damselfly を見よ　北米での英名
civil bluet			blue damselfly を見よ
civil rustic moth		*Caradrina montana* (Bremer)	(チョウ目、ヤガ科) 全北区
cixiid planthoppers	ヒシウンカ科	Cixiidae	(カメムシ目) の昆虫の総称
cizara hawk-moth		*Cizara ardeniae* (Lewin)	(チョウ目、スズメガ科) 豪州区
Claassen's skolly		*Thestor claassensi* Heath et Pringle	(チョウ目、シジミチョウ科) エチオピア区
claddagh			burren green (1) を見よ
clambid beetles			fringe-winged beetles を見よ
clambids			fringe-winged beetles を見よ
Clancy's rustic		*Platyperigea kadenii* (Freyer)	(チョウ目、ヤガ科) 旧北区
clapping cicada		*Amphipsalta cingulata* (Fabricius)	(カメムシ目、セミ科) 豪州区
clara satin moth		*Thalaina clara* Walker	(チョウ目、シャクガ科) 豪州区
Clare's sailer		*Neptis clarei* Neave	(チョウ目、タテハチョウ科) エチオピア区
claret dun		*Leptophlebia vespertina* (Linnaeus)	(カゲロウ目、トビイロカゲロウ科) 旧北区
Clark's day sphinx		*Proserpinus clarkiae* (Boisduval)	(チョウ目、スズメガ科) 新北区。Clark's day sphinx moth ともいう
Clark's rocksitter		*Durbaniella clarki* (van Son)	(チョウ目、シジミチョウ科) エチオピア区
Clark's sorrel copper		*Lycaena clarki* Dickson	(チョウ目、シジミチョウ科) エチオピア区
Clark's widow		*Dingana clarki* van Son	(チョウ目、タテハチョウ科) エチオピア区
claspertail		*Onychogomphus supinus* Hagen	(トンボ目、サナエトンボ科) エチオピア区
Claudina crescent			apricot crescent を見よ
Claudina's Tegosa			apricot crescent を見よ
clavate scale			herculean scale を見よ
clavicorn			clavicorn beetles を見よ

英名	和名	学名	所属、分布、ほか
clavicorn beetles	球角群	Clavicornia	（コウチュウ目）の昆虫の総称
clavigerid beetles			ant-loving beetles (2) を見よ
clavipes sphinx moth		*Aellopos clavipes* (Rothschild et Jordan)	（チョウ目、スズメガ科）新北区、新熱帯区。clavipes sphinx ともいう
claw-tipped bluet		*Enallagma semicirculare* Selys	（トンボ目、イトトンボ科）新北区、新熱帯区
clawler			ヘビトンボの幼虫、カイガラムシの最初の幼虫期
clay			clay moth を見よ
clay-backed cutworm		*Feltia gladiaria* Morrison	（チョウ目、ヤガ科）新北区。トウモロコシ害虫
clay-backed cutworm moth			clay-backed cutworm を見よ
clay-colored billbug		*Calendra aequalis* (Gyllenhal)	（コウチュウ目、ゾウムシ科）新北区。トウモロコシ害虫
claycolored leaf beetle		*Anomoea laticlavia* (Forster)	（コウチュウ目、ハムシ科）新北区
clay-coloured groundling		*Nothris verbascella* (Brahm)	（チョウ目、キバガ科）旧北区
clay-coloured root weevil			clay-coloured weevil を見よ　カナダでの英名
clay-coloured weevil		*Otiorhynchus singularis* (Linnaeus)	（コウチュウ目、ゾウムシ科）旧北区
clay fanfoot		*Paracolax derivalis* (Hübner)	（チョウ目、ヤガ科）旧北区
clay fanfoot		*Paracolax glaucinalis* Schiffermüller	（チョウ目、ヤガ科）旧北区
clay fan-foot (1)	クルマアツバ	*Paracolax tristalis* (Fabricius)	（チョウ目、ヤガ科）日本、旧北区
clay moth		*Mythimna ferrago* (Fabricius)	（チョウ目、ヤガ科）旧北区
clay triple-lines		*Cyclophora linearia* (Hübner)	（チョウ目、シャクガ科）旧北区
clay wainscot			clay moth を見よ
clean mimic white		*Pseudopieris nehemia* (Boisduval)	（チョウ目、シロチョウ科）新熱帯区
clear dagger moth		*Acronicta clarescens* Guenée	（チョウ目、ヤガ科）新北区
Clear Lake gnat		*Chaoborus astictopus* Dyar et Shannon	（ハエ目、ケヨソイカ科）新北区
clear oakworm moth		*Anisota pellucida* (J. E. Smith)	（チョウ目、ヤママユガ科）新北区
clear sailer		*Neptis nandina* Moore	（チョウ目、タテハチョウ科）東洋区
clear sailer (1)	ホソオビミスジ	*Neptis clinia* Moore	（チョウ目、タテハチョウ科）東洋区
clear-spotted Acraea		*Acraea aglaonice* Westwood	（チョウ目、タテハチョウ科）エチオピア区
clear-spotted blue		*Zetona delospila* (Waterhouse)	（チョウ目、シジミチョウ科）豪州区
clear-tipped amberwing		*Perithemis parzefalli* Hoffmann	（トンボ目、トンボ科）新熱帯区

英名	和名	学名	所属、分布、ほか
clear underwing			dusky clearwing moth を見よ
clear wing Acraea		*Acraea rabbaiae* Ward	(チョウ目、タテハチョウ科) エチオピア区
Clearista metalmark		*Pseudonymphidia clearista* (Butler)	(チョウ目、シジミタテハ科) 新熱帯区
Clearlake clubtail		*Gomphus australis* (Needham)	(トンボ目、サナエトンボ科) 新北区
clearwing borer moth		*Melittia snowii* Edwards	(チョウ目、スカシバガ科) 新北区
clearwing chestnut moth			chestnut borer moth を見よ
clearwing hawkmoth			larger pellucid hawk moth を見よ
clearwing mimic-white		*Dismorphia theucharila fortunata* (Lucas)	(チョウ目、シロチョウ科) 新熱帯区
clearwing moths	スカシバガ科	Sesiidae	(チョウ目) の昆虫の総称　翅の全部や一部が透明であることに由来。世界に 1,000 種以上
clearwing swallowtail			big greasy を見よ
clear-winged bugs			gnat bugs を見よ
clear-winged forest glory		*Vestalis gracilis gracilis* (Rambur)	(トンボ目、イトトンボ科) 東洋区
clear-winged grasshopper		*Camnula pellucida* (Scudder)	(バッタ目、バッタ科) 新北区。トウモロコシ害虫
clear-winged moths			clearwing moths を見よ
clearwing-mimic queen		*Lycorea ilione* (Cramer)	(チョウ目、タテハチョウ科) 新熱帯区
clearwings			clearwing moths を見よ
cleftfooted minnow mayflies		Metretopodidae	(カゲロウ目) の昆虫の総称
cleg	アブ		英国の方言
cleg (1)		*Haematopota pluvialis* (Linnaeus)	(ハエ目、アブ科) 旧北区
cleg			long-horned cleg を見よ
cleg fly			cleg (1) を見よ
cleg-flies			horse flies (1) を見よ
clegs		*Haematopota*	(ハエ目、アブ科) の昆虫の総称
clegs			horse flies (1) を見よ
clematis aphid	ボタンヅルフトオヒゲナガアブラムシ	*Macrosiphum clematifoliae* Shinji	(カメムシ目、アブラムシ科) 日本、東洋区
clematis blister beetle			gray blister beetle を見よ
clematis borer moth		*Alcathoe carolinensis* Engelhardt	(チョウ目、スカシバガ科) 新北区
clematis leaf miner		*Phytomyza vitalbae* Kaltenbach	(ハエ目、ハモグリバエ科) 旧北区
clematis looper			pretty chalk carpet を見よ
clematis woolly aphid	ハンショウヅルオオワタムシ	*Eriosoma clematis* (Shinji)	(カメムシ目、アブラムシ科) 日本、東洋区
Clemens' bark moth		*Xylesthia pruniarmiella* Clemens	(チョウ目、ヒロズコガ科) 新北区

英名	和名	学名	所属、分布、ほか
Clemens' Clepsis moth		*Clepsis clemensiana* (Fernald)	（チョウ目、ハマキガ科）新北区
Clemens' Cosmopterix moth		*Cosmopterix clemensella* Stainton	（チョウ目、カザリバガ科）新北区
Clemens' false skeletonizer moth		*Acoloithus falsarius* Clemens	（チョウ目、マダラガ科）新北区
Clemens' grass tubeworm moth		*Acrolophus popeanella* (Clemens)	（チョウ目、Acrolophidae）新北区
Clemens' hawk-moth		*Sphinx luscitiosa* Clemens	（チョウ目、スズメガ科）新北区
Clemens' sphinx			Clemens' hawk-moth を見よ　Clemens' sphinx moth ともいう
Clementine's albatross		*Appias clementina* (Felder)	（チョウ目、シロチョウ科）東洋区
Clench's greenstreak		*Cyanophrys miserabilis* (Clench)	（チョウ目、シジミチョウ科）新北区
Clench's greenstreak		*Cyanophrys pseudolongula* Clench	（チョウ目、シジミチョウ科）新熱帯区
Clench's hairstreak		*Ocaria clenchi* (Johnson)	（チョウ目、シジミチョウ科）新熱帯区
Clench's skipper		*Virga clenchi* Miller	（チョウ目、セセリチョウ科）新熱帯区
Cleon ministreak		*Ministrymon cleon* Fabricius	（チョウ目、シジミチョウ科）新熱帯区
Cleonus metalmark		*Detritivora cleonus* (Stoll)	（チョウ目、シジミタテハ科）新熱帯区
Cleopatra	クレオパトラヤマキチョウ	*Gonepteryx cleopatra* (Linnaeus)	（チョウ目、シロチョウ科）旧北区
Cleopatra wood nymph		*Taygetis cleopatra* C. et R. Felder	（チョウ目、タテハチョウ科）新熱帯区
Cleophes satyr		*Cissia cleophes* (Godman et Salvin)	（チョウ目、タテハチョウ科）新熱帯区
clerid beetles			checkered beetles を見よ
clerids			checkered beetles を見よ
Clerk's snowy bentwing			peach leafminer を見よ
clerodendron thrips			flower thrips (3) を見よ
cleroid longicorn	カッコウカミキリ	*Micolamia cleroides* Bates	（コウチュウ目、カミキリムシ科）日本
Cleta tufted-skipper		*Polyctor cleta* Evans	（チョウ目、セセリチョウ科）新北区、新熱帯区
cleyera whitefly	サカキコナジラミ	*Rusostigma tokyonis* (Kuwana)	（カメムシ目、コナジラミ科）日本
click beetle (1)		*Agriotes litigiosus* Rossi	（コウチュウ目、コメツキムシ科）旧北区
click beetle (2)		*Agriotes sputator* (Linnaeus)	（コウチュウ目、コメツキムシ科）旧北区
click beetle			Pacific coast wireworm を見よ
click beetle			obscure click beetle を見よ
click beetle			common click beetle (1) を見よ
click beetle			garden click beetle を見よ
click beetle			upland click beetle (1) を見よ
click beetles (1)	コメツキムシ科	Elateridae	（コウチュウ目）の昆虫の総称　snap beetles ともいわれる

英名	和名	学名	所属、分布、ほか
click beetles (2)		*Pyrophorus*	(コウチュウ目、コメツキムシ科) の昆虫の総称
click beetles		*Agriotes*	(コウチュウ目、コメツキムシ科) の昆虫の総称
clicker round-winged katydid		*Amblycorypha alexanderi* Walker	(バッタ目、キリギリス科) 新北区
clicks			calicoes を見よ
clidemia leafroller		*Blepharomastix ebulealis* (Guenée)	(チョウ目、メイガ科) 新北区
clidemia thrips		*Liothrips urichi* Karny	(アザミウマ目、クダアザミウマ科) 新北区
clifden nonpareil		*Catocala fraxini* (Linnaeus)	(チョウ目、ヤガ科) 旧北区。日本亜種は *C. f. jezoensis* Matsumura ムラサキシタバ
cliff stiletto		*Thereva strigata* (Fabricius)	(ハエ目、ツルギアブ科) 旧北区
cliff swallowtail			Indra swallowtail を見よ
climbing cutworm		*Abagratis barnesi* Benjamin	(チョウ目、ヤガ科) 新北区
climbing cutworm			common cutworm を見よ　幼虫の英名
climbing locust			speckled bush-cricket を見よ
Climena crow		*Euploea climena* (Stoll)	(チョウ目、タテハチョウ科) 豪州区
Clinton's underwing moth		*Catocala clintonii* Grote	(チョウ目、ヤガ科) 新北区
clio crescent			common crescent を見よ
clio moth		*Ectypia clio* (Packard)	(チョウ目、ヒトリガ科) 新北区
clio tiger moth			clio moth を見よ
clip-thears			common earwig を見よ
clipper	トラフタテハ	*Parthenos sylvia* (Cramer)	(チョウ目、タテハチョウ科) 東洋区。*P. s. lilacinus* Butler も同英名。clipper butterfly ともいう
clippers	トラフタテハ属	*Parthenos*	(チョウ目、タテハチョウ科) の昆虫の総称
Clitus duskywing			Arizona duskywing を見よ
cloaked carpet		*Euphyia biangulata* (Haworth)	(チョウ目、シャクガ科) 旧北区
cloaked carpet		*Euphyia picata* Hübner	(チョウ目、シャクガ科) 旧北区
cloaked knot-horn			Indian meal moth (1) を見よ　cloaked knot ともいう
cloaked marvel moth		*Chytonix palliatricula* (Guenée)	(チョウ目、ヤガ科) 新北区、新熱帯区
cloaked minor	ヨコスジヨトウ	*Mesoligia furuncula* (Denis et Schiffermüller)	(チョウ目、ヤガ科) 日本、旧北区。イネ科牧草の害虫
cloaked Pococera moth		*Pococera subcanalis* (Walker)	(チョウ目、メイガ科) 新北区
cloaked pug		*Eupithecia abietaria* (Goeze)	(チョウ目、シャクガ科) 旧北区。日本亜種は *E. a. debrunneata* Staudinger　オオクロテンカバナミシャク。
cloaked pug		*Eupithecia pini* (Retzius)	(チョウ目、シャクガ科) 旧北区
cloaked pug			spruce cone looper moth を見よ
clock			dor beetle (1) を見よ

英名	和名	学名	所属、分布、ほか
Clodius Apollo	オオアメリカウスバアゲハ	*Parnassius clodius* Ménétriès	（チョウ目、アゲハチョウ科）新北区
Clodius Parnassian			clodius Apollo を見よ
cloistered sister			Miller's sister を見よ
Clonia metalmark		*Esthemopsis clonia* C. et R. Felder	（チョウ目、シジミタテハ科）新熱帯区
clorinde			ghost brimstone を見よ
closebanded yellowhorn moth		*Colocasia propinquilinea* (Grote)	（チョウ目、ヤガ科）新北区。closebanded yellowhorn ともいう
close-wings			grass moths (1) を見よ
close-wings			sod webworm moths を見よ
clothes and scavenger moths			clothes moths を見よ
clothes louse			body louse を見よ
clothes moth (1)	コイガ	*Tineola bisselliella* (Hummel)	（チョウ目、ヒロズコガ科）日本、汎世界。衣類害虫
clothes moth		*Myrmecozela tineoides* Walsingham	（チョウ目、ヒロズコガ科）東洋区
clothes moth			pale corn clothes を見よ
clothes moth			case-making clothes moth (2) を見よ
clothes moths	ヒロズコガ科	Tineidae	（チョウ目）の昆虫の総称
cloud-bordered brindle			clouded-bordered brindle を見よ
cloud copper		*Aloeides nubilus* Henning et Henning	（チョウ目、シジミチョウ科）エチオピア区
cloudforest Argus		*Oxeoschistus simplex* Butler	（チョウ目、タテハチョウ科）新熱帯区
cloud-forest beauty	オオアカタテハ	*Pycina zamba zelys* Godman et Salvin	（チョウ目、タテハチョウ科）新熱帯区
cloudforest crescent		*Anthanassa acesas* (Hewitson)	（チョウ目、タテハチョウ科）新熱帯区
cloud-forest crescent		*Anthanassa otanes fulviplaga* (Butler)	（チョウ目、タテハチョウ科）新熱帯区
cloudforest deadleaf		*Consul panariste* (Hewitson)	（チョウ目、タテハチョウ科）新熱帯区
cloud-forest hoary-skipper		*Carrhenes callipetes* Godman et Salvin	（チョウ目、セセリチョウ科）新熱帯区
cloud-forest king	ニシキヒョウモンダマシ	*Anetia thirza thirza* Geyer	（チョウ目、タテハチョウ科）新熱帯区
cloud-forest monarch			cloud-forest king を見よ
cloudforest sister		*Adelpha aricia* (Hewitson)	（チョウ目、タテハチョウ科）新熱帯区
cloud-forest skipper		*Tigasis nausiphanes* (Schaus)	（チョウ目、セセリチョウ科）新熱帯区
clouded alder bell		*Epinotia sordidana* (Hübner)	（チョウ目、ハマキガ科）旧北区
clouded Apollo	クロホシウスバアゲハ	*Parnassius mnemosyne* (Linnaeus)	（チョウ目、アゲハチョウ科）旧北区

英名	和名	学名	所属、分布、ほか
clouded Argus			gatekeeper を見よ
clouded August thorn		*Ennomos quercaria* (Hübner)	(チョウ目、シャクガ科) 旧北区
clouded border		*Lomaspilis marginata* (Linnaeus)	(チョウ目、シャクガ科) 旧北区。日本亜種は *L. m. amurensis* (Hedemann) シロオビヒメエダシャク (北), *L. m. opis* Butler (本)
clouded border moth			clouded border を見よ
clouded-bordered brindle	カドモンヨトウ	*Apamea crenata* (Hufnagel)	(チョウ目、ヤガ科) 日本、旧北区
clouded brindle		*Apamea epomidion* (Haworth)	(チョウ目、ヤガ科) 旧北区
clouded brown moth		*Mecyna maorialis* Felder	(チョウ目、メイガ科) 豪州区
clouded buff	モンヘリアカヒトリ	*Diacrisia sannio* (Linnaeus)	(チョウ目、ヒトリガ科) 日本、旧北区
clouded centurion		*Sargus cuprarius* (Linnaeus)	(ハエ目、ミズアブ科) 旧北区
clouded cosmet		*Mompha langiella* Hübner	(チョウ目、カザリバガ科) 旧北区
clouded drab		*Orthosia incerta* (Hufnagel)	(チョウ目、ヤガ科) 旧北区。日本亜種は *O. i. incognita* Sugi ミヤマカバキリガ。clouded drab moth ともいう
clouded ermine			clouded buff を見よ
clouded flat		*Tagiades flesus* (Fabricius)	(チョウ目、セセリチョウ科) エチオピア区
clouded forester			clouded flat を見よ
clouded magpie		*Abraxas sylvata* (Scopoli)	(チョウ目、シャクガ科) 旧北区、東洋区。日本亜種は *A. s. microtate* Wehli キタマダラエダシャク
clouded mother-of-pearl	トガリシンジュタテハ	*Salamis anacardii* (Linnaeus)	(チョウ目、タテハチョウ科) エチオピア区
clouded mother-of-pearl		*Protogoniomorpha anacardii* (Linnaeus)	(チョウ目、タテハチョウ科) エチオピア区
clouded oblique-barred conch		*Piercea manniana* Fischer von Röslerstamm	(チョウ目、ハマキガ科) 旧北区
clouded peach stem aphid		*Pterochloroides persicae* (Cholodkovsky)	(カメムシ目、アブラムシ科) 旧北区
clouded pigweed bug		*Sphragisticus nebulosus* (Fallén)	(カメムシ目、ナガカメムシ科) 旧北区
clouded plant bug		*Neurocolpus nubilus* (Say)	(カメムシ目、カスミカメムシ科) 新北区
clouded shield beetle			beet tortoise beetle (1) を見よ
clouded silver	バラシロエダシャク	*Lomographa temerata* (Denis et Schiffermüller)	(チョウ目、シャクガ科) 日本、旧北区。モモ、サクラ害虫
clouded silverline		*Cigaritis nubilus* (Moore)	(チョウ目、シジミチョウ科) 東洋区
clouded skipper		*Lerema accius* Smith et Abbot	(チョウ目、セセリチョウ科) 新北区、新熱帯区
clouded skipper			clouded flat を見よ
clouded slender		*Caloptilia populetorum* (Zeller)	(チョウ目、ホソガ科) 旧北区
clouded sulphur	アメリカモンキチョウ	*Colias philodice philodice* Godart	(チョウ目、シロチョウ科) 新北区

英名	和名	学名	所属、分布、ほか
clouded tortoise beetle			beet tortoise beetle (1) を見よ
clouded underwing moth		*Catocala nebulosa* Hy. Edwards	(チョウ目、ヤガ科) 新北区。clouded underwing ともいう
clouded veneer moth		*Prionapteryx nebulifera* Stephens	(チョウ目、メイガ科) 新北区
clouded yellow	クロケアモンキチョウ	*Colias croceus* (Geoffroy)	(チョウ目、シロチョウ科) 旧北区。clouded yellow butterfly ともいう
clouded yellows	モンキチョウ属	*Colias*	(チョウ目、シロチョウ科) の昆虫の総称
cloudless giant sulphur	ワタリオオキチョウ	*Phoebis sennae* (Linnaeus)	(チョウ目、シロチョウ科) 新北区、新熱帯区。cloudless sulphur ともいう
cloudy hog			large skipper (1) を見よ
cloudy sword-grass			sword-grass を見よ
cloudy white marble			grey bud moth を見よ
cloudy winged white fly			woolly whitefly を見よ　cloudy-winged whitefly とも記す
cloudy-winged whitefly			citrus whitefly (2) を見よ
cloudywings		*Thorybes*	(チョウ目、セセリチョウ科) の昆虫の総称
cloudy wormwood pearl			European corn borer を見よ
clover-and-alflalfa weevils			alfalfa weevils を見よ
clover aphid	ツメクサベニマルアブラムシ	*Nearctaphis bakeri* (Cowen)	(カメムシ目、アブラムシ科) 日本、全北区。ダイズ、マメ科牧草の害虫
clover aphid parasite		*Aphelinus lapisligni* Howard	(ハチ目、ツヤコバチ科) 新北区
clover blue		*Actizera stellata* (Trimen)	(チョウ目、シジミチョウ科) エチオピア区
clover blue			common grass-blue (1) を見よ
clover bud weevil			clover leaf weevil (2) を見よ
clover budworm			clover leaf weevil (2) を見よ
clover callipterus			yellow clover aphid を見よ
clover casebearer			knapweed green case (1) を見よ
clover cutworm			nutmeg を見よ
clover flea beetle			corn flea beetle を見よ
clover flower midge			clover seed midge を見よ
clover gelechiid	コフサキバガ	*Dichomeris acuminata* (Staudinger)	(チョウ目、キバガ科) 日本、全北区、東洋区、エチオピア区、大洋区。マメ科牧草害虫
clover hayworm			gold fringe を見よ
clover head caterpillar		*Grapholitha interstictana* (Clemens)	(チョウ目、ハマキガ科) 新北区。クローバー害虫。clover head caterpillar moth ともいう
clover head midge			clover seed midge を見よ
clover head weevil		*Hypera meles* (Fabricius)	(コウチュウ目、ゾウムシ科) 全北区
clover leaf midge		*Dasineura trifolii* (Loew)	(ハエ目、タマバエ科) 新北区
clover leaf tiger		*Ancylis angulifasciana* Zeller	(チョウ目、ハマキガ科) 新北区

英名	和名	学名	所属、分布、ほか
clover leaf weevil		*Phytonomus zoilus* (Scopoli)	（コウチュウ目、ゾウムシ科）旧北区。ダイズ、マメ科牧草害虫
clover leaf weevil (1)	オオタコゾウムシ	*Hypera punctata* (Fabricius)	（コウチュウ目、ゾウムシ科）日本、全北区。マメ科牧草の害虫
clover leaf weevil (2)	ツメクサタコゾウムシ	*Hypera nigricostris* (Fabricius)	（コウチュウ目、ゾウムシ科）日本、旧北区
clover leaf weevil (3)		*Hypera variabilis* (Herbst)	（コウチュウ目、ゾウムシ科）旧北区
clover leaf weevil			lucerne weevil を見よ
clover leafhopper		*Aceratagallia sanguinolenta* (Provancher)	（カメムシ目、オオヨコバイ科）全北区
clover leaf-roller			fen marble を見よ
clover looper		*Caenurgina crassiuscula* (Haworth)	（チョウ目、ヤガ科）全北区。clover looper moth ともいう
clover mamestra			nutmeg を見よ　米国での英名
clover midge			clover seed midge を見よ
clover Nomophila			lucerne moth を見よ
clover pear-shaped weevil			clover seed weevil (1) を見よ
clover plant-louse		*Metopolophium cryptobium* (Hille Ris Lambers)	（カメムシ目、アブラムシ科）旧北区
clover red midge			clover seed midge を見よ
clover-root bark beetle			clover root borer を見よ
clover root borer		*Hylastinus obscurus* (Marsham)	（コウチュウ目、キクイムシ科）新北区。クローバー害虫
clover root curculio			clover weevil (1) を見よ
clover root scale	ツメクサシロカイガラムシ	*Aulacaspis trifolium* Takagi	（カメムシ目、マルカイガラムシ科）日本。マメ科牧草害虫
clover root weevil			clover weevil (1) を見よ
clover seed caterpillar		*Enarmonia conversana* Walsingham	（チョウ目、ハマキガ科）新北区
clover seed caterpillar		*Enarmonia interstinctana* Clemens	（チョウ目、ハマキガ科）新北区
clover seed chalcid (1)	クローバータネコバチ	*Bruchophagus gibbus* (Goheman)	（ハチ目、カタビロコバチ科）日本、旧北区。マメ科牧草害虫
clover seed chalcid (2)		*Bruchophagus platypterus* (Walker)	（ハチ目、カタビロコバチ科）新北区
clover seed fly			clover seed midge を見よ
clover seed midge		*Dasineura leguminicola* (Lintner)	（ハエ目、タマバエ科）全北区
clover seed moth		*Coleophora frischella* Linnaeus	（チョウ目、ツツミノガ科）全北区、豪州区
clover seed moth			knapweed green case (1) を見よ　豪州での英名
clover seed weevil		*Apion assimile* Kirby	（コウチュウ目、ホソクチゾウムシ科）旧北区
clover seed weevil (1)		*Apion apricans* Herbst	（コウチュウ目、ホソクチゾウムシ科）旧北区

英名	和名	学名	所属、分布、ほか
clover seed weevil (2)		*Apion aestivum* Germar	（コウチュウ目、ホソクチゾウムシ科）旧北区
clover seed weevil (3)		*Tychius picirostris* (Fabricius)	（コウチュウ目、ゾウムシ科）新北区
clover seed worm			clover head caterpillar を見よ
clover springtail			lucerne flea を見よ
clover stem borer			clover stem erotylid を見よ
clover stem erotylid		*Languria mozardi* Latreille	（コウチュウ目、オオキノコムシ科）新北区。レタス、アルファルファなどの害虫
clover thrips	コヌマアザミウマ	*Odontothrips biuncus* John	（アザミウマ目、アザミウマ科）旧北区
clover thrips			red clover thrips を見よ　カナダでの英名
clover webworm			gold fringe を見よ
clover weevil		*Sitona sullcifrons* Thunberg	（コウチュウ目、ゾウムシ科）旧北区
clover weevil (1)	ケチビコフキゾウムシ	*Sitona hispidula* (Fabricius)	（コウチュウ目、ゾウムシ科）日本、全北区。アルファルファ、クローバー害虫
clover weevil			clover seed weevil (1)(2) を見よ
clover weevils			alfalfa weevils を見よ
clovers			budworm moths を見よ
clown beetles			hister beetles を見よ
club beak		*Libythea myrrha* (Godart)	（チョウ目、タテハチョウ科）東洋区
club fleas		Rhopalopsyllidae	（ノミ目）の昆虫の総称
club-legged grasshopper		*Gomphocerus sibiricus* (Linnaeus)	（バッタ目、バッタ科）旧北区
club silverline	ミツボシフタオツバメ	*Cigaritis syama* (Horsfield)	（チョウ目、シジミチョウ科）東洋区。*C. s. terana* (Fruhstorfer) も同英名
clubbed Cyphonia			ant-mimicking treehopper を見よ
clubbed general			soldier fly (1) を見よ
clubbed ringlet		*Ypthima antennata* van Son	（チョウ目、タテハチョウ科）エチオピア区
clubbed shutwing		*Cordulephya divergens* Tillyard	（トンボ目、エゾトンボ科）豪州区
clubhorned beetles			clavicorn beetles を見よ
clubhorned grasshopper		*Aeropedellus clavatus* (Thomas)	（バッタ目、バッタ科）新北区
club-tailed Charaxes	トガリフタオチョウ	*Charaxes zoolina* (Westwood)	（チョウ目、タテハチョウ科）エチオピア区
club-tailed dragonflies		*Gomphus*	（トンボ目、サナエトンボ科）の昆虫の総称
club-tailed dragonflies			clubtails (1) を見よ
club-tailed dragonfly			common clubtail (1) を見よ
club-tailed emperor			club-tailed Charaxes を見よ
clubtails (1)	サナエトンボ科	Gomphidae	（トンボ目）の昆虫の総称　club-tailed dragonflies ともいう

英名	和名	学名	所属、分布、ほか
clubtails		*Octogomphus, Anisogomphus, Burmagomphus, Cyclogomphus* の各属	（トンボ目、サナエトンボ科）の昆虫の総称
Cluentius sphinx		*Neococytius cluentius* (Cramer)	（チョウ目、スズメガ科）新北区、新熱帯区
clusiid flies	クチキバエ科	Clusiidae	（ハエ目）の昆虫の総称
clusiids			clusiid flies を見よ
cluster caterpillar			common cutworm を見よ　豪州での幼虫の英名
cluster flies			blow flies を見よ
cluster fly		*Pollenia rudis* (Fabricius)	（ハエ目、クロバエ科）全北区。幼虫はミミズに寄生
clymene		*Haploa clymene* (Brown)	（チョウ目、ヒトリガ科）新北区。北米での英名。clymene haploa moth, clymene moth ともいう
clymene moth			clymene を見よ
clysonymus longwing			montane longwing を見よ
clytia purplewing			sandbar purplewing を見よ
Clytie hairstreak		*Ministrymon clytie* (Edwards)	（チョウ目、シジミチョウ科）新北区、新熱帯区
Clytie ministreak			Clytie hairstreak を見よ
clytrine beetles			case-bearing leaf beetles (1) を見よ
CMR beetles			blister beetles (1) を見よ　Meloidae の1種が Cape Mounted Riflemen (CMR) に似ることに由来
CMR beetles			blister beetles (3) を見よ　南アフリカでの英名
coach-horse			dragonflies (1) を見よ
Coahuila giant skipper		*Agathymus remingtoni* (Stallings et Turner)	（チョウ目、セセリチョウ科）新北区、新熱帯区
coarse wainscot flat-body			carrot moth (2) を見よ
coarse wainscot flat-body			carrot moth (1) を見よ　南アフリカでの英名
coast Acraea			Chrinda Acraea を見よ
coast Charaxes		*Charaxes ethalion* Boisduval	（チョウ目、タテハチョウ科）エチオピア区
coast copper			Natal opal を見よ
coast dart		*Euxoa cursoria* (Hufnagel)	（チョウ目、ヤガ科）旧北区
coast glider			blond glider を見よ
coast grey shade		*Cnephasia conspersana* Douglas	（チョウ目、ハマキガ科）旧北区
coast knot-horn	ハマベホソメイガ	*Anerastia lotella* (Hübner)	（チョウ目、メイガ科）日本、旧北区
coast leaf-cutter		*Megachile martima* (Kirby)	（ハチ目、ハキリバチ科）旧北区
coast lined June beetle		*Polyphylla crinita* LeConte	（コウチュウ目、コガネムシ科）新北区

英名	和名	学名	所属、分布、ほか
coast policeman	キオビアオバセセリ	*Coeliades sejuncta* (Mabille et Vuillot)	(チョウ目、セセリチョウ科) エチオピア区
coast purple-tip	ツマムラサキシロチョウ	*Colotis erone* (Angas)	(チョウ目、シロチョウ科) エチオピア区
coast purple-tip			scarlet-tip を見よ
coast scarlet		*Axiocerses styx* Rebel	(チョウ目、シジミチョウ科) エチオピア区
coast spurge bell		*Epinotia subsequana* (Haworth)	(チョウ目、ハマキガ科) 旧北区
coast spurge shoot		*Acroclita subsequana* (Herrich-Schäffer)	(チョウ目、ハマキガ科) 旧北区
coastal blue		*Lepidochrysops littoralis* Swanepoel et Vári	(チョウ目、シジミチョウ科) エチオピア区
coastal brown ant			big-headed ant (1) を見よ
coastal fly		*Fannia femoralis* (Stein)	(ハエ目、ヒメイエバエ科) 日本、新北区、新熱帯区
coastal hairstreak		*Harkenclenus titus* (Fabricius)	(チョウ目、シジミチョウ科) 新北区
coastal hairstreak		*Hypolycaena lochmophila* Tite	(チョウ目、シジミチョウ科) エチオピア区
coastal Mellana		*Quasimellana agnesae* (Bell)	(チョウ目、セセリチョウ科) 新熱帯区
coastal pennant		*Macrodiplax cora* (Kaup)	(トンボ目、トンボ科) 東洋区、エチオピア区
coastal petaltail		*Petalura litorea* Theischinger	(トンボ目、ムカシヤンマ科) 豪州区。coastal petaltail dragonfly ともいう
coastal Phaneta moth		*Phaneta argutipunctana* Blanchard et Knudson	(チョウ目、ハマキガ科) 新北区
coastal plain Meganola moth		*Meganola phylla* (Dyar)	(チョウ目、コブガ科) 新北区
coastal silver-stiletto		*Acrosathe annulata* (Fabricius)	(ハエ目、ツルギアブ科) 旧北区
coastal stalk borer		*Chilo orichalcociliella* (Strand)	(チョウ目、メイガ科) エチオピア区
coastal swordtail		*Graphium porthaon* (Hewitson)	(チョウ目、アゲハチョウ科) エチオピア区
coastal tiger beetle		*Laphyridia augulata* (Fabricius)	(コウチュウ目、ハンミョウ科) 日本
coax moth			greater wax moth を見よ
cob borer			maize ear borer を見よ
cobalt fairy playboy		*Paradeudorix cobaltina* (Stempffer)	(チョウ目、シジミチョウ科) エチオピア区
cobalt milkweed beetle		*Chrysochus cobaltinus* LeConte	(コウチュウ目、ハムシ科) 新北区
cobalt pea-blue		*Catochrysops amasea* Waterhouse et Lyell	(チョウ目、シジミチョウ科) 豪州区
cobalt playboy		*Pilodeudorix ula* (Karsch)	(チョウ目、シジミチョウ科) エチオピア区
cobbler moth		*Condica sutor* (Guenée)	(チョウ目、ヤガ科) 新北区
cobra clubtail		*Gomphus vastus* Walsh	(トンボ目、サナエトンボ科) 新北区
cobra inchworm			cross-lined wave moth を見よ

英名	和名	学名	所属、分布、ほか
cobweb skipper		*Hesperia metea* Scudder	(チョウ目、セセリチョウ科) 新北区
Cocala sister		*Adelpha cocala* (Cramer)	(チョウ目、タテハチョウ科) 新熱帯区
coccid-eating pyralid		*Laetilia coccidivora* Comstock	(チョウ目、メイガ科) 新北区。幼虫がカイガラムシ類の卵、幼虫を捕食
coccids			scale insects を見よ
coccinellid adult parasite			ladybird parasitoid を見よ
coccinellid flea beetle	ヘリグロテントウノミハムシ	*Argropistes coccineliformis* Csiki	(コウチュウ目、ハムシ科) 日本、東洋区
coccinellid larval parasites		*Homalotylus*	(ハチ目、トビコバチ科) の昆虫の総称
coccinellid parasitoid			ladybird parasitoid を見よ
coccinellids			ladybird beetles を見よ
coccus			coccids を見よ
cochineal			cochineal insect を見よ
cochineal bug		*Fulgora lanternaria* Donovan	(カメムシ目、テングスケバ科) 新北区
cochineal insect	コチニールカイガラムシ	*Dactylopius coccus* Costa	(カメムシ目、コチニールカイガラムシ科) 新北区、新熱帯区。中南米でサボテンにつくカイガラムシでコチニールえんじ虫といわれる。雌を乾燥したものから赤色顔料や染料をとる
cochineal insects	コチニールカイガラムシ科	Dactylopiidae	(カメムシ目) の昆虫の総称
cochineal insects			soft scales を見よ　米国での英名
cochineal scale			cochineal insect を見よ
cochineal scale insects			cochineal insects を見よ
cochineal scales		*Dactylopius*	(カメムシ目、コチニールカイガラムシ科) の昆虫の総称
cochineal scales			cochineal insects を見よ
cochylid moths			phaloniid moths を見よ
cochylis moth			grape cochylid を見よ
cock tail			common earwig を見よ
cockchafer		*Macrodactylus affinis* Castelnau	(コウチュウ目、コガネムシ科) 新熱帯区。カンキツ害虫
cockchafer		*Macrodactylus suturalis* Mannerheim	(コウチュウ目、コガネムシ科) 新熱帯区。カンキツ害虫
cockchafer			June beetle (1) を見よ
cockchafers		*Melolontha*	(コウチュウ目、コガネムシ科) の昆虫の総称
cockchafers			chafer beetles を見よ
cockchafers			May beetles (3) を見よ
Cockerell scale	シロナガカイガラムシ	*Lopholeucaspis cockerell* (De Charmoy)	(カメムシ目、マルカイガラムシ科) 日本、汎熱帯。カンキツ類、イスノキ害虫
Cockerell scale			false oleander scale を見よ
Cockerell's blue-winged grasshopper		*Leprus cyaneous* Cockerell	(バッタ目、バッタ科) 新北区

英名	和名	学名	所属、分布、ほか
Cockerell's green goldenglow aphid		*Macrosiphum rudbeckiarum* (Cockerell)	(カメムシ目、アブラムシ科) 新北区
Cockerell's moth		*Diedra cockerellana* (Kearfott)	(チョウ目、ハマキガ科) 新北区
cocklebur weevil		*Rhodobaenus tredecimpunctatus* (Iliger)	(コウチュウ目、ゾウムシ科) 新北区
cocklebur weevil		*Rhodobaenus quinquedecimpunctatus* (Say)	(コウチュウ目、ゾウムシ科) 新北区
cockroach hunting wasps	セナガアナバチ亜科	Ampulicinae	(ハチ目、アナバチ科) の昆虫の総称
cockroaches (1)	ゴキブリ科	Blattidae	(ゴミブリ目) の昆虫の総称　世界に4000種。多くは熱帯。英国では American cockroaches という
cockroaches (2)	ゴキブリ目	Blattaria	の昆虫の総称　ゴキブリ目は Blattiformia, Blattopteroidea, Dictyoptera, Oothecaria などの異名がある。世界に約5,500種
cockroaches (3)	ゴキブリ亜目	Blattoidea	(ゴキブリ目) の昆虫の総称
cockroaches (4)	ブラベルスゴキブリ科	Blaberidae	(ゴキブリ目) の昆虫の総称
cocksfoot aphid		*Hyalopteroides humilis* (Walker)	(カメムシ目、アブラムシ科) 旧北区
cocksfoot grub		*Rhopaea verreauxii* Blanchard	(コウチュウ目、コガネムシ科) 豪州区
cocksfoot midge		*Contarinia dactylidis* (Loew)	(ハエ目、タマバエ科) 旧北区
cocksfoot midge		*Dasineura dactylidis* Metcalf	(ハエ目、タマバエ科) 旧北区
cocksfoot midge		*Sitodiplosis dactylidis* Barnes	(ハエ目、タマバエ科) 旧北区
cocksfoot midge			foxtail midge (1) を見よ
cocksfoot moth		*Glyphipterix simplicella* Christoph	(チョウ目、ホソハマキモドキガ科) 旧北区
cocksfoot moth			allied fanner (1)(2)を見よ　cocks foot moth　とも記す
cocksfoot thrips			timothy thrips を見よ
cocktail ant		*Crematogaster laeviceps chasei* Forel	(ハチ目、アリ科) 豪州区
cock-tail beetle			devil's coach-horse (2) を見よ
cocoa ant		*Azteca chartifex* Forel	(ハチ目、アリ科) 新熱帯区
cocoa-bean moth			tobacco moth (1) を見よ
cocoa capsid			cacao capsid (1) を見よ
cocoa clubtail		*Gomphus hybridus* Williamson	(トンボ目、サナエトンボ科) 新北区
cocoa louse		*Pseudococcus hispidus* Morrison	(カメムシ目、コナカイガラムシ科) 東洋区。カンキツ害虫
cocoa mealybug	タイワンコナカイガラムシ	*Planococcus lilacinus* (Cockerell)	(カメムシ目、コナカイガラムシ科) 日本、東洋区。カンキツ、マンゴー、ハイビスカスなどの害虫

英名	和名	学名	所属、分布、ほか
cocoa mort bleu	ナミモンフクロチョウ	*Caligo teucer insularis* (Linnaeus)	(チョウ目、タテハチョウ科) 新熱帯区
cocoa moth		*Conopomorpha cramerella* (Snellen)	(チョウ目、ホソガ科) 東洋区、豪州区
cocoa moth			tobacco moth (1) を見よ
cocoa moth bug		*Colobesthes falcata* (Guérin-Méneville)	(カメムシ目、アオバハゴロモ科) 東洋区
cocoa pod borer		*Characoma stictigrapta* Hampson	(チョウ目、シャクガ科) エチオピア区。カカオ害虫
cocoa pod borer			cocoa moth を見よ
cocoa pod borer			tobacco moth (1) を見よ
cocoa stem borer		*Eulophonotus myrmeleon* Felder	(チョウ目、ボクトウガ科) エチオピア区
cocoa thrips			red-banded thrips を見よ
coconino shieldback		*Ateloplus coconino* Hebard	(バッタ目、キリギリス科) 新北区
coconut black-headed caterpillar			black-headed caterpillar を見よ
coconut bug		*Pseudotheraptus wayi* Brown	(カメムシ目、ヘリカメムシ科) エチオピア区
coconut case caterpillar		*Mahasena corbetti* Tams	(チョウ目、ミノガ科) 東洋区
coconut caterpillar		*Nephantis serinopa* Meyrick	(チョウ目、ヒロバキバガ科) 東洋区。ヤシ害虫
coconut flat moth			coconut leaf moth (3) を見よ
coconut flat moth			coconut leafminer (1) を見よ
coconut fly		*Scholastes bimaculatus* Hendel	(ハエ目、ヒロクチバエ科)。大洋区
coconut grasshopper		*Sexava nubila* (Stål)	(バッタ目、キリギリス科) 東洋区
coconut grasshopper		*Sexava novaeguineae* Branc	(バッタ目、キリギリス科) 東洋区
coconut hispid			palm leaf beetle を見よ
coconut hispine beetle			palm leaf beetle を見よ
coconut leaf-eating caterpillar		*Opisina arenosella* Walker	(チョウ目、ヒロバキバガ科) 東洋区
coconut leafminer		*Promecotheca cumingi* Baly	(コウチュウ目、ハムシ科) 東洋区。ヤシ害虫
coconut leafminer (1)		*Agonoxena argaula* Meyrick	(チョウ目、Agonoxenidae) 新北区
coconut leaf miner			palm leaf beetle を見よ
coconut leafmining beetle		*Promecotheca reichei* Baly	(コウチュウ目、ハムシ科) 東洋区、豪州区、大洋区
coconut leaf moth		*Artona catoxantha* Hampson	(チョウ目、マダラガ科) 東洋区。ヤシ害虫
coconut leaf moth (1)		*Levuana iridescens* Bethune-Baker	(チョウ目、マダラガ科) 大洋区
coconut leaf moth (2)		*Charixena iridoxa* Meyrick	(チョウ目、スガ科) 豪州区

英名	和名	学名	所属、分布、ほか
coconut leaf moth (3)		*Agonoxena pyrogramma* Meyrick	(チョウ目、Agonoxenidae) 東洋区、大洋区、新北区
coconut leafroller		*Hedylepta blackburni* (Butler)	(チョウ目、メイガ科) 新北区、東洋区、大洋区
coconut mealybug		*Pseudococcus nipae* (Maskell)	(カメムシ目、コナカイガラムシ科) 東洋区、全北区、新熱帯区、エチオピア区。温室害虫。中北米での英名
coconut mealybug parasite		*Pseudaphycus utilis* Timberlake	(ハチ目、トビコバチ科)。大洋区
coconut moth (1)		*Atheloca subrufella* Hulst	(チョウ目、メイガ科) 新北区、新熱帯区
coconut moth (2)		*Batrachedra arenosella* Walker	(チョウ目、カザリバガ科) 大洋区
coconut moth			coconut leaf moth (1) を見よ
coconut palm beetle			rhinoceros beetle (1) を見よ
coconut palm borer		*Melittomma insulae* Fairmaire	(コウチュウ目、ツツシンクイ科) エチオピア区
coconut palm scale			coconut scale (1) を見よ
coconut rhinoceros beetle		*Strategus quadrifoveatus* (Palisot et Beauvois)	(コウチュウ目、コガネムシ科) 新熱帯区
coconut rhinoceros beetle			rhinoceros beetle (1) を見よ
coconut scale (1)	ウスイロマルカイガラムシ	*Aspidiotus destructor* Signoret	(カメムシ目、マルカイガラムシ科) 日本、新北区、豪州区、汎熱帯。パパイア、バナナ、チャなどの害虫
coconut scale (2)	タコノキナガカイガラムシ	*Pinnaspis buxi* (Bouché)	(カメムシ目、マルカイガラムシ科) 日本、東洋区、豪州区、大洋区
coconut skipper		*Hidari irava* (Moore)	(チョウ目、セセリチョウ科) 東洋区
coconut slug caterpillar		*Thosea unifascia* Walker	(チョウ目、メイガ科) 東洋区
coconut-spike moth (1)		*Tirathaba rufivena* (Walker)	(チョウ目、メイガ科) 東洋区、豪州区、大洋区
coconut spike moth		*Tirathaba trichogramma* Meyrick	(チョウ目、メイガ科) 東洋区、豪州区
coconut termite		*Neotermes rainbowi* (Hill)	(シロアリ目、レイビシロアリ科) 大洋区
coconut whitefly		*Aleurodicus destructor* Mackie	(カメムシ目、コナジラミ科) 東洋区
coconut zygaenid			coconut leaf moth (1) を見よ
codlin maggot			codling moth を見よ　幼虫の英名
codling moth		*Cydia pomonella* (Linnaeus)	(チョウ目、ハマキガ科) 全北区、豪州区。リンゴの大害虫。codlin moth と記されることあり。細長いリンゴを codling という
codling piercer			codling moth を見よ
Coega copper		*Aloeides clarki* Tite et Dickson	(チョウ目、シジミチョウ科) エチオピア区
Coelestis blue ringlet		*Caeruleuptychia coelestis* Butler	(チョウ目、タテハチョウ科) 新熱帯区
coenagriid damselflies			narrow-winged damselflies を見よ　イトトンボ科の異名 Coenagriidae に由来

英名	和名	学名	所属、分布、ほか
coerulescens flea beetle	ヒメカミナリハムシ	*Altica caerulescens* (Baly)	(コウチュウ目、ハムシ科) 日本、旧北区、東洋区
Coetzer's daisy copper			donkey daisy copper を見よ
Cofaqui giant-skipper		*Megathymus cofaqui* (Strecker)	(チョウ目、セセリチョウ科) 新北区
coffee bean weevil			coffee weevil を見よ　coffee-bean weevil とも記される
coffee beetle			coffee berry borer を見よ
coffee-berry beetle borer			coffee berry borer を見よ
coffee berry borer		*Hypothenemus hampei* (Ferrari)	(コウチュウ目、キクイムシ科) 東洋区、旧北区、エチオピア区、新熱帯区。コーヒー害虫
coffee berry moth		*Prophantis smaragdina* (Butler)	(チョウ目、メイガ科) エチオピア区
coffee berry moth		*Prophantis octoguttalis* Felder et Rogenhofer	(チョウ目、メイガ科) エチオピア区
coffee borer			red coffee borer を見よ
coffee capsid		*Lamprocapsidea coffeae* (China)	(カメムシ目、カスミカメムシ科) エチオピア区
coffee clearwing moth			larger pellucid hawk moth を見よ　アジアでの英名
coffee comma scale			black thread scale を見よ
coffee flies			black flies (2) を見よ　中米での英名
coffee fruit fly		*Ceratitis coffeae* (Bezzi)	(ハエ目、ミバエ科) エチオピア区、
coffee green bug			green scale を見よ
coffee green scale			green scale を見よ
coffee hairy caterpillar		*Eupterote fabia* Cramer	(チョウ目、オビガ科) 東洋区
coffee hawk moth			larger pellucid hawk moth を見よ
coffee lace bug		*Corythucha gossypii* (Fabricius)	(カメムシ目、グンバイムシ科) 新北区、新熱帯区
coffee lace bugs		*Habrochila*	(カメムシ目、グンバイムシ科) の昆虫の総称
coffee leaf miner		*Leucoptera coffeina* Washburn	(チョウ目、ハモグリガ科) 新熱帯区、エチオピア区。コーヒー害虫
coffee leaf miner		*Leucoptera meyricki* Chesquiere	(チョウ目、ハモグリガ科) エチオピア区
coffee leaf miner (1)		*Leucoptera coffeella* (Guérin-Méneville)	(チョウ目、ハモグリガ科) 新熱帯区、エチオピア区。コーヒー害虫
coffee leaf miners		*Leucoptera*	(チョウ目、ハモグリガ科) の昆虫の総称
coffee leaf-rolling thrips			leaf-rolling thrips を見よ
coffee leaf skeletonizer		*Leucophema dohertyi* (Warren)	(チョウ目、フタオガ科) エチオピア区
coffee-loving Pyrausta moth		*Pyrausta tyralis* (Guenée)	(チョウ目、メイガ科) 新北区
coffee mealybug (1)		*Planococcus kenyae* (Le Pelley)	(カメムシ目、コナカイガラムシ科) エチオピア区。コーヒー害虫

英名	和名	学名	所属、分布、ほか
coffee mealybug (2)		*Pseudococcus lilacinus* Cockerell	(カメムシ目、コナカイガラムシ科) エチオピア区。コーヒー害虫
coffee mealybug			citrus mealybug (1) を見よ
coffee playboy		*Deudorix lorisona* (Hewitson)	(チョウ目、シジミチョウ科) エチオピア区
coffee root mealy bug		*Pseudococcus deceptor* Betrem	(カメムシ目、コナカイガラムシ科) 東洋区、エチオピア区。コーヒー害虫
coffee root mealy bug		*Pseudococcus gahani* Green	(カメムシ目、コナカイガラムシ科) 新北区。コーヒー害虫
coffee root mealybug		*Geococcus coffeae* Green	(カメムシ目、コナカイガラムシ科) 大洋区
coffee root mealybug			citrus mealybug (1) を見よ
coffee root mealy bug			coffee mealybug (2) を見よ
coffee seed borer			coffee berry borer を見よ
coffee sphinx			half-blind sphinx moth を見よ
coffee stem borer			coffee white borer を見よ
coffee thrips			African coffee thrips を見よ
coffee thrips			greenhouse thrips を見よ
coffee tip borer		*Eucosma nereidopa* Meyrick	(チョウ目、ハマキガ科) エチオピア区
coffee tortrix			tea tortrix を見よ
coffee twig borer		*Xyleborus compactus* Eichhoff	(コウチュウ目、キクイムシ科) 東洋区、エチオピア区。コーヒー害虫
coffee weevil	ワタミヒゲナガゾウムシ	*Araecerus fasciculatus* (De Geer)	(コウチュウ目、ヒゲナガゾウムシ科) 日本、汎世界。ワタ他の害虫
coffee white borer		*Xylotrechus quadripes* Chevrolat	(コウチュウ目、カミキリムシ科) 東洋区。コーヒー害虫
coffee white stem borer			coffee white borer を見よ
coffin flies			humpbacked flies を見よ
coffin fly		*Ephemera guttulata* Pictet	(カゲロウ目、モンカゲロウ科) 新北区
coffin fly		*Conicera tibialis* Schmitz	(ハエ目、ノミバエ科) 旧北区
coiled rose slug			banded rose sawfly を見よ　幼虫の英名
cola nut gall wasp		*Andricus lignicola* (Hartig)	(ハチ目、タマバチ科) 旧北区
coleopter			beetles を見よ
coleopterans			beetles を見よ
coleopteroid insects		Coleopteroidea	(コウチュウ目) の昆虫の総称
coleopteron			beetles を見よ
Coleotechnites flower moth		*Coleotechnites florae* (Freeman)	(チョウ目、キバガ科) 新北区
colewort bug			cabbage bug (2) を見よ
coliguacho		*Scaptia lata* (Guérin-Méneville)	(ハエ目、アブ科) 新熱帯区
Colima hairstreak		*Symbiopsis* sp.	(チョウ目、シジミチョウ科) 新熱帯区
Coliman Zobera		*Zobera albopunctata* Freeman	(チョウ目、セセリチョウ科) 新熱帯区

英名	和名	学名	所属、分布、ほか
collared dart moth		*Agnorisma bugrai* Kocak	（チョウ目、ヤガ科）新北区。collared dart ともいう
collared ground cricket		*Neonemobius* sp.	（バッタ目、コオロギ科）新北区
collemboles			springtails を見よ
collembolous insects			springtails を見よ
Collenette's variegated		*Euproctis hemicyclia* Collenette	（チョウ目、ドクガ科）豪州区
colletid bees			yellow-faced bees (1) を見よ
colletids			yellow-faced bees (1) を見よ
collier			bean aphid を見よ
Collin's Epitolina		*Epitolina collinsi* Libert	（チョウ目、シジミチョウ科）エチオピア区
Collins' Protea butterfly		*Capys collinsi* Henning et Henning	（チョウ目、シジミチョウ科）エチオピア区
Colombian admiral		*Hypanartia fassli* (Lamas, Willmott et Hall)	（チョウ目、タテハチョウ科）新熱帯区
Colombian banded matalmark		*Crocozona pheretima* C. et R. Felder	（チョウ目、シジミタテハ科）新熱帯区
Colombian blackstreak		*Ocaria aholiba* Hewitson	（チョウ目、シジミチョウ科）新熱帯区
Colombian chequered skipper		*Pyrgus adepta* Plötz	（チョウ目、セセリチョウ科）新熱帯区
Colombian crescent		*Janatella fellula* (Schaus)	（チョウ目、タテハチョウ科）新熱帯区
Colombian red patch		*Castilia castilla* (Felder)	（チョウ目、タテハチョウ科）新熱帯区
colon checkerspot		*Euphydryas colon* Edwards	（チョウ目、タテハチョウ科）新北区。*E. chalcedona* (Doubleday) の亜種とされることもある
colon swift	アトムモンセセリ	*Caltoris cahira austeni* (Moore)	（チョウ目、セセリチョウ科）日本、東洋区。*C. cahira* (Moore) の英名
colona		*Haploa colona* (Hübner)	（チョウ目、ヒトリガ科）新北区。colona moth ともいう
colonid beetles		Colonidae	（コウチュウ目）の昆虫の総称
Colorado alpine	イワベニヒカゲ	*Erebia callias* Edwards	（チョウ目、タテハチョウ科）全北区
Colorado anglewing		*Polygonia hylas* (Edwards)	（チョウ目、タテハチョウ科）新北区
Colorado giant skipper		*Megathymus coloradensis* Riley	（チョウ目、セセリチョウ科）新北区
Colorado hairstreak	コロラドルリアカシジミ	*Hypaurotis crysalus* (Edwards)	（チョウ目、シジミチョウ科）新北区
Colorado potato beetle	コロラドハムシ	*Leptinotarsa decemlineata* (Say)	（コウチュウ目、ハムシ科）全北区。ジャガイモの大害虫。Corolado beetle ともいう。トスジハムシの和名もある
Colorado white			spring white を見よ
colorful caddisflies			northern caddisflies を見よ
colorful foliage ground beetles		Lebiini	（コウチュウ目、オサムシ科）の昆虫の総称
colorful Zale moth		*Zale minerea* (Guenée)	（チョウ目、ヤガ科）新北区

英名	和名	学名	所属、分布、ほか
Colossal flasher		*Narcosius colossus colossus* (Herrich-Schäffer)	（チョウ目、セセリチョウ科）新熱帯区
colour sergeant		*Athyma inara* Westwood	（チョウ目、タテハチョウ科）東洋区
colour sergeant	ネフテミスジ	*Athyma nefte* (Cramer)	（チョウ目、タテハチョウ科）東洋区。*A. n. subrata* Moore ネッタイオオミスジ も同英名
colourful bluetail		*Ischnura pruinescens* (Tillyard)	（トンボ目、イトトンボ科）豪州区
coltsfoot aphid			pear Anuraphis を見よ
Columbacz gnat			Kolumbacs' fly を見よ　Danub 渓谷の地名に由来
Columbatsch fly			Kolumbacs' fly を見よ
Columbia basin wireworm		*Limonius subauratus* LeConte	（コウチュウ目、コメツキムシ科）新北区
Columbia clubtail		*Gomphus lynnae* Paulson	（トンボ目、サナエトンボ科）新北区
Columbia River tiger beetle		*Cicindela columbiana* Casey	（コウチュウ目、ハンミョウ科）新北区。絶滅危惧種
Columbia silk moth			larch silkworm を見よ　北米での英名
Columbian trig		*Cyrtoxipha columbiana* Caudell	（バッタ目、クサヒバリ科）新北区
Columbian defoliator		*Oxydia trychiata* (Guenée)	（チョウ目、シャクガ科）新北区
Columbian emerald moth			Darwin's green を見よ
Columbian shield mantis		*Choeradodis columbica* Stoll	（カマキリ目、カマキリ科）新熱帯区
Columbian skipper		*Hesperia columbia* Scudder	（チョウ目、セセリチョウ科）新北区。Columbia skipper ともいう
Columbian timber beetle		*Corthylus columbianus* Hopkins	（コウチュウ目、キクイムシ科）新北区
Columbine		*Stiboges nymphidia* Butler	（チョウ目、シジミタテハ科）東洋区
Columbine aphid	ハマナスオナガアブラムシ	*Longicaudus trirhodus* (Walker)	（カメムシ目、アブラムシ科）旧北区
Columbine borer		*Papaipema purpurifascia* (Grote et Robinson)	（チョウ目、ヤガ科）新北区
Columbine borer moth		*Papaipema leucostigma* (Harris)	（チョウ目、ヤガ科）新北区
Columbine duskywing		*Erynnis lucilius* (Scudder et Burgess)	（チョウ目、セセリチョウ科）新北区
Columbine leaf miner		*Phytomyza aquilegiae* (Hardy)	（ハエ目、ハモグリバエ科）旧北区
Columbine leafminer	カラマツスジハモグリバエ	*Phytomyza minuscula* Goureau	（ハエ目、ハモグリバエ科）日本、旧北区
Columbine sawfly		*Pristiphora alnivora* (Hartig)	（ハチ目、ハバチ科）旧北区
Columbus swordtail	コロンブスオナガタイマイ	*Eurytides columbus* (Kollar)	（チョウ目、アゲハチョウ科）新熱帯区
Columella hairstreak		*Strymon columella* (Fabricius)	（チョウ目、シジミチョウ科）新熱帯区
column-raiding army ant			driver ant を見よ

英名	和名	学名	所属、分布、ほか
colydiids			cylindrical bark beetles を見よ
colza flower midge			brassica flower midge を見よ
Comanche dancer		*Argia barretti* Calvert	（トンボ目、イトトンボ科）新北区。コマンチ族に由来
Comanche lacewing		*Chrysopa comanche* Banks	（アミメカゲロウ目、クサカゲロウ科）新北区
comb-claw beetles			comb-clawed beetles (1) を見よ
comb-clawed back beetles			comb-clawed beetles (1) を見よ
comb-clawed beetles		*Isomira*	（コウチュウ目、クチキムシ科）の昆虫の総称
comb-clawed beetles		*Mycetochara*	（コウチュウ目、クチキムシ科）の昆虫の総称
comb-clawed beetles (1)	クチキムシ科	Alleculidae	（コウチュウ目）の昆虫の総称
comblipped casemaker caddisflies	アシエダトビケラ科	Calamoceratidae	（トビケラ目）の昆虫の総称　complipped casemake caddisflies　ともいう
comet darner		*Anax longipes* Hagen	（トンボ目、ヤンマ科）新北区
comet moth			moon moth (1) を見よ
comfrey aphid		*Aphis symphyti* Schrank	（カメムシ目、アブラムシ科）旧北区
comic oakblue		*Arhopala comica* de Nicéville	（チョウ目、シジミチョウ科）東洋区
comic oakblue (1)		*Arhopala paramuta* (de Nicéville)	（チョウ目、シジミチョウ科）東洋区
Comitana skipper		*Niconiades comitana* Freeman	（チョウ目、セセリチョウ科）新熱帯区
comma (1)	シータテハ	*Polygonia c-album hamigera* (Butler)	（チョウ目、タテハチョウ科）日本、旧北区。*P. c-album* (Linnaeus) の英名。後翅裏面の C 字紋に由来。comma butterfly ともいう。
comma (2)		*Polygonia comma* Harris	（チョウ目、タテハチョウ科）新北区
comma			angelwings を見よ
comma			green comma を見よ
comma-mark cutworm moth		*Ariathisa comma* Walker	（チョウ目、ヤガ科）豪州区
commander		*Noduza procris* (Cramer)	（チョウ目、タテハチョウ科）東洋区
commander		*Moduza procris milonia* Fruhstorfer	（チョウ目、タテハチョウ科）東洋区。*M. p.* (Cramer) も同英名
commodore	ウスグロイチモンジ	*Auzakia danava* (Moore)	（チョウ目、タテハチョウ科）東洋区
commodore (1)		*Junonia artaxia* Hewitson	（チョウ目、タテハチョウ科）エチオピア区
commodores			buckeyes を見よ
commodores			pansies を見よ
common		*Graphium sarpedon connectens* (Fruhstorfer)	（チョウ目、アゲハチョウ科）、東洋区。common bluebottle を参照
common acacia blue		*Surendra quercetorum* (Moore)	（チョウ目、シジミチョウ科）東洋区
common Acraea			white-barred Acraea を見よ
common Acraea skipper		*Fresna netopha* (Hewitson)	（チョウ目、セセリチョウ科）エチオピア区

英名	和名	学名	所属、分布、ほか
common Actinote		*Actinote anteas* (Doubleday)	(チョウ目、タテハチョウ科) 新熱帯区
common Adreppus		*Adreppus fallax* Sjöstedt	(バッタ目、バッタ科) 豪州区
common Aenetus		*Aenetus eximius* (Scott)	(チョウ目、コウモリガ科) 新北区
common Aenetus			puriri moth を見よ
common aeroplane		*Phaedyma shepherdi* (Moore)	(チョウ目、タテハチョウ科) 豪州区
common Aeschna			common hawker を見よ
common albatross	カワカミシロチョウ	*Appias albina semperi* (Moore)	(チョウ目、シロチョウ科) 日本、東洋区、豪州区。*A. albina* Boisduval の英名
common alder pigmy		*Stigmella alnetella* (Stainton)	(チョウ目、モグリチビガ科) 旧北区
common alpine	アメリカベニヒカゲ	*Erebia epipsodea* Butler	(チョウ目、タテハチョウ科) 新北区
common amberwing		*Perithemis tenera* (Say)	(トンボ目、トンボ科) 新北区
common ambrosia beetle		*Platypus parallelus* (Fabricius)	(コウチュウ目、ナガキクイムシ科) 豪州区
common American walking-stick			walking stick を見よ
common Anastrus		*Anastrus sempiternus sempiternus* (Butler et Druce)	(チョウ目、セセリチョウ科) 新熱帯区
common angle moth		*Macaria aemulataria* Walker	(チョウ目、シャクガ科) 新北区
common annulet		*Gnophos obscurarius* Hübner	(チョウ目、シャクガ科) 旧北区
common anopheles			spotted Anopheles を見よ
common anopheles mosquito		*Anopheles maculipennis* Meigen	(ハエ目、カ科) 旧北区。新北区の亜種 *A. m. treeborni* Aitken は freeborn's malarial mosquito という
common anthelid		*Anthela acuta* (Walker)	(チョウ目、Anthelidae) 豪州区
common antlion		*Myrmeleon acer* Walker	(アミメカゲロウ目、ウスバカゲロウ科) 豪州区
common antlions			ant-lions (2) を見よ
common Apamea moth		*Apamea vulgaris* (Grote et Robinson)	(チョウ目、ヤガ科) 新北区。common Apamea ともいう
common apple leaf roller	アトボシハマキ	*Hoshinoa longicellana* (Walsingham)	(チョウ目、ハマキガ科) 日本、旧北区。リンゴなど果樹、バラ害虫
common apple leafhopper			apple leafhopper (3) を見よ
common Argus		*Callerebia nirmala* (Moore)	(チョウ目、タテハチョウ科) 東洋区
common armyworm		*Leucania convecta* (Walker)	(チョウ目、ヤガ科) 豪州区
common armyworm		*Pseudaletia unipuncta* (Haworth)	(チョウ目、ヤガ科) 全北区、新熱帯区
common armyworm moth		*Pseudaletia separata* Walker	(チョウ目、ヤガ科) 豪州区
common Arugisa moth		*Arugisa lutea* (Smith)	(チョウ目、ヤガ科) 新北区

英名	和名	学名	所属、分布、ほか
common ash bark beetle			ash bark beetle (1) を見よ
common asparagus beetle			asparagus beetle を見よ
common aspen leaf miner			aspen serpentine leafminer moth を見よ
common auger beetle		*Xylopsocus gibbicollis* (Macleay)	（コウチュウ目、ナガシンクイムシ科）豪州区
common Australian anopheline		*Anopheles annulipes* Walker	（ハエ目、カ科）豪州区
common Australian crow			common Indian crow を見よ
common Australian lady beetle			variable ladybird を見よ
common awl	テツイロビロウドセセリ	*Hasora badra badra* (Moore)	（チョウ目、セセリチョウ科）日本、東洋区
common awl robberfly		*Neoitamus cyanurus* (Loew)	（ハエ目、ムシヒキアブ科）旧北区
common awl-fly		*Xylophagus ater* (Panzer)	（ハエ目、キアブ科）旧北区
common backswimmer		*Notonecta undulata* Say	（カメムシ目、マツモムシ科）新北区。米国の普通種
common backswimmer			water boatman (1) を見よ
common bag moth		*Liothula omnivora* Fereday	（チョウ目、ミノガ科）豪州区
common bagworm			live oak bagworm を見よ
common bagworm moth			shining sweep を見よ
common banded awl	オキナワビロウドセセリ	*Hasora chromus inermis* Elwes et Edwards	（チョウ目、セセリチョウ科）日本、東洋区。*H. chromus* (Cramer) クロムスビロウドセセリも同英名
common banded demon		*Notocrypta paralysos* (Wood-Mason et de Nicéville)	（チョウ目、セセリチョウ科）東洋区
common banded hoverfly			hover-fly (2) を見よ
common banded mosquito		*Culex annulirostris* Skuse	（ハエ目、カ科）豪州区
common banded peacock	ホソオビクジャクアゲハ	*Papilio crino* Fabricius	（チョウ目、アゲハチョウ科）東洋区
common banded skipper		*Autochton neis* (Geyer)	（チョウ目、セセリチョウ科）新熱帯区
common banner		*Epiphile adrasta* Hewitson	（チョウ目、タテハチョウ科）新北区
common barkbug		*Aneurus laevis* (Fabricius)	（カメムシ目、Aneuridae）旧北区
common barklice	チャタテ科	Psocidae	（チャタテムシ目）の昆虫の総称
common barklouse		*Loensia maculosa* (Banks)	（チャタテムシ目、チャタテ科）新北区
common baron			baron を見よ
common baskettail		*Epitheca cynosura* (Say)	（トンボ目、エゾトンボ科）新北区
common batwing	ムモンアケボノアゲハ	*Atrophaneura varuna* (White)	（チョウ目、アゲハチョウ科）東洋区

英名	和名	学名	所属、分布、ほか
common beach midget		*Phyllonorycter maestingella* (Müller)	(チョウ目、ホソガ科) 全北区
common beak			nettle-tree butterfly を見よ
common bean weevil			bean weevil を見よ
common bedbug			bed bug (1) を見よ
common bedstraw carpet			common carpet を見よ
common bee fly		*Systoechus oreas* Osten Sacken	(ハエ目、ツリアブ科) 新北区
common beech midget		*Lithocolletis faginella* Zeller	(チョウ目、ホソガ科) 旧北区
common Bematistes		*Acraea epaea* (Cramer)	(チョウ目、タテハチョウ科) エチオピア区
common bent-skipper			marbled duskywing を見よ
common birch pigmy		*Stigmella betulicola* (Stainton)	(チョウ目、モグリチビガ科) 日本、旧北区
common bird-dropping skipper		*Milanion pilumnus* Mabille et Boullet	(チョウ目、セセリチョウ科) 新熱帯区
common birdwing	ヘレナキシタアゲハ	*Troides helena* (Linnaeus)	(チョウ目、アゲハチョウ科) 東洋区。*T. h. cerberus* (C. et R. Felder) も同英名
common biting flies		*Symphoromyia*	(ハエ目、シギアブ科) の昆虫の総称
common biting louse			dog chewing louse を見よ
common biting louse of the dog			dog chewing louse を見よ
common black ant			black ant (1) を見よ
common black beetle			common cockroach を見よ
common black Calosoma			black calosoma を見よ
common black-eye		*Gonatomyrina gorgias* (Stoll)	(チョウ目、シジミチョウ科) エチオピア区
common black eye			black eye を見よ
common black ground beetles			common carabids を見よ
common black vine thrips		*Retithrips syriacus* (Mayet)	(アザミウマ目、アザミウマ科) 旧北区、新熱帯区
common blossom thrips			yellow flower thrips を見よ
common blow fly		*Eucalliphora lilaea* Walker	(ハエ目、クロバエ科) 新北区、新熱帯区
common blue		*Cyclyrius pirithous* (Linnaeus)	(チョウ目、シジミチョウ科) エチオピア区
common blue (1)		*Aricia icarioides* (Boisduval)	(チョウ目、シジミチョウ科) 新北区
common blue (2)	ウスルリシジミ (イカルスヒメシジミ)	*Polyommatus icarus* (Rottemburg)	(チョウ目、シジミチョウ科) 旧北区。common blue butterfly ともいう
common blue			spring azure を見よ
common blue			Lang's short-tailed blue を見よ
common blue			common grass-blue (1) を見よ

英名	和名	学名	所属、分布、ほか
common blue Apollo	ヒマラヤウスバ	*Parnassius hardwickii* Gray	(チョウ目、アゲハチョウ科) 旧北区、東洋区
common blue-banded bush brown		*Bicyclus ephorus* Weymer	(チョウ目、タテハチョウ科) エチオピア区
common blue-banded forester		*Euphaedra harpalyce* Cramer	(チョウ目、タテハチョウ科) エチオピア区
common blue Charaxes		*Charaxes tiridates* (Cramer)	(チョウ目、タテハチョウ科) エチオピア区
common blue damselfly		*Enallagma cyathigera* (Charpentier)	(トンボ目、イトトンボ科) 旧北区。common bluet ともいう
common blue damselfly			blue damselfly を見よ
common blue darner		*Aeschna multicolor* Hagen	(トンボ目、ヤンマ科) 新北区
common blue Morpho			common Morpho を見よ
common bluebottle		*Graphium sarpedon* Linnaeus	(チョウ目、アゲハチョウ科) 旧北区、東洋区、豪州区。*G. s. luctatius* (Fruhstorfer) 、日本亜種 *G. s. nipponum* (Fruhstorfer) アオスジアゲハも同英名
common bluebottle fly		*Calliphora erythrocephala* (Meigen)	(ハエ目、クロバエ科) 全北区
common bluebottle fly			bluebottle fly を見よ
common blue-skipper		*Quadrus cerialis* (Stoll)	(チョウ目、セセリチョウ科) 新熱帯区
common bluet		*Enallagma ebrium* Hagen	(トンボ目、イトトンボ科) 新北区
common bluetail		*Ischnura heterosticta* (Burmeister)	(トンボ目、イトトンボ科) 豪州区
common bluetail (1)	アオモンイトトンボ	*Ischnura senegalensis* (Rambur)	(トンボ目、イトトンボ科) 日本、エチオピア区、東洋区
common bluetail			blue-tailed damselfly を見よ
common body louse			chicken louse を見よ
common bone beetle	ルリホシカムシ	*Necrobia violacea* (Linnaeus)	(コウチュウ目、カッコウムシ科) 日本、汎世界
common booklouse			booklouse (1) を見よ
common bordered-beauty			bordered beauty を見よ
common botfly			horse bot fly を見よ
common braconids		*Apanteles*	(ハチ目、コマユバチ科) の昆虫の総称
common branded skipper			silver-spotted skipper (1) を見よ
common branded skipper			western branded skipper を見よ
common Brangas		*Brangas neora* (Hewitson)	(チョウ目、シジミチョウ科) 新熱帯区
common brassy ringlet		*Erebia cassioides* (Reiner et Hohenwarth)	(チョウ目、タテハチョウ科) 旧北区
common brimstone			brimstone を見よ
common bristletails	内顎綱	*Ectognatha*	の昆虫の総称

英名	和名	学名	所属、分布、ほか
common brown	ミナミヒカゲ	*Heteronympha merope* (Fabricius)	（チョウ目、タテハチョウ科）豪州区。common brown butterfly ともいう
common brown earwig		*Labidura truncata* Kirby	（ハサミムシ目、オオハサミムシ科）豪州区
common brown earwig (1)		*Labidura riparia* (Pallas)	（ハサミムシ目、オオハサミムシ科）豪州区
common brown grass moth		*Crambus vitellus* Doubleday	（チョウ目、メイガ科）豪州区
common brown leafhopper		*Orosius argentatus* (Evans)	（カメムシ目、オオヨコバイ科）豪州区
common brown plume moth		*Platyptilia falcatalis* Walker	（チョウ目、トリバガ科）豪州区
common brown ringlet		*Hypocysta metirius* Butler	（チョウ目、タテハチョウ科）豪州区
common brownie		*Miletus chinensis* C. Felder	（チョウ目、シジミチョウ科）旧北区、東洋区。*M. c. learchus* C. et R. Felder も同英名
common brownie			Boisduval's Miletus を見よ
common buckeye			buckeye を見よ
common buff		*Baliochila aslanga* (Trimen)	（チョウ目、シジミチョウ科）エチオピア区
common burnished brass			burnished brass を見よ
common burrower mayflies			burrowing mayflies を見よ
common burrowing bees			andrenas を見よ
common burying beetle		*Necrophorus vespillo* (Linnaeus)	（コウチュウ目、シデムシ科）旧北区
common bush blue			bush blue を見よ
common bush brown (1)	ウスイロヒトツメコジャノメ	*Mycalesis janardana* Moore	（チョウ目、タテハチョウ科）東洋区。*M. j. sagittigera* Fruhstorfer も同英名
common bush brown (2)		*Bicyclus safitza* (Westwood)	（チョウ目、タテハチョウ科）エチオピア区
common bushbrown (3)	ヒメヒトツメジャノメ	*Mycalesis perseus* (Fabricius)	（チョウ目、タテハチョウ科）旧北区、東洋区、豪州区。*M. p. cepheus* Butler も同英名
common cabbage butterfly			common white を見よ　幼虫は common cabbage worm
common carabids		*Holciophorus, Pterostichus*	（コウチュウ目、オサムシ科）の昆虫の総称
common carder bee		*Bombus agrorum* Fabricius	（ハチ目、ミツバチ科）旧北区
common carder bee			carder bee を見よ
common carpenter bee		*Xylocopa (Xylocopoides) virginica* (Linnaeus)	（ハチ目、ミツバチ科）新北区
common carpet		*Epirrhoe alternata* (Müller)	（チョウ目、シャクガ科）旧北区
common carpet beetle			buffalo carpet beetle を見よ
common carrion beetle		*Silpha lapponica* Herbst	（コウチュウ目、シデムシ科）新北区

英名	和名	学名	所属、分布、ほか
common carrion dermestid		*Dermestes marmoratus* Say	（コウチュウ目、カツオブシムシ科）新北区
common castor		*Ariadne merione* (Cramer)	（チョウ目、タテハチョウ科）東洋区。*A. m. ginosa* Fruhstorfer も同英名
common cattle grub			cattle warble fly (2) を見よ　幼虫の英名
common cattle grub fly			cattle warble fly (2) を見よ
common cattle maggot			cattle warble fly (2) を見よ
common cave cricket		*Hadenoecus subterraneus* (Scudder)	（バッタ目、コロギス科）新北区
common celebean		*Bletogona mycalesis* C. et R. Felder	（チョウ目、タテハチョウ科）東洋区
common Cerulean	コシロウラナミシジミ	*Jamides celeno* (Cramer)	（チョウ目、シジミチョウ科）東洋区。*J. c. aelianus* Fabricius も同英名
common chafer			June beetle (1) を見よ
common checkered beetle			ornate checkered beetle を見よ
common checkered skipper	シロチャマダラセセリ	*Pyrgus communis* (Grote)	（チョウ目、セセリチョウ科）新北区、新熱帯区
common chestnut			chestnut moth を見よ
common chevron			chevron を見よ
common chicken louse			chicken louse を見よ
common ciliate blue			common hairtail を見よ
common citril		*Ceriagrion glabrum* (Burmeister)	（トンボ目、イトトンボ科）エチオピア区
common citrus whitefly			citrus whitefly (1) を見よ
common clearwing			hummingbird moth を見よ　前後翅とも透明部が多い。北米での英名
common click beetle (1)		*Agriotes lineatus* (Linnaeus)	（コウチュウ目、コメツキムシ科）全北区
common click beetle (2)		*Conoderus exsul* (Sharp)	（コウチュウ目、コメツキムシ科）豪州区
common click beetles		*Limonius*	（コウチュウ目、コメツキムシ科）の昆虫の総称
common click beetles		*Ctenicera*	（コウチュウ目、コメツキムシ科）の昆虫の総称
common clothes moth			clothes moth (1) を見よ
common clubtail	ホソバジャコウアゲハ	*Losaria coon* (Fabricius)	（チョウ目、アゲハチョウ科）旧北区、東洋区。*L. c. doubledayi* Wallace も同英名
common clubtail (1)		*Gomphus vulgatissimum* (Linnaeus)	（トンボ目、サナエトンボ科）旧北区
common cockchafer			June beetle (1) を見よ
common cockroach	トウヨウゴキブリ	*Blatta orientalis* Linnaeus	（ゴキブリ目、ゴキブリ科）汎世界
common cockroach			German cockroach を見よ
common Coenagrion			azure damselfly を見よ

英名	和名	学名	所属、分布、ほか
common commander		*Euryphura chalcis* (Felder et Felder)	（チョウ目、タテハチョウ科）エチオピア区
common coneheads		*Neoconocephalus*	（バッタ目、キリギリス科）の昆虫の総称
common copper		*Lycaena salustius* (Fabricius)	（チョウ目、シジミチョウ科）豪州区
common copper			small copper (1) を見よ
common copper butterfly			common copper を見よ
common corixa		*Corixa punctata* (Illiger)	（カメムシ目、ミズムシ科）旧北区
common cotton looper		*Anomis planalis* (Swinhoe)	（チョウ目、ヤガ科）豪州区
common cotton thrips			onion thrips を見よ　北アフリカでの英名
common crane flies		*Leptotarsus*	（ハエ目、ガガンボ科）の昆虫の総称
common crane flies			crane flies (2) を見よ
common cranefly (1)		*Tipula paludosa* Meigen	（ハエ目、ガガンボ科）旧北区
common cranefly (2)		*Tipula oleracea* Linnaeus	（ハエ目、ガガンボ科）旧北区
common crescent		*Eresia clio* (Linnaeus)	（チョウ目、タテハチョウ科）新熱帯区
common crimson-and-gold		*Pyrausta purpuralis* Linnaeus	（チョウ目、メイガ科）旧北区
common crow		*Euploea core* (Cramer)	（チョウ目、タテハチョウ科）東洋区、豪州区
common cruiser			cruiser butterfly を見よ
common currant tortrix			currant twist を見よ　米国での英名
common cutworm	ハスモンヨトウ	*Spodoptera litura* (Fabricius)	（チョウ目、ヤガ科）日本、旧北区、東洋区、大洋区、豪州区、エチオピア区。作物、果樹などの大害虫
common cutworm			Bogong moth を見よ
common cutworm			cutworm (2) を見よ
common cutworm moth			cutworm (2) を見よ
common cutworm moth			saltern ear を見よ
common cyclops		*Erites talcipennis* Wood-Mason et de Nicéville	（チョウ目、タテハチョウ科）東洋区
common dampwood termite		*Zootermopsis angusticollis* (Hagen)	（シロアリ目、オオシロアリ科）新北区、新熱帯区
common damsel bug		*Nabis americoferus* Carayon	（カメムシ目、マキバサシガメ科）新北区。米国で普通種
common damsel bug (1)		*Nabis rugosus* (Linnaeus)	（カメムシ目、マキバサシガメ科）旧北区
common damsel bug			field damsel bug をみよ
common dance flies		*Rhamphomyia*	（ハエ目、オドリバエ科）の昆虫の総称
common darkie		*Allotinus horsfieldi* (Moore)	（チョウ目、シジミチョウ科）東洋区
common darners			darners を見よ

英名	和名	学名	所属、分布、ほか
common dart (1)		*Andronymus neander neander* Plötz	（チョウ目、セセリチョウ科）エチオピア区
common dart		*Ocybadistes flavovittatus* (Latreille)	（チョウ目、セセリチョウ科）豪州区
common dart			Indian dart を見よ
common dart			white dart を見よ
common dart moth			cutworm (2) を見よ
common darter	タイリクアカネ	*Sympetrum striolatum* (Charpentier)	（トンボ目、トンボ科）旧北区。日本亜種 *S. s. imitoides* Bartenef も同英名
common darter dragonfly			common darter を見よ
common dartlet		*Oriens gola* (Moore)	（チョウ目、セセリチョウ科）東洋区。*O. g. pseudolus* Mabille も同英名
common dartlet		*Oriens goloides* (Moore)	（チョウ目、セセリチョウ科）東洋区
common diadem			danaid eggfly を見よ
common disc oakblue		*Arhopala epimuta epiala* Corbet	（チョウ目、シジミチョウ科）東洋区
common dotted border	アフリカヘリグロシロチョウ（ヘリグロネキシロチョウ）	*Mylothris chloris* (Fabricius)	（チョウ目、シロチョウ科）エチオピア区
common dotted wainscot moth		*Leucania semivittata* Walker	（チョウ目、ヤガ科）豪州区
common dry-wood termite		*Kalotermes minor* Hagen	（シロアリ目、レイビシロアリ科）新北区
common duck louse		*Anatoecus dentatus* (Scopoli)	（ハジラミ目、チョウカクハジラミ科）豪州区
common duffer		*Discophora sondaica despoliata* Stichel	（チョウ目、タテハチョウ科）東洋区。large faum を参照
common duffer		*Faunis eumeus* (Drury)	（チョウ目、タテハチョウ科）東洋区
common dusky blue		*Candalides hyacinthinus* (Semper)	（チョウ目、シジミチョウ科）豪州区。common dusky blue butterfly ともいう
common dustywings		*Coniopteryx*	（アミメカゲロウ目、コナカゲロウ科）の昆虫の総称
common ear			ear moth を見よ
common earl	ミヤマイナズマ	*Tanaecia julii* (Lesson)	（チョウ目、タテハチョウ科）東洋区。*T. j. bougainvillei* Corbet も同英名
common earwig		*Forficula auricularia* Linnaeus	（ハサミムシ目、クギヌキハサミムシ科）全北区
common earwigs	クギヌキハサミムシ科	Forficulidae	（ハサミムシ目）の昆虫の総称
common eastern bumble bee		*Bombus impatiens* Cresson	（ハチ目、ミツバチ科）新北区
common eastern firefly		*Photinus pyralis* (Linnaeus)	（コウチュウ目、ホタル科）新北区
common eastern velvet ant		*Dasymutilla occidentalis* (Linnaeus)	（ハチ目、アリバチ科）新北区、大洋区。マルハナバチに寄生
common eggfly	リュウキュウムラサキ	*Hypolimnas bolina* (Linnaeus)	（チョウ目、タテハチョウ科）東洋区、エチオピア区
common elfin		*Sarangesa thecla* (Plötz)	（チョウ目、セセリチョウ科）エチオピア区

英名	和名	学名	所属、分布、ほか
common emerald	キバラヒメアオシャク	*Hemithea aestivaria* (Hübner)	(チョウ目、シャクガ科) 日本、全北区
common emerald moth			common emerald を見よ
common emigrant			lemon migrant を見よ
common emperor		*Bunaea alcinoe* (Stoll)	(チョウ目、ヤママユガ科) 旧北区
common Epicoma moth		*Epicoma melanosticta* (Donovan)	(チョウ目、シャチホコガ科) 豪州区
common Epitola		*Stempfferia cercenoides* (Holland)	(チョウ目、シジミチョウ科) エチオピア区
common Epitolina		*Epitolina dispar* (Kirby)	(チョウ目、シジミチョウ科) エチオピア区
common eucalypt longicorn		*Phoracantha semipunctata* (Fabricius)	(コウチュウ目、カミキリムシ科) 旧北区、豪州区、大洋区
common eumenid wasp		*Euodynerus annulatum* Say	(ハチ目、スズメバチ科) 新北区
common Eupithecia			common pug (1) を見よ　common Eupithecia moth ともいう
common Eupselia moth		*Eupselia carpocapsella* Walker	(チョウ目、マルハキバガ科) 豪州区
common European cockchafer			June beetle (1) を見よ
common European glowworm			glow-worm (2) を見よ　米国での英名
common European mayfly			brown mayfly を見よ
common European white ant		*Reticulitermes lucifugus* Rossi	(シロアリ目、ミゾガシラシロアリ科) 旧北区
common evening brown			evening brown を見よ
common fairytail			spined fairytail を見よ
common fanfoot		*Herminia barbalis* Hübner	(チョウ目、ヤガ科) 旧北区
common fan-foot	カシワアツバ	*Pechipogo strigilata* (Linnaeus)	(チョウ目、ヤガ科) 日本、旧北区
common fantasy		*Pseudaletis clymenus* (Druce)	(チョウ目、シジミチョウ科) エチオピア区
common faun		*Faunis arcesilaus* (Fabricius)	(チョウ目、タテハチョウ科) 東洋区
common faun	マルバネワモン	*Faunis canens* Hübner	(チョウ目、タテハチョウ科) 東洋区。F. c. arcesilas Stichel も同英名
common fern moth		*Sestra flexata* Walker	(チョウ目、シャクガ科) 豪州区
common field grasshopper			field grasshopper を見よ　英国での英名
common five-ring butterfly	コウラナミジャノメ	*Ypthima baldus* (Fabricius)	(チョウ目、タテハチョウ科) 旧北区。*Y. b. newboldi* Distant, *Y. b. zodina* (Fruhstorfer) も同英名。common five-ring ともいう
common flash	ワタナベシジミ	*Rapala nissa* (Kollar)	(チョウ目、シジミチョウ科) 東洋区
common flat-body		*Agonopterix heracliana* (Linnaeus)	(チョウ目、マルハキバガ科) 旧北区

英名	和名	学名	所属、分布、ほか
common flatbug		*Aradus depressus* (Fabricius)	（カメムシ目、ヒラタカメムシ科）旧北区
common flat bugs			flat back bugs を見よ
common fleas	ヒトノミ科	Pulicidae	（ノミ目）の昆虫の総称
common flesh flies		*Sarcophaga*	（ハエ目、ニクバエ科）の昆虫の総称
common flower bug		*Anthocoris nemorum* (Linnaeus)	（カメムシ目、ハナカメムシ科）旧北区
common footman		*Eilema lurideola* (Zincken)	（チョウ目、ヒトリガ科）旧北区
common forest blue		*Aethiopana honorius* (Fabricius)	（チョウ目、シジミチョウ科）エチオピア区
common forest looper		*Pseudocoremia suavis* (Butler)	（チョウ目、シャクガ科）豪州区
common forest queen		*Charaxes eurinome* (Cramer)	（チョウ目、タテハチョウ科）エチオピア区
common forest swift		*Melphina unistriga* (Holland)	（チョウ目、セセリチョウ科）エチオピア区
common forester		*Procis statices* (Linnaeus)	（チョウ目、マダラガ科）旧北区
common forester	ミヤマシロオビヒカゲ	*Lethe insana* (Kollar)	（チョウ目、タテハチョウ科）旧北区、東洋区
common forktail		*Ischnura verticalis* (Say)	（トンボ目、イトトンボ科）新北区
common four ring		*Ypthima huebneri* Kirby	（チョウ目、タテハチョウ科）旧北区、東洋区
common four-ring	カノウラナミジャノメ	*Ypthima praenubilia* Leech	（チョウ目、タテハチョウ科）東洋区
common free-living caddisflies		*Rhyacophila*	（トビケラ目、ナガレトビケラ科）の昆虫の総称
common froghopper			cuckoo spit insect を見よ
common fruit fly			pomace fly を見よ
common fungus moth		*Metalectra discalis* (Grote)	（チョウ目、ヤガ科）新北区
common furniture beetle		*Anobium punctatum* De Geer	（コウチュウ目、シバンムシ科）旧北区。common house borer ともいう
common furniture beetle			furniture beetle (1) を見よ
common garden ant			black ant (1) を見よ
common garden katydid		*Caedicia simplex* (Walker)	（バッタ目、キリギリス科）豪州区。カンキツ害虫
common gem		*Poritia hewitsoni* Moore	（チョウ目、シジミチョウ科）東洋区
common gem			gem (1) を見よ
common Gesonula		*Gesonula mundata* (Walker)	（バッタ目、バッタ科）豪州区
common ginger white		*Oboronia punctatus* (Dewitz)	（チョウ目、シジミチョウ科）エチオピア区
common Glaphyria moth		*Glaphyria glaphyralis* (Guenée)	（チョウ目、メイガ科）新北区
common glassy Acraea	ナカベニホソチョウ	*Acraea quirina* Fabricius	（チョウ目、タテハチョウ科）エチオピア区
common glassy tiger			Euclea tigerwing を見よ

英名	和名	学名	所属、分布、ほか
common glassywing			Pompeius skipper を見よ
common glider		*Trapezostigma loewii* (Kaup)	(トンボ目、トンボ科) 東洋区、豪州区
common glider (1)		*Tramea transmarina* (Kaup)	(トンボ目、トンボ科) 豪州区
common glider	コミスジ	*Neptis sappho intermedia* W. B. Pryer	(チョウ目、タテハチョウ科) 日本、旧北区。Pallas' sailer を参照
common glider			Pallas' sailer を見よ
common glow-worm			glow-worm (2) を見よ
common Gluphisia moth		*Gluphisia septentrionalis* Walker	(チョウ目、シャチホコガ科) 新北区
common gnat			common mosquito を見よ
common goat biting louse			goat louse を見よ
common goat louse			goat louse を見よ
common goat louse			sucking goat louse を見よ
common goat moth			goat moth (1) を見よ
common gold-dot bentwing			laburnum leaf miner を見よ
common gold wasp			ruby-tailed wasp (1) を見よ
common goldenring			golden-ringed dragonfly を見よ
common gooseberry sawfly			gooseberry sawfly (1) を見よ
common Grammoptera		*Grammoptera ruficornis* (Fabricius)	(コウチュウ目、カミキリムシ科) 旧北区
common Graphium			veined swordtail を見よ
common grass-blue (1)	ウスモンシルビアシジミ	*Zizina labradus* (Godart)	(チョウ目、シジミチョウ科) 豪州区
common grass-blue (2)	シルビアシジミ	*Zizina otis* (Fabricius)	(チョウ目、シジミチョウ科) 日本、旧北区、東洋区、豪州区
common grass blue			lesser grass blue を見よ
common grass dart		*Taractrocera maevius* (Fabricius)	(チョウ目、セセリチョウ科) 東洋区
common grass moth		*Crambus flexuosellus* Doubleday	(チョウ目、メイガ科) 豪州区
common grass moth			forage looper を見よ
common grass thrips			foxtail millet thrips を見よ
common grass-veneer moth		*Crambus praefectellus* (Zincken)	(チョウ目、メイガ科) 新北区
common grass yellow			grass yellow を見よ *E. h. contubernalis* Moore, *E. h. solifera* (Butler) も同英名
common gray blowfly			flesh fly (1) を見よ
common gray moth			cranberry spanworm を見よ
common green aphid			corn leaf aphid を見よ
common green birdwing			Priam's birdwing を見よ

英名	和名	学名	所属、分布、ほか
common green capsid (1)	ナガミドリカスミカメムシ	*Lygocoris pabulinus* (Linnaeus)	(カメムシ目、カスミカメムシ科) 日本、旧北区。英国での英名。common green capsid bug ともいう
common green-capsid		*Calocoris sexguttatus* Fabricius	(カメムシ目、カスミカメムシ科) 旧北区
common green Charaxes	アオフタオチョウ	*Charaxes eupale* (Drury)	(チョウ目、タテハチョウ科) エチオピア区
common green colonel		*Oplodontha viridula* (Fabricius)	(ハエ目、ミズアブ科) 旧北区
common green corn aphid			corn leaf aphid を見よ
common green cutworm moth		*Melanchra plena* Walker	(チョウ目、ヤガ科) 豪州区
common green darner		*Anax junius* (Drury)	(トンボ目、ヤンマ科) 日本、全北区、大洋区。green darner ともいう
common green grasshopper		*Omocestus viridulus* (Linnaeus)	(バッタ目、バッタ科) 旧北区
common green lacewing	ニッポンクサカゲロウ	*Chrysoperla carnea* (Stephens)	(アミメカゲロウ目、クサカゲロウ科) 日本、全北区、東洋区
common green mantis			African mantis (1) を見よ
common green planthopper		*Siphanta hebes* (Walker)	(カメムシ目、アオバハゴロモ科) 豪州区
common green shieldbug		*Palomena prasina* (Linnaeus)	(カメムシ目、カメムシ科) 旧北区
common green stink bug			green stink bug (1) を見よ
common grey looper moth		*Xanthorhoe semisignata* Walker	(チョウ目、シャクガ科) 豪州区
common groundhopper		*Tetrix undulata* (Sowerby)	(バッタ目、ヒシバッタ科) 旧北区
common ground-squirrel flea			ground squirrel flea を見よ
common guava blue		*Deudorix isocrates* (Fabricius)	(チョウ目、シジミチョウ科) 東洋区
common gull	タイワンスジグロシロチョウ	*Cepora nerissa* (Fabricius)	(チョウ目、シロチョウ科) 東洋区。*C. n. dapha* Moore も同英名。common gullwing ともいう。タイワンスジグロチョウの和名もあり
common hairstreak			gray hairstreak (1) を見よ
common hairtail		*Anthene definita* (Butler)	(チョウ目、シジミチョウ科) エチオピア区。
common harvestman			dog-day cicada (2) を見よ
common hawker	ルリボシヤンマ	*Aeschna juncea* (Linnaeus)	(トンボ目、ヤンマ科) 日本、全北区
common hawthorn ermel			orchard ermine を見よ
common heath		*Ematurga atomaria* (Linnaeus)	(チョウ目、シャクガ科) 旧北区
common heath beauty			common heath を見よ
common Hebrew character			Hebrew character を見よ

英名	和名	学名	所属、分布、ほか
common hedge blue		*Celastrina puspa* (Horsfield)	（チョウ目、シジミチョウ科）東洋区。日本亜種は *C. p. ishigakiana* (Matsumura) ヤクシマルリシジミ。*C. p. lambi* (Distant) も同英名
common hen louse			chicken louse を見よ
common hide beetle			hide beetle を見よ
common highflier		*Aphnaeus orcas* (Drury)	（チョウ目、シジミチョウ科）エチオピア区
common honey			lesser wax moth を見よ
common hopper		*Platylesches mortili* (Wallengren)	（チョウ目、セセリチョウ科）エチオピア区
common horse biting louse			horse biting louse を見よ
common horse bot			horse bot fly を見よ　幼虫の英名
common horse botfly			horse bot fly を見よ
common horse fly			horse bot fly を見よ
common Hottentot			Latreille's skipper を見よ
common Hottentot skipper		*Gegenes niso* (Linnaeus)	（チョウ目、セセリチョウ科）エチオピア区
common house borer			common furniture beetle を見よ
common house fly			house fly を見よ
common housefly of the Orient		*Musca vicina* Macquart	（ハエ目、イエバエ科）東洋区
common house mosquito			common mosquito を見よ
common hover flies		*Eupeodes*	（ハエ目、ハナアブ科）の昆虫の総称
common hover flies		*Syrphus*	（ハエ目、ハナアブ科）の昆虫の総称
common hover flies		*Scaeva*	（ハエ目、ハナアブ科）の昆虫の総称
common hover flies		*Allograpta*	（ハエ目、ハナアブ科）の昆虫の総称
common hover flies		*Metasyrphus*	（ハエ目、ハナアブ科）の昆虫の総称
common hover fly		*Ischiodon scutellaris* (Fabricius)	（ハエ目、ハナアブ科）豪州区
common hover fly		*Melangyna viridiceps* (Macquart)	（ハエ目、ハナアブ科）豪州区
common hover fly		*Simosyrphus grandicornis* (Macquart)	（ハエ目、ハナアブ科）豪州区
common hoverfly		*Toxomerus marginatus* (Meigen)	（ハエ目、ハナアブ科）新北区
common Hyppa moth		*Hyppa xylinoides* (Guenée)	（チョウ目、ヤガ科）新北区
common ichneumonids		*Ophion*	（ハチ目、ヒメバチ科）の昆虫の総称
common Idia			waved tabby を見よ　common Idia moth ともいう
common imperial		*Cheritra freja* Fabricius	（チョウ目、シジミチョウ科）東洋区。*C. f. frigga* (Fruhstorfer) も同英名
common imperial blue			imperial blue を見よ

英名	和名	学名	所属、分布、ほか
common Indian crow	ガランピマダラ	*Euploea core godartii* Lucas	（チョウ目、タテハチョウ科）日本、東洋区
common indigo ciliate blue		*Anthene sylvanus* (Drury)	（チョウ目、シジミチョウ科）エチオピア区
common ingrailed clay			small square-spot を見よ
common ischnura			blue-tailed damselfly を見よ
common Japanese alder maculated aphid	カバブチアブラムシ	*Symydobius kabae* (Matsumura)	（カメムシ目、アブラムシ科）日本、旧北区
common jay		*Graphium doson* (C. et R. Felder)	（チョウ目、アゲハチョウ科）東洋区。日本の本土亜種 *G. d. albidum* (Wileman), 八重山亜種 *G. d. perillus* (Fruhstorfer) はミカドアゲハ。東洋区の亜種 *G. d. evemonides* Honrath も同英名
common jester		*Symbrenthia hippoclus* (Cramer)	（チョウ目、タテハチョウ科）旧北区、東洋区
common jester	キミスジ	*Symbrenthia lilaea formosanus* Fruhstorfer	（チョウ目、タテハチョウ科）東洋区。*S. l. luciana* (Hewitson) も同英名
common Jezebel	スジグロベニモンシロチョウ	*Delias eucharis* (Drury)	（チョウ目、シロチョウ科）東洋区
common Jezebel (1)		*Delias nigrina* (Fabricius)	（チョウ目、シロチョウ科）豪州区
common joker		*Byblia acheloi acheloia* (Wallengren)	（チョウ目、タテハチョウ科）エチオピア区
common joker (1)	ヘリグロキマダラタテハ	*Byblia anvantara* Boisduval	（チョウ目、タテハチョウ科）エチオピア区
common June beetles			May beetles (2) を見よ
common lace bugs		*Corythucha*	（カメムシ目、グンバイムシ科）の昆虫の総称
common lacewing			common green lacewing を見よ
common lacewings			green lacewings (1) を見よ
common lackey			lackey を見よ
common lancewing		*Epermenia chaerophyllella* (Goeze)	（チョウ目、ササベリガ科）旧北区
common lappet			lappet を見よ
common lappet moth			lappet を見よ
common large fox		*Rhabdomantis sosia* (Mabille)	（チョウ目、セセリチョウ科）エチオピア区
common lascar	キンミスジ	*Pantoporia hordonia* (Stoll)	（チョウ目、タテハチョウ科）東洋区
common lead-belle		*Scotopteryx mucronata* (Scopoli)	（チョウ目、シャクガ科）旧北区。lead belle ともいう
common leaf-cutter bee		*Megachile centuncularis* (Linnaeus)	（ハチ目、ハキリバチ科）全北区
common leafcutter bees			leaf-cutting bees (2) を見よ
common leaf sitter		*Gorgyra sara* Evans	（チョウ目、セセリチョウ科）エチオピア区
common leaf weevil		*Phyllobius pyri* (Linnaeus)	（コウチュウ目、ゾウムシ科）旧北区

英名	和名	学名	所属、分布、ほか
common leopard	ウラベニヒョウモン	*Phalanta phalantha* (Drury)	(チョウ目、タテハチョウ科) 日本、東洋区、豪州区。*P. p. aethiopia* (Rothschild et Jordan) も同英名
common light arches			light arches を見よ
common lime butterfly			chequered swallowtail を見よ
common lineblue	ヒメウラナミシジミ	*Prosotas nora kanoi* (Omoto)	(チョウ目、シジミチョウ科) 日本、東洋区。*P. nora* (Felder) の英名。*P. n. superdates* Fruhstorfer も同英名
common long-cloaked marble	シベチャツマジロヒメハマキ	*Hedya ochroleucana* (Frölich)	(チョウ目、ハマキガ科) 日本、全北区
common longwing	アカスジドクチョウ	*Heliconius erato* (Linnaeus)	(チョウ目、タテハチョウ科) 新北区、新熱帯区
common looper moth		*Xanthorhoe rosearia* Doubleday	(チョウ目、シャクガ科) 豪州区
common looper moth		*Autographa precationis* (Guenée)	(チョウ目、ヤガ科) 新北区、大洋区
common louse of cattle			cattle chewing louse を見よ
common lutestring	フタテントガリバ	*Ochropacha duplaris* (Linnaeus)	(チョウ目、トガリバガ科) 日本、旧北区
common magpie			magpie を見よ
common malaria mosquito			malaria mosquito (1) を見よ
common Malayan powderpost beetle			hairy powderpost beetle (1) を見よ
common manuka moth		*Declana floccosa* Walker	(チョウ目、シャクガ科) 豪州区。ニュージーランドの Manuka という植物に由来
common map	イシガキチョウ	*Cyrestis thyodamas mabella* Fruhstorfer	(チョウ目、タテハチョウ科) 日本。*C. thyodamas* Boisduval 東洋区の英名
common maplet	リサチビイシガケ	*Chersonesia risa* (Doubleday)	(チョウ目、タテハチョウ科) 東洋区
common marbled carpet			marbled carpet moth を見よ
common marbled carpet moth			marbled carpet moth を見よ
common marbled coronet		*Hadena conspersa* Schiffermüller	(チョウ目、ヤガ科) 旧北区
common mayfly			brown mayfly を見よ
common may-tree pigmy		*Stigmella oxyacanthella* (Stainton)	(チョウ目、モグリチビガ科) 旧北区
common meadow blue			eros blue を見よ
common meadow blue			maeadow blue を見よ
common meadow katydid		*Orchelimum vulgare* Harris	(バッタ目、キリギリス科) 新北区
common meal tabby			meal moth を見よ
common mealybug			citrus mealybug (1) を見よ
common mechanitis		*Mechanitis isthmia* Bates	(チョウ目、タテハチョウ科) 新熱帯区

英名	和名	学名	所属、分布、ほか
common Mellana		*Quasimellana eulogius* (Plötz)	（チョウ目、セセリチョウ科）新北区
common Merveille-du-jour		*Griposia aprilina* Linnaeus	（チョウ目、ヤガ科）旧北区
common Mestra			Amymone を見よ
common Metarranthis moth		*Metarranthis hypochraria* (Herrich-Schäffer)	（チョウ目、シャクガ科）新北区。common Metarranthis ともいう
common midge		*Chironomus attenuatus* Walker	（ハエ目、ユスリカ科）新北区。北米最普通種のユスリカ
common midges			midges を見よ　豪州での英名
common midges			bloodworms を見よ
common migrant			mottled migrant を見よ　豪州での英名
common migrant (1)	アオトガリシロチョウ	*Appias celestina* (Boisduval)	（チョウ目、シロチョウ科）東洋区、豪州区
common migrant (2)		*Catopsilia pyranthe crokera* (Macleay)	（チョウ目、シロチョウ科）豪州区
common milkweed bug			small milkweed bug を見よ
common mime	マネシアゲハ（キベリアゲハ）	*Chilasa clytia* (Linnaeus)	（チョウ目、アゲハチョウ科）東洋区
common mimic white		*Enantia lina* (Herbst)	（チョウ目、シロチョウ科）新熱帯区。*E. l. marion* Godman et Salvin, *E. l. virna* Lamas は white mimic white
common mole cricket			northern mole cricket を見よ
common mole cricket			mole cricket (3) を見よ
common moonbeam	ウラギンルリシジミ	*Philiris innotatus* (Miskin)	（チョウ目、シジミチョウ科）豪州区
common Mormon	シロオビアゲハ	*Papilio polytes polytes* Linnaeus	（チョウ目、アゲハチョウ科）日本。東洋区の亜種 *P. p. romulus* Cramer も同英名
common Morpho		*Morpho helenor* (Cramer)	（チョウ目、タテハチョウ科）新熱帯区。*M. h. montezuma* Guenee, *M. h. octavia* Bates, *M. h. guerrerensis* Le Moult et Real も同英名
common Morpho			Peleides blue Morpho を見よ
common mosquito		*Culex pipiens* Linnaeus	（ハエ目、カ科）全北区。日本亜種は *C. p. pallens* Coquillett アカイエカ
common mosquito			spotted Anopheles を見よ
common mother-of-pearl			forest mother-of-pearl (1) を見よ
common mottled brown moth		*Melanchra morosa* Butler	（チョウ目、ヤガ科）豪州区
common mud dauber			black borer wasp を見よ
common Mylon		*Mylon maimon* (Scudder et Burgess)	（チョウ目、セセリチョウ科）新熱帯区
common nawab	オビモンフタオチョウ	*Polyura athamas* (Drury)	（チョウ目、タテハチョウ科）東洋区
common negro bugs		*Corimelaena*	（カメムシ目、ツチカメムシ科）の昆虫の総称
common Nephele		*Nephele comma* Hopffer	（チョウ目、スズメガ科）エチオピア区

英名	和名	学名	所属、分布、ほか
common net-spinning caddisflies			net-spinning caddisflies を見よ
common Nymphidium		*Nymphidium caricae* (Linnaeus)	（チョウ目、シジミタテハ科）新熱帯区
common oak blue		*Arhopala micale* Blanchard	（チョウ目、シジミチョウ科）豪州区。*A. m. amphis* Waterhouse も同英名
common oak midget		*Phyllonorycter quercifoliella* (Zeller)	（チョウ目、ホソガ科）旧北区
common oak moth		*Phoberia atomaris* Hübner	（チョウ目、ヤガ科）新北区
common oblique hoverfly		*Aliograpta obliqua* Say	（ハエ目、ハナアブ科）新北区
common onyx	ミツオシジミ	*Horaga onyx* (Moore)	（チョウ目、シジミチョウ科）東洋区。*H. o. sardonyx* Fruhstorfer も同英名
common orange			common citril を見よ
common orange forester		*Euphaedra ruspina* (Hewitson)	（チョウ目、タテハチョウ科）エチオピア区
common orange legionnaire		*Beris vallata* (Forster)	（ハエ目、ミズアブ科）旧北区
common orange tip			orange tip (1) を見よ　米国での英名
common orange-underwing		*Archiearias parthenias* (Linnaeus)	（チョウ目、ヤガ科）旧北区
common orchard moth		*Tortrix excessana* Walker	（チョウ目、ハマキガ科）豪州区
common Oressinoma		*Oressinoma typhla* Doubleday	（チョウ目、タテハチョウ科）新熱帯区
common pale bell moth		*Tortrix leucaniana* Walker	（チョウ目、ハマキガ科）豪州区
common palm dart			pale darter を見よ
common palmfly	ルリモンジャノメ	*Elymnias hypermnestra* (Linnaeus)	（チョウ目、タテハチョウ科）旧北区、東洋区。*E. h. agina* Fruhstorfer も同英名
common paper wasp			paper wasp (2) を見よ
common paperwasp			Australian paper wasp を見よ
common pasture moth		*Leptomeris rubraria* Doubleday	（チョウ目、シャクガ科）豪州区
common pathfinder skipper		*Pardaleodes edipus* (Stoll)	（チョウ目、セセリチョウ科）エチオピア区
common peacock	クジャクアゲハ	*Achillides polyctor* (Boisduval)	（チョウ目、アゲハチョウ科）東洋区
common pearl white		*Elodina angulipennis* (Lucas)	（チョウ目、シロチョウ科）豪州区
common pearly		*Eresiomera isca* (Hewitson)	（チョウ目、シジミチョウ科）エチオピア区
common pencil-blue			pencilled blue を見よ
common Petrophora moth		*Petrophora divisata* Hübner	（チョウ目、シャクガ科）新北区
common Phanus		*Phanus marshallii* (Kirby)	（チョウ目、セセリチョウ科）新熱帯区
common picture wing		*Rhyothemis variegata* (Linnaeus)	（トンボ目、トンボ科）東洋区

英名	和名	学名	所属、分布、ほか
common pierrot		*Castalius rosimon* (Fabricius)	(チョウ目、シジミチョウ科) 東洋区
common pine aphid			pine adelgid を見よ
common pinkband moth		*Ogdoconta cinereola* (Guenée)	(チョウ目、ヤガ科) 新北区、新熱帯区
common pink-barred			rose-banded wave を見よ
common plant louse			dock aphid (1) を見よ
common plant louse			bean aphid を見よ
common plasterer bee			plaster bee を見よ
common plume		*Emmelina monodactyla* (Linnaeus)	(チョウ目、トリバガ科) 旧北区。common plume moth ともいう
common plume moths		*Platyptilia*	(チョウ目、トリバガ科) の昆虫の総称
common pond damsel			common citril を見よ
common pond skater		*Gerris lacustris* (Linnaeus)	(カメムシ目、アメンボ科) 旧北区
common pondhawk			eastern pondhawk を見よ
common pond-skaters		*Gerris*	(カメムシ目、アメンボ科) の昆虫の総称
common porina moth			spring porina moth を見よ
common posy	フシギノモリノオナガシジミ	*Drupadia ravindra moorei* (Distant)	(チョウ目、シジミチョウ科) 東洋区
common powderpost beetle			powderpost beetle (1) を見よ
common Ptichodis moth		*Ptichodis herbarum* (Guenée)	(チョウ目、ヤガ科) 新北区、新熱帯区
common pug		*Eupithecia vulgata* (Haworth)	(チョウ目、シャクガ科) 旧北区
common pug (1)		*Eupithecia miserulata* Grote	(チョウ目、シャクガ科) 新北区
common punch		*Dodona durga* (Kollar)	(チョウ目、シジミタテハ科) 旧北区、東洋区
common Punchinello		*Zemeros flegyas allica* (Fabricius)	(チョウ目、シジミタテハ科) 東洋区
common Pyrgomorph		*Monistria discrepans* (Walker)	(バッタ目、オンブバッタ科) 豪州区
common quaker		*Orthosia cerasi* (Fabricius)	(チョウ目、ヤガ科) 旧北区
common quaker (1)		*Orthosia stabilis* Denis et Schiffermüller	(チョウ目、ヤガ科) 旧北区
common quaker moth			common quaker (1) を見よ
common rain-barrel mosquito			common mosquito を見よ
common raphidian		*Agulla adnixa* (Hagen)	(アミメカゲロウ目、キスジラクダムシ科) 新北区
common raspberry fruitworm			raspberry fruitworm を見よ

英名	和名	学名	所属、分布、ほか
common Raven	オナシモンキアゲハ	*Papilio castor* Westwood	(チョウ目、アゲハチョウ科) 東洋区
common recluse		*Caenides dacela* (Hewitson)	(チョウ目、セセリチョウ科) エチオピア区
common red ant		*Formica rufa* Linnaeus	(ハチ目、アリ科) 全北区
common red Apollo	テンジクウスバアゲハ	*Parnassius epaphus* (Oberthür)	(チョウ目、アゲハチョウ科) 旧北区、東洋区
common red flash	イアルバスシジミ	*Rapala iarbus* (Fabricius)	(チョウ目、シジミチョウ科) 東洋区
common red forester		*Lethe mekara* (Moore)	(チョウ目、タテハチョウ科) 旧北区、東洋区。*L. m. gopaka* Fruhstorfer も同英名
common red harlequin			Malay red harlequin (1) を見よ
common redeye	アカメセセリ	*Matapa aria* (Moore)	(チョウ目、セセリチョウ科) 東洋区
common red-eye (1)		*Chaetocneme beata* (Hewitson)	(チョウ目、セセリチョウ科) 豪州区
common red-legged robberfly		*Dioctria rufipes* (De Geer)	(ハエ目、ムシヒキアブ科) 旧北区
common redling			double-banded banner を見よ
common ringlet			large heath を見よ　米国での英名
common ringlet		*Ypthima doleta* Kirby	(チョウ目、タテハチョウ科) エチオピア区
common roadside skipper			roadside skipper を見よ
common robber flies		*Stenopogon*	(ハエ目、ムシヒキアブ科) の昆虫の総称
common robber flies		*Machimus*	(ハエ目、ムシヒキアブ科) の昆虫の総称
common robber flies		*Efferia*	(ハエ目、ムシヒキアブ科) の昆虫の総称
common robber-fly		*Neoitamus melanopogon* Schiner	(ハエ目、ムシヒキアブ科) 豪州区
common rose			common rose swallowtail を見よ
common rose pigmy		*Stigmella anomelella* (Goeze)	(チョウ目、モグリチビガ科) 旧北区
common rose swallowtail	ベニモンアゲハ	*Pachliopta aristolochiae* (Fabricius)	(チョウ目、アゲハチョウ科) 東洋区。*P. a. interpositus* (Fruhstorfer), *P. a. asteris* Rothschild も同英名
common rough-winged button			garden rose tortrix を見よ
common rowan-berry argent			apple fruit moth を見よ
common rowan pigmy		*Stigmella sorbi* (Stainton)	(チョウ目、モグリチビガ科) 旧北区
common ruby spot			American ruby-spot を見よ
common rustic		*Mesapamea secalis* (Linnaeus)	(チョウ目、ヤガ科) 旧北区。common rustic moth ともいう
common sailer		*Neptis laeta* Overlaet	(チョウ目、タテハチョウ科) エチオピア区
common sailer			common aeroplane を見よ
common sailor		*Neptis hylas* (Linnaeus)	(チョウ目、タテハチョウ科) 東洋区。日本亜種は *N. h. luculenta* Fruhstorfer　リュウキュウミスジ。*N. h. papaya* Moore も同英名
common sallow		*Cirrhia icteritia* Hufnagel	(チョウ目、ヤガ科) 旧北区

英名	和名	学名	所属、分布、ほか
common sand wasp		*Bembix americana spinolae* Lepeletier	（ハチ目、アナバチ科）新北区
common sanddragon		*Progomphus obscurus* (Rambur)	（トンボ目、サナエトンボ科）新北区
common sandman		*Spialia diomus* (Hopffer)	（チョウ目、セセリチョウ科）エチオピア区
common satyr		*Aulocera swaha* (Kollar)	（チョウ目、タテハチョウ科）東洋区
common savanna bush brown			common bush brown (2) を見よ
common sawflies		*Tenthredo*	（ハチ目、ハバチ科）の昆虫の総称
common sawflies			sawflies (1) を見よ
common scaly crickets		*Cycloptilum*	（バッタ目、コオロギ科）の昆虫の総称
common scarlet		*Axiocerses bambana* Grose-Smith	（チョウ目、シジミチョウ科）エチオピア区
common scarlet		*Axiocerses harpax ugandana* Clench	（チョウ目、シジミチョウ科）エチオピア区
common scarlet (1)		*Axiocerses tjoane* (Wallengren)	（チョウ目、シジミチョウ科）エチオピア区
common scarlet-darter			scarlet darter を見よ
common scarlet-eye		*Nascus phocus* (Cramer)	（チョウ目、セセリチョウ科）新熱帯区
common scavenger water beetles		*Thopisternus*	（コウチュウ目、ガムシ科）の昆虫の総称
common scorpion fly		*Panorpa communis* Linnaeus	（シリアゲムシ目、シリアゲムシ科）旧北区
common scorpionflies			scorpion flies (1) を見よ
common sergeant	シロミスジ	*Athyma perius perius* (Linnaeus)	（チョウ目、タテハチョウ科）日本、東洋区
common sexton beetle			common burying beetle を見よ
common shark			shark moth を見よ
common sheep moth		*Hemileuca (Pseudohazis) eglanterina* (Boisduval)	（チョウ目、ヤママユガ科）新北区。elegant sheepmoth ともいう
common shield bugs		*Eurygaster*	（カメムシ目、キンカメムシ科）の昆虫の総称
common shining cockroach		*Drymaplaneta communis* Tepper	（ゴキブリ目、ゴキブリ科）豪州区
common shore bug		*Saldula saltatoria* (Linnaeus)	（カメムシ目、ミズギワカメムシ科）日本、旧北区
common shore tiger beetle			bronze tiger beetle を見よ
common short-tailed cricket		*Anurogryllus arboreus* Walker	（バッタ目、コオロギ科）新北区
common short-winged katydid		*Dichopetala brevihastata* Morse	（バッタ目、キリギリス科）新北区
common shot-hole borer			shothole borer (1) を見よ
common shot silverline		*Cigaritis ictis* (Hewitson)	（チョウ目、シジミチョウ科）東洋区

英名	和名	学名	所属、分布、ほか
common shutwing		*Cordulephya pygmaea* Selys	(トンボ目、エゾトンボ科) 豪州区
common silver spot			common highflier を見よ
common silver Xenica		*Oreixenica lathoniella* (Westwood)	(チョウ目、タテハチョウ科) 豪州区
common silver Y			silver Y moth (1)(2) を見よ
common silverfish			silverfish (1) を見よ
common silverline		*Spindasis vulcanus* (Fabricius)	(チョウ目、シジミチョウ科) 東洋区
common silver-line		*Cigaritis siphax* (Lucas)	(チョウ目、シジミチョウ科) エチオピア区
common silverstreak		*Theclopsis lydus* Godman et Salvin	(チョウ目、シジミチョウ科) 新熱帯区
common silverstripe		*Fabriciana kamala* (Moore)	(チョウ目、タテハチョウ科) 旧北区、東洋区
common silvery moth			silver Y moth (1) を見よ
common single-spotted clothes moth			casemaking clothes moth (1) を見よ
common six ring		*Ypthima fasciata torone* Fruhstorfer	(チョウ目、タテハチョウ科) 東洋区
common skimmers	トンボ科	Libellulidae	(トンボ目) の昆虫の総称　トンボ目で最多種を含む科
common skip beetle			common click beetle (1) を見よ
common small ermine moth			orchard ermine を見よ
common small ermine moth			apple ermine moth を見よ　英国での英名
common small flat		*Sarangesa dasahara* Moore	(チョウ目、セセリチョウ科) 東洋区
common smoky blue			smoky blue を見よ
common snakeflies		*Agulla*	(アミメカゲロウ目、ラクダムシ科) の昆虫の総称
common snipefly			snipe fly (2) を見よ
common snout			snout (1) を見よ
common snout butterfly			snout butterfly (1) を見よ
common snout butterfly			American snout を見よ
common snow flat	ウラジロシロシタセセリ	*Tagiades japetus* (Stoll)	(チョウ目、セセリチョウ科) 東洋区、豪州区。*T. j. atticus* (Fabricius) も同英名
common snow mosquito			snow mosquito を見よ
common sootywing	アカザマルバネセセリ	*Pholisora catullus* (Fabricius)	(チョウ目、セセリチョウ科) 新北区
common spangle			spangle を見よ
common spangle gall wasp		*Neuroterus quercusbaccarum* (Linnaeus)	(ハチ目、タマバチ科) 旧北区。本種がカシ類の葉の下面につくる虫えいを common spangle gall という。次世代は尾状花序に currant gall をつくる

英名	和名	学名	所属、分布、ほか
common spittlebug	シロオビアワフキ	*Aphrophora intermedia* Uhler	（カメムシ目、アワフキムシ科）日本、旧北区。ダイズ、果樹、クワ、バラなどの害虫
common splendid ghost moth		*Aenetus ligniveren* (Lewin)	（チョウ目、コウモリガ科）豪州区
common spongefly		*Sisyra vicaria* Walker	（アミメカゲロウ目、ミズカゲロウ科）新北区。湖に普通
common spongilla fly		*Sisyra fuscata* (Fabricius)	（アミメカゲロウ目、ミズカゲロウ科）旧北区
common spotted flat		*Celaenorrhinus leucocera* (Kollar)	（チョウ目、セセリチョウ科）東洋区
common spotted flat		*Celaenorrhinus putra* (Moore)	（チョウ目、セセリチョウ科）東洋区
common spotted ladybird		*Harmonia conformis* (Boisduval)	（コウチュウ目、テントウムシ科）豪州区
common Spragueia moth		*Spragueia leo* (Guenée)	（チョウ目、ヤガ科）新北区
common sprawler			sprawler を見よ
common spreadwing		*Lestes plagiatus* (Burmeister)	（トンボ目、アオイトトンボ科）エチオピア区
common spreadwing			emerald damselfly を見よ
common spring moth		*Heliomata cycladata* Grote	（チョウ目、シャクガ科）新北区
common spurwing			powdered grey spurwing を見よ
common stalk borer		*Papaipema nebris* (Guenée)	（チョウ目、ヤガ科）新北区。トウモロコシ害虫
common stick insect		*Clitarchus hookeri* (White)	（ナナフシ目、ナナフシ科）旧北区、豪州区
common stiletto fly		*Thereva nobilitata* (Fabricius)	（ハエ目、ツルギアブ科）旧北区。common stiletto ともいう
common stiletto fly (1)		*Thereva plebeia* (Linnaeus)	（ハエ目、ツルギアブ科）旧北区
common stilt bug		*Neides muticus* (Say)	（カメムシ目、イトカメムシ科）新北区
common stoneflies		*Acroneuria*	（カワゲラ目、カワゲラ科）の昆虫の総称
common stoneflies (1)	カワゲラ科	Perlidae	（カワゲラ目）の昆虫の総称
common straight swift			rice skipper を見よ
common streaky skipper		*Celotes nessus* Edwards	（チョウ目、セセリチョウ科）新北区
common striped hawk		*Hippotion eson* (Cramer)	（チョウ目、スズメガ科）エチオピア区
common sugarcane leafhopper		*Hortensia similis* (Walker)	（カメムシ目、オオヨコバイ科）新北区
common sulphur			clouded sulphur を見よ
common swallowtail butterfly			common yellow swallowtail を見よ
common swift (1)		*Korscheltellus lupulina* (Linnaeus)	（チョウ目、コウモリガ科）全北区
common swift		*Pelopidas agna dingo* Evans	（チョウ目、セセリチョウ科）豪州区

英名	和名	学名	所属、分布、ほか
common swift moth			swift moth を見よ
common swordtail	アゲシラウスオナガタイマイ	*Neographium agesilaus* Guérin-Méneville et Percheron	(チョウ目、アゲハチョウ科) 新熱帯区
common swordtail (1)	ヒメオナガコモンタイマイ (コオナガコモンタイマイ)	*Graphium policenes* (Cramer)	(チョウ目、アゲハチョウ科) エチオピア区
common sylph		*Metisella orientalis* (Aurivillius)	(チョウ目、セセリチョウ科) エチオピア区
common sympetrum			common darter を見よ
common tan wave moth		*Pleuroprucha insulsaria* (Guenée)	(チョウ目、シャクガ科) 新北区
common Telipna		*Telipna acraea* (Westwood)	(チョウ目、シジミチョウ科) エチオピア区
common thick-headed fly		*Thecophora occidensis* (Walker)	(ハエ目、メバエ科) 新北区。幼虫はハナバチ類に寄生
common thorntail		*Ceratogomphus pictus* Hagen	(トンボ目、サナエトンボ科) エチオピア区
common threadtail		*Elattoneura glauca* (Selys)	(トンボ目、ミナミイトトンボ科) エチオピア区
common thread-waisted wasp		*Ammophila procera* Dahlbom	(ハチ目、アナバチ科) 新北区、大洋区
common three ring		*Ypthima pandocus corticaria* Butler	(チョウ目、タテハチョウ科) 東洋区
common three-ring			African ringlet を見よ
common thrips			thrips (1) を見よ
common Ticlear			Euclea tigerwing を見よ
common tiger (1)	シロシタカバマダラ (スジグロシロマダラ)	*Danaus melanippus* (Cramer)	(チョウ目、タテハチョウ科) 日本、東洋区
common tiger (2)		*Mechanitis polymnia* (Linnaeus)	(チョウ目、タテハチョウ科) 新北区、新熱帯区
common tiger (3)	スジグロカバマダラ	*Danaus genutia genutia* (Cramer)	(チョウ目、タテハチョウ科) 日本、東洋区、豪州区
common tiger		*Ictinogomphus ferox* (Rambur)	(トンボ目、サナエトンボ科) エチオピア区。common tigertail ともいう
common tiger			monarch butterfly を見よ
common tiger			African monarch を見よ
common tiger beetle		*Neocicindela tuberculata* (Fabricius)	(コウチュウ目、ハンミョウ科) 豪州区
common tiger beetle			green tiger beetle を見よ
common tiger blue			blue tiger (1) を見よ
common tiger moth			black woolly-bear を見よ
common tigertail		*Ictinogomphus ferox* (Rambur)	(トンボ目、サナエトンボ科) エチオピア区。common tiger ともいう
common Tinsel		*Catapaecilma elegans* (Druce)	(チョウ目、シジミチョウ科) 東洋区
common Tinsel		*Catapaecilma major* Druce	(チョウ目、シジミチョウ科) 東洋区
common tissue			tissue を見よ

英名	和名	学名	所属、分布、ほか
common tit		*Hypolycaena erylus* (Godart)	（チョウ目、シジミチョウ科）東洋区。*H. e. teatus* Fruhstorfer も同英名
common tit (1)		*Hypolycaena phorbas* (Fabricius)	（チョウ目、シジミチョウ科）東洋区、豪州区
common treebrown	シロオビヒカゲ	*Lethe rohria* (Fabricius)	（チョウ目、タテハチョウ科）東洋区
common tree crickets			tree crickets (1) を見よ
common tree nymph	ホソバオオゴマダラ	*Idaea stolli logani* Moore	（チョウ目、タテハチョウ科）東洋区
common true katydid			true katydid を見よ
common twig girdler			twig girdler を見よ
common vagrant			African migrant を見よ
common vapourer		*Orgyia recens* Hübner	（チョウ目、ドクガ科）旧北区、東洋区
common virtuoso katydid		*Amblycorypha longinicta* Walker	（バッタ目、キリギリス科）新北区
common wainscot	タンボキヨトウ	*Mythimna pallens* (Linnaeus)	（チョウ目、ヤガ科）日本、全北区。牧草害虫
common walkingstick			walking stick を見よ
common walking-sticks		Heteronemiidae	（ナナフシ目）の昆虫の総称
common wall butterfly		*Pararge schakra* Kollar	（チョウ目、タテハチョウ科）東洋区。common wall ともいう
common wanderer		*Leptophobia tovaria* (C. et R. Felder)	（チョウ目、シロチョウ科）新熱帯区
common wanderer		*Pareronia anais* (Lesson)	（チョウ目、シロチョウ科）東洋区
common wanderer (1)		*Pareronia hippia* (Fabricius)	（チョウ目、シロチョウ科）東洋区
common wanderer			wanderer (1) を見よ
common wasp	キオビクロスズメバチ	*Vespula vulgaris* (Linnaeus)	（ハチ目、スズメバチ科）日本、全北区、豪州区
common wasps (1)		*Vespa*	（ハチ目、スズメバチ科）の昆虫の総称
common wasps (2)		*Vespula*	（ハチ目、スズメバチ科）の昆虫の総称
common water boatman			water boatman (1) を見よ
common water boatmen		*Corisella*	（カメムシ目、ミズムシ科）の昆虫の総称
common water damsel			brown emerald damselfly を見よ
common water scorpion		*Ranatra brevicollis* Montandon	（カメムシ目、タイコウチ科）新北区
common water strider		*Gerris remigis* Say	（カメムシ目、アメンボ科）新北区
common water strider			common pond skater を見よ
common wave	ミスジコナフエダシャク	*Cabera exanthemata* (Scopoli)	（チョウ目、シャクガ科）日本、旧北区

英名	和名	学名	所属、分布、ほか
common waved silver			common wave を見よ
common web-spinning caddisflies		*Hydropsyche*	（トビケラ目、シマトビケラ科）の昆虫の総称
common whirligig beetles		*Gyrinus*	（コウチュウ目、ミズスマシ科）の昆虫の総称
common whistling moth		*Hecatesia fenestrata* Boisduval	（チョウ目、ヤガ科）豪州区
common white	モンシロチョウ	*Pieris rapae crucivora* Boisduval	（チョウ目、シロチョウ科）日本、旧北区、東洋区
common white			southern cabbage worm を見よ
common white-banded swallowtail	ケブカイチモンジアゲハ	*Papilio cyproeofila* Butler	（チョウ目、アゲハチョウ科）エチオピア区。*P. c. praecyola* Suffert も同英名
common white scale			oleander scale を見よ
common white-spot skipper		*Trapezites petalia* (Hewitson)	（チョウ目、セセリチョウ科）豪州区
common white-spots		*Osmodes thora* (Plötz)	（チョウ目、セセリチョウ科）エチオピア区
common white-streaked grass moth		*Crambus simplex* Butler	（チョウ目、メイガ科）豪州区
common whitetail			white tail を見よ　common whitetail skimmer ともいう
common white wave		*Cabera pusaria* (Linnaeus)	（チョウ目、シャクガ科）旧北区
common whitefly			citrus whitefly (1) を見よ
common wight	シロシタチャバネセセリ	*Iton semamora* (Moore)	（チョウ目、セセリチョウ科）東洋区
common windmill			great windmill (1) を見よ
common winter			winter moth を見よ
common woodbrown		*Lethe sidonis* (Hewitson)	（チョウ目、タテハチョウ科）東洋区
common wood nymph			large wood nymph を見よ
common woolly legs			woolly legs (1) を見よ
common woolly legs			western woolly legs (1) を見よ
common Xenica			marbled Xenica を見よ
common yellow-breast flat		*Gerosis bhagava* (Moore)	（チョウ目、セセリチョウ科）東洋区
common yellow dung-fly			yellow dung fly を見よ
common yellow jacket			common wasp を見よ
common yellow moth		*Epiphryne undosata* Felder	（チョウ目、シャクガ科）豪州区
common yellow Sally		*Isoperla grammatica* (Poda)	（カワゲラ目、Isoperlidae）旧北区
common yellow swallowtail	キアゲハ	*Papilio machaon hippocrates* C. et R. Felder	（チョウ目、アゲハチョウ科）日本。old world swallowtail (1) を参照
common yellow underwing			common yellow underwing moth を見よ

英名	和名	学名	所属、分布、ほか
common yellow underwing moth		*Noctua pronuba* (Linnaeus)	（チョウ目、ヤガ科）全北区。英国では本種の幼虫を単に cutworm という
common yellowjacket			aerial yellowjacket を見よ
common Yeomen	ウスイロネッタイヒョウモン	*Cirrochroa tyche* C. et R. Felder	（チョウ目、タテハチョウ科）東洋区。*C. t. rotundata* Butler も同英名
common zebra blue			Lang's short-tailed blue を見よ
commona hairstreak			purple-brown hairstreak を見よ
compas termite			magnetic termite を見よ
Compassberg skolly		*Thestor compassbergae* Quickelberge et McMaster	（チョウ目、シジミチョウ科）エチオピア区
Complanula skipper		*Turesis complanula* (Herrich-Schäffer)	（チョウ目、セセリチョウ科）新熱帯区
complete red-ling			green-spotted banner を見よ
complex-trilling trig		*Cyrtoxipha confusa* Walker	（バッタ目、クサヒバリ科）新北区
compliant skipper		*Eutychide complana* (Herrich-Schäffer)	（チョウ目、セセリチョウ科）新熱帯区
composite thrips	コスモスアザミウマ	*Microcephalothrips abdominalis* (Crawford)	（アザミウマ目、アザミウマ科）日本、東洋区、大洋区、豪州区、新北区、新熱帯区。キク、コスモス、バラなどの害虫
Compton tortoiseshell			false comma を見よ　米国での英名
comrade midget		*Phyllonorycter comparella* (Duponchel)	（チョウ目、ホソガ科）旧北区
cosmopolitan			painted lady を見よ　米国での英名
Comstock mealybug	クワコナカイガラムシ	*Pseudococcus comstocki* (Kuwana)	（カメムシ目、コナカイガラムシ科）日本、全北区。クワ、チャ、果樹、サクラなどの害虫。アメリカの昆虫学者 Comstock に由来。Comstock's mealybug とも記す
Comstock's giant-skipper		*Agathymus comstocki* (Harbison)	（チョウ目、セセリチョウ科）新熱帯区
Comstock's net-winged midge		*Agathon comstocki* (Kellog)	（ハエ目、アミカ科）新北区
Comstock's Retinia moth			pitch twig moth を見よ
concealer moths	マルハキバガ科	Oecophoridae	（チョウ目）の昆虫の総称
conchuela		*Chlorochroa ligata* (Say)	（カメムシ目、カメムシ科）新北区
concolored skimmer		*Orthemis concolor* Ris	（トンボ目、トンボ科）新熱帯区
concolorous		*Photedes extrema* (Hübner)	（チョウ目、ヤガ科）旧北区
concolorous moth			concolorous を見よ
concolorous wainscot			concolorous を見よ
Condamin's bush brown		*Bicyclus sangmelinae* Condamin	（チョウ目、タテハチョウ科）エチオピア区
Condamin's ringlet		*Ypthima condamini* Kielland	（チョウ目、タテハチョウ科）エチオピア区
conehead darner		*Austroaeschna subapicalis* Theischinger	（トンボ目、ヤンマ科）豪州区

英名	和名	学名	所属、分布、ほか
cone-headed grasshoppers		Copiphorinae	（バッタ目、キリギリス科）の昆虫の総称
coneheaded katydids			cone-headed grasshoppers を見よ
cone-headed locusts			meadow grasshoppers (1) を見よ
cone heads			cone-headed grasshoppers を見よ　cone-heads　とも記す
cone-heads			meadow grasshoppers (1) を見よ
cone moth		*Eucosma rescissoriana* Heinrich	（チョウ目、ハマキガ科）新北区
cone moth			peach moth を見よ
cone moth			pine cone moth を見よ
cone moth			spruce bell moth を見よ
cone moth			spruce seed moth (1) を見よ
cone-nosed grasshoppers			meadow grasshoppers (1) を見よ
cone-noses			assassin bugs (1) を見よ
cone pitch moth		*Blastesthia turionella* Linnaeus	（チョウ目、ハマキガ科）旧北区
cone pyralid			pine knot-horn を見よ
cone resin midge		*Asynapta hopkinsi* Felt	（ハエ目、タマバエ科）新北区
cone tortrix			spruce seed moth (1) を見よ
cone worm			pine knot-horn を見よ
coneflower borer moth		*Papaipema nelita* (Strecker)	（チョウ目、ヤガ科）新北区
conflua skipper		*Tirynthia conflua* (Herrich-Schäffer)	（チョウ目、セセリチョウ科）新熱帯区
confluent-barred slender		*Gracilaria anastomosis* Haworth	（チョウ目、ホソガ科）旧北区
conformist		*Lithophane furcifera* (Hufnagel)	（チョウ目、ヤガ科）旧北区
conformist moth			conformist を見よ
confucian dart	タイワンキマダラセセリ	*Potanthus confucius* (C. et R. Felder)	（チョウ目、セセリチョウ科）東洋区
confusa tigerwing			giant glasswing を見よ
confused		*Apamea furva* (Denis et Schiffermüller)	（チョウ目、ヤガ科）旧北区
confused brindle			confused を見よ
confused cosmet		*Mompha conturbatella* (Hübner)	（チョウ目、カザリバガ科）旧北区
confused dart moth		*Feltia tricosa* (Lintner)	（チョウ目、ヤガ科）新北区
confused Eusarca moth		*Eusarca confusaria* (Hübner)	（チョウ目、シャクガ科）新北区。confused Eusarca ともいう
confused flour beetle	ヒラタコクヌストモドキ	*Tribolium confusum* Jacquelin du Duval	（コウチュウ目、ゴミムシダマシ科）日本、汎世界。著名な貯穀害虫
confused ground cricket		*Eunemobius confusus* (Blatchley)	（バッタ目、コオロギ科）新北区

英名	和名	学名	所属、分布、ほか
confused Haploa			Lyman's Haploa を見よ　confused Haploa moth ともいう
confused Meganola moth		*Meganola minuscula* (Zeller)	（チョウ目、コブガ科）新北区
confused moth			confused を見よ
confused owlet		*Bia actorion* (Stoll)	（チョウ目、タテハチョウ科）新熱帯区
confused Pellicia			confused skipper を見よ
confused recluse		*Leona stoehri* (Karsch)	（チョウ目、セセリチョウ科）エチオピア区
confused satyr		*Cissia confusa* (Staudinger)	（チョウ目、タテハチョウ科）新熱帯区
confused skipper		*Pellicia angra* Evans	（チョウ目、セセリチョウ科）新北区、新熱帯区
confused tigerwing			Lysimnia tigerwing を見よ
confused wheat beetle			confused flour beetle を見よ
confusing ace		*Halpe wantona* Swinhoe	（チョウ目、セセリチョウ科）東洋区
confusing Petrophila moth		*Petrophila confusalis* Walker	（チョウ目、メイガ科）新北区
confusing Phanus		*Phanus confusis* Austin	（チョウ目、セセリチョウ科）新熱帯区
confusing sandman		*Spialia confusa* (Higgins)	（チョウ目、セセリチョウ科）エチオピア区
confusing sister			Iphicleola sister を見よ
Congo floor fly		*Auchmeromyia luteola* (Fabricius)	（ハエ目、クロバエ科）エチオピア区。幼虫は Congo floor maggot
Congo green mantis		*Sphodromantis aurea* Giglio-Tos	（カマキリ目、カマキリ科）エチオピア区
Congo mantis		*Sphodromantis congica* Beier	（カマキリ目、カマキリ科）エチオピア区
Congo white		*Appias phaola* (Doubleday)	（チョウ目、シロチョウ科）エチオピア区
conifer aphids	カサアブラムシ科	Adelgidae	（カメムシ目）の昆虫の総称　英国での英名
conifer Coleotechnites moth		*Coleotechnites coniferella* (Kearfott)	（チョウ目、キバガ科）新北区
conifer needle-binder			larch tortrix（1）を見よ
conifer root aphid		*Pachypappa vesicalis* Koch	（カメムシ目、アブラムシ科）旧北区
conifer sawflies	マツハバチ科	Diprionidae	（ハチ目）の昆虫の総称
conifer sawyer	ヒゲナガカミキリ	*Monochamus grandis* Waterhouse	（コウチュウ目、カミキリムシ科）日本、旧北区。モミ類、マツ類の害虫
conifer scale	ニッポンコノハカイガラムシ	*Fiorinia japonica* Kuwana	（カメムシ目、マルカイガラムシ科）日本、東洋区、新北区。モミ類、トウヒ類の害虫
conifer swift moth		*Korscheltellus gracilis* (Grote)	（チョウ目、コウモリガ科）新北区
conifer weevils	キボシゾウムシ亜科	Pissodinae	（コウチュウ目、ゾウムシ科）の昆虫の総称
conifer woolly aphids			conifer aphids を見よ
coniferous sawfly			introduced pine sawfly を見よ
coniferous tussock moth		*Callitaera abietis* Denis et Schiffermüller	（チョウ目、ドクガ科）日本、旧北区
coniopterygids			dustywings を見よ

英名	和名	学名	所属、分布、ほか
conjoined swift	タイワンオオチャバネセセリ	*Pelopidas conjuncta* (Herrich-Schäffer)	(チョウ目、セセリチョウ科) 東洋区
connected dagger moth		*Acronicta connecta* Grote	(チョウ目、ヤガ科) 新北区
connected looper moth		*Plusia contexta* Grote	(チョウ目、ヤガ科) 新北区
conniption bugs			dobsonflies (1) を見よ　幼虫の英名
connubial underwing moth		*Catocala connubialis* Guenée	(チョウ目、ヤガ科) 新北区
conopid fly		*Conops quadrifasciata* De Geer	(ハエ目、メバエ科) 旧北区
conopid fly		*Physocephala affinis* Williston	(ハエ目、メバエ科) 新北区
conopid-fly		*Myopa testacea* Linnaeus	(ハエ目、メバエ科) 旧北区
Conoveira hairstreak		*Celmia conoveria* (Schaus)	(チョウ目、シジミチョウ科) 新熱帯区
Conrad's copper		*Aloeides conradsi* (Aurivillius)	(チョウ目、シジミチョウ科) エチオピア区
Consobrina canegrub		*Lepidiota consobrina* Girault	(コウチュウ目、コガネムシ科) 豪州区
Consobrina darkwing moth		*Gondysia consobrina* (Guenée)	(チョウ目、ヤガ科) 新北区。Consobrina's darkwing ともいう
consort underwing moth		*Catocala consors* (Smith)	(チョウ目、ヤガ科) 新北区
consperse stink bug		*Eucchistus conspersus* Uhler	(カメムシ目、ヘリカメムシ科) 新北区
conspicuous malachite		*Chlorolestes conspicuus* Hagen	(トンボ目、Synlestidae) エチオピア区
constable	スミナガシ	*Dichorragia nesimachus nesiotes* Fruhstorfer	(チョウ目、タテハチョウ科) 日本、旧北区
Constance's sailer		*Neptis constantiae* Carcasson	(チョウ目、タテハチョウ科) エチオピア区
Constantia evening brown		*Melanitis constantia* (Cramer)	(チョウ目、タテハチョウ科) 東洋区
Constantine's swallowtail		*Papilio constantinus* Ward	(チョウ目、アゲハチョウ科) エチオピア区
constricted longicorn beetle	ドウボソカミキリ	*Pseudocalamobius japonicus japonicus* (Bates)	(コウチュウ目、カミキリムシ科) 日本、旧北区
constricted refuse beetle	コゴモクムシ	*Harpalus tridens* Morawitz	(コウチュウ目、オサムシ科) 日本、旧北区
constricted Sonia moth		*Sonia constrictana* (Zeller)	(チョウ目、ハマキガ科) 新北区
constricted spined beetle	カタビロトゲハムシ	*Dactylispa subquadrata* (Baly)	(コウチュウ目、ハムシ科) 日本、旧北区
constricted thread-legged katydid		*Arethaea constricta* Brunner	(バッタ目、キリギリス科) 新北区
consular oakworm moth		*Anisota consularis* Dyar	(チョウ目、ヤママユガ科) 新北区
contiguous swift		*Polytremis lubricans* (Herrich-Schäffer)	(チョウ目、セセリチョウ科) 旧北区、東洋区

英名	和名	学名	所属、分布、ほか
continental swift		*Parnara ganga* Evans	（チョウ目、セセリチョウ科）東洋区
contorted delphacid		*Phrictopyga contorta* (Muir)	（カメムシ目、ウンカ科）新熱帯区
contracted Datana moth		*Datana contracta* Walker	（チョウ目、シャチホコガ科）新北区。contracted Datana ともいう
contractile scarabs		Acanthocerinae	（コウチュウ目、Acanthoceridae）の昆虫の総称
contrary Charaxes		*Charaxes contrarius* Weymer	（チョウ目、タテハチョウ科）エチオピア区
contrary polyptychus		*Andriasa contraria* Walker	（チョウ目、スズメガ科）エチオピア区
contrasted fen cosmet	ヨシウスオビカザリバ	*Cosmopterix lienigiella* Zeller	（チョウ目、カザリバガ科）日本、旧北区
contrasting Henricus moth		*Henricus contrastana* (Kearfott)	（チョウ目、ハマキガ科）新北区
convergent lady beetle		*Hippodamia convergens* Guérin-Méneville	（コウチュウ目、テントウムシ科）新北区
convergent ladybird			convergent lady beetle を見よ
convex shieldback		*Neduba convexa* Caudell	（バッタ目、キリギリス科）新北区
convolvulus hawk			sweetpotato hawk-moth を見よ
convolvulus hawk moth			sweetpotato hawk-moth を見よ
convolvulus leaf-miner			morning-glory leaf miner を見よ
cook			Fatima を見よ
Cook Strait giant weta		*Deinacrida rugosa* Butler	（バッタ目、Stenopelmatidae）豪州区
Cook's mountain marble		*Olethreutes obsoletana* Stephens	（チョウ目、ハマキガ科）旧北区
Cookson's bush brown		*Bicyclus cooksoni* (Druce)	（チョウ目、タテハチョウ科）エチオピア区
Cooktown azure		*Ogyris aenone* Waterhouse	（チョウ目、シジミチョウ科）豪州区
cool-weather mosquito		*Culiseta incidens* (Thomson)	（ハエ目、カ科）新北区
Cooley spruce gall adelgid			Douglas fir adelges を見よ
Cooley spruce gall aphid			Douglas fir adelges を見よ　北米での英名
Cooley's gall louse			Douglas fir adelges を見よ
coolie			scarlet peacock を見よ
Cooloola monster		*Cooloola propator* Rentz	（バッタ目、Cooloolidae）豪州区
coon		*Psolos fuligo* (Mabille)	（チョウ目、セセリチョウ科）東洋区
coon bug		*Oxycarenus arctatus* (Walker)	（カメムシ目、ナガカメムシ科）豪州区
Coorg forest hopper		*Arnetta mercara* (Evans)	（チョウ目、セセリチョウ科）東洋区
Cootamundra wattle psyllid		*Acizzia acaciaebaileyanae* (Froggatt)	（カメムシ目、キジラミ科）豪州区。豪州南東部の地名に由来
cootie			body louse を見よ　米国での英名

英名	和名	学名	所属、分布、ほか
cooties			sucking lice を見よ
cooties			human lice を見よ
Copa Dichomeris moth		*Dichomeris copa* Hodges	（チョウ目、キバガ科）新北区
copper ant-blue			Cuprea ant-blue を見よ
copper bramble pigmy		*Stigmella splendidissimella* (Herrich-Schäffer)	（チョウ目、モグリチビガ科）旧北区
copper butterflies			coppers (1) を見よ　米国での英名
copper flash		*Rapala pheretima* (Hewitson)	（チョウ目、シジミチョウ科）東洋区。*R. p. sequeira* Distant も同英名
copper-gold longhorn		*Nemophora cupriacella* (Hübner)	（チョウ目、マガリガ科）旧北区
copper-headed sootywing		*Bolla cupreiceps* (Mabille)	（チョウ目、セセリチョウ科）新熱帯区
copper-japan longhorn		*Nemophora fasciella* (Fabricius)	（チョウ目、マガリガ科）旧北区
copper jewel		*Hypochrysops apelles* (Fabricius)	（チョウ目、シジミチョウ科）豪州区
copper looper moth			unspotted looper moth を見よ
copper pencil-blue			Cyprotus blue を見よ
copper sphinx		*Dovania poeccila* Rothschild et Jordan	（チョウ目、スズメガ科）エチオピア区
copper underwing (1)		*Amphipyra pyramidoides* Guenée	（チョウ目、ヤガ科）新北区。copper underwing moth ともいう
copper underwing (2)		*Amphipyra pyramidea* (Linnaeus)	（チョウ目、ヤガ科）旧北区。日本亜種は *A. p. obscura* Oberthür シマカラスヨトウ
copper underwing moth			copper underwing (1)(2) を見よ
copper winged polyptychus		*Polyptychus andosa* (Walker)	（チョウ目、スズメガ科）エチオピア区
coppers (1)	シジミチョウ科	Lycaenidae	（チョウ目）の昆虫の総称　世界に 3,000 種以上
coppers (2)	ベニシジミ亜科	Lycaeninae	（チョウ目、シジミチョウ科）の昆虫の総称
coppers and hairstreaks			coppers (1) を見よ
coppery dancer		*Argia cuprea* (Hagen)	（トンボ目、イトトンボ科）新北区
coppery Dysphania		*Dysphania cuprina* Felder	（チョウ目、シャクガ科）東洋区
coppery ground beetle		*Notiobia cupreola* Bates	（コウチュウ目、オサムシ科）新熱帯区
coppery groundstreak		*Ziegleria denarius* (Butler et Druce)	（チョウ目、シジミチョウ科）新熱帯区
coppery swordtail	オビモンタイマイ	*Graphium latreillianus* (Godart)	（チョウ目、アゲハチョウ科）エチオピア区
copra beetle			red-legged ham beetle を見よ
copromorphid moths			tropical fruitworm moths を見よ
Coprosoma hawk moth			scrofa hawk moth を見よ

英名	和名	学名	所属、分布、ほか
Coptops long-horned beetle		*Coptops aedificator* (Fabricius)	(コウチュウ目、カミキリムシ科) 東洋区、大洋区、エチオピア区
Coracara checkerspot		*Texola coracara* (Dyar)	(チョウ目、タテハチョウ科) 新熱帯区
coral-fronted threadtail		*Neoneura aaroni* Calvert	(トンボ目、Protoneuridae) 新北区
coral hairstreak		*Satyrium titus* (Fabricius)	(チョウ目、シジミチョウ科) 新北区
coral jewel			Miskin's jewel を見よ
coral pink sand dunes tiger beetle		*Cicindela albissima* Rumpp	(コウチュウ目、ハンミョウ科) 新北区
coral-tailed cloudwing			twister を見よ
coral treaders	サンゴアメンボ科	Hermatobatidae	(カメムシ目) の昆虫の総称
Cora's pennant			coastal pennant を見よ
corbie		*Oncopera intricata* Walker	(チョウ目、コウモリガ科) 豪州区
Corcyra dartwhite	オオシロタカネマダラシロチョウ	*Catasticta corcyra* (C. et R. Felder)	(チョウ目、シロチョウ科) 新熱帯区
Cordelia crow	コルデリイルリマダラ	*Euploea cordelia* Martin	(チョウ目、タテハチョウ科) 東洋区。Cordelia's crow ともいう
Cordenillo			Provence hairstreak を見よ
Cordovan pyralid moth		*Acrobasis exsulella* (Zeller)	(チョウ目、メイガ科) 新北区
cordulid dragonflies			green-eyed skimmers を見よ
corduliids			green-eyed skimmers を見よ
coreid bug		*Phthia lunata* Fabricius	(カメムシ目、ヘリカメムシ科) 新熱帯区。カンキツ害虫
coreid bugs	ヘリカメムシ科	Coreidae	(カメムシ目) の昆虫の総称
coreids			coreid bugs を見よ
Corilla scarlet-eye		*Nascus solon corilla* Evans	(チョウ目、セセリチョウ科) 新熱帯区
Corinna daggerwing	ヒメムラサキツルギタテハ	*Marpesia corinna* (Latreille)	(チョウ目、タテハチョウ科) 新熱帯区
corixids			water boatman bugs を見よ
corizid bugs			grass bugs を見よ
cork moth			dark-mottled clothes moth を見よ
corkscrew hooktail		*Paragomphus elpidius* (Ris)	(トンボ目、サナエトンボ科) エチオピア区
corn aphid			corn leaf aphid を見よ
corn aphis			corn leaf aphid を見よ
corn beetle			Sylvanus 属 などヒラタムシの類の昆虫
corn billbugs			grain weevils を見よ
corn blotch leaf fly			corn blotch miner を見よ
corn blotch leafminer			corn blotch miner を見よ
corn blotch miner		*Agromyza parvicornis* Loew	(ハエ目、ハモグリバエ科) 新北区

英名	和名	学名	所属、分布、ほか
corn borer		*Ostrinia narynensis* Mutuura et Munroe	（チョウ目、メイガ科）旧北区
corn borer			Oriental corn borer を見よ
corn budworm			southern armyworm (1) を見よ
corn delphacid			corn hopper を見よ　米国での英名
corn delphacid egg parasite			corn leafhopper egg parasite を見よ
corn delphacid predator		*Cyrtorhinus lividipennis* Reuter	（カメムシ目、カスミカメムシ科）大洋区
corn earworm (1)	アメリカタバコガ	*Heliothis zea* (Boddie)	（チョウ目、ヤガ科）新北区。トウモロコシ、ワタその他多くの作物の害虫
corn earworm (2)	オオタバコガ	*Helicoverpa armigera* (Hübner)	（チョウ目、ヤガ科）日本、東洋区、全北区、エチオピア区、豪州区、大洋区。タバコ、ワタなど多くの作物、草花の害虫
corn earworm moth			corn earworm (1) を見よ
corn emperor			Io moth を見よ
corn-feeding syrphid fly		*Mesogramma polita* (Say)	（ハエ目、ハナアブ科）新北区
corn flea beetle		*Chaetocnema hortensis* Melsheimer	（コウチュウ目、ハムシ科）新北区
corn flea beetle			brassy flea beetle (1)(2) を見よ
corn fulgorid			corn hopper を見よ
corn ground beetle		*Zabrus tenebrioides* Goeze	（コウチュウ目、オサムシ科）旧北区
corn hopper	トウモロコシウンカ	*Peregrinus maidis* (Ashmead)	（カメムシ目、ウンカ科）日本、新北区、汎熱帯。トウモロコシ、イネ害虫
corn lantern fly			corn hopper を見よ
corn leaf aphid	トウモロコシアブラムシ	*Rhopalosiphum maidis* (Fitch)	（カメムシ目、アブラムシ科）日本、汎世界。トウモロコシ害虫
corn leaf aphid			greenbug を見よ
corn leaf miner	ムギスジハモグリバエ	*Phytomyza nigra* Meigen	（ハエ目、ハモグリバエ科）日本、全北区
corn leafhopper		*Dalbulus maidis* (DeLong et Wolcott)	（カメムシ目、オオヨコバイ科）新北区
corn leafhopper			corn hopper を見よ
corn leafhopper egg parasite		*Anagrus osborni* Fullaway	（ハチ目、ホソハネコバチ科）大洋区
corn leaf-tier			clouded skipper を見よ
corn-linear leaf-miner			corn leaf miner を見よ
corn moth			European grain moth を見よ
corn planthopper			corn hopper を見よ
corn root aphid		*Anuraphis maidiradicis* (Forbes)	（カメムシ目、アブラムシ科）新北区。トウモロコシ害虫
corn root aphis			corn root aphid を見よ
corn root aphis			greenbug を見よ
corn root webworm			tobacco webworm を見よ　corn root webworm moth ともいう

英名	和名	学名	所属、分布、ほか
corn sap beetle			dried-fruit beetle (1)(2) を見よ
corn sawfly			wheat stem sawfly (1) を見よ
corn seed maggot			seed-corn maggot (1) を見よ
corn seed maggot			bean seed fly を見よ
corn seedling maggot			rice stem fly (2) を見よ
corn silk beetle		*Metrioidea brunneus* (Crotch)	(コウチュウ目、ハムシ科) 新北区
corn stack bug		*Scolopostethus pictus* (Schilling)	(カメムシ目、ナガカメムシ科) 旧北区
corn stalk borer			southern corn stalk borer (1) を見よ
corn stalk-borer			common stalk borer を見よ
corn stem weevil		*Hyperodes humilis* (Gyllenhal)	(コウチュウ目、ゾウムシ科) 新北区
corn thrips		*Frankliniella williamsi* Hood	(アザミウマ目、アザミウマ科) 新北区
corn thrips			grain thrips を見よ
corn webworm		*Pediasia caliginosella* Clemens	(チョウ目、メイガ科) 新北区
corn webworm		*Sameodes cancellalis* Zeller	(チョウ目、メイガ科) エチオピア区
corn weevil			maize weevil を見よ　米国での英名
corn weevil			granary weevil を見よ
corn wire-worm		*Melanotus communis* (Gyllenhal)	(コウチュウ目、コメツキムシ科) 新北区
corn worm			black armyworm を見よ
Cornelian	ヒイロシジミ	*Deudorix epijarbas* (Moore)	(チョウ目、シジミチョウ科) 東洋区、豪州区。*D. e. cinnabarus* Fruhstorfer も同英名
Cornelians	ヒイロシジミ属	*Deudorix*	(チョウ目、シジミチョウ科) の昆虫の総称
cornfield ant		*Lasius neoniger* Emery	(ハチ目、アリ科) 新北区
cornfield ant (1)	ヒメトビイロケアリ	*Lasius alienus* (Forster)	(ハチ目、アリ科) 日本、全北区。北米で corn root aphid と共生。*Lasius* 属の種に広く使用される
cornfield ant			black ant (1) を見よ　北米亜種 *L. niger americanus* Emery の英名
cornflower aphid			knapweed aphid を見よ
coromandel marsh dart			yellow waxtail を見よ
coronet	イボタケンモン	*Craniophora ligustri* (Denis et Schiffermüller)	(チョウ目、ヤガ科) 日本、旧北区
coronet moth			coronet を見よ
coronets	ヨトウガ亜科	Hadeninae	(チョウ目、ヤガ科) の昆虫の総称
Coronis fritillary	コロニスギンボシヒョウモン	*Speyeria coronis* (Behr)	(チョウ目、タテハチョウ科) 新北区。*S. c. semiramis* (Edwards) オオウスイロギンボシヒョウモンも同英名
coronopus borer moth		*Carmenta minuli* (Edwards)	(チョウ目、スカシバガ科) 新北区。coronopus borer ともいう
Corope skipper		*Cynea corope* (Herrich-Schäffer)	(チョウ目、セセリチョウ科) 新熱帯区

英名	和名	学名	所属、分布、ほか
Correa brown		*Oreixenica correae* (Olliff)	(チョウ目、タテハチョウ科) 豪州区
corroboree cicada			green whizzer を見よ
corsairs		*Rasahus biguttatus* (Say)	(カメムシ目、サシガメ科) 新北区
Corsican dappled white		*Euchloe insularis* (Staudinger)	(チョウ目、シロチョウ科) 旧北区
Corsican fritillary		*Fabriciana elisa* (Godart)	(チョウ目、タテハチョウ科) 旧北区
Corsican grayling		*Hipparchia neomiris* (Godart)	(チョウ目、タテハチョウ科) 旧北区
Corsican heath		*Coenonympha corinna* (Hübner)	(チョウ目、タテハチョウ科) 旧北区
Corsican swallowtail	コルシカアゲハ	*Papilio hospiton* Gene	(チョウ目、アゲハチョウ科) 旧北区
Corusca skipper		*Oxynthes corusca* (Herrich-Schäffer)	(チョウ目、セセリチョウ科) 新熱帯区
corydalids			dobsonflies (1) を見よ
corylus aphid	カバウスブチアブラムシ	*Mesocallis pteleae* Matsumura	(カメムシ目、アブラムシ科) 日本、旧北区
Coryna skipper		*Vettius coryna* (Hewitson)	(チョウ目、セセリチョウ科) 新熱帯区
Coryndon's polyptychus		*Polyptychus coryndoni* Rothschild et Jordan	(チョウ目、スズメガ科) エチオピア区
Cosinga hawker		*Metardaris cosinga* (Hewitson)	(チョウ目、セセリチョウ科) 新熱帯区
cosmet moths	カザリバガ科	Cosmopterigidae	(チョウ目) の昆虫の総称
cosmets			cosmet moths を見よ
cosmopolitan			Lorey leafworm を見よ
cosmopolitan ambrosia beetle			shot-hole borer (3) を見よ
cosmopolitan ambrosia beetle			Saxena ambrosia beetle を見よ
cosmopolitan blue bone beetle			blacklegged ham beetle を見よ
cosmopolitan diving beetle	ヒメゲンゴロウ	*Rhantus pulverosus* (Stephens)	(コウチュウ目、ゲンゴロウ科) 日本、全北区、東洋区、豪州区、大洋区。豪州での英名
cosmopolitan grain barklouse	ヒメチャタテ	*Lachesilla pediculuria* (Linnaeus)	(チャタテムシ目、ヒメチャタテ科) 日本、汎世界
cosmopolitan grain psocid			cosmopolitan grain barklouse を見よ
cosmopolitan ground beetle		*Laemostenus complanatus* (Dejean)	(コウチュウ目、オサムシ科) 豪州区
cosmopolitan moth			Lorey leafworm を見よ
cosmopolitan powder post beetle			oak lyctid を見よ　カナダでの英名
cosmopolitan tea and olive scale			tea scale (1) を見よ
cosmopolite			painted lady を見よ　米国での英名
cosmopterigid moths			cosmet moths を見よ
cosmos thrips			composite thrips を見よ
cossids			carpenter moths (1) を見よ

英名	和名	学名	所属、分布、ほか
Costa Rican Calephelis		*Calephelis costaricicola* Strand	(チョウ目、シジミタテハ科) 新熱帯区
Costa Rican curculio		*Phyrdenus muriceus* (Germar)	(コウチュウ目、ゾウムシ科) 新熱帯区
Costa Rican grass mantis		*Thesprotia insolita* Rehn	(カマキリ目、カマキリ科) 新熱帯区
Costa Rican skipper		*Halotus rica* (Bell)	(チョウ目、セセリチョウ科) 新熱帯区
Costa Rican white		*Hesperocharis costaricensis pasion* (Reakirt)	(チョウ目、シロチョウ科) 新熱帯区
costa spotted mimic white		*Enantia albania* (Bates)	(チョウ目、シロチョウ科) 新北区
costa-spotted metalmark		*Ancyluris jurgensenii jurgensenii* (Saunders)	(チョウ目、シジミタテハ科) 新熱帯区
costa-spotted mimic white		*Enantia albania* (Bates)	(チョウ目、シロチョウ科) 新熱帯区
costal-dotted button			broad-barred marble (2) を見よ
Cosyra firetip		*Yanguna cosyra* Druce	(チョウ目、セセリチョウ科) 新熱帯区
cotoneaster webworm		*Athrips rancidella* Herrich-Schäffer	(チョウ目、キバガ科) 全北区。cotoneaster webworm moth ともいう
cotton aphid	ワタアブラムシ	*Aphis gossypii* Glover	(カメムシ目、アブラムシ科) 日本、汎世界。ワタ他多くの作物、草花などの大害虫
cotton aphis			cotton aphid を見よ
cotton boll weevil			boll weevil を見よ
cotton bollworm		*Helicoverpa stombleri* Okumura et Bauer	(チョウ目、ヤガ科) 新北区
cotton bollworm			corn earworm (1)(2) を見よ
cotton bollworm parasitoid	ヨトウオオサムライコマユバチ	*Microplitis mediator* (Haliday)	(ハチ目、コマユバチ科) 日本、全北区、東洋区
cotton bud caterpillar		*Phycita infusella* Meyrick	(チョウ目、メイガ科) 旧北区、東洋区、豪州区
cotton bud thrips			yellow flower thrips を見よ
cotton bug	アカホシカメムシ	*Dysdercus cingulatus* (Fabricius)	(カメムシ目、ホシカメムシ科) 日本、東洋区。ワタ害虫
cotton bugs			lygus bugs を見よ
cotton caterpillar (1)	ワタヘリクロノメイガ	*Diaphania indica* (Saunders)	(チョウ目、メイガ科) 日本、全北区、東洋区、大洋区、新熱帯区、エチオピア区。ワタ、ウリ類などの害虫
cotton caterpillar			cucumber moth を見よ
cotton cutworm			yellow-striped armyworm moth を見よ
cotton fleahopper		*Pseudatomoscelis seriatus* (Reuter)	(カメムシ目、カスミカメムシ科) 新北区。ワタその他多くの植物につく
cotton flower lint maggot		*Contarinia gossypii* Felt	(ハエ目、タマバエ科) 新熱帯区
cotton gall		*Andricus quercusramuli* (Linnaeus)	(ハチ目、タマバチ科) がカシにつくる綿毛状の虫えいで 20 mm。旧北区
cotton harlequin bug			harlequin bug (2) を見よ
cotton helopeltis		*Helopeltis shoutedeni* Reuter	(カメムシ目、カスミカメムシ科) エチオピア区
cotton jassids		*Empoasca*	(カメムシ目、オオヨコバイ科) の昆虫の総称

英名	和名	学名	所属、分布、ほか
cotton lace bug			coffee lace bug を見よ
cotton leaf caterpillar (1)		*Anomis erosa* Hübner	（チョウ目、ヤガ科）東洋区、豪州区、新北区
cotton leaf caterpillar (2)		*Ennomos erosaria* (Denis et Schiffermülller)	（チョウ目、シャクガ科）旧北区
cotton leaf caterpillar			cotton semi-looper を見よ
cotton leaf miner		*Stigmella gossipii* (Forbes et Leonard)	（チョウ目、モグリチビガ科）新北区、新熱帯区。cotton leafminer moth ともいう
cotton leaf perforator		*Bucculatrix thurberiella* Busck	（チョウ目、チビガ科）新北区、新熱帯区、大洋区
cotton leaf perforator		*Bucculatrix gossypii* Turner	（チョウ目、チビガ科）豪州区
cotton leafhopper		*Amrasca terraereginae* (Paoli)	（カメムシ目、オオヨコバイ科）豪州区
cotton leafhopper			green leafhopper (2) を見よ
cotton leafroller	ワタノメイガ	*Notarcha derogata* (Fabricius)	（チョウ目、メイガ科）日本、東洋区、豪州区、大洋区、エチオピア区。ワタ、オクラ、ハイビスカスなどの害虫
cotton leafworm			common cutworm を見よ
cotton leafworm			Egyptian cotton leafworm を見よ
cotton leafworm			cotton leafworm moth を見よ
cotton leafworm moth		*Alabama argillacea* (Hübner)	（チョウ目、ヤガ科）新北区。成虫は往々南から北に移動する
cotton looper			cotton semi-looper を見よ
cotton lygus		*Taylorilygus vosseleri* (Poppius)	（カメムシ目、カスミカメムシ科）エチオピア区
cotton melon aphid			cotton aphid を見よ
cotton moth			cotton leafworm moth を見よ
cotton plant bug		*Aulacosternus nigrorubrum* Dallas	（カメムシ目、ヘリカメムシ科）豪州区
cotton rough bollworm			spotted bollworm (2) を見よ
cotton scale			striped mealybug を見よ　南アフリカでの英名
cotton seed bug		*Oxycarenus hyalipennis* Costa	（カメムシ目、ナガカメムシ科）旧北区、エチオピア区、東洋区
cotton seed bugs		*Oxycarenus*	（カメムシ目、ナガカメムシ科）の昆虫の総称
cotton seedling thrips			onion thrips を見よ
cotton seedworm	ワタミガ	*Promalactis enopisema* (Butler)	（チョウ目、マルハキバガ科）日本、旧北区。シロスジベニマルハキバガともいう
cotton semi-looper	ワタアカキリバ	*Anomis flava* (Fabricius)	（チョウ目、ヤガ科）日本、全北区、東洋区、豪州区、大洋区、エチオピア区。カンキツ類、ナシなどの害虫
cotton semi-looper			okra leafworm moth を見よ
cotton semi-looper			cotton leaf caterpillar (1)(2) を見よ
cotton sharp-shooter		*Homalodisca triquetra* Fabricius	（カメムシ目、オオヨコバイ科）新熱帯区。サイザル害虫
cotton spotted bollworm			spotted bollworm (1) を見よ

英名	和名	学名	所属、分布、ほか
cotton springtail		*Entomobrya unostrigata* Stach	（トビムシ目、アヤトビムシ科）豪洲区
cotton square borer			gray hairstreak (1) を見よ
cotton stainer		*Dysdercus longirostris* Stål	（カメムシ目、ホシカメムシ科）新熱帯区。カンキツ害虫
cotton stainer		*Dysdercus suturellus* (Herrich-Schäffer)	（カメムシ目、ホシカメムシ科）新北区。ワタ、カンキツ害虫
cotton stainer		*Dysdercus andreae* (Linnaeus)	（カメムシ目、ホシカメムシ科）新北区
cotton stainer (1)		*Dysdercus sidae* Montrouzier	（カメムシ目、ホシカメムシ科）豪州区
cotton stainer			cotton bug を見よ
cotton stainers			red bugs (1)(2) を見よ
cotton stainers			red cotton bugs を見よ
cotton stalkgirdling beetle		*Rhyparida australis* (Boheman)	（コウチュウ目、ハムシ科）豪州区
cotton stem borer			coffee white borer を見よ
cotton stem borers		*Sphenoptera*	（コウチュウ目、タマムシ科）の昆虫の総称
cotton stem moth			large thatch groundling (1) を見よ
cotton thrips		*Frankliniella gossypii* Morgan	（アザミウマ目、アザミウマ科）新北区。ワタ害虫
cotton tipworm			hollyhock moth を見よ。cotton tipworm moth ともいう
cotton webspinner			weed web moth を見よ
cotton weevil			coffee weevil を見よ
cotton white scale		*Pinnaspis minor* (Maskell)	（カメムシ目、マルカイガラムシ科）新熱帯区
cotton white stem borer			coffee white borer を見よ
cotton whitefly			tobacco whitefly を見よ
cotton worm		*Aletia argillacea* Hübner	（チョウ目、ヤガ科）新北区。幼虫の英名
cotton worm			Egyptian cotton leafworm を見よ　幼虫の英名
cottongrass moth			Haworth's fanner を見よ
cotton-peach scale			large cottony scale (1) を見よ
cottonseed bug		*Oxycarenus luctuosus* (Montrouzier)	（カメムシ目、ナガカメムシ科）豪州区
cottonwood borer		*Plectrodera scalator* (Fabricius)	（コウチュウ目、カミキリムシ科）新北区
cottonwood borer			carpenter moth (1) を見よ
cottonwood crown-borer			American hornet moth を見よ
cottonwood dagger moth		*Acronicta lepusculina* (Guenée)	（チョウ目、ヤガ科）新北区。幼虫はポプラ害虫
cottonwood leaf beetle		*Chrysomela scripta* Fabricius	（コウチュウ目、ハムシ科）新北区
cottonwood leaf-miner		*Bucculatrix staintonella* Chambers	（チョウ目、チビガ科）新北区

英名	和名	学名	所属、分布、ほか
cottonwood leafminer moth		*Leucoptera albella* (Chambers)	(チョウ目、ハモグリガ科) 新北区
cottonwood leaf-mining beetle			poplar leafminer (2) を見よ
cotton-wood maple scale			woolly currant scale (1) を見よ　米国での英名
cottonwood scale			willow scale (1) を見よ　米国での英名
cottonwood twig borer		*Gypsonoma haimbachiana* (Kearfott)	(チョウ目、ハマキガ科) 新北区。cottonwood twig borer moth ともいう
cottony alder psyllid			woolly psyllid を見よ
cottony bamboo scale			bamboo scale (2) を見よ
cottony camellia scale			camellia scale (1) を見よ
cottony citrus scale (1)	ミカンワタカイガラムシ	*Pulvinaria aurantii* Cockerell	(カメムシ目、カタカイガラムシ科) 日本、旧北区。カンキツ害虫
cottony citrus scale (2)	ミカンヒメワタカイガラムシ	*Pulvinaria citricola* Kuwana	(カメムシ目、カタカイガラムシ科) 日本、新北区。カンキツ、カキなどの害虫
cottony citrus scale (3)	オキツワタカイガラムシ	*Pulvinaria okitsuensis* Kuwana	(カメムシ目、カタカイガラムシ科) 日本。カンキツ、チャなどの害虫
cottony citrus scale			pulvinaria scale (1)(2) を見よ
cottony cushion scale	イセリアカイガラムシ	*Icerya purchasi* Maskell	(カメムシ目、ワタフキカイガラムシ科) 日本、全北区、豪州区。カンキツ他果樹の大害虫。ベダリアテントウによる生物的防除で有名
cottony cushion scale killer		*Cryptochetum iceryae* (Williston)	(ハエ目、ヒゲブトコバエ科) 新北区
cottony flavicans scale		*Pulvinaria flavicans* Maskell	(カメムシ目、カタカイガラムシ科) 豪州区
cottony grape scale			cottony maple scale (1) を見よ
cottony grass scale		*Pulvinaria elongata* Newstead	(カメムシ目、カタカイガラムシ科) 豪州区
cottony hopbush scale		*Pulvinaria dodonaeae* Maskell	(カメムシ目、カタカイガラムシ科) 豪州区
cottony hydrangea scale			hydrangea scale を見よ
cottony loranthi scale		*Tectopulvinaria loranthi* Froggatt	(カメムシ目、カタカイガラムシ科) 豪州区
cottony maple leaf scale		*Pulvinaria acericola* (Walsh et Riley)	(カメムシ目、カタカイガラムシ科) 新北区
cottony maple scale		*Pulvinaria innumerabilis* (Rathvon)	(カメムシ目、カタカイガラムシ科) 新北区
cottony maple scale (1)		*Pulvinaria vitis* (Linnaeus)	(カメムシ目、カタカイガラムシ科) 全北区
cottony maple scale (2)	モミジワタカイガラムシ	*Lecanium horii* (Kuwana)	(カメムシ目、カタカイガラムシ科) 日本。果樹、樹木害虫
cottony maple scale			woolly currant scale (1) を見よ
cottony mulberry scale	クワワタカイガラムシ	*Pulvinaria kuwacola* Kuwana	(カメムシ目、カタカイガラムシ科) 日本。クワ、カキ、サクラなどの害虫
cottony peach scale			large cottony scale (1) を見よ

英名	和名	学名	所属、分布、ほか
cottony pigface scale		*Pulvinariella mesembryanthemi* (Vallot)	（カメムシ目、カタカイガラムシ科）豪州区
cottony salicornia scale		*Pulvinaria salicorniae* Froggatt	（カメムシ目、カタカイガラムシ科）豪州区
cottony saltbush scale		*Pulvinaria maskelli* Olliff	（カメムシ目、カタカイガラムシ科）豪州区
cottony scale of peach			woolly currant scale（2）を見よ
cottony sweetpotato scale			cottony urbicola scale を見よ
cottony Sydney scale		*Pulvinaria decorata* Borchsenius	（カメムシ目、カタカイガラムシ科）豪州区
cottony Thompson scale		*Pulvinaria thompsoni* Maskell	（カメムシ目、カタカイガラムシ科）豪州区
cottony urbicola scale		*Pulvinaria urbicola* Cockerell	（カメムシ目、カタカイガラムシ科）豪州区、大洋区
Cottrell's brown blue		*Lepidochrysops variabilis* Cottrell	（チョウ目、シジミチョウ科）エチオピア区
Cotundra blue		*Plebejus cotundra* Scott	（チョウ目、シジミチョウ科）新北区
Cotytto clearwing			broad-tipped clearwing を見よ
couch flea beetle		*Chaetocnema australica* (Baly)	（コウチュウ目、ハムシ科）豪州区
couchgrass scale			Ruth grass scale を見よ
couchgrass webworm		*Sclerobia tritalis* (Walker)	（チョウ目、メイガ科）豪州区
couchtip maggot		*Delia urbana* (Malloch)	（ハエ目、ハナバエ科）豪州区
coulee cricket		*Peranabrus scabricollis* (Thomas)	（バッタ目、キリギリス科）新北区
counts			barons を見よ
courtesan		*Euripus nyctelius* (Doubleday)	（チョウ目、タテハチョウ科）東洋区
courtesans		*Euripus*	（チョウ目、タテハチョウ科）の昆虫の総称
courtiers	キゴマダラ属	*Sephisa*	（チョウ目、タテハチョウ科）の昆虫の総称
cousin german		*Paradiarsia sobrina* (Duponchel)	（チョウ目、ヤガ科）旧北区
Coverdale's Anacampsis moth		*Anacampsis coverdalella* Kearfott	（チョウ目、キバガ科）新北区
cow dung sarcophagid		*Ravinia lherminieri* (Robineau-Desvoidy)	（ハエ目、ニクバエ科）大洋区
cow hornfly			horn fly (1) を見よ　米国での英名
cow killer			dragonflies (1) を見よ
cow killer			common eastern velvet ant を見よ
cow killers			velvet ants (1) を見よ
cowboy beetle		*Diaphonia dorsalis* (Donovan)	（コウチュウ目、コガネムシ科）豪州区
cowboy gemmed-satyr		*Cyllopsis caballeroi* Beutelspacher	（チョウ目、タテハチョウ科）新熱帯区
cowdung fly			yellow dung fly を見よ

英名	和名	学名	所属、分布、ほか
cow parsnip borer moth			Heracleum stem borer moth を見よ
cow-parsnip flat-body			parsnip moth (2) を見よ
cowpea aphid	マメアブラムシ	*Aphis craccivora* Koch	（カメムシ目、アブラムシ科）日本、汎世界。ダイズなどマメ類害虫
cowpea beetle			adzuki weevil を見よ
cowpea beetles			cowpea bruchids を見よ
cowpea bruchid			bean weevil を見よ　米国での英名
cowpea bruchid			adzuki weevil を見よ
cowpea bruchids		*Callosobruchus*	（コウチュウ目、ハムシ科）の昆虫の総称
cowpea curculio		*Chaleodermus aeneus* Boheman	（コウチュウ目、ゾウムシ科）新北区。マメ類害虫
cowpea seed beetle			cowpea weevil を見よ
cowpea weevil	ヨツモンマメゾウムシ	*Callosobruchus maculatus* (Fabricius)	（コウチュウ目、ハムシ科）日本、汎世界。貯蔵マメ類害虫
cowpea weevil			adzuki weevil を見よ
cowpea weevils			cowpea bruchids を見よ
coxcomb prominent		*Ptilodon capucina* (Linnaeus)	（チョウ目、シャチホコガ科）旧北区
coyote brush borer plume moth		*Hellinsia grandis* (Fish)	（チョウ目、トリバガ科）新北区
coyote brush gall moth		*Gnorimoschema baccharisella* Busck	（チョウ目、キバガ科）新北区
coyote brush twig borer moth		*Coleotechnites bacchariella* (Keifer)	（チョウ目、キバガ科）新北区
coyote cloudywing		*Achalarus toxeus* (Plötz)	（チョウ目、セセリチョウ科）新北区、新熱帯区
coyotero thread-legged katydid		*Arethaea coyotero* Hebard	（バッタ目、キリギリス科）新北区
crab leaf miner			apple pygmy を見よ
crab louse	ケジラミ	*Pthirus pubis* (Linnaeus)	（シラミ目、ケジラミ科）日本、全北区、豪州区。ヒトに寄生
crabgrass leaf beetle		*Oulema rufotincta* (Clark)	（コウチュウ目、ハムシ科）豪州区
crab-hole mosquito		*Deinocerites cancer* Theobald	（ハエ目、カ科）新北区
crablike rove beetles	シリボソハネカクシ亜科	Tachyporinae	（コウチュウ目、ハネカクシ科）の昆虫の総称
crabronid wasps		Crabronidae	（ハチ目）の昆虫の総称　Sphecidae の異名
crabs			crab louse を見よ
crab's-claw skipper		*Gorgythion vox* Evans	（チョウ目、セセリチョウ科）新熱帯区
crabweed case		*Coleophora saxicolella* (Duponchel)	（チョウ目、ツツミノガ科）旧北区
cracker grasshopper		*Trimerotropis vermiculatus* (Kirby)	（バッタ目、バッタ科）新北区
Cramer's blue Morpho			blue Morpho (1) を見よ
Cramer's eighty-eight			eighty eight butterfly (2) を見よ

英名	和名	学名	所属、分布、ほか
Cramer's green hairstreak			Janias greenstreak を見よ
Cramer's Mesene		*Mesene phareus* Cramer	（チョウ目、シジミチョウ科）新熱帯区
Cramer's metalmark		*Menander pretus picta* (Godman et Salvin)	（チョウ目、シジミタテハ科）新熱帯区
Cramer's midget		*Phyllonorycter harrisella* (Linnaeus)	（チョウ目、ホソガ科）旧北区
Cramer's midget		*Charis anius* (Cramer)	（チョウ目、シジミタテハ科）新熱帯区
Cramer's nightfighter			common scarlet-eye を見よ
Cramer's owlet		*Eryphanis automedon* (Cramer)	（チョウ目、タテハチョウ科）新熱帯区
Cramer's paradise skipper			mangrove skipper (2) を見よ
Cramer's proboscis skipper		*Saliana salius* (Cramer)	（チョウ目、セセリチョウ科）新熱帯区
Cramer's redwing		*Pyrrhogyra crameri* Aurivillius	（チョウ目、セセリチョウ科）新熱帯区
Cramer's ringlet		*Cissia myncea* (Cramer)	（チョウ目、タテハチョウ科）新熱帯区
Cramer's satyr		*Magneuptychia antonoe* (Cramer)	（チョウ目、タテハチョウ科）新熱帯区
Cramer's shoemaker		*Archaeoprepona licomedes* (Cramer)	（チョウ目、タテハチョウ科）新熱帯区
Cramer's sphinx moth		*Erinnyis crameri* (Schaus)	（チョウ目、スズメガ科）新北区、新熱帯区。Cramer's sphinx ともいう
Cramer's swallowtail		*Battus lycidas* (Cramer)	（チョウ目、アゲハチョウ科）新熱帯区
cranberry blackhead			black-headed fireworm を見よ
cranberry black-head fireworm			marbled single-dot bell を見よ　幼虫の英名
cranberry blossom worm			pointed sallow moth を見よ
cranberry blue	カラフトルリシジミ	*Vaccinina optilete daisetsuzana* (Matsumura)	（チョウ目、シジミチョウ科）日本。原名亜種の英名
cranberry blue			Yukon blue を見よ
cranberry fireworm		*Acrobasis vaccinii* Riley	（チョウ目、メイガ科）新北区
cranberry fireworm			black-headed fireworm を見よ　幼虫の英名
cranberry fireworm			holly tortrix を見よ
cranberry fritillary	コケモモヒョウモン	*Boloria aquilonaris* (Stichel)	（チョウ目、タテハチョウ科）旧北区
cranberry fruitworm			cranberry fireworm を見よ　cranberry fruitworm moth ともいう
cranberry girdle worm			garden grass-veneer を見よ
cranberry girdler		*Chrysoteuchia topiaria* (Zeller)	（チョウ目、メイガ科）新北区
cranberry girdler			garden grass-veneer を見よ
cranberry moth			chestnut moth を見よ
cranberry rootworm		*Rhabdopterus picipes* (Olivier)	（コウチュウ目、ハムシ科）新北区

英名	和名	学名	所属、分布、ほか
cranberry spanworm		*Anavitrinelia pampinaria* (Guenée)	（チョウ目、シャクガ科）新北区
cranberry span-worm			black-banded orange moth を見よ
cranberry spanworm moth		*Ematurga amitaria* (Guenée)	（チョウ目、シャクガ科）新北区
cranberry tree leaf beetle		*Galerucella viburni* Paykull	（コウチュウ目、ハムシ科）全北区
cranberry weevil		*Anthonomus musculus* Say	（コウチュウ目、ゾウムシ科）新北区
crane flies (1)	ガガンボ科	Tipulidae	（ハエ目）の昆虫の総称
crane flies	ガガンボ亜科	Tipulinae	（ハエ目、ガガンボ科）の昆虫の総称
crane flies (2)		*Tipula*	（ハエ目、ガガンボ科）の昆虫の総称
crane fly		*Tipula illustris* Doane	（ハエ目、ガガンボ科）新北区
crane fly		*Tipula longicornis* Schummel	（ハエ目、ガガンボ科）新北区
crane fly		*Erioptera caliptera* Say	（ハエ目、ガガンボ科）新北区、新熱帯区
crape myrtle aphid	サルスベリヒゲマダラアブラムシ	*Sarucallis kahawaluokalani* (Kirkaldy)	（カメムシ目,アブラムシ科）日本、全北区、東洋区、大洋区。サルスベリ害虫
Crassus swallowtail	マエオビアオジャコウ	*Battus crassus* (Cramer)	（チョウ目、アゲハチョウ科）新熱帯区
crataegus leaf aphid	サンザシハマキワタムシ	*Prociphilus crataegicola* Shinji	（カメムシ目、アブラムシ科）日本
crater termite		*Odontotermes badius* (Haviland)	（シロアリ目、シロアリ科）エチオピア区
Crathis tigerwing		*Olyras crathis staudingeri* Godman et Salvin	（チョウ目、タテハチョウ科）新熱帯区
Craw mealybug			bamboo scale (2) を見よ
crawler	シラミ		米国での俗称
crawling water beetle		*Haliplus fulvus* (Fabricius)	（コウチュウ目、コガシラミズムシ科）旧北区
crawling water beetles	コガシラミズムシ科	Haliplidae	（コウチュウ目）の昆虫の総称
Crawshay's caper white		*Belenois crawshayi* Butler	（チョウ目、シロチョウ科）エチオピア区
Crawshay's ciliate blue		*Anthene crawshayi* (Butler)	（チョウ目、シジミチョウ科）エチオピア区
Crawshay's sapphire		*Iolaus crawshayi littoralis* (Stempffer et Bennett)	（チョウ目、シジミチョウ科）エチオピア区
crazy ant	アシナガキアリ	*Anoplolepis longipes* (Jerdon)	（ハチ目、アリ科）日本、汎亜熱帯、汎熱帯
crazy ant		*Paratrechina fulva* (Mayr)	（ハチ目、アリ科）新熱帯区
crazy ant			hairy ant を見よ
crazy-spotted skipper		*Paratrytone rhexenor* Godman	（チョウ目、セセリチョウ科）新熱帯区
cream-banded Charaxes		*Charaxes hansali* C. et R. Felder	（チョウ目、タテハチョウ科）エチオピア区
cream-banded checkerspot			black checkerspot を見よ

英名	和名	学名	所属、分布、ほか
cream-banded dusky emperor		*Asterocampa idyja argus* (H. Bates)	(チョウ目、タテハチョウ科) 新北区
cream-banded nymph		*Manerebia inderena* (Adams)	(チョウ目、タテハチョウ科) 新熱帯区
cream-banded swallowtail		*Papilio leucotaenia* Rothschild	(チョウ目、アゲハチョウ科) エチオピア区
cream-banded swordtail	ヒトスジタイマイ	*Graphium illyris* (Hewitson)	(チョウ目、アゲハチョウ科) エチオピア区
cream-blotched cosmet		*Mompha propinquella* (Stainton)	(チョウ目、カザリバガ科) 旧北区
cream-bordered green			cream-bordered green pea を見よ
cream-bordered green pea		*Earias clorana* (Linnaeus)	(チョウ目、ヤガ科) 旧北区
cream-bordered green pea moth			cream-bordered green pea を見よ
cream-cloak apple shoot			eyespotted bud moth を見よ
cream-coloured argent		*Argyresthia curvella* (Linnaeus)	(チョウ目、メムシガ科) 旧北区
cream-edged Dichomeris moth		*Dichomeris flavocostella* (Clemens)	(チョウ目、キバガ科) 新北区
cream-lined swallowtail			Delalande's butterfly を見よ
cream pierid blue		*Larinopoda lircaea* (Hewitson)	(チョウ目、シジミチョウ科) エチオピア区
cream-shouldered groundling		*Teleiodes luculella* (Hübner)	(チョウ目、キバガ科) 旧北区
cream spot ladybird	シロジュウシホシテントウ	*Calvia quattuordecimguttata* (Linnaeus)	(コウチュウ目、テントウムシ科) 日本、旧北区
cream spot tiger		*Arctia villica* Linnaeus	(チョウ目、ヒトリガ科) 旧北区。cream spot tiger moth ともいう
creamspotted ichneumon		*Echthromorpha intricatoria* (Fabricius)	(ハチ目、ヒメバチ科) 豪州区
cream-spotted pigmy		*Stigmella trimaculella* (Haworth)	(チョウ目、モグリチビガ科) 旧北区
cream-spotted tigerwing		*Tithorea tarricina duenna* Bates	(チョウ目、タテハチョウ科) 新熱帯区
cream-spotted tigerwing			Tarricina tiger を見よ
cream-streaked lady beetle		*Harmonia quadripunctata* (Pontoppidan)	(コウチュウ目、テントウムシ科) 旧北区
cream striped hawk		*Theretra cajus* (Cramer)	(チョウ目、スズメガ科) エチオピア区
cream striped swordtail			coastal swordtail を見よ
cream-tipped swampdamsel		*Leptobasis melinogaster* Gonzalez-Soriano	(トンボ目、イトトンボ科) 新北区
cream wave		*Scopura floslactata* (Haworth)	(チョウ目、シャクガ科) 旧北区。日本亜種は *S. f. claudata* Prout ヤスジマルバヒメシャク

英名	和名	学名	所属、分布、ほか
creamy crescent			common crescent を見よ
creamy Graphium		*Graphium ucalegon* (Hewitson)	（チョウ目、アゲハチョウ科）エチオピア区
creamy hairstreak			jade-blue hairstreak を見よ
creamy hawk		*Praedora leucophaea* Rothschild et Jordan	（チョウ目、スズメガ科）エチオピア区
creamy marblewing		*Euchloe ausonides* (Lucas)	（チョウ目、シロチョウ科）新北区
creamy metalmark			firestreak Nymphidium を見よ
creamy silver spot		*Aphnaeus flavescens* Stempffer	（チョウ目、シジミチョウ科）エチオピア区
creamy stripe-streak			jade-blue hairstreak を見よ creamy stripestreak とも記される
creamy white		*Melete lycimnia isandra* (Boisduval)	（チョウ目、シロチョウ科）新熱帯区。
creek grasshopper		*Bermius odontocercus* Stål	（バッタ目、バッタ科）豪州区
creek pygmy grasshopper		*Paratettix* sp.	（バッタ目、ヒシバッタ科）豪州区
creeper			這うことから昆虫を指す
creeping pine woolly aphid	キタマツカサアブラムシ	*Pineus cembrae* (Cholodkovsky)	（カメムシ目、カサアブラムシ科）日本、旧北区。エゾマツ類害虫
creeping thistle lacebug			thistle lace bug を見よ
creeping water bug		*Ambrysus femoratus* Palisot de Beauvois	（カメムシ目、コバンムシ科）新北区
creeping water bugs	コバンムシ科	Naucoridae	（カメムシ目）の昆虫の総称
creepy-crawly			這う虫、昆虫。creeper を見よ
crenulate bark beetles	カワノキクイムシ亜科	Hylesininae	（コウチュウ目、キクイムシ科）の昆虫の総称
crenulate darkie		*Allotinus drumila* (Moore)	（チョウ目、シジミチョウ科）東洋区
crenulate moths	フタオガ科	Epiplemidae	（チョウ目）の昆虫の総称
crenulate mottle			crenulate darkie を見よ
crenulate oakblue		*Apporasa atkinsoni* (Hewitson)	（チョウ目、シジミチョウ科）東洋区
crenulate Temnora		*Temnora crenulata* (Holland)	（チョウ目、スズメガ科）エチオピア区
Creole			Creole pearly eye を見よ
Creole pearly eye		*Enodia creola* (Skinner)	（チョウ目、タテハチョウ科）新北区
creon skipper		*Creonpyge creon* (Druce)	（チョウ目、セセリチョウ科）新熱帯区
creosote bush bagworm		*Thyridopteryx mcadi* H. Edwards	（チョウ目、ミノガ科）新北区
creosote bush beetle		*Pyrota postica* LeConte	（コウチュウ目、ツチハンミョウ科）新北区
creosote bush grasshopper		*Bootettix punctatus* (Scudder)	（バッタ目、バッタ科）新北区
creosote bush katydid		*Insara covilleae* Rehn et Hebard	（バッタ目、キリギリス科）新北区
creosote moth		*Digrammia colorata* (Grote)	（チョウ目、シャクガ科）新北区

英名	和名	学名	所属、分布、ほか
creosote shieldback		*Eremopedes covilleae* Hebard	（バッタ目、キリギリス科）新北区
crepuscular rock-rose moth		*Teleiodes sequax* (Haworth)	（チョウ目、キバガ科）新北区
crescent	ショウブオオヨトウ	*Celaena leucostigma* (Hübner)	（チョウ目、ヤガ科）日本。イネ、ムギ、トウモロコシ害虫
crescent bluet		*Coenagrion lunulatum* (Charpentier)	（トンボ目、イトトンボ科）旧北区
crescent dart		*Agrotis trux* (Hübner)	（チョウ目、ヤガ科）旧北区
crescent groundling		*Calyocolum tricolorella* (Haworth)	（チョウ目、キバガ科）旧北区
crescent-marked lily aphid			mottled arum aphid を見よ　カナダでの英名
crescent metalmark		*Apodemia phyciodoides* Barnes et Benjamin	（チョウ目、シジミタテハ科）新北区
crescent moth			crescent を見よ
crescentspots	ミカズキタテハ属	*Phyciodes*	（チョウ目、タテハチョウ科）の昆虫の総称
crescent striped			crescent striped moth を見よ
crescent striped moth	イシカリヨトウ	*Apamea oblonga* (Haworth)	（チョウ目、ヤガ科）日本、旧北区
crescent-winged Caudellia moth		*Caudellia apyrella* Dyar	（チョウ目、メイガ科）新北区
Cresson's grass tubeworm moth		*Acrolophus cressoni* (Walsingham)	（チョウ目、Acrolophidae）新北区
crested bentwing		*Bucculatrix cristatella* (Zeller)	（チョウ目、チビガ科）旧北区
crested katydid		*Alectoria superba* Brunner von Wattenwyl	（バッタ目、キリギリス科）豪州区
Cretan Argus		*Plebejus psylorita* (Freyer)	（チョウ目、シジミチョウ科）旧北区
Cretan grayling		*Hipparchia cretica* (Rebel)	（チョウ目、タテハチョウ科）旧北区
Cretan heath		*Coenonympha thyrsis* (Freyer)	（チョウ目、タテハチョウ科）旧北区
Cretan small heath			Cretan heath を見よ
cribage hawk		*Hoplistopus penricei* Rothschild et Jordan	（チョウ目、スズメガ科）エチオピア区
cribrate weevil		*Otiorhynchus citricollis* Gyllenhal	（コウチュウ目、ゾウムシ科）新北区
cricket			house cricket を見よ
cricket-cockroaches			rock crawlers (1) を見よ
cricket grasshoppers			gryllacridoids を見よ
crickets (1)	コオロギ科	Gryllidae	（バッタ目）の昆虫の総称
crickets (2)		Saltatoria	（バッタ目）の昆虫の総称
crickets (3)	キリギリス亜科	Tettigoniinae	（バッタ目、キリギリス科）の昆虫の総称
crickets			long-horned grasshoppers (3) を見よ
crickets			bush crickets (1) を見よ
Cricula silkmoth		*Cricula trifenestrata* (Helfer)	（チョウ目、ヤママユガ科）東洋区

英名	和名	学名	所属、分布、ほか
crimson Acraea			blood-red Acraea をみよ
crimson-banded black	ヘリアカタテハ	*Biblis hyperia* (Cramer)	(チョウ目、タテハチョウ科) 新北区、新熱帯区
crimson-bodied lichen moth		*Lerina incarnata* Walker	(チョウ目、ヒトリガ科) 新北区
crimson darter			scarlet skimmer を見よ
crimson dropwing			crimson marsh glider を見よ
crimson Hawaiian damselfly		*Megalagrion leptodemas* (Perkins)	(トンボ目、イトトンボ科) 大洋区
crimson marsh glider	ベニトンボ	*Trithemis aurora* (Burmeister)	(トンボ目、トンボ科) 東洋区
crimson patch		*Chlosyne janais* (Drury)	(チョウ目、タテハチョウ科) 新熱帯区
crimson patch checkerspot			crimson patch を見よ
crimson-patched longwing		*Heliconius cerato cruentus* Lamas	(チョウ目、タテハチョウ科) 新熱帯区。*H. c. petiverana* (Doubleday) も同英名
crimson-patched longwing			common longwing を見よ
crimson ramble			bed bug (1) を見よ
crimson-ringed butterfly			Apollo を見よ
crimson rose	ヘクトールベニモンアゲハ	*Pachliopta hector* (Linnaeus)	(チョウ目、アゲハチョウ科) 東洋区
crimson speckled		*Utetheisa pulchella* (Linnaeus)	(チョウ目、ヒトリガ科) 旧北区
crimson speckled flunkey			crimson-speckled を見よ
crimson-speckled footman			crimson speckled を見よ
crimson-speckled moth			crimson speckled を見よ　米国での英名
crimson-spot hairstreak		*Thereus orasus* (Godman et Salvin)	(チョウ目、シジミチョウ科) 新熱帯区
crimson-spotted swallowtail		*Papilio menatius* Hübner	(チョウ目、アゲハチョウ科) 新熱帯区
crimson tip		*Colotis hetaera* (Gerstaecker)	(チョウ目、シロチョウ科) エチオピア区
crimson-tip			scarlet-tip を見よ
crinan ear		*Amphipoea crinanensis* (Burrows)	(チョウ目、ヤガ科) 旧北区
crinkled flannel moth		*Megalopyge crispata* (Packard)	(チョウ目、Megalopygidae) 新北区。幼虫はベリー類、リンゴを加害
crinkled flannel moth			yellow flannel moth (1) を見よ
crinkled leafwing	ウスベニキノハタテハ	*Anaea glycerium* Doubleday	(チョウ目、タテハチョウ科) 新北区、新熱帯区
crinum borer			lily borer を見よ　アフリカでの英名
criocerine leaf-beetles			asparagus beetles を見よ

英名	和名	学名	所属、分布、ほか
crisp skipper		*Mictris crispus* (Herrich-Schäffer)	（チョウ目、セセリチョウ科）新熱帯区
Crithona crescent		*Anthanassa crithona* (Salvin)	（チョウ目、タテハチョウ科）新熱帯区
crocea skipper		*Croitana croites* (Hewitson)	（チョウ目、セセリチョウ科）豪州区
crochet-hooked stiletto			common stiletto fly (1) を見よ
crocus geometer moth		*Xanthotype sospeta* (Drury)	（チョウ目、シャクガ科）新北区。北米での英名。crocus geometer ともいう
Croesus eyemark		*Semomesia croesus* (Fabricius)	（チョウ目、シジミタテハ科）新熱帯区
Cresus prominent		*Chliara cresus* (Cramer)	（チョウ目、シャチホコガ科）旧北区
Croker's Frother		*Rhodogastria crokeri* Macleay	（チョウ目、ヒトリガ科）豪州区
Crolinus hairstreak		*Tmolus crolinus* Butler et Druce	（チョウ目、シジミチョウ科）新熱帯区
crop mirid		*Sidnia kimbergi* (Stål)	（カメムシ目、カスミカメムシ科）豪州区
cross-line skipper		*Polites origenes* (Fabricius)	（チョウ目、セセリチョウ科）新北区
cross-lined wave moth		*Timandra amaturaria* Walker	（チョウ目、シャクガ科）新北区
cross-striped cabbage worm		*Evergestis rimosalis* (Guenée)	（チョウ目、メイガ科）新北区、豪州区。キャベツ害虫。cross-striped cabbageworm moth ともいう
cross-toothed rove beetles	オオキバハネカクシ亜科	Oxyporinae	（コウチュウ目、ハネカクシ科）の昆虫の総称
cross-winged tortoise beetle		*Helocassis crucipennis* (Boheman)	（コウチュウ目、ハムシ科）新熱帯区
Crossley's forest queen	ウスコモンマルバネタテハ	*Euxantha crossleyi* (Ward)	（チョウ目、タテハチョウ科）エチオピア区
Cross's wave moth		*Leptostales crossii* (Hulst)	（チョウ目、シャクガ科）新北区
crotalaria moth	タイワンベニゴマダラヒトリ	*Utetheisa lotrix* (Cramer)	（チョウ目、ヒトリガ科）日本、東洋区、豪州区、エチオピア区
crotalaria pod borer			orange sun moth を見よ
croton bug			German cockroach を見よ
croton bugs			cockroaches (2) を見よ
croton caterpillar		*Achaea janata* (Linnaeus)	（チョウ目、ヤガ科）新北区、豪州区、大洋区
croton chaff scale		*Parlatoria crotonis* (Douglas)	（カメムシ目、マルカイガラムシ科）大洋区
croton mussel scale	クロトンカキカイガラムシ	*Lepidosaphes tokionis* (Kuwana)	（カメムシ目、マルカイガラムシ科）日本(小笠原)、新北区、汎熱帯。クロトン害虫
croton weed crown borer		*Acalolepta argentata* (Aurivillius)	（コウチュウ目、カミキリムシ科）豪州区
croton weed gall fly			pamakani gall fly を見よ
croton whitefly		*Orchamoplatus mammaeferus* (Quaintancer et Baker)	（カメムシ目、コナジラミ科）大洋区
crow butterflies			crows を見よ

英名	和名	学名	所属、分布、ほか
crow egg-fly			Malayan egg-fly を見よ
crow-feather case		*Metriotes lutarea* (Haworth)	(チョウ目、ツツミノガ科) 旧北区
crow louse		*Myrsidea cornicis* (De Geer)	(ハジラミ目、タンカクトリハジラミ科) 旧北区
Crowley's Epitola		*Cerautola crowleyi* (Sharpe)	(チョウ目、シジミチョウ科) エチオピア区
Crowley's Euptera		*Euptera crowleyi centralis* Libert	(チョウ目、タテハチョウ科) エチオピア区
crown			coronet を見よ
crown girdler			strawberry root weevil (1) を見よ
crown silver spot		*Aphnaeus coronae* Talbot	(チョウ目、シジミチョウ科) エチオピア区
crown stick insect		*Onchestus rentzi* Brock et Hasenpusch	(ナナフシ目、ナナフシ科) 豪州区
crown whitefly		*Aleuroplatus coronata* (Quaintance)	(カメムシ目、コナジラミ科) 新北区
crowned hairstreak		*Evenus coronata* (Hewitson)	(チョウ目、シジミチョウ科) 新熱帯区
crowned Phlyctaenia moth			garden elder pearl を見よ
crowned slug moth		*Isa textula* (Herrich-Schäffer)	(チョウ目、イラガ科) 新北区
crows	ルリマダラ属	*Euploea*	(チョウ目、タテハチョウ科) の昆虫の総称
crucifer borer		*Baris chlorizans* Germar	(コウチュウ目、ゾウムシ科) 旧北区
crucifer borer		*Baris laticollis* (Marsham)	(コウチュウ目、ゾウムシ科) 旧北区
crucifer caterpillar	ナノメイガ	*Evergestis forficalis* (Linnaeus)	(チョウ目、メイガ科) 日本、旧北区、東洋区、エチオピア区。アブラナ科野菜害虫
crucifer flea beetle			turnip flea beetle (2) を見よ
crucifer leafminer	アブラナハモグリバエ	*Liriomyza brassicae* (Riley)	(ハエ目、ハモグリバエ科) 日本、東洋区、新北区。アブラナ科野菜害虫
crucifer midge			cabbage gall midge を見よ
crucifer shield bug			cabbage bug (2) を見よ
cruciferous looper		*Xanthorhoe saturata* (Guenée)	(チョウ目、シャクガ科) 日本、東洋区。アブラナ科野菜害虫
crucifex ground beetle		*Panagaeus crux-major* (Linnaeus)	(コウチュウ目、オサムシ科) 旧北区
cruel Toxonprucha moth		*Toxonprucha crudelis* (Grote)	(チョウ目、ヤガ科) 新北区
cruiser	ヒメチャイロタテハ	*Vindula arsinoe* (Cramer)	(チョウ目、タテハチョウ科) 豪州区。*V. a. ada* (Butler) も同英名
cruiser		*Vindula dejone erotella* (Butler)	(チョウ目、タテハチョウ科) 東洋区。Erichson's cruiser を参照
cruiser			cruiser butterfly を見よ
cruiser butterfly	チャイロタテハ	*Vindula erota* (Fabricius)	(チョウ目、タテハチョウ科) 東洋区
cruisers			river skimmers (2) を見よ
crusader bug		*Mictis profana* (Fabricius)	(カメムシ目、ヘリカメムシ科) 豪州区。カンキツ害虫

英名	和名	学名	所属、分布、ほか
crustwax whitefly			orange whitefly を見よ
Cryphia moth		*Cryphia cuerva* (Barnes)	（チョウ目、ヤガ科）新北区
cryptic fritillary			Pleistocene fritillary を見よ
cryptic hairstreak		*Ostrinotes* sp.	（チョウ目、シジミチョウ科）新熱帯区
cryptic Mylon		*Mylon cajus* Plötz	（チョウ目、セセリチョウ科）新熱帯区
cryptic Remella		*Remella vopiscus* (Herrich-Schäffer)	（チョウ目、セセリチョウ科）新北区、新熱帯区
cryptic scarlet-eye		*Venada* sp.	（チョウ目、セセリチョウ科）新熱帯区
cryptic skipper		*Noctuana lactifera bipuncta* (Plötz)	（チョウ目、セセリチョウ科）新熱帯区
cryptic spreadwing		*Lestes dissimulans* Fraser	（トンボ目、アオイトトンボ科）エチオピア区
cryptic-winged shieldback		*Eremopedes cryptoptera* (Rehn et Hebard)	（バッタ目、キリギリス科）新北区
cryptochetid flies	ヒゲブトコバエ科	Cryptochetidae	（ハエ目）の昆虫の総称
cryptochetids			cryptochetid flies を見よ
cryptomeria bark borer	スギカミキリ	*Semanotus japonicus* (Lacordaire)	（コウチュウ目、カミキリムシ科）日本、東洋区。スギ、ヒノキなどの害虫
cryptomeria bark midge	スギザイノタマバエ	*Resseliella odai* (Inouye)	（ハエ目、タマバエ科）日本。スギ害虫
cryptomeria bud miner	スギメムシガ	*Argyresthia anthocephala* Meyrick	（チョウ目、メムシガ科）日本。スギ害虫
cryptomeria cone moth	スギカサガ	*Cydia cryptomeriae* (Issiki)	（チョウ目、ハマキガ科）日本。スギ害虫
cryptomeria needle gall midge	スギタマバエ	*Contarinia inouei* Mani	（ハエ目、タマバエ科）日本。スギ害虫
cryptomeria torymid	スギモンオナガコバチ	*Megastigmus cryptomeriae* Yano	（ハチ目、オナガコバチ科）日本
cryptomeria twig borer	スギノアカネトラカミキリ	*Anaglyptus subfasciatus* Pic	（コウチュウ目、カミキリムシ科）日本。スギ、ヒノキなどの害虫
cryptomeria webworm	スギハマキ	*Homona issikii* Yasuda	（チョウ目、ハマキガ科）日本、東洋区。スギ害虫
cryptophagid beetles			silken fungus beetles を見よ
cryptorufes dotted blue		*Euphilotes enoptes cryptorufes* Pratt et Emmel	（チョウ目、シジミチョウ科）新熱帯区
crypts		*Cryptolaemus*	（コウチュウ目、テントウムシ科）の昆虫の総称
crystal-winged skipper		*Onenses hyalophora* (Felder)	（チョウ目、セセリチョウ科）新北区、新熱帯区
Ctemene dartwhite	タカネマダラシロチョウ	*Catasticta ctemene* Hewitson	（チョウ目、シロチョウ科）新熱帯区
ctenophorine craneflies		Ctenophorinae	（ハエ目、ガガンボ科）の昆虫の総称
ctenuchas			wasp moths を見よ
ctenuchid moths			wasp moths を見よ
ctenuchids			wasp moths を見よ
Cuban bluet		*Enallagma truncatum* (Gundlach)	（トンボ目、イトトンボ科）新熱帯区

英名	和名	学名	所属、分布、ほか
Cuban cockroach		*Panchlora nivea* (Linnaeus)	（ゴキブリ目、ブラベルスゴキブリ科）新北区。バナナに多し
Cuban crescent	ホソオビミカズキタテハ	*Anthanassa frisia* (Poey)	（チョウ目、タテハチョウ科）新北区、新熱帯区
Cuban crescentspot			Cuban crescent を見よ
Cuban fly		*Lixophaga diatraeae* (Townsend)	（ハエ目、ヤドリバエ科）新北区、新熱帯区
Cuban ground cricket		*Neonemobius cubensis* (Saussure)	（バッタ目、コオロギ科）新北区
Cuban kite swallowtail		*Eurytides celadon* (Lucas)	（チョウ目、アゲハチョウ科）新北区
Cuban laurel thrips	ガジュマルクダアザミウマ	*Gynaikothrips ficorum* (Marchal)	（アザミウマ目、クダアザミウマ科）日本、汎世界
Cuban laurel thrips predator		*Montandoniola moraguesi* (Puton)	（カメムシ目、ハナカメムシ科）大洋区
Cuban leaf cutting ant		*Atta insularis* Guérin	（ハチ目、アリ科）新熱帯区。カンキツ害虫
Cuban peacock		*Anartia chrysopelea* (Hübner)	（チョウ目、タテハチョウ科）新北区、新熱帯区
Cuban snout		*Libytheana motya* (Hübner)	（チョウ目、タテハチョウ科）新北区、新熱帯区
Cuban spanworm moth		*Oxydia cubana* (Warren)	（チョウ目、シャクガ科）新北区
Cuban sphinx			Sagra sphinx moth を見よ
Cuban sphinx moth		*Manduca brontes cubensis* (Kitching et Cadiou)	（チョウ目、スズメガ科）新北区、新熱帯区
Cuban trig		*Anaxipha imitator* (Saussure)	（バッタ目、クサヒバリ科）新北区
Cuban white-backed rice planthopper		*Sogatodes cubanus* (Crawford)	（カメムシ目、ウンカ科）新北区。イネ害虫
cuckoo bees		*Coelioxys*	（ハチ目、ハキリバチ科）の昆虫の総称
cuckoo bees		*Sphecodes*	（ハチ目、コハナバチ科）の昆虫の総称
cuckoo bees (1)		Nomadinae	（ハチ目、コシブトハナバチ科）の昆虫の総称
cuckoo bees (2)		Psithyrinae	（ハチ目、ミツバチ科）の昆虫の総称　とくに *Psithyrus* の種
cuckoo bees			long-tongued bees を見よ
cuckoo bees			spotted cuckoo bees を見よ
cuckoo bees			digger bees (1) を見よ。主に Nomadini, Epeolini の種
cuckoo bumblebee		*Psithyrus vestalis* (Geoffroy)	（ハチ目、ミツバチ科）旧北区
cuckoo-feather slender		*Caloptilia ligustrinella* (Zeller)	（チョウ目、ホソガ科）旧北区
cuckoo-feather slender (1)		*Caloptilia cuculipennella* (Hübner)	（チョウ目、ホソガ科）日本、旧北区
cuckoo-flies			inquiline gall wasps を見よ
cuckoo spit			spittle bugs (2) を見よ

英名	和名	学名	所属、分布、ほか
cuckoo spit			cuckoo spit insect を見よ
cuckoo spit insect	ホソアワフキ	*Philaenus spumarius* (Linnaeus)	(カメムシ目、アワフキムシ科) 日本、全北区。イネ科牧草の害虫
cuckoo spit insects			spittle bugs (1) を見よ
cuckoo spittle			spittle bugs (2) を見よ
cuckoo wasp		*Trichrysis tridens* Lepeletier	(ハチ目、セイボウ科) 新北区
cuckoo wasp	ミツバセイボウ	*Chrysis cyanea* Linnaeus	(ハチ目、セイボウ科) 日本、東洋区、旧北区
cuckoo wasp (1)	ヤドリスズメバチ	*Vespula austriaca* (Panzer)	(ハチ目、スズメバチ科) 日本、全北区
cuckoo wasps		Ceropalidae	(ハチ目) の昆虫の総称
cuckoo wasps (1)	セイボウ科	Chrysididae	(ハチ目) の昆虫の総称
cuckoo wasps			chrysidoid wasps を見よ
cuckoospit			spittle bugs (2) を見よ
cuckoospit			cuckoo spit insect を見よ
cucubano		*Pyrophorus luminosa* Illiger	(コウチュウ目、コメツキムシ科) 新熱帯区。成虫は発光する
cucujid beetles			flat bark beetles (1) を見よ
cucujo		*Pyrophorus noctilucus* Linnaeus	(コウチュウ目、コメツキムシ科) 新熱帯区。成幼虫とも発光する
cucullina ringlet		*Hermeuptychia cucullina* (Weymer)	(チョウ目、タテハチョウ科) 新熱帯区
cucumber beetles	ヒゲナガハムシ亜科	Galerucinae	(コウチュウ目、ハムシ科) の昆虫の総称
cucumber-feeder ladybeetle	トホシテントウ	*Epilachna admirabilis* Crotch	(コウチュウ目、テントウムシ科) 日本、東洋区
cucumber fly		*Dacus cucumis* French	(ハエ目、ミバエ科) 豪州区
cucumber looper	ウリキンウワバ	*Anadevidia peponis* (Fabricius)	(チョウ目、ヤガ科) 日本、旧北区、東洋区、豪州区
cucumber moth		*Diaphania bivitralis* (Guenée)	(チョウ目、メイガ科) 東洋区
cucumber moth			cotton caterpillar (1) を見よ
cucumber root maggots			mushroom flies を見よ
cucurbit fly			Ethiopian fruit fly を見よ
cucurbit ladybird		*Epilachna cucurbitae* Richards	(コウチュウ目、テントウムシ科) 豪州区
cucurbit ladybird		*Epilachna argus* (Geoffroy)	(コウチュウ目、テントウムシ科) 旧北区、エチオピア区。南アフリカでの英名
cucurbit leaf beetle	ウリハムシモドキ	*Aulacophora femoralis* (Motschulsky)	(コウチュウ目、ハムシ科) 日本、旧北区。トウモロコシ、ダイズ、ウリ類、アブラナ科野菜などの害虫
cucurbit longicorn		*Apomecyna saltator* (Fabricius)	(コウチュウ目、カミキリムシ科) 新北区、大洋区
cucurbit longhorn			cucurbit longicorn を見よ
cucurbit looper			cucumber looper を見よ
cucurbit midge		*Prodiplosis citrulli* (Felt)	(ハエ目、タマバエ科) 新北区

英名	和名	学名	所属、分布、ほか
cucurbit shield bug		*Megymenjum affine* Boisduval	（カメムシ目、Dinidoridae）豪州区
cucurbit stem borer	ヨツスジシラホシサビカミキリ	*Apomecyna histrio* (Fabricius)	（コウチュウ目、カミキリムシ科）日本、東洋区、豪州区
cucurbit thrips			honeysuckle thrips を見よ
cudweed		*Cucullia gnaphalii* (Hübner)	（チョウ目、ヤガ科）旧北区。cudweed moth ともいう
cudweed grasshopper		*Hypochlora alba* (Dudge)	（バッタ目、バッタ科）新北区
cudweed shark			cudweed を見よ
culex			house mosquitoes を見よ　複数形は culices
culicid			カ科に属する虫、カを指す
cullum's humble bee		*Bombus cullumanus* Kirby	（ハチ目、ミツバチ科）旧北区
Culminicola hairstreak		*Penaincisalia culminicola* (Staudinger)	（チョウ目、シジミチョウ科）新熱帯区
cultivated silk worm			silkworm を見よ
culture louse			white snow flea を見よ
Culver's root borer moth		*Papaipema sciata* Bird	（チョウ目、ヤガ科）新北区
Cumberland gem	マダラムラサキヨトウ	*Eucarta amethystina* (Hübner)	（チョウ目、ヤガ科）旧北区。イングランド北西部の地名に由来
cup moth			mottled cup moth を見よ
cup moths			slug caterpillar moths を見よ
cup moths			slug moths を見よ
cup plant borer moth		*Papaipema polymniae* Bird	（チョウ目、ヤガ科）新北区
Cupavia skipperling		*Dalla cupavia* (Mabille)	（チョウ目、セセリチョウ科）新熱帯区
cupedids			reticulated beetles を見よ
Cupenthus hairstreak		*Megathecla cupentus* (Stoll)	（チョウ目、シジミチョウ科）新熱帯区
cupid		*Euchrysops cnejus cnidus* Waterhouse	（チョウ目、シジミチョウ科）豪州区
cupid			gram blue を見よ　豪州での英名
cupid dart moth		*Abagrotis cupida* (Grote)	（チョウ目、ヤガ科）新北区。cupid dart ともいう
cupped spangle			oak leaf cupped-gall cynipid がカシ類の葉につくる有毛の虫えい
cuppery click beetle	コガネコメツキ	*Apholistus puncticolis* (Motschulsky)	（コウチュウ目、コメツキムシ科）日本
Cuprea ant-blue		*Acrodipsas cuprea* (Sands)	（チョウ目、シジミチョウ科）豪州区
cupreous blue		*Eicochrysops messapus* (Godart)	（チョウ目、シジミチョウ科）エチオピア区。*E. m. nandiana* (Bethune-Baker) も同英名
cupreous chafer	ドウガネブイブイ	*Anomala cuprea* (Hope)	（コウチュウ目、コガネムシ科）日本。多くの作物、果樹、花木などの害虫
cupreous flattened click beetle	ドウガネヒラタコメツキ	*Corymbitodes gratus* (Lewis)	（コウチュウ目、コメツキムシ科）日本
cupreous hairtail		*Anthene princeps* (Butler)	（チョウ目、シジミチョウ科）エチオピア区

英名	和名	学名	所属、分布、ほか
cupreous leaf beetle	ドウガネツヤハムシ	*Oomorphoides cupreatus* (Baly)	（コウチュウ目、ハムシ科）日本、旧北区
cupreous polished chafer	リュウキュウツヤハナムグリ	*Protaetia pryeri* (Janson)	（コウチュウ目、コガネムシ科）日本
cupreous polished chafer			drone beetle を見よ
cupreous Protea butterfly		*Capys cupreus* Henning et Henning	（チョウ目、シジミチョウ科）エチオピア区
cupressus aphid			cypress aphid (1) を見よ
curculio	ゾウムシ		（コウチュウ目、ゾウムシ科）の昆虫。weevils (1) を見よ
curculio beetle			apple weevil (1) を見よ
curculios			weevils (1) を見よ
curenulate-winged grasshopper		*Cordillacris crenulata* (Bruner)	（バッタ目、バッタ科）新北区
curious ciliate blue		*Anthene locuples* (Grose-Smith)	（チョウ目、シジミチョウ科）エチオピア区
curl grub			African black beetle を見よ
curled dock leaf-miner		*Pegomyia bicolor* Wiedemann	（ハエ目、ハナバエ科）旧北区
curled rose sawfly			banded rose sawfly を見よ　北米での英名
curled rose slug			banded rose sawfly を見よ　米国での幼虫の英名
curler flat-body		*Exaeretia ciniflonella* (Lienig et Zeller)	（チョウ目、マルハキバガ科）旧北区
Curle's brown		*Stygionympha curlei* Henning et Henning	（チョウ目、タテハチョウ科）エチオピア区
curlewbug			southern corn billbug を見よ　curlew-bug とも記す
currant and gooseberry borer			currant clearwing を見よ
currant and gooseberry fruit fly			currant fruit fly を見よ
currant and gooseberry moth			magpie を見よ
currant and gooseberry sawfly			gooseberry sawfly (1) を見よ
currant and lettuce aphid		*Dactynotus sonchi* (Linnaeus)	（カメムシ目、アブラムシ科）旧北区
currant and lettuce aphid			currant-lettuce aphid を見よ
currant aphid	スグリトックリアブラムシ	*Cryptomyzus ribis* (Linnaeus)	（カメムシ目、アブラムシ科）日本、新北区。currant スグリ に由来
currant aphid			black currant aphid (2) を見よ
currant aphid			currant-lettuce aphid を見よ
currant aphid			permanent currant aphid を見よ
currant blister aphid			currant aphid を見よ
currant blister aphid			black currant aphid (2) を見よ

英名	和名	学名	所属、分布、ほか
currant borer			spotted bronze bright を見よ
currant borer			currant clearwing を見よ　米国での英名
currant borer moth			currant clearwing を見よ
currant bud moth		*Stathmopoda chalcotypa* Meyrick	（チョウ目、ニセマイコガ科）豪州区
currant clearwing		*Synanthedon tipuliformis* (Clerck)	（チョウ目、スカシバガ科）新北区。幼虫がスグリの幹に食入
currant clearwing moth		*Synanthedon salmachus* Linnaeus	（チョウ目、スカシバガ科）全北区、豪州区
currant clearwing moth			currant clearwing を見よ
currant fig moth			almond moth を見よ
currant fruit fly		*Epochra canadensis* (Loew)	（ハエ目、ミバエ科）新北区
currant fruit moth			greater brown cloaked bell を見よ　米国での英名
currant fruit weevil		*Pseudanthonomus validus* Dietz	（コウチュウ目、ゾウムシ科）新北区
currant fruitworm		*Carposina fernaldana* Busck	（チョウ目、シンクイガ科）新北区。スグリ害虫。currant fruitworm moth ともいう
currant-gall			lenticular gall wasp の虫えいの英名
currant-lettuce aphid	チシャミドリアブラムシ	*Hyperomyzus lactucae* (Linnaeus)	（カメムシ目、アブラムシ科）日本、全北区、豪州区、エチオピア区。レタス害虫
currant-lettuce aphid			lettuce aphid (1) を見よ
currant-lettuce aphid			gooseberry aphid (2) を見よ
currant mealybug			citrophilus mealybug (1) を見よ
currant midge			gooseberry midge を見よ
currant moth			magpie を見よ
currant moth			tobacco moth (1) を見よ
currant plant-louse			currant aphid を見よ
currant pug		*Eupithecia assimilata* Doubleday	（チョウ目、シャクガ科）旧北区
currant root aphid			elm leaf aphid (1) を見よ
currant sawfly			gooseberry sawfly (1) を見よ
currant shoot borer			spotted bronze bright を見よ　currant shoot borer moth ともいう
currant-snowthistle aphid			currant-lettuce aphid を見よ
currant spanworm		*Hame ribearia* (Fitch)	（チョウ目、シャクガ科）新北区。currant spanworm moth ともいう
currant spinach			spinach moth を見よ
currant stem aphid			black currant aphid (1) を見よ
currant stem girdler		*Janus integer* (Norton)	（ハチ目、クキバチ科）新北区
currant stem girdler			social pear sawfly (1) を見よ
currant twist	サクラトビハマキ	*Pandemis ribeana* Hübner	（チョウ目、ハマキガ科）旧北区。currant twist moth ともいう
currant worm			magpie を見よ　幼虫の英名

英名	和名	学名	所属、分布、ほか
currant worm			gooseberry sawfly（1）を見よ　幼虫の英名
currant-yellow rattle aphid			northern currant aphid を見よ
Curtis's ash-bud ermel			ash bud moth を見よ
Curtis's plain plume		*Leioptilus tephradactyla* Hübner	（チョウ目、トリバガ科）旧北区
Curtomerus long-horned beetle		*Curtomerus flavus* (Fabricius)	（コウチュウ目、カミキリムシ科）大洋区
curtonotid flies		Curtonotidae	（ハエ目）の昆虫の総称
curve-dentated bark beetle	キョクシキクイムシ	*Pityokteines curvidens* (German)	（コウチュウ目、キクイムシ科）日本、旧北区
curve-horned chafer	ヒゲコガネ	*Polyphylla laticollis* Lewis	（コウチュウ目、コガネムシ科）日本
curve-lined angle moth		*Digrammia continuata* Walker	（チョウ目、シャクガ科）新北区
curve-lined Argyria moth		*Vaxi auratella* (Clemens)	（チョウ目、メイガ科）新北区
curve-lined bird-dropping moth		*Tarache terminimaculata* (Grote)	（チョウ目、ヤガ科）新北区
curve-lined Cydosia moth		*Cydosia nobilitella* Cramer	（チョウ目、ヤガ科）新北区
curve-lined looper moth			hemlock looper (2) を見よ
curve-lined owlet moth		*Phyprosopus callitrichoides* Grote	（チョウ目、ヤガ科）新北区。curve-lined owlet ともいう
curve-lined Theope		*Theope bacenis* Schaus	（チョウ目、シジミタテハ科）新熱帯区
curve-lined Vaxi			curve-lined Argyria moth を見よ
curve-tailed bush cricket			bush katydid を見よ
curve-tailed earwig		*Labia curvicauda* (Motschulsky)	（ハサミムシ目、チビハサミムシ科）旧北区
curve-toothed geometer		*Eutrapela clemataria* J. E. Smith	（チョウ目、シャクガ科）新北区。curve-toothed geometer moth ともいう
curve-winged cotton moth		*Pyroderces pyrrhodes* Meyrick	（チョウ目、カザリバガ科）豪州区
curve-winged Emesis			curve-winged metalmark を見よ
curve-winged metalmark		*Emesis emesia* (Hewitson)	（チョウ目、シジミタテハ科）新北区
curved-bar grass yellow			fairy yellow を見よ
curved halter moth		*Capis curvata* Grote	（チョウ目、ヤガ科）新北区
curved-line Agonopterix moth		*Agonopterix curvilinella* (Beutenmüller)	（チョウ目、マルハキバガ科）新北区
cushion scale			camellia scale (1) を見よ
cushion scale			pear mealybug を見よ
cushow lady			seven-spot ladybird を見よ
Cutter's forester		*Bebearia cutteri* (Hewitson)	（チョウ目、タテハチョウ科）エチオピア区

英名	和名	学名	所属、分布、ほか
cutworm		*Euxoa henrietta* (Smith)	(チョウ目、ヤガ科) 新北区
cutworm		*Parastichtis purpurea* var. *crispa* Harvey	(チョウ目、ヤガ科) 新北区。カンキツ害虫
cutworm (1)		*Xylomyges curialis* Grote	(チョウ目、ヤガ科) 新北区。カンキツ害虫
cutworm (2)	カブラヤガ	*Agrotis segetum* (Denis et Schiffermüller)	(チョウ目、ヤガ科) 日本、旧北区。多くの作物、タバコ、チャ、観賞用植物の害虫
cutworm (3)	オオカブラヤガ	*Agrotis tokionis* Butler	(チョウ目、ヤガ科) 日本、旧北区。タマネギ、トウモロコシの害虫
cutworm			spotted cutworm moth (1) を見よ
cutworm moths			owlet moths を見よ
cutworm wasp		*Podalonia luctuosa* (Smith)	(ハチ目、アナバチ科) 新北区
cutworms (1)		*Agrotis, Euxoa, Feltia*	(チョウ目、ヤガ科) の昆虫の総称　幼虫の英名。ネキリムシ、ヨトウムシといわれる
cutworms			ligustrum moth を見よ
cutworms			darts and rustics を見よ　幼虫の英名
cutworms			owlet moths を見よ　幼虫の英名
cutworms			surface caterpillars を見よ　幼虫の英名
Cuvier's purplewing		*Eunica cuvieri* (Godart)	(チョウ目、タテハチョウ科) 新熱帯区
cyan-banded Perisama		*Perisama dorbignyi jurinei* (Guenée)	(チョウ目、タテハチョウ科) 新熱帯区
Cyan emperor		*Doxocopa cyane* (Latreille)	(チョウ目、タテハチョウ科) 新熱帯区
Cyan leafwing		*Memphis praxias* (Hopffer)	(チョウ目、タテハチョウ科) 新熱帯区
Cyananthe blue wing		*Myscelia cyananthe streckeri* Skinner	(チョウ目、タテハチョウ科) 新北区。dark wave を参照
Cyane jewel		*Hypochrysops cyane* (Waterhouse et Lyell)	(チョウ目、シジミチョウ科) 豪州区
cyanophyllum scale	シュロマルカイガラムシ	*Abgrallaspis cyanophylli* (Signoret)	(カメムシ目、マルカイガラムシ科) 日本、新北区、汎熱帯。ラン害虫
cycad blue	キヤムラシジミ	*Chilades kiamurae* Matsumura	(チョウ目、シジミチョウ科) 日本
cycad blue			plains cupid を見よ
cycad blue		*Theclinesthes onycha* (Hewitson)	(チョウ目、シジミチョウ科) 豪州区
cycad butterfly	マルバネカラスシジミ	*Eumaeus minyas* (Hübner)	(チョウ目、シジミチョウ科) 新熱帯区
cyclamen borer			black vine weevil を見よ
cyclamen tortrix		*Clepsis spectrana* (Treitschke)	(チョウ目、ハマキガ科) 旧北区。シクラメン、イチゴ他の害虫
cyclamen weevil			black vine weevil を見よ
cyclbalanopsis scale	クリシロカイガラムシ	*Pseudaulacaspis kiushiuensis* (Kuwana)	(カメムシ目、マルカイガラムシ科) 日本、東洋区
Cyclops sootywing		*Bolla cyclops* (Mabille)	(チョウ目、セセリチョウ科) 新熱帯区
cyclorrhaphous flies			circular-seamed flies を見よ
Cyda hairstreak		*Micandra cyda* (Godman et Salvin)	(チョウ目、シジミチョウ科) 新熱帯区

英名	和名	学名	所属、分布、ほか
Cyda mottled-skipper		*Codatractus cyda* (Godman)	（チョウ目、セセリチョウ科）新熱帯区
cydnid bugs			ground bugs を見よ
Cydno longwing		*Heliconius cydno* (Doubleday)	（チョウ目、タテハチョウ科）新熱帯区。cydno ともいう
Cydonia metalmark		*Chamaelimnas cydonia* Stichel	（チョウ目、シジミタテハ科）新熱帯区
Cydrara hairstreak		*Tmolus cydrara* (Hewitson)	（チョウ目、シジミチョウ科）新熱帯区
Cyledis mottled-skipper		*Codatractus cyledis* (Dyar)	（チョウ目、セセリチョウ科）新熱帯区
cylindrical auger beetle		*Xylion cylindricus* Macleay	（コウチュウ目、ナガシンクイムシ科）豪州区
cylindrical bark beetles	ホソカタムシ科	Colydiidae	（コウチュウ目）の昆虫の総称
cylindrical hardwood borer			red-headed ash borer を見よ
cylindrical leaf beetles			case-bearing leaf beetles (2) を見よ
cylindrical weevils	カツオゾウムシ亜科	Cleoninae	（コウチュウ目、ゾウムシ科）の昆虫の総称
cylindrotomid flies	シリブトガガンボ亜科	Cylindrotominae	（ハエ目、ガガンボ科）の昆虫の総称
cymbalaria aphid		*Myzus cymbalariae* Stroyan	（カメムシ目、アブラムシ科）豪州区
cymbidium scale	タブカキカイガラムシ	*Lepidosaphes machili* (Maskell)	（カメムシ目、マルカイガラムシ科）日本、旧北区。シキミなどの害虫
cymbidium scale			orchid scale を見よ
cymbidium thrips			dendrobium thrips を見よ
cyna blue		*Zizula cyna* (Edwards)	（チョウ目、シジミチョウ科）新北区、新熱帯区
Cynea skipper		*Cynea cynea* (Hewitson)	（チョウ目、セセリチョウ科）新熱帯区
Cyneas checkerspot		*Chlosyne cyneas* (Godman et Salvin)	（チョウ目、タテハチョウ科）新北区
cynical quaker moth		*Orthodes cynica* Guenée	（チョウ目、ヤガ科）新北区
cynipid gall wasps			gall wasps (1) を見よ
cynipid wasps			gall wasps (2) を見よ
cynipids			gall wasps (1) を見よ
cynone skipper		*Anisynta cynone* Hewitson	（チョウ目、セセリチョウ科）豪州区
cynophyllum scale		*Aspidiotus cyanophylli* Signoret	（カメムシ目、マルカイガラムシ科）新北区。温室害虫
Cynosura eighty-eight			BD butterfly を見よ
cynthea moth			eri-silkworm moth を見よ
cynthia moth		*Samia cynthia* (Drury)	（チョウ目、ヤママユガ科）全北区、東洋区、豪州区。日本亜種は *S. c. preyeri* (Butler) シンジュサン
cynthia silkmoth			cynthia moth を見よ

英名	和名	学名	所属、分布、ほか
Cynthia's fritillary		*Euphydryas cynthia* (Schiffermüller)	（チョウ目、タテハチョウ科）旧北区
cypress aphid (1)		*Cinara cupressi* (Buckton)	（カメムシ目、アブラムシ科）旧北区
cypress aphid (2)		*Cinara fresia* (Blanchard)	（カメムシ目、アブラムシ科）全北区、大洋区
cypress bark beetle		*Phloeosinus cupressi* Hopkins	（コウチュウ目、キクイムシ科）豪州区
cypress bark beetle	ヒノキノキクイムシ	*Phloeosinus rudis* Blandford	（コウチュウ目、キクイムシ科）日本、旧北区。スギ、ヒノキなどの害虫
cypress bark borer		*Physocnemum andreae* (Haldemann)	（コウチュウ目、カミキリムシ科）新北区
cypress bark moth	ヒノキカワムグリガ	*Epinotia granitalis* (Butler)	（チョウ目、ハマキガ科）日本。スギ、ヒノキ害虫
cypress bark weevil		*Aesiotes leucurus* Pascoe	（コウチュウ目、ゾウムシ科）豪州区
cypress carpet moth		*Thera cupressata* (Geyer)	（チョウ目、シャクガ科）旧北区。cypress carpet ともいう
cypress clubtail		*Gomphus minutus* Rambur	（トンボ目、サナエトンボ科）新北区
cypress cockroach			Pacific beetle cockroach を見よ
cypress cone moth		*Laspeyresia cupressana* (Kearfott)	（チョウ目、ハマキガ科）新北区
cypress emerald moth		*Nemoria elfa* Ferguson	（チョウ目、シャクガ科）新北区
cypress flower gall midge		*Taxodiomyia cupressi* (Schweinitz)	（ハエ目、タマバエ科）新北区
cypress gall midge		*Contarinia rugosa* Gagné	（ハエ目、タマバエ科）大洋区
cypress girdler			beetle roach を見よ
cypress jewel beetle		*Diadoxus regius* Peterson	（コウチュウ目、タマムシ科）豪州区
cypress katydids		*Inscudderia*	（バッタ目、キリギリス科）の昆虫の総称
cypress leafminer	ヒノキハモグリガ	*Argyresthia chamaecypariae* Moriuti	（チョウ目、メムシガ科）日本。ヒノキ害虫
cypress longicorn		*Tritocosmia latecostata* Fairmaire	（コウチュウ目、カミキリムシ科）豪州区
cypress looper		*Anacamptodes pergracilis* Hulst	（チョウ目、シャクガ科）新北区。cypress looper moth ともいう
cypress moth		*Argyresthia cupressella* Walsingham	（チョウ目、メムシガ科）新北区
cypress moths			shiny head-standing moths を見よ
cypress pine aphid			arborvitae aphid を見よ
cypress pine sawfly		*Zenarge turneri* Rohwer	（ハチ目、ミフシハバチ科）豪州区
cypress pug		*Eupithecia phoeniceata* (Rambur)	（チョウ目、シャクガ科）旧北区
cypress sawfly	サワラハバチ	*Monoctenus itoi* Okutani	（ハチ目、マツハバチ科）日本。ヒノキ害虫
cypress seed chalcid		*Megastigmus wachtii* Mayr	（ハチ目、オナガコバチ科）旧北区
cypress sphinx			bald cypress sphinx moth を見よ
cypress tip moth			cypress moth を見よ
cypress twig gall midge		*Taxodiomyia cupressiananassa* (Osten Sacken)	（ハエ目、タマバエ科）新北区

英名	和名	学名	所属、分布、ほか
Cypria metalmark			orange-barred Emesis を見よ
Cyprian honey bee		*Apis mellifica cyprica* Pollman	（ハチ目、ミツバチ科）．旧北区。キプロスに由来
Cypris Morpho	キプリスモルフォ	*Morpho cypris* Westwood	（チョウ目、タテハチョウ科）新熱帯区
Cyprotus blue	アカガネミナミシジミ	*Candalides cyprotus* (Olliff)	（チョウ目、シジミチョウ科）豪州区。Cyprotus blue-butterfly ともいう
cypselosomatidae flies		Cypselosomatidae	（ハエ目）の昆虫の総称
cyrano darner		*Nasiaeschna pentacantha* (Rambur)	（トンボ目、ヤンマ科）新北区
Cyril's brown		*Argynnina cyrila* Waterhouse et Lyell	（チョウ目、タテハチョウ科）豪州区
Cytherea sister			smooth-banded sister を見よ

英名	和名	学名	所属、分布、ほか
D			
d'Abrera's pigmy		*Baeotis staudingeri* d'Abrera	（チョウ目、シジミタテハ科）新熱帯区
d'Abrera's tiger	デブレライアサギマダラ	*Parantica dabrerai* (Miller et Miller)	（チョウ目、タテハチョウ科）東洋区
dactylopid scales			cochineal insects を見よ
dadap borer			Erythrina borer moth を見よ　dadap はデイゴのインドネシア語名。
daddy long water legs			crane flies (1) を見よ
daddy long-legs			hanging flies を見よ
daddy long-legs			crane flies (1) を見よ　ザトウムシ目の動物の総称でもある
Daedalus shieldback		*Pediodectes daedalus* Rehn et Hebard	（バッタ目、キリギリス科）新北区
daffodil fly			narcissus bulb fly を見よ　daffodil ラッパスイセンに由来
dagger-marked cutworm moth		*Persectania atristirga* Walker	（チョウ目、ヤガ科）豪州区
dagger-marked scoparia moth		*Scoparia rotuella* Felder	（チョウ目、メイガ科）豪州区
dagger moth (1)		*Acronicta americana* (Harris)	（チョウ目、ヤガ科）新北区
dagger moth (2)		*Acronicta auricoma* (Denis et Schiffermüller)	（チョウ目、ヤガ科）全北区
dagger moths	ケンモンヤガ亜科	Acronictinae	（チョウ目、ヤガ科）の昆虫の総称
dagger-tails			daggerwings を見よ
daggerwings	ツルギタテハ属	*Marpesia*	（チョウ目、タテハチョウ科）の昆虫の総称
daggers		Apatelinae	（チョウ目、ヤガ科）の昆虫の総称
dahuria bark beetle	シラカバノキクイムシ	*Scolytus dahuricus* (Chapuis)	（コウチュウ目、キクイムシ科）日本、旧北区
daikon flea beetle			cabbage flea beetle (2) を見よ
daikon leaf beetle			brassica leaf beetle を見よ
daikon weevil	ダイコンサルゾウムシ	*Ceuthorhynchidius albosuturalis* (Roelofs)	（コウチュウ目、ゾウムシ科）日本、旧北区
daimio ambrosia beetle	ダイミョウキクイムシ	*Scolytoplatypus daimio* Blandford	（コウチュウ目、キクイムシ科）日本、旧北区。大名に由来
daimyo bark beetle			daimio ambrosia beetle を見よ
daimyo oak aphid			oak spined aphid を見よ
daimyo oak bark beetle	カシワノキクイムシ	*Trypodendron signatum* (Fabricius)	（コウチュウ目、キクイムシ科）日本、旧北区
daimyo oak curculio	ハイイロチョッキリ	*Mechoris ursulus* Roelofs	（コウチュウ目、オトシブミ科）日本、旧北区
daimyo oak cutworm	カシワオビキリガ	*Conistra ardescens* (Butler)	（チョウ目、ヤガ科）日本
daimyo oak leaf beetle	カシワツツハムシ	*Cryptocephalus scitulus* Baly	（コウチュウ目、ハムシ科）日本
daimyo oak leaf roller	カバカギバヒメハマキ	*Ancylis partifana* (Christoph)	（チョウ目、ハマキガ科）日本、旧北区

英名	和名	学名	所属、分布、ほか
daimyo oak leaf worm			Hebrew character を見よ
daimyo oak leafhopper	カシハトムネヨコバイ	*Macropsis quercus* (Matsumura)	(カメムシ目、オオヨコバイ科) 日本
daimyo oak sucker			lacquer psylla を見よ
daimyo oak tussock moth			oak tussock moth を見よ
daimyo oak yellow-dotted noctuid	カシワキボシキリガ	*Lithophane pruinosa* (Butler)	(チョウ目、ヤガ科) 日本、旧北区
dainty coenagrion			dainty damselfly を見よ
dainty damselfly		*Coenagrion scitulum* (Rambur)	(トンボ目、イトトンボ科) 旧北区
dainty egg white		*Dismorphia lysis peruana* (Hewitson)	(チョウ目、シロチョウ科) 新熱帯区
dainty grass-blue			tiny grass blue (1) を見よ
dainty gray moth		*Glena plumosaria* (Packard)	(チョウ目、シャクガ科) 新北区
dainty sulphur	チビキチョウ	*Nathalis iole* Boisduval	(チョウ目、シロチョウ科) 新北区
dainty swallowtail	ホシボシアゲハ	*Papilio anactus* (Macleay)	(チョウ目、アゲハチョウ科) 豪州区
dainty white sailor		*Dynamine agacles* Dalman	(チョウ目、タテハチョウ科) 新熱帯区
daisy thrips			red clover thrips を見よ
Dakota skipper		*Hesperia dacotae* (Skinner)	(チョウ目、セセリチョウ科) 新北区
Dakotah dash		*Polites dakotah* (Edwards)	(チョウ目、セセリチョウ科) 新北区
dalcerid moths			tropical slug caterpillars をみよ
Dale's oak clearwing		*Synanthedon conopiformis* (Esper)	(チョウ目、スカシバガ科) 旧北区
Dalmatian ringlet		*Proterebia afra* (Fabricius)	(チョウ目、タテハチョウ科) 旧北区
d'Almeida's glasswing		*Ithomia arduinna* d'Almeida	(チョウ目、タテハチョウ科) 新熱帯区
Damara copper		*Aloeides damarensis* (Trimen)	(チョウ目、シジミチョウ科) エチオピア区
Damel's blue		*Jalmenus daemeli* (Semper)	(チョウ目、シジミチョウ科) 豪州区。Damel's blue butterfly ともいう
Dame's violet moth			grey-streaked smudge を見よ
dammar bees			stingless bees (1) を見よ　dammar 樹脂 に由来
Dammer's dotted blue		*Euphilotes enoptes dammersi* (Comstock et Henne)	(チョウ目、シジミチョウ科) 新熱帯区。Dammer's blue ともいう
Dammer's planthopper		*Loxophora dammersi* Van Duzee	(カメムシ目、テングスケバ科) 新北区
damnacanthus sawfly	アリドウシハバチ	*Nematus damnacanti* (Takeuchi)	(ハチ目、ハバチ科) 日本
Damo hairstreak			sky-blue hairstreak を見よ
damoiselle			demoiselle Agrion を見よ
damoiselle Agrion			demoiselle Agrion を見よ
damoiselle flies			damselflies を見よ

英名	和名	学名	所属、分布、ほか
Damon blue		*Polyommatus damon* (Schiffermüller)	（チョウ目、シジミチョウ科）旧北区
damp barklice		Epipsocidae	（チャタテムシ目）の昆虫の総称
damp barklouse		*Propsocus pulchripennis* (Perkins)	（チャタテムシ目、ケチャタテモドキ科）大洋区、豪州区
damp-loving field cricket		*Gryllus alogus* Rehn	（バッタ目、コオロギ科）新北区
dampwood borer		*Hadrobregmus australiensis* Pic	（コウチュウ目、シバンムシ科）豪州区
dampwood termite		*Porotermes adamsoni* (Froggatt)	（シロアリ目、オオシロアリ科）豪州区
dampwood termite	オオシロアリ	*Hodotermopsis japonica* Holmgren	（シロアリ目、オオシロアリ科）日本。腐木などに営巣
dampwood termites		*Prohinotermes*	（シロアリ目、ミゾガシラシロアリ科）の昆虫の総称
dampwood termites		*Zootermopsis*	（シロアリ目、シュウカクシロアリ科）の昆虫の総称
damp-wood termites		*Neotermes, Paraneotermes*	（シロアリ目、レイビシロアリ科）の昆虫の総称
dampwood termites			drywood termites (1) を見よ
dampwood termites			subterranean termites (1) を見よ
damsel bug			Pacific damsel bug を見よ
damsel bug			common damsel bug (1) を見よ
damsel bug			field damsel bug を見よ
damsel bug			marsh damsel bug を見よ
damsel bugs	マキバサシガメ科	Nabidae	（カメムシ目）の昆虫の総称　捕食性
damselflies	イトトンボ亜目	Zygoptera	（トンボ目）の昆虫の総称　damsel fly といわれるトンボ
damselflies			broad-winged damselflies を見よ
damselfly		*Megalagrion mauka* Daigle	（トンボ目、イトトンボ科）大洋区
damson aphis			damson-hop aphid を見よ
damson-hop aphid		*Phorodon humuli* (Schrank)	（カメムシ目、アブラムシ科）全北区、エチオピア区。プラム害虫
Dana blue			dingy lineblue を見よ
danaid eggfly	メスアカムラサキ	*Hypolimnas misippus* (Linnaeus)	（チョウ目、タテハチョウ科）日本、東洋区、豪州区、新熱帯区、エチオピア区。斑紋が多型なので Danaid eggflies と複数形にすることもある
Danaids			monarchs を見よ
dance flies			long-legged flies (2) を見よ　水面でおどるように飛翔することに由来
dance fly		*Empis tesselata* Fabricius	（ハエ目、オドリバエ科）旧北区
dance fly		*Hilara maura* Fabricius	（ハエ目、オドリバエ科）旧北区
dancer flies			long-legged flies (2) を見よ
dancers		*Onychargia*	（トンボ目、イトトンボ科）の昆虫の総称
dancers (1)		*Argia*	（トンボ目、イトトンボ科）の昆虫の総称
dancing Acraea			small orange Acraea (2) を見よ

英名	和名	学名	所属、分布、ほか
dancing dropwing		*Trithemis pallidinervis* (Kirby)	(トンボ目、トンボ科) 東洋区
dancing jewel		*Platycypha caligata* (Selys)	(トンボ目、Chlorocyphidae) エチオピア区
dancing moth		*Dryadaula terpsichorella* (Busck)	(チョウ目、ヒロズコガ科) 新北区、大洋区
dancing swordtail		*Graphium polistratus* (Grose-Smith)	(チョウ目、アゲハチョウ科) エチオピア区
dandelion gall wasp		*Phanacis taraxaci* (Ashmead)	(ハチ目、タマバチ科) 新北区。タンポポに由来
dandelion thrips		*Ceratothrips frici* (Uzel)	(アザミウマ目、アザミウマ科) 豪州区
Dannfelt's monarch		*Amauris dannfelti* Aurivillius	(チョウ目、タテハチョウ科) エチオピア区
Danube clouded yellow	ミルミドーネモンキチョウ	*Colias myrmidone* (Esper)	(チョウ目、シロチョウ科) 旧北区
dapdap caterpillar		*Streblote dorsalis* Walker	(チョウ目、カレハガ科) 東洋区
Daphne's opal		*Chrysoritis daphne* (Dickson)	(チョウ目、シジミチョウ科) エチオピア区
daphniphyllus bark beetle	ユズリハノキクイムシ	*Xyleborus volvulus* (Fabricius)	(コウチュウ目、キクイムシ科) 日本、東洋区
dappled marble	アウソニアツマキチョウ	*Euchloe ausonia* (Hübner)	(チョウ目、シロチョウ科) 旧北区
dappled monarch		*Danaus limniace petiverana* Doubleday	(チョウ目、タテハチョウ科) エチオピア区
dappled monarch			African blue tiger を見よ
dappled white		*Euchloe simplonia* (Freyer)	(チョウ目、シロチョウ科) 旧北区
dappled white			dappled marble を見よ
Dardaris skipperling		*Dardarina dardaris* (Hewitson)	(チョウ目、セセリチョウ科) 新熱帯区
Dardus Eurybia			Dardus underleaf を見よ
Dardus underleaf		*Eurybia dardus* (Fabricius)	(チョウ目、シジミタテハ科) 新熱帯区
Dares skipper		*Pompeius dares* (Plötz)	(チョウ目、セセリチョウ科) 新熱帯区
daring owl-butterfly		*Dynastor darius stygianus* Butler	(チョウ目、タテハチョウ科) 新熱帯区
dark Apamea			black-dashed Apamea moth を見よ
dark apple redbug		*Heterocordylus malinus* Reuter	(カメムシ目、カスミカメムシ科) 新北区。リンゴ他害虫
dark arches		*Apamea monoglypha* (Hufnagel)	(チョウ目、ヤガ科) 旧北区
dark banded ace		*Halpe ormenes vilasina* Fruhstorfer	(チョウ目、セセリチョウ科) 東洋区
dark-banded carpet moth		*Cidaria deltoidata* Walker	(チョウ目、シャクガ科) 豪州区
dark-banded fireworm			broken-banded leafroller moth を見よ
dark-banded geometer moth		*Ecliptopera atricolorata* (Grote et Robinson)	(チョウ目、シャクガ科) 新北区。dark-banded geometer ともいう
dark-banded owlet moth		*Phalaenophana pyramusalis* (Walker)	(チョウ目、ヤガ科) 新北区。dark-banded owlet ともいう

英名	和名	学名	所属、分布、ほか
dark-banded Pyrausta moth		*Pyrausta insignitalis* (Guenée)	（チョウ目、メイガ科）新北区
dark-banded scarlet		*Axiocerses croesus* (Trimen)	（チョウ目、シジミチョウ科）エチオピア区
dark banded Temnora		*Temnora atrofasciata* (Holland)	（チョウ目、スズメガ科）エチオピア区
dark bark-borer moth		*Gymnobathra omphalota* Meyrick	（チョウ目、マルハキバガ科）豪州区
dark-barred grey marble		*Endothenia pullana* (Haworth)	（チョウ目、ハマキガ科）旧北区
dark-barred twin-spot carpet			dark twin-spot carpet を見よ
dark blue jungle glory	ルリワモンチョウ	*Thaumantis klugius lucipor* Westwood	（チョウ目、タテハチョウ科）東洋区
dark blue lady beetle		*Curinus coeruleus* Mulsant	（コウチュウ目、テントウムシ科）大洋区
dark blue pansy			blue pansy (1) を見よ
dark blue policeman		*Coeliades bixana* Evans	（チョウ目、セセリチョウ科）エチオピア区
dark blue royal		*Pratapa icetas* (Hewitson)	（チョウ目、シジミチョウ科）東洋区
dark blue tiger	コモンマダラ	*Tirumala septentrionis* (Butler)	（チョウ目、タテハチョウ科）東洋区
dark blue tiger (1)	ミナミコモンマダラ	*Tirumala hamata* (Macleay)	（チョウ目、タテハチョウ科）旧北区、東洋区、豪州区
dark bluet			Norfolk damselfly を見よ
dark bordered beauty		*Epione parallelaria* Schiffermüller	（チョウ目、シャクガ科）旧北区
dark bordered beauty		*Epione vespertaria* (Linnaeus)	（チョウ目、シャクガ科）旧北区
dark-bordered granite moth		*Digrammia neptaria* (Guenée)	（チョウ目、シャクガ科）新北区
dark bordered straw			bordered straw を見よ
dark-brand bushbrown	マルバヒトツメコジャノメ	*Mycalesis mineus* (Linnaeus)	（チョウ目、タテハチョウ科）旧北区、東洋区。*M. m. macromalayana* Fruhstorfer も同英名
dark branded swift		*Caltoris brunnea* (Snellen)	（チョウ目、セセリチョウ科）東洋区
dark brindled clothes moth		*Monopis rusticella* (Hübner)	（チョウ目、ヒロズコガ科）旧北区
dark brindled-brown groundling		*Bryotropha affinis* (Haworth)	（チョウ目、キバガ科）旧北区
dark brocade		*Blepharita adusta* (Esper)	（チョウ目、ヤガ科）旧北区
dark brown ant			black ant (1) を見よ　米国での英名
dark brown banded leaf-cut weevil	ウスモンオトシブミ	*Apoderus balteatus* Roelofs	（コウチュウ目、オトシブミ科）日本
dark brown-headed rice stem borer		*Chilo diffusillioneus* (De Joannis)	（チョウ目、メイガ科）エチオピア区
dark brown forester		*Euphaedra losinga losinga* (Hewitson)	（チョウ目、タテハチョウ科）エチオピア区
dark brown missile		*Meza mabea* (Holland)	（チョウ目、セセリチョウ科）エチオピア区
dark-brown Scoparia moth		*Scoparia penumbralis* Dyar	（チョウ目、メイガ科）新北区

英名	和名	学名	所属、分布、ほか
dark buckeye (1)		*Junonia genoveva nigrosuffusa* Barnes et McDunnough	(チョウ目、タテハチョウ科) 新北区
dark buckeye (2)	ナンベイタテハモドキ	*Junonia evarete nigrosuffusa* Barnes et McDunnough	(チョウ目、タテハチョウ科) 新北区、新熱帯区。West Indian buckeye を参照
dark bush-cricket			bush cricket を見よ
dark Calephelis		*Calephelis velutina* (Godman et Salvin)	(チョウ目、シジミタテハ科) 新熱帯区
dark catseye		*Zipaetis scylax* Hewitson	(チョウ目、タテハチョウ科) 東洋区。dark catseye butterfly ともいう
dark Cerulean (1)		*Jamides bochus* (Stoll)	(チョウ目、シジミチョウ科) 東洋区。日本のルリウラナミシジミ *J. b. ishigakianus* Shirozu, *J. b. nabonassar* Fruhstorfer も同英名
dark Cerulean (2)		*Jamides phaseli* (Mathew)	(チョウ目、シジミチョウ科) 豪州区
dark Ceylon 6-lineblue		*Nacaduba calauria* (Felder)	(チョウ目、シジミチョウ科) 東洋区、豪州区
dark-cheeked scarlet-eye		*Bungalotis astylos* (Cramer)	(チョウ目、セセリチョウ科) 新熱帯区
dark chestnut		*Conistra ligula* (Esper)	(チョウ目、ヤガ科) 旧北区
dark chopper		*Gonometa postica* (Walker)	(チョウ目、カレハガ科) エチオピア区
dark ciliate blue		*Anthene seltuttus* (Rober)	(チョウ目、シジミチョウ科) 豪州区。*A. s. affinis* (Waterhouse et Turner) も同英名
dark clay case		*Coleophora saturatella* Stainton	(チョウ目、ツツミノガ科) 旧北区
dark clouded yellow			clouded yellow を見よ
dark Cornelian			Cornelian を見よ
dark crescent piercer		*Cydia internana* Guenée	(チョウ目、ハマキガ科) 旧北区
dark crimson porina moth		*Oxycanus enysii* Butler	(チョウ目、コウモリガ科) 豪州区
dark crimson underwing		*Catocala sponsa* (Linnaeus)	(チョウ目、ヤガ科) 旧北区
dark cupid		*Tongeia potanini glycon* Corbet	(チョウ目、シジミチョウ科) 東洋区
dark dagger	ヒメリンゴケンモン	*Triaena sugii* Kinoshita	(チョウ目、ヤガ科) 日本
dark dagger (1)		*Acronicta tridens* (Denis et Schiffermüller)	(チョウ目、ヤガ科) 旧北区
dark dagger moth			dark dagger (1) を見よ
dark daggerwing		*Marpesia themistocles* (Fabricius)	(チョウ目、タテハチョウ科) 新熱帯区
dark dart			black cutworm を見よ
dark darter	タケアカセセリ	*Telicota ohara* (Plötz)	(チョウ目、セセリチョウ科) 豪州区
dark Doberes		*Doberes anticus* (Plötz)	(チョウ目、セセリチョウ科) 新熱帯区
dark-edged bee-fly			bee fly (1) を見よ
dark-edged Eusarca moth		*Eusarca fundaria* (Guenée)	(チョウ目、シャクガ科) 新北区

英名	和名	学名	所属、分布、ほか
dark edged snow flat		*Tagiades menaka manis* Evans	（チョウ目、セセリチョウ科）東洋区。spotted snow flat を参照
dark elaterid beetle			obscure click beetle を見よ
dark elfin			dusted elfin を見よ
dark emerald damselfly		*Lestes macrostigma* (Eversmann)	（トンボ目、アオイトトンボ科）旧北区
dark Emesis		*Emesis ocypore aethalia* H. Bates	（チョウ目、シジミタテハ科）新熱帯区。dingy Emesis を参照
dark emperor		*Anax chloromelas* Ris	（トンボ目、ヤンマ科）エチオピア区
dark evehawker		*Anaciaeschna donaldi* Fraser	（トンボ目、ヤンマ科）東洋区
dark evening brown	クロコノマチョウ	*Melanitis phedima oitensis* Matsumura	（チョウ目、タテハチョウ科）日本。*M. phedima* (Cramer) の英名。*M. p. abdullae* Distant も同英名
dark falcate Emesis		*Emesis tegula* Godman et Salvin	（チョウ目、シジミタテハ科）新熱帯区
dark fishfly		*Nigronia serricornis* (Say)	（アミメカゲロウ目、ヘビトンボ科）新北区。翅は黒色に白の斑紋
dark flour beetle		*Tribolium destructor* Uyttenboogaart	（コウチュウ目、ゴミムシダマシ科）旧北区
dark flower scarab			spangled flower beetle を見よ
dark forest-blue			Eone blue を見よ
dark forestwraith		*Platysticta apicalis* Kirby	（トンボ目、Platystictidae）東洋区
dark-fringed mournful duskywing		*Erynnis tristis pattersoni* Burns	（チョウ目、セセリチョウ科）新熱帯区
dark fruit fly	ウスグロミバエ	*Anastrepha serpentina* (Wiedemann)	（ハエ目、ミバエ科）新北区、新熱帯区
dark fruit-tree tortrix			brown tortrix を見よ
dark fruit tree tortrix moth			brown tortrix を見よ
dark gemmed-satyr		*Cyllopsis whiteorum* Miller et Maza	（チョウ目、タテハチョウ科）新熱帯区
dark giant horsefly			horse fly (2) を見よ
dark glassy tiger		*Parantica agleoides* (C. et R. Felder)	（チョウ目、タテハチョウ科）東洋区
dark-glittered threadtail		*Elattoneura centralis* (Hagen)	（トンボ目、ミナミイトトンボ科）東洋区
dark grass blue		*Zizina antanossa* (Mabille)	（チョウ目、シジミチョウ科）エチオピア区。clover blue ともいう
dark grass blue (1)	ハマヤマトシジミ	*Zizeeria karsandra* (Moore)	（チョウ目、シジミチョウ科）日本、旧北区、東洋区、豪州区
dark grass blue			African grass blue を見よ
dark gray comma			dark grey anglewing を見よ
dark gray lichen moth		*Crambidia lithosioides* Dyar	（チョウ目、ヒトリガ科）新北区
dark grayling		*Pseudochazara mniszechii* Herrich-Schäffer	（チョウ目、タテハチョウ科）旧北区
dark green apple capsid		*Orthotylus marginalis* Reiter	（カメムシ目、カスミカメムシ科）旧北区

英名	和名	学名	所属、分布、ほか
dark green-banded blue		*Nothodanis schaeffera* (Eschscholtz)	(チョウ目、シジミチョウ科) 豪州区
dark green fritillary		*Speyeria aglaja* (Linnaeus)	(チョウ目、タテハチョウ科) 旧北区。日本亜種は *S. a. basaluis* (Matsumura) ギンボシヒョウモン
dark green grain aphid			corn leaf aphid を見よ
dark grey anglewing	シベリアキタテハ	*Polygonia progne* (Cramer)	(チョウ目、タテハチョウ科) 新北区
dark grey carpet moth		*Microdes squamulata* Guenée	(チョウ目、シャクガ科) 豪州区
dark grey cutworm			cutworm (3) を見よ
dark-grey virgin smoke		*Luffia ferchaultella* Stephens	(チョウ目、ミノガ科) 旧北区
dark-headed aspen leafroller moth		*Anacampsis innocuella* (Zeller)	(チョウ目、キバガ科) 新北区
dark-headed rice borer		*Chilo polychrysus* (Meyrick)	(チョウ目、メイガ科) 東洋区、豪州区。イネ害虫
dark-headed stem borer			dark-headed rice borer を見よ
dark headed striped borer			dark-headed rice borer を見よ
dark Himalayan oakblue	ラマムラサキシジミ	*Arhopala rama* (Kollar)	(チョウ目、シジミチョウ科) 日本、東洋区。
dark-hindwinged prominent	クロシタシャチホコ	*Mesophalera sigmata* (Butler)	(チョウ目、シャチホコガ科) 日本、旧北区
dark Hottentot			pigmy skipper を見よ　dark Hottentot skipper ともいう
dark hunter		*Austrogomphus bifurcatus* Tillyard	(トンボ目、サナエトンボ科) 豪州区
dark-inlaid grass-veneer		*Crambus lathoniellus* Zincken	(チョウ目、メイガ科) 旧北区
dark-inlaid grass-veneer			hook-streak grass veneer を見よ
dark Jerusalem cricket		*Stenopelmatus nigrocapitatus* Tinkham et Rentz	(バッタ目、Stenopelmatidae) 新北区
dark Jezebel		*Delias berinda* (Moore)	(チョウ目、シロチョウ科) 東洋区
dark Judy	ホシボシシジミタテハ	*Abisara fylla* (Westwood)	(チョウ目、シジミタテハ科) 旧北区、東洋区
dark jungle glory		*Thaumantis noureddin noureddin* Westwood	(チョウ目、タテハチョウ科) 東洋区
dark kite swallowtail			dark zebra swallowtail を見よ
dark knot-horn	シロスジマダラメイガ	*Assara terebrella* (Zincken)	(チョウ目、メイガ科) 日本、旧北区
dark leaf sitter		*Gorgyra diversata* Evans	(チョウ目、セセリチョウ科) エチオピア区
dark Lestes		*Lestes congener* Hagen	(トンボ目、アオイトトンボ科) 新北区
dark Leuconycta			marbled-green Leuconycta moth を見よ
dark lilac nymph		*Sallya amulia intermedia* Carcasson	(チョウ目、タテハチョウ科) エチオピア区

英名	和名	学名	所属、分布、ほか
dark line blue		*Pseudonacaduba aethiops* (Mabille)	(チョウ目、シジミチョウ科) エチオピア区
dark long-cloaked marble		*Apotomis sauciana* (Frölich)	(チョウ目、ハマキガ科) 旧北区
dark manuka moth		*Declana junctilinea* Walker	(チョウ目、シャクガ科) 豪州区
dark Marathyssa moth		*Marathyssa inficita* (Walker)	(チョウ目、ヤガ科) 新北区
dark marbled carpet	ツマキナカジロナミシャク	*Dysstroma citratum* (Linnaeus)	(チョウ目、シャクガ科) 日本。dark marbled carpet moth ともいう
dark mealworm			dark mealworm beetle を見よ
dark mealworm beetle	コメノゴミムシダマシ	*Tenebrio obscurus* Fabricius	(コウチュウ目、ゴミムシダマシ科) 日本、汎世界。貯穀害虫
dark Metanema moth		*Metanema determinata* Walker	(チョウ目、シャクガ科) 新北区
dark mimic-white		*Lieinix lala lala* (Godman et Salvin)	(チョウ目、シロチョウ科) 新熱帯区。*L. l. turrenti* (Maza et Maza) も同英名
dark mottle		*Logania distanti* Semper	(チョウ目、シジミチョウ科) 東洋区
dark-mottled clothes moth		*Nemapogon cloacella* (Haworth)	(チョウ目、ヒロズコガ科) 全北区
dark mottled peach aphid			peach aphid を見よ
dark mottled willow		*Spodoptera cilium* Guenée	(チョウ目、ヤガ科) 旧北区
dark mountain moth		*Aletia griseipennis* Felder	(チョウ目、ヤガ科) 豪州区
dark-night mole cricket		*Gryllotalpa monanka* Otte et Alexander	(バッタ目、コオロギ科) 豪州区
dark northern stiletto		*Thereva valida* Loew	(ハエ目、ツルギアブ科) 旧北区
dark oblique-barred moth		*Eucymatoge anguligera* Butler	(チョウ目、シャクガ科) 豪州区
dark oblique-barred twist			brown tortrix を見よ
dark opal		*Chrysoritis nigricans* (Aurivillius)	(チョウ目、シジミチョウ科) エチオピア区
dark opal		*Nesolycaena medicta* (Miskin)	(チョウ目、シジミチョウ科) 豪州区
dark orange-dart		*Ocybadistes ardea* Bethune-Baker	(チョウ目、セセリチョウ科) 豪州区
dark oystershell scale	クロカキカイガラムシ	*Lepidosaphes tubulorum* Ferris	(カメムシ目、マルカイガラムシ科) 日本、東洋区
dark palm dart		*Telicota bambusae* (Moore)	(チョウ目、セセリチョウ科) 東洋区
dark palm dart		*Telicota pythias* Mabille	(チョウ目、セセリチョウ科) 東洋区
dark palm dart			dark darter を見よ
dark pear pyralid	ナシモンクロマダラメイガ	*Acrobasis bellulella* (Ragonot)	(チョウ目、メイガ科) 日本、東洋区
dark pellucid-spotted saturnid	クロウスタビガ	*Rhodinia jankowskii hattoris* Inoue	(チョウ目、ヤママユガ科) 日本
dark pencil-blue		*Candalides consimilis* Waterhouse	(チョウ目、シジミチョウ科) 豪州区

英名	和名	学名	所属、分布、ほか
dark Phalaenostola moth		*Phalaenostola eumelusalis* (Walker)	（チョウ目、ヤガ科）新北区
dark Phanus		*Phanus obscurior obscurior* Kaye	（チョウ目、セセリチョウ科）新熱帯区
dark pied pierrot			black pie を見よ
dark pierrot		*Tarucus ananda* (de Nicéville)	（チョウ目、シジミチョウ科）東洋区
dark pincertail		*Onychogomphus assimilis* Schneider	（トンボ目、サナエトンボ科）旧北区
dark posy		*Drupadia theda* (C. et R. Felder)	（チョウ目、シジミチョウ科）東洋区。*D. t. thesmia* (Hewitson) も同英名
dark purple azure		*Ogyris abrota* Westwood	（チョウ目、シジミチョウ科）豪州区
dark purple-shaded piercer		*Cydia gallicana* Guenée	（チョウ目、ハマキガ科）旧北区
dark purplewing	アイイロタテハ	*Eunica alcmena alcmena* Doubleday	（チョウ目、タテハチョウ科）新熱帯区
dark ranger		*Kedestes niveostriga* (Trimen)	（チョウ目、セセリチョウ科）エチオピア区
dark red veronica moth		*Mecyna daiclealis* Meyrick	（チョウ目、メイガ科）豪州区
dark-ribboned wave moth		*Leptostales rubromarginaria* (Packard)	（チョウ目、シャクガ科）新北区
dark rockbrown		*Chazara persephone* (Hübner)	（チョウ目、タテハチョウ科）旧北区、東洋区
dark roller	コゲチャカギバヒメハマキ	*Ancylis upupana* (Treitschke)	（チョウ目、ハマキガ科）日本、旧北区
dark round-spot		*Pilodeudorix leonina* (Bethune-Baker)	（チョウ目、シジミチョウ科）エチオピア区
dark sage		*Psoltoda macallumi* Moulds	（カメムシ目、セミ科）豪州区
dark Saliana		*Saliana severus* (Mabille)	（チョウ目、セセリチョウ科）新熱帯区
dark scallop			dark umber を見よ
dark scallop moth		*Cepphis decoloraria* (Hulst)	（チョウ目、シャクガ科）新北区
dark Scoparia moth		*Scoparia philerga* Meyrick	（チョウ目、メイガ科）豪州区
dark serrate-margined geometrid			scalloped hazel を見よ
dark shieldback		*Idiostatus fuscus* Caudell	（バッタ目、キリギリス科）新北区
dark shoulder			flame shoulder を見よ
darkshouldered cornuted shadowdamsel		*Drepanosticta tropica* (Hagen)	（トンボ目、Platystictidae）東洋区
dark-shouldered ringtail		*Erpetogomphus leptophis* Garrison	（トンボ目、サナエトンボ科）新熱帯区
dark-sided cutworm		*Euxoa messoria* (Harris)	（チョウ目、ヤガ科）新北区。タバコ害虫
dark-sided cutworm moth			dark-sided cutworm を見よ
dark silver-striped piercer	カシワギンオビヒメハマキ	*Strophedra nitidana* (Fabricius)	（チョウ目、ハマキガ科）日本、旧北区

英名	和名	学名	所属、分布、ほか
dark skipper		*Semalea sextilis* (Plötz)	（チョウ目、セセリチョウ科）エチオピア区
dark skipper			dark ranger を見よ
dark small-branded swift			small branded swift を見よ
dark spectacle		*Abrostola triplasia* (Linnaeus)	（チョウ目、ヤガ科）日本、旧北区
dark spectacle			spectacle (1) を見よ
dark spinach	チャイロナミシャク	*Pelurga comitata* (Linnaeus)	（チョウ目、シャクガ科）日本、旧北区
dark-spotted looper moth		*Diachrysia aereoides* (Grote)	（チョウ目、ヤガ科）新北区。dark-spotted looper ともいう
dark-spotted Palthis moth		*Palthis angulalis* (Hübner)	（チョウ目、ヤガ科）新北区
dark-spotted tiger moth		*Spilosoma canescens* (Butler)	（チョウ目、ヒトリガ科）豪州区
dark spreadwing			dark emerald damselfly を見よ
dark spruce bell			coast spruce bell を見よ
dark stoneroot borer moth		*Papaipema duplicatus* Bird	（チョウ目、ヤガ科）新北区
dark straw ace		*Pithauria murdava* (Moore)	（チョウ目、セセリチョウ科）東洋区
dark strawberry tortrix moth		*Olethreutes lacunana* (Denis et Schiffermüller)	（チョウ目、ハマキガ科）全北区
dark swift			colon swift を見よ
dark sword grass			black cutworm を見よ　米国での英名
dark sword grass moth			black cutworm を見よ
dark swordtail			coastal swordtail を見よ
dark-tail wasps		*Cameronella*	（ハチ目、コガネコバチ科）の昆虫の総称
dark tawny wave		*Cyclophora nanaria* (Walker)	（チョウ目、シャクガ科）新北区
dark Tegosa			Nigrella crescent を見よ
dark Telemiades		*Telemiades nicomedes* (Möschler)	（チョウ目、セセリチョウ科）新熱帯区
dark three-spot missile		*Meza banda* (Evans)	（チョウ目、セセリチョウ科）エチオピア区
dark Tinsel		*Acupicta delicatum* (de Nicéville)	（チョウ目、シジミチョウ科）東洋区
dark tropical buckeye			dark buckeye (1) を見よ
dark tufted-skipper		*Nisoniades laurentina* (Williams et Bell)	（チョウ目、セセリチョウ科）新熱帯区
dark tussock		*Dicallomera fascelina* (Linnaeus)	（チョウ目、ドクガ科）旧北区
dark twin-spot carpet		*Xanthorhoe ferrugata* (Clerck)	（チョウ目、シャクガ科）日本、旧北区
dark umber		*Philereme transversata* (Hufnagel)	（チョウ目、シャクガ科）旧北区。日本亜種は *P. t. japonaria* (Leech) トビスジヤエナミシャク
dark underwing			heath fritillary を見よ

英名	和名	学名	所属、分布、ほか
dark underwing wainscot moth		*Leucania sulcana* Fereday	(チョウ目、ヤガ科) 豪州区
dark-veined white		*Pieris bryoniae* (Hübner)	(チョウ目、シロチョウ科) 全北区
dark velvet bob		*Koruthaialos butleri* (de Nicéville)	(チョウ目、セセリチョウ科) 東洋区
dark velvety maculated longicorn	ヤハズカミキリ	*Uraecha bimaculata bimaculata* Thomson	(コウチュウ目、カミキリムシ科) 日本
dark vulgar bush brown		*Bicyclus sandace* (Hewitson)	(チョウ目、タテハチョウ科) エチオピア区
dark wall		*Lasiommata menava* Moore	(チョウ目、タテハチョウ科) 旧北区
dark wanderer		*Pareronia ceylanica* (C. et R. Felder)	(チョウ目、シロチョウ科) 東洋区
dark wave		*Myscelia cyananthe* C. et R. Felder	(チョウ目、タテハチョウ科) 新北区
dark-waved bell		*Epinotia brunnichiana* (Linnaeus)	(チョウ目、ハマキガ科) 旧北区
dark wavy-striped geometrid	ウスグロオオナミシャク	*Triphosa dubitata amblychiles* Prout	(チョウ目、シャクガ科) 日本、旧北区
dark-webbed ringlet		*Physaeneura panda* (Boisduval)	(チョウ目、タテハチョウ科) エチオピア区
dark whiteface			white-faced darter (2) を見よ
dark-winged black		*Pachygaster atra* (Panzer)	(ハエ目、ミズアブ科) 旧北区
dark-winged fungus flies	クロバネキノコバエ科	Sciaridae	(ハエ目) の昆虫の総称
dark-winged fungus gnats		*Lycoria*	(ハエ目、クロバネキノコバエ科) の昆虫の総称
darkwinged fungus gnats			dark-winged fungus flies を見よ
dark-winged groundling		*Brachythemis fuscopalliata* (Selys)	(トンボ目、トンボ科) 旧北区
dark-winged horse bot fly	アカウマバエ	*Gasterophilus pecorum* (Fabricius)	(ハエ目、ウマバエ科) 日本、旧北区、東洋区
dark-winged quaker moth		*Eremobina claudens* (Walker)	(チョウ目、ヤガ科) 新北区。dark-winged quaker ともいう
dark-winged soldier		*Oxycera analis* Wiedemann	(ハエ目、ミズアブ科) 旧北区
dark wood nymph		*Cercyonis oetus* (Boisduval)	(チョウ目、タテハチョウ科) 新北区
dark yellow-banded flat	キンオビキコモンセセリ	*Celaenorrhinus aurivittata* (Moore)	(チョウ目、セセリチョウ科) 東洋区。*C. a. cameroni* Distant も同英名
dark yellow-banded flat		*Celaenorrhinus cameroni* (Distant)	(チョウ目、セセリチョウ科) 東洋区
dark yellow underwing		*Anarta cordigera* (Thunberg)	(チョウ目、ヤガ科) 旧北区
dark Zanclognatha moth		*Zanclognatha obscuripennis* (Grote)	(チョウ目、ヤガ科) 新北区。dark Zanclognatha ともいう
dark zebra swallowtail		*Protographium (Neographium) philolaus* (Boisduval)	(チョウ目、アゲハチョウ科) 新熱帯区

英名	和名	学名	所属、分布、ほか
darkened longwing		*Eueides procula* Doubleday	（チョウ目、タテハチョウ科）新熱帯区
darkened rusty clearwing		*Greta morgane oto* (Hewitson)	（チョウ目、タテハチョウ科）新熱帯区。rusty clearwing を参照
darkened white		*Pereute charops* (Boisduval)	（チョウ目、シロチョウ科）新北区
darkened yellow-haired skipper		*Cogia cajeta eluina* Godman et Salvin	（チョウ目、セセリチョウ科）新熱帯区
darkening midget		*Phyllonorycter nigrescentella* (Logan)	（チョウ目、ホソガ科）旧北区
darker commodore	カモシカタテハモドキ	*Precis antilope* (Feisthamel)	（チョウ目、タテハチョウ科）エチオピア区
darker Diacme moth		*Diacme adipaloides* (Grote et Robinson)	（チョウ目、メイガ科）新北区
darker Moodna moth		*Moodna ostrinella* (Clemens)	（チョウ目、メイガ科）新北区
darkies	ヒカゲシジミ属	*Allotinus*	（チョウ目、シジミチョウ科）の昆虫の総称
darkling beetle		*Gondwanocrypticus plalensis* (Fairmaire)	（コウチュウ目、ゴミムシダマシ科）新熱帯区
darkling beetles		*Alobates*	（コウチュウ目、ゴミムシダマシ科）の昆虫の総称
darkling beetles		*Bolitotherus*	（コウチュウ目、ゴミムシダマシ科）の昆虫の総称
darkling beetles		*Uloma*	（コウチュウ目、ゴミムシダマシ科）の昆虫の総称
darkling beetles (1)	ゴミムシダマシ科	Tenebrionidae	（コウチュウ目）の昆虫の総称
darkling bog button		*Acleris lorquiniana* (Duponchel)	（チョウ目、ハマキガ科）旧北区
darkling ground beetles			darkling beetles (1) を見よ
darling underwing moth			bronze underwing を見よ
darners	ヤンマ科	Aeschnidae	（トンボ目）の昆虫の総称
darning needle			dragonflies (1) を見よ
dart dagger moth			speared dagger moth を見よ
dart moth			cutworm (2) を見よ
dart moths			cutworms (1) を見よ
dart tail wasps			dark-tail wasps を見よ
darter			common darter を見よ
darter dragonflies		*Sympetrum*	（トンボ目、トンボ科）の昆虫の総称　darters ともいう。英国での名で、北米では meadowhawks という
darter dragonflies			common skimmers を見よ
darting cruiser		*Phyllomacromia picta* (Hagen)	（トンボ目、ヤンマ科）エチオピア区
dartlets		*Oriens*	（チョウ目、セセリチョウ科）の昆虫の総称
darts		*Potanthus*	（チョウ目、セセリチョウ科）の昆虫の総称
darts and rustics		Euxoinae	（チョウ目、ヤガ科）の昆虫の総称

英名	和名	学名	所属、分布、ほか
darts and swifts			intermediate skippers を見よ
Darwin brown crow		*Euploea darchia* (Macleay)	(チョウ目、タテハチョウ科) 豪州区。Darwin brown-crow butterfly ともいう
Darwin ringlet		*Hypocysta adiante antirius* Butler	(チョウ目、タテハチョウ科) 豪州区
Darwin ringlet			orange ringlet を見よ
Darwin's beetle		*Chiasognathus grantii* Stephens	(コウチュウ目、クワガタムシ科) 新熱帯区
Darwin's green		*Nemoria darwiniata* (Dyar)	(チョウ目、シャクガ科) 新北区
Darwin's heath		*Coenonympha darwiniana* Staudinger	(チョウ目、タテハチョウ科) 旧北区
dashed ringtail		*Erpetogomphus heterodon* Garrison	(トンボ目、サナエトンボ科) 新北区、新熱帯区
dashed slender robberfly		*Leptogaster guttiventris* Zetterstedt	(ハエ目、ムシヒキアブ科) 旧北区
dasheen horn worm	イッポンセスジスズメ	*Theretra silhetensis* Walker	(チョウ目、スズメガ科) 日本、東洋区。サトイモ害虫。dasheen タロイモ に由来
dasheen tree borer	シロテンコウモリ	*Palpifer sexnotata* (Moore)	(チョウ目、コウモリガ科) 日本、東洋区。サトイモ、コンニャクなどの害虫
dashwing			cadmus を見よ
dasytid beetles		Dasytidae	(コウチュウ目) の昆虫の総称 ジョウカイモドキ科に入れられる
date bug		*Asarcopus palmarum* Horvath	(カメムシ目、マルウンカ科) 新北区
date embiid			black webspinner を見よ
date moth			Mediterranean flour moth を見よ
date palm scale		*Parlatoria blanchardi* (Targioni-Tozzetti)	(カメムシ目、マルカイガラムシ科) 東洋区、全北区。ナツメヤシ害虫
date stone beetle			button weevil を見よ
date stone borer			button weevil を見よ
Datis crescent			Hewitson's mimic を見よ
Datura leaf beetle			three lined potato beetle (1) を見よ
daubers			cicada killers を見よ 豪州での英名
Davara scrub-hairstreak		*Strymon davara* (Hewitson)	(チョウ目、シジミチョウ科) 新熱帯区
Davidson's beetle		*Dascillus davidsoni* LeConte	(コウチュウ目、Dascillidae) 新北区
Davies's Colletes		*Colletes daviesanus* Smith	(ハチ目、ムカシハナバチ科) 旧北区
Davis' tussock moth		*Halysidota davisii* H. Edwards	(チョウ目、ヒトリガ科) 新北区
Davis's conehead		*Belocephalus davisi* Rehn et Hebard	(バッタ目、キリギリス科) 新北区
Davis's tree cricket		*Oecanthus exclamationis* Davis	(バッタ目、カンタン科) 新北区
daw bugs			white grubs を見よ daw コクマルガラスに由来
daw bugs			May beetles (3) を見よ 米国での英名

英名	和名	学名	所属、分布、ほか
dawn bluetail			aurora bluetail を見よ
dawn clouded yellow	コベニモンキチョウ	*Colias aurorina* Herrich-Schäffer	（チョウ目、シロチョウ科）旧北区
dawnflies		*Capila*	（チョウ目、セセリチョウ科）の昆虫の総称
Dawson grasshopper		*Melanoplus dawsoni* (Scudder)	（バッタ目、バッタ科）新北区
Dawson River black fly		*Austrosimulium pestilens* Mackerras et Mackerras	（ハエ目、ブユ科）豪州区。豪州の川の名に由来
Dawson's burrowing bee		*Amegilla dawsoni* (Rayment)	（ハチ目、ミツバチ科）豪州区
Dawson's giant-skipper		*Agathymus dawsoni* Comstock	（チョウ目、セセリチョウ科）新熱帯区
day emerald moth		*Mesothea incertata* (Walker)	（チョウ目、シャクガ科）新北区
day flies			burrowing mayflies を見よ
day flies			mayflies を見よ
day flying moths			giant silkworm moths を見よ
dayfeeding armyworm			African armyworm を見よ
dayfly			mayflies を見よ
dayflying moths			forester moths を見よ
dayflying moths			leaf skeletonizer moths (2) を見よ　南米での英名
day-flying moths			giant silkworm moths を見よ
day moths			Saturnioidea（チョウ目）の昆虫の総称
dazzling glasswing	オオスカシマダラ	*Godyris duillia* (Hewitson)	（チョウ目、タテハチョウ科）新熱帯区
dazzling nightfighter		*Porphyrogenes omphale* Butler	（チョウ目、セセリチョウ科）新熱帯区
de la Maza's mimic-white		*Enantia mazai mazai* Llorente	（チョウ目、シロチョウ科）新熱帯区。*E. m. diazi* Llorente も同英名
de la Maza's skipper			glowing skipper を見よ
de Lesse's brassy ringlet		*Erebia nivalis* Lorkovic et de Lesse	（チョウ目、タテハチョウ科）旧北区
de Niceville's dull oakblue		*Arhopala agrata* de Nicéville	（チョウ目、シジミチョウ科）東洋区。de Niceville's oakblue ともいう
de Niceville's spotted flat		*Celaenorrhinus sumitra* (Moore)	（チョウ目、セセリチョウ科）東洋区
de Niceville's windmill		*Byasa polla* (de Nicéville)	（チョウ目、アゲハチョウ科）東洋区
de Prunner's ringlet		*Erebia triaria* (de Prunner)	（チョウ目、タテハチョウ科）旧北区
dead leaf mantis		*Deroplatys desicata* Beier	（カマキリ目、カマキリ科）東洋区
dead leaf mantis			giant dead leaf mantis を見よ
dead-leaf wood nymph		*Taygetis angulosa* Weymer	（チョウ目、タテハチョウ科）新熱帯区
dead-nettle leaf beetle		*Chrysolina fastuosa* (Scopoli)	（コウチュウ目、ハムシ科）旧北区
deadly fire ant			imported fire ant (1) を見よ

英名	和名	学名	所属、分布、ほか
deadwood borer moth		*Scolecocampa liburna* (Geyer)	(チョウ目、ヤガ科) 新北区。dead-wood borer ともいう
death watch			booklouse (2) を見よ
death watch			furniture beetle (2) を見よ
death watch			common furniture beetle を見よ
death watch			death watch beetle (1) を見よ
death watch			death watch beetles を見よ
deathwatch and drugstore beetles			death watch beetles を見よ
death watch beetle (1)		*Xestobium rufovillosum* (De Geer)	(コウチュウ目、シバンムシ科) 旧北区
death watch beetle (2)		*Vrilleta decorata* Van Dyke	(コウチュウ目、ヒョウホンムシ科) 新北区
death watch beetle			common furniture beetle を見よ
death watch beetles	シバンムシ科	Anobiidae	(コウチュウ目) の昆虫の総称
death watches			death watch beetles を見よ
deathwatches			booklice (1) を見よ
death's head	メンガタスズメ	*Acherontia styx crathis* Rothschild et Jordan	(チョウ目、スズメガ科) 日本、旧北区、東洋区
death's head			death's head hawkmoth (1) を見よ
death's head cockroach		*Blaberus craniifer* Burmeister	(ゴキブリ目、ブラベルスゴキブリ科) 新北区。胸部背面の斑に由来。科内で最大種
death's-head hawk			death's head hawkmoth (1) を見よ
death's head hawkmoth (1)	クロメンガタスズメ	*Acherontia lachesis* (Fabricius)	(チョウ目、スズメガ科) 日本、東洋区
death's head hawkmoth (2)		*Acherontia atropos* (Linnaeus)	(チョウ目、スズメガ科) 旧北区
death's head moth			death's head hawkmoth (1) を見よ
death's head moths		*Acherontia*	(チョウ目、スズメガ科) の昆虫の総称
Debora's harlequin		*Mimeresia debora* (Kirby)	(チョウ目、シジミチョウ科) エチオピア区
debris bug		*Lyctocoris campestris* (Fabricius)	(カメムシ目、ハナカメムシ科) 旧北区、豪州区
debris bugs			flower bugs を見よ
decay moths		*Barea*	(チョウ目、マルハキバガ科) の昆虫の総称
Deccan grasshopper			Jola grasshopper を見よ
December eggar		*Poecilocampa populi* (Linnaeus)	(チョウ目、カレハガ科) 旧北区
December moth			December eggar を見よ
deceptive Apotomis moth		*Apotomis deceptana* (Kearfott)	(チョウ目、ハマキガ科) 新北区
deceptive Bomolocha moth		*Hypena deceptalis* Walker	(チョウ目、ヤガ科) 新北区
deceptive diadem			deceptive mimic を見よ
deceptive eggfly			deceptive mimic を見よ
deceptive mimic		*Hypolimnas deceptor* (Trimen)	(チョウ目、タテハチョウ科) エチオピア区

英名	和名	学名	所属、分布、ほか
deceptive widow		*Palpopleura deceptor* (Calvert)	（トンボ目、トンボ科）エチオピア区
deceptive widow		*Tarsocera imitator* Vári	（チョウ目、タテハチョウ科）エチオピア区
Decim			アメリカの周期セミのうち *Magicicada septendecim* (Linnaeus)(17年セミ) と *M. tredecim* (Walsh et Riley)(13年セミ) の俗称
decima ant-blue		*Acrodipsas decima* Miller et Lane	（チョウ目、シジミチョウ科）豪州区
Decora skipper			yellow-veined skipper を見よ
decorated granite moth		*Digrammia decorata* (Grossbeck)	（チョウ目、シャクガ科）新北区
decorated longhorn beetle		*Xylotribus decorator* (Fabricius)	（コウチュウ目、カミキリムシ科）新熱帯区
decorated owlet moth		*Pangrapta decoralis* Hübner	（チョウ目、ヤガ科）新北区。decorated owlet ともいう
Decula			アメリカの周期セミのうち *Magicicada septemdecula* Alexander et Moore (17年セミ) と *M. tredecula* Alexander et Moore (13年セミ) の俗称
Dedecora skipper		*Styriodes dedecora* (Plötz)	（チョウ目、セセリチョウ科）新熱帯区
deep-blue eyed-metalmark		*Mesosemia asa asa* Hewitson	（チョウ目、シジミタテハ科）新熱帯区。*M. a. asopis* Stichel も同英名
deep-brown dart		*Aporophyla lutulenta* (Denis et Schiffermüller)	（チョウ目、ヤガ科）旧北区
deep brown rustic			deep-brown dart を見よ
deep brown shade		*Neosphaleroptera nubilana* (Hübner)	（チョウ目、ハマキガ科）旧北区
deep-green hairstreak		*Theritas mavors* Hübner	（チョウ目、シジミチョウ科）新熱帯区
deep mountain stag beetle			miyama stag beetle を見よ
deep yellow Euchlaena moth		*Euchlaena amoenaria* (Guenée)	（チョウ目、シャクガ科）新北区。deep yellow Euchlaena ともいう
deer and horse flies			horse flies (1) を見よ
deer biting louse		*Rhabdopelidon longicorne* (Nitzsch)	（ハジラミ目、ケモノハジラミ科）旧北区
deer biting louse			roe biting louse を見よ
deer flies			horse flies (1)（3）を見よ
deer fly		*Chrysops vittatus* Wiedemann	（ハエ目、アブ科）新北区
deer fly			deer ked を見よ
deer fly			callidus deer fly を見よ
deer ked		*Lipoptena cervi* Linnaeus	（ハエ目、シラミバエ科）旧北区
deer ked			deer louse (1)(2)(3) を見よ
deer ked		*Lipoptena (Lipoptennella) depressa* (Say)	（ハエ目、シラミバエ科）新北区
deer keds			louse flies を見よ

英名	和名	学名	所属、分布、ほか
deer louse (1)	ヒメシカシラミバエ	*Lipoptena fortisetosa* Bequaert	(ハエ目、シラミバエ科) 日本
deer louse (2)		*Lipoptena japonica* Maa	(ハエ目、シラミバエ科) 日本
deer louse (3)		*Lipoptena sikae* Mogi	(ハエ目、シラミバエ科) 日本
deer louse fly		*Lipoptena depressa* (Say)	(ハエ目、シラミバエ科) 新北区
deer nostril fly		*Cephenomyia auribarbis* (Meigen)	(ハエ目、ヒツジバエ科) 旧北区
deer warble		*Hypoderma diana* Brauer	(ハエ目、ウシバエ科) 旧北区
deer warble fly			deer warble を見よ
deerfly			deer warble を見よ
definite patch		*Chlosyne definita* (Aaron)	(チョウ目、タテハチョウ科) 新北区
definite tussock moth		*Orgyia definita* Packard	(チョウ目、ドクガ科) 新北区。definite-marked tussock moth ともいう
defoliating hemlock moth			Alstroemer's flat-body を見よ
defoliating leaf bugs		*Campylomma*	(カメムシ目、カスミカメムシ科) の昆虫の総称
DeGeer's long-horn		*Adela degeerella* Linnaeus	(チョウ目、マガリガ科) 旧北区
degenerate scale	サカキマルカイガラムシ	*Dynaspidiotus degeneratus* (Leonardi)	(カメムシ目、マルカイガラムシ科) 日本、旧北区。ツバキ、ヒサカキなどの害虫
Deidamia Morpho	デイダミアモルフォ	*Morpho deidamia* (Hübner)	(チョウ目、タテハチョウ科) 新熱帯区
Dejean's Omus		*Omus dejani* Reiche	(コウチュウ目、ハンミョウ科) 新北区
dejected satyr		*Pedaliodes dejecta dejecta* (Bates)	(チョウ目、タテハチョウ科) 新熱帯区
dejected underwing moth		*Catocala dejecta* Strecker	(チョウ目、ヤガ科) 新北区
Delagoa grizzled skipper		*Spialia delagoae* (Trimen)	(チョウ目、セセリチョウ科) エチオピア区
Delagoa sandman			Delagoa grizzled skipper を見よ
Delalande skipper		*Telemiades delalande* (Latreille)	(チョウ目、セセリチョウ科) 新熱帯区
Delalande's butterfly		*Papilio delalandei* Godart	(チョウ目、アゲハチョウ科) エチオピア区
Delattin's grayling		*Hipparchia volgensis* (Mazochin-Porshnyakov)	(チョウ目、タテハチョウ科) 旧北区
Delaware skipper		*Anatrytone logan* (Edwards)	(チョウ目、セセリチョウ科) 新北区
delicate		*Mythimna vitellina* (Hübner)	(チョウ目、ヤガ科) 旧北区
delicate apple capsid		*Malacocoris chlorizans* (Panzer)	(カメムシ目、カスミカメムシ科) 旧北区
delicate blue			Australian hedge blue を見よ
delicate Cycnia		*Cycnia tenera* Hübner	(チョウ目、ヒトリガ科) 新北区。delicate Cycnia moth ともいう
delicate meadow katydid		*Orchelimum delicatum* Brunner	(バッタ目、キリギリス科) 新北区
delicate metalmark		*Astraeodes areuta* (Westwood)	(チョウ目、シジミタテハ科) 新熱帯区

英名	和名	学名	所属、分布、ほか
delicate moth			delicate を見よ
delicate pearl-white			northern pearl white を見よ
delicate skipper		*Apaustus gracilis* (Felder et Felder)	（チョウ目、セセリチョウ科）新熱帯区
delicate soldier		*Oxycera nigricornis* Olivier	（ハエ目、ミズアブ科）旧北区
delicate strawberry roller			strawberry leaf folder を見よ
delicate wainscot			delicate を見よ
delightful bird-dropping moth		*Tarache delecta* Walker	（チョウ目、ヤガ科）新北区
delightful dagger moth		*Acronicta vinnula* Grote	（チョウ目、ヤガ科）新北区
delilah underwing moth		*Catocala delilah* Strecker	（チョウ目、ヤガ科）新北区。delilah underwing ともいう
delineated sister			Fruhstorfer's sister を見よ
delphacid planthoppers			planthoppers (1) を見よ
delphacids			planthoppers (1) を見よ
delphinium cutworm			silver eight を見よ　カナダでの幼虫の英名
delphinium moth			silver eight を見よ
delta green ground beetle		*Elaphrus viridis* Horn	（コウチュウ目、オサムシ科）新北区
delta scale		*Lecanium deltae* Lizer	（カメムシ目、カタカイガラムシ科）新熱帯区。カンキツ害虫
delta-spotted spiketail		*Cordulegaster diastatops* (Selys)	（トンボ目、オニヤンマ科）新北区
deltoid moths			owlet moths を見よ
deltoid moths			snouts (1) を見よ
deltoid moths			gems (1) を見よ
Demea silverstreak		*Theclopsis demea* (Hewitson)	（チョウ目、シジミチョウ科）新熱帯区
Demeter longwing		*Heliconius demeter* Staudinger	（チョウ目、タテハチョウ科）新北区、新熱帯区
demoiselle Agrion		*Calopteryx virgo* (Linnaeus)	（トンボ目、イトトンボ科）旧北区
demoiselles			broad-winged damselflies を見よ　細い形を少女にたとえた英名。demoiselle ともいわれる
demon Charaxes	マルスミフタオチョウ	*Charaxes etheocles* (Cramer)	（チョウ目、タテハチョウ科）エチオピア区
demon emperor		*Charaxes phaeus* Hewitson	（チョウ目、タテハチョウ科）エチオピア区
demons		*Indothemis*	（トンボ目、トンボ科）の昆虫の総称
demons	クロセセリ属	*Notocrypta*	（チョウ目、セセリチョウ科）の昆虫の総称
Demophon shoemaker			one-spotted Prepona を見よ　silver-king shoemaker ともいう
dendrobium mealybug		*Pseudococcus dendrobiorum* Williams	（カメムシ目、コナカイガラムシ科）豪州区

英名	和名	学名	所属、分布、ほか
dendrobium thrips	ランノアザミウマ	*Helionothrips errans* (Williams)	(アザミウマ目、アザミウマ科) 日本、全北区、東洋区、エチオピア区
dengue-fever mosquito			yellow fever mosquito を見よ
dengue mosquito			yellow fever mosquito を見よ　豪州での英名
dentate bush brown		*Bicyclus dentata* (Sharpe)	(チョウ目、タテハチョウ科) エチオピア区
dentate sister			Godman's sister (1) を見よ
dentate stink beetle		*Eleodes dentipes* Eschscholtz	(コウチュウ目、ゴミムシダマシ科) 新北区
dentated pug	トラノオナミシャク	*Anticollix sparsata* (Treitschke)	(チョウ目、シャクガ科) 日本、旧北区
dented beetle		*Uloma tenebrionodes* White	(コウチュウ目、ゴミムシダマシ科) 豪州区
deodar aphid		*Cinara curvipes* (Patch)	(カメムシ目、アブラムシ科) 新北区。ヒマラヤスギに由来
deodar weevil		*Pissodes nemorensis* Hopkins	(コウチュウ目、ゾウムシ科) 新北区、エチオピア区
deomuga silkworm			gorse silkworm を見よ
Depicta copper		*Aloeides depicta* Tite et Dickson	(チョウ目、シジミチョウ科) エチオピア区
deploring Desmia moth		*Desmia deploralis* Hamspon	(チョウ目、メイガ科) 新北区
depluming louse			chicken wing louse を見よ
depressed flour beetle	コヒメコクヌストモドキ	*Palorus subdepressus* (Wollaston)	(コウチュウ目、ゴミムシダマシ科) 日本、汎世界。貯穀害虫
deprived grizzled skipper		*Spialia depauperata* (Strand)	(チョウ目、セセリチョウ科) エチオピア区
deprived grizzled skipper			wandering sandman を見よ
derbid planthoppers	ハネナガウンカ科	Derbidae	(カメムシ目) の昆虫の総称
derelict Eucosma moth		*Eucosma derelicta* Heinrich	(チョウ目、ハマキガ科) 新北区
dermestid beetle			Frisch's carrion beetle を見よ
dermestid beetles			larder beetles を見よ
dermestids			larder beetles を見よ
Dero clearwing		*Dircenna dero* (Hübner)	(チョウ目、タテハチョウ科) 新熱帯区
derodontid beetles			tooth-necked fungus beetles を見よ
Descimon's elfin		*Penaincisalia descimoni* Johnson	(チョウ目、シジミチョウ科) 新熱帯区
Descimon's ringlet		*Erebia serotina* Descimon et de Lesse	(チョウ目、タテハチョウ科) 旧北区
desert ant			harvester ant (1) を見よ
desert Apollo	イランアゲハ	*Hypermnestra helios* (Nickerl)	(チョウ目、アゲハチョウ科) 旧北区
desert Babul blue		*Azanus ubaldus* (Stoll)	(チョウ目、シジミチョウ科) 東洋区
desert bath white		*Pontia glauconome* (Klug)	(チョウ目、シロチョウ科) 旧北区、エチオピア区

英名	和名	学名	所属、分布、ほか
desert bean cupid		*Euchrysops nilotica* (Aurivillius)	（チョウ目、シジミチョウ科）エチオピア区
desert blister beetle		*Cysteodemus wislizeni* LeConte	（コウチュウ目、ツチハンミョウ科）新北区
desert blister beetle (1)		*Lytta magister* Horn	（コウチュウ目、ツチハンミョウ科）新北区
desert blue			inland hairstreak を見よ
desert bluet		*Enallagma deserti* Selys	（トンボ目、イトトンボ科）旧北区
desert buckwheat blue			Rita blue を見よ
desert checkered skipper		*Pyrgus philetas* Edwards	（チョウ目、セセリチョウ科）新北区、新熱帯区
desert cloudywing			Mexican hoary edge を見よ
desert cockroach			Boll's sand cockroach を見よ
desert cockroaches		Polyphagidae	（ゴキブリ目）の昆虫の総称
desert corn flea beetle		*Chaetocnema ectypa* Horn	（コウチュウ目、ハムシ科）新北区
desert cricket		*Gryllus desertus* Pallas	（バッタ目、コオロギ科）旧北区
desert cutworm		*Euxoa ridingsiana* Grote	（チョウ目、ヤガ科）新北区
desert dampwood termite		*Paraneotermes simplicicornis* (Banks)	（シロアリ目、レイビシロアリ科）新北区
desert darkling beetle		*Glyptasida costipennis* (LeConte)	（コウチュウ目、ゴミムシダマシ科）新北区
desert elfin	アリゾナコツバメ	*Callophrys fotis* (Strecker)	（チョウ目、シジミチョウ科）新北区
desert false underwing		*Drasteria tejonica* (Behr)	（チョウ目、ヤガ科）新北区
desert fire ant		*Solenopsis aurea* Wheeler	（ハチ目、アリ科）新北区
desert firetail		*Telebasis salva* (Hagen)	（トンボ目、イトトンボ科）新北区
desert forktail		*Ischnura barberi* Currie	（トンボ目、イトトンボ科）新北区
desert fourring		*Ypthima bolanica* Marshall	（チョウ目、タテハチョウ科）東洋区
desert fritillary		*Melitaea deserticola* Oberthür	（チョウ目、タテハチョウ科）旧北区、エチオピア区
desert fritillary			lesser spotted fritillary を見よ
desert gray skipper			Carus skipper を見よ
desert green hairstreak		*Callophrys sheridanii comstocki* Henne	（チョウ目、シジミチョウ科）新北区
desert grizzled skipper			Aden skipper を見よ
desert hackberry butterfly			empress Leilia を見よ
desert harvester ant			harvester ant (3) を見よ
desert leopard	ヒョウモンヒメキマダラツバメ	*Apharitis myrmecophilia* Dumont	（チョウ目、シジミチョウ科）エチオピア区
desert locust	サバクトビバッタ	*Schistocera gregaria* (Forskal)	（バッタ目、バッタ科）旧北区、東洋区、エチオピア区。集団で移動する大害虫
desert long-horned grasshoppers			monkey grasshoppers (1)(2) を見よ

英名	和名	学名	所属、分布、ほか
desert marble		*Euchloe lotta* Beutenmüller	(チョウ目、シロチョウ科) 新北区
desert marigold moth		*Schinia miniana* (Grote)	(チョウ目、ヤガ科) 新北区、新熱帯区
desert mealybug		*Spirococcus larreae* Ferris	(カメムシ目、コナカイガラムシ科) 新北区
desert Mexican metalmark		*Apodemia mejicanus deserti* Barnes et McDunnough	(チョウ目、シジミタテハ科) 新熱帯区
desert net-winged beetle		*Lycus fernandezi* Duges	(コウチュウ目、ベニボタル科) 新北区。成虫幼虫ともに捕食性
desert orange-tip (1)	サバクツマキチョウ	*Anthocharis cethura* (C. et R. Felder)	(チョウ目、シロチョウ科) 新北区、新熱帯区。*A. c. pima* Edwards は Pima orange-tip という
desert orange tip (2)	モロッコツマアカシロチョウ	*Colotis evagore* (Klug)	(チョウ目、シロチョウ科) 旧北区、エチオピア区
desert orange tip			Pima orange-tip を見よ
desert orange-tip			small orange-tip を見よ
desert pebble mantis		*Eremiaphila zetterstedti* Beier	(カマキリ目、Eremiaphilidae) エチオピア区
desert Pixie		*Melanis leucophlegma* (Stichel)	(チョウ目、シジミタテハ科) 新熱帯区
desert scorpionfly		*Boreus notoperatus* Fitch	(シリアゲムシ目、ユキシリアゲムシ科) 新北区
desert seed-harvester ant			harvester ant (1)(3) を見よ
desert skunk beetle			armored stink beetle を見よ
desert spider beetle		*Cysteodemus armatus* LeConte	(コウチュウ目、ツチハンミョウ科) 新北区
desert subterranean termite		*Heterodermes aureus* (Snyder)	(シロアリ目、ミゾガシラシロアリ科) 新北区。
desert swallowtail		*Papilio rudkini* F. et R. Chermock	(チョウ目、アゲハチョウ科) 新北区
desert swallowtail (1)		*Papilio saharae* Oberthür	(チョウ目、アゲハチョウ科) エチオピア区
desert termites		*Amitermes, Gnathamitermes*	(シロアリ目、シロアリ科) の昆虫の総称
desert termites			soldierless termites (1) を見よ
desert water scorpion			water scorpion (2) を見よ
desert white			desert bath white を見よ
Desjardins' beetle	モンセマルホソヒラタムシ	*Cryptomorpha disjardinsii* (Guérin-Méneville)	(コウチュウ目、ホソヒラタムシ科) 日本、旧北区
desmergate	職兵蟻型		職蟻と兵蟻の間に介在する型のアリを指す
Desmond's giant cupid		*Lepidochrysops desmondi* Stempffer	(チョウ目、シジミチョウ科) エチオピア区
Desmond's green-banded swallowtail		*Papilio desmondi* van Someren	(チョウ目、アゲハチョウ科) エチオピア区
destroyer ant	ミゾヒメアリ	*Monomorium destructor* (Jerdon)	(ハチ目、アリ科) 日本、汎亜熱帯、汎熱帯
destroyer clothes moth			clothes moth (1) を見よ

英名	和名	学名	所属、分布、ほか
destructive mealybug			citrus mealybug (1) を見よ
destructive prune worm			tricolored Acrobasis moth を見よ destructive pruneworm moth ともいう
detached dart		*Potanthus trachala* (Mabille)	(チョウ目、セセリチョウ科) 東洋区
detached white-spots		*Osmodes lux* Holland	(チョウ目、セセリチョウ科) エチオピア区
detracted owlet moth		*Lesmone detrahens* (Walker)	(チョウ目、ヤガ科) 新北区
detritus moth		*Opogona omoscopa* (Meyrick)	(チョウ目、ヒロズコガ科) 豪州区
Deucalion beauty			leopard-spotted beauty を見よ
Deva skipper		*Atrytonopsis deva* (Edwards)	(チョウ目、セセリチョウ科) 新北区、新熱帯区
devastating grasshopper			devastating locust を見よ
devastating locust		*Melanoplus devastator* Scudder	(バッタ目、バッタ科) 新北区
devil horses			mantids (2) を見よ
devil rider			two-striped walking-stick を見よ
devilhoppers			treehoppers (1) を見よ
Devilliers swallowtail		*Battus devilliersii* (Godart)	(チョウ目、アゲハチョウ科) 新北区
devil's blue			quaker (1) を見よ
devil's coach horse (1)		*Creophilus erythrocephalus* (Fabricius)	(コウチュウ目、ハネカクシ科) 豪州区
devil's coach horse (2)		*Staphylinus olens* Müller	(コウチュウ目、ハネカクシ科) 全北区
devil's coachhorse beetle		*Creophilus oculatus* (Fabricius)	(コウチュウ目、ハネカクシ科) 豪州区
devil's coach-horses			rove beetles を見よ
devil's coachman			common earwig を見よ
devil's darning needles			dragonflies (1) を見よ
devil's flower mantis (1)		*Blepharopsis mendica* Fabricius	(カマキリ目、Empusidae) エチオピア区
devil's flower mantis (2)		*Idolomantis diabolica* (Saussure)	(カマキリ目、Empusidae) エチオピア区
devil's needle			dragonflies (1) を見よ
devil's riding horse			dragonflies (1) を見よ
devil's rope pear cochineal		*Dactylopius tomentosus* (Lamarck)	(カメムシ目、コチニールカイガラムシ科) 豪州区
Devon carpet	チビアトクロナミシャク	*Lampropteryx otregiata* Metcalf	(チョウ目、シャクガ科) 日本、旧北区。イングランドの地名に由来
Devon red-legged robberfly		*Neomochtherus pallipes* (Meigen)	(ハエ目、ムシヒキアブ科) 旧北区
Devonshire wainscot		*Mythimna putrescens* (Hübner)	(チョウ目、ヤガ科) 旧北区
Devonshire woody gall			marble gall wasp の虫えい

英名	和名	学名	所属、分布、ほか
DeVries' Theope		*Theope devriesi* Hall et Willmott	（チョウ目、シジミタテハ科）新熱帯区
dew footman		*Philea irrorella* Linnaeus	（チョウ目、マルハキバガ科）旧北区
dew footman			dew moth を見よ
dew moth		*Setina irrorella* (Linnaeus)	（チョウ目、ヒトリガ科）旧北区
Dewick's plusia			chrysanthemum silver phytometra を見よ
Dewick's silver spangle			chrysanthemum silver phytometra を見よ
dewy ringlet	ウスグロベニヒカゲ（ハイイロベニヒカゲ）	*Erebia pandrose* (Borkhausen)	（チョウ目、タテハチョウ科）旧北区
dexid flies	アシナガヤドリバエ科	Dexiidae	（ハエ目）の昆虫の総称　ホソカ科 Dixidae とは別。ヤドリバエ科に入れられる
Deyrolle's blue skipper		*Quadrus deyrollei* (Mabille)	（チョウ目、セセリチョウ科）新熱帯区
Dia leafwing		*Memphis dia dia* (Godman et Salvin)	（チョウ目、タテハチョウ科）新熱帯区
diabolic shieldback		*Neduba diabolica* (Scudder)	（バッタ目、キリギリス科）新北区
diabolical fungus moth		*Metalectra diabolica* Barnes et Benjamin	（チョウ目、ヤガ科）新北区
diabrotica beetle		*Diabrotica speciosa* (Germar)	（コウチュウ目、ハムシ科）新熱帯区。カンキツ害虫
diadem			danaid eggfly を見よ　南アでの英名
diadems			eggflies を見よ
diamond beetle		*Chrysolopus spectabilis* (Fabricius)	（コウチュウ目、ゾウムシ科）豪州区
diamond beetles		*Entimus*	（コウチュウ目、ゾウムシ科）の昆虫の総称
diamond spot pearl			beet webworm を見よ
diamondback			diamondback moth を見よ
diamondback cabbage moth			diamondback moth を見よ
diamondback Epinotia moth		*Epinotia lindana* (Fernald)	（チョウ目、ハマキガ科）新北区
diamondback moth	コナガ	*Plutella xylostella* (Linnaeus)	（チョウ目、スガ科）日本、汎世界。キャベツなどアブラナ科野菜の大害虫
diamondback moths			ermine moths (1) を見よ
diamondbacked spittlebug		*Lepyronia quadrangularis* (Say)	（カメムシ目、アワフキムシ科）新北区
diamond-barred pigmy		*Stigmella marginicolella* Stainton	（チョウ目、モグリチビガ科）旧北区
diamond-spot pearl	クロミャクノメイガ	*Sitochroa verticalis* (Linnaeus)	（チョウ目、メイガ科）日本、旧北区。テンサイ、マメ科牧草害虫
diamond-spot pearl (1)	ヘリキスジノメイガ	*Margaritia sticticalis* (Linnaeus)	（チョウ目、メイガ科）日本、全北区。幼虫は多食性
Diana	ダイアナギンボシヒョウモン	*Speyeria diana* (Cramer)	（チョウ目、タテハチョウ科）新北区。ローマ神話のダイアナ女神に由来。Diana butterfly ともいう

英名	和名	学名	所属、分布、ほか
Diana moonbeam		*Philiris diana* Waterhouse et Lyell	(チョウ目、シジミチョウ科) 豪州区
Diana's Choreutis moth		*Choreutis diana* (Hübner)	(チョウ目、ハマキモドキガ科) 新北区
diapriid wasps	ハエヤドリクロバチ科	Diapriidae	(ハチ目) の昆虫の総称
diapriids			diapriid wasps を見よ
diastatid flies	ホソショウジョウバエ科	Diastatidae	(ハエ目) の昆虫の総称
Diaz's gemmed-satyr		*Cyllopsis diazi* Miller	(チョウ目、タテハチョウ科) 新熱帯区
Diaz's sister		*Adelpha diazi* Beutelspacher	(チョウ目、タテハチョウ科) 新熱帯区
Dicaea hairstreak		*Dicya dicaea* (Hewitson)	(チョウ目、シジミチョウ科) 新熱帯区
dichthadiigyne	雌職蟻型		生殖職蟻型のさらに発達した型。dichthadia ともいう
Dickinson's horse			dragonflies (1) を見よ
Dickinson's mare			dragonflies (1) を見よ
Dickson's brown		*Stygionympha dicksoni* (Riley)	(チョウ目、タテハチョウ科) エチオピア区
Dickson's copper		*Aloeides dicksoni* Henning	(チョウ目、シジミチョウ科) エチオピア区
Dickson's copper (1)		*Chrysoritis dicksoni* (Gabriel)	(チョウ目、シジミチョウ科) エチオピア区
Dickson's geranium bronze		*Cacyreus dicksoni* Pennington	(チョウ目、シジミチョウ科) エチオピア区
Dickson's skolly		*Thestor dicksoni* Riley	(チョウ目、シジミチョウ科) エチオピア区
Dickson's strandveld copper			Dickson's copper (1) を見よ
Dickson's sylph		*Tsitana dicksoni* Evans	(チョウ目、セセリチョウ科) エチオピア区
Dickson's widow		*Tarsocera dicksoni* (van Son)	(チョウ目、タテハチョウ科) エチオピア区
dicky rice weevil			spinelegged citrus weevil を見よ
dictyophard planthoppers	テングスケバ科	Dictyopharidae	(カメムシ目) の昆虫の総称
dictyospermum scale	オンシツマルカイガラムシ	*Chrysomphalus dictyospermi* (Morgan)	(カメムシ目、マルカイガラムシ科) 日本、新北区、汎熱帯。温室害虫。米国での英名
dicyrtomid springtails		Dicyrtomidae	(トビムシ目) の昆虫の総称
Dido longwing			bamboo page を見よ
differenciate sawfly	メスアカハバチ	*Nematinus japonicus* (Marlatt)	(ハチ目、ハバチ科) 日本
different grasshopper			differential grasshopper を見よ
different-horned tree cricket		*Oecanthus varicornis* Walker	(バッタ目、カンタン科) 新北区
differential grasshopper		*Melanoplus differentialis* (Thomas)	(バッタ目、バッタ科) 新北区。米国で普通
differentiated grasshopper			Oriental long-headed locust (1) を見よ
difficult skipper		*Zera difficilis* (Weeks)	(チョウ目、セセリチョウ科) 新熱帯区

英名	和名	学名	所属、分布、ほか
digger bees		*Ptilothrix*	（ハチ目、ミツバチ科）の昆虫の総称
digger bees (1)		Anthophorinae	（ハチ目、コシブトハナバチ科）の昆虫の総称
digger bees			long-tongued bees を見よ
digger bees			hairy-footed bees を見よ
digger beetle		*Cicindela scutellaris* Say	（コウチュウ目、ハンミョウ科）新北区
digger wasp		*Podalonia affinis* (Kirby)	（ハチ目、アナバチ科）旧北区
digger wasp			blue-winged wasp を見よ
digger wasps			cicada killers を見よ
digger wasps			crabronid wasps を見よ
digger wasps			sphecoid wasps を見よ
digging grasshopper		*Acrotylus insubricus* (Scopoli)	（バッタ目、バッタ科）旧北区
digging wasps			spider wasps を見よ
Diggle's blue	ウラジロニシキシジミ	*Hypochrysops digglesii* (Hewitson)	（チョウ目、シジミチョウ科）豪州区
Dilecta banner		*Epiphile dilecta* Rober	（チョウ目、タテハチョウ科）新熱帯区
Dil's grayling		*Pseudochazara orestes* de Prins et Poorten	（チョウ目、タテハチョウ科）旧北区
dilute green Charaxes		*Charaxes dilutus* Rothschild	（チョウ目、タテハチョウ科）エチオピア区
diluted skipper		*Cynea diluta* (Herrich-Schäffer)	（チョウ目、セセリチョウ科）新熱帯区
diminutive clubtail		*Gomphus diminutus* Needham	（トンボ目、サナエトンボ科）新北区
diminutive grass-veneer moth		*Raphiptera argillaceellus* (Packard)	（チョウ目、メイガ科）新北区
diminutive shieldback		*Aglaothorax diminutiva* (Rentz et Birchim)	（バッタ目、キリギリス科）新北区
diminutive wave moth		*Idaea scintillularia* (Hulst)	（チョウ目、シャクガ科）新北区
dimorphic admiral		*Antanartia dimorphica* Howarth	（チョウ目、タテハチョウ科）エチオピア区
dimorphic Bomolocha moth		*Hypena bijugatis* Walker	（チョウ目、ヤガ科）新北区。dimorphic Bomolocha, dimorphic Hypena ともいう
dimorphic Eulithis moth		*Eulithis molliculata* (Walker)	（チョウ目、シャクガ科）新北区
dimorphic flower longhorn		*Anastrangalia laetifica* (LeConte)	（コウチュウ目、カミキリムシ科）新北区
dimorphic grass skipper		*Callimormus corades* (Felder)	（チョウ目、セセリチョウ科）新熱帯区
dimorphic grass skipper (1)		*Anthoptus epictetus* (Fabricius)	（チョウ目、セセリチョウ科）新熱帯区
dimorphic gray moth		*Tornos scolopacinaria* Guenée	（チョウ目、シャクガ科）新北区
dimorphic leafcutter moth		*Neargyractis slossonalis* (Dyar)	（チョウ目、メイガ科）新北区、新熱帯区
dimorphic Macalla moth		*Epipaschia superatalis* Clemens	（チョウ目、メイガ科）新北区

英名	和名	学名	所属、分布、ほか
dimorphic pinion moth		*Lithophane patefacta* Walker	(チョウ目、ヤガ科) 新北区
dimorphic Sitochroa moth		*Sitochroa chortalis* (Grote)	(チョウ目、メイガ科) 新北区
dimorphic skipper		*Cabirus procas* (Cramer)	(チョウ目、セセリチョウ科) 新熱帯区
dimorphic skipper		*Hyalothyrus neleus pemphigargyra* (Mabille)	(チョウ目、セセリチョウ科) 新熱帯区
dimorphic Tosale moth		*Tosale oviplagalis* (Walker)	(チョウ目、メイガ科) 新北区
dimpling bug			apple dimpling bug を見よ
Dina yellow		*Eurema dina* (Poey)	(チョウ目、シロチョウ科) 新北区。新熱帯区の *E. d. westwoodi* (Boisduval) も同英名
dinergate	兵蟻		大きな頭部と大腮をもつアリ
Dingaan's widow		*Dingana dingana* (Trimen)	(チョウ目、タテハチョウ科) エチオピア区
dingiest sailer			chocolate sailor を見よ
dingy angle		*Macaria bicolorata* (Fabricius)	(チョウ目、シャクガ科) 旧北区。*M. b. praeatomata* Haworth も同英名
dingy Antinephele		*Antinephele achlora* Holland	(チョウ目、スズメガ科) エチオピア区
dingy arctic fritillary			dusky-winged fritillary を見よ
dingy bushbrown			common bushbrown (2) を見よ
dingy cutworm		*Feltia ducens* Walker	(チョウ目、ヤガ科) 新北区
dingy cutworm			dingy cutworm moth (1) を見よ　北米での英名
dingy cutworm moth		*Feltia jaculifera* (Guenée)	(チョウ目、ヤガ科) 新北区
dingy cutworm moth (1)		*Feltia subgothica* (Haworth)	(チョウ目、ヤガ科) 新北区。トウモロコシ害虫
dingy cutworms		*Feltia*	(チョウ目、ヤガ科) の昆虫の総称　cutworms を参照
dingy dart		*Suniana lascivia* (Rosenstock)	(チョウ目、セセリチョウ科) 豪州区
dingy darter		*Telicota eurotas* (Felder)	(チョウ目、セセリチョウ科) 豪州区
dingy drill		*Dichrorampha sedatana* Busck	(チョウ目、ハマキガ科) 旧北区
dingy duskflyer		*Zyxomma petiolatum* Rambur	(トンボ目、トンボ科) 東洋区
dingy Emesis		*Emesis ocypore* (Geyer)	(チョウ目、シジミタテハ科) 新熱帯区
dingy footman		*Eilema griseola* (Hübner)	(チョウ目、ヒトリガ科) 旧北区。日本亜種は *E. g. aegrota* (Butler) キシタホソバ
dingy fritillary			dusky-winged fritillary を見よ
dingy grass dart		*Taractrocera dolon* (Plötz)	(チョウ目、セセリチョウ科) 豪州区
dingy jewel		*Hypochrysops ignitus ollifi* Miskin	(チョウ目、シジミチョウ科) 豪州区。fiery jewel を参照
dingy larch bell			larch tortrix (1) を見よ
dingy lineblue		*Petrelaea dana* (de Nicéville)	(チョウ目、シジミチョウ科) 東洋区

英名	和名	学名	所属、分布、ほか
dingy mocha		*Cyclophora pendularia* (Linnaeus)	(チョウ目、シャクガ科) 旧北区
dingy mocha		*Cyclophora orbicularia* Hübner	(チョウ目、シャクガ科). 旧北区
dingy purplewing	モニマアイイロタテハ	*Eunica monima* (Stoll)	(チョウ目、タテハチョウ科) 新北区、新熱帯区
dingy ring	ミナミウラナミジャノメ	*Ypthima arctoa* (Fabricius)	(チョウ目、タテハチョウ科) 豪州区
dingy ringlet		*Hypocysta pseudirius* Butler	(チョウ目、タテハチョウ科) 豪州区
dingy roseate conch	ナカハスジベニホソハマキ	*Cochylidia subroseana* (Haworth)	(チョウ目、ホソハマキガ科) 日本、旧北区
dingy sailer		*Neptis vikasi* Horsfield	(チョウ目、タテハチョウ科) 東洋区
dingy scrub hopper		*Acromachus dubius* (Elwes et Edwards)	(チョウ目、セセリチョウ科) 旧北区、東洋区
dingy shears		*Parastichtis ypsillon* (Denis et Schiffermüller)	(チョウ目、ヤガ科) 旧北区
dingy shears		*Enargia fissipuncta* (Haworth)	(チョウ目、ヤガ科) 旧北区
dingy shell	ハンノナミシャク	*Euchoeca nebulata* (Scopoli)	(チョウ目、シャクガ科) 日本、旧北区
dingy shield skipper		*Signeta tymbophora* (Meyrick et Lower)	(チョウ目、セセリチョウ科) 豪州区
dingy skipper	ヒメミヤマセセリ	*Erynnis tages* (Linnaeus)	(チョウ目、セセリチョウ科) 旧北区。dingy skipper butterfly ともいう
dingy smudge		*Acrolepia autumnitella* Curtis	(チョウ目、アトヒゲコガ科) 旧北区
dingy spruce bell		*Epinotia pygmaeana* (Hübner)	(チョウ目、ハマキガ科) 旧北区
dingy straw flat-body			gorse tip moth を見よ
dingy straw neb		*Metzneria lappella* (Linnaeus)	(チョウ目、キバガ科) 全北区
dingy swallowtail			chequered swallowtail を見よ
dingy swift			Mediterranean skipper を見よ
dingy white plume		*Pterophorus baliodactylus* (Zeller)	(チョウ目、トリバガ科) 旧北区
Diomus grizzled skipper			common sandman を見よ
Dion skipper			sedge skipper を見よ
dioptid moths			oak moths を見よ
dioscorea leaf beetle	ヤマイモハムシ	*Lema honorata* Baly	(コウチュウ目、ハムシ科) 日本、旧北区、東洋区。ナガイモ害虫
Diotima jewel	エグリウラスジタテハ	*Orophila diotima* Hewitson	(チョウ目、タテハチョウ科) 新熱帯区
diplurans			entotrophs を見よ
diplurans			anajapygid entotrophs を見よ
diplurans			slender entotrophs を見よ
diprionid sawflies			conifer sawflies を見よ

英名	和名	学名	所属、分布、ほか
dipterans			flies を見よ
dipteroid insects		Dipteroidea	の昆虫の総称
dipteron			flies を見よ
dipterous insects			flies を見よ
dipterous leaf miners			leafminer flies を見よ
dirce beauty			zebra (2) を見よ
Dirphia skipper		*Motasingha dirphia* (Hewitson)	(チョウ目、セセリチョウ科) 豪州区
dirty albatross			Congo white を見よ
dirty-blue satyr		*Cepheuptychia glaucina* Bates	(チョウ目、タテハチョウ科) 新熱帯区
dirty sailer	タイワンミスジ	*Neptis nata* Moore	(チョウ目、タテハチョウ科) 東洋区
Disa alpine			spruce bog alpine を見よ
discoid cockroach		*Blaberus discoidalis* Audinet-Serville	(ゴキブリ目、ブラベルスゴキブリ科) 新北区、新熱帯区
discolored Renia moth		*Renia discoloralis* Guenée	(チョウ目、ヤガ科) 新北区。discolored Renia ともいう
disguised scrub-hairstreak			Limenia hairstreak を見よ
dishclout frit			marsh fritillary を見よ
disk-headed stink bug		*Cataulax pudens* (Distant)	(カメムシ目、カメムシ科) 新熱帯区
disk-tailed carrion beetle		*Oxelytrum discicolle* (Brulle)	(コウチュウ目、シデムシ科) 新熱帯区
dismal mystic	ウスモンヒカゲ	*Lethe ocellata* (Poujade)	(チョウ目、タテハチョウ科) 東洋区
dismal satyr		*Steremnia umbracina* Butler	(チョウ目、タテハチョウ科) 新熱帯区
dismal sylph		*Tsitana tsita* (Trimen)	(チョウ目、セセリチョウ科) エチオピア区
dispar skipper			barred skipper を見よ
disparaged arches moth		*Orthodes detracta* (Walker)	(チョウ目、ヤガ科) 新北区
dissimilar bark-beetle			shot-hole borer (3) を見よ
dissimilar chicken louse			brown chicken louse を見よ
distinct fritillary		*Clossiana distincta* (Gibson)	(チョウ目、タテハチョウ科) 新北区
distinct Sparganothis moth		*Sparganothis distincta* (Walsingham)	(チョウ目、ハマキガ科) 新北区
distinctive scaly cricket		*Cycloptilum distinctum* Hebard	(バッタ目、コオロギ科) 新北区
distinguished Colymychus moth		*Colymychus talis* (Grote)	(チョウ目、メイガ科) 新北区
distinguished cypress owlet moth		*Cutina distincta* (Grote)	(チョウ目、ヤガ科) 新北区
disturbed tigerwing			common tiger (2) を見よ

英名	和名	学名	所属、分布、ほか
distylium fruit gall midge	イスノキハリオタマバエ	*Asphondylia itoi* Uechi et Yukawa	（ハエ目、タマバエ科）日本。イスノキ害虫。イスノキの属名 *Distylium* に由来
distylium gall aphid	イスオオムネアブラムシ	*Nipponaphis distychii* Pergande	（カメムシ目、アブラムシ科）日本
distylium scale	イスシロカイガラムシ	*Aulacaspis latissima* (Cockerell)	（カメムシ目、マルカイガラムシ科）日本、旧北区
distylium whitefly	イスノキコナジラミ	*Acanthobemisia distylii* Takahashi	（カメムシ目、コナジラミ科）日本。イスノキ害虫
ditch dun		*Habrophlebia fusca* (Curtis)	（カゲロウ目、トビイロカゲロウ科）旧北区
ditch jewel		*Brachythemis contaminata* (Fabricius)	（トンボ目、トンボ科）東洋区
diurnal Lepidoptera		Diurna	（チョウ目）の昆虫の総称　昼飛性のチョウ目の昆虫
diurnal predatory katydid		*Austrophlugis malidupa* Rentz	（バッタ目、キリギリス科）豪州区
diva Hemerophila moth		*Hemerophila diva* (Riley)	（チョウ目、ハマキモドキガ科）新北区
diva moth		*Divana diva* (Butler)	（チョウ目、カストニアガ科）旧北区、新熱帯区。diva プリマドンナ に由来
divebomber Charaxes			blue Charaxes を見よ
divergent dart moth		*Euxoa divergens* (Walker)	（チョウ目、ヤガ科）新北区。divergent dart ともいう
divergent metallic wood borer		*Dicerca divaricata* Say	（コウチュウ目、タマムシ科）新北区
divergent woolly legs		*Lachnocnema divergens* Gaede	（チョウ目、シジミチョウ科）エチオピア区
diverse cloudy wing			western cloudywing を見よ
diverse damselflies	ムカシトンボ亜目	Anisozygoptera	（トンボ目）の昆虫の総称
diverse white		*Appias epaphia* (Cramer)	（チョウ目、シロチョウ科）エチオピア区。*A. e. contracta* (Butler) も同英名
diversicorn beetles		Diversicornia	（コウチュウ目）の昆虫の総称
divided Olethreutes moth		*Olethreutes bipartitana* (Clemens)	（チョウ目、ハマキガ科）新北区
diving beetles (1)	ゲンゴロウ科	Dytiscidae	（コウチュウ目）の昆虫の総称　幼虫は water tigers といわれる
diving beetles (2)		*Dytiscus*	（コウチュウ目、ゲンゴロウ科）の昆虫の総称
diving grasshopper		*Bermiella acuta* (Stål)	（バッタ目、バッタ科）豪州区
diving water beetle			great black water beetle を見よ
Divisa scarlet		*Crocothemis divisa* Baumann	（トンボ目、トンボ科）エチオピア区
Dixa gnats			dixid midges を見よ
Dixa midges			dixid midges を見よ
dixid midges	ホソカ科	Dixidae	（ハエ目）の昆虫の総称
dixie skipper		*Hesperia meskei* (Edwards)	（チョウ目、セセリチョウ科）新北区
Djah paradise skipper		*Abantis ja* Druce	（チョウ目、セセリチョウ科）エチオピア区

英名	和名	学名	所属、分布、ほか
dobson			アミメカゲロウ目、ヘビトンボ科の幼虫を指す。dobson fly を見よ
dobsonflies (1)	ヘビトンボ科	Corydalidae	(アミメカゲロウ目) の昆虫の総称　dobson fly ヘビトンボといわれる昆虫。幼虫は dobsons, hellgrammites, conniption bugs, crawlers, arnly, toe-biters といわれ、川釣りの餌とされる
dobsonflies (2)		*Corydalus*	(アミメカゲロウ目、ヘビトンボ科) の昆虫の総称
dobsonflies (3)		*Dysmicohermes*	(アミメカゲロウ目、ヘビトンボ科) の昆虫の総称
dobsonfly		*Corydalus cornutus* (Linnaeus)	(アミメカゲロウ目、ヘビトンボ科) 新北区。幼虫は渓流釣りの餌
dock aphid		*Brachycaudus rumexicolens* (Patch)	(カメムシ目、アブラムシ科) 豪州区
dock aphid (1)	ギシギシアブラムシ	*Aphis rumicis* Linnaeus	(カメムシ目、アブラムシ科) 日本、汎世界
dock false worm			dock sawfly を見よ　米国での英名
dock rustic moth		*Resapamea passer* (Guenée)	(チョウ目、ヤガ科) 新北区
dock sawfly		*Ametastegia glabrata* (Fallén)	(ハチ目、ハバチ科) 全北区
docker		*Antichloris caca* Hübner	(チョウ目、ヒトリガ科) 旧北区、新熱帯区
Dodava hairstreak		*Nesiostrymon dodava* (Hewitson)	(チョウ目、シジミチョウ科) 新熱帯区
dodder gall weevil		*Smicronyx sculpticollis* Casey	(コウチュウ目、ゾウムシ科) 新北区
Dodd's azure		*Ogyris iphis* Waterhouse et Lyell	(チョウ目、シジミチョウ科) 豪州区
Dodd's bunyip		*Tamasa doddi* (Goding et Froggatt)	(カメムシ目、セミ科) 豪州区
dodonaea long-horned beetle			Hawaiian aalii long-horned beetle を見よ
dog biting louse			dog chewing louse を見よ
dog chewing louse	イヌハジラミ	*Trichodectes canis* (De Geer)	(ハジラミ目、ケモノハジラミ科) 日本、汎世界。イヌに寄生
dog-day cicada (1)		*Tibicen canicularis* Harris	(カメムシ目、セミ科) 新北区
dog-day cicada (2)		*Tibicen linnei* (Smith et Grossbeck)	(カメムシ目、セミ科) 新北区
dog-day cicadas		*Tibicen*	(カメムシ目、セミ科) の昆虫の総称　本属に *Lyristes* の使用が提案されている
dogday harvestfly			dog-day cicada (1) を見よ
dog dung fly			bazar fly を見よ　米国での英名
dog flea	イヌノミ	*Ctenocephalides canis* (Curtis)	(ノミ目、ヒトノミ科) 日本、汎世界。イヌに寄生
dog fly	イヌシラミバエ	*Hippobosca longipennis* (Fabricius)	(ハエ目、シラミバエ科) 日本、旧北区
dog fly			stable fly を見よ　米国での英名
dog ked			dog fly を見よ
dog louse		*Heterodoxus spiniger* (Enderlein)	(シラミ目、ミナミケモノハジラミ科) 日本、汎世界

英名	和名	学名	所属、分布、ほか
dog louse			dog chewing louse を見よ
dog sucking louse	イヌジラミ	*Linognathus setosus* (von Olfers)	(シラミ目、ケモノホソジラミ科) 日本、汎世界。イヌに寄生
dogbane beetle		*Chrysochus auratus* Fabricius	(コウチュウ目、ハムシ科) 新北区。dogbane leaf beetle ともいう
dogbane hawk moth		*Hyles chamyla* Denso	(チョウ目、スズメガ科) 旧北区
dogbane Saucrobotys moth		*Saucrobotys futilalis* (Lederer)	(チョウ目、メイガ科) 新北区
dogbane tiger moth			delicate Cycnia を見よ
dogberry aphid		*Nasonovia cynosbati* (Oestlund)	(カメムシ目、アブラムシ科) 新北区
dogface butterfly	イヌモンキチョウ	*Zerene cesonia* Stoll	(チョウ目、シロチョウ科) 新北区、新熱帯区
dogface butterfly			southern dogface を見よ
dog's head	ホソアメリカオオトラフトンボ	*Tetragoneuria canis* MacLachlan	(トンボ目、エゾトンボ科) 新北区
dog's tooth		*Mamestra suasa* (Denis et Schiffermüller)	(チョウ目、ヤガ科) 旧北区
dogtail dragonfly		*Tetragoneuria cynosura* (Say)	(トンボ目、エゾトンボ科) 新北区
dogwood aphid	ミズキヒラタアブラムシ	*Anoecia corni* (Fabricius)	(カメムシ目、アブラムシ科) 日本、汎世界
dogwood borer			dogwood moth を見よ　dogwood borer moth ともいう
dogwood calligrapha		*Calligrapha philadelphica* Linnaeus	(コウチュウ目、ハムシ科) 新北区
dogwood clubgall midge		*Resseliella clavata* (Beutenmüller)	(ハエ目、タマバエ科) 新北区
dogwood moth		*Synanthedon scitula* (Harris)	(チョウ目、スカシバガ科) 新北区
dogwood scale		*Chionaspis corni* Cooley	(カメムシ目、マルカイガラムシ科) 新北区
dogwood spittle bug		*Clastoptera proteus* Fitch	(カメムシ目、コガシラアワフキ科) 新北区
dogwood thyatirid moth		*Euthyatira pudens* (Guenée)	(チョウ目、カギバガ科) 新北区
dogwood twig borer		*Oberea tripunctata* (Swederus)	(コウチュウ目、カミキリムシ科) 新北区
Doherty's dull oakblue			Khamti oakblue を見よ
Doherty's green oakbue		*Arhopala hellenore* Doherty	(チョウ目、シジミチョウ科) 東洋区
Doherty's longtail		*Himantopterus dohertyi* (Elwes)	(チョウ目、マダラガ科) 東洋区
Doherty's oakblue		*Arhopala dohertyi* Bethune-Baker	(チョウ目、シジミチョウ科) 東洋区
Dolicaon swordtail	ドリカオンオナガタイマイ	*Eurytides dolicaon* (Cramer)	(チョウ目、アゲハチョウ科) 新熱帯区
dolichopodids			long-legged flies (1) を見よ
Dolium hairstreak		*Nicolaea dolium* (Druce)	(チョウ目、シジミチョウ科) 新熱帯区

英名	和名	学名	所属、分布、ほか
Doll's clearwing moth		*Paranthrene dollii* (Neumoegen)	（チョウ目、スカシバガ科）新北区
Doll's sphinx		*Sphinx dollii* Neumoegen	（チョウ目、スズメガ科）新北区。Doll's sphinx moth ともいう
Dolor skipper		*Methionopsis dolor* Evans	（チョウ目、セセリチョウ科）新熱帯区
dome-backed spiny ant		*Polyrhachis australis* Mayr	（ハチ目、アリ科）豪州区
domestic capricorn beetle			house longhorn を見よ
domestic container mosquito			striped mosquito を見よ
domestic groundling		*Bryotropha domestica* (Haworth)	（チョウ目、キバガ科）旧北区
domesticated silkworm			silkworm を見よ
Dominula skipper		*Anisynta dominula* (Plötz)	（チョウ目、セセリチョウ科）豪州区
Domophon shoemaker			one-spotted Prepona を見よ
Donaldson's dropwing		*Trithemis donaldsoni* (Calvert)	（トンボ目、トンボ科）エチオピア区
Don-Dwala		*Bradinopyga cornuta* Ris	（トンボ目、トンボ科）エチオピア区
donkey daisy copper		*Chrysoritis zonarius* (Riley)	（チョウ目、シジミチョウ科）エチオピア区
Donnysa skipper		*Hesperilla donnysa* Hewitson	（チョウ目、セセリチョウ科）豪州区
Donovan's amsacta		*Amsacta marginata* Donovan	（チョウ目、ヒトリガ科）豪州区
Donysa sister		*Adelpha donysa donysa* (Hewitson)	（チョウ目、タテハチョウ科）新熱帯区
Donzel's silver-line	ヒョウモンツバメ	*Cigaritis zohra* Donzel	（チョウ目、シジミチョウ科）旧北区
doodle bug			antlion を見よ　幼虫の英名
doodlebug			June beetle (1) を見よ
doodle bugs			antlions (1) を見よ　ウスバカゲロウ科の幼虫の俗称
dor			chafer beetles を見よ
dor beetle		*Geotrupes mutator* (Marsham)	（コウチュウ目、センチコガネ科）旧北区
dor beetle		*Geotrupes vernalis* (Linnaeus)	（コウチュウ目、センチコガネ科）旧北区
dor beetle (1)		*Geotrupes stereorarius* (Linnaeus)	（コウチュウ目、センチコガネ科）旧北区。人畜の糞に生息
dorbeetle (2)		*Geotrupes balyi* Jekel	（コウチュウ目、センチコガネ科）新北区
dor-beetle			June beetle (1) を見よ
dor beetles		*Geotrupes*	（コウチュウ目、センチコガネ科）の昆虫の総称
dor beetles			dung beetles (1) を見よ
dor-fly			June beetle (1) を見よ
Dorantes longtail			lilac-banded longtail を見よ

英名	和名	学名	所属、分布、ほか
Dorantes skipper			lilac-banded longtail を見よ
dorbug		*Lachnosterna fusca* (Fröhlich)	（コウチュウ目、コガネムシ科）新北区
Dorcas copper	アラスカベニシジミ	*Lycaena dorcas* Kirby	（チョウ目、シジミチョウ科）新北区
Dorippus tiger		*Danaus dorippus* (Klug)	（チョウ目、タテハチョウ科）エチオピア区
Doris	ドリスドクチョウ	*Laparus doris* (Linnaeus)	（チョウ目、タテハチョウ科）新熱帯区。Doris butterfly ともいう
Doris heliconian			Doris を見よ
Doris longwing			Doris を見よ　*L. d. viridis* (Staudinger) も同英名
Doris' pinemoth		*Coloradia doris* Barnes	（チョウ目、ヤママユガ科）新北区
Doris tiger moth		*Grammia doris* (Boisduval)	（チョウ目、ヒトリガ科）新北区
dormouse louse		*Schizophthirus pleurophaeus* (Burmeister)	（シラミ目、フトゲシラミ科）旧北区
dorr			June beetle（1）を見よ
dorsal dropwing		*Trithemis dorsalis* (Rambur)	（トンボ目、トンボ科）エチオピア区
Dorset clothes		*Eudarcia richardsoni* (Walsingham)	（チョウ目、ヒロズコガ科）旧北区。イングランド南西部の地名
Dorset cream wave		*Stegania trimaculata* (De Villers)	（チョウ目、シャクガ科）旧北区
Dorset straw			eastern bordered straw を見よ
Dorylas hairstreak			mottled hairstreak を見よ
doryline ants			army ants（1）を見よ
dos Passos' gemmed-satyr		*Cyllopsis dospassosi* Miller	（チョウ目、タテハチョウ科）新熱帯区
Dos Passos' hairstreak		*Callophrys dospassosi dospassosi* Clench	（チョウ目、シジミチョウ科）新熱帯区。*C. d. searsi* Clench も同英名
dot			white dot を見よ
dot-and-dash swordgrass moth		*Xylena curvimacula* (Morrison)	（チョウ目、ヤガ科）新北区
dot-banded satyr		*Oxeoschistus hilara* Bates	（チョウ目、タテハチョウ科）新熱帯区
dot-bordered grey		*Exoplisia cadmeis* (Hewitson)	（チョウ目、シジミタテハ科）新熱帯区
dot-bordered heliconian			Doris を見よ
dot-bordered wave			dotted border wave を見よ
dot-collared sabre-wing		*Jemadia pseudognetus* (Mabille)	（チョウ目、セセリチョウ科）新熱帯区
dot-collared skipper			dot-collared sabre-wing を見よ
dot-dash sergeant	マレーミヤマオオミスジ	*Athyma kanwa* Moore	（チョウ目、タテハチョウ科）東洋区
dot-lined angle moth		*Psamatodes abydata* (Guenée)	（チョウ目、シャクガ科）新北区
dot-lined wave moth		*Idaea tacturata* (Walker)	（チョウ目、シャクガ科）新北区

英名	和名	学名	所属、分布、ほか
dot-lined white moth		*Artace cribrarius* (Ljungh)	（チョウ目、カレハガ科）新北区
dot moth			white dot を見よ
dot moths		*Mamestra*	（チョウ目、ヤガ科）の昆虫の総称
dot-winged baskettail		*Epitheca petechialis* (Muttkowski)	（トンボ目、エゾトンボ科）新北区
dotted Antaeotricha moth		*Antaeotricha humilis* (Zeller)	（チョウ目、マルハキバガ科）新北区
dotted beauty			dotted carpet を見よ
dotted bee-fly		*Bombylius discolor* Mikan	（ハエ目、ツリアブ科）旧北区
dotted blue		*Euphilotes enoptes* (Boisduval)	（チョウ目、シジミチョウ科）新北区
dotted blue		*Tarucus sybaris sybaris* (Hopffer)	（チョウ目、シジミチョウ科）エチオピア区
dotted border		*Mylothris bernice* (Hewitson)	（チョウ目、シロチョウ科）エチオピア区
dotted border		*Agriopis marginaria* (Fabricius)	（チョウ目、シャクガ科）旧北区
dotted border moth			dotted carpet を見よ
dotted-border moth		*Graphiphora compta* Walker	（チョウ目、ヤガ科）豪州区
dotted border wave		*Idaea sylvestris* Linnaeus	（チョウ目、シャクガ科）旧北区
dotted borders		*Mylothris*	（チョウ目、シロチョウ科）の昆虫の総称
dotted carpet		*Alcis jubata* (Thunberg)	（チョウ目、シャクガ科）旧北区。日本亜種は *A. j. melanonota* Prout コケエダシャク
dotted Carteris moth		*Carteris oculatalis* (Möschler)	（チョウ目、ヤガ科）新北区
dotted checkerspot		*Poladryas minuta* (Edwards)	（チョウ目、タテハチョウ科）新北区
dotted chestnut		*Conistra rubiginea* (Denis et Schiffermüller)	（チョウ目、ヤガ科）旧北区
dotted clay		*Xestia baja* (Denis et Schiffermüller)	（チョウ目、ヤガ科）旧北区
dotted cloaked marble		*Hedya atropunctana* (Zetterstedt)	（チョウ目、ハマキガ科）旧北区
dotted cloaked marble (1)	シロモンヒメハマキ	*Hedya dimidiana* (Clerck)	（チョウ目、ハマキガ科）日本、旧北区。ウメ、スモモ、サクラなどの害虫
dotted crambus	ゴマフツトガ	*Chilo pulveratus* (Wileman et South)	（チョウ目、メイガ科）日本
dotted Ecdytolopha moth		*Gymnandrosoma punctidiscanum* (Dyar)	（チョウ目、ハマキガ科）新北区
dotted fanfoot		*Macrochilo cribrumalis* (Hübner)	（チョウ目、ヤガ科）旧北区
dotted flat		*Celaenorrhinus monartus* (Plötz)	（チョウ目、セセリチョウ科）新熱帯区
dotted footman		*Pelosia muscerda* (Hufnagel)	（チョウ目、ヒトリガ科）旧北区。日本亜種は *P. m. tetrasticta* (Hampson) ホシホソバ
dotted glory		*Asterope markii* (Hewitson)	（チョウ目、タテハチョウ科）新熱帯区

英名	和名	学名	所属、分布、ほか
dotted grain moth			willow tail prominent を見よ
dotted gray moth		*Glena cribrataria* (Guenée)	(チョウ目、シャクガ科) 新北区。dotted gray ともいう
dotted grey groundling		*Athrips mouffetella* (Linnaeus)	(チョウ目、キバガ科) 旧北区
dotted graylet moth		*Hyperstrotia perverens* (Barnes et McDunnough)	(チョウ目、ヤガ科) 新北区。dotted graylet ともいう
dotted knot-horn		*Phycita spissicella* Fabricius	(チョウ目、メイガ科) 旧北区
dotted leaftier moth		*Psilocorsis reflexella* Clemens	(チョウ目、マルハキバガ科) 新北区
dotted margin smoke		*Diplodoma herminata* Geoffroy	(チョウ目、ミノガ科) 旧北区
dotted metalmark		*Argyrogrammana stilbe holosticta* (Godman et Salvin)	(チョウ目、シジミタテハ科) 新熱帯区
dotted prince		*Stalachtis phlegia* (Cramer)	(チョウ目、シジミタテハ科) 新熱帯区
dotted roadside skipper		*Amblyscirtes eos* (Edwards)	(チョウ目、セセリチョウ科) 新北区、新熱帯区
dotted rustic		*Rhyacia simulans* (Hufnagel)	(チョウ目、ヤガ科) 旧北区
dotted sable	シロモンノメイガ	*Bocchoris inspersalis* (Zeller)	(チョウ目、メイガ科) 日本、旧北区、東洋区、豪州区、エチオピア区
dotted sallow moth		*Anathix ralla* (Grote et Robinson)	(チョウ目、ヤガ科) 新北区
dotted shade			Eana grass Tortrix moth を見よ
dotted skipper		*Hesperia attalus* (Edwards)	(チョウ目、セセリチョウ科) 新北区
dotted white geometrid	ホシシャク	*Naxa seriaria* (Guenée)	(チョウ目、シャクガ科) 日本、旧北区
double-arched roller	ナミモンカギバヒメハマキ	*Anclylis geminana* (Donovan)	(チョウ目、ハマキガ科) 日本、旧北区
double-banded banner		*Pyrrhogyra otolais otolais* Bates	(チョウ目、タテハチョウ科) 新熱帯区
double-banded carpet moth		*Spargania magnoliata* Guenée	(チョウ目、シャクガ科) 新北区
double-banded grass-veneer moth			grass webworm (1) を見よ
double-banded groundstreak		*Lamprospilus tarpa* (Godman et Salvin)	(チョウ目、シジミチョウ科) 新熱帯区
double-banded orange			African golden Arab を見よ
double-banded Zale moth		*Zale calycanthata* (Smith et Abbott)	(チョウ目、ヤガ科) 新北区。double-banded zale ともいう
double-barred			clover looper を見よ
double-barred birch pigmy		*Stigmella continuella* (Stainton)	(チョウ目、モグリチビガ科) 旧北区
double-barred nettle-tap			apple leaf skeletonizer (1) を見よ

英名	和名	学名	所属、分布、ほか
double-barred orange		*Olethreutes bifasciana* (Haworth)	(チョウ目、ハマキガ科) 旧北区
double-barred smudge		*Ypsolophus alpellus* (Denis et Schiffermüller)	(チョウ目、クチブサガ科) 旧北区
double bay-streaked button		*Acleris rufana* (Denis et Schiffermüller)	(チョウ目、ハマキガ科) 旧北区
double-branded blue crow	ルリマダラ	*Euploea sylvester lactifica* Butler	(チョウ目、タテハチョウ科) 日本、東洋区。double-banded crow ともいう。豪州区の *E. sylvester* (Fabricius) two-branded crow にも本英名が使われる
double-brown groundling		*Aroga velocella* (Duponchel)	(チョウ目、キバガ科) 旧北区
double-cross ladybird			variable ladybird を見よ
double dart		*Graphiphora augur* (Fabricius)	(チョウ目、ヤガ科) 日本、旧北区
double dart moth			double dart を見よ
double-dotted skipper			double-spotted skipper を見よ
double drummer		*Thopha saccata* (Fabricius)	(カメムシ目、セミ科) 豪州区
double-eye moths		Amphitheridae	(チョウ目) の昆虫の総称
double-headed hawk moth		*Coequosa triangularis* (Donovan)	(チョウ目、スズメガ科) 豪州区
double kidney	ヤナギキリガ	*Ipimorpha retusa* (Linnaeus)	(チョウ目、ヤガ科) 旧北区
double line	フタオビキヨトウ	*Mythimna turca* (Linnaeus)	(チョウ目、ヤガ科) 日本。イネ科牧草害虫
double-lined brown moth		*Hormoschista latipalpis* (Walker)	(チョウ目、ヤガ科) 新北区
double-lined Doryodes moth		*Doryodes bistrialis* (Geyer)	(チョウ目、ヤガ科) 新北区
double-lined Furcula			western Furcula moth を見よ
double-lined gray moth		*Cleora sublunaria* (Guenée)	(チョウ目、シャクガ科) 新北区
double-lined prominent moth		*Lochmaeus bilineata* (Packard)	(チョウ目、シャチホコガ科) 新北区
double-lined wainscot			double line を見よ
double lobed moth		*Lateroligia ophiogramma* Esper	(チョウ目、ヤガ科) 全北区。double lobed ともいう
double orange-spot piercer		*Cydia aurana* (Fabricius)	(チョウ目、ハマキガ科) 旧北区
double shoot		*Rhyacionia logaea* Durrant	(チョウ目、ハマキガ科) 旧北区
double shoot (1)	マツツマアカシンムシ	*Rhyacionia duplana simulata* Heinrich	(チョウ目、ハマキガ科) 日本、旧北区。マツ類害虫
double silver-barred cosmet		*Mompha subbistrigella* (Haworth)	(チョウ目、カザリバガ科) 旧北区
double spined bark beetle	オオハキクイムシ	*Ips duplicatus* (Sahlberg)	(コウチュウ目、キクイムシ科) 日本、旧北区

英名	和名	学名	所属、分布、ほか
double-spined spruce bark beetle			double spined bark beetle を見よ
double-spot brocade		*Meganephria bimaculosa* (Linnaeus)	（チョウ目、ヤガ科）旧北区
double-spotted cicada		*Cicadetta labeculata* (Distant)	（カメムシ目、セミ科）豪州区
double spotted flat		*Celaenorrhinus pyrrha* de Nicéville	（チョウ目、セセリチョウ科）東洋区
double-spotted lineblue		*Nacaduba biocellata* (C. et R. Felder)	（チョウ目、シジミチョウ科）豪州区
double-spotted owl-butterfly		*Eryphanis aesacus* (Herrich-Schäffer)	（チョウ目、タテハチョウ科）新熱帯区
double-spotted skipper		*Decinea percosius* (Godman)	（チョウ目、セセリチョウ科）新北区、新熱帯区
double-spotted spangle		*Megalographa bimaculata* (Stephens)	（チョウ目、ヤガ科）旧北区
double-spotted spangle			two-spotted looper moth を見よ
double square-spot		*Xestia triangulum* (Hufnagel)	（チョウ目、ヤガ科）旧北区
double-striped bluet		*Enallagma basidens* Calvert	（トンボ目、イトトンボ科）新北区
double-striped longtail			Bell's longtail を見よ
double-striped pug		*Gymnoscelis rufifasciata* (Haworth)	（チョウ目、シャクガ科）旧北区
double-striped pug		*Gymnoscelis pumilata* Hübner	（チョウ目、シャクガ科）旧北区
double-striped red knot-horn		*Cryptoblabes bistriga* Haworth	（チョウ目、メイガ科）旧北区
double-striped Scoparia moth		*Scoparia biplagialis* Walker	（チョウ目、メイガ科）新北区
double-striped tabby		*Hypsopygia glaucinalis* Linnaeus	（チョウ目、メイガ科）旧北区
double swallowtail			two-tailed swallowtail を見よ
double-toothed prominent			elm leaf caterpillar を見よ
double-toothed prominent moth			elm leaf caterpillar を見よ
double-tufted royal		*Dacalana penicilligera* (de Nicéville)	（チョウ目、シジミチョウ科）東洋区
double-tufted royal		*Dacalana vidura* (Horsfield)	（チョウ目、シジミチョウ科）東洋区
double-tufted wasp moth		*Didasys belae* Grote	（チョウ目、ヒトリガ科）新北区
Doubleday's Baileya moth		*Baileya doubledayi* (Guenée)	（チョウ目、ヤガ科）新北区
Doubleday's Notocelia moth			rose eucosmid を見よ

英名	和名	学名	所属、分布、ほか
Doubleday's orange		*Colotis doubledayi angolanus* Talbot	（チョウ目、シロチョウ科）エチオピア区
Doubleday's pug		*Eupithecia cauchiata* (Duponchel)	（チョウ目、シャクガ科）旧北区
Doubleday's sailor		*Dynamine setabis* Doubleday	（チョウ目、タテハチョウ科）新熱帯区
Doubleday's skipper		*Toxidia doubledayi* (Felder)	（チョウ目、セセリチョウ科）豪州区
Doubleday's untailed Charaxes		*Charaxes doubledayi* Aurivillius	（チョウ目、タテハチョウ科）エチオピア区
doubtful conch		*Cochylis dubitana* (Hübner)	（チョウ目、ホソハマキガ科）旧北区
doubtful Apamea moth		*Apamea dubitans* Walker	（チョウ目、ヤガ科）新北区。doubtful Apamea ともいう
Douglas fir adelges		*Adelges cooleyi* (Gillette)	（カメムシ目、カサアブラムシ科）全北区
Douglas-fir beetle		*Dendroctonus pseudotsugae* Hopkins	（コウチュウ目、キクイムシ科）新北区
Douglas fir borer		*Clytus blaisdelli* Van Dyke	（コウチュウ目、カミキリムシ科）新北区
Douglas fir chermes			Douglas fir adelges を見よ　米国での英名
Douglas fir cone moth		*Barbara colfaxina* (Kearfott)	（チョウ目、ハマキガ科）新北区。Douglas-fir conemoth とも記す
Douglas fir cone moth			larch tortrix (2) を見よ
Douglas fir engraver		*Scolytus unispinosus* LeConte	（コウチュウ目、キクイムシ科）新北区
Douglas fir needle midge		*Contarinia pseudotsugae* Condrashoff	（ハエ目、タマバエ科）新北区
Douglas-fir pitch moth		*Synanthedon novaroensis* (Edwards)	（チョウ目、スカシバガ科）新北区
Douglas fir seed fly		*Megastigmus spermatotrophus* Dalla Torre	（ハチ目、オナガコバチ科）旧北区
Douglas fir tussock moth		*Orgyia pseudotsugata* (McDunnough)	（チョウ目、ドクガ科）新北区。雌は無翅
Douglas fir twig weevil		*Cylindrocopturus furnissi* Buchanan	（コウチュウ目、ゾウムシ科）新北区
Douglas moths		Douglasiidae	（チョウ目）の昆虫の総称
douglasiid moths			Douglas moths を見よ
dove-colored knot-horn		*Acrobasis advenella* Zincken	（チョウ目、メイガ科）旧北区
Dover lazy conch			orange-barred conch を見よ　イングランド南部の地名に由来
Dover twist		*Periclepsis cinctana* (Denis et Schiffermüller)	（チョウ目、ハマキガ科）旧北区
dowdy plume	キキョウトリバ	*Stenoptilia zophodactyla* (Duponchel)	（チョウ目、トリバガ科）日本、全北区、豪州区、東洋区、新熱帯区、エチオピア区。リンドウ、トルコキキョウの害虫
down dwarf		*Elachista bedellella* (Sircom)	（チョウ目、クサモグリガ科）旧北区
downland horsefly		*Tabanus glaucopsis* Meigen	（ハエ目、アブ科）旧北区

英名	和名	学名	所属、分布、ほか
downland robberfly		*Machimus rusticus* (Meigen)	（ハエ目、ムシヒキアブ科）旧北区
downland villa		*Villa cingulata* (Meigen)	（ハエ目、ツリアブ科）旧北区
downlooker fly			snipe fly (1) (2) を見よ
downlooker snipefly			snipe fly (1) (2) を見よ
downy birch aphid			silver birch aphid を見よ
downy emerald		*Cordulia aenea* (Linnaeus)	（トンボ目、エゾトンボ科）旧北区。日本亜種は *C. a. amurensis* Selys　カラカネトンボ
downy leather-winged beetle		*Podabrus pruinosus* LeConte	（コウチュウ目、ジョウカイボン科）新北区
drab angle moth		*Speranza evagaria* (Hulst)	（チョウ目、シャクガ科）新北区
drab brown wave moth		*Lobocleta ossularia* (Geyer)	（チョウ目、シャクガ科）新北区
drab carpet		*Minoa murinata* (Scopoli)	（チョウ目、シャクガ科）旧北区
drab carpet moth		*Phrissogonus testulatus* Guenée	（チョウ目、シャクガ科）豪州区
drab Condylolomia moth		*Condylolomia participalis* Grote	（チョウ目、メイガ科）新北区
drab cutworm moth		*Melanchra homoscia* Meyrick	（チョウ目、ヤガ科）豪州区
drab looper			drab carpet を見よ
drab prominent moth		*Misogada unicolor* (Packard)	（チョウ目、シャチホコガ科）新北区。drab prominent ともいう
drab red moth		*Melanchra infensa* Walker	（チョウ目、ヤガ科）豪州区
drab three-spot missile		*Meza cybeutes* (Holland)	（チョウ目、セセリチョウ科）エチオピア区
drab wood-soldierfly		*Solva marginata* (Meigen)	（ハエ目、キアブモドキ科）旧北区
drabs and quakers			green fruitworms を見よ
dracaena thrips			banded-winged palm thrips を見よ
Draco skipper		*Polites draco* (Edwards)	（チョウ目、セセリチョウ科）新北区
dragon lubber		*Dracotettix monstrosus* Bruner	（バッタ目、バッタ科）新北区
dragon skipper			Draco skipper を見よ
dragonflies (1)	トンボ目	Odonata	の昆虫の総称　世界に約 5900 種。一般に dragon fly といわれる。damselflies を参照
dragonflies (2)	不均翅亜目	Anisoptera	（トンボ目）の昆虫の総称
dragonfly shaped geometrid		*Cystidia stratonice* (Stoll)	（チョウ目、シャクガ科）日本
dragonhunter			black clubtail を見よ
dragontails	スソビキアゲハ属	*Lamproptera*	（チョウ目、アゲハチョウ科）の昆虫の総称
drain fly			bathroom fly を見よ
Drake Mackerel mayfly			brown mayfly を見よ
Drakensberg brown		*Pseudonympha poetula* Trimen	（チョウ目、タテハチョウ科）エチオピア区
Drakensberg copper		*Chrysoritis oreas* (Trimen)	（チョウ目、シジミチョウ科）エチオピア区

英名	和名	学名	所属、分布、ほか
Drakensberg daisy copper			Drakensberg copper を見よ
Drakensberg malachite		*Chlorolestes draconicus* Balinsky	(トンボ目、Synlestidae) エチオピア区
drakes			mayflies を見よ
Draudt's hairstreak		*Lamasina draudti* (Lathy)	(チョウ目、シジミチョウ科) 新熱帯区
Draudti hairstreak			Draudt's hairstreak を見よ
dreaded lily borer			lily borer を見よ
dreamy duskywing	マメミヤマセセリ	*Erynnis icelus* Scudder et Burgess	(チョウ目、セセリチョウ科) 新北区
Dreisbach's Calephelis			Nogales metalmark を見よ
drepanosiphonine plantlice		Drepanosiphinae	(カメムシ目、アブラムシ科) の昆虫の総称
Drexel's Datana moth		*Datana drexelii* Hy. Edwards	(チョウ目、シャチホコガ科) 新北区
dried apple beetle		*Doticus pestilens* Olliff	(コウチュウ目、ヒゲナガゾウムシ科) 豪州区。カンキツ害虫
driedapple beetle		*Araecerus palmaris* (Pascoe)	(コウチュウ目、ヒゲナガゾウムシ科) 豪州区
dried bean beetle			bean weevil を見よ
dried currant figmoth			almond moth を見よ
dried currant moth			almond moth を見よ　英国での英名
dried fig and chocolate moth			almond moth を見よ
dried fig moth			almond moth を見よ
dried fish book louse			grain psocid を見よ
dried fruit beetle (1)	ガイマイデオキスイ	*Carpophilus dimidiatus* (Fabricius)	(コウチュウ目、ケシキスイ科) 日本、新北区、亜熱帯、熱帯。貯穀害虫
dried-fruit beetle (2)	クリヤケシキスイ	*Carpophilus hemipterus* (Linnaeus)	(コウチュウ目、ケシキスイ科) 日本、汎世界。貯穀害虫
dried fruit beetles		*Carpophilus*	(コウチュウ目、ケシキスイ科) の昆虫の総称
dried fruit beetles			sap beetles (1) を見よ　豪州での英名
dried-fruit knot-horn			almond moth を見よ
dried-fruit moth		*Vitula edmandsae serratilinella* Ragonot	(チョウ目、メイガ科) 新北区。幼虫は貯蔵食品とマルハナバチの巣につく
dried fruit moth			almond moth を見よ
dried fruit moth			Indian meal moth (1) を見よ
dried fruit moth			carob moth (1) を見よ
dried-fruit moth			American wax moth を見よ
dried fruit moths			cereal moths を見よ
drilid beetles		Drilidae	(コウチュウ目) の昆虫の総称
drinker		*Euthrix potatoria* (Linnaeus)	(チョウ目、カレハガ科) 旧北区。日本亜種は *E. p. bergmani* (Bryk) ヨシカレハ
drinker moth			drinker を見よ

英名	和名	学名	所属、分布、ほか
drive beetles			riffle beetles を見よ
driver ant		*Eciton hamatum* (Fabricius)	(ハチ目、アリ科) 新北区、新熱帯区
driver ants			safari ants を見よ
driver ants			army ants (1) を見よ
Dromus grizzled skipper			forest sandman を見よ
drone			ミツバチの雄
drone bee			drone fly を見よ
drone beetle	カナブン	*Rhomborrhina japonica* Hope	(コウチュウ目、コガネムシ科) 日本。ナシ、モモなど果樹害虫
drone flies		*Eristalis*	(ハエ目、ハナアブ科) の昆虫の総称
drone flies			flower flies を見よ
drone fly		*Eristalis tenax* (Linnaeus)	(ハエ目、ハナアブ科) 日本、汎世界。成虫はミツバチに似るので drone bee ともいわれる。幼虫は汚水に生息
drooping shadowdamsel		*Drepanosticta lankanensis* (Fraser)	(トンボ目、Platystictidae) 東洋区
dropwings		*Trithemis*	(トンボ目、トンボ科) の昆虫の総称
drosophila			pomace fly を見よ
Drucei hairstreak		*Theritas drucei* (Lathy)	(チョウ目、シジミチョウ科) 新熱帯区
Druce's pearly hairstreak		*Strephonota purpurantes* (Druce)	(チョウ目、シジミチョウ科) 新熱帯区
Druce's underleaf		*Eurybia caerulescens* Druce	(チョウ目、シジミタテハ科) 新熱帯区
drug darkling beetle	キュウリュウチュウ	*Palenbus dermestoides* (Chevrolat)	(コウチュウ目、ゴミムシダマシ科) 日本、汎世界
drugstore beetle			biscuit beetle を見よ
drugstore beetles			death watch beetles を見よ
drugstore weevil			biscuit beetle を見よ　米国での英名
drummer			giant drummer を見よ
drummer			discoid cockroach を見よ
drummers			ovoviviparous cockroaches を見よ
drumming katydid			oak bush-cricket を見よ
Drury's delight		*Mesoxantha ethosea* (Drury)	(チョウ目、タテハチョウ科) エチオピア区
Drury's egg white		*Dismorphia crisia* (Drury)	(チョウ目、シロチョウ科) 新熱帯区。*D. c. alvarez* J. et R. G. Maza, *D. c. virgo* (Bates) も同英名
Drury's owl moth		*Dactyloceras lucina* (Drury)	(チョウ目、イボタガ科) エチオピア区
Drusilla crescent			orange-patched crescent を見よ
Drusilla groundstreak		*Calycopis drusilla* Field	(チョウ目、シジミチョウ科) 新熱帯区
Drusius cloudywing		*Thorybes drusius* (Edwards)	(チョウ目、セセリチョウ科) 新北区、新熱帯区
dry-fungus beetles	ヒメキノコムシ科	Sphindidae	(コウチュウ目) の昆虫の総称

英名	和名	学名	所属、分布、ほか
dry-land wireworm		*Ctenicera glauca* (Germar)	(コウチュウ目、コメツキムシ科) 新北区
dry leaf	コノハタテハモドキ	*Junonia tugela* Trimen	(チョウ目、タテハチョウ科) エチオピア区
dry weather flies			dog-day cicadas を見よ
Dryad		*Minois dryas* Scopoli	(チョウ目、タテハチョウ科) 旧北区。ギリシャ神話の森の妖精 Dryas に由来。日本亜種は *M. d. bipunctata* (Motschulsky) ジャノメチョウで two-spotted Dryad、Dryad butterfly という
dryandra moth		*Carthaea saturnioides* Walker	(チョウ目、Carthaeidae) 豪州区
Dryburgh's skolly		*Thestor dryburghi* van Son	(チョウ目、シジミチョウ科) エチオピア区
dryinid wasps			pincher wasps を見よ
dryinids			pincher wasps を見よ
dryomyzid flies	ベッコウバエ科	Dryomyzidae	(ハエ目) の昆虫の総称
dryomyzids			dryomyzid flies を見よ
Dryope hairstreak		*Satyrium dryope* Edwards	(チョウ目、シジミチョウ科) 新北区
dryopid beetles			long-toed water beetles を見よ
dryopids			long-toed water beetles を見よ
drywood termites		*Cryptotermes*	(シロアリ目、レイビシロアリ科) の昆虫の総称 豪州での英名
drywood termites (1)	レイビシロアリ科	Kalotermitidae	(シロアリ目) の昆虫の総称
drywood termites (2)		*Incistermes*	(シロアリ目、レイビシロアリ科) の昆虫の総称
dry-wood termites (3)		*Kaltotermes, Procryptotermes*	(シロアリ目、レイビシロアリ科) の昆虫の総称
dry-wooden longicorn			dry-wooden longicorn beetle を見よ
dry-wooden longicorn beetle	イエカミキリ	*Stromatium longicorne* (Newman)	(コウチュウ目、カミキリムシ科) 日本、東洋区。クチナシ、モミ類、乾材の害虫
dubia roach			South American cockroach を見よ
dubious fairy hairstreak		*Hypolycaena dubia* Aurivillius	(チョウ目、シジミチョウ科) エチオピア区
dubious flitter	ムモンニセクロセセリ	*Quedara monteithi* (Wood-Mason et de Nicéville)	(チョウ目、セセリチョウ科) 東洋区
dubious scalloped bush brown		*Bicyclus dubia* (Aurivillius)	(チョウ目、タテハチョウ科) エチオピア区
dubious tiger moth		*Spilosoma dubia* (Walker)	(チョウ目、ヒトリガ科) 新北区
duboisia flea beetle		*Psylliodes paritis* Weise	(コウチュウ目、ハムシ科) 豪州区
duboisia leaf beetle			duboisia flea beetle を見よ
duck louse		*Lipeurus anatis* (Fabricius)	(ハジラミ目、チョウカクハジラミ科) 旧北区
duck louse			slender duck louse を見よ
Duckweed firetail		*Telebasis byersi* Westfall	(トンボ目、イトトンボ科) 新北区
Dudgeon carpenter worm moths		Dudgeoneidae	(チョウ目) の昆虫の総称
Duellona metalmark		*Necyria larunda* Godman et Salvin	(チョウ目、シジミタテハ科) 新熱帯区
Duke of Burgundy fritillary	セイヨウシジミタテハ	*Hamearis lucina* (Linnaeus)	(チョウ目、シジミタテハ科) 旧北区。Duke of Burgundy ともいう

英名	和名	学名	所属、分布、ほか
Duke's blue		*Lepidochrysops dukei* Cottrell	(チョウ目、シジミチョウ科) エチオピア区
Dukes' skipper			scarce swamp skipper を見よ
dull Astraptes		*Astraptes anaphus* (Cramer)	(チョウ目、セセリチョウ科) 新北区、新熱帯区
dull Babul blue			Indian Babul blue を見よ
dull brown job		*Nastra chao* (Mabille)	(チョウ目、セセリチョウ科) 新熱帯区
dull brown looper moth		*Xanthorhoe aegrota* Butler	(チョウ目、シャクガ科) 豪州区
dull copper		*Paralucia pyrodiscus* (Doubleday)	(チョウ目、シジミチョウ科) 豪州区
dull copper			streaked copper を見よ
dull Doryodes moth		*Doryodes spadaria* Guenée	(チョウ目、ヤガ科) 新北区
dull firetip			Araxes skipper を見よ
dull firetip			Arizona Araxes skipper を見よ
dull forester		*Lethe gulnihal* de Nicéville	(チョウ目、タテハチョウ科) 東洋区
dull four-spined legionnaire		*Chorisops tibialis* (Meigen)	(ハエ目、ミズアブ科) 旧北区
dull green hairstreak	ムラサキミドリシジミ	*Esakiozephyrus icana* (Moore)	(チョウ目、シジミチョウ科) 東洋区
dull grey mountain moth		*Notoreas omichlias* Meyrick	(チョウ目、シャクガ科) 豪州区
dull heath-blue			Mathew's blue を見よ
dull jewel		*Hypochrysops epicurus* Miskin	(チョウ目、シジミチョウ科) 豪州区
dull oak blue	オオムラサキツバメ	*Arhopala centaurus* (Fabricius)	(チョウ目、シジミチョウ科) 豪州区。東洋区の *A. c. nakula* C. et R. Felder も同英名
dull reddish dart moth		*Xestia dilucida* Morrison	(チョウ目、ヤガ科) 新北区。dull reddish dart ともいう
dull skipper		*Mnaseas bicolor* (Mabille)	(チョウ目、セセリチョウ科) 新熱帯区
dull spotted moth		*Leptocroca scholaea* Meyrick	(チョウ目、マルハキバガ科) 豪州区
duller broom midget		*Phyllonorycter scopariella* (Duponchel)	(チョウ目、ホソガ科) 旧北区
dumble-dor			dor beetle (1) を見よ
Dumeril skipper		*Drephalys dumeril* (Latreille)	(チョウ目、セセリチョウ科) 新熱帯区
Dumeril's luperina			Dumeril's rustic を見よ
Dumeril's rustic		*Luperina dumerilii* (Duponchel)	(チョウ目、ヤガ科) 旧北区
dump flies			black garbage flies を見よ
dump fly		*Ophyra aenescens* (Wiedemann)	(ハエ目、ハナバエ科) 新北区、大洋区、新熱帯区
dumpling bug		*Niastama punctaticollis* Reuter	(カメムシ目、カスミカメムシ科) 豪州区
dun bar			dun-bar moth を見よ　dun-bar の使用例もあり

英名	和名	学名	所属、分布、ほか
dun-bar moth		*Cosmia trapezina* (Linnaeus)	(チョウ目、ヤガ科) 旧北区
dun fly			cleg (1) を見よ
Dun sedge skipper			Dun skipper (1) を見よ
Dun skipper		*Euphyes ruricola melacomet* Harris	(チョウ目、セセリチョウ科) 新北区
Dun skipper (1)		*Euphyes vestris* (Boisduval)	(チョウ目、セセリチョウ科) 新北区。*E. v. harbisoni* Brown et McGuire も同英名
dune crickets		Schizodactylidae	(バッタ目) の昆虫の総称
dune robberfly	シロズヒメムシヒキ	*Philonicus albiceps* (Meigen)	(ハエ目、ムシヒキアブ科) 日本、旧北区
dune villa		*Villa modesta* (Meigen)	(ハエ目、ツリアブ科) 旧北区
dunflies			horse flies (2) を見よ
dunflies			clegs を見よ
dung beetle		*Aphodius fossor* (Linnaeus)	(コウチュウ目、コガネムシ科) 旧北区
dung beetle		*Aphodius merdarius* (Fabricius)	(コウチュウ目、コガネムシ科) 旧北区
dung beetle (1)		*Aphodius fimetarius* Linnaeus	(コウチュウ目、コガネムシ科) 全北区
dung beetles		*Phanaeus*	(コウチュウ目、コガネムシ科) の昆虫の総称
dung beetles (1)	センチコガネ科	Geotrupidae	(コウチュウ目) の昆虫の総称
dung beetles (2)		*Aphodius*	(コウチュウ目、コガネムシ科) の昆虫の総称
dung beetles (3)	ダイコクコガネ亜科	Scarabaeinae	(コウチュウ目、コガネムシ科) の昆虫の総称
dung beetles			chafer beetles を見よ
dung breeding fly			Oriental face fly を見よ
dung flies	フンバエ科	Scatophagidae	(ハエ目) の昆虫の総称
dung flies			yellow dung-flies を見よ　米国での英名
dung fly			yellow dung fly を見よ
dung maggots			mushroom flies を見よ
dung midges			dung flies を見よ
dung roller beetle			tumble-bugs (1) を見よ
dung rollers			chafer beetles を見よ
duns		*Leptophlebia*	(カゲロウ目、トビイロカゲロウ科) の昆虫の総称
duns			mayflies を見よ
duns			caddisflies (1) を見よ　南アフリカでの英名
dupla scale			camellia mining scale を見よ
Duponchel's sphinx		*Amphonyx duponchel* (Poey)	(チョウ目、スズメガ科) 新北区、新熱帯区。Dupochel's sphinx moth ともいう
d'Urban's brown		*Neita durbani* (Trimen)	(チョウ目、タテハチョウ科) エチオピア区
d'Urban's woolly legs		*Lechnocnema durbani* Trimen	(チョウ目、シジミチョウ科) エチオピア区
Durdham-down crest		*Telephila schmidtiellus* (Heyden)	(チョウ目、キバガ科) 旧北区

英名	和名	学名	所属、分布、ほか
Durham Argus			brown Argus (1) を見よ
durian fruit borer		*Mudaria magniplaga* Walker	（チョウ目、ヤガ科）東洋区
durra stalk borer		*Sesamia cretica* Lederer	（チョウ目、ヤガ科）旧北区、エチオピア区。イネ害虫。durra アズキモロコシ に由来
durra stem borer		*Sesamia vuteria* Stoll	（チョウ目、ヤガ科）旧北区
durra stem borer			pink borer (2) を見よ
durra stem borer			durra stalk borer を見よ
Dury's metalmark		*Apodemia duryi* (W. H. Edwards)	（チョウ目、シジミタテハ科）新熱帯区
duskfliers		*Zyxomma*	（トンボ目、トンボ科）の昆虫の総称
duskhawkers		*Gynacantha*	（トンボ目、ヤンマ科）の昆虫の総称
dusky Acraea		*Acraea esebria* Hewitson	（チョウ目、タテハチョウ科）エチオピア区
dusky Apamea moth		*Apamea plutonia* Grote	（チョウ目、ヤガ科）新北区。dusky Apamea, dusky quaker ともいう
dusky azure		*Celastrina nigra* (Forbes)	（チョウ目、シジミチョウ科）新北区
dusky azure			sooty azure を見よ
duskyback leafroller		*Archips mortuana* Kearfott	（チョウ目、ハマキガ科）新北区
dusky-backed Filatima moth		*Filatima pseudacaciella* (Chambers)	（チョウ目、キバガ科）新北区
dusky birch sawfly		*Croesus latitarsus* Norton	（ハチ目、ハバチ科）新北区
dusky black-vein		*Aporia nabellica* (Boisduval)	（チョウ目、シロチョウ科）旧北区、東洋区
dusky blue			African line blue を見よ
dusky blue cupid		*Celastrina huegelii dipora* (Moore)	（チョウ目、シジミチョウ科）東洋区。large hedge blue を参照
dusky blue cupid		*Celastrina dipora* (Moore)	（チョウ目、シジミチョウ科）東洋区
dusky blue groundstreak			dusky blue hairstreak を見よ
dusky blue hairstreak		*Calycopis isobeon* (Butler et Druce)	（チョウ目、シジミチョウ科）新北区。dusky-blue hairstreak とも記す
dusky brindled pearl		*Pyrausta prunalis* (Denis et Schiffermüller)	（チョウ目、メイガ科）旧北区
dusky brindled pearl		*Udea nivalis* (Drury)	（チョウ目、メイガ科）旧北区
dusky brocade	マツバラシラクモヨトウ	*Apamea remissa* (Hübner)	（チョウ目、ヤガ科）日本、旧北区
dusky brocade		*Apamea obscura* (Barnes et McDunnough)	（チョウ目、ヤガ科）新北区
dusky bushblue		*Arhopala paraganesa* (de Nicéville)	（チョウ目、シジミチョウ科）東洋区
dusky bush-brown			nigger を見よ
dusky carpet		*Tephronia sepiaria* (Hufnagel)	（チョウ目、シャクガ科）旧北区
dusky Charaxes			demon emperor を見よ
dusky clearwing			dusky clearwing moth を見よ
dusky clearwing moth		*Paranthrene tabaniformis* (Rottemburg)	（チョウ目、スカシバガ科）全北区。ポプラ、ヤナギなどの害虫

英名	和名	学名	所属、分布、ほか
dusky clubtail		*Gomphus spicatus* Hagen	(トンボ目、サナエトンボ科) 新北区
dusky cockroach			tawny cockroach (1) を見よ
dusky copper			African copper を見よ
dusky cotton stainers			cotton seed bugs を見よ
dusky cutworm moth		*Agrotis venerabilis* Walker	(チョウ目、ヤガ科) 新北区。幼虫は多くの作物、園芸植物を加害
dusky danaid			black friar を見よ
dusky dancer		*Argia translata* Hagen	(トンボ目、イトトンボ科) 新北区、新熱帯区
dusky diadem		*Ethope himachala* (Moore)	(チョウ目、タテハチョウ科) 東洋区
dusky dotted border		*Mylothris sagata* Grose-Smith	(チョウ目、シロチョウ科) エチオピア区
dusky emperor		*Asterocampa idyja* (Geyer)	(チョウ目、タテハチョウ科) 新北区
dusky emperor			cream-banded dusky emperor を見よ
dusky Euselasia		*Euselasia eubule* (R. Felder)	(チョウ目、シジミタテハ科) 新熱帯区
dusky evening brown		*Gnophodes chelys* (Fabricius)	(チョウ目、タテハチョウ科) エチオピア区
dusky-faced meadow katydid		*Orchelimum campestre* Blatchley	(バッタ目、キリギリス科) 新北区
dusky fanfoot		*Herminia zelleralis* (Wocke)	(チョウ目、ヤガ科) 旧北区
dusky giant owl butterfly			dusky owl-butterfly を見よ
dusky grasshopper		*Encoptolophus costalis* (Scudder)	(バッタ目、バッタ科) 新北区
dusky green underwing		*Albulina omphisa* Moore	(チョウ目、シジミチョウ科) 東洋区。
dusky grizzled skipper		*Pyrgus cacaliae* (Rambur)	(チョウ目、セセリチョウ科) 旧北区
dusky ground cricket		*Allonemobius funeralis* (Hart)	(バッタ目、コオロギ科) 新北区
dusky heath	ウスグロヒメヒカゲ	*Coenonympha dorus* (Esper)	(チョウ目、タテハチョウ科) 旧北区
dusky hedge blue		*Oreolyce vardhana* (Moore)	(チョウ目、シジミチョウ科) 東洋区
dusky Herpetogramma moth			tropical sod webworm を見よ
dusky knight			dingy ring を見よ
dusky knot-grass dagger	ナシケンモン	*Apatele rumicis* (Linnaeus)	(チョウ目、ヤガ科) 旧北区。ナシなど果樹、ダイズ、バラ、サクラなどの害虫
dusky labyrinth		*Neope yama* (Moore)	(チョウ目、タテハチョウ科) 東洋区
dusky lady beetles		Scymnini	(コウチュウ目、テントウムシ科) の昆虫の総称
dusky ladybird beetles			dusky lady beetles を見よ
dusky large blue		*Maculinea nausithous* (Bergstrasser)	(チョウ目、シジミチョウ科) 旧北区
dusky leafroller moth			urtica leaf roller を見よ

英名	和名	学名	所属、分布、ほか
dusky lemon sallow		*Xanthia gilvago* (Denis et Schiffermüller)	(チョウ目、ヤガ科) 旧北区
dusky long-horned groundling		*Lita sexpunctella* (Fabricius)	(チョウ目、キバガ科) 旧北区
dusky Madoryx		*Madoryx plutonius* (Hübner)	(チョウ目、スズメガ科) 新熱帯区
dusky marbled brown		*Gluphisia crenata vertunea* (Bray)	(チョウ目、シャチホコガ科) 旧北区
dusky meadow blue		*Tiora devanica* (Moore)	(チョウ目、シジミチョウ科) 東洋区
dusky meadow brown		*Hyponephele lycaon* Rottemburg	(チョウ目、タテハチョウ科) 旧北区
dusky metalmark			fatal metalmark を見よ
dusky-orange sallow		*Cirrhia gilvago palleago* Hübner	(チョウ目、ヤガ科) 旧北区
dusky owl-butterfly		*Caligo illioneus oberon* Butler	(チョウ目、タテハチョウ科) 新熱帯区
dusky pasture scarab		*Sericesthis nigrolineata* Boisduval	(コウチュウ目、コガネムシ科) 豪州区
dusky peacock		*Semiothisa signaria* (Hübner)	(チョウ目、シャクガ科) 旧北区
dusky plume	オオカマトリバ	*Oidaematophorus lithodactylus* (Treitschke)	(チョウ目、トリバガ科) 日本、旧北区。幼虫はオグルマなどを食う
dusky roadside skipper			blue-dusted roadside skipper を見よ
dusky royal		*Tajuria thyia* de Niceville	(チョウ目、シジミチョウ科) 東洋区
dusky sallow		*Eremobia ochroleuca* (Denis et Schiffermüller)	(チョウ目、ヤが科) 旧北区
dusky sallow rustic			dusky sallow を見よ
dusky sap beetle		*Carpophilus lugubris* Murray	(コウチュウ目、ケシキスイ科) 新北区
dusky sapphire		*Iolaus subinfuscata* (Grunberg)	(チョウ目、シジミチョウ科) エチオピア区
dusky Saucrobotys moth		*Saucrobotys fumoferalis* Hulst	(チョウ目、メイガ科) 新北区
dusky shieldback		*Clinopleura infuscata* Caudell	(バッタ目、キリギリス科) 新北区
dusky silver Y moth		*Syngrapha octoscripta* (Grote)	(チョウ目、ヤガ科) 新北区。dusky silver Y ともいう
dusky stink bug		*Euschistus tristigmus* (Say)	(カメムシ目、カメムシ科) 新北区
dusky sweep		*Pachytelia opacella* Herrich-Schäffer	(チョウ目、ミノガ科) 旧北区
dusky Temnora		*Temnora sardanus* (Walker)	(チョウ目、スズメガ科) エチオピア区
dusky thorn		*Ennomos fuscantaria* Haworth	(チョウ目、シャクガ科) 旧北区
dusky-veined Acraea		*Acraea igola* Trimen	(チョウ目、タテハチョウ科) エチオピア区
dusky veined walnut aphid			large walnut aphid を見よ

英名	和名	学名	所属、分布、ほか
duskywing moth		*Euclidia ardita* Franclemont	（チョウ目、ヤガ科）新北区
dusky-winged fritillary		*Clossiana improba* (Butler)	（チョウ目、タテハチョウ科）全北区
duskywings	ミヤマセセリ属	*Erynnis*	（チョウ目、セセリチョウ科）の昆虫の総称
dusky wireworm			obscure click beetle を見よ　カナダでの英名
dusky yellow-breast flat		*Gerosis phisara* (Moore)	（チョウ目、セセリチョウ科）東洋区
dusky yellowstreak		*Electrogena lateralis* (Curtis)	（カゲロウ目、ヒラタカゲロウ科）旧北区
dust beetles			death watch beetles を見よ
dust beetles			lyctids を見よ
dust lice		Atropidae	（チャタテムシ目）の昆虫の総称
dust lice			booklice (2) を見よ
dust-lice			booklice (3) を見よ
dust louse			booklouse (2)(3) を見よ
dust louse			cosmopolitan grain barklouse を見よ
dusted apple pigmy		*Bohemannia pulverosella* (Stainton)	（チョウ目、モグリチビガ科）旧北区
dusted black groundling		*Gelechia nigra* (Haworth)	（チョウ目、キバガ科）旧北区
dusted elfin		*Sarangesa seineri* Strand	（チョウ目、セセリチョウ科）エチオピア区
dusted skipper		*Atrytonopsis hianna* Scudder	（チョウ目、セセリチョウ科）新北区、新熱帯区
dusty			cotton seed bugs を見よ
dusty brown beetle			rice ground beetle を見よ
dusty brown beetles		*Gonocephalum*	（コウチュウ目、ゴミムシダマシ科）の昆虫の総称
dusty flat-body		*Agonopterix assimilella* (Treitschke)	（チョウ目、マルハキバガ科）旧北区
dusty hooktip	オビカギバ	*Drepana curvatula* (Borkhausen)	（チョウ目、カギバガ科）日本、旧北区
dusty June beetle		*Ambionoxia palpalis* (Horn)	（コウチュウ目、コガネムシ科）新北区
dusty lacewings			dustywings を見よ
dusty lined Matigramma moth		*Matigramma pulverilinea* (Grote)	（チョウ目、ヤガ科）新北区
dusty marble		*Bactra lancealana* (Hübner)	（チョウ目、ハマキガ科）旧北区
dusty millers			owlet moths を見よ　米国での英名
dusty peacock			pale-marked angle moth を見よ
dusty skimmer			cardinal meadowhawk を見よ
dusty sweep		*Acanthopsyche atra* Fuss	（チョウ目、ミノガ科）旧北区
dusty-tail roach			Surinam cockroach を見よ
dusty-wing		*Conwentzia hageni* Banks	（アミメカゲロウ目、コナカゲロウ科）新北区

英名	和名	学名	所属、分布、ほか
dustywings	コナカゲロウ科	Coniopterygidae	（アミメカゲロウ目）の昆虫の総称　小型で世界に240種
Duwweltjie caterpillar			geometrician を見よ　幼虫の英名
dwarf bee			lesser honey bee を見よ
dwarf blue		*Oraidium barberae* (Trimen)	（チョウ目、シジミチョウ科）エチオピア区
dwarf blue			eastern pygmy blue を見よ
dwarf bush-cricket			four-spot bush-cricket を見よ
dwarf clouded yellow		*Colias dubia* Elwes	（チョウ目、シロチョウ科）東洋区
dwarf cream wave		*Idaea fuscovenosa* (Goeze)	（チョウ目、シャクガ科）旧北区
dwarf cricket		*Metioche vittaticollis* (Stål)	（バッタ目、コオロギ科）大洋区
dwarf crow	ホリシャルリマダラ	*Euploea tulliolus* (Fabricius)	（チョウ目、タテハチョウ科）東洋区、豪州区。*E. t. ledereri* Felder et Felder も同英名
dwarf grass-veneer		*Platytes cerussella* Schiffermüller	（チョウ目、メイガ科）旧北区
dwarf honey bee			lesser honey bee を見よ
dwarf leafhoppers			cotton jassids を見よ
dwarf percher		*Diplacodes pumila* Dijkstra	（トンボ目、トンボ科）エチオピア区
dwarf pine bell		*Epinotia nanana* (Treitschke)	（チョウ目、ハマキガ科）旧北区
dwarf pug		*Eupithecia tantillaria* Boisduval	（チョウ目、シャクガ科）旧北区
dwarf sandman		*Spialia nanus* (Trimen)	（チョウ目、セセリチョウ科）エチオピア区
dwarf siberian pine caterpillar			Takamuku lasiocampid を見よ
dwarf skimmer		*Nannothemis bella* (Uhler)	（トンボ目、トンボ科）新北区
dwarf yellow			dainty sulphur を見よ
dwarfs	チベットシロチョウ属	*Baltia*	（チョウ目、シロチョウ科）の昆虫の総称
dwarfs			grass miner moths を見よ
Dyar's duskywing			pacuvius duskywing を見よ
Dyar's gemmed-satyr		*Cyllopsis suivalens suivalens* (Dyar)	（チョウ目、タテハチョウ科）新熱帯区。*C. s. escalantei* Miller も同英名
Dyar's lichen mooth		*Afrida ydatodes* Dyar	（チョウ目、ヒトリガ科）新北区
Dyar's metalmark		*Pheles eulesca* (Dyar)	（チョウ目、シジミタテハ科）新熱帯区
Dyar's Myscelus		*Myscelus perissodora* (Dyar)	（チョウ目、セセリチョウ科）新熱帯区
Dyar's ringlet		*Euptychoides hotchkissi* Dyar	（チョウ目、タテハチョウ科）新熱帯区
Dyar's skipperling		*Dalla dividuum* (Dyar)	（チョウ目、セセリチョウ科）新熱帯区
Dyar's swallowtail		*Battus ingenuus* (Dyar)	（チョウ目、アゲハチョウ科）新熱帯区
Dymas checkerspot	コビトヒョウモンモドキ	*Dymasia dymas* (Edwards)	（チョウ目、タテハチョウ科）新北区
Dyonis checkerspot		*Dynamine dyonis* Geyer	（チョウ目、タテハチョウ科）新北区

英名	和名	学名	所属、分布、ほか
Dyson's blue doctor	ツバメシジミタテハ	*Rhetus dysonii* Saunders	（チョウ目、シジミタテハ科）新熱帯区
Dyson's silverpatch		*Aides dysoni* Godman	（チョウ目、セセリチョウ科）新熱帯区
dytiscids			diving beetles (1) を見よ

英名	和名	学名	所属、分布、ほか
E			
Eana grass Tortrix moth		*Eana osseana* (Scopoli)	（チョウ目、ハマキガ科）全北区
Eanes longwing			small flambeau を見よ
ear borer			maize ear borer を見よ
ear-cutting caterpillar			armyworm (2) を見よ
ear-eating caterpillar			armyworm (2) を見よ
ear marked geometrid	ミミモンエダシャク	*Eilicrinia wehrlii* Djakonov	（チョウ目、シャクガ科）日本、旧北区
ear miner moth		*Apamea basilinea* (Denis et Schiffermüller)	（チョウ目、ヤガ科）旧北区
ear miner moth			wheat earworm を見よ
ear moth		*Amphipoea oculea* (Linnaeus)	（チョウ目、ヤガ科）旧北区
eared commodore	トガリタテハモドキ	*Precis tugela* Trimen	（チョウ目、タテハチョウ科）エチオピア区
earhead bug			rice bug (2) を見よ
ear-head shield bug		*Menida histrio* (Fabricius)	（カメムシ目、カメムシ科）東洋区。イネ害虫
ear-piercer			common earwig を見よ
early arctic		*Oeneis rosovi* (Kurentzov)	（チョウ目、タテハチョウ科）新北区
early brown stoneflies		*Capnia*	（カワゲラ目、クロカワゲラ科）の昆虫の総称
early bumblebee		*Bombus pratorum* (Linnaeus)	（ハチ目、ミツバチ科）旧北区
early button slug moth		*Tortricidia testacea* Packard	（チョウ目、イラガ科）新北区
early cattle grub			cattle warble fly (2) を見よ　幼虫の英名
early elfin		*Incisalia fotis* (Strecker)	（チョウ目、シジミチョウ科）新北区
early elfin			moss elfin を見よ
early grey		*Xylocampa areola* (Esper)	（チョウ目、ヤガ科）旧北区
early grey moth			early grey を見よ
early hairstreak	ハツハルカラスシジミ	*Erora laeta* (Hewitson)	（チョウ目、シジミチョウ科）新北区
early juniper carpet moth		*Thera contractata* (Packard)	（チョウ目、シャクガ科）新北区
early moth		*Thera primaria* (Haworth)	（チョウ目、シャクガ科）旧北区
early moth			early umber を見よ
early nesting humble bee			early bumblebee を見よ
early shoot borer		*Chilo infuscatellus* (Snellen)	（チョウ目、メイガ科）旧北区、東洋区。イネ害虫
early spring millers		*Xylomania*	（チョウ目、ヤガ科）の昆虫の総称
early spring millers		*Xylomyges*	（チョウ目、ヤガ科）の昆虫の総称
early thorn		*Selenia dentaria* (Fabricius)	（チョウ目、シャクガ科）旧北区

英名	和名	学名	所属、分布、ほか
early thorn moth		*Cosmotriche lunigera* Esper	（チョウ目、カレハガ科）旧北区
early thorn moth			early thorn を見よ
early tooth-striped		*Trichopteryx carpinata* (Borkhausen)	（チョウ目、シャクガ科）旧北区
early umber		*Theria rupicapraria* (Denis et Schiffermüller)	（チョウ目、シャクガ科）旧北区
early Zanclognatha moth		*Zanclognatha cruralis* (Guenée)	（チョウ目、ヤガ科）新北区。early Zanclognatha ともいう
earth-boring dung beetles		Geotrupinae	（コウチュウ目、コガネムシ科）の昆虫の総称
earth-boring dung beetles			dung beetles (1) を見よ
earth-boring scarab beetle			Japanese dung beetle (1) を見よ
earth crab			mole cricket (3) を見よ
earth fleas			springtails を見よ　南アフリカでの英名
earthworm flies		*Pollenia*	（ハエ目、クロバエ科）の昆虫の総称
earwig			common earwig を見よ
earwig flies		Meropidae	（シリアゲムシ目）の昆虫の総称
earwig scorpionflies			earwig flies を見よ　米国での英名
earwigfly			シリアゲムシ目の種
earwiglike entotrophs	ハサミコムシ科	Japygidae	（コムシ目）の昆虫の総称
earwigs	ハサミムシ目	Dermaptera	の昆虫の総称　世界に約 1,500 種。英名は耳と虫を意味する古英語、あるいは耳から入って脳に産卵するという昔話に由来するともいわれる
earwigs			common earwigs を見よ
East African forest Acraea		*Acraea pharsalus* Ward	（チョウ目、タテハチョウ科）エチオピア区
East Indian harvesting ant		*Pheidologeton diversus* (Jerdon)	（ハチ目、アリ科）東洋区
East-Mexican acacia skipper		*Cogia hippalus hiska* Evans	（チョウ目、セセリチョウ科）新熱帯区。acacia skipper を参照
East-Mexican banded-skipper		*Autochton siermadror* Burns	（チョウ目、セセリチョウ科）新熱帯区
East-Mexican banner		*Catonephele mexicana* Jenkins et de la Maza	（チョウ目、タテハチョウ科）新熱帯区
East-Mexican white-skipper		*Heliopyrgus sublinea* (Schaus)	（チョウ目、セセリチョウ科）新熱帯区
east slope ringlet			northwest ringlet を見よ
eastern alchemist			sweetpotato leaf worm を見よ
eastern amberwing			common amberwing を見よ
eastern ant cricket		*Myrmecophilus pergandei* (Bruner)	（バッタ目、アリヅカコオロギ科）新北区
eastern azure sapphire		*Heliophorus moorei saphir* Blanchard	（チョウ目、シジミチョウ科）旧北区。azure sapphire を参照

英名	和名	学名	所属、分布、ほか
eastern baron blue		*Pseudophilotes vicrama* (Moore)	(チョウ目、シジミチョウ科) 旧北区
eastern bath white		*Pontia edusa* (Fabricius)	(チョウ目、シロチョウ科) 旧北区
eastern beach tiger beetle		*Cicindela dorsalis* Say	(コウチュウ目、ハンミョウ科) 新北区
eastern bentwinged cicada		*Froggatoides typicus* Distant	(カメムシ目、セミ科) 豪州区
eastern black acorn weevil		*Curculio baculi* Chittenden	(コウチュウ目、ゾウムシ科) 新北区
eastern black swallowtail			black swallowtail を見よ
eastern blackheaded budworm			black-headed budworm を見よ　eastern black-headed budworm moth ともいう
eastern blacktail		*Nesciothemis farinosa* (Förster)	(トンボ目、トンボ科) エチオピア区
eastern blotched leopard			blotched leopard を見よ
eastern blue beauty			blue mother-of-pearl を見よ
eastern blue darner		*Aeschna verticalis* Hagen	(トンボ目、ヤンマ科) 新北区。開帳 10 cm
eastern blue sapphire		*Heliophorus oda* (Hewitson)	(チョウ目、シジミチョウ科) 東洋区
eastern bordered straw		*Heliothis nubigera* Herrich-Schäffer	(チョウ目、ヤガ科) 旧北区
eastern boxelder twig borer moth			boxelder twig borer を見よ
eastern brown Argus		*Plebejus carmon* (Gethard)	(チョウ目、シジミチョウ科) 旧北区
eastern brown Argus			Grecian Argus を見よ
eastern brown crow	マサキルリマダラ	*Euploea tulliolus koxinga* Fruhstorfer	(チョウ目、タテハチョウ科) 日本、東洋区。*E. tulliolus* (Fabricius) の英名。eastern brown-crow butterfly ともいう
eastern brown crow			dwarf crow を見よ
eastern buck moth			buck moth を見よ
eastern bush blue		*Cacyreus virilis* Stempffer	(チョウ目、シジミチョウ科) エチオピア区
eastern cactus-boring moth		*Melitara prodenialis* Walker	(チョウ目、メイガ科) 新北区
eastern carpenter bee			common carpenter bee をみよ
eastern cloudywing		*Thorybes confusis* Bell	(チョウ目、セセリチョウ科) 新北区
eastern comma	ウスイロシータテハ (ヨーロッパキタテハ)	*Polygonia egea* (Cramer)	(チョウ目、タテハチョウ科) 旧北区、東洋区
eastern comma			comma (2) を見よ
eastern common tiger			common tiger (1) をみよ
eastern courtier	キゴマダラ	*Sephisa chandra* (Moore)	(チョウ目、タテハチョウ科) 旧北区、東洋区
Eastern Covadonga skipper		*Pheraeus covadonga covadonga* Freeman	(チョウ目、セセリチョウ科) 新熱帯区
eastern crane fly		*Tipula abdominalis* (Say)	(ハエ目、ガガンボ科) 新北区

英名	和名	学名	所属、分布、ほか
eastern cricket		*Modicogryllus frontalis* (Fieber)	（バッタ目、コオロギ科）旧北区
eastern cypress katydid		*Inscudderia walkeri* Hebard	（バッタ目、キリギリス科）新北区
eastern dappled white			dappled marble を見よ
eastern death-watch beetle		*Hemicoelus carinatus* (Say)	（コウチュウ目、シバンムシ科）新北区
eastern digger bee		*Ptilothrix bombiformis* (Cresson)	（ハチ目、コシブトハナバチ科）新北区
eastern dingy lineblue		*Petrelaea tombugensis* (Rober)	（チョウ目、シジミチョウ科）東洋区、豪州区
eastern dobsonfly			dobsonfly を見よ
eastern dotted border		*Mylothris agathina* (Cramer)	（チョウ目、シロチョウ科）エチオピア区
eastern dusk-flat			common red-eye (1) を見よ
eastern elf			forest-watcher を見よ
eastern elm piercer		*Laspeyresia leguminana* (Lienig et Zeller)	（チョウ目、ハマキガ科）旧北区
eastern eyed click beetle		*Alaus oculatus* (Linnaeus)	（コウチュウ目、コメツキムシ科）新北区
eastern false wireworm		*Pterohelaeus darlingensis* Carter	（コウチュウ目、ゴミムシダマシ科）豪州区
eastern festoon	シロタイスアゲハ	*Allancastria cerisyi* (Godart)	（チョウ目、アゲハチョウ科）旧北区
eastern field wireworm		*Limonius agonus* (Say)	（コウチュウ目、コメツキムシ科）新北区
eastern five ring		*Ypthima similis* Elwes et Edwards	（チョウ目、タテハチョウ科）東洋区
eastern fivering		*Ypthima persimilis* Elwes et Edwards	（チョウ目、タテハチョウ科）東洋区
eastern five-spined ips			five-spined bark beetle を見よ
eastern flat		*Netrocoryne repanda* Felder	（チョウ目、セセリチョウ科）豪州区
eastern forktail			common forktail を見よ
eastern fungus weevil		*Euparius marmoreus* (Olivier)	（コウチュウ目、ヒゲナガゾウムシ科）新北区、新熱帯区
eastern giant ichneumon			giant ichneumon wasp を見よ
eastern golden haired blowfly			brown blowfly を見よ
eastern grape leafhopper		*Erythroneura comes* (Say)	（カメムシ目、オオヨコバイ科）新北区
eastern grapeleaf skeletonizer			grapeleaf skeletonizer を見よ
eastern Graphium			white-dappled swallowtail を見よ
eastern grass blue		*Chilades putli* (Kollar)	（チョウ目、シジミチョウ科）東洋区、豪州区

英名	和名	学名	所属、分布、ほか
eastern grass tubeworm moth		*Acrolophus plumifrontella* (Clemens)	（チョウ目、Acrolophidae）新北区
eastern grass-veneer moth		*Crambus laqueatellus* Clemens	（チョウ目、メイガ科）新北区
eastern grayback			gray petaltail を見よ
eastern green bush cricket		*Tettigonia caudata* (Charpentier)	（バッタ目、キリギリス科）旧北区
eastern green fruit beetle			green June beetle を見よ
eastern green-streaked playboy		*Pilodeudorix congoana* (Aurivillius)	（チョウ目、シジミチョウ科）エチオピア区
eastern greenish black-tip		*Elphinstonia penia* (Freyer)	（チョウ目、シロチョウ科）旧北区
eastern Hallelesis		*Hallelesis asochis* (Hewitson)	（チョウ目、タテハチョウ科）エチオピア区
eastern Hercules beetle			unicorn beetle を見よ
eastern hillside brown		*Stygionympha scotia* Quickelberge	（チョウ目、タテハチョウ科）エチオピア区
eastern horn lerp		*Creiis corniculata* (Froggatt)	（カメムシ目、キジラミ科）豪州区
eastern hornet			Oriental hornet を見よ
eastern lady beetle		*Coccinella transversoguttata* Falderman	（コウチュウ目、テントウムシ科）新北区。亜種 *C. t. richardsoni* Brown は transverse lady beetle という
eastern larch beetle		*Dendroctonus simplex* LeConte	（コウチュウ目、キクイムシ科）新北区
eastern larch borer		*Tetropium cinnamopterum* Kirby	（コウチュウ目、カミキリムシ科）新北区
eastern larch borer			larch longhorn beetle を見よ
eastern large heath		*Coenonympha rhodopensis* Elwes	（チョウ目、タテハチョウ科）旧北区
eastern leaf-footed bug			leaf-footed bug (1) を見よ
eastern least clubtail		*Stylogomphus albistylus* (Hagen)	（トンボ目、サナエトンボ科）新北区
eastern Lineodes moth		*Lineodes fontella* Walsingham	（チョウ目、メイガ科）新北区
eastern long-winged grasshopper		*Epacromius tergestinus* (Charpentier)	（バッタ目、バッタ科）旧北区
eastern lubber grasshopper		*Romalea guttata* (Latreille)	（バッタ目、Romaleidae）新北区。体長１０cm。米国では教材としてよく使用
eastern meadow fritillary		*Clossiana bellona toddi* (Holland)	（チョウ目、タテハチョウ科）新北区。meadow fritillary (1) を参照
eastern Nycteoline			black-striped bark-like moth を見よ
eastern olive triangle	コドルスタイマイ	*Graphium codrus* (Cramer)	（チョウ目、アゲハチョウ科）東洋区
eastern opal		*Chrysoritis orientalis* (Swanepoel)	（チョウ目、シジミチョウ科）エチオピア区

英名	和名	学名	所属、分布、ほか
eastern orange albatross			flame albatross を見よ
eastern orange tip	キイロクモマツマキチョウ	*Anthocharis damone* (Boisduval)	(チョウ目、シロチョウ科) 旧北区
eastern pale clouded yellow			Oriental clouded yellow を見よ
eastern Panthea moth		*Panthea furcilla* Packard	(チョウ目、ヤガ科) 新北区。eastern Panthea ともいう
eastern peach borer			peach tree borer を見よ
eastern pine catkin borer moth		*Calosima dianella* Dietz	(チョウ目、ネマルハキバガ科) 新北区
eastern pine elfin	ヒガシコツバメ	*Callophrys niphon* (Hübner)	(チョウ目、シジミチョウ科) 新北区
eastern pine looper		*Lambdina pellucidaria* (Grote et Robinson)	(チョウ目、シャクガ科) 新北区
eastern pine seedworm		*Cydia toreuta* Grote	(チョウ目、ハマキガ科) 新北区。eastern pine seedworm moth ともいう
eastern pine shoot borer			pine shoot borer を見よ
eastern pine weevil			deodar weevil を見よ
eastern pistol casebearer			pistol casebearer (1) を見よ
eastern pondhawk		*Erythemis simplicicollis* (Say)	(トンボ目、トンボ科) 新北区、大洋区。common pondhawk ともいう
eastern pygmy blue		*Brephidium isophthalma* (Herrich-Schäffer)	(チョウ目、シジミチョウ科) 新北区
eastern raspberry fruitworm		*Byturus rubi* Barber	(コウチュウ目、キスイモドキ科) 新北区。キイチゴ害虫
eastern red damsel		*Amphiagrion saucium* (Burmeister)	(トンボ目、イトトンボ科) 新北区
eastern red lacewing	アカハレギチョウ	*Cethosia cydippe* (Linnaeus)	(チョウ目、タテハチョウ科) 豪州区。東洋区のアカハレギチョウ *C. c. chrysippe* (Fabricius) も red lacewing という
eastern ringed Xenica		*Geitoneura acantha* (Donovan)	(チョウ目、タテハチョウ科) 豪州区。*Xenica* は本種の旧属名
eastern ringtail		*Erpetogomphus designatus* Hagen	(トンボ目、サナエトンボ科) 新北区
eastern rock grayling		*Hipparchia syriaca* (Staudinger)	(チョウ目、タテハチョウ科) 旧北区
eastern saltmarsh mosquito			golden saltmarsh mosquito を見よ
eastern sand wasp			common sand wasp を見よ
eastern sandgrinder		*Arenopsaltria nibivena* (Walker)	(カメムシ目、セミ科) 豪州区
eastern saw-tailed bush cricket		*Barbitistes constrictus* Brunner von Wattenwyl	(バッタ目、キリギリス科) 旧北区
eastern sawyer			southern pine sawyer を見よ
eastern scarlet			common scarlet (1) を見よ
eastern shieldbacks		*Atlanticus*	(バッタ目、キリギリス科) の昆虫の総称

英名	和名	学名	所属、分布、ほか
eastern short-tailed blue		Cupido decoloratus (Staudinger)	(チョウ目、シジミチョウ科) 旧北区
eastern silverstripe			Pallas' fritillary を見よ
eastern smoke fly		Microsania imperfecta (Zoew)	(ハエ目、ヒラタアシバエ科) 新北区
eastern spider beetle		Ptinus raptor Sturm	(コウチュウ目、ヒョウホンムシ科) 旧北区
eastern spruce budworm			spruce bud moth (1) を見よ
eastern spruce-gall adelgid			spruce pineapple gall adelges を見よ 北米での英名
eastern spruce gall aphid			spruce pineapple gall adelges を見よ 北米での英名
eastern striped albatross	オルフェルナトガリシロチョウ	Appias olferna Swinhoe	(チョウ目、シロチョウ科) 東洋区、豪州区
eastern striped cricket		Miogryllus saussurei (Scudder)	(バッタ目、コオロギ科) 新北区
eastern subterranean termite		Reticulitermes flavipes (Kollar)	(シロアリ目、ミゾガシラシロアリ科) 全北区
eastern swamp dotted border		Mylothris rubricosta (Mabille)	(チョウ目、シロチョウ科) エチオピア区
eastern swamp emerald		Procordulia jacksoniensis (Rambur)	(トンボ目、エゾトンボ科) 豪州区
eastern tailed-blue	アメリカツバメシジミ	Cupido comyntas (Godart)	(チョウ目、シジミチョウ科) 新北区、新熱帯区
eastern tent caterpillar			American tent caterpillar を見よ
eastern tent caterpillar moth			American tent caterpillar を見よ
eastern thread-legged katydid		Arethaea phalangium (Scudder)	(バッタ目、キリギリス科) 新北区
eastern tiger swallowtail			tiger swallowtail を見よ
eastern tree yellow	ブチローサムモンキチョウ	Gandaca butyrosa (Butler)	(チョウ目、シロチョウ科) 東洋区
eastern treehole mosquito		Ochlerotatus triseriatus (Say)	(ハエ目、カ科) 新北区
eastern tumblebug		Canthon pilularius Linnaeus	(コウチュウ目、コガネムシ科) 新北区。糞で巣を作り、産卵して地中にうめる
eastern velvet ant			common eastern velvet ant を見よ
eastern white-faced darter			white-faced darter (2) を見よ
eastern white-lady swordtail			white-dappled swallowtail を見よ
eastern willow clearwing moth		Synanthedon proxima (Edwards)	(チョウ目、スカシバガ科) 新北区
eastern wood white		Leptidea duponcheli (Staudinger)	(チョウ目、シロチョウ科) 旧北区
eastern yellow glassy tiger	クレオナアサギマダラ	Parantica cleona (Stoll)	(チョウ目、タテハチョウ科) 東洋区

英名	和名	学名	所属、分布、ほか
eastern yellowjacket		*Vespula maculifrons* (Buysson)	（ハチ目、スズメバチ科）新北区
eboneus skipper		*Zera eboneus* (Bell)	（チョウ目、セセリチョウ科）新熱帯区
ebony grasshopper		*Boopedon nubilum* (Say)	（バッタ目、バッタ科）新北区
ebony jewelwing		*Calopteryx maculata* (Beauvois)	（トンボ目、イトトンボ科）新北区
ebony scale			citrus scale を見よ
Ebor grassgrub		*Oncopera alboguttata* Tindale	（チョウ目、コウモリガ科）豪州区
Ebusus skipper		*Ebusus ebusus nigrior* Miller	（チョウ目、セセリチョウ科）新熱帯区
Ecaille rose		*Eusharia festiva* (Hufnagel)	（チョウ目、ヒトリガ科）旧北区
Echina skipper		*Aethilla echina echina* Hewitson	（チョウ目、セセリチョウ科）新熱帯区
echo moth		*Seirarctia echo* (Smith)	（チョウ目、ヒトリガ科）新北区
Ection army ant		*Ection burchellii* Westwood	（ハチ目、アリ科）新熱帯区
ectobiid cockroaches		Ectobiidae	（ゴキブリ目）の昆虫の総称
ectognathous insects			common bristletails を見よ
Edessa glasswing		*Heterosais edessa* (Hewitson)	（チョウ目、タテハチョウ科）新熱帯区
edge leafwing			Oenomais leafwing を見よ
edible grasshopper		*Homorocoryphus nitidulus vicinus* Walker	（バッタ目、キリギリス科）エチオピア区
Edinburgh pug			white-spotted pug (1) を見よ
Edith's checker butterfly			Edith's checkerspot を見よ
Edith's checkerspot	ニシシマヒョウモンモドキ	*Euphydryas editha* (Boisduval)	（チョウ目、タテハチョウ科）新北区。マルバネヒョウモンモドキの別和名あり。Edith's checker butterfly ともいう
Edith's copper	ネバダヤマベニシジミ	*Lycaena editha* (Mead)	（チョウ目、シジミチョウ科）新北区
Edith's Fabiola moth		*Fabiola edithella* Busck	（チョウ目、マルハキバガ科）新北区
Edmund's snaketail		*Ophiogomphus edmundo* Needham	（トンボ目、サナエトンボ科）新北区
Edocla redring			green-spotted banner を見よ
Edocla redwing		*Pyrrhogyra edocla* Doubleday	（チョウ目、セセリチョウ科）新熱帯区
Edwards' anthophora		*Anthophora edwardsii* Cresson	（ハチ目、コシブトハナバチ科）新北区
Edwards' azure		*Celastrina ladon violacea* (W. H. Edwards)	（チョウ目、シジミチョウ科）新北区。spring azure を参照
Edwards' cuckoo bee			Edwards' cuckoo wasp を見よ
Edward's cuckoo wasp		*Parnopes edwardsii* (Cresson)	（ハチ目、セイボウ科）新北区
Edward's false underwing		*Drasteria edwardsii* (Behr)	（チョウ目、ヤガ科）新北区

英名	和名	学名	所属、分布、ほか
Edward's fritillary	ウスミドリギンボシヒョウモン	*Speyeria edwardsii* (Reakirt)	（チョウ目、タテハチョウ科）新北区
Edwards' glasywing		*Hemihyalea edwardsii* (Packard)	（チョウ目、ヒトリガ科）新北区。Edwards' glassywing moth ともいう
Edward's hairstreak	エドワードカラスシジミ	*Satyrium edwardsii* (Grote et Robinson)	（チョウ目、シジミチョウ科）新北区
Edward's skipperling		*Oarisma edwardsii* (Barnes)	（チョウ目、セセリチョウ科）新北区、新熱帯区
Edwards spring azure			Edwards' azure を見よ
Edwards' wasp moth		*Lymire edwardsii* (Grote)	（チョウ目、ヒトリガ科）新北区
Edwards' water penny		*Eubrianax edwardsii* (LeConte)	（コウチュウ目、ヒラタドロムシ科）新北区
Edwards's forester		*Euphaedra edwardsii* (van der Hoeven)	（チョウ目、タテハチョウ科）エチオピア区
effective Euchlaena moth		*Euchlaena effecta* (Walker)	（チョウ目、シャクガ科）新北区
Effusa spreadwing		*Potamanaxas effusa* (Draudt)	（チョウ目、セセリチョウ科）新熱帯区
egg-case insects		Oothecaria	卵鞘を作る昆虫。cockroaches (2) を見よ
eggflies	リュウキュウムラサキ属	*Hypolimnas*	（チョウ目、タテハチョウ科）の昆虫の総称
eggfruit caterpillar			buff-tip moth を見よ
eggfruit caterpillar			buff tip を見よ
egg-parasite wasps			pointed-tailed wasps を見よ
egg parasites			scelionid wasps を見よ
egg-pod insects			egg-case insects を見よ
egger			oak egger を見よ　カレハガ科の種にも使われる
egger moth			oak egger を見よ
eggers			tent caterpillar moths (1) を見よ　eggars とも記す。
eggers and lappets			tent caterpillar moths (1) を見よ
eggplant flea beetle		*Epitrix fuscula* Crotch	（コウチュウ目、ハムシ科）新北区。野菜害虫
eggplant fruit-borer	ナスノメイガ	*Leucinodea orbonalis* Guenée	（チョウ目、メイガ科）日本、東洋区、エチオピア区。ナス害虫
eggplant lace bug		*Gargaphia solani* Heidemann	（カメムシ目、グンバイムシ科）新北区
eggplant leaf miner (1)		*Tildenia glochinella* (Zeller)	（チョウ目、キバガ科）新北区、新熱帯区。eggplant leafminer moth ともいう
eggplant leaf miner (2)		*Tildenia inconspicuella* (Murtfeldt)	（チョウ目、キバガ科）新北区
eggplant leafroller moth		*Lineodes integra* (Zeller)	（チョウ目、メイガ科）新北区。eggplant leafroller ともいう
eggplant stem borer			Indian meal moth (2) を見よ
eggplant tortoise beetle		*Cassida pallidula* Boheman	（コウチュウ目、ハムシ科）新北区
eggplant webworm moth		*Rhectocraspeda periusalis* (Walker)	（チョウ目、メイガ科）新北区

英名	和名	学名	所属、分布、ほか
Egina daggerwing		*Marpesia egina* (Bates)	（チョウ目、タテハチョウ科）新熱帯区
Egleis fritillary		*Speyeria egleis* (Behr)	（チョウ目、タテハチョウ科）新北区
Egyptian alfalfa weevil		*Hypera brunneipennis* (Boheman)	（コウチュウ目、ゾウムシ科）新北区
Egyptian beetle		*Blaps polychresta* Forskal	（コウチュウ目、ゴミムシダマシ科）豪州区
Egyptian bollworm	ミスジアオリンガ	*Earias insulana* (Boisduval)	（チョウ目、ヤガ科）日本、旧北区、東洋区。ワタ害虫
Egyptian bollworm moth			Egyptian bollworm を見よ
Egyptian cotton bollworm			Egyptian bollworm を見よ
Egyptian cotton leafworm		*Spodoptera littoralis* (Boisduval)	（チョウ目、ヤガ科）旧北区、エチオピア区
Egyptian cotton worm			common cutworm を見よ
Egyptian cottony-cussion scale	エジプトワタフキカイガラムシ	*Icerya aegyptiaca* (Douglas)	（カメムシ目、ワタフキカイガラムシ科）日本、東洋区、エチオピア区。トベラ、シャリンバイの害虫
Egyptian flower mantis			devil's flower mantis (1) を見よ
Egyptian fluted scale			Egyptian cottony-cussion scale を見よ
Egyptian grasshopper		*Anacridium aegyptium* (Linnaeus)	（バッタ目、バッタ科）旧北区
Egyptian hibiscus mealybug			hibiscus mealybug (1) を見よ
Egyptian house fly			common housefly of the Orient を見よ
Egyptian icerya			Egyptian cottony-cussion scale を見よ
Egyptian mealybug			Egyptian cottony-cussion scale を見よ
Egyptian praying mantis		*Miomantis abyssinica* Giglio-Tos	（カマキリ目、カマキリ科）旧北区
Egyptian praying mantis (1)		*Miomantis paykullii* (Stål)	（カマキリ目、カマキリ科）エチオピア区
Egyptian pygmy mantis			Egyptian praying mantis (1) を見よ
Egyptian sacred scarab			scarab を見よ
Eichhorn's crow		*Euploea eichhorni* Staudinger	（チョウ目、タテハチョウ科）豪州区
eight-barred Lygropia moth		*Lygropia octonalis* (Zeller)	（チョウ目、メイガ科）新北区
eight-dentated bark beetle			eight-spined ips を見よ
eight-eyed longicorn			eightspotted lichen longicorn beetle を見よ
eight-spined ips		*Ips typographus* (Linnaeus)	（コウチュウ目、キクイムシ科）旧北区。エゾマツ類の害虫。日本亜種は *I. t. japonicus* Niijima ヤツバキクイムシ
eight-spot		*Amyna axis* (Guenée)	（チョウ目、ヤガ科）東洋区、豪州区、エチオピア区、新北区。eight-spot moth ともいう

英名	和名	学名	所属、分布、ほか
eight spot numberwing		*Callicore felderi* (Hewitson)	(チョウ目、タテハチョウ科) 新熱帯区
eight-spot short-tail skipper			eight-spotted longtail を見よ
eight-spotted forester		*Alypia octomaculata* (Fabricius)	(チョウ目、ヤガ科) 新北区。ブドウ他害虫。eight-spotted forester moth ともいう
eight-spotted leaf beetle	ヤツボシツツハムシ	*Cryptocephalus japanus* Baly	(コウチュウ目、ハムシ科) 日本、旧北区
eightspotted lichen longicorn beetle	ヤツメカミキリ	*Eutetrapha ocelota* (Bates)	(コウチュウ目、カミキリムシ科) 日本。ウメ、サクラの害虫。eightspotted lichen longicorn ともいう
eight-spotted longtail		*Polythrix octomaculata* (Sepp)	(チョウ目、セセリチョウ科) 新北区、新熱帯区
eight-striped prominent	ヤスジシャチホコ	*Epodonta lineata* (Oberthür)	(チョウ目、シャチホコガ科) 日本、旧北区
eight-toothed bark beetle			eight-spined ips を見よ
eight-toothed engraver beetle			eight-spined ips を見よ
eight-toothed spruce beetle			eight-spined ips を見よ
eighteen-spotted ladybird		*Myrrha octodecimguttata* (Linnaeus)	(コウチュウ目、テントウムシ科) 旧北区
eighteen-spotted ladybird			common spotted ladybird を見よ
eighty-eight butterfly	マイムナウズマキタテハ	*Callicore maimuna* (Hewitson)	(チョウ目、タテハチョウ科) 新熱帯区
eighty-eight butterfly (1)	アンナウラモジタテハ	*Diaethria anna* (Guérin-Méneville)	(チョウ目、タテハチョウ科) 新北区。後翅裏面の斑紋 88 に由来。*D. a. mixteca* Maza, *D. a. salvadorensis* (Franz et Schroder) も同英名
eighty eight butterfly (2)	クリメナウラスジタテハ	*Diaethria clymena* (Cramer)	(チョウ目、タテハチョウ科) 新北区、新熱帯区
eighty nine butterfly	ウラモジタテハ	*Diaethria neglecta* (Salvin)	(チョウ目、タテハチョウ科) 新熱帯区
El Hierro grayling		*Hipparchia bacchus* Higgins	(チョウ目、タテハチョウ科) 旧北区
El Segundo blue		*Euphilotes battoides allyni* Shields	(チョウ目、シジミチョウ科) 新北区
elachistid moths			grass miner moths を見よ
Elada checkerspot		*Texola elada* (Hewitson)	(チョウ目、タテハチョウ科) 新北区
elasmids		Elasmidae	(ハチ目) の昆虫の総称
elater			click beetles (1) を見よ
elaterid			click beetle (1) を見よ
Elban heath		*Coenonympha elbana* Staudinger	(チョウ目、タテハチョウ科) 旧北区
elbowed pierrot		*Pycnophallium elna* (Hewitson)	(チョウ目、シジミチョウ科) 東洋区
elbowed red ant		*Myrmica scabrinodis* Nylander	(ハチ目、アリ科) 旧北区
elder aphid	ニワトコアブラムシ	*Aphis sambuci* Linnaeus	(カメムシ目、アブラムシ科) 日本、全北区

英名	和名	学名	所属、分布、ほか
elder moth		*Zotheca tranquilla* Grote	(チョウ目、ヤガ科) 新北区
elder shoot borer			elder shoot borer moth を見よ
elder shoot borer moth		*Achatodes zeae* (Harris)	(チョウ目、ヤガ科) 新北区
elderberry longhorn		*Desmocerus palliatus* (Forster)	(コウチュウ目、カミキリムシ科) 新北区。elderberry borer ともいう
electra buck moth		*Hemileuca electra* Wright	(チョウ目、ヤママユガ科) 新北区
electric green swordtail			green-spotted swallowtail を見よ
electric light bug			giant water bug (1) を見よ
electric light bugs			giant water bugs (2) を見よ　灯火によく飛来することに由来
electron Pixie		*Melanis electron* (Fabricius)	(チョウ目、シジミタテハ科) 新熱帯区
elegant Acraea		*Acraea egina* (Cramer)	(チョウ目、タテハチョウ科) エチオピア区
elegant Anteros		*Anteros chrysoprasta roratus* Godman et Salvin	(チョウ目、シジミタテハ科) 新熱帯区
elegant Aristotelia moth		*Aristotelia elegantella* (Chambers)	(チョウ目、キバガ科) 新北区
elegant bush katydid		*Insara elegans* (Scudder)	(バッタ目、キリギリス科) 新北区
elegant flat spreadwing		*Lestes inaequalis* Walsh	(トンボ目、アオイトトンボ科) 新熱帯区。elegant spreadwing ともいう
elegant grasshoppers		*Zonocerus*	(バッタ目、バッタ科) の昆虫の総称
elegant grass-veneer moth		*Microcrambus elegans* (Clemens)	(チョウ目、メイガ科) 新北区
elegant looper		*Philtraea elegantaria* H. Edwards	(チョウ目、シャクガ科) 新北区
elegant malachite		*Chlorolestes elegans* Pinhey	(トンボ目、Synlestidae) エチオピア区
elegant midget		*Phyllonorycter lautella* (Zeller)	(チョウ目、ホソガ科) 旧北区
elegant prominent moth		*Odontosia elegans* (Strecker)	(チョウ目、シャチホコガ科) 新北区
elegant Pygarctia		*Euchaetes elegans* Stretch	(チョウ目、ヒトリガ科) 新北区、新熱帯区
elegant sheepmoth			common sheep moth を見よ
elegant shieldback		*Idiostatus elegans* Caudell	(バッタ目、キリギリス科) 新北区
elegant sphinx		*Sphinx perelegans* Hy. Edwards	(チョウ目、スズメガ科) 新北区。elegant sphinx moth ともいう
elegant tailed slug moth		*Packardia elegans* (Packard)	(チョウ目、イラガ科) 新北区
elegant Temnora		*Temnora elegans* (Rothschild)	(チョウ目、スズメガ科) エチオピア区
elegant wave		*Acidalia decorata* (Denis et Schiffermüller)	(チョウ目、シャクガ科) 旧北区
elenchids	エダヒゲネジレバネ科	Elenchidae	(ネジレバネ目) の昆虫の総称
elephant beetle		*Xylotrupes gideon* (Linnaeus)	(コウチュウ目、コガネムシ科) 豪州区
elephant beetle			strawberry blossom weevil (2) を見よ

英名	和名	学名	所属、分布、ほか
elephant beetles		*Strategeus*	（コウチュウ目、コガネムシ科）の昆虫の総称
elephant beetles			Hercules beetles を見よ
elephant emperor		*Anax indicus* Lieftinck	（トンボ目、ヤンマ科）東洋区
elephant flies			horse flies (2) を見よ
elephant hawk			elephant hawk moth を見よ
elephant hawk moth		*Deilephila elpenor* (Linnaeus)	（チョウ目、スズメガ科）旧北区。日本亜種は *D. e. lewisii* (Butler)　ベニスズメ
elephant louse		*Haematomyzus elephatis* Piaget	（ハジラミ目、ケモノタンカクハジラミ科）エチオピア区
elephant mosquito		*Toxorhynchites rutilus* (Coquillett)	（ハエ目、カ科）新北区
elephant stag beetle			American stag beetle を見よ　米国での英名
elephant weevil		*Orthorhinus cylindrirostris* (Fabricius)	（コウチュウ目、ゾウムシ科）豪州区
elephant weevil		*Rhyncodes ursus* White	（コウチュウ目、オトシブミ科）豪州区
elephant weta		*Motuweta isolata* Johns	（バッタ目、Stenopelmatidae）豪州区
Eleuchia heliconian		*Heliconius eleuchia eleuchia* Hewitson	（チョウ目、タテハチョウ科）新熱帯区
Eleuchia longwing			Eleuchia heliconian を見よ
eleven-spot ladybird		*Coccinella undecimpunctata* Linnaeus	（コウチュウ目、テントウムシ科）旧北区、豪州区。eleven-spotted ladybird ともいう
elf	ハスオビチビタテハ	*Microtia elva* Bates	（チョウ目、タテハチョウ科）新熱帯区。elf butterfly ともいう
elfin butterflies			elfins (1) を見よ
elfin skipper		*Sarangesa motozi* (Wallengren)	（チョウ目、セセリチョウ科）エチオピア区
elfins		*Incisalia*	（チョウ目、シジミチョウ科）の昆虫の総称
elfins (1)	コツバメ属	*Callophrys*	（チョウ目、シジミチョウ科）の昆虫の総称
elfs		*Tetrathemis*	（トンボ目、トンボ科）の昆虫の総称
elgin shoot moth			double shoot (1) を見よ
Elgner's jewel		*Hypochrysops elgneri* (Waterhouse)	（チョウ目、シジミチョウ科）豪州区
Elgon crimson tip		*Colotis elgonensis* (Sharpe)	（チョウ目、シロチョウ科）エチオピア区
Elgon giant cupid		*Lepidochrysops elgonae* Stempffer	（チョウ目、シジミチョウ科）エチオピア区
Elia brown		*Nesoxenica leprea elia* Waterhouse et Lyell	（チョウ目、タテハチョウ科）豪州区
Eliena skipper		*Trapezites eliena* (Hewitson)	（チョウ目、セセリチョウ科）豪州区
Eliot's Cerulean		*Jamides elioti* Hirowatari et Lassidy	（チョウ目、シジミチョウ科）東洋区
Eliot's hawker		*Zosteraeschna elioti* (Kirby)	（トンボ目、ヤンマ科）エチオピア区
Elis emperor	アカオビアメリカコムラサキ	*Doxocopa elis* (C. et R. Felder)	（チョウ目、タテハチョウ科）新熱帯区
Elisa skimmer		*Celithemis elisa* (Hagen)	（トンボ目、トンボ科）新北区

英名	和名	学名	所属、分布、ほか
Elissa page		*Metamorpha elissa* Hübner	（チョウ目、タテハチョウ科）新熱帯区
Elissa roadside skipper		*Amblyscirtes elissa* Godman	（チョウ目、セセリチョウ科）新北区、新熱帯区
elk nostril fly		*Cephenomyia ulrichi* Brauer	（ハエ目、ヒツジバエ科）旧北区
Ella's barred blue		*Spindasis ella* Hewitson	（チョウ目、シジミチョウ科）エチオピア区
Ella's barred-blue			Ella's silverline を見よ
Ella's silverline		*Cigaritis ella* (Hewitson)	（チョウ目、シジミチョウ科）エチオピア区
elliptical barklice			damp barklice を見よ
Ellis's bushblue		*Arhopala ellisi* Evans	（チョウ目、シジミチョウ科）東洋区
Ellis's bushblue			pale bushblue を見よ
ello sphinx		*Erynnyis ello* (Linnaeus)	（チョウ目、スズメガ科）新北区、新熱帯区。北米での英名。ello sphinx moth ともいう
elm aphid		*Tinocallis platani* (Kaltenbach)	（カメムシ目、アブラムシ科）全北区
elm aphid			elm leaf aphid (1) を見よ
elm aphis			elm leaf aphid (1) を見よ
elm balloon-gall aphid			woolly pear aphid を見よ
elm bark beetle (1)	ヨーロッパニレノキクイムシ	*Scolytus scolytus* (Fabricius)	（コウチュウ目、キクイムシ科）旧北区。ニレ害虫
elm bark beetle (2)	セスジキクイムシ	*Scolytus multistriatus* (Marsham)	（コウチュウ目、キクイムシ科）全北区。ニレ害虫
elm bark beetle (3)	ニレカワノキクイムシ	*Scolytus frontalis* Blandford	（コウチュウ目、キクイムシ科）日本、東洋区。ニレ害虫
elm bark beetle			large elm bark beetle を見よ
elm bark beetles			bark beetles (2) を見よ
elm bark borer		*Physocnemum brevilineum* Say	（コウチュウ目、カミキリムシ科）新北区
elm bark louse			European elm scale を見よ
elm bark-louse			purple scale を見よ
elm bark weevil			elm bark beetle (2) を見よ
elm borer		*Saperda tridentata* Olivier	（コウチュウ目、カミキリムシ科）新北区
elm butterfly			large tortoiseshell を見よ
elm calligrapha		*Calligrapha scalaris* (LeConte)	（コウチュウ目、ハムシ科）新北区。ニレの葉を食害。calligrapha を見よ
elm casebearer		*Coleophora ulmifoliella* McDunnough	（チョウ目、ツツミノガ科）新北区
elm casebearer			mud-feather case を見よ　elm casebearer moth ともいう
elm cockscomb-gall aphid		*Colopha ulmicola* (Fitch)	（カメムシ目、アブラムシ科）全北区。ニレに虫えい (elm cockscomb gall) をつくる
elm-currant aphid			elm leaf aphid (2) を見よ
elm flea beetle		*Altica ulmi* Woods	（コウチュウ目、ハムシ科）新北区
elm flea beetle		*Altica calinata* Germar	（コウチュウ目、ハムシ科）新北区

英名	和名	学名	所属、分布、ほか
elm flea weevil		*Rhynchaenus saltator* Hustache	(コウチュウ目、ゾウムシ科) 旧北区
elm gall aphid	ニレイボフシ	*Kaltenbachiella nirecola* (Matsumura)	(カメムシ目、アブラムシ科) 日本、旧北区
elm gall bug		*Anthocoris gallarumulmi* (De Geer)	(カメムシ目、ハナカメムシ科) 旧北区
elm grey		*Scoparia ancipitella* La Harpe	(チョウ目、メイガ科) 旧北区
elm lace bug		*Corythucha ulmi* Osborn et Drake	(カメムシ目、グンバイムシ科) 新北区
elm leaf aphid (1)		*Tinocallis ulmifolii* (Monell)	(カメムシ目、アブラムシ科) 新北区
elm leaf aphid (2)	ニレヨスジワタムシ	*Tetraneura ulmi* (Linnaeus)	(カメムシ目、アブラムシ科) 日本、旧北区
elm leaf beetle	ニレハムシ	*Galerucella xanthomelaena* (Schrank)	(コウチュウ目、ハムシ科) 全北区。ニレ害虫
elm leaf beetle		*Galerucella luteola* (Muller)	(コウチュウ目、ハムシ科) 旧北区。elm leaf-beetle とも記す
elm leaf beetle		*Galerucella calmariensis* (Linnaeus)	(コウチュウ目、ハムシ科) 旧北区
elm leaf caterpillar		*Nerice bidentata* (Walker)	(チョウ目、シャチホコガ科) 新北区
elm leaf gall aphid			elm leaf aphid (2) を見よ
elm leafhopper		*Ribautiana ulmi* (Linnaeus)	(カメムシ目、オオヨコバイ科) 豪州区
elm leafminer		*Fenusa ulmi* Sundevall	(ハチ目、ハバチ科) 全北区。北米での英名
elm leaf-mining sawfly			elm leafminer を見よ
elm leaf-mining weevil		*Rhynchaenus alni* (Linnaeus)	(コウチュウ目、ゾウムシ科) 旧北区
elm leaftier moth		*Canarsia ulmiarrosorella* (Clemens)	(チョウ目、メイガ科) 新北区
elm prominent		*Heterocampa bilineata* Packard	(チョウ目、シャチホコガ科) 新北区
elm rosette aphid			woolly apple aphid を見よ
elm sawfly		*Cimbex americana* Leach	(ハチ目、コンボウハバチ科) 新北区
elm sawfly			alder leafminer を見よ
elm scale			European elm scale を見よ
elm scurly scale		*Chionaspis americana* Johnson	(カメムシ目、マルカイガラムシ科) 新北区。ニレ害虫
elm spanworm moth		*Ennomos subsignaria* (Hübner)	(チョウ目、シャクガ科) 新北区。幼虫はリンゴ、ニレなどにつく。elm spanworm ともいう
elm sphinx		*Ceratomia amyntor* (Geyer)	(チョウ目、スズメガ科) 新北区。幼虫の形態から four horned sphinx ともいわれる。elm sphinx moth ともいう
elm sucker		*Psylla ulmi* Förster	(カメムシ目、キジラミ科) 旧北区
elm tree beetle		*Galeruca caprae* Linnaeus	(コウチュウ目、ハムシ科) 旧北区
elm tree beetle			yellow willow leaf beetle を見よ
elm-tree white scale			oystershell scale (1) を見よ

英名	和名	学名	所属、分布、ほか
elmids			riffle beetles を見よ
elongate-bodied springtail			water springtail を見よ
elongate-bodied springtails			hypogastrurid springtails を見よ
elongate-bodied springtails			slender springtails を見よ
elongate-bodied springtails			snowfleas を見よ
elongate brown flower click beetle	ホソサビキコリ	*Agrypnus fuliginosus* (Candèze)	（コウチュウ目、コメツキムシ科）日本
elongate cicada	ツクツクボウシ	*Meimuna opalifera* (Walker)	（カメムシ目、セミ科）日本、旧北区、東洋区
elongate coccus			long brown scale (1) を見よ
elongate cottony scale	オオワタコナカイガラムシ	*Phenacoccus pergandei* Cockerell	（カメムシ目、コナカイガラムシ科）日本、旧北区
elongate flea beetle		*Systena elongata* (Fabricius)	（コウチュウ目、ハムシ科）新北区
elongate hemlock scale			hemlock scale (1) を見よ　米国での英名
elongate long-necked leaf beetle	ホソクビナガハムシ	*Lilioceris parvicollis* (Baly)	（コウチュウ目、ハムシ科）日本
elongate longicorn beetle	ホソカミキリ	*Distenia graciis graciis* (Blessig)	（コウチュウ目、カミキリムシ科）日本、旧北区
elongate-necked blister beetle	クビナガツチハンミョウ	*Meloe violaceus semenovi* Jakowlew	（コウチュウ目、ツチハンミョウ科）日本、旧北区
elongate parlatoria scale			cattleya scale を見よ
elongate platypodid	カギナガキクイムシ	*Platypus hamatus* Blandford	（コウチュウ目、ナガキクイムシ科）日本。木材害虫
elongate twig ant			slender Pseudomyrmex を見よ
elongate weevil	カツオゾウムシ	*Lixus impressiventris* Roelofs	（コウチュウ目、ゾウムシ科）日本、旧北区
elongate white-marmorated longicorn			longwinged oak longicorn beetle を見よ
elongate whitefly		*Dialeurodes elongata* Dozier	（カメムシ目、コナジラミ科）東洋区。カンキツ害虫
elongated blackstreak		*Ocaria elongata* Hewitson	（チョウ目、シジミチョウ科）新熱帯区
Elsa hairstreak		*Magnastigma elsa* (Hewitson)	（チョウ目、シジミチョウ科）新熱帯区
Elsa sphinx		*Sagenosoma elsa* (Strecker)	（チョウ目、スズメガ科）新北区、新熱帯区。Elsa sphinx moth ともいう
Elthamm copper		*Pralucia pyrodiscus lucida* Crosby	（チョウ目、シジミチョウ科）豪州区
elusive adjutant		*Aethriamanta brevipennis* (Rambur)	（トンボ目、トンボ科）東洋区
elusive clubtail		*Stylurus notatus* (Rambur)	（トンボ目、サナエトンボ科）新北区

英名	和名	学名	所属、分布、ほか
elusive sanddragon		*Progomphus zephyrus* Needham	(トンボ目、サナエトンボ科) 新熱帯区
Elvina Eurybia			blind Eurybia を見よ
Elwes' pied flat		*Coladenia agnioides* Elwes et Edwards	(チョウ目、セセリチョウ科) 東洋区
Elwes's silverline		*Cigaritis elwesi* Evans	(チョウ目、シジミチョウ科) 東洋区
emarginate short-winged katydid		*Dichopetala emarginata* Brunner von Wattenwyl	(バッタ目、キリギリス科) 新北区
embiids			webspinners (1) を見よ
embiopterans			webspinners (1) を見よ
embolemid wasps	アリモドキバチ科	Embolemidae	(ハチ目) の昆虫の総称
Emelina grass blue		*Zizina emelina* de l'Orza	(チョウ目、シジミチョウ科) 旧北区
emerald		*Cyaniriodes libna andersonii* (Moore)	(チョウ目、シジミチョウ科) 東洋区
emerald Aguna		*Aguna claxon* Evans	(チョウ目、セセリチョウ科) 新北区
emerald ash borer		*Agrilus planipennis* Fairmaire	(コウチュウ目、タマムシ科) 新北区
emerald bug			southern green stink bug を見よ
emerald carpet moth		*Chloroclystis muscosata* Walker	(チョウ目、シャクガ科) 豪州区
emerald cockroach wasp		*Ampulex compressa* (Fabricius)	(ハチ目、アナバチ科) 新北区
emerald cutworm moth		*Erana graminosa* Walker	(チョウ目、ヤガ科) 豪州区
emerald damselflies	アオイトトンボ科	Lestidae	(トンボ目) の昆虫の総称
emerald damselfly	アオイトトンボ	*Lestes sponsa* (Hansemann)	(トンボ目、アオイトトンボ科) 日本、旧北区、エチオピア区
emerald dragonflies			green-eyed skimmers を見よ
emerald fruit chafer		*Rhabdotis aulica* (Fabricius)	(コウチュウ目、コガネムシ科) エチオピア区
emerald hairstreak		*Erora subflorens* Schaus	(チョウ目、シジミチョウ科) 新熱帯区
emerald hairstreak			Damel's blue を見よ
emerald moth		*Eucrostes disparata* (Walker)	(チョウ目、シャクガ科) 日本、東洋区、豪州区、エチオピア区
emerald moth			common emerald を見よ
emerald-patched cattleheart		*Parides sesostris zestos* (Gray)	(チョウ目、アゲハチョウ科) 新熱帯区。sesostris cattle-heart を参照
emerald spreadwing			scarce emerald を見よ
emerald Sri Lanka spreadwing		*Sinhalestes orientalis* (Hagen)	(トンボ目、アオイトトンボ科) 東洋区
emerald-striped slim		*Aciagrion pimheyi* Samways	(トンボ目、イトトンボ科) エチオピア区
emerald-studded blue			Acmon blue を見よ
emeralds	アオシャク亜科	Geometrinae	(チョウ目、シャクガ科) の昆虫の総称
emeralds			green-eyed skimmers を見よ

英名	和名	学名	所属、分布、ほか
Emerantia crescent		*Eresia emerantia* (Hewitson)	（チョウ目、タテハチョウ科）新熱帯区
Emessa hairstreak		*Gargina emessa* Hewitson	（チョウ目、シジミチョウ科）新熱帯区
emex weevil			South African emex weevil を見よ
emigrants	ウスキシロチョウ属	*Catopsilia*	（チョウ目、シロチョウ科）の昆虫の総称
Emilia banner	シロモントンボマダラモドキ	*Vila emilia* (Cramer)	（チョウ目、タテハチョウ科）新熱帯区
emma cricket		*Teleogryllus emma* (Ohmachi et Matsuura)	（バッタ目、コオロギ科）日本
emma field cricket			emma cricket を見よ
emma's dancer		*Argia emma* Kennedy	（トンボ目、イトトンボ科）新北区
emmets			アリ科の一部の階級に対する英名
emperor	ペレイデスモルフォ	*Morpho peleides insularis* Fruhstorfer	（チョウ目、タテハチョウ科）新熱帯区
emperor (1)		*Opodiphthera eucalypti* (Scott)	（チョウ目、ヤママユガ科）豪州区
emperor (2)		*Saturnia pavonia* (Linnaeus)	（チョウ目、ヤママユガ科）旧北区
emperor			emperor dragonfly をみよ
emperor			Peleides blue Morpho を見よ
emperor cicada			giant cicada を見よ
emperor dragonfly		*Anax imperator* Leach	（トンボ目、ヤンマ科）旧北区、エチオピア区。emperor ともいう
emperor gum moth			emperor (1) を見よ
emperor moth		*Saturnia carpini* Schiffermüller	（チョウ目、ヤママユガ科）旧北区
emperor moth		*Syntherata janetta* (White)	（チョウ目、ヤママユガ科）豪州区
emperor moth			emperor (1)(2) を見よ
emperor moth			cecropia moth を見よ
emperor moths			giant silkworm moths を見よ
emperor of Morocco			purple emperor (2) を見よ
emperor swallowtail	オオシロモンアゲハ	*Papilio hesperus* Westwood	（チョウ目、アゲハチョウ科）エチオピア区
emperor swallowtail	サカハチアゲハ	*Papilio ophidicephalus* Oberthür	（チョウ目、アゲハチョウ科）エチオピア区
emperors		*Anax*	（トンボ目、ヤンマ科）の昆虫の総称
emperors	コムラサキ属	*Apatura*	（チョウ目、タテハチョウ科）の昆虫の総称
emperors		*Hemianax*	（トンボ目、ヤンマ科）の昆虫の総称
emperors (1)		Apaturinae	（チョウ目、タテハチョウ科）の昆虫の総称 emperor butterfly ともいわれる
emperors			darners を見よ
emperors			giant silkworm moths を見よ
empid fly		*Empis livida* Linnaeus	（ハエ目、オドリバエ科）旧北区
empids			long-legged flies (2) を見よ
empress			emperor (2) を見よ

英名	和名	学名	所属、分布、ほか
empress		*Sasakia funebris* (Leech)	（チョウ目、タテハチョウ科）東洋区
empress Alicia	エノキタテハ	*Asterocampa celtis alicia* (Edwards)	（チョウ目、タテハチョウ科）新北区
empress Antonia		*Asterocampa antonia* (Edwards)	（チョウ目、タテハチョウ科）新北区
empress cicada			giant cicada を見よ
Empress Eugenie Morpho		*Morpho eugenia* Deyrolle	（チョウ目、タテハチョウ科）新熱帯区
empress flora		*Asterocampa flora* Edwards	（チョウ目、タテハチョウ科）新北区
empress Leilia		*Asterocampa leilia* (Edwards)	（チョウ目、タテハチョウ科）新北区、新熱帯区
empress Louisa		*Asterocampa clyton louisa* D. Stallings et Turner	（チョウ目、タテハチョウ科）新北区。tawny emperor (1) を参照
empress tree granulated weevil	コブコブゾウムシ	*Styanax rugosus* (Roelofs)	（コウチュウ目、ゾウムシ科）日本
empress tree weevil			paulownia leaf weevil を見よ
Emylius gem		*Calospila emylius* (Cramer)	（チョウ目、シジミタテハ科）新熱帯区
Enania glasswing		*Oleria enania* Haensch	（チョウ目、タテハチョウ科）新熱帯区
Encedon Acraea			white-barred Acraea を見よ
enchanter's cosmet		*Mompha terminella* (Humphreys et Westwood)	（チョウ目、カザリバガ科）旧北区
encyrtid parasites			encyrtid wasps を見よ
encyrtid wasps	トビコバチ科	Encyrtidae	（ハチ目）の昆虫の総称
encyrtids			encyrtid wasps を見よ
endromid moths			glory moths を見よ
endrosid moths		Endrosidae	（チョウ目）の昆虫の総称
Endymion hairstreak		*Electrostrymon endymion* (Fabricius)	（チョウ目、シジミチョウ科）新北区、新熱帯区
Endymion opal		*Chrysoritis endymion* (Pennington)	（チョウ目、シジミチョウ科）エチオピア区
Engelmann spruce weevil			white pine weevil を見よ
Engel's Salebriaria moth		*Salebriaria engeli* (Dyar)	（チョウ目、メイガ科）新北区
English grain aphid	ムギヒゲナガアブラムシ	*Sitobion akebiae* Shinji	（カメムシ目、アブラムシ科）日本
English grain aphid		*Sitobion granarium* (Kirby)	（カメムシ目、アブラムシ科）新北区
English grain aphid			grain aphid (1) を見よ 北米での英名
English grain aphis			greenbug を見よ
English grain-louse			grain aphid (1) を見よ 米国での英名
English privet hawk moth			privet hawk を見よ
English scarab			tumblebug (1) を見よ
English wasp			common wasp を見よ 豪州での英名

英名	和名	学名	所属、分布、ほか
engrailed (1)		*Ectropis biundulata* de Villeneuve	（チョウ目、シャクガ科）旧北区
engrailed (2)	フトフタオビエダシャク	*Ectropis bistortata* (Goeze)	（チョウ目、シャクガ科）日本、旧北区
engrailed moth			engrailed (1)(2) を見よ
engraved big weevil			olive weevil を見よ
engraved thick weevil	フトアナアキゾウムシ	*Dyscerus gigas* (Kono)	（コウチュウ目、ゾウムシ科）日本
engraver beetle			eight-spined ips を見よ
engraver beetle			pine ips を見よ
engraver beetles			bark beetles (1) を見よ
Enops tufted-skipper		*Polyctor enops* (Godman et Salvin)	（チョウ目、セセリチョウ科）新北区、新熱帯区
enormous giant stick insect			spiny stick insect を見よ
ensete root mealybug		*Cataenococcus ensete* Williams et Matile-Ferrero	（カメムシ目、コナカイガラムシ科）エチオピア区
ensign coccid		*Orthezia urticae* (Linnaeus)	（カメムシ目、ハカマカイガラムシ科）旧北区
ensign coccids			ensighn scales を見よ
ensign flies			ensighn wasps を見よ
ensign scales	ハマカイガラムシ科	Ortheziidae	（カメムシ目）の昆虫の総称
ensign wasps	ヤセバチ科	Evaniidae	（ハチ目）の昆虫の総称
entomobryid springtails			slender springtails を見よ
entomolite			化石の昆虫
entotrophs	コムシ目	Diplura	の昆虫の総称　2～50mm、通常10mm以下。世界で約660種。コムシ目は Entotrophi といい、英名はそれに由来
Enyo falcon		*Corades enyo* (Hewitson)	（チョウ目、タテハチョウ科）新熱帯区
Eolian grayling		*Hipparchia leighebi* Kudma	（チョウ目、タテハチョウ科）旧北区
Eone blue		*Pseudodipsas eone* (C. et R. Felder)	（チョウ目、シジミチョウ科）豪州区
Epallia grasshopper		*Epallia exigua* Sjostedt	（バッタ目、バッタ科）豪州区
epaulet skimmer		*Orthetrum chrysostigma* (Burmeister)	（トンボ目、トンボ科）エチオピア区
epauletted pitcher-plant moth		*Exyra fax* Grote	（チョウ目、ヤガ科）新北区。pitcher plant moth ともいう
epermeniid moths			fringe-tufted moths を見よ
ephemera	カゲロウ		複数形は ephemeras, ephemerae。mayflies を見よ
ephemeral flies			mayflies を見よ
ephemerans			mayflies を見よ
ephemerid			mayflies を見よ
ephemeron			ephemera を見よ

英名	和名	学名	所属、分布、ほか
Ephora tufted skipper		*Nisoniades ephora* (Herrich-Schäffer)	（チョウ目、セセリチョウ科）新熱帯区
Epigaea looper moth		*Syngrapha epigaea* (Grote)	（チョウ目、ヤガ科）新北区
Epilachna beetles		*Epilachna*	（コウチュウ目、テントウムシ科）の昆虫の総称
Epimachia glory		*Myscelus epimachia* Herrich-Schäffer	（チョウ目、セセリチョウ科）新熱帯区
Epione underwing moth		*Catocala epione* Drury	（チョウ目、ヤガ科）新北区
epiplemid moths			crenulate moths を見よ
epipsocids			damp barklice を見よ
epipyropid moths			planthopper parasites を見よ
Epistrophus white Morpho		*Morpho epistrophus* (Fabricius)	（チョウ目、タテハチョウ科）新熱帯区
epsilon		*Scotia ypsilon* Hufnagel	（チョウ目、ヤガ科）旧北区
equatorial mountain blue		*Harpendyreus aequatorialis* (Sharpe)	（チョウ目、シジミチョウ科）エチオピア区
equine sucking louse			horse sucking louse を見よ　南アフリカでの英名
Erasa chocolate moth		*Argyrostrotis erasa* (Guenée)	（チョウ目、ヤガ科）新北区
Erato heliconian	エラートドクチョウ	*Heliconius erato petiveranus* Doubleday	（チョウ目、タテハチョウ科）新北区。common longwing を参照
erebid moth		*Bulia deducta* Morrison	（チョウ目、ヤガ科）新北区
Erebus moth		*Erebus terminitincta* (Gaede)	（チョウ目、ヤガ科）豪州区
Erechthias clothes moth			Caribbean scavenger moth を見よ
ergate			職蟻に対する英名
eri silkworm			cynthia moth を見よ
eri-silkworm moth	エリサン	*Samia ricini* (Jones)	（チョウ目、ヤママユガ科）汎世界
Erichson's cruiser		*Vindula dejone* (Erichson)	（チョウ目、タテハチョウ科）東洋区
Erichson's white-skipper		*Heliopyrgus domicella domicella* (Erichson)	（チョウ目、セセリチョウ科）新熱帯区
Erichson's white skipper			white-banded skipper を見よ
erigeron crabronid	ナミギングチ	*Ectemius continuus* (Fabricius)	（ハチ目、アナバチ科）日本、全北区
erigeron flower moth		*Schinia obscurata* Strecker	（チョウ目、ヤガ科）新北区
erigeron root aphid		*Aphis middletonii* (Thomas)	（カメムシ目、アブラムシ科）新北区
eriococcid mealybugs			mealybugs (3) を見よ
eriococcid scales			mealybugs (2)(3) を見よ
eriococcus caterpillar		*Stathmopoda melanochra* Meyrick	（チョウ目、マルハキバガ科）豪州区
eriocraniid moths			primitive moths　(1) を見よ

英名	和名	学名	所属、分布、ほか
Eriogonum mealybug		*Spirococcus eriogoni* (Ehrhorn)	（カメムシ目、コナカイガラムシ科）新北区
Eriphyle ringlet	ヒメアルプスベニヒカゲ	*Erebia eriphyle* (Freyer)	（チョウ目、タテハチョウ科）旧北区
ermel			orchard ermine を見よ
ermels			ethmiid moths を見よ
ermels			ermine moths (1) を見よ
ermine			tiger moths (1) を見よ
ermine moth			orchard ermine を見よ
ermine moth			apple ermine moth を見よ 米国での英名
ermine moths (1)	スガ科	Yponomeutidae	（チョウ目）の昆虫の総称
ermine moths (2)		*Yponomeuta*	（チョウ目、スガ科）の昆虫の総称
ermine moths			tiger moths (1) を見よ
Ernestine's moth		*Phytometra ernestinana* (Blanchard)	（チョウ目、ヤガ科）新北区、新熱帯区
Erodyle checkerspot	クロヒョウモンモドキ	*Chlosyne erodyle erodyle* (H. Bates)	（チョウ目、タテハチョウ科）新熱帯区
eros blue	チョウセンヒメシジミ	*Polyommatus eros* (Ochsenheimer)	（チョウ目、シジミチョウ科）旧北区、東洋区
Erostratus swallowtail		*Papilio erostratus erostratus* Westwood	（チョウ目、アゲハチョウ科）新熱帯区
Erota metalmark		*Notheme erota* (Cramer)	（チョウ目、シジミタテハ科）新熱帯区
Erotia sister	キオビイチモンジ	*Adelpha erotia* (Hewitson)	（チョウ目、タテハチョウ科）新熱帯区
erotylids			pleasing fungus beetles (1) を見よ
erratic ant		*Tapinoma erraticum* (Latreille)	（ハチ目、アリ科）旧北区
erratic army worm			black army cutworm moth を見よ
erratic roach		*Arenivaga erratica* (Rehn)	（ゴキブリ目、Corydiidae）新北区
Erybathis hairstreak		*Laothus erybathis* (Hewitson)	（チョウ目、シジミチョウ科）新熱帯区
Erymanthis sister			Godman's sister (1) を見よ
Erythrina borer moth		*Terastia meticulosalis* Guenée	（チョウ目、メイガ科）東洋区、新北区、新熱帯区。Erythrina borer, Erythrina twigborer ともいう
Erythrina gall wasp		*Quadrastichus erythrinae* Kim	（ハチ目、ヒメコバチ科）旧北区
Erythrina seed beetle		*Specularius impressithorax* (Pic)	（コウチュウ目、ハムシ科）新北区、エチオピア区
Escalante's giant-skipper		*Agathymus escalantei* Stallings, Turner et Stallings	（チョウ目、セセリチョウ科）新熱帯区
Escalante's ruby-eye		*Carystoides escalantei* Freeman	（チョウ目、セセリチョウ科）新熱帯区
Escher's blue		*Polyommatus escheri* (Hübner)	（チョウ目、シジミチョウ科）旧北区
Eskimo alpine		*Erebia occulta* Roos et Kimmich	（チョウ目、タテハチョウ科）新北区
Eskimo arctic			Alberta arctic を見よ

英名	和名	学名	所属、分布、ほか
Eskimo arctic			sentinel arctic (1) を見よ
Esmeralda		*Cithaerias esmeralda* (Salazar et Constantino)	(チョウ目、タテハチョウ科) 新熱帯区
Esmeralda longtail		*Urbanus esmeraldus* (Butler)	(チョウ目、セセリチョウ科) 新北区、新熱帯区
Esperanza eyed-metalmark		*Mesosemia esperanza* Schaus	(チョウ目、シジミタテハ科) 新熱帯区
Esperanza sister		*Adelpha erymanthis esperanza* Balcazar et Willmott	(チョウ目、タテハチョウ科) 新熱帯区
Esperanza swallowtail		*Papilio esperanza* Beutelspacher	(チョウ目、アゲハチョウ科) 新熱帯区
Esper's marbled white	オオシロジャノメ	*Melanargia russiae* (Esper)	(チョウ目、タテハチョウ科) 旧北区
Essex emerald		*Thetidea smaragdaria* (Fabricius)	(チョウ目、シャクガ科) 旧北区
Essex Phaneta moth		*Phaneta essexana* (Kearfott)	(チョウ目、ハマキガ科) 新北区
Essex skipper			European skipper を見よ　Essex skipper butterfly ともいう
Essex Y		*Syngrapha circumflexa* (Linnaeus)	(チョウ目、ヤガ科) 旧北区、東洋区、エチオピア区
Esta longtail		*Urbanus esta* Evans	(チョウ目、セセリチョウ科) 新熱帯区
Estcourt blue		*Lepidochrysops pephredo* (Trimen)	(チョウ目、シジミチョウ科) エチオピア区
Estela hairstreak		*Callophrys estela* Clench	(チョウ目、シジミチョウ科) 新熱帯区
Estelle's giant-skipper			brown bullet を見よ
esther moth		*Hypagyrtis esther* (Barnes)	(チョウ目、シャクガ科) 新北区
Ethelda sister		*Adelpha ethelda* (Hewitson)	(チョウ目、タテハチョウ科) 新熱帯区
ether's mon			dragonflies (1) を見よ
ether's nild			dragonflies (1) を見よ
Ethilla longwing	エチラドクチョウ	*Heliconius ethilla* (Godart)	(チョウ目、タテハチョウ科) 新北区、新熱帯区
Ethiopian fruit fly		*Didacus ciliatus* Loew	(ハエ目、ミバエ科) エチオピア区
Ethiopian mantis		*Sphodromantis aethiopica* La Greca et Lombardo	(カマキリ目、カマキリ科) エチオピア区
ethmiid moths	スエヒロキバガ科	Ethmiidae	(チョウ目) の昆虫の総称
Etias doctor	ワオオビシジミタテハ	*Ancyluris etias* (Saunders)	(チョウ目、シジミタテハ科) 新熱帯区
etiella moth			lucerne seed web moth を見よ
Eublemma moth		*Eublemma cochylioides* (Guenée)	(チョウ目、ヤガ科) 日本、旧北区、東洋区、大洋区、豪州区
Eubolina moth		*Eubolina impartialis* Harvey	(チョウ目、ヤガ科) 新北区
eucalypt-defoliating sawfly		*Pergagrapta bella* (Newman)	(ハチ目、Pergidae) 豪州区
eucalypt keyhole borer		*Xyleborus truncatus* (Erichson)	(コウチュウ目、キクイムシ科) 豪州区

英名	和名	学名	所属、分布、ほか
eucalypt leafgall scale		*Opisthoscelis subrotunda* Schrader	（カメムシ目、フクロカイガラムシ科）豪州区
eucalypt pinworm		*Atractocerus kreuslerae* Pascoe	（コウチュウ目、ツツシンクイ科）豪州区
eucalypt psyllid			redgum sugar lerp を見よ
eucalypt ringbarking longicorn		*Tryphocaria mastersi* Pascoe	（コウチュウ目、カミキリムシ科）豪州区
eucalypt shoot psyllid			eucalyptus psyllid を見よ
eucalyptus borer			common eucalypt longicorn を見よ
eucalyptus brown looper		*Thyrinteina arnobia* Stoll	（チョウ目、シャクガ科）新熱帯区
eucalyptus flies		Fergusoninidae	（ハエ目）の昆虫の総称
eucalyptus lappet		*Porela vetusta* (Walker)	（チョウ目、シャクガ科）豪州区
eucalyptus leaf beetles (1)		*Paropsis*	（コウチュウ目、ハムシ科）の昆虫の総称
eucalyptus leaf beetles (2)		*Chrysophtharta*	（コウチュウ目、ハムシ科）の昆虫の総称
eucalyptus leafroller		*Strepsicrates rhothia* Meyrick	（チョウ目、ハマキガ科）東洋区
eucalyptus long-horned beetle			common eucalypt longicorn を見よ
eucalyptus longhorned borer			common eucalypt longicorn を見よ
eucalyptus psyllid		*Blastopsylla occidentalis* Taylor	（カメムシ目、キジラミ科）豪州区
eucalyptus snout beetle			eucalyptus weevil を見よ
eucalyptus sucker		*Ctenarytaina eucalypti* (Maskell)	（カメムシ目、キジラミ科）旧北区、豪州区
eucalyptus thrips		*Thrips australis* (Bagnall)	（アザミウマ目、アザミウマ科）豪州区
eucalyptus tortoise beetle		*Paropsis charybidis* Stål	（コウチュウ目、ハムシ科）豪州区
eucalyptus tortoise beetles			eucalyptus leaf beetles (1) (2) を見よ
eucalyptus weevil		*Gonipterus scutellatus* (Gyllenhal)	（コウチュウ目、ゾウムシ科）エチオピア区、豪州区
eucharitid wasps	アリヤドリコバチ科	Eucharitidae	（ハチ目）の昆虫の総称
eucharitids			eucharitid wasps を見よ
Euclea tigerwing		*Hypothyris euclea* (Godart)	（チョウ目、タテハチョウ科）新熱帯区。*H. e. valora* (Haensch) も同英名
eucnemid beetles			false click beetles を見よ
eucnemids			false click beetles を見よ
eucoilid wasps	ツヤヤドリタマバチ科	Eucoilidae	（ハチ目）の昆虫の総称
Eudoxus Charaxes		*Charaxes eudoxus* (Drury)	（チョウ目、タテハチョウ科）エチオピア区
Eufala skipper			rice leaffolder を見よ

英名	和名	学名	所属、分布、ほか
Eugene's blue		*Polyommatus menelaos* Brown	(チョウ目、シジミチョウ科) 旧北区
Eugenia caterpillar		*Targalla delatrix* Guenée	(チョウ目、ヤガ科) 新北区、大洋区
Eugenia psyllid		*Trioza eugeniae* Froggatt	(カメムシ目、トガリキジラミ科) 豪州区、新北区
eulophid lady beetle parasite		*Tetrastichus coccinellae* Kurdjumov	(ハチ目、ヒメコバチ科) 旧北区
eulophid parasites			eulophid wasps を見よ
eulophid wasps	ヒメコバチ科	Eulophidae	(ハチ目) の昆虫の総称
eulophids			eulophid wasps を見よ
eumastacids			monkey grasshoppers (1) を見よ
Eumeda checkerspot		*Chlosyne eumeda* (Godman et Salvin)	(チョウ目、タテハチョウ科) 新熱帯区
Eungella darner		*Austroaeschna eungella* Theischinger	(トンボ目、ヤンマ科) 豪州区
Eunice crescent			tiger crescent を見よ
Eunoe mimic-white		*Dismorphia eunoe eunoe* (Doubelday)	(チョウ目、シロチョウ科) 新熱帯区。*D. e. chamula* Llorente et Luis, *D. e. populuca* Llorente et Luis も同英名
Eunomia eighty-eight			Eunomia numberwing を見よ
Eunomia numberwing		*Callicore eunomia* (Hewitson)	(チョウ目、タテハチョウ科) 新熱帯区
euonymous defoliator moth			pellucid zygaenid を見よ
euonymus alatus scale			willow oystershell scale を見よ
euonymus aphid		*Aphis euonymi* (Fabricius)	(カメムシ目、アブラムシ科) 旧北区
euonymus ermine moth	マサキスガ	*Yponomeuta meguronis* Matsumura	(チョウ目、スガ科) 日本。マサキ害虫
euonymus gall midge	マサキタマバエ	*Masakimyia pustulae* Yukawa et Sunose	(ハエ目、タマバエ科) 日本。マサキ害虫
euonymus scale	マサキナガカイガラムシ	*Unaspis euonymi* (Comstock)	(カメムシ目、マルカイガラムシ科) 日本、汎世界。イヌツゲ、マサキ害虫
eupatorium borer moth		*Papaipema eupatorii* (Lyman)	(チョウ目、ヤガ科) 新北区
eupatorium borer moth			ironweed clearwing moth を見よ
eupatorium gall fly			pamakani gall fly を見よ
eupatorium plume moth		*Oidaematophorus eupatorii* (Fernald)	(チョウ目、トリバガ科) 新北区
eupelmid wasps	ナガコバチ科	Eupelmidae	(ハチ目) の昆虫の総称
eupelmids			eupelmid wasps を見よ
Euphonious hairstreak		*Terenthina terentia* Hewitson	(チョウ目、シジミチョウ科) 新熱帯区
Euphorbus giant owl		*Caligo euphorbus* (C. et R. Felder)	(チョウ目、タテハチョウ科) 新熱帯区
Euphrates swordtail		*Graphium euphrates* (C. et R. Felder)	(チョウ目、アゲハチョウ科) 東洋区

英名	和名	学名	所属、分布、ほか
Eurasian baskettail		*Epitheca bimaculata* (Charpentier)	(トンボ目、エゾトンボ科) 旧北区
Eurasian beak			nettle-tree butterfly を見よ
Eurasian comma			comma (1) を見よ
Eurasian hemp moth			hemp stemborer を見よ
Eurasian meadow katydid			short-winged cone-head を見よ
Eurasian pine adelgid			pine adelgid を見よ
Euribates skipper		*Dyscophellus euribates* (Stoll)	(チョウ目、セセリチョウ科) 新北区
European alder leafminer			alder leafminer を見よ　米国での英名
European alder leaf-miner			elm sawfly を見よ　北米での英名
European Amazon ant			Amazon ant を見よ
European apple sawfly			apple fruit sawfly を見よ　北米での英名
European apple sucker			apple sucker を見よ　米国での英名
European beak			nettle-tree butterfly を見よ
European bean weevil			broadbean weevil を見よ
European birch aphid			silver birch aphid を見よ
European birch sawfly			betula argid sawfly を見よ
European black willow aphid			black willow aphid を見よ　米国での英名
European blackberry flower midge			blackberry flower midge を見よ
European blackcurrant aphid			black currant aphid (2) を見よ
European blowfly			blue blowfly を見よ
European blue bottle			blue blowfly を見よ　豪州での英名
European blue bottle fly			blue blowfly を見よ
European boxwood leaf-miner			box leaf-mining midge を見よ　米国での英名
European brown scale			European fruit lecanium を見よ　米国での英名
European bumble bee			buff-tailed bumble bee を見よ
European cabbage worm			common white を見よ　*P. rapae* Linnaeus の英名
European calosoma beetle			forest caterpillar-hunter を見よ
European chafer		*Rhizotrogus majalis* (Razoumowsky)	(コウチュウ目、コガネムシ科) 全北区。芝害虫

英名	和名	学名	所属、分布、ほか
European cherry fruit fly			cherry fruit fly (2) を見よ
European chestnut weevil			chestnut weevil を見よ
European chicken flea			chicken flea を見よ
European chinchbug		*Ischnodemus sabuleti* (Fallén)	(カメムシ目、ナガカメムシ科) 旧北区.
European cockchafer			June beetle (1) を見よ　米国での英名
European codling moth			splended piercer を見よ　米国での英名
European common blue			common blue (2) を見よ
European common cockchafer			June beetle (1) を見よ
European corn borer		*Ostrinia nubilalis* (Hübner)	(チョウ目、メイガ科) 全北区。1909 年に欧州から米国に入ったトウモロコシ害虫
European corn borer moth			European corn borer を見よ
European crane fly			marsh crane fly を見よ　米国での英名
European dung beetle			dung beetle (1) を見よ
European earwig			common earwig を見よ　米国、豪州での英名
European elm bark beetle			elm bark beetle (2) を見よ
European elm case bearer			mud-feather case を見よ
European elm leaf-curl aphid			elm leaf aphid (2) を見よ
European elm scale		*Gossyparia spuria* (Modeer)	(カメムシ目、フクロカイガラムシ科) 全北区。米国での英名
European field cricket		*Liogryllus campestris* (Linnaeus)	(バッタ目、コオロギ科) 旧北区
European field cricket			field cricket (1) を見よ　米国での英名
European firefly		*Luciola lusitanica* (Charpentier)	(コウチュウ目、ホタル科) 旧北区
European flesh fly	シリアカニクバエ	*Parasarcophaga crassipalpis* (Macquart)	(ハエ目、ニクバエ科) 日本、全北区、東洋区、豪州区、大洋区、エチオピア区。豪州での英名
European fruit lecanium	ミズキカタカイガラムシ	*Lecanium corni* Bouché	(カメムシ目、カタカイガラムシ科) 日本、全北区。果樹、クワ、樹木害虫
European fruit lecanium			vine scale (1) を見よ
European fruit scale			oystershell scale (1) を見よ　北米での英名
European fruit scale			pear scale (1) を見よ
European furniture beetle			common furniture beetle を見よ
European gall aphid			poplar-spiral-gall aphid を見よ

英名	和名	学名	所属、分布、ほか
European giant ant			herculean ant (2) を見よ
European giant peacock			emperor (2) を見よ
European glow worm			glow-worm (2) を見よ
European goatmoth			goat moth (1) を見よ　米国での英名
European gold tail			browntail moth (2) を見よ
European golden-Y moth			golden Y-moth を見よ
European gooseberry aphid			gooseberry aphid (1) を見よ
European grain aphid			oat bird-cherry aphid を見よ　米国での英名
European grain louse			grain aphid (1) を見よ　北アフリカでの英名
European grain moth	コクガ	*Nemapogon granella* (Linnaeus)	（チョウ目、ヒロズコガ科）日本、汎世界。シイタケ、貯穀害虫。米国での英名
European grapevine moth			grapevine moth (1) を見よ
European green botfly			sheep bot fly を見よ　豪州での英名
European ground beetle			forest caterpillar-hunter を見よ
European gypsy moth			gypsy moth を見よ
European hardwood ambrosia beetle		*Trypodendron domesticum* (Linnaeus)	（コウチュウ目、キクイムシ科）全北区
European harvester ant			harvester ant (2) を見よ
European honey bee			honey bee を見よ
European honeysuckle leaf roller			honeysuckle moth を見よ
European honeysuckle leaf roller			tooth-streaked hooked smudge を見よ　米国での英名
European honeysuckle moth			honeysuckle moth を見よ
European hop flea beetle			hop flea beetle を見よ　米国での英名
European hornet			hornet を見よ
European hornet moth			hornet moth を見よ
European horse biting louse			pilose biting horse louse を見よ
European horse chewing louse			pilose biting horse louse を見よ　米国での英名
European horseradish webworm			chequered straw pearl を見よ　幼虫の英名

英名	和名	学名	所属、分布、ほか
European house borer			house longhorn を見よ
European house cricket			house cricket を見よ
European house moth			brown-dotted clothes moth を見よ
European June beetle			summer chafer を見よ　米国での英名
European June bug			summer chafer を見よ
European lackey moth			lackey を見よ
European lanternfly		*Dictyophara europaea* (Linnaeus)	(カメムシ目、テングスケバ科)旧北区
European large tortoiseshell			large tortoiseshell を見よ
European leaf beetle			willow leaf beetle (1) を見よ
European leaf weevil		*Polydrosus impressifrons* Gyllenhal	(コウチュウ目、ゾウムシ科)旧北区
European leafroller			rose tortrix moth を見よ　北米での英名
European lilac leaf miner		*Gracilaria syringella* (Fabricius)	(チョウ目、ホソガ科) 全北区
European lilac leaf miner			confluent-barred slender を見よ　北米での英名
European lyctus beetle			oak lyctid を見よ
European maize borer			European corn borer を見よ
European mantid			mantid を見よ　米国での英名
European mantis			mantid を見よ　カナダでの英名
European map butterfly			map butterfly を見よ
European maple aphid			black hairy aphid を見よ
European mole cricket			mole cricket (3) を見よ
European mouse flea		*Leptopsylla segnis* (Schönherr)	(ノミ目、ホソノミ科) 日本、汎世界
European musk beetle			musk beetle を見よ　米国での英名
European oak bark beetle			oak bark beetle (1) を見よ
European oak borer		*Agrilus sulcicollis* Lacordaire	(コウチュウ目、タマムシ科)旧北区
European orange-tip			orange tip (1) を見よ　米国での英名
European owl moth			Hartig's brahmaea を見よ
European paper wasp			yellow paper wasp を見よ
European pea moth			pea moth を見よ
European peach aphid			black peach aphid (1) を見よ

英名	和名	学名	所属、分布、ほか
European peach scale			European fruit lecanium を見よ 米国での英名
European peach scale			vine scale (2) を見よ
European pear scale			Italian pear scale を見よ 米国での英名
European permanent currant aphid			permanent currant aphid を見よ
European pigeon bug			pigeon bug を見よ 米国での英名
European pigeon louse		*Goniocotes bidentotus* Scopoli	(ハジラミ目、チョウカクハジラミ科) 旧北区
European pigeon louse			small pigeon louse (1) を見よ 米国での英名
European pine engraver		*Ips mannsfeldi* (Wachtl)	(コウチュウ目、キクイムシ科) 旧北区
European pine moth			pine-tree lappet を見よ 米国での英名
European pine resin midge		*Cecidomyia pini* (De Geer)	(ハエ目、タマバエ科) 旧北区。
European pine sawfly			fox colored sawfly を見よ 米国での英名
European pine sawfly			introduced pine sawfly を見よ 米国での英名
European pine shoot moth			pine shoot moth (1) を見よ 北米での英名
European pineshoot moth			pine twig moth を見よ
European poplar clearwing moth			dusky clearwing moth を見よ
European poplar shoot borer moth			poplar cloaked bell を見よ
European potter wasp		*Ancistrocerus gazella* (Panzer)	(ハチ目、スズメバチ科) 旧北区、豪州区
European powder post beetle			powderpost beetle (1) を見よ
European predacious diving beetle			water beetle (1) を見よ
European privet sawfly		*Macrophya punctumalbum* (Linnaeus)	(ハチ目、ハバチ科) 全北区。カナダでの英名
European processionary moths		*Thaumetopoea*	(チョウ目、シャチホコガ科) の昆虫の総称
European rabbit flea		*Spilopsyllus cuniculi* (Dale)	(ノミ目、ヒトノミ科) 旧北区、豪州区
European raphidian		*Raphidia ratzeburgi* Brauer	(アミメカゲロウ目、キスジラクダムシ科) 旧北区
European raspberry aphid			rubus aphid (2) を見よ 米国での英名
European raspberry beetle			raspberry beetle (1) を見よ
European rat flea			northern rat flea を見よ 豪州での英名
European red-bellied clerid			ant beetle を見よ
European rhinocerus beetle			rhinoceros beetle (3) を見よ

英名	和名	学名	所属、分布、ほか
European rose sawfly			banded rose sawfly を見よ　米国での英名
European rose slug			rose slug sawfly を見よ
European rosette willow gall midge	ヤナギシントメタマバエ	*Rhabdophaga rosaria* (H. Loew)	（ハエ目、タマバエ科）日本、全北区。ヤナギ害虫
European shot-hole borer			shot-hole borer (3) を見よ　米国での英名
European skipper	カラフトセセリ	*Thymelicus lineola* Ochsenheimer	（チョウ目、セセリチョウ科）全北区
European skipperling			European skipper を見よ
European small tortoiseshell			small tortoiseshell を見よ　米国での英名
European spanishfly			blister beetle を見よ
European spiny mason wasp		*Hoplomerus spinipes* (Linnaeus)	（ハチ目、セイボウ科）旧北区
European spittle insect			cuckoo spit insect を見よ
European spruce bark beetle			eight-spined ips を見よ
European spruce bark beetle			European spruce beetle を見よ
European spruce beetle	エゾマツオオキクイムシ	*Dendroctonus micans* (Kugelmann)	（コウチュウ目、キクイムシ科）日本、全北区
European spruce longhorn beetle			black spruce beetle を見よ
European spruce needleminer moth			dwarf pine bell を見よ
European spruce sawfly	シマトウヒハバチ	*Gilpinia polytoma* (Hartig)	（ハチ目、マツハバチ科）日本、新北区
European spruce sawfly			spruce sawfly (1) を見よ
European stag beetle			stag beetle を見よ
European strawberry weevil			black vine weevil を見よ
European tarnished plantbug			tarnished plant bug (2) を見よ
European thrips		*Thrips physapus* Linnaeus	（アザミウマ目、アザミウマ科）全北区。米国での英名
European tortoise bug			Hottentot bug (1) を見よ
European tree cricket			tree cricket (1) を見よ
European turtle bug		*Podops inuncta* (Fabricius)	（カメムシ目、カメムシ科）旧北区
European tussock moth			common vapourer を見よ　米国での英名
European velvet ant			velvet ant を見よ
European vine moth			grapevine moth (1) を見よ
European vine weevil			black vine weevil を見よ　米国での英名
European walnut aphid			walnut aphid を見よ　米国での英名

英名	和名	学名	所属、分布、ほか
European wasp			German wasp を見よ　豪州での英名
European wasp beetle			wasp beetle (1) を見よ
European water cricket		*Corixa geoffroyi* Leach	（カメムシ目、ミズムシ科）全北区
European water cricket			common corixa を見よ　米国での英名
European wheat sawfly			wheat stem sawfly (1) を見よ
European wheat stem maggot		*Meromyza saltatrix* (Linnaeus)	（ハエ目、キモグリバエ科）旧北区
European wheat stem sawfly			wheat stem sawfly (1) を見よ
European winter moth			winter moth (1) を見よ　米国での英名
European wood wasp			horntail (2) を見よ
eurya whitefly	ヒサカキコナジラミ	*Aleurotuberculatus euryae* (Kuwana)	（カメムシ目、コナジラミ科）日本。ヒサカキ、イヌツゲなどの害虫
Eurynome		*Speyeria mormonia eurynome* (Edwards)	（チョウ目、タテハチョウ科）新北区。Mormon fritillary を参照
eurystethid beetles		Eurystethidae	（コウチュウ目）の昆虫の総称
eurytomid wasps			jointworms (2) を見よ
eurytomids			jointworms (2) を見よ
eutrichosomatids		Eutrichosomatidae	（ハチ目）の昆虫の総称　Encyrtidae トビコバチ科の異名
Evadnes skipper		*Panoquina evadnes* (Stoll)	（チョウ目、セセリチョウ科）新熱帯区
evanioid flies			evanioid wasps を見よ
evanioid wasps	ヤセバチ上科	Evanioidea	（ハチ目）の昆虫の総称
Evans' Acraea skipper		*Fresna carlo* Evans	（チョウ目、セセリチョウ科）エチオピア区
Evans' Mylon		*Mylon salvia* Evans	（チョウ目、セセリチョウ科）新熱帯区
Evans' orange sprite		*Celaenorrhinus ovalis* Evans	（チョウ目、セセリチョウ科）エチオピア区
Evans' recluse		*Leona halma* Evans	（チョウ目、セセリチョウ科）エチオピア区
Evans' ruby-eye		*Perichares aurina* Evans	（チョウ目、セセリチョウ科）新熱帯区
Evan's skipper		*Panoquina evansi* (Freeman)	（チョウ目、セセリチョウ科）新北区、新熱帯区
Evans' tufted skipper		*Nisoniades evansi* Steinhauser	（チョウ目、セセリチョウ科）新熱帯区
evechurr			field cricket (1) を見よ
eve-churrs			mole crickets (1) を見よ
evehawkers		*Anaciaeschna*	（トンボ目、ヤンマ科）の昆虫の総称
even-lined Renia			yellow-spotted Renia moth を見よ
even-lined sallow moth		*Ipimorpha pleonectusa* Grote	（チョウ目、ヤガ科）新北区
even-spurred ground crickets		*Eunemobius*	（バッタ目、コオロギ科）の昆虫の総称

英名	和名	学名	所属、分布、ほか
evening brown	ウスイロコノマチョウ	*Melanitis leda* (Linnaeus)	(チョウ目、タテハチョウ科) 日本、東洋区、豪州区、エチオピア区。日本亜種 *M. l. ismene* (Cramer) も同英名
evening brown			zebra caterpillar moth を見よ　南米での英名
evening brown butterfly		*Melanitis leda bankia* (Fabricius)	(チョウ目、タテハチョウ科) 豪州区
evening browns		*Cyllogenes*	(チョウ目、タテハチョウ科) の昆虫の総称
evening browns		*Gnophodes*	(チョウ目、タテハチョウ科) の昆虫の総称
evening browns			browns (2) を見よ
evening cicada	ヒグラシ	*Tanna japonensis japonensis* (Distant)	(カメムシ目、セミ科) 日本、旧北区
evening hawker		*Anaciaeschna triangulifera* McLachlan	(トンボ目、ヤンマ科) エチオピア区
evening moth			hummingbird hawkmoth を見よ
evening skimmer		*Tholymis citrina* Hagen	(トンボ目、トンボ科) 新熱帯区
everchur			mole cricket (3) を見よ
Everest clouded yellow	セイカイウスベニモンキチョウ	*Colias berylla* Fawcett	(チョウ目、シロチョウ科) 東洋区
everglades conehead		*Neoconocephalus pahavokee* Whitesell	(バッタ目、キリギリス科) 新北区
evergreen bagworm			bagworm (1) を見よ　北米での英名
evergreen bagworm moth			bagworm (1) を見よ　北米での英名
evergreen clover leaf-cut weevil			blue leaf-cut weevil を見よ
evergreen coneworm moth		*Dioryctria abietivorella* (Grote)	(チョウ目、メイガ科) 新北区
evergreen oak bark beetle			Saxena ambrosia beetle を見よ
evergreen oak cottony scale			cyclbalanopsis scale を見よ
evergreen oak elongate scolytid			oak borer (2) を見よ
evergreen oak hairy aphid	カシケクダアブラムシ	*Allotrichosiphum kashicola* (Kurisaki)	(カメムシ目、アブラムシ科) 日本
evergreen oak horn worm	クチバスズメ	*Marumba sperchius* (Ménétriès)	(チョウ目、スズメガ科) 日本、旧北区、東洋区
evergreen oak leafhopper			oak leafhopper (2) を見よ
evergreen oak long-leg weevil	カシアシナガゾウムシ	*Mecysolobus piceus* (Roelofs)	(コウチュウ目、ゾウムシ科) 日本、旧北区
evergreen oak red scale			oak scale (1) を見よ
evergreen oak scale	シイマルカイガラムシ	*Hypaspidiotus jordani* (Kuwana)	(カメムシ目、マルカイガラムシ科) 日本
evergreen oak shoot sawfly			oak shoot sawfly を見よ
evergreen oak thyridid	マダラマドガ	*Rhodoneura vittula* Guenée	(チョウ目、マドガ科) 日本、旧北区、東洋区

英名	和名	学名	所属、分布、ほか
evergreen oak woolly aphid	シラカシムネアブラムシ	*Xenothoracaphis kashifoliae* (Uye)	（カメムシ目、アブラムシ科）日本
evergreen spanworm			false hemlock looper を見よ
everlasting bud moth		*Eublemma minima* (Guenée)	（チョウ目、ヤガ科）新北区、新熱帯区、エチオピア区
everlasting Tebenna moth		*Tebenna gnaphaliella* (Kearfott)	（チョウ目、ハマキモドキガ科）新北区
Evershed's ace		*Thoressa evershedi* (Evans)	（チョウ目、セセリチョウ科）東洋区
Eversmann's Parnassian		*Parnassius eversmanni* Ménétriès	（チョウ目、アゲハチョウ科）新北区。日本亜種は *P. e. daisetsuzanus* Matsumura ウスバキチョウ
Eversmann's rustic			black army cutworm moth を見よ
exasperating Platynota moth		*Platynota exasperatana* (Zeller)	（チョウ目、ハマキガ科）新北区
excelsior eighty-eight			superb numberwing を見よ
exclamation moth		*Homaledra heptathalama* Busck	（チョウ目、ツツミノガ科）新北区
exhausted brocade moth		*Neoligia exhausta* (Smith)	（チョウ目、ヤガ科）新北区
exile		*Apamea maillardi* (Geyer)	（チョウ目、ヤガ科）旧北区。exile moth ともいう
exile		*Apamea marmorata* (Zetterstedt)	（チョウ目、ヤガ科）旧北区
exiled dagger moth		*Acronicta exilis* Grote	（チョウ目、ヤガ科）新北区
exotic birch-leafmining sawfly		*Profenus thomsoni* (Konow)	（ハチ目、ハバチ科）日本、旧北区
exotic greenhouse invasive		*Duponchelia fovealis* Zeller	（チョウ目、メイガ科）全北区
exotic leafroller moth			apple tortrix を見よ
exotic pumpkin caterpillar moth			cotton caterpillar (1) を見よ
exposed bird dropping moth			nun (1) を見よ
exposed Lipocosma moth		*Lipocosma septa* Munroe	（チョウ目、メイガ科）新北区
exquisite sailor		*Dynamine athemon* (Linnaeus)	（チョウ目、タテハチョウ科）新熱帯区
extinct shieldback		*Neduba extincta* Rentz	（バッタ目、キリギリス科）新北区
extra lascar	サンダカキンミスジ	*Pantoporia sandaka* (Butler)	（チョウ目、タテハチョウ科）東洋区
extra-striped snaketail		*Ophiogomphus anomalus* Harvey	（トンボ目、サナエトンボ科）新北区
eye-cap moths	ヒラタモグリガ科	Opostegidae	（チョウ目）の昆虫の総称
eye flies			frit flies を見よ
eye gnat		*Hippelates pusio* (Loew)	（ハエ目、キモグリバエ科）新北区
eye gnat		*Hippelates collusor* Loew	（ハエ目、キモグリバエ科）新北区
eye gnats		*Hippelates*	（ハエ目、キモグリバエ科）の昆虫の総称

英名	和名	学名	所属、分布、ほか
eye gnats			frit flies を見よ
eyegnats			cryptochetid flies を見よ
eye-marked drepanid	ヒトツメカギバ	*Auzata superba* (Butler)	（チョウ目、カギバガ科）日本、旧北区
eyeringed Chionodes moth		*Chionodes discoocellella* (Chambers)	（チョウ目、キバガ科）新北区
eyespot anthelid		*Anthela ocellata* (Walker)	（チョウ目、Anthelidae）豪州区
eyespotted bud moth	リンゴシロヒメハマキ	*Spilonota ocellana* (Denis et Schiffermüller)	（チョウ目、ハマキガ科）日本、全北区。リンゴ、ナシなど果樹害虫。北米での英名
eye-spotted lady beetle			eyed ladybird を見よ　カナダでの英名
eyed Baileya moth		*Baileya ophthalmica* (Guenée)	（チョウ目、ヤガ科）新北区
eyed brown		*Satyrodes eurydice* (Linnaeus)	（チョウ目、タテハチョウ科）新北区
eyed bush brown		*Henotesia perspicua* (Trimen)	（チョウ目、タテハチョウ科）エチオピア区
eyed click beetle			eastern eyed click beetle を見よ
eyed Dysodia moth		*Dysodia oculatana* Clemens	（チョウ目、マドガ科）新北区
eyed elater		*Alaus melanops* LeConte	（コウチュウ目、コメツキムシ科）新北区
eyed elater			eastern eyed click beetle を見よ
eyed hawk			willow hawk moth (1) を見よ　eyed hawk moth ともいう
eyed hawk moth			willow hawk moth (1) を見よ
eyed ladybird		*Anatis ocellata* (Linnaeus)	（コウチュウ目、テントウムシ科）旧北区
eyed ladybird			ashy grey lady beetle を見よ
eyed Paectes moth		*Paectes oculatrix* (Guenée)	（チョウ目、ヤガ科）新北区
eyed pansy		*Junonia orithya madagascariensis* Guenée	（チョウ目、タテハチョウ科）エチオピア区。blue pansy (2) を参照
eyed sister			Paroeca sister を見よ
eyed sphinx			willow hawk moth (2) を見よ
eyed spreadwing		*Cyclosemia leppa* Evans	（チョウ目、セセリチョウ科）新熱帯区
eyed stink bug			two-spotted stink bug を見よ
eyed tiger-moth			great leopard moth を見よ
eyed X poplar hawk		*Smerinthus hybridus* Stephens	（チョウ目、スズメガ科）旧北区
eyed Zulu		*Alaena interposita hauttecoeuri* Oberthür	（チョウ目、シジミチョウ科）エチオピア区
eyelet sober		*Thiotricha subocellea* (Stephens)	（チョウ目、キバガ科）旧北区

英名	和名	学名	所属、分布、ほか
F			
F sharp			human flea を見よ
Fabricius angel		*Chorinea octauius* (Fabricius)	（チョウ目、シジミタテハ科）新熱帯区
Fabricius sabre-wing		*Jemadia gnetus* (Fabricius)	（チョウ目、セセリチョウ科）新熱帯区
Fabricius's clothes moth		*Haplotinea insectella* (Fabricius)	（チョウ目、ヒロズコガ科）旧北区
Fabricius's nettle tap	イラクサハマキモドキ	*Anthophila fabriciana* (Linnaeus)	（チョウ目、ハマキモドキガ科）日本、旧北区
fabulous green sphinx of Kauai			green sphinx を見よ
face fly		*Musca autumnalis* De Geer	（ハエ目、イエバエ科）全北区
faceted skipper		*Synapte syracea* (Godman)	（チョウ目、セセリチョウ科）新北区
facilis skipper		*Eutocus facilis* (Plötz)	（チョウ目、セセリチョウ科）新熱帯区
faded eighty-eight			Astala eighty-eight を見よ
faded faceted-skipper		*Synapte shiva* (Evans)	（チョウ目、セセリチョウ科）新熱帯区
Fadus sphinx moth		*Aellopos fadus* (Cramer)	（チョウ目、スズメガ科）新北区、新熱帯区。Fadus sphinx ともいう
faga blue		*Nabokovia faga* Dognin	（チョウ目、シジミチョウ科）新熱帯区
fagaceae bark beetle	アカガシノキクイムシ	*Xyleborus concisus* Blandford	（コウチュウ目、キクイムシ科）日本、東洋区
faggot bagworm			faggot case moth を見よ
faggot case moth		*Clania ignobilis* (Walker)	（チョウ目、ミノガ科）豪州区
faggot worm		*Clania crameri* Westwood	（チョウ目、ミノガ科）東洋区
faggot-worm (1)		*Kotochalia doubledayi* Westwood	（チョウ目、ミノガ科）東洋区
fagus bivalve-shaped gall midge	ブナカイガラタマバエ	*Hartigiola faggalli* (Monzen)	（ハエ目、タマバエ科）日本。ブナノキ害虫
faint silver-striped bell			coast spurge bell を見よ
faint-spotted angle moth		*Digrammia ocellinata* (Guenée)	（チョウ目、シャクガ科）新北区
faint-spotted Palthis moth		*Palthis asopialis* (Guenée)	（チョウ目、ヤガ科）新北区
fairy flies	ホソハネコバチ科	Mymaridae	（ハチ目）の昆虫の総称　体長1mm以内の卵寄生蜂
fairyflies			minute egg parasites を見よ
fairy hairstreak		*Hypolycaena lebona* (Hewitson)	（チョウ目、シジミチョウ科）エチオピア区
fairy moths		*Adela*	（チョウ目、マガリガ科）の昆虫の総称
fairy moths			yucca moths (1) を見よ
fairy-ring longhorn beetle		*Pseudovadonia livida* (Fabricius)	（コウチュウ目、カミキリムシ科）旧北区
fairy wasps			fairy flies を見よ

英名	和名	学名	所属、分布、ほか
fairy white		*Argyrocheila undifera* Staudinger	(チョウ目、シジミチョウ科) エチオピア区
fairy yellow		*Eurema daira* (Godart)	(チョウ目、シロチョウ科) 新北区、新熱帯区
faithful beauty moth		*Composia fidelissima* Herrich-Schäffer	(チョウ目、ヒトリガ科) 新北区、新熱帯区。faithful beauty ともいう
falcate Acraea	カギバホソチョウ	*Acraea perenna* Doubleday	(チョウ目、タテハチョウ科) エチオピア区
falcate Dismorphia		*Lieinix nemesis* (Latreille)	(チョウ目、シロチョウ科) 新熱帯区
falcate Emesis			falcate metalmark (1) を見よ
falcate gemmed-satyr		*Cyllopsis clinas* (Godman et Salvin)	(チョウ目、タテハチョウ科) 新熱帯区
falcate metalmark		*Apodemia hypoglauca hypoglauca* (Godman et Salvin)	(チョウ目、シジミタテハ科) *A. h. wellingi* Ferris も同英名
falcate metalmark (1)		*Emesis tenedia* C. et R. Felder	(チョウ目、シジミタテハ科) 新北区、新熱帯区
falcate metalmark			curve-winged metalmark を見よ
falcate oakblue		*Mahathala ameria* (Hewitson)	(チョウ目、シジミチョウ科) 東洋区
falcate orange-tip	ミデイアツマキチョウ	*Anthocharis midea* (Hübner)	(チョウ目、シロチョウ科) 新北区
falcate polyptychus		*Falcatula falcatus* (Rothschild et Jordan)	(チョウ目、スズメガ科) エチオピア区
falcate Prepona		*Archaeoprepona phaedra aelia* (Godman et Salvin)	(チョウ目、タテハチョウ科) 新熱帯区
falcate red Charaxes		*Charaxes paphianus* Ward	(チョウ目、タテハチョウ科) エチオピア区
falcate skipper		*Spathilepia clonius* (Cramer)	(チョウ目、セセリチョウ科) 新北区、新熱帯区
falcate Theope		*Theope phaeo* Prittwitz	(チョウ目、シジミタテハ科) 新熱帯区
falcate yellow			plain sulphur を見よ
Falcifera sphinx moth		*Callionima falcifera* (Gehlen)	(チョウ目、スズメガ科) 新北区、新熱帯区
falcon sphinx moth		*Xylophanes falco* (Walker)	(チョウ目、スズメガ科) 新北区、新熱帯区
Falerina hairstreak		*Enos falerina* (Hewitson)	(チョウ目、シジミチョウ科) 新熱帯区
fall armyworm			black armyworm を見よ
fall armyworm			common cutworm を見よ
fall armyworm moth			black armyworm を見よ
fall canker moth			fall cankerworm を見よ
fall cankerworm		*Alsophila pometaria* (Harris)	(チョウ目、シャクガ科) 新北区。雌は無翅
fall cankerworm moth			fall cankerworm を見よ
fall field cricket			northern field cricket を見よ
fall spruce needle moth		*Argyrotaenia occultana* Freeman	(チョウ目、ハマキガ科) 新北区
fall webworm	アメリカシロヒトリ	*Hyphantria cunea* (Drury)	(チョウ目、ヒトリガ科) 日本、全北区。ダイズ、果樹、クワ、チャなど多くの植物害虫
fall webworm moth			fall webworm を見よ

英名	和名	学名	所属、分布、ほか
fallax sabre-wing		*Jemadia fallax* (Mabille)	(チョウ目、セセリチョウ科) 新熱帯区
fallow beetle			hide beetle を見よ
fallow deer nostril fly		*Cephenomyia multispinosa* Ulrich	(ハエ目、ヒツジバエ科) 旧北区
false acacia sawfly		*Nematus tibialis* Newman	(ハチ目、ハバチ科) 旧北区
false Acraea			sweet potato Acraea を見よ
false Altinote		*Gnathotriche mundina* Druce	(チョウ目、タテハチョウ科) 新熱帯区
false antlike flower beetles	クビボソムシ亜科	Pedilinae	(コウチュウ目、アカハネムシ科) の昆虫の総称
false antlike flower beetles			cardinal beetles を見よ
false Apollo	シリアアゲハ	*Archon apollinus* (Herbst)	(チョウ目、アゲハチョウ科) 旧北区
false apple leaf miner			apple leaf miner (1) を見よ
false armored scales		Conchaspididae	(カメムシ目) の昆虫の総称
false Baton blue		*Pseudophilotes abencerragus* (Pierret)	(チョウ目、シジミチョウ科) 旧北区
false black flour beetle			dark flour beetle を見よ
false blister beetle		*Oxycopis mcdonaldi* (Arnett)	(コウチュウ目、カミキリモドキ科) 新北区
false blister beetles	カミキリモドキ科	Oedemeridae	(コウチュウ目) の昆虫の総称
false bombardier beetle		*Galerita bicolor* Drury	(コウチュウ目、オサムシ科) 大洋区、新北区
false bombardier beetles		Galeritini	(コウチュウ目、オサムシ科) の昆虫の総称
false brown bamboo scale			penicillate scale を見よ
false brown scale			orchid soft scale を見よ
false budworm			corn earworm (2) を見よ 幼虫の英名。false bud-worm とも記す
false bumble-bees			cuckoo bees (2) を見よ
false burnet moth		*Euclodea glyphica* (Linnaeus)	(チョウ目、ヤガ科) 旧北区
false burnet moths		Urodidae	(チョウ目) の昆虫の総称
false cabbage aphid			turnip aphid (1) を見よ
false cabbage sawfly	オスグロハバチ	*Dolerus similis japonicus* Kirby	(ハチ目、ハバチ科) 日本
false caterpillars			sawflies (1) を見よ 幼虫の英名
false celery leaftier		*Udea profundalis* (Packard)	(チョウ目、メイガ科) 新北区
false chestnut leaved oak bug	サジクヌギカメムシ	*Urostylis striicornis* Scott	(カメムシ目、クヌギカメムシ科) 日本、旧北区
false chief	イチモンジホソチョウモドキ	*Pseudacraea lucretia* (Cramer)	(チョウ目、タテハチョウ科) エチオピア区
false chinch bug		*Nysius raphanus* Howard	(カメムシ目、ナガカメムシ科) 新北区

英名	和名	学名	所属、分布、ほか
false chinch bug		*Nysius ericae* (Schilling)	（カメムシ目、ナガカメムシ科）旧北区。カンキツ、ブドウ害虫
false chinch bugs			big-eyed bugs (1) を見よ
false click beetles	コメツキダマシ科	Eucnemidae	（コウチュウ目）の昆虫の総称
false clown beetles	エンマムシダマシ科	Sphaeritidae	（コウチュウ目）の昆虫の総称
false codling moth		*Cryptophlebia batrachopa* Meyrick	（チョウ目、ハマキガ科）エチオピア区
false codling moth (1)		*Argyroploce leucotreta* Meyrick	（チョウ目、ハマキガ科）旧北区、エチオピア区。カンキツ害虫
false coffee playboy		*Deudorix kayonza* Stempffer	（チョウ目、シジミチョウ科）エチオピア区
false comma	エルタテハ	*Nymphalis vaualbum samurai* (Fruhstorfer)	（チョウ目、タテハチョウ科）日本。全北区の *N. vaualbum* (Denis et Schiffermüller) の英名
false crane flies		*Bittacomorpha*	（シリアゲムシ目、ガガンボモドキ科）の昆虫の総称
false crane flies			phantom crane flies を見よ
false craneflies			wood gnats を見よ
false crocus geometer moth		*Xanthotype urticaria* (Swett)	（チョウ目、シャクガ科）新北区
false daggerwing		*Hypanartia dione* Latreille	（チョウ目、タテハチョウ科）新熱帯区
false daimyo oak cutworm			daimyo oak yellow-dotted noctuid を見よ
false darkling beetles	ナガクチキムシ科	Melandryidae	（コウチュウ目）の昆虫の総称
false death's head			discoid cockroach を見よ　false death's head cockroack ともいう
false dewy ringlet	ステニオベニヒカゲ（ニセハイイロベニヒカゲ）	*Erebia sthennyo* Graslin	（チョウ目、タテハチョウ科）旧北区
false diadem			false chief を見よ
false diamondback moths	アトヒゲコガ科	Acrolepiidae	（チョウ目）の昆虫の総称
false diving beele	ゲンゴロウモドキ	*Dytiscus dauricus* Gebler	（コウチュウ目、ゲンゴロウ科）日本、全北区
false Doris	マエモンシロチョウ	*Archonias brassolis* (Fabricius)	（チョウ目、シロチョウ科）新熱帯区
false dotted border		*Belenois thysa thysa* (Hopffer)	（チョウ目、シロチョウ科）エチオピア区
false duskywing		*Gesta invisus* (Butler et Druce)	（チョウ目、セセリチョウ科）新北区
false Eros blue	シベリアヒメシジミ	*Polyommatus eroides* (Frivaldszky)	（チョウ目、シジミチョウ科）旧北区
false evergreen oak scale	カシニセタマカイガラムシ	*Lecanodiaspis quercus* Cockerell	（カメムシ目、ニセタマカイガラムシ科）日本、旧北区
false-eye leafhopper		*Empoasca vitis* (Goethe)	（カメムシ目、ヒメヨコバイ科）日本、旧北区
false flower beetle		*Cyphon variabilis* (Thunberg)	（コウチュウ目、マルハナノミ科）旧北区
false fritillary		*Pseudargynnis hegemone* (Godart)	（チョウ目、タテハチョウ科）エチオピア区

英名	和名	学名	所属、分布、ほか
false garden mantid		*Pseudomantis albofimbriata* (Stål)	（カマキリ目、カマキリ科）豪州区
false German cockroach	ヒメチャバネゴキブリ	*Blattella lituricollis* (Walker)	（ゴキブリ目、チャバネゴキブリ科）日本、新北区
false golden-banded skipper		*Autochton pseudocellus* (Coolidge et Clemence)	（チョウ目、セセリチョウ科）新北区、新熱帯区
false grayling		*Arethusana arethusa* (Schiffermüller)	（チョウ目、タテハチョウ科）旧北区
false greenhouse leaftier			false celery leaftier を見よ
false heath fritillary	ウスイロヒョウモンモドキ	*Melitaea protomedia* Ménétriès	（チョウ目、タテハチョウ科）日本、旧北区
false hemlock looper		*Nepytia canosaria* (Walker)	（チョウ目、シャクガ科）新北区。成虫は false hemlock moth という
false hemlock looper			western false hemlock looper を見よ
false hornworm			false sphinx を見よ　北米での英名
false ilex hairstreak		*Satyrium esculi* (Hübner)	（チョウ目、シジミチョウ科）旧北区
false jumping bush cricket		*Orocharis luteolira* Walker	（バッタ目、コオロギ科）新北区
false katydids			bush katydids (1) を見よ
false ladybird beetles			darkling beetles (1) を見よ
false lesser bulb fly			onion bulb fly を見よ
false lizard barklice	ニセケチャタテ科	Pseudocaeciliidae	（チャタテムシ目）の昆虫の総称
false longhorn beetles	クビナガムシ科	Cephaloidae	（コウチュウ目）の昆虫の総称
false long-horned beetles			false longhorn beetles を見よ
false mallow skipper		*Carcharodes tripolinus* Verity	（チョウ目、セセリチョウ科）旧北区
false mantids	カマキリモドキ科	Mantispidae	（アミメカゲロウ目）の昆虫の総称　過変態でクモの巣に寄生。地中のコガネムシ、ヤガ幼虫を捕食
false March brown		*Ecdyonurus venosus* (Fabricius)	（カゲロウ目、ヒラタカゲロウ科）旧北区
false marmorated grasshopper	クルマバッタモドキ	*Oedaleus infernalis* Saussure	（バッタ目、バッタ科）日本、旧北区
false meadow brown		*Pyronia janiroides* (Herrich-Schäffer)	（チョウ目、タテハチョウ科）旧北区
false mealworm		*Alobates pennsylvanica* (De Geer)	（コウチュウ目、ゴミムシダマシ科）新北区。false mealworm beetle ともいう
false melon beetle	ウリハムシモドキ	*Atrachya menetriesi* (Faldermann)	（コウチュウ目、ハムシ科）日本、旧北区。トウモロコシ、ダイズ、アブラナ科野菜など多くの作物害虫
false metallic wood-boring beetles			throscid beetles を見よ
false Methona	トンボシロチョウ	*Patia orise* (Boisduval)	（チョウ目、シロチョウ科）新熱帯区
false Mnestra ringlet		*Erebia aethiopella* (Hoffmannsegg)	（チョウ目、タテハチョウ科）旧北区

英名	和名	学名	所属、分布、ほか
false mocha		*Cyclophora porata* (Linnaeus)	（チョウ目、シャクガ科）旧北区
false numberwing		*Paulogramma pyracmon* (Godart)	（チョウ目、タテハチョウ科）新熱帯区
false Nymphidium		*Synargis calyce* (C. et R. Felder)	（チョウ目、シジミタテハ科）新熱帯区
false oleander scale	アオキシロカイガラムシ	*Pseudaulacaspis cockerelli* (Cooley)	（カメムシ目、マルカイガラムシ科）日本、東洋区、新北区、汎熱帯。バナナ、マンゴー、キウイなどの害虫
false oriental fruit fly		*Dacus opiliae* Drew et Hardy	（ハエ目、ミバエ科）豪州区
false owlet moths	トガリバガ科	Thyatiridae	（チョウ目）の昆虫の総称
false paddy borer	ヒトスジオオメイガ	*Scirpophaga lineata* (Butler)	（チョウ目、メイガ科）日本、東洋区
false parlatoria scale	ラヌスマルカイガラムシ	*Pseudoparlatoria parlatorioides* (Comstock)	（カメムシ目、マルカイガラムシ科）日本、汎熱帯
false pathfinder skipper		*Xanthodisca astrape* (Holland)	（チョウ目、セセリチョウ科）エチオピア区
false pit scale insects	ニセタマカイガラムシ科	Lecanodiaspididae	（カメムシ目）の昆虫の総称　世界に約70種
false pit scales			false pit scale insects を見よ
false plain tiger			danaid eggfly を見よ
false plume moths		Tineodidae	（チョウ目）の昆虫の総称
false potato beetle			horse-nettle beetle を見よ
false powderpost beetles	ナガシンクイムシ科	Bostrichidae	（コウチュウ目）の昆虫の総称
false red-striped delicate geometrid	コベニスジヒメシャク	*Timandra comptaria* Walker	（チョウ目、シャクガ科）日本、旧北区
false rice grasshopper	イナゴモドキ	*Parapleurus alliaceus* (Germar)	（バッタ目、バッタ科）日本、旧北区
false rice stink bug		*Aelia fieberi* Scott	（カメムシ目、カメムシ科）日本、旧北区
false ringlet	ヒメヒカゲ	*Coenonympha aedippus annulifer* Butler	（チョウ目、タテハチョウ科）日本。*C. aedippus* Pionneau の英名
false roadside-skipper		*Repens florus* (Godman)	（チョウ目、セセリチョウ科）新熱帯区
false robust conehead		*Neoconocephalus bivocatus* Walker	（バッタ目、キリギリス科）新北区
false rose stem sawfly			rose stem sawfly (1)(2) を見よ
false sago scale	ソテツマルカイガラムシ	*Aonidiella sotetsu* (Takahashi)	（カメムシ目、マルカイガラムシ科）日本、東洋区
false Saliana		*Justinia norda* Evans	（チョウ目、セセリチョウ科）新熱帯区
false sealed bell		*Pelochrista caecimaculana* (Hübner)	（チョウ目、ハマキガ科）旧北区
false silver berry scale	グミシロカイガラムシ	*Aulacaspis difficilis* (Cockerell)	（カメムシ目、マルカイガラムシ科）日本、旧北区
false silver-bottom brown		*Pseudonympha magoides* van Son	（チョウ目、タテハチョウ科）エチオピア区
false skin beetles	ムクゲキノコムシ科	Biphyllidae	（コウチュウ目）の昆虫の総称

英名	和名	学名	所属、分布、ほか
false slender-footed robberfly		*Leptarthus vitripennis* (Meigen)	（ハエ目、ムシヒキアブ科）旧北区
false sphinx		*Pheosia rimosa* Packard	（チョウ目、シャチホコガ科）新北区。日本亜種は *P. r. fusiformis* Matsumura シロジマシャチホコ
false spruce webworm	カラマツヒラタハバチ	*Cephalcia abietis* (Linnaeus)	（ハチ目、ヒラタハバチ科）日本、旧北区
false stable fly	オオイエバエ	*Muscina stabulans* (Fallén)	（ハエ目、イエバエ科）日本、新北区、豪州区
false stainer			cotton plant bug を見よ
false strawberry sawfly			rose sawfly (2) を見よ
false sweet flag aphid	ショウブアブラムシ	*Schizaphis acori* (Shinji)	（カメムシ目、アブラムシ科）日本
false sweet flag cutworm			crescent を見よ
false swift		*Borbo fallax* (Gaede)	（チョウ目、セセリチョウ科）エチオピア区
false thistle conch	トガリホソハマキ	*Cochylis hybridella* (Hübner)	（チョウ目、ホソハマキガ科）日本、旧北区
false tiger beetles (1)	クチキムシダマシ科	Othniidae	（コウチュウ目）の昆虫の総称
false tiger beetles	クチキムシダマシ亜科	Othniinae	（コウチュウ目、クチキムシダマシ科）の昆虫の総称
false tiger moth	サラサヒトリ	*Camptoloma interiorata* (Walker)	（チョウ目、ヒトリガ科）日本、旧北区。カシ類の害虫
false underwing moth		*Allotria elonympha* (Hübner)	（チョウ目、ヤガ科）新北区
false underwings		*Drasteria*	（チョウ目、ヤガ科）の昆虫の総称
false wanderer		*Pseudacraea eurytus* (Linnaeus)	（チョウ目、タテハチョウ科）エチオピア区
false water Betony		*Shargacucullia prenanthis* (Boisduval)	（チョウ目、ヤガ科）旧北区
false water penny			water penny beetles を見よ
false webworm		*Acantholyda maculiventris* Norton	（ハチ目、ヒラタハバチ科）新北区
false webworm			lucerne moth を見よ
false white-tipped long-leg weevil	ホウジロアシナガゾウムシ	*Mecysolobus erro* Pascoe	（コウチュウ目、ゾウムシ科）日本、東洋区
false wild silkworm moth	クワゴモドキシャチホコ	*Gonoclostera timoniorum* (Bremer)	（チョウ目、シャチホコガ科）日本、旧北区
false-windowed sphinx moth		*Madoryx pseudothyreus* (Grote)	（チョウ目、スズメガ科）新北区、新熱帯区。false-windowed sphinx ともいう
false wireworm		*Eleodes opaca* (Say)	（コウチュウ目、ゴミムシダマシ科）新北区
false wireworm (1)		*Gonocephalum walkeri* (Champion)	（コウチュウ目、ゴミムシダマシ科）豪州区
false wireworm (2)		*Isopteron punctatissimus* (Pascoe)	（コウチュウ目、ゴミムシダマシ科）豪州区
false wireworm (3)		*Gonocephalum misellum* (Blackburn)	（コウチュウ目、ゴミムシダマシ科）豪州区
false wireworms			olive beetles を見よ
false wireworms			comb-clawed beetles (1) を見よ

英名	和名	学名	所属、分布、ほか
false wireworms			darkling beetles (1) を見よ　幼虫の英名
false Y moth			pale Y moth を見よ
false yellow-headed vineworm			Sparganothis fruitworm moth を見よ
false yellow-hindwinged noctuid	キシタアシブトクチバ	*Ophiusa coronata* (Fabricius)	(チョウ目、ヤガ科) 日本、東洋区、豪州区、大洋区
false yellow scale		*Aonidiella comperei* McKenzie	(カメムシ目、マルカイガラムシ科) 東洋区。カンキツ害虫
false yucca moths			yucca moths (3) を見よ
false zebra longwing			Athis longwing を見よ
familiar bluet			blue damselfly を見よ　北米での英名
fan-bearing wood-borer			beech-furniture borer を見よ
fan-bristled robberfly		*Dysmachus trigonus* (Meigen)	(ハエ目、ムシヒキアブ科) 旧北区
fanfoot		*Herminia tarsipennalis* (Treitschke)	(チョウ目、ヤガ科) 旧北区
fanfoot moth			fanfoot を見よ
fan-footed wave			small fan-footed wave を見よ
fan moths			many plume moths を見よ　南アフリカでの英名
fanners and nettle-taps			sedge moths を見よ
fantastic skipper		*Vettius fantasos* (Cramer)	(チョウ目、セセリチョウ科) 新北区
Far East rusty longicorn beetle	サビカミキリ	*Arhopalus coreanus* (Sharp)	(コウチュウ目、カミキリムシ科)日本,旧北区。マツ、スギや木材の害虫。Far East rusty longicorn ともいう
Far Eastern knotweed borer	ウスジロキノメイガ	*Ostrinia latipennis* (Warren)	(チョウ目、メイガ科) 日本、旧北区
Far Eastern urticating moth		*Euproctis flava* Bremer	(チョウ目、ドクガ科) 旧北区
farmer skipper			rural skipper を見よ
Farren's lancewing		*Cataplectica farreni* Walsingham	(チョウ目、ササベリガ科) 旧北区
fasciate sallow pigmy		*Nepticula salicis* Stainton	(チョウ目、モグリチビガ科)旧北区
fasciolated Melipotis moth		*Melipotis fasciolaris* (Hübner)	(チョウ目、ヤガ科) 新北区
fashion commodore	ニジタテハモドキ	*Precis pelarga pelarga* (Fabricius)	(チョウ目、タテハチョウ科) エチオピア区
fast-calling tree cricket		*Oecanthus celerinictus* Walker	(バッタ目、カンタン科) 新北区
fast-tinkling trig		*Anaxipha* sp.	(バッタ目、クサヒバリ科) 新北区
fat flies			small-headed flies を見よ
fatal metalmark		*Calephelis nemesis* (Edwards)	(チョウ目、シジミタテハ科) 新北区
fateful barklice	ヒメチャタテ科	Lachesillidae	(チャタテムシ目) の昆虫の総称
Fatima	ベニモンシロオビタテハ	*Anartia fatima* (Fabricius)	(チョウ目、タテハチョウ科) 新北区、新熱帯区。Fatima butterfly ともいう

英名	和名	学名	所属、分布、ほか
Fatimella Emesis			noble Emesis を見よ
fatsia psylla	ヤツデキジラミ	*Cacopsylla fatsiae* (Jensen)	(カメムシ目、キジラミ科) 日本
fatsia scale			Fukaya pyriform scale を見よ
Faunus anglewing			green comma を見よ
faust leaf-roller weevil	ファウストハマキチョッキリ	*Byctiscus fausti* Sharp	(コウチュウ目、オトシブミ科) 日本
Fawcett's clouded yellow		*Colias nina* Fawcett	(チョウ目、シロチョウ科) 東洋区
fawn-coloured lawn moth		*Crambus bonifatellus* (Hulst)	(チョウ目、メイガ科) 新北区
fawn darner		*Boyeria vinosa* (Say)	(トンボ目、ヤンマ科) 新北区
fawn hairstreak	シャクナゲミドリシジミ	*Chrysozephyrus birupa* (Moore)	(チョウ目、シジミチョウ科) 東洋区
fawn lettuce bell		*Eucosma conterminana* (Guenée)	(チョウ目、ハマキガ科) 旧北区
fawn sallow		*Copipanolis styracis* Guenée	(チョウ目、ヤガ科) 旧北区
fawn-spotted skipper		*Cymaenes odilia* (Burmeister)	(チョウ目、セセリチョウ科) 新北区
fawn-spotted skipper		*Cymaenes trebius* (Mabille)	(チョウ目、セセリチョウ科) 新熱帯区
feather chewing lice			bird lice (1) を見よ
feather legs			platycnemid damselflies を見よ
featherlegs		*Copera*	(トンボ目、モノサシトンボ科) の昆虫の総称
feather lice			chewing lice を見よ
feather-winged beetles	ムクゲキノコムシ科	Ptiliidae	(コウチュウ目) の昆虫の総称
feather-winged moths			many plume moths を見よ　米国での英名
feathered beauty		*Peribatodes secundaria* (Denis et Schiffermüller)	(チョウ目、シャクガ科) 旧北区
feathered brindle		*Aporophyla australis* (Boisduval)	(チョウ目、ヤガ科) 旧北区。亜種 *A. a. pascuea* (Humphreys et Westwood) も同英名
feathered diamond-back bright		*Incurvaria muscalella* (Denis et Schiffermüller)	(チョウ目、マガリガ科) 旧北区
feathered ear		*Pachetra sagittigera* Hufnagel	(チョウ目、ヤガ科) 旧北区。亜種 *P. s. britannica* (Turner) も同英名
feathered flunkey		*Coscinia striata* (Linnaeus)	(チョウ目、ヒトリガ科) 旧北区
feathered footman		*Spiris striata* (Linnaeus)	(チョウ目、ヒトリガ科) 旧北区
feathered gothic		*Tholera decimalis* (Poda)	(チョウ目、ヤガ科) 旧北区
feathered moth		*Phrissogonus laticostatus* (Walker)	(チョウ目、シャクガ科) 豪州区
feathered moth			November moth (1) を見よ
feathered ranunculus		*Polymixis lichenea* (Hübner)	(チョウ目、ヤガ科) 旧北区
feathered slender			cuckoo-feather slender (1) を見よ

英名	和名	学名	所属、分布、ほか
feathered thorn			November moth (1) を見よ　英国での英名。feathered thorn moth ともいう
February miller		*Feralia februalis* Grote	(チョウ目、ヤガ科) 新北区
February red		*Taeniopteryx nebulosa* (Linnaeus)	(カワゲラ目、ミジカオカワゲラ科) 旧北区
feeble grass moth		*Amolita fessa* Grote	(チョウ目、ヤガ科) 新北区
feeble grasshopper			lesser field grasshopper を見よ
Felder's bee skipper		*Oxynetra semihyalina* Felder et Felder	(チョウ目、セセリチョウ科) 新熱帯区
Felder's crescent		*Telenassa jana* (Felder et Felder)	(チョウ目、タテハチョウ科) 新熱帯区
Felder's emperor		*Doxocopa zunilda* (Godart)	(チョウ目、タテハチョウ科) 新熱帯区
Felders' hairstreak		*Paiwarria antinous* (C. et R. Felder)	(チョウ目、シジミチョウ科) 新熱帯区
Felder's line blue		*Prosotas felderi* (Murray)	(チョウ目、シジミチョウ科) 豪州区
Felder's lineblue (1)	フェルダーウラナミシジミ	*Catopyrops ancyra* (Felder)	(チョウ目、シジミチョウ科) 東洋区、豪州区
Felder's pigmy		*Baeotis zonata* R. Felder	(チョウ目、シジミタテハ科) 新熱帯区
Felder's ringlet		*Euptychoides griphe* (Felder et Felder)	(チョウ目、タテハチョウ科) 新熱帯区
Felder's royal	マントラヤドリギツバメ	*Tajuria mantra* (C. et R. Felder)	(チョウ目、シジミチョウ科) 東洋区
Felder's sister		*Adelpha malea* Felder et Felder	(チョウ目、タテハチョウ科) 新熱帯区
Felder's sister (1)		*Adelpha felderi* (Boisduval)	(チョウ目、タテハチョウ科) 新熱帯区
Felder's sister			Boeotia sister を見よ
Felder's skipper		*Potamanaxas unifasciata* (C. et R. Felder)	(チョウ目、セセリチョウ科) 新熱帯区
felt clothes moth		*Monopis imella* (Hübner)	(チョウ目、ヒロズコガ科) 旧北区
felt louse			crab louse を見よ
felt scale insect			mealybugs (3) を見よ
felt scales			mealybugs (3) を見よ
felted beech coccus			beech scale (1) を見よ
felted grass coccid			Rhodes grass scale を見よ
felted pine coccid			araucaria scale を見よ
Feltham's opal		*Chrysoritis felthami* (Trimen)	(チョウ目、シジミチョウ科) エチオピア区
fen cricket			field cricket (1) を見よ
fen marble	クローバヒメハマキ	*Olethreutes doubledayana* (Barret)	(チョウ目、ハマキガ科) 日本、旧北区。マメ科牧草の害虫
fen neb		*Aristotelia subdecurtella* (Stainton)	(チョウ目、キバガ科) 旧北区
fen snout		*Nemotelus pantherinus* (Linnaeus)	(ハエ目、ミズアブ科) 旧北区
fen square-spot		*Diarsia florida* (Schmidt)	(チョウ目、ヤガ科) 旧北区

英名	和名	学名	所属、分布、ほか
fen wainscot		*Arenostola phragmitidis* (Hübner)	(チョウ目、ヤガ科) 旧北区
Fender's blue butterfly		*Icaricia icarioides fenderi* Macy	(チョウ目、シジミチョウ科) 新北区。Fender's blue ともいう。common blue (1) を参照
fenestrated drepanid	スカシカギバ	*Macrauzata maxima* Inoue	(チョウ目、カギバガ科) 日本、旧北区
fennel aphid		*Dysaphis foeniculus* (Theobald)	(カメムシ目、アブラムシ科) 豪州区
Fenn's wainscot	スジグロウスキヨトウ	*Photedes brevilinea* (Fenn)	(チョウ目、ヤガ科) 日本、旧北区
Fenton's wood white	エゾヒメシロチョウ	*Leptidea morsei* (Fenton)	(チョウ目、シロチョウ科) 日本、旧北区
Ferentina calico			gray cracker を見よ
ferment flies			vinegar flies (1)(2) を見よ
fern		*Horisme tersata* (Denis et Schiffermüller)	(チョウ目、シャクガ科) 旧北区。日本亜種は *H. t. tetricata* (Guenee) アトシロナミシャク
fern aphid (1)	シダフクレアブラムシ	*Amphorophora ampullata* Buckton	(カメムシ目、アブラムシ科) 日本、全北区
fern aphid (2)		*Idiopterus nephrelepidis* Davis	(カメムシ目、アブラムシ科) 全北区、豪州区
fern broad-headed leafhopper	シダヨコバイ	*Japanagallia pteridis* (Matsumura)	(カメムシ目、オオヨコバイ科) 日本、旧北区
fernbug		*Bryocoris pteridis* (Fallén)	(カメムシ目、カスミカメムシ科) 旧北区
fern bug			bracken bug を見よ
fern carpet			fern を見よ
fern caterpillar		*Callopistria* sp.	(ハエ目、Pterocallidae) 大洋区
fern-chafer			June beetle (1) を見よ
fern flies		Teratomyzidae	(ハエ目) の昆虫の総称
fern looper			brown silver-line を見よ
fern moth			fern を見よ
fern petiole sawflies		Blasticotomidae	(ハチ目) の昆虫の総称
fern planthopper	ヒシウンカモドキ	*Cixiopsis punctatus* Matsumura	(カメムシ目、グンバイウンカ科) 日本
fern sawfly	トガリシダハバチ	*Hemitaxonus japonicus* Rohwer	(ハチ目、ハバチ科) 日本
fern sawfly	ゼンマイハバチ	*Strongylogaster osmundae* (Takeuchi)	(ハチ目、ハバチ科) 日本
fern scale		*Pinnaspis caricis* Ferris	(カメムシ目、マルカイガラムシ科) 豪州区
fern scale (1)	ハランナガカイガラムシ	*Pinnaspis aspidistrae* (Signoret)	(カメムシ目、マルカイガラムシ科) 日本、全北区、エチオピア区
fern stem borer		*Blasticotoma filiceti* Klug	(ハチ目、Blastomidae) 旧北区
fern stem sawfly		*Heptamelus ochroleucus* (Stephens)	(ハチ目、ハバチ科) 旧北区
fern tortrix			cyclamen tortrix を見よ
fern weevil		*Syagrius intrudens* Waterhouse	(コウチュウ目、ゾウムシ科) 旧北区
fern weevil (1)		*Syagrius fulvitarsis* Pascoe	(コウチュウ目、ゾウムシ科) 豪州区

英名	和名	学名	所属、分布、ほか
Fernald's Cosmopterix moth		*Cosmopterix fernaldella* Walsingham	(チョウ目、カザリバガ科) 新北区
Fernald's Eucosma moth		*Eucosma fernaldana* (Grote)	(チョウ目、ハマキガ科) 新北区
Fernald's Helcystogramma moth		*Helcystogramma fernaldella* (Busck)	(チョウ目、キバガ科) 新北区
Feronia cracker			variable cracker を見よ
Ferrar's Cerulean		*Jamides ferrari* Evans	(チョウ目、シジミチョウ科) 東洋区
Ferris'copper		*Chalceria ferrisi* Johnson et Balogh	(チョウ目、シジミチョウ科) 新北区
Ferris's copper			white mountains copper を見よ
Ferris's Antaeotricha moth		*Antaeotricha arizonensis* Ferris	(チョウ目、マルハキバガ科) 新北区
ferruginous Eulia moth	ボカシハマキ	*Eulia ministrana* (Linnaeus)	(チョウ目、ハマキガ科) 日本、全北区
ferruginous flour beetle			red flour beetle を見よ
ferruginous glider		*Tramea limbata* (Desjardins)	(トンボ目、トンボ科) エチオピア区、東洋区
ferruginous hairstreak		*Thecla leechii* de Nicéville	(チョウ目、シジミチョウ科) 東洋区
ferruginous Phaneta moth		*Phaneta ferruginana* (Fernald)	(チョウ目、ハマキガ科) 新北区
ferruginous skipper		*Enosis achelous* (Plötz)	(チョウ目、セセリチョウ科) 新熱帯区
ferruginous swift		*Borbo ferruginea* (Aurivillius)	(チョウ目、セセリチョウ科) エチオピア区
fervid Plagodis moth		*Plagodis fervidaria* (Herrich-Schäffer)	(チョウ目、シャクガ科) 新北区
fescue aphid			grass aphid を見よ
Festiva skipper		*Sostrata festiva* Erichson	(チョウ目、セセリチョウ科) 新熱帯区
festive dung beetle		*Oxysternon festivum* (Linnaeus)	(コウチュウ目、コガネムシ科) 新熱帯区
festoon		*Apoda limacodes* Hufnagel	(チョウ目、イラガ科) 旧北区
festoon moth		*Apoda avellana* (Linnaeus)	(チョウ目、イラガ科) 旧北区。festoon ともいう
festoon moth			festoon を見よ
festooned roller		*Ancylis diminutana* Haworth	(チョウ目、ハマキガ科) 旧北区
festuca scale	トボシガラフクロカイガラムシ	*Eriococcus festucarum* Lindinger	(カメムシ目、フクロカイガラムシ科) 日本
fever flies			march flies を見よ
fever fly		*Dilophus febrilis* (Linnaeus)	(ハエ目、ケバエ科) 旧北区
feverfly			spotted Anopheles を見よ
few-spotted ermel		*Hyponomeuta rorellus* Hübner	(チョウ目、スガ科) 旧北区
fibertworm			Catalina cherry moth を見よ
ficus agaonid	イタビコバチ	*Blastophaga callida* Grandi	(ハチ目、イチジクコバチ科) 日本

英名	和名	学名	所属、分布、ほか
ficus gall midge		*Horidiplosis ficifolii* Harris	(ハエ目、タマバエ科) 全北区、東洋区
ficus oystershell scale	イタビカキカイガラムシ	*Lepidosaphes buzenensis* (Kuwana)	(カメムシ目、マルカイガラムシ科) 日本
ficus psylla	ヒゲブトキジラミ	*Caenohomotoma radiata* (Kuwayama)	(カメムシ目、ヒゲブトキジラミ科) 日本
ficus whitefly		*Singhiella simplex* (Singh)	(カメムシ目、コナジラミ科) 東洋区、新北区
fiddle beetle	バイオリンムシ	*Mormolyce phyllodes* Hagenbach	(コウチュウ目、オサムシ科) 東洋区。バイオリン状の扁平なオサムシ
fiddle-shaped planthopper	オヌキグンバイウンカ	*Mesepora onukii* (Matsumura)	(カメムシ目、グンバイウンカ科) 日本、東洋区
fiddler beetle		*Eupoecila australasiae* (Donovan)	(コウチュウ目、コガネムシ科) 豪州区
field and house crickets			field crickets (1) を見よ
field ant	クロクサアリ	*Lasius (Dendrolasius) fuliginosus* (Latreille)	(ハチ目、アリ科) 日本、旧北区。強い臭気のあるアリ
field anthocorid			debris bug を見よ
field anthocoris			debris bug を見よ　米国での英名
field bindweed moth			four-spotted moth を見よ
field bugs			stink bugs を見よ
field cockroach		*Blattella vaga* Hebard	(ゴキブリ目、チャバネゴキブリ科) 新北区。湿地に生息
field crescent		*Phyciodes pratensis* (Behr)	(チョウ目、タテハチョウ科) 新北区
field crescent			field crescentspot (1) をみよ
field crescentspot		*Phyciodes pulchella* (Boisduval)	(チョウ目、タテハチョウ科) 新北区
field crescentspot (1)		*Phyciodes campestris* (Behr)	(チョウ目、タテハチョウ科) 新北区
field cricket (1)		*Gryllus campestris* Linnaeus	(バッタ目、コオロギ科) 旧北区
field cricket	タイワンエンマコオロギ	*Teleogryllus taiwanemma* (Ohmachi et Matsuura)	(バッタ目、コオロギ科) 日本、東洋区
field cricket	エゾエンマコオロギ	*Teleogryllus ezoemma* (Ohmachi et Matsuura)	(バッタ目、コオロギ科) 日本
field cricket			emma cricket を見よ
field cricket			Jamaican field cricket を見よ
field cricket			northern field cricket を見よ
field cricket			large brown cricket を見よ
field crickets		*Gryllus*	(バッタ目、コオロギ科) の昆虫の総称
field crickets		*Acheta*	(バッタ目、コオロギ科) の昆虫の総称
field crickets (1)		Gryllinae	(バッタ目、コオロギ科) の昆虫の総称
field cuckoobee		*Psithyrus campestris* (Panzer)	(ハチ目、ミツバチ科) 旧北区
field damsel bug	オオマキバサシガメ	*Nabis ferus* (Linnaeus)	(カメムシ目、マキバサシガメ科) 日本、新北区
field digger wasp		*Melinus arvensis* (Linnaeus)	(ハチ目、アナバチ科) 旧北区

英名	和名	学名	所属、分布、ほか
field grasshopper	ヒナバッタ	*Chorthippus brunneus* (Thunberg)	(バッタ目、バッタ科) 日本、旧北区。ダイズ、牧草の害虫
field grasshoppers			long-horned grasshoppers (2) を見よ
field ground beetle		*Carabus granulatus* Linnaeus	(コウチュウ目、オサムシ科) 旧北区。日本亜種は *C. g. yezoensis* Bates エゾアカガネオサムシ
field maybeetle			horse chestnut cockchafer を見よ 米国での英名
field mosquitoes		*Aedes*	(ハエ目、カ科) の昆虫の総称
field moths			cereal stem moths を見よ
field rush case		*Coleophora alticolella* Zeller	(チョウ目、ツツミノガ科) 旧北区
field skipper		*Atalopedes campestris* (Boisduval)	(チョウ目、セセリチョウ科) 新北区、新熱帯区
field tiger beetle			green tiger beetle を見よ
fields			cereal stem moths を見よ
Field's giant-skipper		*Agathymus fieldi* Freeman	(チョウ目、セセリチョウ科) 新熱帯区
Field's Mellana		*Quasimellana fieldi* (Bell)	(チョウ目、セセリチョウ科) 新熱帯区
fiery Acraea		*Acraea acrita* Hewitson	(チョウ目、タテハチョウ科) エチオピア区
fiery Campylotes		*Campylotes desgodinsi* (Oberthür)	(チョウ目、マダラガ科) 東洋区
fiery clearwing		*Bembecia chrysidiformis* (Esper)	(チョウ目、スカシバガ科) 旧北区
fiery clouded yellow		*Colias eogene* C. Felder	(チョウ目、シロチョウ科) 東洋区
fiery copper		*Thersamonia thetis* (Klug)	(チョウ目、シジミチョウ科) 旧北区
fiery copper			dull copper を見よ
fiery emperor		*Anax immaculifrons* Rambur	(トンボ目、ヤンマ科) 東洋区
fiery Euselasia		*Euselasia aurantiaca aurantiaca* (Salvin et Godman)	(チョウ目、シジミタテハ科) 新熱帯区。*E. a. aurum* Callaghan, Llorente et Luis も同英名
fiery-eyed dancer		*Argia oenea* Hagen	(トンボ目、イトトンボ科) 新北区、新熱帯区
fiery hunter		*Calosoma calidum* Fabricius	(コウチュウ目、オサムシ科) 新北区。ヨトウムシなどを捕食
fiery jewel		*Hypochrysops ignitus* (Leach)	(チョウ目、シジミチョウ科) 豪州区
fiery Myscelus		*Myscelus assaricus michaeli* Nicolay	(チョウ目、セセリチョウ科) 新熱帯区
fiery satyr		*Lasiophila orbifera* Butler	(チョウ目、タテハチョウ科) 新熱帯区
fiery scarlet-eye			Amazon nightfighter を見よ
fiery searcher			caterpillar hunter を見よ
fiery skipper	カンシャクアカセセリ	*Hylephila phyleus* (Drury)	(チョウ目、セセリチョウ科) 新北区、新熱帯区、大洋区
fiery small fox		*Teniorhinus ignita* (Mabille)	(チョウ目、セセリチョウ科) エチオピア区
fiery tussock		*Dasychira pyrosoma* Hampson	(チョウ目、ドクガ科) 旧北区

英名	和名	学名	所属、分布、ほか
fiery yellow glider		*Cymothoe lucasii* (Doumet)	（チョウ、タテハチョウ科）エチオピア区
fig and orange borer			large auger beetle を見よ
fig bark beetle		*Aricerus eichhoffi* (Blandford)	（コウチュウ目、ゾウムシ科）豪州区
fig bark beetle		*Hypoborus ficus* (Erichson)	（コウチュウ目、キクイムシ科）旧北区。イチジク害虫。
fig bark beetle	チビコキクイムシ	*Hypothenemus eruditus* Westwood	（コウチュウ目、キクイムシ科）日本、全北区、エチオピア区、新熱帯区。イチジク害虫
fig beetle		*Cotinis texana* Casey	（コウチュウ目、コガネムシ科）新北区。イチジク、ヤシ害虫
fig black fly		*Lonchaea aristella* Becker	（ハエ目、クロツヤバエ科）旧北区
fig caprifier			fig wasp を見よ
fig eaters			foresters (1) を見よ
fig fruit borer	モンキシロノメイガ	*Cirrhochrista brizoalis* (Walker)	（チョウ目、メイガ科）日本、東洋区、豪州区
fig fruitborer		*Phycomorpha prasinochroa* (Meyrick)	（チョウ目、Copromorphidae）豪州区
fig insect			fig wasp を見よ
fig-insects			fig wasps (1) を見よ
fig leaf beetle		*Poneridia australis* (Boheman)	（コウチュウ目、ハムシ科）豪州区
fig leaf beetle (1)		*Poneridia semipullata* (Clark)	（コウチュウ目、ハムシ科）豪州区
fig leaf miner	イチジクホソガ	*Melanocereops ficuvorella* (Yazaki)	（チョウ目、ハモグリガ科）日本
fig leaf moth		*Talanga tolumnialis* (Walker)	（チョウ目、メイガ科）豪州区
fig leaf moth (1)		*Anthophila nemorana* (Hübner)	（チョウ目、ハマキモドキガ科）旧北区
fig leafhopper		*Dialecticopteryx australica* Kirkaldy	（カメムシ目、オオヨコバイ科）豪州区
fig longicorn		*Acalolepta vastator* (Newman)	（コウチュウ目、カミキリムシ科）豪州区
fig longicorn beetle	イチジクカミキリ	*Batocera rubus* (Linnaeus)	（コウチュウ目、カミキリムシ科）日本、東洋区
fig looper			silver Y moth (2) を見よ
fig moth			almond moth を見よ
fig moth			Mediterranean flour moth を見よ
fig moth			fig leaf moth (1) を見よ
fig mussel scale		*Lepidosaphes conchyformis* (Gmelin)	（カメムシ目、マルカイガラムシ科）全北区。イチジク害虫
fig pyralid		*Ghesquierellana hirtusalis* Walker	（チョウ目、メイガ科）エチオピア区
fig scale			fig mussel scale を見よ
fig scale			Mediterranean fig scale を見よ　米国での英名
fig shoot caterpillar		*Azochis gripusalis* Walker	（チョウ目、メイガ科）新熱帯区。fig borer ともいう

英名	和名	学名	所属、分布、ほか
fig sphinx moth		*Pachylia ficus* (Linnaeus)	(チョウ目、スズメガ科) 新北区、新熱帯区。fig sphinx ともいう
fig sucker		*Homotoma ficus* (Linnaeus)	(カメムシ目、ネッタイキジラミ科) 旧北区
fig tree bark borer		*Hylesinus oleiperda* (Fabricius)	(コウチュウ目、キクイムシ科) 旧北区
figtree blue	イチジクシジミ	*Myrina silenus* (Fabricius)	(チョウ目、シジミチョウ科) エチオピア区
fig tree blues		*Myrina*	(チョウ目、シジミチョウ科) の昆虫の総称
fig tree leaf beetle			fig leaf beetle (1) を見よ
fig-tree moth		*Ocnerogyia amanda* Staudinger	(チョウ目、ドクガ科) 旧北区
fig tussock moth			transparent tussock を見よ
fig wasp		*Blastophaga (Blastophaga) psenes* (Linnaeus)	(ハチ目、イチジクコバチ科) 新北区、豪州区。米国に移入
fig wasps		*Blastoppa*	(ハチ目、イチジクコバチ科) の昆虫の総称
fig wasps (1)	イチジクコバチ科	Agaonidae	(ハチ目) の昆虫の総称
fig wax scale			wax scale を見よ
fig whitefly		*Aleyrodes elevatus* Silvestri	(カメムシ目、コナジラミ科) 旧北区
figeater			green June beetle を見よ 米国南部での呼称
figeater beetle			fig beetle を見よ
figeater beetle			green fruit beetle を見よ fig beetle ともいう
figitid wasps	ヤドリタマバチ科	Figitidae	(ハチ目) の昆虫の総称
figitids			figitid wasps を見よ
figure-eight looper moth			dusky silver Y moth を見よ
figure-eight sallow moth		*Psaphida resumens* Walker	(チョウ目、ヤガ科) 新北区
figure of eight		*Diloba caeruleocephala* (Linnaeus)	(チョウ目、ヤガ科) 旧北区
figure of eight moth			figure of eight を見よ
figure-of-eight swift		*Baoris pagana* de Nicéville	(チョウ目、セセリチョウ科) 東洋区
figure of eighty		*Tethea ocularis* (Linnaeus)	(チョウ目、トガリバガ科) 旧北区。日本亜種は *T. o. tanakai* Inoue チョウセントガリバ
figure-seven moth			great kidney を見よ
figured tiger moth		*Grammia figurata* (Drury)	(チョウ目、ヒトリガ科) 新北区
figwort stem borer moth		*Papaipema sauzalitae* (Grote)	(チョウ目、ヤガ科) 新北区
figwort weevil		*Cionus scrophulariae* (Linnaeus)	(コウチュウ目、ゾウムシ科) 旧北区
Fijian ginger weevil		*Elytroteinus subtruncatus* (Fairmaire)	(コウチュウ目、ゾウムシ科) 新北区
Fijian leaf holder		*Susumia exigua* (Butler)	(チョウ目、メイガ科) 豪州区
filament bearer			bordered chequer を見よ
filament borer		*Nematocampa filamenfaria* Guenée	(チョウ目、シャクガ科) 新北区。ホップその他の草本につく
filaria mosquito			southern house mosquito を見よ

英名	和名	学名	所属、分布、ほか
filbert aphid	サワシバヒゲマダラアブラムシ	*Myzocallis coryli* (Goetze)	（カメムシ目、アブラムシ科）日本、全北区、豪州区
filbert weevil		*Curculio occidentis* (Casey)	（コウチュウ目、ゾウムシ科）新北区
filbert worm			Catalina cherry moth を見よ　filbert worm moth ともいう
filigree skimmer		*Pseudoleon superbus* (Hagen)	（トンボ目、トンボ科）新熱帯区
filigreed moth		*Chimoptesis pennsylvaniana* (Kearfott)	（チョウ目、ハマキガ科）新北区
fillet dart moth		*Euxoa redimicula* (Morrison)	（チョウ目、ヤガ科）新北区
filter fly			bathroom fly を見よ
filth flies			jackal flies を見よ
fin-tipped sootywing		*Bolla subapicatus* (Schaus)	（チョウ目、セセリチョウ科）新熱帯区
fine-banded amberwing		*Perithemis lais* (Perty)	（トンボ目、トンボ科）新熱帯区
fine-lined gray moth		*Exelis pyrolaria* Guenée	（チョウ目、シャクガ科）新北区
fine-lined hairstreak		*Arawacus sito* (Boisduval)	（チョウ目、シジミチョウ科）新熱帯区
fine-lined sallow moth		*Catabena lineolata* Walker	（チョウ目、ヤガ科）新北区
fine-lined stripestreak			fine-lined hairstreak を見よ
fine-spotted skipperling		*Piruna sina* Freeman	（チョウ目、セセリチョウ科）新熱帯区
fine streaked hawk		*Pantophaea favillacea* (Walker)	（チョウ目、スズメガ科）エチオピア区
finger-net caddisflies	カワトビケラ科	Philopotanidae	（トビケラ目）の昆虫の総称
fingered dagger moth		*Acronicta dactylina* Grote	（チョウ目、ヤガ科）新北区
fingered Lemmeria moth		*Lemmeria digitalis* (Grote)	（チョウ目、ヤガ科）新北区
fingered lerp		*Cardiaspina maniformis* Taylor	（カメムシ目、キジラミ科）　豪州区
Finlayson's oakworm moth		*Anisota finlaysoni* Riotte	（チョウ目、ヤママユガ科）新北区
finned grasshopper		*Trachyrhachys aspera* Scudder	（バッタ目、バッタ科）新北区
finned-willow prominent moth		*Notodonta scitipennis* Walker	（チョウ目、シャチホコガ科）新北区
finnish dart			black army cutworm moth を見よ
Fiona's Charaxes		*Charaxes fionae* Henning	（チョウ目、タテハチョウ科）エチオピア区
Fiorinia scale			avocado scale を見よ
fir aphid	ハネナガオオアブラムシ	*Cinara longipennis* (Matsumura)	（カメムシ目、アブラムシ科）日本、旧北区
fir bark beetle	トドマツノキクイムシ	*Phloeosinus proximus* Blandford	（コウチュウ目、キクイムシ科）日本、旧北区。モミ、ツガ、マツ類の害虫
fir bark beetle (1)	トドマツアトマルキクイムシ	*Dryocoetes striatus* Eggers	（コウチュウ目、キクイムシ科）日本、旧北区。トドマツ害虫

英名	和名	学名	所属、分布、ほか
fir bark beetle (2)	アカマツノコキクイムシ	*Hypothenemus oblongus* Niijima	(コウチュウ目、キクイムシ科) 日本
fir budmoth	トドマツメムシガ	*Argyresthia nemorivaga* Moriuti	(チョウ目、メムシガ科) 日本。トドマツ類害虫
fir budmoth (1)	アカトドマツヒメハマキ	*Epinotia apiculana* Falkovitsh	(チョウ目、ハマキガ科) 日本、旧北区。トドマツ、エゾマツ害虫
fir budmoth		*Zeiraphera rufimitrana* Herrich-Schäffer	(チョウ目、ハマキガ科) 旧北区
fir budworm		*Choristoneura murinana* Hübner	(チョウ目、ハマキガ科) 旧北区
fir cone borer moth		*Eucosma siskiyouana* (Kearfott)	(チョウ目、ハマキガ科) 新北区
fir cone looper		*Eupithecia spermaphaga* (Dyar)	(チョウ目、シャクガ科) 新北区。幼虫は種々の針葉樹につく
fir cone moth	トドマツカサガ	*Cydia kamijoi* (Oku)	(チョウ目、ハマキガ科) 日本。トドマツ害虫
fir coneworm			evergreen coneworm moth を見よ
fir crescent piercer		*Cydia coniferana* (Saxesen)	(チョウ目、ハマキガ科) 旧北区
fir engraver		*Scolytus ventralis* LeConte	(コウチュウ目、キクイムシ科) 新北区
fir-needle beetle		*Polydrosus pilosus* Gredler	(コウチュウ目、ゾウムシ科) 旧北区
fir seed chalcid		*Megastigmus pinus* Parfitt	(ハチ目、オナガコバチ科) 旧北区
fir seed chalcid		*Megastigmus strobilobius* Ratzeburg	(ハチ目、オナガコバチ科) 旧北区
fir seed moth		*Cydia bracteatana* (Fernald)	(チョウ目、ハマキガ科) 新北区
fir-shoot tortricid			fir budworm を見よ
fir tree borer		*Semanotus litigiosus* (Casey)	(コウチュウ目、カミキリムシ科) 新北区
fir-tree weevil			pine weevil (1) を見よ
fir tussock moth		*Orgyia detrita* Guérin	(チョウ目、ドクガ科) 新北区
fir tussock moth (1)	ハラアカマイマイ	*Lymantria fumida* Butler	(チョウ目、ドクガ科) 日本。モミ類害虫
fir weevil			silver fir weevil を見よ
fire ant (1)		*Solenopsis molesta* (Say)	(ハチ目、アリ科) 新北区
fire ant (2)	アカカミアリ	*Solenopsis geminata* (Fabricius)	(ハチ目、アリ科) 日本、全北区、汎亜熱帯、汎熱帯。南北アメリカの著名種
fire ant			imported fire ant (1) を見よ
fire ants		*Solenopsis*	(ハチ目、アリ科) の昆虫の総称　ヒアリといわれる
fire-banded metalmark		*Panaropsis elegans* (Schaus)	(チョウ目、シジミタテハ科) 新熱帯区
fire bee		*Trigona melliicolor* Packard	(ハチ目、ミツバチ科) 新熱帯区
fire beetle		*Merimna atrata* (Laporte et Gory)	(コウチュウ目、タマムシ科) 豪州区
fire beetle (1)		*Pyrochroa coccinea* Linnaeus	(コウチュウ目、アカハネムシ科) 旧北区
fire beetles			click beetles (2) を見よ
fire beetles			cardinal beetles を見よ

英名	和名	学名	所属、分布、ほか
fireblight beetle		*Pyrgoides orphana* (Erichson)	（コウチュウ目、ハムシ科）豪州区
fire blister beetle		*Pyrota palpalis* Champion	（コウチュウ目、ツチハンミョウ科）新北区
fire-bordered Eurybia		*Eurybia cyclopia* Stichel	（チョウ目、シジミタテハ科）新熱帯区
firebrat		*Lepismodes inquilinus* Newman	（シミ目、シミ科）豪州区
firebrat (1)	マダラシミ	*Thermobia domestica* (Packard)	（シミ目、シミ科）日本、新北区
firebrats			silverfish (4) を見よ　シミ科の英名に firebrats を使用するのは稀
fire bug			red bug (1) を見よ　firebug とも記す
fire bug			harlequin bug (1) を見よ　北米での英名
fire bugs			red bugs (2) を見よ
firebush scale			willow oystershell scale を見よ
fire-colored beetle		*Dendroides concolor* (Newman)	（コウチュウ目、アカハネムシ科）新北区
fire-colored beetle		*Dendroides bicolor* Newman	（コウチュウ目、アカハネムシ科）新北区
fire-colored beetles			cardinal beetles を見よ
firecracker skimmer			red skimmer を見よ
fire flee			dragonflies (1) を見よ
fireflies (1)	ホタル科	Lampyridae	（コウチュウ目）の昆虫の総称
fireflies (2)		*Ellychnia*	（コウチュウ目、ホタル科）の昆虫の総称
fire-flies			click beetles (2) を見よ
firefly			glow-worm (2) を見よ
firefly			European firefly を見よ
firefly beetles			fireflies (1) を見よ
firefly-shaped longicorn	ホタルカミキリ	*Dere thoracica* White	（コウチュウ目、カミキリムシ科）日本、旧北区、東洋区
fire-grid burnet			fire-grid burnet moth を見よ
fire-grid burnet moth		*Arniocera erythropyga* Wallengren	（チョウ目、マダラガ科）エチオピア区
fire-spotted satyr		*Pseudomaniola phaselis* (Hewitson)	（チョウ目、タテハチョウ科）新熱帯区
firestreak Nymphidium		*Nymphidium ascolia* Hewitson	（チョウ目、シジミタテハ科）新熱帯区
firetail		*Cicadetta denisoni* (Distant)	（カメムシ目、セミ科）豪州区
firethorn leaf miner		*Phyllonorycter leucographella* (Zeller)	（チョウ目、ホソガ科）旧北区
fireweed caterpillar		*Alypia langtonii* Couper	（チョウ目、ヤガ科）新北区
fireweed clearwing moth		*Albuna pyramidalis* (Walker)	（チョウ目、スカシバガ科）新北区
fireweed Mompha moth			confused cosmet を見よ

英名	和名	学名	所属、分布、ほか
firewood borer		*Neoclytus conjuntus* (LeConte)	(コウチュウ目、カミキリムシ科) 新北区
fireworm		*Lithophane georgii* Grote	(チョウ目、ヤガ科) 新北区
fireworm			glow-worm (2) を見よ
fireworm			black-headed fireworm を見よ
first-born geometer			infant を見よ
Fischer's bush cricket		*Barbitistes fischeri* (Yersin)	(バッタ目、キリギリス科) 旧北区
fish killers			giant water bugs (2) を見よ
fish moth			silverfish (3) を見よ
fish moths			silverfish (2)(4) を見よ　fish-moths とも記す
Fisher's estuarine moth		*Gortyna borelii lunata* (Freyer)	(チョウ目、ヤガ科)　旧北区
fishflies		*Chauliodes*	(アミメカゲロウ目、ヘビトンボ科) の昆虫の総称
fishflies		*Rambur*	(アミメカゲロウ目、ヘビトンボ科) の昆虫の総称
fishflies			dobsonflies (1) を見よ
fishflies			alderflies (1) を見よ
fishfly		*Chauliodes rastiicornis* (Rambur)	(アミメカゲロウ目、ヘビトンボ科) 新北区。
fissured prominent			false sphinx を見よ
Fitzsimon's jewel		*Platycypha fitzsimonsi* Pinhey	(トンボ目、Chlorocyphidae) エチオピア区
five-banded tiphiid wasp		*Myzinum quinquecinctum* (Fabricius)	(ハチ目、コツチバチ科) 新北区
five-bar swordtail (1)		*Graphium aristeus* (Stoll)	(チョウ目、アゲハチョウ科) 東洋区、豪州区。five-barred swordtail ともいう
fivebar swordtail		*Graphium antiphates* (Cramer)	(チョウ目、アゲハチョウ科) 旧北区、東洋区
five bar swordtail	オナガタイマイ	*Pathysa antiphates itamputi* Butler	(チョウ目、アゲハチョウ科) 東洋区
five continent	ウスアオイチモンジ	*Adelpha cytherea* (Linnaeus)	(チョウ目、タテハチョウ科) 新熱帯区
five-lined gray moth		*Glena quinquelinearia* (Packard)	(チョウ目、シャクガ科) 新北区
five-spined bark beetle		*Ips grandicollis* (Eichhoff)	(コウチュウ目、キクイムシ科) 新北区、豪州区
five-spot burnet		*Zygaena trifolii* (Esper)	(チョウ目、マダラガ科) 旧北区
five-spot ladybird		*Coccinella quinquepunctata* Linnaeus	(コウチュウ目、テントウムシ科) 旧北区
five-spotted asparagus beetle		*Crioceris quinquepunctata* (Scopoli)	(コウチュウ目、ハムシ科) 旧北区
five-spotted Eucosma moth		*Eucosma quinquemaculana* (Robinson)	(チョウ目、ハマキガ科) 新北区
five-spotted Glyphidocera moth		*Glyphidocera latiflosella* (Chambers)	(チョウ目、Glyphidoceridae) 新北区
five-spotted hawk-moth			tobacco hornworm を見よ　北米での英名。five spotted hawk moth とも記す

英名	和名	学名	所属、分布、ほか
five-spotted ladybird		*Hippodamia quinquesignata* (Kirby)	(コウチュウ目、テントウムシ科) 新北区
five-spotted longwing		*Heliconius hecalesia* Hewitson	(チョウ目、タテハチョウ科) 新熱帯区。*H. h. octavia* Bates も同英名
five-striped kite-swallowtail		*Eurytides macrosilaus penthesilaus* (Felder et Felder)	(チョウ目、アゲハチョウ科) 新熱帯区
five-striped leaftail		*Phyllogomphoides albrighti* (Needham)	(トンボ目、サナエトンボ科) 新北区
flabellate grasshopper		*Melanoplus occidentalis* (Thomas)	(バッタ目、バッタ科) 新北区
flag skipper		*Moeris striga stroma* Evans	(チョウ目、セセリチョウ科) 新熱帯区
flagstaff shieldback		*Eremopedes cylindricerca* Rentz	(バッタ目、キリギリス科) 新北区
flag-tailed spinyleg		*Dromogomphus spoliatus* (Hagen)	(トンボ目、サナエトンボ科) 新北区
flambeau			Julia を見よ
flambeau			ruddy daggerwing を見よ
flambeau daggerwing			ruddy daggerwing を見よ
flamboyant flower beetle			striped love beetle を見よ
flame	モクメヨトウ	*Axylia putris* (Linnaeus)	(チョウ目、ヤガ科) 日本、旧北区。マメ科牧草の害虫
flame			Julia を見よ
flame albatross		*Appias zarinda* (Boisduval)	(チョウ目、シロチョウ科) 東洋区
flame-bordered Charaxes	カバシタフタオチョウ	*Charaxes protoclea* Feisthamel	(チョウ目、タテハチョウ科) エチオピア区
flame brocade		*Trigonophora flammea* (Esper)	(チョウ目、ヤガ科) 旧北区
flame carpet		*Xanthorhoe designata* (Hufnagel)	(チョウ目、シャクガ科) 旧北区。日本亜種は *X. d. rectantemediana* (Wehrli) トビスジシロナミシャク
flame copper			small copper (1) を見よ　米国での英名
flame longhorn		*Adela flammeusella* Chambers	(チョウ目、マガリガ科) 新北区
flame moth			flame を見よ
flame rustic			flame を見よ
flame shoulder		*Ochropleura plecta* (Linnaeus)	(チョウ目、ヤガ科) 旧北区。日本亜種は *O. p. glaucimacula* (Graeser) マエジロヤガ
flame skimmer			red skimmer を見よ
flame skipper		*Hesperilla idothea* (Miskin)	(チョウ目、セセリチョウ科) 豪州区
flame-tailed pondhawk		*Erythemis peruviana* (Rambur)	(トンボ目、トンボ科) 新熱帯区
flame-tipped hunter		*Austroepigomphus turneri* (Martin)	(トンボ目、サナエトンボ科) 豪州区
flame tree looper			poinciana looper を見よ

英名	和名	学名	所属、分布、ほか
flame wainscot		*Mythimna flammea* (Curtis)	（チョウ目、ヤガ科）旧北区。イグサ、マメ科牧草の害虫。日本亜種は *M. f. stenoptera* (Staudinger) ナカスジキヨトウ
flamingo leafwing		*Fountainea ryphea* (Cramer)	（チョウ目、タテハチョウ科）新熱帯区
flanged looper			white-fringed emerald moth を見よ
flangetails		*Ictinogomphus*	（トンボ目、サナエトンボ科）の昆虫の総称
flannel moth			yellow flannel moth (2) を見よ
flannel moths		Megalopygidae	（チョウ目）の昆虫の総称
flap case		*Coleophora kuehnella* (Goeze)	（チョウ目、ツツミノガ科）旧北区
flasher ruby-eye		*Perichares deceptus* (Butler et Druce)	（チョウ目、セセリチョウ科）新熱帯区
flashes	トラフシジミ属	*Rapala*	（チョウ目、シジミチョウ科）の昆虫の総称
flashing Astraptes		*Astraptes fulgerator* (Watch)	（チョウ目、セセリチョウ科）新北区、新熱帯区
flashwings		*Vestalis*	（トンボ目、イトトンボ科）の昆虫の総称
flat		*Tagiades cohaerens* Mabille	（チョウ目、セセリチョウ科）東洋区
flat back bugs		*Aradus*	（カメムシ目、ヒラタカメムシ科）の昆虫の総称
flat-backed kelp fly		*Coelopa vanduzeei* Cresson	（ハエ目、ハマベバエ科）新北区
flat bark beetles		*Brontes*	（コウチュウ目、ヒラタムシ科）の昆虫の総称
flat bark beetles (1)	ヒラタムシ科	Cucujidae	（コウチュウ目）の昆虫の総称
flat beetles			net-winged beetles (1) を見よ　中南米での英名
flat-bodies		*Agonopterix*	（チョウ目、マルハキバガ科）の昆虫の総称
flat-bodies		*Depressaria*	（チョウ目、マルハキバガ科）の昆虫の総称
flat body			bordered sallow を見よ
flat body			striped hawk を見よ
flat bug		*Aradus robustus* Uhler	（カメムシ目、ヒラタカメムシ科）新北区。樹皮下に生息
flat bugs	ヒラタカメムシ科	Aradidae	（カメムシ目）の昆虫の総称
flat-celled shot-borer			Saxena ambrosia beetle を見よ
flat-celled shot-hole borer			Saxena ambrosia beetle を見よ
flat clown beetle		*Hololepta populnea* LeConte	（コウチュウ目、エンマムシ科）新北区。幼虫は他の甲虫の幼虫を捕食
flat-faced longhorned beetle		*Onychocerus scorpio* Fabricius	（コウチュウ目、カミキリムシ科）新熱帯区
flat-faced long-horns	フトカミキリ亜科	Lamiinae	（コウチュウ目、カミキリムシ科）の昆虫の総称
flat flies			louse flies を見よ
flat-footed ambrosia beetles			pinhole beetles を見よ
flat-footed flies	ヒラタアシバエ科	Platypezidae	（ハエ目）の昆虫の総称　一部の種は smoke flies と呼ばれる
flat grain beetle	アフリカヒラタキクイムシ	*Cryptolestes minutus* Fabricius	（コウチュウ目、ヒラタムシ科）エチオピア区

英名	和名	学名	所属、分布、ほか
flat grain beetle (1)	カクムネチビヒラタムシ	*Cryptolestes pusillus* (Schoenherr)	(コウチュウ目、ヒラタムシ科) 日本、汎世界。貯穀害虫
flat grain beetle			rusty grain beetle を見よ　米国での英名
flat grain beetles		*Cryptolestes*	(コウチュウ目、ヒラタムシ科) の昆虫の総称
flat grain beetles	ホソヒラタムシ科	Silvanidae	(コウチュウ目) の昆虫の総称
flatheaded apple tree borer	リンゴムツボシタマムシ	*Chrysobothris femorata* (Olivier)	(コウチュウ目、タマムシ科) 新北区
flat-headed baldcypress sapwood borer		*Acmaeodera pulchella* (Herbst)	(コウチュウ目、タマムシ科) 新北区
flat-headed beetles	タマムシ科	Buprestidae	(コウチュウ目) の昆虫の総称　幼虫は flat-headed borers という
flat-headed borer		*Chrysobothris lepida* Castalnau et Gory	(コウチュウ目、タマムシ科) 新熱帯区。カンキツ害虫
flat-headed borer			oak borer (1) を見よ
flat-headed borer			jewel beetle (1) を見よ
flatheaded borers		*Chrysobothris*	(コウチュウ目、タマムシ科) の昆虫の総称
flatheaded citrus borer	ミカンナガタマムシ	*Agrilus auriventris* E. Saunders	(コウチュウ目、タマムシ科) 日本、旧北区。カンキツ害虫
flatheaded citrus borer			Ales' flat-headed citrus borer を見よ
flatheaded cone borer		*Chrysophana placida* LeConte	(コウチュウ目、タマムシ科) 新北区
flat-headed evergreen oak borer	クロナガタマムシ	*Agrilus cyaneoniger* E. Saunders	(コウチュウ目、タマムシ科) 日本
flatheaded fir borer		*Melanophila drummondi* Kirby	(コウチュウ目、タマムシ科) 新北区
flat-headed grape borer	ブドウナガタマムシ	*Agrilus marginicollis* E. Saunders	(コウチュウ目、タマムシ科) 日本。ブドウ害虫
flat-headed mayflies			stream mayflies を見よ
flat-headed mulberry borer	クワナガタマムシ	*Agrilus komareki* Obenberger	(コウチュウ目、タマムシ科) 日本。クワ害虫
flatheaded pasture webworm		*Oncopera mitocera* (Turner)	(チョウ目、コウモリガ科) 豪州区
flat-headed pear borer			sinuate pear tree borer を見よ
flat-headed wood-borer			beech borer を見よ
flat-headed wood borers			flat-headed beetles を見よ
flat-headed zelkova borer	ケヤキナガタマムシ	*Agrilus spinipennis* Lewis	(コウチュウ目、タマムシ科) 日本。ケヤキ害虫
flat-horned ground beetles		Helluonini	(コウチュウ目、オサムシ科) の昆虫の総称
flat-legged horntail	ヒラアシキバチ	*Tremex longicollis* Konow	(ハチ目、キバチ科) 日本、旧北区、東洋区
flat-notcher moth			flax-notcher moth を見よ
flat nymph mayflies		*Epeorus*	(カゲロウ目、ヒラタカゲロウ科) の昆虫の総称
flat nymph mayflies		*Heptagenia*	(カゲロウ目、ヒラタカゲロウ科) の昆虫の総称

英名	和名	学名	所属、分布、ほか
flat nymph mayflies			brown stream mayflies を見よ
flat rove beetles	ヒラタハネカクシ亜科	Piestinae	(コウチュウ目、ハネカクシ科) の昆虫の総称
flat scale		*Paralecanium expansum* (Green)	(カメムシ目、カタカイガラムシ科) 豪州区。カンキツ害虫
flat white spotted leaf roller			pear leaf roller (3) を見よ
flatid planthoppers	アオバハゴロモ科	Flatidae	(カメムシ目) の昆虫の総称
flats			bed bugs を見よ
flats			bed bug (1) を見よ
flattened flower chafer	ヒラタハナムグリ	*Nipponovalgus angusticollis angusticollis* (Waterhouse)	(コウチュウ目、コガネムシ科) 日本
flattened rice leafhopper	イネヒラタヨコバイ	*Stroggylocephalus agrestis* (Fallén)	(カメムシ目、オオヨコバイ科) 日本、全北区
flattened sandy ground beetle	オオスナゴミムシダマシ	*Gonocephalum pubeus* Marseul	(コウチュウ目、ゴミムシダマシ科) 日本、旧北区、東洋区
flatwoods ground cricket		*Pictonemobius uliginosus* Mays et Gross	(バッタ目、コオロギ科) 新北区
Flaveola clouded yellow		*Colias flaveola* Blanchard	(チョウ目、シロチョウ科) 新熱帯区
flavescent clover weevil	ナガチビコフキゾウムシ	*Sitona flavescens* (Marsham)	(コウチュウ目、ゾウムシ科) 旧北区。マメ科牧草害虫
flax arctid	アマヒトリ	*Phragmatobia amurensis* Seitz	(チョウ目,ヒトリガ科)日本　旧北区。アマ、ダイズ、リンゴなどの害虫
flax bollworm		*Heliothis ononia* (Fabricius)	(チョウ目、ヤガ科) 全北区。flax bollworm moth ともいう
flax budworm		*Heliothis maritima* Graslin	(チョウ目、ヤガ科) 旧北区。日本亜種は *H. m. adaucta* Butler ツメクサガ。ダイズ、マメ科牧草、園芸花類の害虫
flax caterpillar			slender burnished brass を見よ
flax flea beetle			small flax flea beetle を見よ
flax grub			white flax moth を見よ
flax-notcher moth		*Persectania steropastis* Meyrick	(チョウ目、ヤガ科) 豪州区
flax seed			wheat midge を見よ　蛹の英名
flax thrips		*Thrips lini* Ladureau	(アザミウマ目、アザミウマ科) 旧北区
flax thrips		*Thrips linarius* Uzel	(アザミウマ目、アザミウマ科) 旧北区
flax tortrix		*Cnephasia interjectaria* Boisduval	(チョウ目、ハマキガ科) 旧北区
flax tortrix moth		*Cnephasia asseclana* (Denis et Schiffermüller)	(チョウ目、ハマキガ科) 旧北区
flax tortrix moth			flax tortrix を見よ
flea			human flea, fleas を見よ　一般にノミのようにとぶ虫を指す
flea bee-fly		*Phthiria pulicaria* (Mikan)	(ハエ目、ツリアブ科) 旧北区
flea beetle		*Phyllotreta nigripes* (Fabricius)	(コウチュウ目、ハムシ科) 旧北区

英名	和名	学名	所属、分布、ほか
flea beetles (1)		Halticinae	（コウチュウ目、ハムシ科）の昆虫の総称　ノミハムシ亜科 Alticinae の異名
flea beetles	ノミハムシ亜科	Alticinae	（コウチュウ目、ハムシ科）の昆虫の総称
flea beetles		*Systena*	（コウチュウ目、ハムシ科）の昆虫の総称
flea beetles		*Epitrix*	（コウチュウ目、ハムシ科）の昆虫の総称
flea beetles			leaf beetles (1) を見よ
flea beetles			cucumber beetles を見よ
flea beetles			cabbage flea beetles を見よ
flea hopper			とびはねるカスミカメムシの類（*Halticus* 属）
flea louse			pear sucker を見よ　幼虫の英名
flea weevils		*Rhynchaenus*	（コウチュウ目、ゾウムシ科）の昆虫の総称
flea weevils		Rhynchaeninae	（コウチュウ目、ゾウムシ科）の昆虫の総称
fleabane cochylid moth		*Eugnosta erigeronana* (Riley)	（チョウ目、ハマキガ科）新北区
fleabane tortoise beetle	ベニカメノコハムシ	*Cassida murraea* Linnaeus	（コウチュウ目、ハムシ科）日本、旧北区
fleahoppers		*Halticus*	（カメムシ目、カスミカメムシ科）の昆虫の総称
flea-hoppers			defoliating leaf bugs を見よ　米国での英名
flealocust			cuckoo spit insect を見よ
fleas	ノミ目	Siphonaptera	の昆虫の総称　世界に約 2,200 種
fleas			giant water bugs (2) を見よ　アメリカ、フロリダでの俗称
flecked camel cricket	マダラカマドウマ	*Diestrammena japonica* Karny	（バッタ目、カマドウマ科）日本、全北区、東洋区
flecked general		*Stratiomys singularior* (Harris)	（ハエ目、ミズアブ科）旧北区
flecked glassy-wing		*Hemihyalea labecula* (Grote)	（チョウ目、ヒトリガ科）新北区
flecked mayflies		*Callibaetus*	（カゲロウ目、コカゲロウ科）の昆虫の総称
flecked snout		*Nemotelus notatus* Zetterstedt	（ハエ目、ミズアブ科）旧北区
fleece worm			green botfly を見よ　幼虫の英名
fleece worm			green blow fly (1) を見よ　幼虫の英名
fleeing ask			dragonflies (1) を見よ
fleeing ether			dragonflies (1) を見よ
fleeing snake			dragonflies (1) を見よ
flesh flies	ニクバエ科	Sarcophagidae	（ハエ目）の昆虫の総称
flesh flies			common flesh flies を見よ
flesh fly	センチニクバエ	*Sarcophaga peregrina* Robineau-Desvoidy	（ハエ目、ニクバエ科）日本、旧北区、東洋区、大洋区。幼虫が便池、糞などに発生する普通種
flesh fly	ナミニクバエ	*Parasarcophaga similis* (Meade)	（ハエ目、ニクバエ科）日本、旧北区。幼虫は動物死体、人畜糞などに発生
flesh fly (1)		*Sarcophaga carnaria* (Linnaeus)	（ハエ目、ニクバエ科）旧北区

英名	和名	学名	所属、分布、ほか
flesh fly (2)		*Sarcophaga haemorrhoidalis* (Fallén)	（ハエ目、ニクバエ科）全北区、東洋区、豪州区
Fletcher scale		*Parthenolecanium fletcheri* Cockerell	（カメムシ目、カタカイガラムシ科）新北区。Fletcher's scale とも記す
Fletcher's Cydia moth		*Cydia fletcherana* (Kearfott)	（チョウ目、ハマキガ科）新北区
Fletcher's pug		*Eupithecia egenaria* Herrich-Schäffer	（チョウ目、シャクガ科）旧北区
flier			flyer を見よ
flies	ハエ目	Diptera	の昆虫の総称　世界に約 150,000 種
flightless bush crickets		*Hapithus*	（バッタ目、コオロギ科）の昆虫の総称
flightless dung beetle		*Circellium bacchus* (Fabricius)	（コウチュウ目、コガネムシ科）エチオピア区
flightless field cricket			Indian house cricket を見よ
Flint's cruiser		*Macromia flinti* Lieftinck	（トンボ目、ヤマトンボ科）東洋区
flip flop			African wood white を見よ
flitters		*Hyarotis*	（チョウ目、セセリチョウ科）の昆虫の総称
floating-heart waterlily moth		*Parapoynx seminealis* (Walker)	（チョウ目、メイガ科）新北区
flocculent white fly			woolly whitefly を見よ
floodwater mosquito	カラフトヤブカ	*Aedes sticticus* (Meigen)	（ハエ目、カ科）日本、全北区。人畜から吸血
flood-water mosquito			brown saltmarsh mosquito (1) を見よ
flood-water mosquito			vexans mosquito を見よ
floor maggot			Congo floor fly を見よ
Flores tiger		*Parantica wegneri* (Nieuwenhuis)	（チョウ目、タテハチョウ科）東洋区
Florida bark mantis			lichen mantid を見よ
Florida baskettail		*Epitheca stella* (Williamson)	（トンボ目、エゾトンボ科）新北区
Florida black scale			Caribbean black scale を見よ
Florida bluet		*Enallagma pollutum* (Hagen)	（トンボ目、イトトンボ科）新北区
Florida buckeye			West Indian buckeye を見よ
Florida Callopistria moth			Florida fern caterpillar を見よ
Florida carpenter ant			red carpenter ant　(2) を見よ
Florida cockroach			Australian cockroach を見よ
Florida damp-wood termite		*Prorhinotermes simplex* (Hagen)	（シロアリ目、ミゾガシラシロアリ科）新北区
Florida damp-wood termites			（シロアリ目、レイビシロアリ科）新北区。*Neotermes castaneus* (Burmeister), *N. jouteli* (Banks), *N. luykxi* Nickle et Collins の総称
Florida duskywing		*Ephyriades brunnea floridensis* Strecker	（チョウ目、セセリチョウ科）新北区

英名	和名	学名	所属、分布、ほか
Florida Eucereon		*Nelphe carolina* H. Edwards	（チョウ目、ヒトリガ科）新北区
Florida Eutelia moth		*Eutelia furcata* (Walker)	（チョウ目、ヤガ科）新北区、新熱帯区
Florida false katydid		*Amblycorypha floridana* Rehn et Hebard	（バッタ目、キリギリス科）新北区
Florida fern caterpillar		*Callopistria floridensis* (Guenée)	（チョウ目、ヤガ科）新北区、新熱帯区。成虫は Florida fern moth
Florida flannel moth		*Lagoa lacyi* Barnes et McDunnough	（チョウ目、Megalopygidae）新北区
Florida flower thrips		*Frankliniella cephalica bispinosa* Morgan	（アザミウマ目、アザミウマ科）新北区。カンキツ害虫
Florida glades mosquito		*Psorophora confinnis* (Lynch Arribalzaga)	（ハエ目、カ科）新北区
Florida goatweed butterfly		*Anaea floridalis* Johnson et Comstock	（チョウ目、タテハチョウ科）新北区
Florida goatweed butterfly			Florida leafwing を見よ
Florida ground mealybug		*Rhizoecus floridanus* Hambleton	（カメムシ目、コナカイガラムシ科）新北区
Florida harvester ant		*Pogonomyrmex badius* (Latreille)	（ハチ目、アリ科）新北区
Florida katydid		*Lea floridensis* (Beutenmüller)	（バッタ目、キリギリス科）新北区
Florida leaf-footed bug		*Acanthocephala femorata* (Fabricius)	（カメムシ目、ヘリカメムシ科）新北区
Florida leafwing		*Anaea troglodyta* (Fabricius)	（チョウ目、タテハチョウ科）新熱帯区
Florida leafwing			Florida goatweed butterfly を見よ
Florida morning-glory leafminer moth		*Bedellia minor* Busck	（チョウ目、Bedelliidae）新北区
Florida oakgall moth		*Synanthedon sapygaeformis* (Walker)	（チョウ目、スカシバガ科）新北区
Florida pink scavenger moth		*Pyroderces badia* (Hodges)	（チョウ目、カザリバガ科）全北区、大洋区
Florida predatory stink bug		*Euthyrhynchus floridanus* (Linnaeus)	（カメムシ目、カメムシ科）新北区、新熱帯区
Florida purplewing		*Eunica tatila* (Herrich-Schäffer)	（チョウ目、タテハチョウ科）新北区、新熱帯区
Florida red scale	アカホシマルカイガラムシ	*Chrysomphalus ficus* Ashmead	（カメムシ目、マルカイガラムシ科）日本、全北区、汎熱帯。カンキツ害虫
Florida scale			avocado scale を見よ
Florida Sciota moth		*Sciota floridensis* (Heinrich)	（チョウ目、メイガ科）新北区
Florida scorpionfly		*Panorpa floridana* Byers	（シリアゲムシ目、シリアゲムシ科）新北区
Florida stink bug tachinid		*Trichopoda pennipes* (Fabricius)	（ハエ目、ヤドリバエ科）大洋区、新北区
Florida swamp skipper		*Euphyes berryi* (Bell)	（チョウ目、セセリチョウ科）新北区

英名	和名	学名	所属、分布、ほか
Florida Tetanolita moth		*Tetanolita floridana* Smith	（チョウ目、ヤガ科）新北区。Florida owlet ともいう
Florida tortoise beetle		*Hemisphaerota cyanea* (Say)	（コウチュウ目、ハムシ科）新北区
Florida true katydid			Florida katydid を見よ
Florida tussock moth		*Halysidota cinctipes* Grote	（チョウ目、ヒトリガ科）新北区
Florida wax scale	フロリダロウムシ	*Ceroplastes floridensis* Comstock	（カメムシ目、カタカイガラムシ科）日本、新北区、汎熱帯。カンキツ、マンゴー、イヌツゲなどの害虫
Florida white		*Appias drusilla neumoegnii* (Skinner)	（チョウ目、シロチョウ科）新北区、新熱帯区。*A. d. tenuis* (Lamas) も同英名
Florida whitefly		*Trialeurodes floridensis* (Quaintance)	（カメムシ目、コナジラミ科）新北区
Florida woods cockroach		*Eurycotis floridana* F. Walker	（ゴキブリ目、ゴキブリ科）新北区
Floridian grass-veneer moth		*Thaumatopsis floridalis* (Barnes et McDunnough)	（チョウ目、メイガ科）新北区
flounced chestnut		*Agrochola helvola* (Linnaeus)	（チョウ目、ヤガ科）旧北区
flounced chestnut		*Agonum scitulum* Dejean	（コウチュウ目、オサムシ科）旧北区
flounced rustic		*Luperina testacea* (Denis et Schiffermüller)	（チョウ目、ヤガ科）旧北区
flounced rustic			cabbage armyworm を見よ
flounced rustic moth			flounced rustic を見よ
flour beetle			confused flour beetle を見よ　小麦粉につく小甲虫にも使われる
flour beetles		*Tribolium*	（コウチュウ目、ゴミムシダマシ科）の昆虫の総称
flour moth			tobacco moth (1) を見よ
flour moth			Mediterranean flour moth を見よ
flour weevil			confused flour beetle を見よ
floury bake			floury miller を見よ
floury baker			floury miller を見よ
floury miller		*Abricta curvicosta* (Germar)	（カメムシ目、セミ科）豪州区
floury scots-fir argent		*Cedestis subfasciella* (Stephens)	（チョウ目、スガ科）旧北区
flow-legged beetle			yellow mealworm を見よ
flower bee		*Anthophora plumipes* Pallas	（ハチ目、コシブトハナバチ科）旧北区
flower bees			hairy-footed bees を見よ
flower bees			sweat bees (1) を見よ
flower beetle		*Trichotinus affinis* (Gory et Percheron)	（コウチュウ目、コガネムシ科）新北区
flower beetle			rose chafer (1) を見よ
flower beetles		*Trichiotinus*	（コウチュウ目、コガネムシ科）の昆虫の総称
flower beetles			dried fruit beetles (3) を見よ

英名	和名	学名	所属、分布、ほか
flower beetles (1)	ハナムグリ亜科	Cetoniinae	(コウチュウ目、コガネムシ科) の昆虫の総称
flower beetles (2)	スジコガネ亜科	Rutelinae	(コウチュウ目、コガネムシ科) の昆虫の総称
flower beetles (3)		*Chremastocheilus*	(コウチュウ目、コガネムシ科) の昆虫の総称
flower beetles			chafers を見よ
flower beetles			soft-winged flower beetles (1) を見よ
flower bug			common flower bug を見よ
flower bugs	ハナカメムシ科	Anthocoridae	(カメムシ目) の昆虫の総称　体長 2～5 mm の小型昆虫
flower chafer	ハナムグリ	*Eucetonia pilifera* (Motschulsky)	(コウチュウ目、コガネムシ科) 日本
flower chafers			chafers を見よ
flower flies	ハナアブ科	Syrphidae	(ハエ目) の昆虫の総称　世界に 5,000 種
flower flies			anthomyiine flies (1) を見よ
flower flies			fruit flies を見よ
flower fly			American flower fly を見よ
flower-girl hopper		*Platylesches neba* (Hewitson)	(チョウ目、セセリチョウ科) エチオピア区
flower lepturine		*Stenocorus vestitus* Haldemann	(コウチュウ目、カミキリムシ科) 新北区
flower longhorned beetle		*Leptura obliterata* Haldemann	(コウチュウ目、カミキリムシ科) 新北区
flower longhorns		*Zorion*	(コウチュウ目、カミキリムシ科) の昆虫の総称
flower long-horns	ハナカミキリ亜科	Lepturinae	(コウチュウ目、カミキリムシ科) の昆虫の総称
flower-loving flies		Apioceridae	(ハエ目) の昆虫の総称
flower-loving fly		*Rhaphiomidas trochilus* (Coquillett)	(ハエ目、Aploceridae) 新北区
flower mantis		*Hymenopus coronatus* (Olivier)	(カマキリ目、カマキリ科) 東洋区。ランの花そっくりの種
flower midge			clover seed midge を見よ
flower moths	キヌバコガ科	Scythrididae	(チョウ目) の昆虫の総称
flower scarab		*Trigonopeltastes delta* (Förster)	(コウチュウ目、コガネムシ科)、新北区。成虫は種々の花に普通
flower thrips		*Frankliniella cestrum* Moulton	(アザミウマ目、アザミウマ科) 新北区
flower thrips		*Kakothrips pisivorus* (Westwood)	(アザミウマ目、アザミウマ科) 旧北区
flower thrips (1)	ハナアザミウマ	*Thrips hawaiiensis* (Morgan)	(アザミウマ目、アザミウマ科) 日本、旧北区、豪州区、大洋区。多くの作物、果樹、園芸植物の害虫
flower thrips (2)		*Frankliniella tritici* (Fitch)	(アザミウマ目、アザミウマ科) 全北区
flower thrips (3)	ヒラズハナアザミウマ	*Frankliniella intonsa* (Trybom)	(アザミウマ目、アザミウマ科) 日本、旧北区、東洋区。多くの作物、果樹、園芸植物の害虫
flower thrips			yellow flower thrips を見よ
flower wasps			tiphiid wasps を見よ　豪州での英名
flower weevils	ヒメゾウムシ亜科	Baridinae	(コウチュウ目、ゾウムシ科) の昆虫の総称

英名	和名	学名	所属、分布、ほか
flowing-line Bomolocha moth		*Hypena manalis* Walker	（チョウ目、ヤガ科）新北区。flowing-line Hypena ともいう
fluff louse	ヒメニワトリハジラミ	*Goniocotes gallinae* (De Geer)	（ハジラミ目、チョウカクハジラミ科）日本、汎世界
fluff louse			brown chicken louse を見よ
fluff louse			lesser chicken louse を見よ
fluffy tit	アマサフタオシジミ	*Zeltus amasa* (Hewitson)	（チョウ目、シジミチョウ科）東洋区。ツメアシフサオシジミの和名もあり。*Z. a. maximinianus* Fruhstorfer も同英名
fluffy tit		*Zeltus etolus* (Fabricius)	（チョウ目、シジミチョウ科）東洋区
fluted scale		*Icerya montserratensis* Riley et Howard	（カメムシ目、ワタフキカイガラムシ科）新熱帯区
fluted scale		*Icerya palmeri* Riley et Howard	（カメムシ目、ワタフキカイガラムシ科）新北区、新熱帯区
fluted scale			cottony cushion scale を見よ
fluted scale			yellow cottony-cushion scale を見よ
fluted scale			cottony cushion scale を見よ
fluted scales			giant coccids を見よ
flutterers		*Rhyothemis*	（トンボ目、トンボ科）の昆虫の総称
fly			ハエ、飛ぶ昆虫。flies を見よ
fly			damson-hop aphid を見よ
fly bug			masked hunter bug を見よ
flyblow			（肉などの上に産みつけた）アオバエ blowfly の卵と幼虫
fly-honeysuckle aphid			honeysuckle aphid を見よ
fly-like insects			dipteroid insects を見よ
flyer			昆虫を指す
flying adder			dragonflies (1) を見よ
flying asp			dragonflies (1) を見よ
flying earwig Hawaiian damselfly		*Megalagrion nesiotes* (Perkins)	（トンボ目、イトトンボ科）大洋区。尾角がハサミムシに似ることから
flying handkerchief			mocker swallowtail を見よ
flying pansy			California dogface を見よ
flyspeck scale			pineapple long scale を見よ
fogged skipper		*Cynea irma* (Möschler)	（チョウ目、セセリチョウ科）新熱帯区
foggy-winged twister			twister を見よ
folivorous flea beetle	アザミカミナリハムシ	*Altica cyanea* (Weber)	（コウチュウ目、ハムシ科）日本、旧北区
Fologne's elm pigmy		*Stigmella ulmivora* (Fologne)	（チョウ目、モグリチビガ科）旧北区
Folsom's campodeid		*Campodea folsomi* Silvestri	（コムシ目、ナガコムシ科）新北区
Fontaine's sapphire		*Iolaus fontainei* (Stempffer)	（チョウ目、シジミチョウ科）エチオピア区

英名	和名	学名	所属、分布、ほか
Fontana grasshopper		*Trimerotropis fontana* Thomas	(バッタ目、バッタ科) 新北区。前翅に帯状紋有り
foodwater mosquito		*Aedes (Ochlerotatus) sinicus* (Meigen)	(ハエ目、カ科) 新北区
foolish swift		*Borbo fatuellus fatuellus* Hepffer	(チョウ目、セセリチョウ科) エチオピア区
foot spinners			webspinners (1) を見よ
footman moths	コケガ亜科	Lithosiinae	(チョウ目、ヒトリガ科) の昆虫の総称
footman moths			tiger moths (1) を見よ
footmen		*Eilema*	(チョウ目、ヒトリガ科) の昆虫の総称
footmen moths			tiger moths (1) を見よ
forage armyworm moth		*Leucania subpunctata* (Harvey)	(チョウ目、ヤガ科) 新北区
forage looper		*Caenurgina erechtea* (Cramer)	(チョウ目、ヤガ科) 新北区
forage looper moth			forage looper を見よ
Forbes' Acleris moth		*Acleris forbesana* (McDunnough)	(チョウ目、ハマキガ科) 新北区
Forbes scale		*Quadraspidiotus forbesi* (Johnson)	(カメムシ目、マルカイガラムシ科) 日本、新北区
Forbes' silkmoth		*Rothschildia forbesi* Benjamin	(チョウ目、ヤママユガ科) 新北区。新熱帯区
Forbes' silkmoth			Lebeau's silkmoth を見よ
Forbes's tree cricket		*Oecanthus forbesi* Titus	(バッタ目、カンタン科) 新北区
forcepfly		*Merope tuber* Newman	(シリアゲムシ目、Meropidae) 新北区
forcepsflies			earwig flies を見よ 米国での英名
forcipate firetail		*Telebasis versicolor* Fraser	(トンボ目、イトトンボ科) 新熱帯区
foreign grain beetle	カドコブホソヒラタムシ	*Ahasver advena* (Waltlus)	(コウチュウ目、ホソヒラタムシ科)。日本、汎世界。貯穀の菌を餌とする
forest beauty			bush beauty を見よ
forest bob		*Scobura isota* (Swinhoe)	(チョウ目、セセリチョウ科) 東洋区
forest bob		*Scobura*	(チョウ目、セセリチョウ科) の昆虫の総称
forest bob		*Scobura cephala* (Hewitson)	(チョウ目、セセリチョウ科) 東洋区
forest bug	アシアカカメムシ	*Pentatoma rufipes* (Linnaeus)	(カメムシ目、カメムシ科) 日本、旧北区
forest bugs			stink bugs を見よ
forest caper white		*Belenois theora* (Doubleday)	(チョウ目、シロチョウ科) エチオピア区
forest caper white			forest white (1) を見よ
forest caterpillar-hunter		*Calosoma sycophanta* (Linnaeus)	(コウチュウ目、オサムシ科) 全北区
forest cockroach		*Ectobius sylvestris* (Poda)	(ゴキブリ目、チャバネゴキブリ科) 全北区
forest cuckoo bumblebee			four-coloured cuckoo bee を見よ
forest darner		*Austroaeschna pulchra* Tillyard	(トンボ目、ヤンマ科) 豪州区

英名	和名	学名	所属、分布、ほか
forest day mosquito			tiger mosquito を見よ 米国での英名
forest elfin			elfin skipper を見よ
forest emperor			forest king Charaxes を見よ
forest false dotted border			woodland white を見よ
forest flies			louse flies を見よ
forest fly	ウマジラミバエ	*Hippobosca equina* Linnaeus	(ハエ目、シラミバエ科) 日本、旧北区、エチオピア区。ウマ、ウシ、イヌ、トリなどから吸血
forest giant owl		*Caligo eurilochus* (Cramer)	(チョウ目、タテハチョウ科) 新熱帯区
forest glade nymph	マルバネモンキタテハ	*Aterica galene* (Brown)	(チョウ目、タテハチョウ科) エチオピア区
forest goblins		Platystictidae	(トンボ目) の昆虫の総称
forest golden piper		*Eurytela alinda* Mabille	(チョウ目、タテハチョウ科) エチオピア区
forest grass yellow		*Eurema senegalensis* Boisduval	(チョウ目、シロチョウ科) エチオピア区
forest grizzled skipper		*Spialia ploetzi* (Aurivillius)	(チョウ目、セセリチョウ科) エチオピア区
forest highflier			common highflier を見よ
forest hopper		*Astictopterus jama* C. et R. Felder	(チョウ目、セセリチョウ科) 東洋区
forest inland floodwater mosquito			vexans mosquito を見よ
forest king Charaxes		*Charaxes xiphares* (Stoll)	(チョウ目、タテハチョウ科) エチオピア区
forest ladies			mantids (1)(2) を見よ 豪州での英名
forest leopard		*Phalanta eurytis* (Doubleday)	(チョウ目、タテハチョウ科) エチオピア区
forest leopard fritillary			forest leopard を見よ
forest malachite		*Chlorolestes tessellatus* Burmeister	(トンボ目、Synlestidae) エチオピア区
forest maybeetle			June beetle (1) を見よ
forest monarch	アカチャウスグロマダラ	*Tirumala formosa* (Godman)	(チョウ目、タテハチョウ科) エチオピア区
forest mort bleu	オオフクロチョウ	*Caligo eurilochus minor* Cramer	(チョウ目、タテハチョウ科) 新熱帯区
forest mother-of-pearl	シンジュタテハ	*Protogoniomorpha parhassus* (Drury)	(チョウ目、タテハチョウ科) エチオピア区
forest mother-of-pearl (1)		*Salamis parhassus* (Druce)	(チョウ目、タテハチョウ科) エチオピア区。mother-of-pearl を参照
forest pearl Charaxes	ネジロフタオチョウ	*Charaxes fulvescens* (Aurivillius)	(チョウ目、タテハチョウ科) エチオピア区
forest pied pierrot		*Tuxentius carana* (Hewitson)	(チョウ目、シジミチョウ科) エチオピア区
forest pierrot	ゴイシシジミ	*Taraka hamada hamada* (H. Druce)	(チョウ目、シジミチョウ科) 日本、旧北区、東洋区。*T. h. mendesia* Fruhstorfer も同英名
forest pygmy grasshopper		*Tetrix* sp.	(バッタ目、ヒシバッタ科) 豪州区

英名	和名	学名	所属、分布、ほか
forest quaker	リュウキュウウラボシシジミ	*Pithecops corvus ryukyuensis* Shirozu	（チョウ目、シジミチョウ科）日本。*P. corvus* Fruhstorfer 東洋区の英名
forest quaker		*Pithecops hylax* Horsfield	（チョウ目、シジミチョウ科）東洋区
forest queen	シロマルバネタテハ	*Euxantha wakefieldi* (Ward)	（チョウ目、タテハチョウ科）エチオピア区
forest ringlet		*Dodonidia helmsii* Butler	（チョウ目、タテハチョウ科）豪州区
forest sandman		*Spialia dromus* (Plötz)	（チョウ目、セセリチョウ科）エチオピア区
forest scaly cricket		*Cycloptilum trigonipalpum* (Rehn et Hebard)	（バッタ目、コオロギ科）新北区
forest shield bug			forest bug を見よ
forest silverfishes		Lepidotrichidae	（シミ目）の昆虫の総称
forest silver-stiletto		*Pandivirilia melaleuca* (Loew)	（ハエ目、ツルギアブ科）旧北区
forestskimmers		*Cratilla*	（トンボ目、トンボ科）の昆虫の総称
forest spider wasp		*Priocnemis oregona* (Banks)	（ハチ目、ベッコウバチ科）新北区
forest swallowtail		*Papilio euphranor* Trimen	（チョウ目、アゲハチョウ科）エチオピア区
forest tent caterpillar		*Malacosoma disstria* Hübner	（チョウ目、カレハガ科）新北区
forest tree caterpillar moth			forest tent caterpillar を見よ
forest tree drywood termite		*Neotermes connexus* (Snyder)	（シロアリ目、レイビシロアリ科）新北区、大洋区。forest tree termite ともいう
forest-watcher		*Notiothemis jonesi* Ris	（トンボ目、トンボ科）エチオピア区
forest white		*Phrissura aegis cynis* (Hewitson)	（チョウ目、シロチョウ科）東洋区
forest white (1)		*Belenois zochalia zochalia* (Boisduval)	（チョウ目、シロチョウ科）エチオピア区
forest windowfly		*Scenopinus niger* (De Geer)	（ハエ目、マドギワアブ科）旧北区
forestwraiths		*Platysticta*	（トンボ目、Platystictidae）の昆虫の総称
forest Zulu			Zulu を見よ
forester		*Adscita statices* (Linnaeus)	（チョウ目、マダラガ科）旧北区
forester			eight-spotted forester を見よ
forester moths	トラガ科	Agaristidae	（チョウ目）の昆虫の総称
foresters (1)		*Euphaedra*	（チョウ目、タテハチョウ科）の昆虫の総称
foresters (2)		Procrinae	（チョウ目、マダラガ科）の昆虫の総称
foresters			leaf skeletonizer moths（2）を見よ
forget-me-not		*Catochrysops strabo strabo* Fabricius	（チョウ目、シジミチョウ科）東洋区
forget-me-not			silver forget-me-not（1）を見よ　forget-me-not はワスレナグサの英名。
forget-me-not aphid		*Myzodes myosotidis* Börner	（カメムシ目、アブラムシ科）旧北区
forget-me-not blue	ムラサキオナガウラナミシジミ	*Catochrysops strabo luzonensis* Tite	（チョウ目、シジミチョウ科）日本、東洋区

英名	和名	学名	所属、分布、ほか
forgotten frigid owlet moth		*Nycteola metaspilella* (Walker)	（チョウ目、ヤガ科）新北区
fork hunter		*Austrogomphus divaricatus* Watson	（トンボ目、サナエトンボ科）豪州区
forktail damselfly			fragile forktail を見よ
fork-tailed bluet		*Proischnura subfurcata* (Selys)	（トンボ目、イトトンボ科）エチオピア区
fork-tailed bush katydid			fork-tailed katydid を見よ
fork-tailed bush katydids		*Scudderia*	（バッタ目、キリギリス科）の昆虫の総称
forktailed damselfly			fragile forktail を見よ
fork-tailed katydid		*Scudderia furcata* Brunner von Wattenwyl	（バッタ目、キリギリス科）新北区。カンキツ害虫
forktails		*Macrogomphus*	（トンボ目、サナエトンボ科）の昆虫の総称
forktails			bluetails を見よ
forked dagger moth			speared dagger moth を見よ
forked Euchlaena moth		*Euchlaena pectinaria* (Denis et Schiffermüller)	（チョウ目、シャクガ科）新北区
forked fungus beetle		*Bolitotherus cornutus* (Panzer)	（コウチュウ目、ゴミムシダマシ科）新北区
forked red-barred twist			apple leafroller (1) を見よ
Formosa carpet moth		*Dysstroma formosa* (Hulst)	（チョウ目、シャクガ科）新北区
Formosa looper moth		*Chrysanympha formosa* (Grote)	（チョウ目、ヤガ科）新北区。Formosa looper ともいう
Formosan hairy aphid	タイワンヒゲナガアブラムシ	*Uroleucon formosanum* (Takahashi)	（カメムシ目、アブラムシ科）日本、旧北区、東洋区
Formosan lettuce aphid			Formosan hairy aphid を見よ
Formosan rice plant weevil			rice curculio を見よ
Formosan subterranean termite	イエシロアリ	*Coptotermes formosanus* Shiraki	（シロアリ目、ミゾガシラシロアリ科）日本、東洋区、新北区、大洋区
Formosan swift			rice swift を見よ　香港での使用
Formosan white ant			Formosan subterranean termite を見よ
Formosus jewelmark		*Anteros formosus* (Cramer)	（チョウ目、シジミタテハ科）新熱帯区
Forrer's leafwing		*Memphis forreri* (Godman et Salvin)	（チョウ目、タテハチョウ科）新熱帯区
Forrer's leafwing			Guatemalan leafwing を見よ
Forsakal's button		*Acleris forsakaleana* (Linnaeus)	（チョウ目、ハマキガ科）全北区
Forsebia moth		*Forsebia perlaeta* Edwards	（チョウ目、ヤガ科）新北区
Forster's furry blue		*Polyommatus ainsae* (Forster)	（チョウ目、シジミチョウ科）旧北区

英名	和名	学名	所属、分布、ほか
Forster's twist		*Lozotaenia forsterana* (Fabricius)	(チョウ目、ハマキガ科) 旧北区
Forsyth's Epipagis moth		*Epipagis forsythae* Munroe	(チョウ目、メイガ科) 新北区
fortified carpet		*Scotopteryx moeniata* (Scopoli)	(チョウ目、シャクガ科) 旧北区
fosted elfin		*Incisalia irus* (Godart)	(チョウ目、シジミチョウ科) 新北区
Foulquier's grizzled skipper		*Pyrgus bellieri* (Oberthür)	(チョウ目、セセリチョウ科) 旧北区
Foulquier's grizzled skipper		*Pyrgus foulquieri* Oberthür	(チョウ目、セセリチョウ科) 旧北区
fourbanded leafroller		*Argyrotaenia quadrifasciana* (Fernald)	(チョウ目、ハマキガ科) 新北区
four-banded longhorn beetle	カラフトヨツスジハナカミキリ	*Leptura quadrifasciata* Linnaeus	(コウチュウ目、カミキリムシ科) 日本、旧北区
four-bar slender		*Parectopa ononidis* (Zeller)	(チョウ目、ホソガ科) 旧北区
fourbar swordtail	アゲテスオナガタイマイ	*Graphium agetes* (Westwood)	(チョウ目、アゲハチョウ科) 東洋区
four-bar swordtail	ムカシタイマイ	*Protographium leosthenes* (Doubleday)	(チョウ目、アゲハチョウ科) 豪州区。four-barred swordtail ともいう
four-barred gray moth		*Aethalura intertexta* (Walker)	(チョウ目、シャクガ科) 新北区
four-barred major		*Oxycera rara* (Scopoli)	(ハエ目、ミズアブ科) 旧北区
four-colour Acraea		*Acraea quadricolor* (Rogenhofer)	(チョウ目、タテハチョウ科) エチオピア区
four-coloured cuckoo bee		*Bombus sylvestris* (Lepeletier)	(ハチ目、ミツバチ科) 旧北区
four-dotted Agonopterix moth		*Agonopterix robiniella* Packard	(チョウ目、マルハキバガ科) 新北区
four-dotted alpine		*Erebia dabanensis* Erschoff	(チョウ目、タテハチョウ科) 新北区
four-dotted footman		*Cybosia mesomella* (Linnaeus)	(チョウ目、ヒトリガ科) 旧北区
four-eyed spruce bark beetle		*Polygraphus rufipennis* (Kirby)	(コウチュウ目、キクイムシ科) 新北区
four-eyed spruce bark beetle			cherry-tree bark beetle を見よ
four-horned sphinx			elm sphinx を見よ　北米での英名
four-lined borer		*Resapamea stipata* (Morrison)	(チョウ目、ヤガ科) 新北区。成虫は four-lined borer moth
four-lined Cabera moth		*Cabera quadrifasciaria* (Packard)	(チョウ目、シャクガ科) 新北区
four-lined chocolate moth		*Argyrostrotis quadrifilaris* (Hübner)	(チョウ目、ヤガ科) 新北区
four-lined geometrid	ヨスジナミシャク	*Xanthorhoe quadrifasciata ignobilis* (Butler)	(チョウ目、シャクガ科) 日本。large twin-spot carpet も参照
four-lined horsefly		*Atylotus rusticus* (Linnaeus)	(ハエ目、アブ科) 旧北区
four-lined leafroller moth			fourbanded leafroller を見よ

英名	和名	学名	所属、分布、ほか
four-lined plant bug		*Poecilocapsus lineatus* (Fabricius)	(カメムシ目、カスミカメムシ科) 新北区。多くの植物を加害
four-lined wave moth		*Scopula quadrilineata* (Packard)	(チョウ目、シャクガ科) 新北区。four-lined wave ともいう
four-marked ash-borer			ivory marked beetle を見よ
four-pair scaly cricket		*Cycloptilum quatrainum* Love et Walker	(バッタ目、コオロギ科) 新北区
four-spot		*Libellula quadrimaculata* Linnaeus	(トンボ目、トンボ科) 全北区。日本亜種は *L. q. asahinai* Schmidt ヨツボシトンボ
four-spot bush-cricket		*Phaneroptera nana* Fieber	(バッタ目、キリギリス科) 全北区
four spot green sailor		*Dynamine postverta* (Cramer)	(チョウ目、タテハチョウ科) 新熱帯区。four-spotted sailor ともいう
four-spot lady bird			cream-streaked lady beetle を見よ
four-spot Pixie		*Melanis hillapana* (Rober)	(チョウ目、シジミタテハ科) 新熱帯区
four-spot sylph		*Metisella quadrisignatus* (Butler)	(チョウ目、セセリチョウ科) エチオピア区
four-spotted angle moth		*Speranza coortaria* (Hulst)	(チョウ目、シャクガ科) 新北区
four-spotted angle moth		*Trigrammia quadrinotaria* Herrich-Schäffer	(チョウ目、シャクガ科) 新北区
four-spotted aphis fly		*Baccha clavata* (Fabricius)	(ハエ目、ハナアブ科) 新北区。four-spotted aphid fly ともいう
four-spotted baboon beetle		*Babia quadriguttata* (Olivier)	(コウチュウ目、ハムシ科) 新北区
four-spotted bean weevil			cowpea weevil を見よ
four-spotted bird-dropping moth		*Tarache tetragona* (Walker)	(チョウ目、ヤガ科) 新北区、新熱帯区
four-spotted chaser			four-spot を見よ
four-spotted coconut weevil			palm weevil borer を見よ
four-spotted conifer longicorn			four-spotted conifer longicorn beetle を見よ
four-spotted conifer longicorn beetle	シラフヨツボシヒゲナガカミキリ	*Monochamus urussovii* (Fischer)	(コウチュウ目、カミキリムシ科) 日本、旧北区。トドマツ、エゾマツ他の害虫
fourspotted cup moth		*Doratifera quadriguttata* (Walker)	(チョウ目、イラガ科) 豪州区
four-spotted egg eater		*Collops quadrimaculatus* Fabricius	(コウチュウ目、ジョウカイモドキ科) 新北区
four-spotted eighty-eight		*Callicore brome* (Doyere)	(チョウ目、タテハチョウ科) 新熱帯区
four-spotted footman	ヨツボシホソバ	*Lithosia quadra* (Linnaeus)	(チョウ目、ヒトリガ科) 日本、旧北区
four-spotted fungus moth		*Metalectra quadrisignata* (Walker)	(チョウ目、ヤガ科) 新北区
four-spotted ghost moth		*Sthenopis purpurascens* (Packard)	(チョウ目、コウモリガ科) 新北区

英名	和名	学名	所属、分布、ほか
four-spotted Gluphisia moth		*Gluphisia avimacula* Hudson	（チョウ目、シャチホコガ科）新北区
four-spotted grasshopper		*Philbostroma quadramaculatum* (Thomas)	（バッタ目、バッタ科）新北区
four-spotted greenish geometrid	ヨツテンアオシャク	*Comibaena diluta* (Warren)	（チョウ目、シャクガ科）日本、東洋区
four-spotted large ant	ヨツボシオオアリ	*Camponotus quadrinotatus* Forel	（ハチ目、アリ科）日本、旧北区
four-spotted leaf beetle	ヨツボシナガサルハムシ	*Clytra arida* Weise	（コウチュウ目、ハムシ科）日本、旧北区
four-spotted leaf beetle		*Clytra quadripunctata* (Linnaeus)	（コウチュウ目、ハムシ科）旧北区
four-spotted leaf bug	ヨツモンカスミカメムシ	*Adelphocoris albonotatus* (Jakovlev)	（カメムシ目、カスミカメムシ科）日本、旧北区
four-spotted leafhopper	ヨツテンヨコバイ	*Macrosteles quadrimaculatus* (Matsumura)	（カメムシ目、オオヨコバイ科）日本、旧北区
four-spotted Libellula			four-spot を見よ
four-spotted lift			yellow-spotted lift を見よ
four-spotted longicorn beetle	ヨツボシカミキリ	*Stenygrinum quadrinatatum* Bates	（コウチュウ目、カミキリムシ科）日本、旧北区、東洋区
four-spotted longtail			short-tail skipper を見よ
four-spotted melon leaf beetle	ヨツボシハムシ	*Paridea quadriplagiata* (Baly)	（コウチュウ目、ハムシ科）日本、旧北区
four-spotted moth		*Tyta luctuosa* (Denis et Schiffermüller)	（チョウ目、ヤガ科）旧北区。four spotted ともいう
four-spotted Palpita moth		*Palpita quadristigmalis* (Guenée)	（チョウ目、メイガ科）新北区
four-spotted skipperling			spotted skipperling を見よ
four-spotted stink bug	ヨツボシカメムシ	*Homalogonia obtusa* (Walker)	（カメムシ目、カメムシ科）日本、旧北区
four-spotted tree cricket		*Oecanthus nigricornis quadripunctatus* Beutenmüller	（バッタ目、カンタン科）新北区
four-spotted yellowneck moth		*Oegoconia novimundi* (Busck)	（チョウ目、Symmocidae）旧北区
four-spotted yellowneck moth		*Oegoconia quadripuncta* (Haworth)	（チョウ目、Symmocidae）新北区。four-spotted yellowneck ともいう
four-spurred assassin bug		*Zelus tetracanthus* Stål	（カメムシ目、サシガメ科）新北区
four-striped flower longicorn	ヨツスジハナカミキリ	*Leptura ochraceofasciata ochraceofasciata* Motschulsky	（コウチュウ目、カミキリムシ科）日本、旧北区
four-striped leaftail		*Phyllogomphoides stigmatus* (Say)	（トンボ目、サナエトンボ科）新北区
four-striped rhombic planthopper	ヨスジヒシウンカ	*Reptalus quadricinctus* Matsumura	（カメムシ目、ヒシウンカ科）日本、旧北区

英名	和名	学名	所属、分布、ほか
four-toothed mason wasp		*Monobia quadridens* (Linnaeus)	(ハチ目、スズメバチ科) 新北区
fourteen-spot ladybird		*Propylea quatuordecimpunctata* (Linnaeus)	(コウチュウ目、テントウムシ科) 旧北区
fourteen-spotted asparagus beetle		*Crioceris quatuordecimpunctata* (Scopoli)	(コウチュウ目、ハムシ科) 旧北区
fourteen spotted lady beetle			cream spot ladybird を見よ
fowl body louse			chicken body louse を見よ　南アフリカでの英名
fowl bug			pigeon bug を見よ
fowl fly	ハトシラミバエ	*Ornithomyia avicularia aobatonis* Matsumura	(ハエ目、シラミバエ科) 日本
fowl fly		*Ornithomyia fringillina* Curtis	(ハエ目、シラミバエ科) 全北区
fowl head louse			chicken head louse を見よ　南アフリカでの英名
fowl louse			chicken louse を見よ
fowl wing louse			chicken wing louse を見よ　南アフリカでの英名
fox			fox moth を見よ
fox biting louse		*Eichlerella vulpis* Denny	(ハジラミ目、ケモノハジラミ科) 旧北区
fox colored sawfly	マツノキハバチ	*Neodiprion sertifer* (Geoffroy)	(ハチ目、マツハバチ科) 日本、全北区。マツ類害虫。fox-coloured sawfly とも記す
fox-face metalmark			Banner metalmark を見よ
fox flea		*Chaetopsylla globiceps* (Taschenberg)	(ノミ目、ケナガノミ科) 旧北区
fox moth		*Macrothylacia rubi* (Linnaeus)	(チョウ目、カレハガ科) 旧北区
foxglove aphid		*Dysaulacorthum antirrhini* (Machiat)	(カメムシ目、アブラムシ科) 旧北区
foxglove aphid		*Myzus convolvuli* (Kaltenbach)	(カメムシ目、アブラムシ科) 新北区
foxglove aphid (1)		*Dysaulacorthum anthrisci* Börner	(カメムシ目、アブラムシ科) 旧北区
foxglove aphid			potato aphid（2）を見よ
foxglove pug		*Eupithecia pulchellata* (Stephens)	(チョウ目、シャクガ科) 旧北区
Fox's blue-banded bush brown		*Bicyclus sweadneri* Fox	(チョウ目、タテハチョウ科) エチオピア区
foxtail midge		*Contarinia merceri* Barnes	(ハエ目、タマバエ科) 旧北区
foxtail midge		*Dasineura alopecuri* (Reuter)	(ハエ目、タマバエ科) 旧北区
foxtail midge (1)		*Stenodiplosis geniculati* (Reuter)	(ハエ目、タマバエ科) 旧北区
foxtail millet thrips	アワキオビアザミウマ	*Anaphothrips sudanensis* Trybom	(アザミウマ目、アザミウマ科) 日本、全北区、東洋区、エチオピア区、新熱帯区

英名	和名	学名	所属、分布、ほか
foxtrot copper		*Phasis thero* (Linnaeus)	(チョウ目、シジミチョウ科) エチオピア区
foxy Charaxes		*Charaxes jasius saturnus* Butler	(チョウ目、タテハチョウ科) エチオピア区
foxy emperor			two-tailed Pasha を見よ
fractured western snout moth		*Diastictis fracturalis* (Zeller)	(チョウ目、メイガ科) 新北区
fragile dagger moth		*Acronicta fragilis* Guenée	(チョウ目、ヤガ科) 新北区
fragile forktail		*Ischnura posita* (Hagen)	(トンボ目、イトトンボ科) 大洋区
fragile gray		*Anacamptodes fragilaria* (Grossbeck)	(チョウ目、シャクガ科) 新北区、大洋区
fragile lacewings		Apochrysidae	(アミメカゲロウ目) の昆虫の総称
fragile whistling cricket			tree cricket (1) を見よ
fragile white carpet moth		*Hydrelia albifera* (Walker)	(チョウ目、シャクガ科) 新北区
Francillon's carrot conch		*Aethes francillana* (Fabricius)	(チョウ目、ホソハマキガ科) 旧北区
Franck's sphinx		*Sphinx franckii* Neumoegen	(チョウ目、スズメガ科) 新北区。Franck's sphinx moth ともいう
Frangipani hornworm			giant grey sphinx を見よ
Fraser's shadowdamsel		*Drepanosticta fraseri* Lieftinck	(トンボ目、Platystictidae) 東洋区
fraternal Renia moth		*Renia fraternalis* Smith	(チョウ目、ヤガ科) 新北区。fraternal Renia ともいう
fraximus elongate scolytid	タブノキナガキクイムシ	*Platypus contaminatus* (Blandford)	(コウチュウ目、ナガキクイムシ科) 日本、東洋区
fraximus looper	ホソバトガリエダシャク	*Planociampa modesta* (Butler)	(チョウ目、シャクガ科) 日本
fraximus sawfly	クロハバチ	*Macrophya ignava* Smith	(ハチ目、ハバチ科) 日本、旧北区
freak	クビワチョウ	*Calinaga buddha* Moore	(チョウ目、タテハチョウ科) 東洋区
freckled groundling		*Mirificarma lentiginosella* (Zeller)	(チョウ目、キバガ科) 旧北区
free beaks	頸吻群	Auchenorrhyncha	(カメムシ目) の昆虫の総称
free-living caddisflies			primitive caddisflies を見よ
free state blue		*Lepidochrysops letsea* (Trimen)	(チョウ目、シジミチョウ科) エチオピア区
freeborn's malarial mosquito		*Anopheles maculipennis freeborni* Aitken	(ハエ目、カ科) 新北区
Freeman's blue-skipper		*Pythonides mundo* Freeman	(チョウ目、セセリチョウ科) 新熱帯区
Freeman's firetip		*Jonaspyge tzotzili* (Freeman)	(チョウ目、セセリチョウ科) 新熱帯区
Freeman's gemmed-satyr		*Cyllopsis gemma freemani* (Stalling et Turner)	(チョウ目、タテハチョウ科) 新熱帯区。gemmed satyr を参照
Freeman's metalmark		*Calephelis freemani* McAlpine	(チョウ目、シジミタテハ科) 新北区

英名	和名	学名	所属、分布、ほか
Freeman's Palpita moth		*Palpita freemanalis* Munroe	（チョウ目、メイガ科）新北区
Freeman's ruby-eye		*Carystoides floresi* Freeman	（チョウ目、セセリチョウ科）新熱帯区
Freeman's skipper		*Vinpeius tinga* (Evans)	（チョウ目、セセリチョウ科）新熱帯区
Freeman's skipperling		*Dalla freemani* Warren	（チョウ目、セセリチョウ科）新熱帯区
French bean fly	インゲンモグリバエ	*Ophiomyia phaseoli* (Tryon)	（ハエ目、ハモグリバエ科）日本、東洋区、大洋区、豪州区。ダイズ、インゲンなどの害虫。幼虫は French bean miner
French bean root aphid			bean root aphid を見よ
French duke	キオビイナズマ	*Euthalia franciae* (Gray)	（チョウ目、タテハチョウ科）東洋区
French wasp		*Polistes gallicus* Linnaeus	（ハチ目、スズメバチ科）全北区
French's canegrub		*Lepidiota frenchi* Blackburn	（コウチュウ目、コガネムシ科）豪州区
Freya's fritillary	アサヒヒョウモン	*Clossiana freija asahidakeana* (Matsumura)	（チョウ目、タテハチョウ科）日本。*C. freija* (Quensel) の英名。Freija's fritillary, Freija fritillary, Freyja's fritillary, Freyja fritillary とも記す
Freyer's dappled white			dappled white を見よ
Freyer's fritillary		*Melitaea arduinna* (Esper)	（チョウ目、タテハチョウ科）旧北区
Freyer's grayling		*Neohipparchia fatua* (Freyer)	（チョウ目、タテハチョウ科）旧北区
Freyer's pug		*Eupithecia intricata arceuthata* (Freyer)	（チョウ目、シャクガ科）旧北区。原名亜種 *E. intricata* Zetterstedt は別に Edinburgh pug ともいう
Freyer's purple emperor			purple emperor (1) を見よ *A. metis* Freyer の英名
friar		*Amauris niavius dominicanus* Trimen	（チョウ目、カノコガ科）新熱帯区
friar	シロモンマダラ	*Amauris niavius* (Linnaeus)	（チョウ目、タテハチョウ科）エチオピア区。英名は修道士の意
Friday's blue		*Plebejus fridayi* Chermock	（チョウ目、シジミチョウ科）新北区
Fridericus spreadwing		*Ouleus fridericus* (Geyer)	（チョウ目、セセリチョウ科）新熱帯区
friendly fly		*Sarcophaga aldrichi* Parker	（ハエ目、ニクバエ科）新北区
friendly hawker		*Zosteraeschna minuscula* (McLachlan)	（トンボ目、ヤンマ科）エチオピア区
friendly Probole moth		*Probole amicaria* (Herrich-Schäffer)	（チョウ目、シャクガ科）新北区
frigate bird fly		*Olfersia spinifera* (Leach)	（ハエ目、シラミバエ科）新北区
frigid owlet moth		*Nycteola frigidana* (Walker)	（チョウ目、ヤガ科）新北区
frigid wave moth		*Scopula frigidaria* (Möschler)	（チョウ目、シャクガ科）全北区
frilly grass tubeworm moth		*Acrolophus mycetophagus* Davis	（チョウ目、Acrolophidae）新北区

英名	和名	学名	所属、分布、ほか
fringe-tree sallow moth		*Sympistis chionanthi* (J. E. Smith)	（チョウ目、ヤガ科）新北区。fringe-tree sallow ともいう
fringe-tufted moths	ササベリガ科	Epermeniidae	（チョウ目）の昆虫の総称
fringe-winged beetles	タマキノコムシモドキ科	Clambidae	（コウチュウ目）の昆虫の総称 体長1mm以下の微小甲虫
fringed azalea whitefly			rhododendron whitefly (1) を見よ
fringed blue		*Neolucia agricola* (Westwood)	（チョウ目、シジミチョウ科）豪州区。fringed blue butterfly ともいう
fringed dawnfly		*Capila pennicillatum* (de Nicéville)	（チョウ目、セセリチョウ科）東洋区
fringed heath-blue			fringed blue を見よ
fringed nettle grub		*Darna nararia* Moore	（チョウ目、イラガ科）東洋区
fringed orchid aphid		*Cerataphis orchidearum* (Westwood)	（カメムシ目、アブラムシ科）全北区
fringed redeye		*Matapa cresta* Evans	（チョウ目、セセリチョウ科）東洋区
Frisch's carrion beetle	フイリカツオブシムシ	*Dermestes frischii* Kugelann	（コウチュウ目、カツオブシムシ科）日本、汎世界。乾燥動植物の害虫
Frisch's case			large clover case を見よ
Frisch's gold long-horn		*Adela fibulella* Fabricius	（チョウ目、マガリガ科）旧北区
frit flies	キモグリバエ科	Chloropidae	（ハエ目）の昆虫の総称
frit fly		*Oscinella frit* (Linnaeus)	（ハエ目、キモグリバエ科）東洋区、新北区。ムギの大害虫
fritillaries	ヒメヒョウモン属	*Boloria*	（チョウ目、タテハチョウ科）の昆虫の総称
fritillaries (1)	ギンボシヒョウモン属	*Speyeria*	（チョウ目、タテハチョウ科）の昆虫の総称
Fritzgaertner's flat		*Celaenorrhinus fritzgaertneri* (Bailey)	（チョウ目、セセリチョウ科）新北区、新熱帯区
frog beetle			kangaroo beetle を見よ
frog cicada		*Venustria superba* Goding et Froggatt	（カメムシ目、セミ科）豪州区
frog-flies			leafhoppers (2) を見よ
Froggatt's buzzer		*Froggattina australis* (Walker)	（バッタ目、バッタ科）豪州区
Froggatt's canegrub		*Lepidiota froggatti* Macleay	（コウチュウ目、コガネムシ科）豪州区
froghopper			アワフキムシ科 Cercopidae の昆虫。とび方がカエルに似る。spittle bugs (2) を見よ
froghopper			cuckoo spit insect を見よ
froghoppers			spittle bugs (1)(2) を見よ
froghoppers			spittle insects を見よ
frosted chafer			Japanese cockchafer を見よ
frosted click beetle	シモフリコメツキ	*Actenicerus pruinosus* (Motschulsky)	（コウチュウ目、コメツキムシ科）日本
frosted dagger moth		*Acronicta hastulifera* (Smith)	（チョウ目、ヤガ科）新北区

英名	和名	学名	所属、分布、ほか
frosted dartwhite		*Catasticta hegemon* Godman et Salvin	（チョウ目、シロチョウ科）新熱帯区
frosted elfin	オナガコツバメ	*Callophrys irus* (Godart)	（チョウ目、シジミチョウ科）新北区
frosted flasher		*Astraptes alardus* Stoll	（チョウ目、セセリチョウ科）新北区、新熱帯区
frosted green		*Polyploca ridens* (Fabricius)	（チョウ目、トガリバガ科）旧北区。frosted green moth ともいう
frosted green lutestring		*Achlya ridens* (Fabricius)	（チョウ目、トガリバガ科）旧北区
frosted greentail		*Arcas tuneta* Hewitson	（チョウ目、シジミチョウ科）新熱帯区
frosted hawk moth	シモフリスズメ	*Psilogramma incerta* (Walker)	（チョウ目、スズメガ科）日本、旧北区
frosted mimic-white		*Lieinix nemesis atthis* (Doubleday)	（チョウ目、シロチョウ科）新熱帯区。*L. n. nayaritensis* Llorente も同英名。
frosted mimic-white			falcate Dismorphia を見よ
frosted orange		*Gortyna ochracea* Hübner	（チョウ目、ヤガ科）旧北区。frosted orange moth ともいう
frosted orange (1)		*Gortyna flavago* Denis et Schiffermüller	（チョウ目、ヤガ科）旧北区
frosted pear aphid			poplar horned aphid を見よ
frosted scale		*Eulecanium pruinosum* (Coquillett)	（カメムシ目、カタカイガラムシ科）新北区、豪州区。クルミ害虫
frosted skipper			hoary-edged skipper を見よ
frosted tufted skipper		*Pellicia costimacula* Herrich-Schäffer	（チョウ目、セセリチョウ科）新熱帯区
frosted yellow		*Isturgia limbaria* (Fabricius)	（チョウ目、シャクガ科）旧北区
frosty-banded skipper		*Cymaenes fraus* (Godman)	（チョウ目、セセリチョウ科）新熱帯区
frosty Hawaiian damselfly		*Megalagrion eudytum* (Perkins)	（トンボ目、イトトンボ科）大洋区
froth-worm			cuckoo spit insect を見よ
frothy moth		*Plagiomimicus spumosum* (Grote)	（チョウ目、ヤガ科）新北区
Fruhstorfer's junglewatcher		*Hylaeothemis fruhstorferi* (Karsch)	（トンボ目、トンボ科）東洋区
Fruhstorfer's owl-butterfly		*Caligo oedipus fruhstorferi* Stichel	（チョウ目、タテハチョウ科）新熱帯区
Fruhstorfer's sister		*Adelpha delinita* Fruhstorfer	（チョウ目、タテハチョウ科）新熱帯区
fruit bark beetle			shothole borer (1) を見よ
fruit beetle		*Carpophilus sexpustulatus* (Fabricius)	（コウチュウ目、ケシキスイ科）旧北区
fruit beetle			dried-fruit beetle (2) を見よ
fruit calpe			fruit piercing moth (7) を見よ
fruit chafers			flower beetles (2) を見よ　中南米での英名
fruit cricket	カネタタキ	*Ornebius kanetataki* (Matsumura)	（バッタ目、カネタタキ科）日本

英名	和名	学名	所属、分布、ほか
fruit flies	ミバエ科	Tephritidae	(ハエ目)の昆虫の総称 世界に4,000種以上
fruit flies			vinegar flies (1) を見よ 英国での英名
fruit fly			citrus fruit fly を見よ
fruit fly		*Monachrosticus citricola* Bezzi	(ハエ目、ミバエ科) 東洋区。カンキツ害虫
fruit fly (1)		*Drosophila funebris* (Fabricius)	(ハエ目、ショウジョウバエ科) 汎世界
fruit fly			pomace fly を見よ
fruit mining tortrix		*Pammene rhediella* (Clerck)	(チョウ目、ハマキガ科) 旧北区
fruit notoxus		*Notoxus constrictus* Fall	(コウチュウ目、アリモドキ科) 新北区
fruit piercing moth	ツキワクチバ	*Artena dotata* (Fabricius)	(チョウ目、ヤガ科) 日本、東洋区。成虫が口吻を果実に刺して吸汁するので吸蛾類という
fruit piercing moth		*Adris suthepensis* Bunziger et Honey	(チョウ目、ヤガ科) 東洋区
fruit-piercing moth		*Gonodonta incurva* Sepp	(チョウ目、ヤガ科) 新熱帯区。カンキツ害虫
fruit-piercing moth		*Othreis ancilla* Cramer	(チョウ目、ヤガ科) 東洋区。カンキツ害虫
fruit-piercing moth		*Othreis apta* Walker	(チョウ目、ヤガ科) 新熱帯区。カンキツ害虫
fruit-piercing moth		*Perigea cupentia* Cramer	(チョウ目、ヤガ科) 新熱帯区。カンキツ害虫
fruit-piercing moth		*Zale fietilis* Guenée	(チョウ目、ヤガ科) 新熱帯区。カンキツ害虫
fruit-piercing moth		*Ophiusa indiscriminata* Hampson	(チョウ目、ヤガ科) 東洋区、大洋区、豪州区
fruit-piercing moth	キタエグリバ	*Calyptra hokkaida* (Wileman)	(チョウ目、ヤガ科) 日本、旧北区。果実吸汁害虫
fruit-piercing moth	キンイロエグリバ	*Calyptra lata* (Butler)	(チョウ目、ヤガ科) 日本、旧北区。果実吸汁害虫
fruit-piercing moth (1)	ヒメアケビコノハ	*Eudocima phalonia* (Linnaeus)	(チョウ目、ヤガ科) 日本、東洋区、豪州区、大洋区、エチオピア区
fruit-piercing moth (2)		*Othreis materna* (Linnaeus)	(チョウ目、ヤガ科) 東洋区、豪州区、エチオピア区。豪州区での英名
fruit-piercing moth (3)	キマエコノハ	*Eudocima salaminia* (Cramer)	(チョウ目、ヤガ科) 日本、東洋区、豪州区。カンキツなどの吸汁加害
fruit-piercing moth (4)		*Othreis fullonia* (Clerck)	(チョウ目、ヤガ科) 日本、全北区、東洋区、エチオピア区、豪州区
fruit-piercing moth (5)	アケビコノハ	*Eudocima tyrannus* (Guenée)	(チョウ目、ヤガ科) 日本、旧北区、東洋区、豪州区。果実吸汁害虫
fruit-piercing moth (6)	オオエグリバ	*Calyptra gruesa* (Draudt)	(チョウ目、ヤガ科) 日本、旧北区。果実吸汁害虫
fruit-piercing moth (7)	ウスエグリバ	*Calyptra thalictri* (Borkhausen)	(チョウ目、ヤガ科) 日本、旧北区。果実吸汁害虫
fruit-piercing moth (8)	ムクゲコノハ	*Lagoptera juno* (Dalman)	(チョウ目、ヤガ科) 日本、旧北区、東洋区。トマト、カンキツなど果樹害虫
fruit-piercing moth (9)	アカエグリバ	*Oraesia excavata* (Butler)	(チョウ目、ヤガ科) 日本、旧北区、豪州区。果実吸汁害虫
fruit-piercing moth (10)	ヒメエグリバ	*Oraesia emarginata* (Fabricius)	(チョウ目、ヤガ科) 日本、東洋区、豪州区。果実吸汁害虫
fruit-piercing moth (11)	シラフクチバ	*Sypnoides picta* (Butler)	(チョウ目、ヤガ科) 日本、旧北区。ナシ吸汁害虫

英名	和名	学名	所属、分布、ほか
fruit-piercing moth			croton caterpillar を見よ
fruit-piercing moth			soapberry noctuid を見よ
fruitspotting bug		*Amblypelta nitida* Stål	(カメムシ目、ヘリカメムシ科) 豪州区
fruitspotting bug		*Dasynus fuscescens* (Distant)	(カメムシ目、ヘリカメムシ科) 豪州区
fruit-sucking moth			fruit-piercing moth (1)(3)を見よ
fruit-tree bark beetle			shothole borer (1) を見よ
fruit-tree borer			cherry tree borer(1)(3)を見よ
fruit tree coccus			calico scale を見よ
fruit-tree katydid			Japanese broad-winged katydid を見よ
fruit tree leaf-folder		*Archips argyrospilus* (Walker)	(チョウ目、ハマキガ科) 新北区
fruit tree leafhopper		*Erythroneura flammigera* (Geoffroy)	(カメムシ目、オオヨコバイ科) 旧北区
fruit tree leafhopper		*Ribautiana debilis* (Douglas)	(カメムシ目、オオヨコバイ科) 旧北区
fruit tree leafhopper		*Typhlocyba avellanae* Edwards	(カメムシ目、オオヨコバイ科) 旧北区
fruit tree leafhopper		*Typhlocyba crataegi* Douglas	(カメムシ目、オオヨコバイ科) 旧北区
fruit tree leafhopper		*Typhlocyba hippocastani* Edwards	(カメムシ目、オオヨコバイ科) 旧北区
fruit-tree leafhopper	ハンノヒメヨコバイ	*Alnella alneti* (Dahlbom)	(カメムシ目、オオヨコバイ科) 日本、旧北区。ハンノキ害虫
fruit tree leafhopper			bramble leafhopper を見よ
fruit tree leafhopper			prune leafhopper を見よ
fruit tree leafhopper			oak leafhopper (1) を見よ
fruit tree leaf-roller			fruit tree leaf-folder を見よ　fruit tree leafroller moth ともいう
fruit-tree looper	シモフリトゲエダシャク	*Phigalia sinuosaria* Leech	(チョウ目、シャクガ科) 日本、旧北区。リンゴなど果樹害虫
fruit-tree pinhole borer			Saxena ambrosia beetle を見よ
fruit-tree root weevil		*Leptopius squalidus* (Boheman)	(コウチュウ目、ゾウムシ科) 豪州区。fruit tree rood weevil とも記す
fruit tree thrips			pear thrips を見よ
fruit tree tortrix		*Ditula angustiorana* (Haworth)	(チョウ目、ハマキガ科) 全北区。リンゴ、ブドウ他害虫
fruit tree tortrix			great brown twist を見よ
fruit tree tortrix			rhomboid tortrix moth を見よ
fruit tree tortrix			brown oak tortrix を見よ
fruit-tree tortrix			marbled orchard tortrix を見よ
fruit-tree tortrix			currant twist を見よ
fruit-tree tortrix			brown tortrix を見よ
fruit tree tortrix moth			great brown twist を見よ

英名	和名	学名	所属、分布、ほか
fruit tree tortrix moth			fruit tree tortrix を見よ
fruit tree tortrix moth			rhomboid tortrix moth を見よ
fruit tree tortrix moth			brown oak tortrix を見よ
fruit tree tortrix moth			grey bud moth を見よ
fruit tree tortrix moths		*Apotomis, Olethreutes*	（チョウ目、ハマキガ科）の昆虫の総称
fruit weevils	ハナゾウムシ亜科	Anthonominae	（コウチュウ目、ゾウムシ科）の昆虫の総称
fruit worm beetles	キスイモドキ科	Byturidae	（コウチュウ目）の昆虫の総称　fruitworm beetles とも記される。
fruitworm moths	シンクイガ科	Carposinidae	（チョウ目）の昆虫の総称、
fruitworms			fruit worm beetles を見よ
fruitlet mining tortrix			fruit mining tortrix を見よ
fuchsia moth		*Izatha epiphanes* Meyrick	（チョウ目、マルハキバガ科）豪州区
Fuelleborn's bottle-tail			Fuelleborn's riverking を見よ
Fuelleborn's riverking		*Zygonoides fuelleborni* (Grunberg)	（トンボ目、トンボ科）エチオピア区
Fukaya pyriform scale	フカヤカタカイガラムシ	*Protopulvinaria fukayai* (Kuwana)	（カメムシ目、カタカイガラムシ科）日本。クチナシなどの害虫。Fukaya coccus ともいう
fulgorid planthoppers			lantern flies を見よ
fulgorines		Fulgorina	（カメムシ目）の昆虫の総称
fulgoroids			lantern bugs を見よ
fuliginous alderfly		*Sialis fuliginosa* Pictet	（アミメカゲロウ目、センブリ科）旧北区
full-spotted ermel		*Hyponomeuta evonymellus* Linnaeus	（チョウ目、スガ科）旧北区
full stop swift		*Caltoris cormasa* (Hewitson)	（チョウ目、セセリチョウ科）東洋区
Fullaway's banana Hedyleptan moth		*Omiodes fullawayi* Swezey	（チョウ目、メイガ科）大洋区
Fuller	ヨーロッパヒゲコガネ	*Polyphylla fullo* (Linnaeus)	（コウチュウ目、コガネムシ科）旧北区
Fuller rose beetle	フラーバラゾウムシ	*Pantomorus godmani* (Crotch)	（コウチュウ目、ゾウムシ科）新北区、大洋区。アボカド害虫。Fuller's rose beetle とも記す
Fuller rose weevil			Fuller rose beetle を見よ
Fuller's rose weevil		*Pantomorus cervinus* (Boheman)	（コウチュウ目、ゾウムシ科）旧北区
Fulton's ground cricket		*Allonemobius fultoni* Howard et Furth	（バッタ目、コオロギ科）新北区
Fulton's trig		*Anaxipha* sp.	（バッタ目、クサヒバリ科）新北区
Fulvia checkerspot		*Chlosyne fulvia* (Edwards)	（チョウ目、タテハチョウ科）新北区、新熱帯区。*C. f. coronado* (Smith et Brock) も同英名

英名	和名	学名	所属、分布、ほか
fulvous-bordered green mantle		*Caria fulvimargo* Lathy	（チョウ目、シジミタテハ科）新熱帯区
fulvous clover		*Chloridea maritima* Graslin	（チョウ目、ヤガ科）旧北区
fulvous dawnfly		*Copila phanaeus* (Hewitson)	（チョウ目、セセリチョウ科）東洋区
fulvous-edged Pyrausta moth		*Pyrausta nexalis* (Hulst)	（チョウ目、メイガ科）新北区
fulvous forest skimmer		*Neurothemis fulva* (Drury)	（トンボ目、トンボ科）東洋区
fulvous hairstreak			Angelic hairstreak を見よ
fulvous pied flat		*Pseudocoladenia dan* (Fabricius)	（チョウ目、セセリチョウ科）旧北区、東洋区。*P. d. dhyana* Fruhstorfer も同英名
fulvous ranger		*Kedestes mohozutza* (Wallengren)	（チョウ目、セセリチョウ科）エチオピア区
fulvous sealed bell		*Eucosma fulvana* Stephens	（チョウ目、ハマキガ科）旧北区
fulvous sphinx		*Coelonia fulvinotata* (Butler)	（チョウ目、スズメガ科）エチオピア区
fun flies	ツヤコバチ科	Aphelinidae	（ハチ目）の昆虫の総称
Fundania sister		*Adelpha malea fundania* Fruhstorfer	（チョウ目、タテハチョウ科）新熱帯区
fundatrices			stem mother を見よ
fundatrigeniae			stem mother を見よ
funerary dagger moth		*Acronicta funeralis* Grote et Robinson	（チョウ目、ヤガ科）新北区
funereal Apotomis moth		*Apotomis funerea* (Meyrick)	（チョウ目、ハマキガ科）新北区
funereal duskywing	モフクセセリ	*Erynnis funeralis* (Scudder et Burgess)	（チョウ目、セセリチョウ科）新北区、新熱帯区
funereal ermel		*Ethmia funerella* Haworth	（チョウ目、スエヒロキバガ科）旧北区
funereal Nephele		*Nephele funebris* (Fabricius)	（チョウ目、スズメガ科）エチオピア区
fungus ants			leafcutter ants を見よ
fungus beetles			hairy fungus beetles を見よ
fungus beetles			handsome fungus beetles を見よ
fungus bugs			flat bugs を見よ
fungus-eating ladybird		*Illeis galbula* (Mulsant)	（コウチュウ目、テントウムシ科）豪州区
fungus flies			mushroom flies を見よ
fungus gardening ant			fungus growing ant (1) を見よ
fungus gnat		*Leia varia* Walker	（ハエ目、キノコバエ科）新北区
fungus gnat maggots			dark-winged fungus gnats を見よ　カナダでの英名
fungus gnats	ツノキノコバエ科	Keroplatidae	（ハエ目）の昆虫の総称
fungus gnats		*Mycetophila*	（ハエ目、キノコバエ科）の昆虫の総称
fungus gnats			mushroom flies を見よ
fungus gnats			dark-winged fungus gnats を見よ

英名	和名	学名	所属、分布、ほか
fungus grain moth			Fabricius's clothes moth を見よ
fungus growers			agricultural ants を見よ
fungus growing ant		*Cyphomyrmex minutus* Mayr	(ハチ目、アリ科) 新北区
fungus growing ant		*Trachymyrmex arizonensis* Wheeler	(ハチ目、アリ科) 新北区
fungus growing ant (1)		*Trachymyrmex septentrionalis* (McCock)	(ハチ目、アリ科) 新北区、新熱帯区
fungus growing ant			Texas leaf cutting ant を見よ
fungus growing termite	オオキノコシロアリ	*Macrotermes bellicosus* (Smeathman)	(シロアリ目、シロアリ科) エチオピア区。最大のシロアリで、女王は5 cm以上、職アリは6 mm
fungus growing termite		*Macrotermes gilvus* (Hagen)	(シロアリ目、シロアリ科) 東洋区
fungus maggots			mushroom flies を見よ
fungus midges			mushroom flies を見よ
fungus moths			clothes moths を見よ
fungus weevils	ヒゲナガゾウムシ科	Anthribidae	(コウチュウ目) の昆虫の総称
funnel ant		*Aphaenogaster pythia* Forel	(ハチ目、アリ科) 豪州区
fur beetle		*Attagenus pellio* (Linnaeus)	(コウチュウ目、カツオブシムシ科) 汎世界
furniture beetle (1)		*Anobium striatum* Olivier	(コウチュウ目、シバンムシ科) 旧北区
furniture beetle (2)		*Anobium pertinax* Linnaeus	(コウチュウ目、シバンムシ科) 旧北区
furniture beetle			common furniture beetle を見よ
furniture beetles			death watch beetles を見よ
furniture beetles			lycids を見よ
furniture borer			common furniture beetle を見よ
furniture borer			furniture beetle (1) を見よ
furniture borers			death watch beetles を見よ
furniture carpet beetle	アカオビカツオブシムシ	*Dermestes vorax* Motschulsky	(コウチュウ目、カツオブシムシ科) 日本、旧北区
furniture carpet beetle		*Anthrenus flavipes* LeConte	(コウチュウ目、カツオブシムシ科) 新北区、豪州区。家具、カーペット害虫
furniture cockroach			brown-banded cockroach (2) を見よ　北アフリカでの英名
furniture moth		*Tineola furciferella* Zagulyaev	(チョウ目、ヒロズコガ科) 旧北区
furniture moth			mottled bran marble を見よ
furry blue		*Polyommatus dolus* (Hübner)	(チョウ目、シジミチョウ科) 旧北区
furry yellow caterpillar			wild cherry moth を見よ
furtive forktail		*Ischnura prognata* (Hagen)	(トンボ目、イトトンボ科) 新北区

英名	和名	学名	所属、分布、ほか
fuscomarginated geometrid			Nitobe looper moth を見よ
fuscous birch pigmy		*Stigmella confusella* (Wood et Walsingham)	（チョウ目、モグリチビガ科）旧北区
fuscous-speckled clothes		*Nemaxera betulinella* (Fabricius)	（チョウ目、ヒロオズコガ科）旧北区
fuscous-speckled clothes moth		*Nemapogon emortuellus* Zeller	（チョウ目、ヒロズコガ科）．旧北区
fuscus swallowtail			yellow Helen (1) を見よ　fuscous swallowtail とも記す。
fusedback mayflies	マダラカゲロウ亜目	Pannota	（カゲロウ目）の昆虫の総称
fussorial wasps			sphecoid wasps を見よ

英名	和名	学名	所属、分布、ほか
G			
Gabb's checkerspot		*Chlosyne gabbii* (Behr)	（チョウ目、タテハチョウ科）新北区
Gabina hairstreak		*Erora gabina* (Godman et Salvin)	（チョウ目、シジミチョウ科）新熱帯区
Gabina skipper		*Orthos gabina* (Godman)	（チョウ目、セセリチョウ科）新熱帯区
gad-bree			cattle warble-fly (1) (2) を見よ
gad flies			horse flies (1)(2) を見よ　米国、豪州での英名
gadflies			bot flies (2) を見よ
gad fly			cattle warble-fly (1)(2) を見よ
gadfly			アブ、アブ科の幼虫、ウマアブ
gad-fly			sheep bot fly を見よ
Gadira hairstreak		*Ignata gadira* (Hewitson)	（チョウ目、シジミチョウ科）新熱帯区
Gaelic groundling		*Bryotropha boreella* (Douglas)	（チョウ目、キバガ科）旧北区
Gaika blue			tiny grass blue (2) を見よ
Gaika brown		*Pseudonympha gaika* Riley	（チョウ目、タテハチョウ科）エチオピア区
Gala skipper		*Paratrytone gala* (Godman)	（チョウ目、セセリチョウ科）新熱帯区
Galapagos blue		*Leptotes parrhasioides* (Wallengren)	（チョウ目、シジミチョウ科）新熱帯区
galaxy satyr		*Pedaliodes phrasicla* Hewitson	（チョウ目、タテハチョウ科）新熱帯区
galerucine beetles			cucumber beetles を見よ
galerucine leaf-beetles			cucumber beetles を見よ
gall			虫こぶ、虫えい
gall aphids		Pemphiginae	（カメムシ目、アブラムシ科）の昆虫の総称
gall aphids			bark aphids を見よ
gall-flies		Cynipinae	（ハチ目、タマバチ科）の昆虫の総称
gall-flies			gall wasps (1)(2) を見よ　gallflies とも記す
gall flies			fruit flies を見よ
gall fly		*Spathegaster baccarum* (Linnaeus)	（ハチ目、タマバチ科）旧北区
gall fly			rice gall midge を見よ
gall gnats			gall midges を見よ
gall-like coccids			mealybugs (3) を見よ
gall-like coccids			gall-like scales を見よ　米国での英名
gall-like coccids			kermes (1) を見よ
gall-like scale insects			gall-like scales を見よ
gall-like scales	タマカイガラムシ科	Kermestidae	（カメムシ目）の昆虫の総称
gall-making aphid			woolly apple aphid を見よ
gall-making aphids			woolly aphids を見よ　南アフリカでの英名

英名	和名	学名	所属、分布、ほか
gallmaking maple borer		*Xylotrechus aceris* Fisher	(コウチュウ目、カミキリムシ科) 新北区
gall-making signal		*Augasma aeratella* (Zeller)	(チョウ目、マイコガ科) 旧北区
gallmidge parasites			platygasterid wasps を見よ 南アフリカでの英名
gall midges	タマバエ科	Cecidomyiidae	(ハエ目) の昆虫の総称
gall moths		Cecidosidae	(チョウ目) の昆虫の総称
gall-nut			gall-nut wasp の虫えい
gall-nut wasp		*Andricus gallaetinctoriae* (Olivier)	(ハチ目、タマバチ科) 旧北区
gall pine scale		*Matsucoccus gallicolus* Morrison	(カメムシ目、ワタフキカイガラムシ科) 新北区
gall wasp			oak apple gall wasp を見よ
gall wasp			mossy rose gall wasp を見よ
gall wasp inquilines			inquiline gall wasps を見よ
gall wasps (1)	タマバチ科	Cynipidae	(ハチ目) の昆虫の総称 フシバチともいう
gall wasps (2)	タマバチ上科	Cynipoidea	(ハチ目) の昆虫の総称
gallicoles			bark aphids を見よ
galling aphid		*Forda formicaria* von Heyden	(カメムシ目、アブラムシ科) 旧北区
gallinipper		*Psorophora ciliata* (Fabricius)	(ハエ目、カ科) 新北区
gallium carpet		*Epirrhoe galiata* (Denis et Schiffermüller)	(チョウ目、シャクガ科) 旧北区
gallium horn worm			bedstraw hawk を見よ
gallium looper	アトクロナミシャク	*Lampropteryx minna* (Butler)	(チョウ目、シャクガ科) 日本、旧北区
gallium sphinx			bedstraw hawk を見よ 米国での英名。gallium sphinx moth ともいう
Gallo's anomalous blue		*Polyommatus galloi* Balletto et Toso	(チョウ目、シジミチョウ科) 旧北区
Gambian spotted-eye mantis		*Pseudoharpax virescens* Serville	(カマキリ目、Hymenopodidae) エチオピア区
Gamble's sprite		*Pseudagrion gamblesi* Pinhey	(トンボ目、イトトンボ科) エチオピア区
gamma			silver Y moth (1) を見よ gamma moth ともいう
Gauthier's striped swordtail		*Graphium biokoensis* Gauthier	(チョウ目、アゲハチョウ科) エチオピア区
garbage cricket			house cricket を見よ
Garcia tiger swallowtail		*Papilio alexiares garcia* Rothschild et Jordan	(チョウ目、アゲハチョウ科) 新熱帯区。Mexican tiger swallowtail を参照
garden Acraea		*Acraea horta* (Linnaeus)	(チョウ目、タテハチョウ科) エチオピア区
garden ant			field ant を見よ
garden ant			black ant (1) を見よ 米国での英名
garden arches moth		*Lacanobia radix* (Walker)	(チョウ目、ヤガ科) 新北区
garden armyworm			velvet armyworm moth を見よ

英名	和名	学名	所属、分布、ほか
garden bagworm			snailcase bagworm moth を見よ
garden Bermius		*Bermius brachycerus* Stål	（バッタ目、バッタ科）豪州区
garden carpet		*Xanthorhoe fluctuata* (Linnaeus)	（チョウ目、シャクガ科）旧北区。日本亜種は *X. f. malleola* Inoue クロモンミヤマナミシャク
garden carpet moth			garden carpet を見よ
garden carrion beetle			garden silphid を見よ
garden chafer		*Phyllopertha horticola* (Linnaeus)	（コウチュウ目、コガネムシ科）旧北区
garden click beetle		*Athous haemorrhoidalis* (Fabricius)	（コウチュウ目、コメツキムシ科）旧北区
garden cutworm			heart and dart を見よ
garden dart			golden dart moth を見よ
garden dart moth		*Euxoa nigricans* Linnaeus	（チョウ目、ヤガ科）全北区
garden elder pearl		*Phlyctaenia coronata* (Hufnagel)	（チョウ目、メイガ科）旧北区
garden flea beetle		*Epitrix pubescens* (Koch)	（コウチュウ目、ハムシ科）旧北区
garden flea hopper		*Halticus citri* (Ashmead)	（カメムシ目、カスミカメムシ科）新北区
garden fleahopper (1)		*Halticus bractatus* (Say)	（カメムシ目、カスミカメムシ科）新北区。体長 3～4 mm。園芸植物、野菜害虫
garden fleas			springtails を見よ
garden flies			march flies を見よ
garden fruit chafer		*Pachnoda sinuata* (Fabricius)	（コウチュウ目、コガネムシ科）エチオピア区
garden grass-veneer		*Crambus hortuellus* Hübner	（チョウ目、メイガ科）旧北区
garden grass veneer			grass moth (1) を見よ
garden ground beetle		*Carabus hortensis* Linnaeus	（コウチュウ目、オサムシ科）旧北区
garden inspector		*Precis archesia* (Cramer)	（チョウ目、タテハチョウ科）エチオピア区
garden leafhopper			garden fleahopper (1) を見よ
garden long-legged fly		*Condylostylus pilicornis* (Aldrich)	（ハエ目、アシナガバエ科）新北区
garden looper	フチベニヒメシャク	*Idaea jakima* (Butler)	（チョウ目、シャクガ科）日本、旧北区
garden maggot		*Bibio imitator* Walker	（ハエ目、ケバエ科）豪州区
garden mantis			green mantid を見よ
garden miller		*Galgula partita* Guenée	（チョウ目、ヤガ科）新北区
garden pea leafminer	ナモグリバエ	*Chromatomyia horticola* (Goureau)	（ハエ目、ハモグリバエ科）日本、東洋区。各種作物、園芸植物害虫
garden pebble			crucifer caterpillar を見よ　garden pebble moth ともいう
garden rose tortrix		*Acleris variegana* (Denis et Schiffermüller)	（チョウ目、ハマキガ科）全北区、エチオピア区。garden rose tortrix moth ともいう
garden silphid		*Silpha ramosa* Say	（コウチュウ目、シデムシ科）新北区
garden soldier fly			blue soldier fly を見よ
garden springtail	キボシマルトビムシ	*Bourletiella hortensis* (Fitch)	（トビムシ目、マルトビムシ科）日本、全北区。各種作物害虫

英名	和名	学名	所属、分布、ほか
garden springtails		*Bourletiella*	（トビムシ目、マルトビムシ科）の昆虫の総称
garden swift moth			common swift (1) を見よ
garden thrips			flower thrips (3) を見よ
garden tiger moth			black woolly-bear を見よ　garden tiger ともいう
garden tortrix		*Clepsis peritana* (Clemens)	（チョウ目、ハマキガ科）全北区、新熱帯区。garden tortrix moth ともいう
garden webworm		*Phlyctaenodes similaris* Guenée	（チョウ目、メイガ科）新北区
garden webworm		*Achyra llaguenalis* Munroe	（チョウ目、メイガ科）新熱帯区
garden webwrom		*Achyra rantalis* (Guenée)	（チョウ目、メイガ科）新北区。garden webworm moth ともいう
garden webworm			beet webworm を見よ　幼虫の英名
garden weevil		*Phlycinus callosus* Boheman	（コウチュウ目、ゾウムシ科）豪州区
garden wireworm			garden click beetle を見よ　幼虫の英名
gardenia budworm	アヤニジュウシトリバ	*Alucita flavofascia* (Inoue)	（チョウ目、ニジュウシトリバガ科）日本。クチナシの害虫
Gargophia hairstreak		*Gargina gargophia* (Hewitson)	（チョウ目、シジミチョウ科）新熱帯区
Garita skipperling		*Oarisma garita* (Reakirt)	（チョウ目、セセリチョウ科）新北区。Garita skipper ともいう
garnet-patched hairstreak		*Nicolaea velina* (Hewitson)	（チョウ目、シジミチョウ科）新熱帯区
gartered Halysidota			Florida tussock moth をみよ
gasteruptionid wasps	コンボウヤセバチ科	Gasteruptionidae	（ハチ目）の昆虫の総称　gasteruptiids ともいう
gastrilegine bees		Gastrilegina	（ハチ目）の昆虫の総称
gastrilegines			gastrilegine bees を見よ
gatekeeper		*Pyronia tithonus* (Linnaeus)	（チョウ目、タテハチョウ科）旧北区。gatekeeper butterfly ともいう
gatekeeper			hedge brown を見よ
Gaudy Altinote		*Altinote negra demonica* (Hopffer)	（チョウ目、タテハチョウ科）新熱帯区
Gaudy baron	アカホシイナズマ	*Euthalia lubentina* (Cramer)	（チョウ目、タテハチョウ科）東洋区
Gaudy checkerspot		*Chlosyne gaudialis gaudialis* (Bates)	（チョウ目、タテハチョウ科）新熱帯区
Gaudy commodore	オクタビアタテハモドキ	*Precis octavia* (Cramer)	（チョウ目、タテハチョウ科）エチオピア区
Gaudy eyed-metalmark		*Mesosemia gaudiolum* Bates	（チョウ目、シジミタテハ科）新熱帯区
Gaudy nightfighter		*Porphyrogenes spadix* Austin et O. Mielke	（チョウ目、セセリチョウ科）新熱帯区
Gaudy sister		*Adelpha boreas* (Butler)	（チョウ目、タテハチョウ科）新熱帯区
Gaudy sphinx		*Eumorpha labruscae* (Linnaeus)	（チョウ目、スズメガ科）新北区。Gaudy sphinx moth ともいう
Gaumeri hairstreak			white-tipped hairstreak を見よ

英名	和名	学名	所属、分布、ほか
Gaura borer moth		*Euhagena emphytiformis* (Walker)	（チョウ目、スカシバガ科）新北区
Gavarnie blue		*Agriades pyrenaicus* (Boisduval)	（チョウ目、シジミチョウ科）旧北区
Gavarnie ringlet	ガバルニーベニヒカゲ	*Erebia gorgone* Boisduval	（チョウ目、タテハチョウ科）旧北区
Gayra skipper		*Vacerra gayra* (Dyar)	（チョウ目、セセリチョウ科）新熱帯区
gazella dung beetle			brown dung beetle を見よ
gela top borer			early shoot borer を見よ
gelastocodids			toad bugs を見よ
gelechiid moths			twirler moths を見よ
gelechiids			twirler moths を見よ
Gelonaetha long-horned beetle		*Gelonaetha hirta* (Fairmaire)	（コウチュウ目、カミキリムシ科）大洋区
gelonus bugs		*Gelonus*	（カメムシ目、ヘリカメムシ科）の昆虫の総称
gem (1)	トビスジヒメナミシャク	*Orthonama obstipata* (Fabricius)	（チョウ目、シャクガ科）日本、旧北区。gem moth ともいう
gem (2)		*Chloroselas pseudozeritis* (Trimen)	（チョウ目、シジミチョウ科）エチオピア区
gem swallowtail			pink spotted swallowtail (1) を見よ
gemmate bush katydid		*Insara gemmicula* Rehn et Hebard	（バッタ目、キリギリス科）新北区
gemmed satyr		*Cyllopsis gemma* (Hübner)	（チョウ目、タテハチョウ科）新北区
gemmed shoot			pine shoot moth (1) を見よ
gemmed sphinx		*Sphinx geminus* (Rothschild et Jordan)	（チョウ目、スズメガ科）新熱帯区
gems		*Libellago*	（トンボ目、Chlorocyphidae）の昆虫の総称
gems		*Chloroselas*	（チョウ目、シジミチョウ科）の昆虫の総称
gems	キララシジミ属	*Poritia*	（チョウ目、シジミチョウ科）の昆虫の総称
gems (1)		Plusiinae	（チョウ目、ヤガ科）の昆虫の総称
general purple and gold			peppermint Pyrausta を見よ
genista aphid		*Aphis genistae* Scopoli	（カメムシ目、アブラムシ科）旧北区
genista bell moth		*Catamacta gavisana* Walker	（チョウ目、ハマキガ科）豪州区
genista broom moth		*Uresiphita reversalis* (Guenée)	（チョウ目、メイガ科）新北区、新熱帯区。genista caterpillar, genista caterpillar moth ともいう
Genji firefly	ゲンジボタル	*Luciola cruciata* Motschulsky	（コウチュウ目、ホタル科）日本。日本の代表的ホタル
Genoveva azure	ミイロヤドリギシジミ	*Ogyris genoveva* Hewitson	（チョウ目、シジミチョウ科）豪州区
Genoveva buckeye			black buckeye を見よ
gentian leafminer	リンドウハモグリバエ	*Chromatomyia gentianae* (Hendel)	（ハエ目、ハモグリバエ科）日本、旧北区
gentilla groundstreak		*Calycopis gentilla* (Schaus)	（チョウ目、シジミチョウ科）新熱帯区
gentles			blow flies を見よ　幼虫の英名

英名	和名	学名	所属、分布、ほか
geometer moths	シャクガ科	Geometridae	(チョウ目)の昆虫の総称 幼虫はシャクトリムシ、成虫はシャクトリガと俗称される。geometers、geometrid moths ともいう
geometrician		*Grammodes stolidus* (Fabricius)	(チョウ目、ヤガ科) 旧北区。geometrician moth ともいう
Geonoma-feeding ruby-eye		*Perichares geonomaphaga* Burns	(チョウ目、セセリチョウ科) 新熱帯区
George's carpet moth		*Plemyria georgii* Hulst	(チョウ目、シャクガ科) 新北区
Georgia river cruiser			swift river cruiser を見よ
Georgia satyr		*Neonympha areolatus* (Smith)	(チョウ目、タテハチョウ科) 新北区
Georgian prominent moth		*Hyperaeschra georgica* (Herrich-Schäffer)	(チョウ目、シャチホコガ科) 新北区
georyssid beetles			minute mud-loving beetles を見よ
geotrupid dung beetles			dor beetles を見よ 南アフリカでの英名
Gerald's brown		*Stygionympha geraldi* Pennington	(チョウ目、タテハチョウ科) エチオピア区
geranium aphid			pelargonium aphid を見よ
geranium Argus		*Aricia eumedon* (Esper)	(チョウ目、シジミチョウ科) 旧北区
geranium blue			geranium bronze を見よ
geranium bronze		*Cacyreus marshalli* Butler	(チョウ目、シジミチョウ科) エチオピア区
geranium mealybug		*Spirococcus geraniae* (Rau)	(カメムシ目、コナカイガラムシ科) 新北区
geranium plume moth		*Amblyptilia pica* Walsingham	(チョウ目、トリバガ科) 全北区
geranium sawfly		*Protemphytus carpini* (Hartig)	(ハチ目、ハバチ科) 旧北区
Gerda's buff		*Toxochitona gerda* (Kirby)	(チョウ目、シジミチョウ科) エチオピア区
Gerina Epitola		*Geritola gerina* (Hewitson)	(チョウ目、シジミチョウ科) エチオピア区
German burying beetle			gravedigger beetle を見よ
German cockroach	チャバネゴキブリ	*Blattella germanica* (Linnaeus)	(ゴキブリ目、チャバネゴキブリ科) 日本、汎世界。著名屋内害虫
German cockroaches	チャバネゴキブリ科	Blattellidae	(ゴキブリ目)の昆虫の総称
German cousin moth		*Sideridis congermana* (Morrison)	(チョウ目、ヤガ科) 新北区
German duck			bed bug (1) を見よ
German grain louse			grain aphid (1) を見よ
German honey bee			honey bee を見よ
German millet stem maggot	モロコシクキイエバエ	*Atherigona biseta* Karl	(ハエ目、イエバエ科) 日本、東洋区。アワ、キビ害虫
German roach			common cockroach を見よ
German wasp		*Vespula germanica* (Fabricius)	(ハチ目、スズメバチ科) 旧北区
German yellow jacket			German wasp を見よ 米国での英名

英名	和名	学名	所属、分布、ほか
Germander speedwell gall		*Jaapiella veronicae* (Vallot)	(ハエ目、タマバエ科) がイヌノフグリ類似種の葉につくる有毛の虫えいで 7～9 mm。旧北区
germanium aphid		*Aulacorthum geranii* Kaltenbach	(カメムシ目、アブラムシ科) 旧北区
Gerthard's underwing			Herodias underwing moth を見よ
geum aphid			potato aphid (1) を見よ 米国での英名
geum leaf-mining sawfly		*Metallus gei* (Brischke)	(ハチ目、ハバチ科) 旧北区
geum sawfly			raspberry sawfly (1) を見よ
geum sawfly			geum leaf-mining sawfly を見よ
Geyer grained moth	フサヤガ	*Eutelia geyeri* (Felder et Rogenhofer)	(チョウ目、ヤガ科) 日本、東洋区
Geyer's blue skipper		*Pythonides herennius* Geyer	(チョウ目、セセリチョウ科) 新熱帯区
ghost and swift moths			ghost moths を見よ
ghost ant	アワテコヌカアリ	*Tapinoma melanocephalum* (Fabricius)	(ハチ目、アリ科) 日本、豪州区、汎亜熱帯、汎熱帯
ghost brimstone	マエモンオオヤマキチョウ	*Anteos clorinde* (Godart)	(チョウ目、シロチョウ科) 新北区、新熱帯区
ghost fly			greenhouse whitefly を見よ
ghost larva		*Chaoborus crystallinus* (De Geer)	(ハエ目、ケヨソイカ科). 旧北区
ghost mantis		*Phyllocrania paradoxa* (Burmeister)	(カマキリ目、Hymenopodidae) 旧北区、エチオピア区
ghost midges			phantom midges (1) を見よ
ghost moth		*Hepialus humuli* (Linnaeus)	(チョウ目、コウモリガ科) 旧北区
ghost moth			puriri moth を見よ
ghost moth			gold swift を見よ
ghost moths	コウモリガ科	Hepialidae	(チョウ目) の昆虫の総称
ghost moths			carpenter moths (1) を見よ
ghost swift			ghost moth を見よ
ghost swift moth			ghost moth を見よ
ghost walker	マルクビバイオリンムシ	*Mormolyce castelnaudi* H. Deyrolle	(コウチュウ目、オサムシ科) 東洋区
ghost yellow		*Eurema albula* (Cramer)	(チョウ目、シロチョウ科) 新熱帯区
giant African fruit beetle		*Mecynorrhina polyphemus* (Fabricius)	(コウチュウ目、コガネムシ科) エチオピア区
giant African longhorn		*Petrognatha gigas* Fabricius	(コウチュウ目、カミキリムシ科) エチオピア区。giant African longhorn beetle ともいう
giant African mantis		*Sphodromantis viridis* Forskal	(カマキリ目、カマキリ科) エチオピア区
giant African mantis			African mantis (2) を見よ
giant African skipper	イフィスセセリ	*Pyrrhochalcia iphis* Drury	(チョウ目、セセリチョウ科) エチオピア区

英名	和名	学名	所属、分布、ほか
giant African swallowtail			African giant swallowtail を見よ
giant agrippa	ナンベイオオヤガ	*Thysania agrippina* (Cramer)	（チョウ目、ヤガ科）新北区、新熱帯区。翅長が最大の蛾
giant Amazonian ant		*Dinoponera gigantea* (Perty)	（ハチ目、アリ科）新熱帯区。最大のアリで雌は4 cm
giant anthelid		*Chelepteryx collesi* Gray	（チョウ目、Anthelidae）豪州区
giant ant-lions		*Palpares*	（アミメカゲロウ目、ウスバカゲロウ科）の昆虫の総称
giant arid-land katydid			greater arid-land katydid を見よ
giant Asian mantis		*Hierodula membranacea* Burmeister	（カマキリ目、カマキリ科）東洋区
giant Asian mantis (1)		*Hierodula patellifera* Serville	（カマキリ目、カマキリ科）日本、東洋区、大洋区。Asian mantis ともいう
giant bagworm	オオミノガ	*Eumeta japonica* Heylaerts	（チョウ目、ミノガ科）日本。各種樹木害虫
giant bamboo mealybug			bamboo mealybug を見よ
giant bark aphid		*Longistigma caryae* (Harris)	（カメムシ目、アブラムシ科）新北区
giant bark feeder			giant bark aphid を見よ
giant Batrachedra moth		*Batrachedra enomis* Meyrick	（チョウ目、カザリバガ科）新北区
giant bee			giant honey bee を見よ
giant bird louse		*Ancistrona vagelli* (Fabricius)	（ハジラミ目、タンカクハジラミ科）新北区
giant black bostrichid			apate borer を見よ
giant black stonefly			giant stonefly (2) を見よ
giant black water beetle		*Hydrophilus triangularis* Say	（コウチュウ目、ガムシ科）新北区
giant blue swallowtail	ザルモクシスオオアゲハ	*Papilio zalmoxis* Hewitson	（チョウ目、アゲハチョウ科）エチオピア区
giant brown phasmid		*Palophus titan* Sjöstedt	（ナナフシ目、Bacteriidae）豪州区
giant burrowing cockroach		*Macropanesthia rhinoceros* Saussure	（ゴキブリ目、オオゴキブリ科）豪州区。体長8 cm
giant butterfly moths			sun moths (1) を見よ
giant carpenter ant		*Camponotus laevigatus* (Smith)	（ハチ目、アリ科）新北区
giant carrion beetle			American burying beetle を見よ
giant casemaker caddisflies			large caddisflies を見よ
giant Charaxes	オオフタオビフタオチョウ	*Charaxes castor* (Cramer)	（チョウ目、タテハチョウ科）エチオピア区
giant cicada	テイオウセミ	*Pomponia imperatoria* (Westwood)	（カメムシ目、セミ科）東洋区。体長7 cm の大型セミ
giant cicada killer			cicada-killer wasp を見よ
giant click beetle		*Chalcolepidius erythroloma* Candèze	（コウチュウ目、コメツキムシ科）大洋区

英名	和名	学名	所属、分布、ほか
giant coccid		*Icerya pulcher* Leonardi	(カメムシ目、ワタフキカイガラムシ科) 東洋区。カンキツ害虫
giant coccids	ワタフキカイガラムシ科	Margarodidae	(カメムシ目) の昆虫の総称
giant coccoids			giant coccids を見よ
giant coccus	オオカタカイガラムシ	*Lecanium glandi* Kuwayama	(カメムシ目、カタカイガラムシ科) 日本
giant cockroaches			cockroaches (4) を見よ
giant cocoa termite		*Neotermes papua* (Desneux)	(シロアリ目、レイビシロアリ科) 東洋区
giant crane fly		*Tipula maxima* Poda	(ハエ目、ガガンボ科) 旧北区
giant crane fly		*Holorusia rubiginosa* Loew	(ハエ目、ガガンボ科) 新北区
giant crepuscular skipper		*Gretna balenge* (Holland)	(チョウ目、セセリチョウ科) エチオピア区
giant crow	パエナレタルリマダラ	*Euploea phaenareta* (Schaller)	(チョウ目、タテハチョウ科) 東洋区
giant cupids	アフリカゴマシジミ属	*Lepidochrysops*	(チョウ目、シジミチョウ科) の昆虫の総称
giant darner		*Anax walsinghami* McLachlan	(トンボ目、ヤンマ科) 新北区
giant dead leaf mantis		*Deroplatys desicata* Beier	(カマキリ目、カマキリ科) 東洋区
giant death's head roach			death's head cockroach を見よ
giant devil's flower mantis			devil's flower mantis (2) を見よ
giant dragonfly			south-eastern petaltail を見よ
giant drummer		*Blaberus giganteus* (Linnaeus)	(ゴキブリ目、ブラベルスゴキブリ科) 新熱帯区。脚で木をたたく習性あり
giant earwig		*Titanolabis colossea* (Dohrn)	(ハサミムシ目、オオハサミムシ科) 豪州区
giant earwig			common brown earwig (1) を見よ
giant emperor		*Saturnia pyri* Denis et Schiffermüller	(チョウ目、ヤママユガ科) 旧北区
giant Entheus		*Entheus matho matho* Godman et Salvin	(チョウ目、セセリチョウ科) 新熱帯区
giant Eucosma moth		*Eucosma giganteana* (Riley)	(チョウ目、ハマキガ科) 新北区。giant Eucosma ともいう
giant eyed-metalmark		*Mesosemia grandis* Druce	(チョウ目、シジミタテハ科) 新熱帯区
giant false powderpost beetle			large bostrichid を見よ
giant fish killer			giant water bug (2) を見よ
giant fishkillers			giant water bugs (1) を見よ
giant flea		*Hystrichopsylla talpae* Curtis	(ノミ目、ケブカノミ科) 旧北区
giant flower-loving flies		*Rhaphiomidas*	(ハエ目、Apioceridae) の昆虫の総称

英名	和名	学名	所属、分布、ほか
giant forest ant		*Camponotus gigas* (Latreille)	(ハチ目、アリ科) 東洋区
giant geometer	トビモンオオエダシャク	*Biston robustum* Butler	(チョウ目、シャクガ科) 日本、旧北区。カンキツ、リンゴ、チャ、ツバキなどの害虫。giant geometrid ともいう
giant glasswing		*Methona confusa* Butler	(チョウ目、タテハチョウ科) 新熱帯区
giant golden stink bug	オオキンカメムシ	*Eucorysses grandis* (Thunberg)	(カメムシ目、キンカメムシ科) 日本、東洋区
giant grasshopper		*Valanga irregularis* (Walker)	(バッタ目、バッタ科) 豪州区
giant gray moth		*Cymatophora approximaria* Hübner	(チョウ目、シャクガ科) 新北区
giant gray sphinx			giant grey sphinx を見よ
giant grayling		*Berberia abdelkader* Pierret	(チョウ目、タテハチョウ科) 旧北区
giant green slantface		*Acrida conica* (Fabricius)	(バッタ目、バッタ科) 豪州区
giant green water beetle		*Dytiscus marginicollis* LeConte	(コウチュウ目、ゲンゴロウ科) 新北区
giant grey sphinx		*Pseudosphinx tetrio* (Linnaeus)	(チョウ目、スズメガ科) 旧北区
giant hairstreak			Marsyas hairstreak を見よ
giant hairy aphid	クワナケクダアブラムシ	*Greenidia kuwanai* (Pergande)	(カメムシ目、アブラムシ科) 日本、旧北区
giant Hawaiian dragonfly		*Anax strenuus* Hagen	(トンボ目、ヤンマ科) 新北区、大洋区。giant Hawaiian darner, pinao ともいう。
giant hawk moth	オオシモフリスズメ	*Langia zenzeroides nawai* Rothschild et Jordan	(チョウ目、スズメガ科) 日本、旧北区。モモ、ウメ、サクラ害虫
giant hawker		*Tetracanthagyna plagiata* (Waterhouse)	(トンボ目、ヤンマ科) 東洋区。開帳 20 cm の最大型種
giant honey bee	オオミツバチ	*Apis dorsata* Fabricius	(ハチ目、ミツバチ科) 東洋区。大型で時に人を襲う
giant hooktip moths		Cyclidiidae	(チョウ目) の昆虫の総称
giant hornet		*Vespa mandarinia* Smith	(ハチ目、スズメバチ科) 旧北区、東洋区。日本亜種 *V. m. japonica* Radoszkowski オオスズメバチも同英名
giant hornet			hornet を見よ
giant horntail			polished horntail を見よ
giant horntail			horntail (1) を見よ
giant ichneumon wasp		*Megarhyssa macrurus* (Linnaeus)	(ハチ目、ヒメバチ科) 新北区。lunar giant ichneumon も参照
giant ichneumons		*Megarhyssa*	(ハチ目、ヒメバチ科) の昆虫の総称
giant jewel beetle		*Stenocera orissa* Buquet	(コウチュウ目、タマムシ科) エチオピア区
giant jumping plant louse	オオガタキジラミ	*Psylla magnifera* Kuwayama	(カメムシ目、キジラミ科) 日本
giant katydid	クツワムシ	*Mecopoda nipponensis* (de Haan)	(バッタ目、キリギリス科) 日本
giant katydid		*Stilpnochlora couloniana* (Saussure)	(バッタ目、キリギリス科) 新北区

英名	和名	学名	所属、分布、ほか
giant lacewing		*Osmylus fulvicephalus* (Scopoli)	(アミメカゲロウ目、ヒロバカゲロウ科) 旧北区
giant lacewings	アメリカオオアミメカゲロウ科	Polystoechotidae	(アミメカゲロウ目) の昆虫の総称 大型で新北区、新熱帯区に分布
giant ladybird	オオテントウ	*Synonycha grandis* (Thunberg)	(コウチュウ目、テントウムシ科) 日本、東洋区、豪州区
giant lappet moths	オビガ科	Eupteropidae	(チョウ目) の昆虫の総称
giant leaf crickets	コノハギス科	Phyllophoridae	(バッタ目) の昆虫の総称
giant leaf grasshopper	オオコノハギス	*Siliquofera grandis* (Blanchard)	(バッタ目、キリギリス科) 豪州区
giant leaf insect		*Phyllium giganteum* Hausleithner	(ナナフシ目、コノハムシ科) 東洋区
giant leaf roach			peppered cockroach を見よ
giant leopard moth			great leopard moth を見よ
giant locust		*Tropidacris latreillei* (Petry)	(バッタ目、バッタ科) 新熱帯区、エチオピア区
giant looper		*Ascotis selenaria reciprocaris* (Walker)	(チョウ目、シャクガ科) 旧北区、東洋区、エチオピア区。原名亜種は本英名と mugwort looper という
giant Malaysian shield mantis		*Rhombodera basalis* de Haan	(カマキリ目、カマキリ科) 東洋区
giant margarodid scale			giant mealybug を見よ
giant mayfly		*Hexagenia limbata* (Serville)	(カゲロウ目、モンカゲロウ科) 新北区。Green Bay fly とも云われる
giant mealybug	オオワラジカイガラムシ	*Drosicha corpulenta* (Kuwana)	(カメムシ目、ワタフキカイガラムシ科) 日本、旧北区。カンキツ、クリ、カシ類の害虫
giant mealybugs	トゲコナカイガラムシ科	Putoidae	(カメムシ目) の昆虫の総称
giant mesquite bug		*Thasus neocalifornicus* Brailovsky et Barrera	(カメムシ目、ヘリカメムシ科) 新北区
giant metallic ceiba borer		*Euchroma gigantea* (Linnaeus)	(コウチュウ目、タマムシ科) 新熱帯区。カポックなどの属 *Ceiba* に由来
giant Mexican metalmark		*Apodemia mejicanus maxima* (Weeks)	(チョウ目、シジミタテハ科) 新熱帯区
giant Mistletoe hairstreak		*Iolaphilus julus* Hewitson	(チョウ目、シジミチョウ科) エチオピア区
giant mole cricket			prairie mole cricket を見よ
giant mole weta		*Deinacrida talpa* Gibbs	(バッタ目、Stenopelmatidae) 豪州区
giant moth borer		*Castnia licoides* Boisduval	(チョウ目、カストニアガ科) 新熱帯区
giant moth borer			giant sugarcane borer を見よ
giant mountain beaver flea		*Hystrichopsylla schefferi* Chapin	(ノミ目、ケブカノミ科) 新北区。ビーバーにつく体長1cmの大型種
giant mountain copper		*Aloeides pallida* Tite et Dickson	(チョウ目、シジミチョウ科) エチオピア区
giant noctuid			black witch を見よ 北米での英名
giant northern termite		*Mastotermes darwiniensis* Froggatt	(シロアリ目、ムカシシロアリ科) 豪州区
giant orange sulphur			orange giant sulphur を見よ

英名	和名	学名	所属、分布、ほか
giant orange-tip		*Colotis lucasi* (Grandidier)	(チョウ目、シロチョウ科) エチオピア区
giant owl			owl butterfly (1) を見よ
giant palm-boring beetle			California palm borer を見よ
giant paper wasp		*Polistes gigas* Kirby	(ハチ目、スズメバチ科) 東洋区
giant petaltail		*Petalura ingentissima* Tillyard	(トンボ目、ムカシヤンマ科) 豪州区
giant pine weevil		*Eurhamphus fasciculatus* Shuckard	(コウチュウ目、ゾウムシ科) 豪州区
giant predatory click beetle		*Paracalais gibboni* (Newman)	(コウチュウ目、コメツキムシ科) 豪州区
giant prickly stick insect		*Extatosoma tiaratum* (Macleay)	(ナナフシ目、ナナフシ科) 東洋区、豪州区
giant purple empress			Japanese emperor を見よ
giant redeye	オオコウモリセセリ	*Gangara thyrsis* (Fabricius)	(チョウ目、セセリチョウ科) 東洋区
giant red eyed African silkmoth		*Aurivillius triamis* Rothschild	(チョウ目、ヤママユガ科) エチオピア区
giant-reed skipper			Yuma skipper を見よ
giant resin bee		*Megachile sculpturalis* Smith	(ハチ目、ハキリバチ科) 日本、全北区
giant roadside-skipper		*Amblyscirtes raphaeli* Freeman	(チョウ目、セセリチョウ科) 新熱帯区
giant robber fly		*Proctacanthus occidentalis* Hine	(ハエ目、ムシヒキアブ科) 新北区
giant Sakhalin fir aphid	トドマツオオアブラムシ	*Cinara todocola* Inouye	(カメムシ目、アブラムシ科) 日本、旧北区
giant salmon fly			giant stonefly (1) を見よ
giant sand treader camel cricket		*Daihinibaenetes giganteus* Tinkham	(バッタ目、Rhaphidophoridae) 新北区
giant saturn	オオムラサキトガリバワモン	*Zeuxidia aurelius aurelius* (Cramer)	(チョウ目、タテハチョウ科) 東洋区
giant scale insects			giant coccids を見よ
giant scales			giant coccids を見よ
giant sicklewing		*Achlyodes busirus heros* Ehrmann	(チョウ目、セセリチョウ科) 新熱帯区
giant silk moth	クスサン	*Caligula japonica* (Moore)	(チョウ目、ヤママユガ科) 日本、旧北区。果樹、サクラ、カエデ、カシ類、クスノキなどの害虫
giant silkmoth			Chinese oak silkmoth を見よ
giant silkmoth			tulip-tree silkmoth を見よ
giant silkworm moth			giant emperor を見よ
giant silkworm moths	ヤママユガ科	Saturniidae	(チョウ目) の昆虫の総称　世界に1,100種
giant silkworms			giant silkworm moths を見よ
giant snail beetle		*Tefflus zanzibaricus alluaudi* (Sternberg)	(コウチュウ目、オサムシ科) 大洋区。

英名	和名	学名	所属、分布、ほか
giant South American grasshopper		*Tropidacris violaceus* Ucksmucks	（バッタ目、バッタ科）新熱帯区。体長 12 cm の大型種
giant sphinx		*Cocytius antaeus* (Drury)	（チョウ目、スズメガ科）旧北区。giant sphinx moth ともいう
giant spider-hunting wasps			tarantula hawks を見よ
giant spiny stick insect			New Guinea spiny stick insect を見よ
giant spreadwing		*Archilestes grandis* (Rambur)	（トンボ目、アオイトトンボ科）新北区、新熱帯区
giant stag beetle			American stag beetle を見よ
giant stick bug		*Phobaeticus serratipes* (Gray)	（ナナフシ目、ナナフシ科）東洋区
giant stick insect		*Argosarchus horridus* (White)	（ナナフシ目、ナナフシ科）豪州区
giant stoneflies		Pteronarcyoidea	（カワゲラ目）の昆虫の総称
giant stoneflies (1)		Pteronarcidae	（カワゲラ目）の昆虫の総称
giant stonefly (1)		*Pteronarcys californica* Newport	（カワゲラ目、Pteronarcidae）新北区。体長 5 cm、開帳 9.5 cm の大型種
giant stonefly (2)		*Pteronarcys dorsalis* Say	（カワゲラ目、Pteronarcidae）新北区。開帳 10.6 cm
giant sugarcane borer		*Castnia licus* (Drury)	（チョウ目、カストニアガ科）新熱帯区
giant sugarcane moth borer			giant moth borer を見よ
giant sulphur	ヤナギモンキチョウ	*Colias gigantea* Strecker	（チョウ目、シロチョウ科）新北区
giant swallowtail	オオタスキアゲハ	*Papilio cresphontes* Cramer	（チョウ目、アゲハチョウ科）新北区。giant swallowtail butterfly ともいう
giant swallowtails		Heraclides	（チョウ目、アゲハチョウ科）の昆虫の総称
giant tadpole killers			giant water bugs (1) を見よ
giant termite			giant northern termite を見よ
giant Texas katydid			greater arid-land katydid を見よ
giant three-horned beetle			Caucasus beetle を見よ
giant thrips		*Acanthinothrips spectrum* (Haliday)	（アザミウマ目、Idolothripidae）豪州区。1836 年にダーウインにより発見された黒色大型種
giant tiger moth		*Platarctia parthenos* (Harris)	（チョウ目、ヒトリガ科）新北区
giant torrent midge		*Edwardsina gigantea* Zwick	（ハエ目、アミカ科）豪州区。絶滅危惧種
giant tropical ant			bullet ant を見よ
giant tropical damselfly			helicopter damselfly を見よ
giant walkingstick		*Megaphasma denticrus* (Stål)	（ナナフシ目、Bacunculidae）新北区。北米最大のナナフシ
giant waraji coccus			giant mealybug を見よ

英名	和名	学名	所属、分布、ほか
giant water bug	タガメ	*Lethocerus deyrollei* (Vuillefroy)	（カメムシ目、コオイムシ科）日本、旧北区、東洋区。日本最大の水生カメムシ目の種。幼虫は水生昆虫、魚などを捕食
giant water bug		*Lethocerus maximus* De Carlo	（カメムシ目、コオイムシ科）新熱帯区
giant water bug (1)		*Lethocerus americanus* (Leidy)	（カメムシ目、コオイムシ科）新北区。体長5〜6 cm
giant water bug (2)	タイワンタガメ	*Lethocerus indicus* (Lepeletier et Serville)	（カメムシ目、コオイムシ科）日本、東洋区
giant water bugs (1)		*Lethocerus*	（カメムシ目、コオイムシ科）の昆虫の総称
giant water bugs (2)	コオイムシ科	Belostomatidae	（カメムシ目）の昆虫の総称　熱帯に10 cmをこえる種あり
giant water bugs (3)		*Belostoma*	（カメムシ目、コオイムシ科）の昆虫の総称
giant water scavenger beetle			giant black water beetle を見よ
giant water scavenger beetles		*Hydrophilus*	（コウチュウ目、ガムシ科）の昆虫の総称
giant western dragonfly			western petaltail を見よ
giant weta			weta-punga を見よ
giant wheel bug			wheel bug を見よ
giant white		*Ganyra josephina* (Godart)	（チョウ目、シロチョウ科）新北区、新熱帯区
giant white butterfly			Siam tree nymph を見よ
giant whitefly		*Aleurodicus dugesii* Cockerell	（カメムシ目、コナジラミ科）新北区、新熱帯区
giant wild silk worm			atlas moth を見よ
giant willow aphid			large willow aphid を見よ
giant wood moth			witjuti grub を見よ
giant wood roach			peppered cockroach を見よ
giant wood wasp			horntail (1) を見よ
Giffard's whitefly	ヒメコナジラミ	*Bemisia giffardi* (Kotinsky)	（カメムシ目、コナジラミ科）日本、東洋区、豪州区、大洋区、新北区。Giffard whitefly ともいう。カンキツ害虫
Gifu sawfly	コシアキハバチ	*Tenthredo gifui* Marlatt	（ハチ目、ハバチ科）日本
gigantic water-veneer		*Schoenobius gigantellus* (Denis et Schiffermüller)	（チョウ目、メイガ科）旧北区
gigas metalmark		*Pachythone gigas gigas* Godman et Salvin	（チョウ目、シジミタテハ科）新熱帯区
Gila shieldback		*Platyoplus gilaensis* Tinkham	（バッタ目、キリギリス科）新北区
Gilbert's agave skipper			West Texas giant skipper を見よ
Gilbert's blue		*Candalides gilberti* Waterhouse	（チョウ目、シジミチョウ科）豪州区。Gilbert's blue butterfly ともいう
Gilbert's flasher		*Astraptes gilberti* Freeman	（チョウ目、セセリチョウ科）新北区
Gilbert's flasher		*Astraptes alector hopfferi* (Plötz)	（チョウ目、セセリチョウ科）新熱帯区

英名	和名	学名	所属、分布、ほか
gilded fly		*Chrysotoxum arcuatum* (Linnaeus)	(ハエ目、ハナアブ科) 旧北区
gilded Presba		*Syncordulia legator* Dijkstra, Samways et Simaika	(トンボ目、エゾトンボ科) エチオピア区
Gillette felt scale		*Eriococcus gillettei* Tinsley	(カメムシ目、フクロカイガラムシ科) 新北区
Gillette's checkerspot		*Euphydryas gillettii* (Barnes)	(チョウ目、タテハチョウ科) 新北区
Gillette's shieldback		*Plagiostira gillettei* Caudell	(バッタ目、キリギリス科) 新北区
gilt ground beetle			golden ground beetle を見よ
gilt May pigmy			glossy hawthorn pigmy を見よ
gingelly borer moth	ホソトガリノメイガ	*Antigastra catalaunalis* (Duponchel)	(チョウ目、メイガ科) 日本、全北区、東洋区、エチオピア区、新熱帯区
ginger maggot		*Eumerus figurans* Walker	(ハエ目、ハナアブ科) 新北区
ginger-rhizome gnat	ショウガクロバネキノコバエ	*Phytosciara zingiberis* Sasakawa	(ハエ目、クロバネキノコバエ科) 日本。ショウガ害虫
ginger robberfly	キンホソイシアブ	*Choerades gilvus* (Linnaeus)	(ハエ目、ムシヒキアブ科) 日本、全北区
ginger stable-flies		*Protohystricia*	(ハエ目、ヤドリバエ科) の昆虫の総称
ginkgo borer	イチョウヒゲビロウドカミキリ	*Acalolepta ginkgovora* Makihara	(コウチュウ目、カミキリムシ科) 日本。イチョウ害虫
gipsy			gypsy moth を見よ
gipsy cuckoobee		*Psithyrus bohemicus* (Seidl)	(ハチ目、ミツバチ科) 旧北区
gipsy moth			gypsy moth を見よ
Girard's grass-veneer moth		*Crambus girardellus* Clemens	(チョウ目、メイガ科) 新北区
girdle worm			garden grass-veneer を見よ
girdler			樹皮を輪状に食う昆虫
girlfriend underwing moth		*Catocala amica* (Hübner)	(チョウ目、ヤガ科) 新北区。girlfriend underwing ともいう
gisborne cockroach		*Drymaplaneta semivitta* (Walker)	(ゴキブリ目、ゴキブリ科) 豪州区
Gisella sailor	ルリアカガネタテハ	*Dynamine gisella* Hewitson	(チョウ目、タテハチョウ科) 新熱帯区
glacier flea		*Isotoma saltans* (Nicolet)	(トビムシ目、ツチトビムシ科) 旧北区
glad-eye bushbrown		*Mycalesis patnia* Moore	(チョウ目、タテハチョウ科) 東洋区
gladiator meadow katydid		*Orchelimum gladiator* Bruner	(バッタ目、キリギリス科) 新北区
gladiator short-winged katydid		*Dichopetala gladiator* Rehn et Hebard	(バッタ目、キリギリス科) 新北区
gladiolus thrips	グラジオラスアザミウマ	*Taeniothrips simplex* (Morrison)	(アザミウマ目、アザミウマ科) 日本、全北区、豪州区。グラジオラス、キク、アマリリスの害虫
gladstone grasshopper		*Melanoplus gladstoni* Scudder	(バッタ目、バッタ科) 新北区
Glandon blue			arctic blue を見よ

英名	和名	学名	所属、分布、ほか
Glanville fritillary	アトグロヒョウモンモドキ	*Melitaea cinxia* (Linnaeus)	(チョウ目、タテハチョウ科) 旧北区。チョウ愛好の Glanville 夫人に由来
Glaser's Dichomeris moth		*Dichomeris bolize* Hodges	(チョウ目、キバガ科) 新北区
glasshouse and potato aphid			potato aphid (2) を見よ
glasshouse aphid			potato aphid (2) を見よ
glasshouse black sciarid		*Bradysia impatiens* (Johannsen)	(ハエ目、クロバネキノコバエ科) 豪州区
glasshouse camel-cricket			greenhouse camel cricket を見よ
glasshouse leafhopper		*Empoasca pallidifrons* Edwards	(カメムシ目、オオヨコバイ科) 旧北区
glasshouse leafhopper		*Hauptidia maroccana* (Melichar)	(カメムシ目、オオヨコバイ科) 旧北区
glasshouse mealybug			tuber mealybug を見よ
glasshouse mealybug			grape mealybug を見よ
glasshouse orthezia			Jacarand bug を見よ
glasshouse potato aphid			potato aphid (2) を見よ
glasshouse striped sciarid		*Bradysia tritici* (Coquillett)	(ハエ目、クロバネキノコバエ科) 豪州区
glasshouse thrips			greenhouse thrips を見よ
glasshouse tomato moth			bright-line brown-eye を見よ　米国での英名
glasshouse whitefly			greenhouse whitefly を見よ
glasshouse whitefly parasite			greenhouse whitefly parasite を見よ
glasshouse wing-spot flies		*Scatella*	(ハエ目、ミギワバエ科) の昆虫の総称
glasswing	クロホシホソチョウ	*Acraea andromacha* (Fabricius)	(チョウ目、タテハチョウ科) 豪州区
glasswing metalmark		*Ithomeis aurantiaca* H. Bates	(チョウ目、シジミタテハ科) 新熱帯区
glasswing moth			glasswing を見よ
glasswings		*Ornipholidotos*	(チョウ目、シジミチョウ科) の昆虫の総称
glassworm			glow-worm (2) を見よ
glassworms			fireflies (1) を見よ
glassy Acraea	ヘリブトホソチョウ	*Acraea sambavae* Ward	(チョウ目、タテハチョウ科) エチオピア区
glassy bluebottle	タイワンタイマイ	*Graphium cloanthus* (Westwood)	(チョウ目、アゲハチョウ科) 東洋区
glassy cutworm			glassy cutworm moth を見よ　北米での英名
glassy cutworm moth		*Apamea devastator* (Brace)	(チョウ目、ヤガ科) 新北区。幼虫は多様な植物につく
glassy Graphium			Westwood's white-lady を見よ
glassy star scale		*Vinsonia stellifera* (Westwood)	(カメムシ目、カタカイガラムシ科) 新熱帯区

英名	和名	学名	所属、分布、ほか
glassy tiger	ヒメコモンアサギマダラ	*Parantica aglea maghaba* (Fruhstorfer)	(チョウ目、タテハチョウ科) 日本、東洋区。*P. aglea* Linnaeus の英名
glassy-winged sharpshooter		*Homalodisca viripennis* (Germar)	(カメムシ目、オオヨコバイ科) 新北区
glassy-winged skipper		*Xenophanes tryxus* (Stoll)	(チョウ目、セセリチョウ科) 新北区、新熱帯区
glassy-winged toothpick grasshopper		*Stenacris vitreipennis* (Marshall)	(バッタ目、バッタ科) 新北区
glassy-wings			clearwing moths を見よ
Glauce leafwing		*Memphis glauce* (C. et R. Felder)	(チョウ目、タテハチョウ科) 新熱帯区
Glaucoma metalmark		*Periplacis glaucoma isthmica* (Godman et Salvin)	(チョウ目、シジミタテハ科) 新熱帯区
glaucous cracker		*Hamadryas glauconome* (Bates)	(チョウ目、タテハチョウ科) 新北区
glaucous leaf weevil		*Phyllobius calcaratus* (Fabricius)	(コウチュウ目、ゾウムシ科) 旧北区
glaucous shears	タカネハイイロヨトウ	*Papestra biren* (Goeze)	(チョウ目、ヤガ科) 日本、旧北区
glazed oakblue		*Arhopala paralea* (Evans)	(チョウ目、シジミチョウ科) 東洋区
glazed Pellicia		*Pellicia arina* Evans	(チョウ目、セセリチョウ科) 新北区、新熱帯区
glazed skipper			glazed Pellicia を見よ
glazed tufted-skipper			glazed Pellicia を見よ
Glehn's spruce aphid			spruce shoot aphid を見よ
Gleichen's dwarf	ギンモンクサモグリガ	*Elachista gleichenella* (Fabricius)	(チョウ目、クサモグリガ科) 日本、旧北区
Glenmore awl-fly		*Xylophagus junki* (Szilady)	(ハエ目、キアブ科) 旧北区
glevum bentwing		*Dryadaula pactolia* Meyrick	(チョウ目、ヒロズコガ科) 旧北区
glided river cruiser		*Macromia pacifica* Hagen	(トンボ目、ヤマトンボ科) 新北区
gliders (1)		*Tramea*	(トンボ目、トンボ科) の昆虫の総称
gliders		*Hydrobasileus*	(トンボ目、トンボ科) の昆虫の総称
gliders		*Pantala*	(トンボ目、トンボ科) の昆虫の総称
gliders			sailers を見よ　英国での英名
glistening Cerulean		*Jamides elpis* (Godart)	(チョウ目、シジミチョウ科) 東洋区。*J. e. pseudelpis* Butler も同英名
glistening Cerulean		*Lampides kankena* (Felder)	(チョウ目、シジミチョウ科) 東洋区
glistening Cerulean (1)	ルリムラサキタテハモドキ (マダガスカルタテハモドキ)	*Jamides caerulea* (Druce)	(チョウ目、シジミチョウ科) 東洋区
glistening demoiselle		*Phaon iridipennis* (Burmeister)	(トンボ目、イトトンボ科) エチオピア区
glistening line-blue		*Sahulana scintillata* (Lucas)	(チョウ目、シジミチョウ科) 豪州区。glistening blue ともいう
glistening pearl-white		*Elodina queenslandica* de Baar et Hancock	(チョウ目、シロチョウ科) 豪州区

英名	和名	学名	所属、分布、ほか
glistening rustic			Miranda moth を見よ
glittering demoiselle		*Calopteryx exul* Selys	(トンボ目、イトトンボ科) エチオピア区
glittering Magdalena moth		*Daulia magdalena* (Fernald)	(チョウ目、メイガ科) 新北区
glittering mantle		*Caria sponsa* (Staudinger)	(チョウ目、シジミタテハ科) 新熱帯区
glittering sapphire		*Lasaia agesilas* (Latreille)	(チョウ目、シジミタテハ科) 新北区、新熱帯区。*L. a. callania* Clench は shining-blue Lasaia という
globe artichoke flower aphid			thistle aphid を見よ
globe artichokes		*Oidaematophorus phoebus* Barnes et Lindsey	(チョウ目、トリバガ科) 新北区
globe-bearing treehopper		*Bocydium globulare* (Fabricius)	(カメムシ目、ツノゼミ科) 新熱帯区
globe flies			beetle flies を見よ
globemallow leaf beetle		*Calligrapha serpentina* (Rogers)	(コウチュウ目、ハムシ科) 新北区
globe skimmer			globe trotter を見よ
globe trotter	ウスバキトンボ	*Pantala flavescens* (Fabricius)	(トンボ目、トンボ科) 日本。汎世界。世界漫遊家という英名は洋上を長距離飛翔することに由来
globose bamboo fringed scale			hemispherical scale (1) を見よ
globose scale		*Sphaerolecanium prunastri* (Fonscolombe)	(カメムシ目、カタカイガラムシ科) 全北区
globose spider beetle			museum beetle (1) を見よ
globular planthopper	マルウンカ	*Gergithus variabilis* (Butler)	(カメムシ目、マルウンカ科) 日本、旧北区、東洋区
globular refuse beetle	マルガタゴミムシ	*Amara chalcites* Dejean	(コウチュウ目、オサムシ科) 日本、旧北区
globular spider beetle		*Trigonogenius globulum* Solier	(コウチュウ目、シバンムシ科) 新北区
globular springtails	マルトビムシ科	Sminthuridae	(トビムシ目) の昆虫の総称
globular stink bug	マルカメムシ	*Megacopta punctatissima* (Montandon)	(カメムシ目、マルカメムシ科) 日本、旧北区
globular stink bug			rice bug (3) を見よ
globular treehopper	マルツノゼミ	*Gargara genistae* (Fabricius)	(カメムシ目、ツノゼミ科) 日本、旧北区
gloomy scale		*Melanaspis tenebricosa* (Comstock)	(カメムシ目、マルカイガラムシ科) 新北区
gloomy Temnora			grey Temnora を見よ
gloomy underwing			Andromeda underwing moth を見よ
glorious beetle		*Chrysina gloriosa* (LeConte)	(コウチュウ目、コガネムシ科) 新北区。glorious scarab ともいう
glorious Begum		*Agatasa calydonia* (Hewitson)	(チョウ目、タテハチョウ科) 東洋区
glorious blue-skipper		*Paches loxus gloriosus* Rober	(チョウ目、セセリチョウ科) 新熱帯区

英名	和名	学名	所属、分布、ほか
glorious checkerspot	ベニモンヒョウモンモドキ	*Chlosyne janais gloriosa* Bauer	（チョウ目、タテハチョウ科）新熱帯区
glorious Habrosyne moth		*Habrosyne gloriosa* (Guenée)	（チョウ目、カギバガ科）新北区
glorious purplewing	ルリフチアイイロタテハ	*Eunica sophonisba* Cramer	（チョウ目、タテハチョウ科）新熱帯区
glorious squash vine borer moth			vine borer を見よ
glorious velvet ant		*Dasymutilla gloriosa* (Saussure)	（ハチ目、アリバチ科）新北区
glory moths		Endromidae	（チョウ目）の昆虫の総称
glossy black Idia moth			dried leaf moth を見よ
glossy-brown case		*Coleophora violacea* (Strøm)	（チョウ目、ツツミノガ科）旧北区
glossy brown moth		*Bityla defigurata* Walker	（チョウ目、ヤガ科）豪州区
glossy hawthorn pigmy		*Stigmella hybnerella* (Hübner)	（チョウ目、モグリチビガ科）旧北区
glossy shield bug			predatory shield bug (1) を見よ
Glover scale	ミカンナガカキカイガラムシ	*Lepidosaphes gloverii* (Packard)	（カメムシ目、マルカイガラムシ科）日本、汎世界。カンキツ害虫
Glover's scale			Glover scale を見よ　豪州での英名
Glover's silk moth		*Hyalophora gloveri* (Strecker)	（チョウ目、ヤママユガ科）新北区
glow-worm	オーストラリアヒカリキノコバエ	*Arachnocampa richardsae* Harrison	（ハエ目、キノコバエ科）豪州区。幼虫が発光
glow-worm (1)	ニュージーランドヒカリキノコバエ	*Arachnocampa luminosa* Skuse	（ハエ目、キノコバエ科）豪州区。幼虫が発光
glow-worm (2)		*Lampyris noctiluca* Linnaeus	（コウチュウ目、ホタル科）旧北区
glowworm (3)		*Zarhipis integripennis* LeConte	（コウチュウ目、Phengodidae）新北区。雌成虫と幼虫は発光
glowworm			glow-worm (2) を見よ
glowworm beetles		Phengodidae	（コウチュウ目）の昆虫の総称　雌は発光する。南北米に約200種
glow-worm beetles		*Zarhipis*	（コウチュウ目、Phengodidae）の昆虫の総称
glow worms		*Arachnocampa*	（ハエ目、キノコバエ科）の昆虫の総称　幼虫が発光するヒカリキノコバエ類
glow-worms			fireflies (1) を見よ　英国での英名
glowing skipper		*Anatrytone mazai* (Freeman)	（チョウ目、セセリチョウ科）新北区、新熱帯区
Glycera glasswing		*Napeogenes glycera* Godman	（チョウ目、タテハチョウ科）新熱帯区
Glycera longwing		*Dione glycera* (C. et R. Felder)	（チョウ目、タテハチョウ科）新熱帯区
glyphipterigid moths			sedge moths を見よ
Gmelin's banded-skipper			twin-spot banded skipper を見よ

英名	和名	学名	所属、分布、ほか
gnat			血を吸う小さな羽虫。米国ではヌカカ、ユスリカまたはブヨを指し、英国ではイエカを指す
gnat			common mosquito を見よ
gnat bugs	クビナガカメムシ科	Enicocephalidae	（カメムシ目）の昆虫の総称　捕食性
gnats			midges を見よ　豪州での英名
gnats			mosquitoes (1) を見よ
gnats			malaria mosquitoes を見よ
gnats and mosquitoes			mosquitoes (1) を見よ
gnawer beetles			hide beetles (2) を見よ
gnawing barklice		Polypsocidae	（チャタテムシ目）の昆虫の総称
gnawing beetles		Temnochilidae	（コウチュウ目）の昆虫の総称
gnawing beetles (1)	コクヌスト科	Trogositidae	（コウチュウ目）の昆虫の総称
Gnidus metalmark			spider-wing cupid を見よ
Gnosia hairstreak		*Gargina gnosia* (Hewitson)	（チョウ目、シジミチョウ科）新熱帯区
goat			goat moth (1) を見よ
goat biting louse			goat louse を見よ　北米での英名
goat biting louse			reddish chewing louse of sheep and goats を見よ
goat biting louse			Angora goat biting louse (1) を見よ　goat-biting louse とも記す
goat-chafers			long-horned beetles を見よ
goat chewing louse			goat louse を見よ
goat louse	ヤギハジラミ	*Damalinia caprae* (Gurlt)	（ハジラミ目、ケモノハジラミ科）日本、汎世界。ヤギに寄生
goat louse			Angora goat biting louse (1) を見よ
goat louse			sucking goat louse を見よ
goat moth (1)		*Cossus cossus* (Linnaeus)	（チョウ目、ボクトウガ科）旧北区。日本亜種は *C. c. orientalis* Gaede オオボクトウ
goat moth (2)	ボクトウガ	*Cossus jezoensis* (Matsumura)	（チョウ目、ボクトウガ科）日本、旧北区。幼虫が果樹、樹木に穿入
goat moth			carpenter moth (1) を見よ
goat moths			carpenter moths (1) を見よ
goat sallow moth		*Homoglaea hircina* Morrison	（チョウ目、ヤガ科）新北区
goat sheard aphid		*Brachycaudus tragopogonia* Kaltenbach	（カメムシ目、アブラムシ科）旧北区
goat sucking louse			sucking goat louse を見よ
goatweed butterfly		*Anaea andria* Scudder	（チョウ目、タテハチョウ科）新北区
goatweed leafwing			goatweed butterfly を見よ
Godart's Agrias	クラウデイアミイロタテハ	*Agrias claudina* (Godart)	（チョウ目、タテハチョウ科）新熱帯区
Godart's Altinote			lamplight Altinote を見よ
Godart's blue ringlet		*Chloreuptychia herseis* (Godart)	（チョウ目、タテハチョウ科）新熱帯区

英名	和名	学名	所属、分布、ほか
Godart's cycadian		*Eumaeus godartii* (Boisduval)	(チョウ目、シジミチョウ科) 新熱帯区
Godart's groundstreak		*Calycopis calus* (Godart)	(チョウ目、シジミチョウ科) 新熱帯区
Godart's metalmark		*Esthemopsis pherephatte pherephatte* (Godart)	(チョウ目、シジミタテハ科) 新熱帯区
Godart's numberwing		*Callicore pygas* (Godart)	(チョウ目、タテハチョウ科) 新熱帯区
Godart's purplewing		*Eunica sydonia* Godart	(チョウ目、タテハチョウ科) 新熱帯区
Godart's sailor		*Dynamine ines* Godart	(チョウ目、タテハチョウ科) 新熱帯区
Godart's white		*Ganyra phaloe* (Godart)	(チョウ目、シロチョウ科) 新熱帯区。*G. p. tiburtia* (Fruhstorfer) も同英名
Godma tufted-skipper		*Nisoniades godma* Evans	(チョウ目、セセリチョウ科) 新熱帯区
Godman's bent-skipper		*Helias godmani* (Mabille et Boullet)	(チョウ目、セセリチョウ科) 新熱帯区
Godman's Euselasia		*Euselasia eucrates leucorrhoa* (Godman et Salvin)	(チョウ目、シジミタテハ科) 新熱帯区
Godman's leaf beetle		*Metaxyonyha godmani* Jacoby	(コウチュウ目、ハムシ科) 新北区
Godman's mapwing		*Hypanartia godmanii* (H. Bates)	(チョウ目、タテハチョウ科) 新熱帯区
Godman's metalmark		*Behemothia godmanii* (Dewitz)	(チョウ目、シジミタテハ科) 新熱帯区
Godman's Sarota		*Sarota myrtea* Godman et Salvin	(チョウ目、シジミタテハ科) 新熱帯区
Godman's sister		*Adelpha seriphia* (C. et R. Felder)	(チョウ目、タテハチョウ科) 新熱帯区。*A. s. godmani* Fruhstorfer も同英名
Godman's sister (1)		*Adelpha erymanthis* Godman et Salvin	(チョウ目、タテハチョウ科) 新熱帯区
Godman's skipper		*Zariaspes mythecus* Godman	(チョウ目、セセリチョウ科) 新熱帯区
Godman's skipperling		*Dalla lalage* (Godman)	(チョウ目、セセリチョウ科) 新熱帯区
Goeden's shieldback		*Idiostatus goedeni* Rentz	(バッタ目、キリギリス科) 新北区
goerids		Goerinae	(トビケラ目、エグリトビケラ科) の昆虫の総称 northern caddisflies を見よ
goggle eye			wood nymph を見よ
gold-banded Cydosia			straight-lined Cydosia moth を見よ
gold-banded Etiella moth			lima-bean pod borer を見よ
gold-banded Euphaedra			gold-banded forester を見よ
gold-banded forester		*Euphaedra neophron* (Hopffer)	(チョウ目、タテハチョウ科) エチオピア区
gold-barred argent		*Argyresthia pygmaeella* Hübner	(チョウ目、メムシガ科) 旧北区
gold-barred basil dwarf		*Stephensia brunnichella* (Linnaeus)	(チョウ目、クサモグリガ科) 旧北区

英名	和名	学名	所属、分布、ほか
gold-barred beech pigmy		*Nepticula hemargyrella* Kollar	（チョウ目、モグリチビガ科）旧北区
gold-barred beech pigmy (1)		*Stigmella tityrella* (Stainton)	（チョウ目、モグリチビガ科）旧北区
gold-barred rose pigmy		*Stigmella centifoliella* (Zeller)	（チョウ目、モグリチビガ科）旧北区
gold beetle			コガネ虫（一般に前翅が金色の甲虫の俗称）
gold-bent midget		*Lithocolletis roboris* Zeller	（チョウ目、ホソガ科）旧北区
gold-bordered hairstreak		*Rekoa palegon* (Cramer)	（チョウ目、シジミチョウ科）新熱帯区
gold-brindled purple	キンマダラスイコバネ	*Eriocrania sparrmannella* (Bosc)	（チョウ目、スイコバネガ科）日本、旧北区
gold bug		*Callichroma holochlora* Bates	（コウチュウ目、カミキリムシ科）新熱帯区
gold bug (1)		*Metriona propinqua* (Boheman)	（コウチュウ目、ハムシ科）新熱帯区
gold bug			beet tortoise beetle (2) を見よ
goldbug			golden tortoise beetle を見よ
gold bugs			sweetpotato beetles を見よ
gold bugs			tortoise beetles (1) を見よ
goldcap moss-eater moth		*Epimartyria auricrinella* Walsingham	（チョウ目、コバネガ科）新北区
gold chafer			rose chafer (1) を見よ
gold-costa skipper			caicus skipper を見よ
gold-dot skipper		*Eutychide paria* (Plötz)	（チョウ目、セセリチョウ科）新熱帯区
gold-dotted case		*Goniodoma limoniella* (Stainton)	（チョウ目、ツツミノガ科）旧北区
gold-dotted slender	タデキボシホソガ	*Calybites phasianipenella* (Issiki)	（チョウ目、ホソガ科）日本、旧北区、東洋区
gold dotted Temnora		*Basiothia aureata* (Karsch)	（チョウ目、スズメガ科）エチオピア区
gold drop Helicopis		*Helicopis cupido* (Linnaeus)	（チョウ目、シジミタテハ科）新熱帯区
gold-dust weevil		*Hypomeces squamosus* (Fabricius)	（コウチュウ目、ゾウムシ科）東洋区
gold-edged giant owl		*Caligo atreus* (Kollar)	（チョウ目、タテハチョウ科）新熱帯区
gold-edged owl-butterfly			owl butterfly (2) を見よ
gold-flecked sootywing		*Bolla atahuallpai* (Lindsey)	（チョウ目、セセリチョウ科）新熱帯区
gold fly			green botfly を見よ
gold-four argent	ズミメムシガ	*Argyresthia ivella* (Haworth)	（チョウ目、メムシガ科）日本、旧北区
gold fringe		*Hypsopygia costalis* (Fabricius)	（チョウ目、メイガ科）全北区。gold fringe moth ともいう
gold-fringed borer			gold-fringed rice borer を見よ
gold-fringed rice borer		*Chilo auricilius* (Dudgeon)	（チョウ目、メイガ科）東洋区。イネ害虫

英名	和名	学名	所属、分布、ほか
gold-fringed snout moth		*Asturodes fimbriauralis* Guenée	(チョウ目、メイガ科) 新熱帯区
gold-fringed tabby		*Pyralis costalis* (Fabricius)	(チョウ目、メイガ科) 旧北区、新北区
gold-fringed tabby			gold fringe を見よ
gold-headed metalmark		*Caria stillaticia* Dyar	(チョウ目、シジミタテハ科) 新熱帯区
gold-hunter's hairstreak	クリイロカラスシジミ	*Satyrium auretorum* (Boisduval)	(チョウ目、シジミチョウ科) 新北区、新熱帯区。*S. a. spadix* (H. Edwards) も同英名
gold juniper argent		*Argyresthia aurulentella* Stainton	(チョウ目、メムシガ科) 旧北区
gold-lined Melanomma moth		*Melanomma auricinctaria* Grote	(チョウ目、ヤガ科) 新北区
gold moth		*Basilodes pepita* Guenée	(チョウ目、ヤガ科) 新北区
gold moths		Axiidae	(チョウ目) の昆虫の総称
gold-rayed skipperling		*Dalla ramirezi* Freeman	(チョウ目、セセリチョウ科) 新熱帯区
gold rowan argent		*Argyresthia sorbiella* (Treitschke)	(チョウ目、メムシガ科) 旧北区
gold spangle		*Autographa bractea* (Denis et Schiffermüller)	(チョウ目、ヤガ科) 旧北区
gold spangle			gold spot を見よ
gold spot	イネキンウワバ	*Plusia festucae* (Linnaeus)	(チョウ目、ヤガ科) 日本、旧北区。イネ害虫。成虫前翅の金白色紋に由来
gold-spot Aguna		*Aguna asander* (Hewitson)	(チョウ目、セセリチョウ科) 新北区。gold-spotted Aguna ともいう
goldspot skipperling		*Dalla cypselus* (C. et R. Felder)	(チョウ目、セセリチョウ科) 新熱帯区
gold-spotted ghost moth		*Sthenopis auratus* (Grote)	(チョウ目、コウモリガ科) 新北区
gold-spotted oak borer	オオスイコバネ	*Agrilus coxalis* Waterhouse	(コウチュウ目、タマムシ科) 新北区
gold spotted sylph		*Metisella metis* (Linnaeus)	(チョウ目、セセリチョウ科) エチオピア区。gold-spotted sylph とも記す
gold-spotted woodrust dwarf			wood-rush dwarf を見よ
gold-sprinkled purple		*Eriocrania semipurpurella* (Stephans)	(チョウ目、スイコバネガ科) 日本、旧北区
gold-stained satyr		*Cissia pseudoconfusa* Singer, DeVries et Ehrlich	(チョウ目、タテハチョウ科) 新熱帯区
gold-stripe grass-veneer moth		*Microcrambus biguttellus* Forbes	(チョウ目、メイガ科) 新北区
gold-striped leaftier moth		*Machimia tentoriferella* Clemens	(チョウ目、マルハキバガ科) 新北区
gold-striped tortoise beetle		*Cassida nobilis* Linnaeus	(コウチュウ目、ハムシ科) 旧北区
gold swift	キンスジコウモリ	*Phymatopus hecta* (Linnaeus)	(チョウ目、コウモリガ科) 日本、旧北区
gold swift			ghost moth を見よ
goldtail		*Allocnemis leucosticta* Selys	(トンボ目、モノサシトンボ科) エチオピア区

英名	和名	学名	所属、分布、ほか
gold-tail			browntail moth (1)(2) を見よ　英国での英名
goldtail moth			browntail moth (2) を見よ　米国での英名。gold-tailed moth ともいう
gold-tail moth		*Porthesia auriflua* Fabricius	（チョウ目、ドクガ科）旧北区
goldtipped tubular thrips		*Heplothrips gowdeyi* (Franklin)	（アザミウマ目、クダアザミウマ科）豪州区
gold triangle			gold fringe を見よ
gold-tufted skipper		*Typhedanus ampyx* (Godman et Salvin)	（チョウ目、セセリチョウ科）新熱帯区
gold-washed skipper		*Vinius tryhana tryhana* (Kaye)	（チョウ目、セセリチョウ科）新熱帯区
gold wasps			cuckoo wasps (1) を見よ
goldwing		*Synthymia fixa* (Fabricius)	（チョウ目、ヤガ科）旧北区
goldwing moth			goldwing を見よ
golden albatross	オウゴントガリシロチョウ	*Appias aurosa* Yata, Chainey et Vane-Wright	（チョウ目、シロチョウ科）東洋区
golden amberwing		*Perithemis electra* Ris	（トンボ目、トンボ科）新熱帯区
golden angle		*Caprona ransonnetii* (Felder)	（チョウ目、セセリチョウ科）東洋区、エチオピア区
golden ant		*Polyrhachis ornata* Mayr	（ハチ目、アリ科）豪州区
golden ant-blue		*Acrodipsas aurata* (Sands)	（チョウ目、シジミチョウ科）豪州区
golden apple beetles			leaf beetles (1) を見よ　この英名の使用は稀
golden Arab		*Colotis chrysonome* (Klug)	（チョウ目、シロチョウ科）エチオピア区
golden aster flower moth		*Schinia tuberculum* Hübner	（チョウ目、ヤガ科）新北区
golden azure			Sydney azure を見よ
golden-banded gem		*Parcella amarynthina* (C. et R. Felder)	（チョウ目、シジミタテハ科）新熱帯区
golden-banded metalmark		*Pirascca tyriotes* (Godman et Salvin)	（チョウ目、シジミタテハ科）新熱帯区
golden-banded ruby-eye		*Lychnuchoides saptine* (Godman et Salvin)	（チョウ目、セセリチョウ科）新熱帯区
golden-banded sister		*Adelpha salmoneus salmonides* Hall	（チョウ目、タテハチョウ科）新熱帯区
golden-banded skipper	キンオビセセリ	*Autochton cellus* (Boisduval et LeConte)	（チョウ目、セセリチョウ科）新北区
golden bandwing		*Cryptobothrus chrysophorus* Rehn	（バッタ目、バッタ科）豪州区
golden birdwing (1)	キシタアゲハ	*Troides aeacus* (C. et R. Felder)	（チョウ目、アゲハチョウ科）東洋区
golden birdwing (2)		*Ornithoptera aesacus* Ney	（チョウ目、アゲハチョウ科）東洋区
golden-bloomed grey longhorn		*Agapanthia villosoviridescens* (De Geer)	（コウチュウ目、カミキリムシ科）旧北区。golden-bloomed grey longhorn beetle ともいう
golden bluet		*Enallagma sulcatum* Williamson	（トンボ目、イトトンボ科）新北区
golden borer moth		*Papaipema cerina* (Grote)	（チョウ目、ヤガ科）新北区

英名	和名	学名	所属、分布、ほか
golden brown fern moth		*Musotima nitidalis* Walker	（チョウ目、メイガ科）豪州区
golden-brown tubic		*Batia unitella* (Hübner)	（チョウ目、マルハキバガ科）旧北区
golden buprestid		*Buprestis aurulenta* Linnaeus	（コウチュウ目、タマムシ科）全北区
golden chalcid			red scale parasite (1) を見よ
golden ciliate blue		*Anthene scintillula* (Holland)	（チョウ目、シジミチョウ科）エチオピア区
golden clearwing		*Albuna oberthuri* (Le Cerf)	（チョウ目、スカシバガ科）豪州区
golden copper		*Chrysoritis chrysaor* (Trimen)	（チョウ目、シジミチョウ科）エチオピア区
golden copper			fiery copper を見よ
golden cricket wasp		*Liris aurulenta* (Fabricius)	（ハチ目、アナバチ科）新北区
golden cricket wasp		*Liris opulenta* (Lepeletier)	（ハチ目、アナバチ科）大洋区
golden danaid			African monarch を見よ
golden danaid			monarch butterfly を見よ
golden dart moth		*Euxoa nigricans* (Linnaeus)	（チョウ目、ヤガ科）全北区
golden dartlet			aurora bluetail を見よ
golden digger wasp		*Sphex ichneumoneus* (Linnaeus)	（ハチ目、アナバチ科）新北区
golden drummer		*Thopha colorata* Distant	（カメムシ目、セミ科）豪州区
golden egg bug		*Phylomorpha laciniata* de Villers	（カメムシ目、ヘリカメムシ科）旧北区
golden-8-moth			silver eight を見よ
golden emperor		*Anapsaltodea pulchra* (Ashton)	（カメムシ目、セミ科）豪州区
golden emperor		*Loepa katinka* (Linnaeus)	（チョウ目、ヤママユガ科）旧北区、東洋区。golden emperor moth ともいう
golden emperor	キンイロマドタテハ	*Dilipa morgiana* (Westwood)	（チョウ目、タテハチョウ科）東洋区
golden Euselasia		*Euselasia chrysippe* (H. Bates)	（チョウ目、シジミタテハ科）新熱帯区
golden eye			golden－eye lacewing (2) を見よ
golden-eye lacewing (1)		*Chrysopa (Chrysopa) oculata* Say	（アミメカゲロウ目、クサカゲロウ科）新北区。成虫は灯火に飛来。卵は長い柄で葉上に産付。goldeneyed lacewing ともいう
golden-eye lacewing (2)		*Chrysopa perla* (Linnaeus)	（アミメカゲロウ目、クサカゲロウ科）旧北区
golden-eyed flies			green lacewings (1) を見よ
golden-eyed fly			golden-eye lacewing (2) を見よ
golden-eyed lacewings			green lacewings (1) を見よ
golden eyes			green lacewings (1) を見よ
golden flash			scarce scarlet を見よ
Golden Gate blue		*Orachrysops montanus* Henning et Henning	（チョウ目、シジミチョウ科）エチオピア区。南アの Golden Gate 高原に因む

英名	和名	学名	所属、分布、ほか
Golden Gate brown		*Pseudonympha paragaika* Vári	（チョウ目、タテハチョウ科）エチオピア区
Golden Gate widow		*Torynesis orangica* Vári	（チョウ目、タテハチョウ科）エチオピア区
golden gem		*Libellago lineata* (Burmeister)	（トンボ目、Chlorocyphidae）東洋区
golden glider	ウスベニイロタテハ	*Cymothoe capella* (Ward)	（チョウ、タテハチョウ科）エチオピア区
goldenglow aphid		*Dactynotus rudbeckiae* (Fitch)	（カメムシ目、アブラムシ科）新北区
golden-green minute leaf beetle	アオバネサルハムシ	*Basilepta fulvipes* (Motschulsky)	（コウチュウ目、ハムシ科）日本、旧北区、東洋区
golden ground beetle		*Carabus auratus* Linnaeus	（コウチュウ目、オサムシ科）旧北区
goldenhaired bark beetle			pine root bark beetle を見よ
golden-haired dung fly			yellow dung fly を見よ
golden haired longhorn beetle		*Leptura aurulenta* Fabricius	（コウチュウ目、カミキリムシ科）旧北区
golden haired mortar bee			teddy bear bee を見よ
golden-haired robberfly		*Choerades marginatus* (Linnaeus)	（ハエ目、ムシヒキアブ科）旧北区
golden hairstreak	アメリカゼフィルス	*Habrodais grunnus* (Boisduval)	（チョウ目、シジミチョウ科）新北区
golden-headed scallopwing			golden-headed sootywing を見よ
golden-headed sootywing		*Staphylus ceos* (Edwards)	（チョウ目、セセリチョウ科）新北区、新熱帯区
golden-headed sootywing			copper-headed sootywing を見よ
golden heath eye			small heath を見よ
golden heliconian			tiger longwing を見よ
golden horsefly		*Atylotus fulvus* (Meigen)	（ハエ目、アブ科）旧北区
golden hoverfly		*Chrysotoxum octomaculatum* Curtis	（ハエ目、ハナアブ科）旧北区
golden hunting wasp		*Sphictostethus nitidus* (Fabricius)	（ハチ目、ベッコウバチ科）豪州区
golden Jezebel		*Delias aruna* (Boisduval)	（チョウ目、シロチョウ科）豪州区
golden Kaiser-i-Hind	オウゴンテングアゲハ	*Teinopalpus aureus* Mell	（チョウ目、アゲハチョウ科）東洋区
golden lady slipper		*Pierella hyceta* Hewitson	（チョウ目、タテハチョウ科）新熱帯区
golden leaf roller			pear leaf roller (3) を見よ
golden leaf roller moth			Stockholm button を見よ
golden longwing			tiger longwing を見よ
golden looper moth		*Argyrogramma verruca* (Fabricius)	（チョウ目、ヤガ科）新北区、新熱帯区。golden looper ともいう
golden mayflies		*Hexagenia*	（カゲロウ目、モンカゲロウ科）の昆虫の総称

英名	和名	学名	所属、分布、ほか
golden mealybug		*Nipaecoccus aurilanatus* (Maskell)	(カメムシ目、コナカイガラムシ科) 豪州区
golden mosquito		*Coquillettidia xanthogaster* (Edwards)	(ハエ目、カ科) 豪州区
golden mottled-skipper		*Codatractus bryaxis* (Hewitson)	(チョウ目、セセリチョウ科) 新熱帯区
golden mountain satyr		*Lymanopoda translucida* Staudinger	(チョウ目、タテハチョウ科) 新熱帯区
golden norhtern bumble bee		*Bombus fervidus* (Fabricius)	(ハチ目、ミツバチ科) 新北区、大洋区
golden oak plutellid		*Abebaea cervella* Walsingham	(チョウ目、スガ科) 新北区
golden oak scale		*Asterolecanium quercicola* (Bouché)	(カメムシ目、フサカイガラムシ科) 旧北区
golden oak scale (1)		*Asterolecanium variolosa* (Ratzeburg)	(カメムシ目、フサカイガラムシ科) 全北区、豪州区
golden opal			Heidelberg copper を見よ
golden-orange bumble bee		*Bombus borealis* Kirby	(ハチ目、ミツバチ科) 新北区
golden pansy		*Junonia chorimene* (Guérin-Méneville)	(チョウ目、タテハチョウ科) エチオピア区
golden paper wasp		*Polistes fuscatus* (Fabricius)	(ハチ目、スズメバチ科) 新北区、新熱帯区。northern paper wasp ともいう。亜種 *P. f. aurifer* Saussure も同英名
golden pigmy		*Stigmella aurella* (Fabricius)	(チョウ目、モグリチビガ科) 旧北区
golden piper		*Eurytela dryope* Cramer	(チョウ目、タテハチョウ科) エチオピア区
golden plusia			silver eight を見よ
golden plusia moth			silver eight を見よ
golden polistes		*Polistes aurifer* Saussure	(ハチ目、スズメバチ科) 新北区
goldenrain tree aphid	モクゲンジニタイケアブラムシ	*Periphyllus koelreuteriae* (Takahashi)	(カメムシ目、アブラムシ科) 日本
goldenrain-tree bug			red-shouldered bug をみよ
goldenrain-tree bug			red-shouldered bug を見よ
golden-rayed blue		*Candalides noelkeri* Braby et Douglas	(チョウ目、シジミチョウ科) 豪州区
golden rim			Polydamas swallowtail を見よ
golden-rim swallowtail			Polydamas swallowtail を見よ
golden-ringed dragonflies			biddies を見よ
golden-ringed dragonfly		*Cordulegaster boltonii* (Donovan)	(トンボ目、オニヤンマ科) 旧北区
golden-ringed spiketail		*Cordulegaster annulatus* (Latreille)	(トンボ目、オニヤンマ科) 旧北区
goldenrod borer plume moth		*Hellinsia kellicottii* (Fish)	(チョウ目、トリバガ科) 新北区。goldenrod borer ともいう
golden-rod brindle	シロスジキリガ	*Lithomoia solidaginis* (Hübner)	(チョウ目、ヤガ科) 日本、旧北区

英名	和名	学名	所属、分布、ほか
goldenrod elliptical gall moth			goldenrod gall moth (1) を見よ
goldenrod flower moth		*Schinia nundina* Drury	（チョウ目、ヤガ科）新北区
goldenrod gall fly		*Eurosta solidaginis* (Fitch)	（ハエ目、ミバエ科）新北区
golden rod gall-maker			goldenrod gall midge を見よ
golden rod gall-maker			goldenrod gall moth (1) を見よ
goldenrod gall midge		*Rhopalomyia solidaginis* (Loew)	（ハエ目、タマバエ科）新北区
goldenrod gall moth		*Epiblema scudderiana* (Clemens)	（チョウ目、ハマキガ科）新北区
goldenrod gall moth (1)		*Gnorimoschema gallaeosolidaginis* Riley	（チョウ目、キバガ科）新北区
goldenrod gall moths		*Gnorimoschema*	（チョウ目、キバガ科）の昆虫の総称
goldenrod gallmaker			goldenrod gall fly を見よ
goldenrod lace-bug			chrysanthemum lace bug を見よ
golden-rod pug		*Eupithecia virgaureata* Doubleday	（チョウ目、シャクガ科）旧北区。日本亜種は *E. v. invisa* Butler アザミカバナミシャク
goldenrod soldier beetle			Pennsylvania leatherwing beetle を見よ
goldenrod spindle gall moth			goldenrod gall moth (1) を見よ
goldenrod stowaway moth		*Cirrhophanus triangulifer* Grote	（チョウ目、ヤガ科）新北区。goldenrod stowaway ともいう
golden royal		*Pseudotajuria donatana donatana* (de Nicéville)	（チョウ目、シジミチョウ科）東洋区
golden saltmarsh mosquito		*Aedes sollicitans* (Walker)	（ハエ目、カ科）新北区
golden sapphire		*Heliophorus brahma* (Moore)	（チョウ目、シジミチョウ科）東洋区
golden Scottish stiletto		*Thereva handlirschi* Kröber	（ハエ目、ツルギアブ科）旧北区
golden skipper		*Poanes taxiles* (Edwards)	（チョウ目、セセリチョウ科）新北区、新熱帯区
golden-snout scallopwing			golden-snouted sootywing を見よ
golden-snouted sootywing		*Staphylus vulgata* (Möschler)	（チョウ目、セセリチョウ科）新熱帯区
golden snowflea		*Onychiurus cocklei* (Folsom)	（トビムシ目、シロトビムシ科）新北区
golden spider beetle		*Niptus hololeucus* (Falderman)	（コウチュウ目、ヒョウホンムシ科）汎世界
golden stag beetle	オウゴンオニクワガタ	*Allotopus moellenkampi moseri* Moellenkamp	（コウチュウ目、クワガタムシ科）東洋区
golden stag beetle		*Lamprima aurata* Latreille	（コウチュウ目、クワガタムシ科）豪州区
golden sulphur			western sulphur を見よ

英名	和名	学名	所属、分布、ほか
golden sun moth		*Synemon plana* Walker	（チョウ目、カストニアガ科）豪州区
golden swift			gold swift を見よ
golden sylph		*Metisella midas* (Butler)	（チョウ目、セセリチョウ科）エチオピア区
golden-tabbed robberfly		*Eutolmus rufibarbis* (Meigen)	（ハエ目、ムシヒキアブ科）日本、旧北区
golden-tailed moth			browntail moth (2) を見よ
golden-tailed spiny ant		*Polyrhachis ammon* (Fabricius)	（ハチ目、アリ科）豪州区
golden target beetle			ringed tortoise beetle を見よ
golden-tip		*Colotis hildebrandti* (Staudinger)	（チョウ目、シロチョウ科）エチオピア区
golden tortoise beetle		*Metriona bicolor* (Fabricius)	（コウチュウ目、ハムシ科）新北区。サツマイモ害虫
golden tree flitter			yellow-base flitter を見よ
golden twin spot		*Chrysodeixis chalcites* (Esper)	（チョウ目、ヤガ科）全北区
golden twin spot moth			golden twin spot を見よ
golden-veined glasswing		*Godyris sappho* (Haensch)	（チョウ目、タテハチョウ科）新熱帯区
golden velvet ant		*Dasymutilla aureola* (Cresson)	（ハチ目、アリバチ科）新北区
golden-winged dancer		*Argia rhoadsi* Calvert	（トンボ目、イトトンボ科）新北区
golden-winged Palyas moth		*Phrygionis auriferaria* (Hulst)	（チョウ目、シャクガ科）新北区
golden wasp		*Vespa auraria* Smith	（ハチ目、スズメバチ科）東洋区
golden white		*Melete polyhymnia florinda* (Butler)	（チョウ目、シロチョウ科）新熱帯区。*M. p. serrana* Maza も同英名
golden Y-moth		*Autographa jota* (Linnaeus)	（チョウ目、ヤガ科）旧北区
golds			mandibulate moths を見よ
Goldsmith beetle		*Cotalpa lanigera* Linnaeus	（コウチュウ目、コガネムシ科）新北区
Goldsmith beetle			rose chafer (1) を見よ　米国での英名
golf-club skipper			Araxes skipper を見よ
Goliath beetle		*Goliathus druryi* Westwood	（コウチュウ目、コガネムシ科）エチオピア区
Goliath beetle		*Goliathus regius* Klug	（コウチュウ目、コガネムシ科）エチオピア区
Goliath beetle (1)	ゴライアスオオツノコガネ	*Goliathus goliatus* (Linnaeus)	（コウチュウ目、コガネムシ科）エチオピア区。アフリカ産の大型カブトムシ
Goliath beetles		*Goliathus*	（コウチュウ目、コガネムシ科）の昆虫の総称　エチオピア区。*G. atlas* Nickered, *G. goliatus* (Linnaeus) が所属し、体長 12.5 cm 幅 5 cm の巨大昆虫
Goliath birdwing	ゴライアストリバネアゲハ	*Ornithoptera goliath* Oberthür	（チョウ目、アゲハチョウ科）東洋区
Goliath spreadwing		*Conognathus platon* Felder et Felder	（チョウ目、セセリチョウ科）新熱帯区

英名	和名	学名	所属、分布、ほか
Goliath stick insect		*Eurycnema goliath* (Gray)	(ナナフシ目、ナナフシ科) 豪州区
Goliathus beetle			Goliath beetle (1) を見よ
Golo bush brown		*Bicyclus golo* (Aurivillius)	(チョウ目、タテハチョウ科) エチオピア区
golubatz fly			Kolumbacs' fly を見よ
gomphids			clubtails (1) を見よ
Gondwanaland moths			miniature ghost moths を見よ
good King Henry			painted lady を見よ
Good's Epitola		*Geritola goodii* (Holland)	(チョウ目、シジミチョウ科) エチオピア区
Goodson's greenstreak			Goodson's hairstreak を見よ
Goodson's hairstreak		*Cyanophrys goodsoni* (Clench)	(チョウ目、シジミチョウ科) 新北区
goon borer			bamboo powderpost beetle を見よ
goose body louse		*Trinoton anserinus* (Fabricius)	(ハジラミ目、タンカクハジラミ科) 全北区、豪州区
goose-feather case		*Coleophora bernoulliella* (Goeze)	(チョウ目、ツツミノガ科) 旧北区
goose-feather case (1)		*Coleophora anatipennella* (Hübner)	(チョウ目、ツツミノガ科) 旧北区
goosefoot leafminer moth			Schaffer's neb (2) を見よ
goosefoot pug		*Eupithecia sinuosaria* (Eversmann)	(チョウ目、シャクガ科) 旧北区
goose wing lice		*Anaticola*	(シラミ目、Philopteridae) の昆虫の総称
goose wing louse			slender goose louse を見よ　豪州での英名
goose wing louse			slender duck louse を見よ　豪州での英名
gooseberry aphid (1)		*Aphis grossulariae* Kaltenbach	(カメムシ目、アブラムシ科) 旧北区
gooseberry aphid (2)		*Hyperomyzus pallidus* Hille Ris Lambers	(カメムシ目、アブラムシ科) 旧北区
gooseberry barkminer moth		*Pseudopostega quadristrigella* (Frey et Boll)	(チョウ目、ヒラタモグリガ科) 新北区
gooseberry borer			currant clearwing を見よ
gooseberry bryobia		*Bryobia ribis* Thomas	(アザミウマ目、アザミウマ科) 旧北区
gooseberry bud midge		*Rhopalomyia grossulariae* Felt	(ハエ目、タマバエ科). 旧北区
gooseberry flower midge		*Contarinia ribis* Kieffer	(ハエ目、タマバエ科) 旧北区
gooseberry fruitworm		*Zophodia convulutella* (Hübner)	(チョウ目、メイガ科) 全北区
gooseberry fruitworm moth		*Zophodia grossulariella* Zincken	(チョウ目、メイガ科) 全北区
gooseberry gall midge			gooseberry bud midge を見よ
gooseberry leaf midge		*Dasineura ribicola* (Kieffer)	(ハエ目、タマバエ科) 旧北区

英名	和名	学名	所属、分布、ほか
gooseberry midge		*Cecidomyia grossulariae* Fitch	（ハエ目、タマバエ科）旧北区
gooseberry moth			magpie を見よ
gooseberry sawflies		*Nematus*	（ハチ目、ハバチ科）の昆虫の総称
gooseberry sawfly		*Nematus consobrinus* Vollenhoven	（ハチ目、ハバチ科）旧北区
gooseberry sawfly (1)		*Nematus ribesii* (Scopoli)	（ハチ目、ハバチ科）汎世界
gooseberry sawfly (2)	スグリハバチ	*Priophorus pallipes* (Lepeletier)	（ハチ目、ハバチ科）日本、旧北区
gooseberry sawfly			gooseberry sawfly (1) を見よ
gooseberry-sowthistle aphid			gooseberry aphid (2) を見よ
gooseberry spanworm			currant spanworm を見よ
gooseberry weevil		*Ecrizothis inaequalis* Blackburn	（コウチュウ目、ゾウムシ科）豪州区
gooseberry weevil (1)	スグリゾウムシ	*Pseudocneorhinus bifasciatus* Roelofs	（コウチュウ目、ゾウムシ科）日本、全北区。カンキツ、リンゴ、イチゴなどの害虫
gooseberry witchbroom aphid		*Kakimia houghtonensis* (Troop)	（カメムシ目、アブラムシ科）新北区
Gopher tortoise moth		*Ceratophaga vicinella* Dietz	（チョウ目、ヒロズコガ科）新北区
goree silkworm			gorse silkworm を見よ
Gorge claspertail			claspertail を見よ
Gorgon copper	ゴルゴンベニシジミ	*Lycaena gorgon* (Boisduval)	（チョウ目、シジミチョウ科）新北区、新熱帯区
Gorgone checkerspot			Gorgone crescentspot を見よ
Gorgone crescentspot	ゴルゴンヒョウモンモドキ	*Chlosyne gorgone* (Hübner)	（チョウ目、タテハチョウ科）新北区
Gorgus skipper		*Tosta gorgus* (Bell)	（チョウ目、セセリチョウ科）新熱帯区
gorse-broom lace bug		*Dictyonota strichnocera* Fieber	（カメムシ目、グンバイムシ科）旧北区
gorse emerald			Jersey emerald を見よ
gorse knot-horn		*Nephopterix genistella* Duponchel	（チョウ目、メイガ科）旧北区
gorse lacebug			gorse-broom lace bug を見よ
gorse seed weevil		*Apion ulicis* (Forster)	（コウチュウ目、ホソクチゾウムシ科）新北区、豪州区
gorse shieldbug			green shield bug (1) を見よ
gorse silkworm		*Theophila religiosa* (Helfer)	（チョウ目、ヤママユガ科）東洋区
gorse soft shoot moth		*Agonopterix umbellana* (Fabricius)	（チョウ目、マルハキバガ科）旧北区、豪州区、大洋区
gorse thrips		*Odontothripiella australis* (Bagnall)	（アザミウマ目、アザミウマ科）豪州区
gorse tip moth		*Agonopterix nervosa* (Haworth)	（チョウ目、マルハキバガ科）全北区
gorse weevil			gorse seed weevil を見よ

英名	和名	学名	所属、分布、ほか
Gossamer damselfly			aurora bluetail を見よ
Gossamer-winged butterflies			coppers (1) を見よ　米国での英名
Gossamerwings		Euphaeidae	（トンボ目）の昆虫の総称
Gossamer-wings			coppers (1) を見よ
gossypium aphid			oleander aphid を見よ
gothic		*Naenia typica* (Linnaeus)	（チョウ目、ヤガ科）旧北区
gothic dart			dingy cutworm moth (1) を見よ
gothic moth			gothic を見よ
gothic type			gothic を見よ
gourd-shaped ground weevil	スナムグリヒョウタンゾウムシ	*Scepticus tigrinus* (Roelofs)	（コウチュウ目、ゾウムシ科）日本
gourd-shaped leaf bug	ヒョウタンカスミカメムシ	*Pilophorus setulosus* Horvath	（カメムシ目、カスミカメムシ科）日本、旧北区
gout fly			chloropid gout fly を見よ
gouty pitch midge		*Cecidomyia piniinopis* Osten Sacken	（ハエ目、タマバエ科）新北区
Gowan's copper		*Aloeides gowani* Tite et Dickson	（チョウ目、シジミチョウ科）エチオピア区
graceful clearwing			slender clearwing moth を見よ
graceful clearwing			slender clearwing moth を見よ
graceful ghost moth		*Hepialus mustelinus* Packard	（チョウ目、コウモリガ科）新北区
graceful grasshopper		*Melanoplus gracilis* (Bruner)	（バッタ目、バッタ科）豪州区
graceful grass-veneer moth		*Parapediasia decorellus* (Zincken)	（チョウ目、メイガ科）新北区。graceful grass-veneer ともいう
graceful meadow katydid		*Conocephalus gracillimus* (Morse)	（バッタ目、キリギリス科）新北区
graceful narrow-winged grasshopper			graceful grasshopper を見よ
graceful slug moth		*Lithacodes gracea* Dyar	（チョウ目、イラガ科）新北区
graceful underwing moth		*Catocala gracilis* Edwards	（チョウ目、ヤガ科）新北区。graceful underwing ともいう
gracile line-blue		*Prosotas gracilis* (Rober)	（チョウ目、シジミチョウ科）豪州区
gracile Palpita moth		*Palpita gracialis* (Hulst)	（チョウ目、メイガ科）新北区
gracious whisp		*Agriocnemis gratiosa* Gerstacker	（トンボ目、イトトンボ科）エチオピア区
Graham bean weevil	アカイロマメゾウムシ	*Callosobruchus analis* (Fabricius)	（コウチュウ目、ハムシ科）旧北区
Graham's ace		*Sovia grahami* (Evans)	（チョウ目、セセリチョウ科）東洋区
Graham's blue		*Lepidochrysops grahami* (Trimen)	（チョウ目、シジミチョウ科）エチオピア区
grain aphid (1)		*Sitobion avenae* (Fabricius)	（カメムシ目、アブラムシ科）全北区
grain aphid (2)		*Sitobion* nr. *fragariae* (Walker)	（カメムシ目、アブラムシ科）豪州区
grain aphid			oat bird-cherry aphid を見よ

英名	和名	学名	所属、分布、ほか
grain aphid			rose grain aphid を見よ
grain aphid			potato aphid (1) を見よ
grain beetle			lesser grain borer を見よ
grain beetle			yellow mealworm を見よ
grain borer		*Calamoblus filum* (Rossi)	（コウチュウ目、カミキリムシ科）旧北区
grain gall midge			wheat blossom midge を見よ
grain-gnawing beetles			gnawing beetles (1) を見よ
grain moth			Angoumois grain moth を見よ
grain psocid	カツブシチャタテ	*Liposcelis entomophilus* (Enderlein)	（チャタテムシ目、コナチャタテ科）日本、汎世界。貯蔵食品などの害虫
grain striped moth	フタスジシマメイガ	*Orthopygia glaucinalis* (Linnaeus)	（チョウ目、メイガ科）日本、旧北区
grain thrips		*Limothrips cerealium* (Haliday)	（アザミウマ目、アザミウマ科）全北区、豪州区
grain toxoptera			greenbug を見よ
grain weevil			granary weevil を見よ
grain weevils		*Sitophilus*	（コウチュウ目、ゾウムシ科）の昆虫の総称
grain weevils			billbugs (1)（2）を見よ
grain worm	コクマルハキバガ	*Martyringa xeraula* (Meyrick)	（チョウ目、マルハキバガ科）日本、東洋区、新北区。貯穀害虫
grained moth			willow prominent を見よ
grained noctuid			flame を見よ
grained prominent			willow prominent を見よ
grains			silkworm を見よ　カイコガ卵の英名
gram aphid			potato aphid（1）を見よ
gram blue	オジロシジミ	*Euchrysops cnejus* (Fabricius)	（チョウ目、シジミチョウ科）日本、東洋区、豪州区、大洋区
gram caterpillar			corn earworm (2) を見よ　南アフリカでの幼虫の英名
gram pod borer			corn earworm (2) を見よ
gramang ant		*Platiolepis longipes* Jerdon	（ハチ目、アリ科）東洋区
Grammy's needle			dragonflies (1) を見よ
Granadilla purple scale			nigra scale を見よ　南アフリカでの英名
granary barklouse		*Cerobasis guestfalica* (Kolbe)	（チャタテムシ目、コチャタテ科）汎世界
granary booklice	コチャタテ科	Trogiidae	（チャタテムシ目）の昆虫の総称
granary booklouse		*Lepinotus inquilinus* Heyden	（チャタテムシ目、コチャタテ科）日本、汎世界
granary moth			southern old lady moth を見よ
granary weevil	グラナリアコクゾウムシ	*Sitophilus granarius* (Linnaeus)	（コウチュウ目、ゾウムシ科）日本、汎世界。貯穀害虫
grand Duchess	パタライナズマ	*Euthalia patala* (Kollar)	（チョウ目、タテハチョウ科）東洋区
grand duke		*Bassarona iva* (Moore)	（チョウ目、タテハチョウ科）東洋区

英名	和名	学名	所属、分布、ほか
grand groundling		*Gelechia turpella* (Denis et Schiffermüller)	（チョウ目、キバガ科）旧北区
grand imperial	オナガツバメシジミ	*Neocheritra amrita* (C. et R. Felder)	（チョウ目、シジミチョウ科）東洋区
grand lecanium			giant coccus を見よ
grand shieldback		*Pediodectes grandis* (Rehn)	（バッタ目、キリギリス科）新北区
grand skipper		*Gamia buchholzi* (Plötz)	（チョウ目、セセリチョウ科）エチオピア区
granite moth		*Macaria granitata* Guenée	（チョウ目、シャクガ科）新北区
granite moth		*Digrammia imparilata* Ferguson	（チョウ目、シャクガ科）新北区
granite moth		*Digrammia pictipennata* (Hulst)	（チョウ目、シャクガ科）新北区
granitose fern moth		*Callopistria granitosa* (Guenée)	（チョウ目、ヤガ科）新北区
grannom	カクスイトビケラ	*Brachycentrus subnubilus* Curtis	（トビケラ目、カクスイトビケラ科）日本、旧北区
grannom	カワゲラ		stoneflies を見よ
granny moth			southern old lady moth を見よ
Granny's cloak moth			northern brown house moth (1) を見よ
Grant's longwing		*Capnobotes granti* Rentz et Birchim	（バッタ目、キリギリス科）新北区
Grant's stag beetle			Darwin's beetle を見よ
granular ringlet		*Ypthima granulosa* Butler	（チョウ目、タテハチョウ科）エチオピア区
granulate cutworm		*Feltia subterranea* (Fabricius)	（チョウ目、ヤガ科）全北区。granulate cutworm moth, granulate cutworm ともいう
granulated grouse locust		*Acrydium granulatum* (Kirby)	（バッタ目、ヒシバッタ科）新北区
granulated grouse locust (1)		*Tetrix subulata* (Linnaeus)	（バッタ目、ヒシバッタ科）新北区
granulated long-horned weevil	シロオビタマゾウムシ	*Stereonychidius galloisi* (Hustache)	（コウチュウ目、ゾウムシ科）日本
grape apple gall		*Schizomyia vitispomum* (Osten Sacken)	（ハエ目、タマバエ科）新北区
grape asteropetes	ヒメトラガ	*Asteropetes noctuina* (Butler)	（チョウ目、トラガ科）日本
grape berry moth (1)	ブドウヒメハマキ	*Endopiza viteana* Clemens	（チョウ目、ハマキガ科）新北区
grape berry moth (2)		*Episimus argutanus* (Clemens)	（チョウ目、ハマキガ科）新北区
grapeberry moth			grape cochylid を見よ
grapeberry moth			grapevine moth (1) を見よ
grapeberry root weevil			black vine weevil を見よ
grape blossom midge		*Contarinia johnsoni* Felt	（ハエ目、タマバエ科）新北区
grape borer	ブドウトラカミキリ	*Xylotrechus pyrrhoderus* Bates	（コウチュウ目、カミキリムシ科）日本、旧北区。ブドウ害虫
grape boring plume moth		*Oxyptilus regulus* Meyrick	（チョウ目、トリバガ科）東洋区。ブドウ害虫

英名	和名	学名	所属、分布、ほか
grape bud beetle		*Glyptoscelis squamulata* Crotch	(コウチュウ目、ハムシ科) 新北区。ブドウ害虫
grape bud moth			grape cochylid を見よ
grape cane borer			apple twig borer (2) を見よ
grape cane gall maker		*Ampeloglypter sesostris* (LeConte)	(コウチュウ目、ゾウムシ科) 新北区
grape cane girdler		*Ampeloglypter ater* LeConte	(コウチュウ目、ゾウムシ科) 新北区
grape clearwing moth	ブドウスカシバ	*Paranthrene regalis* (Butler)	(チョウ目、スカシバガ科) 日本、旧北区。幼虫がブドウのつるに食入して虫えいを作る
grape cochylid	ブドウホソハマキ	*Eupoecillia ambiguella* (Hübner)	(チョウ目、ホソハマキガ科) 日本、旧北区。ブドウ害虫
grape colaspis (1)		*Colaspis brunnea* (Fabricius)	(コウチュウ目、ハムシ科) 新北区
grape colaspis (2)		*Colaspis flavida* (Say)	(コウチュウ目、ハムシ科) 新北区。ブドウ、イネほか多くの作物害虫
grape curculio		*Craponius inaequalis* (Say)	(コウチュウ目、ゾウムシ科) 新北区
grape curculio			pear leaf roller (1) を見よ
grape flea beetle		*Altica chalybea* (Illiger)	(コウチュウ目、ハムシ科) 新北区
grape fruit fly			pomace fly を見よ
grape fruit moth			grapevine moth (1) を見よ
grape gall midge	ブドウタマバエ	*Janetiella oenophila* (Haimhoffen)	(ハエ目、タマバエ科) 日本、旧北区
grape hawkmoth		*Pholus vitis* (Linnaeus)	(チョウ目、スズメガ科) 新熱帯区
grape hawkmoth		*Pholus achemon* Drury	(チョウ目、スズメガ科) 新北区
grape horn worm	ブドウスズメ	*Acosmeryx castanea* Rothschild et Jordan	(チョウ目、スズメガ科) 日本、東洋区。ブドウ害虫
grape horn worm	クルマスズメ	*Ampelophaga rubiginosa* Bremer et Grey	(チョウ目、スズメガ科) 日本、旧北区。ブドウ害虫
grape leaf beetle	ドウガネサルハムシ	*Scelodonta lewisii* Baly	(コウチュウ目、ハムシ科) 日本、東洋区
grape leaf folder			grape leaf roller を見よ　grape leaffolder moth ともいう
grapeleaf louse	ブドウネアブラムシ	*Viterus vitifolii* (Fitch)	(カメムシ目、ネアブラムシ科) 日本、汎世界。ブドウ害虫
grapeleaf skeletonizer		*Harrisina americana* Guérin-Méneville	(チョウ目、マダラガ科) 新北区。grapeleaf skeletonizer moth、grape leaf skeletonizer ともいう
grape leaf roller		*Desmia funeralis* (Hübner)	(チョウ目、メイガ科) 新北区。幼虫はブドウの葉を食う
grape leaf worm	ブドウスカシクロバ	*Illiberis tenuis* (Butler)	(チョウ目、マダラガ科) 日本、旧北区、東洋区。ブドウ害虫
grape leafhopper	フタテンヒメヨコバイ	*Arboridia apicalis* (Nawa)	(カメムシ目、オオヨコバイ科) 日本、旧北区。ブドウ害虫
grape leafhopper	フタテンオオヨコバイ	*Epiacanthus stramineus* (Motschulsky)	(カメムシ目、オオヨコバイ科) 日本、旧北区。カンキツ、ブドウ害虫
grape leafhopper		*Arboridia kermansbat* Dlabola	(カメムシ目、オオヨコバイ科) 旧北区
grape leafhopper		*Eurythroneura comes* (Say)	(カメムシ目、オオヨコバイ科) 新北区

英名	和名	学名	所属、分布、ほか
grape leafhopper (1)		*Erythroneura elegantula* Osborn	（カメムシ目、オオヨコバイ科）新北区。ブドウ害虫
grape leafhopper			false-eye leafhopper を見よ
grape leafminer	ブドウハムグリガ	*Phyllocnistis toparcha* Meyrick	（チョウ目、コハモグリガ科）日本。ブドウ害虫
grape leafroller	モンキクロノメイガ	*Herpetogramma luctuosalis zelleri* (Bremer)	（チョウ目、メイガ科）日本、旧北区、東洋区。ブドウ害虫
grape leaf-roller weevil	ブドウハマキチョッキリ	*Aspidobyctiscus lacunipennis* (Jekel)	（コウチュウ目、オトシブミ科）日本、旧北区、東洋区
grape mealybug		*Pseudococcus maritimus* (Ehrhorn)	（カメムシ目、コナカイガラムシ科）全北区。ブドウ害虫
grape mealybug			citrus mealybug (1) を見よ
grape midge			grapevine midge を見よ
grape pelidnota			spotted grape beetle を見よ
grape phylloxera			grapeleaf louse を見よ　豪州での英名
grape plume moth		*Pterophorus walsinghami* (Fernald)	（チョウ目、トリバガ科）新北区
grape plume moth		*Pterophorus periscelidactylus* Fitch	（チョウ目、トリバガ科）新北区
grape plume moth (1)	ブドウトリバ	*Nippoptilia vitis* (Sasaki)	（チョウ目、トリバガ科）日本、旧北区、東洋区。幼虫はブドウの花蕾を加害
grape root borer		*Vitacea polistiformis* (Harris)	（チョウ目、スカシバガ科）全北区。grape root borer moth ともいう
grape rootworm		*Fidia viticida* Walsh	（コウチュウ目、ハムシ科）新北区
grape sawfly		*Erythraspides vitis* (Harris)	（ハチ目、ハバチ科）新北区
grape scale		*Diaspidiotus uvae* (Comstock)	（カメムシ目、マルカイガラムシ科）新北区
grape seed chalcid		*Evoxysoma vitis* (Saunders)	（ハチ目、カタビロコバチ科）新北区
grape seed weevil			grape curculio を見よ
grape sphinx moth			achemon sphinx を見よ
grape spittlebug	クロスジアワフキ	*Aphrophora vitis* (Matsumura)	（カメムシ目、アワフキムシ科）日本
grape thrips		*Drepanothrips reuteri* Uzel	（アザミウマ目、アザミウマ科）日本、全北区、新熱帯区。ブドウ害虫
grape tiger longicorn			grape borer を見よ
grape tiger moth		*Grammia arge* Drury	（チョウ目、ヒトリガ科）新北区
grape tiger moth			grape asteropetes を見よ
grape tree borer	キマダラコウモリ	*Endoclita sinensis* (Moore)	（チョウ目、コウモリガ科）日本、旧北区、東洋区。カンキツ、スギなど果樹、樹木害虫
grape trunk borer		*Cyltoleptus albofasciatus* Laporte et Gory	（コウチュウ目、カミキリムシ科）新北区
grape tube gallmaker		*Schizomyia viticola* (Osten Sacken)	（ハエ目、タマバエ科）新北区
grape tussock moth	ブドウドクガ	*Neocifuna eurydice* (Butler)	（チョウ目、ドクガ科）日本、旧北区
grape Vesperus		*Vesperus xatarti* Dufour	（コウチュウ目、カミキリムシ科）旧北区

英名	和名	学名	所属、分布、ほか
grape-vine aphid		*Aphis illinoisensis* Shimer	（カメムシ目、アブラムシ科）新北区
grapevine beetle			spotted grape beetle を見よ
grape vine bug		*Capsodes sulcatus* (Fieber)	（カメムシ目、カスミカメムシ科）旧北区
grapevine Epimenis moth		*Psychomorpha epimenis* (Drury)	（チョウ目、ヤガ科）新北区。grapevine Epimenis ともいう
grapevine hawk moth			vine hawk-moth (1) を見よ
grapevine hawk-moth			elephant hawk moth を見よ
grapevine Hoplia		*Hoplia callipyge* LeConte	（コウチュウ目、コガネムシ科）新北区。ブドウ害虫
grape-vine leafhopper		*Acia lineatifrons* (Naude)	（カメムシ目、オオヨコバイ科）エチオピア区
grapevine looper			lesser grapevine looper moth を見よ
grapevine midge		*Contarinia viticola* Rubsaamen	（ハエ目、タマバエ科）旧北区
grapevine moth (1)	ホソバヒメハマキ	*Lobesia botrana* (Denis et Schiffermüller)	（チョウ目、ハマキガ科）日本、旧北区
grapevine moth (2)		*Phalaenoides glycinae* Lewin	（チョウ目、ヤガ科）豪州区
grape vine phylloxera			grapeleaf louse を見よ
grapevine scale			European fruit lecanium を見よ
grape-vine sphinx			Virginia-creeper sphinx moth を見よ
grape vine thrips		*Rhipiphorothrips cruentatus* Hood	（アザミウマ目、アザミウマ科）東洋区
grapevine thrips			grape thrips を見よ
grape whitefly		*Trialeurodes vittata* (Quaintance)	（カメムシ目、コナジラミ科）新北区
grape whitefly	ブドウコナジラミ	*Aleurolobus taonabae* (Kuwana)	（カメムシ目、コナジラミ科）日本、旧北区。ブドウ害虫
graphic beauty			Baeotus beauty を見よ
graphic crescent			Vesta crescentspot (1) を見よ
graphic flutterer		*Rhyothemis graphiptera* (Rambur)	（トンボ目、トンボ科）豪州区
graphic moth		*Drasteria graphica* Hübner	（チョウ目、ヤガ科）新北区
grappletail		*Octogomphus specularis* (Hagen)	（トンボ目、サナエトンボ科）新北区
grappletails		*Heliogomphus*	（トンボ目、サナエトンボ科）の昆虫の総称
grass and cereal fly		*Geomyza tripunctata* Fallen	（ハエ目、ヒメコバエ科）旧北区
grass and cereal fly		*Opomyza florum* Fabricius	（ハエ目、ヒメコバエ科）旧北区
grass and cereal fly		*Opomyza germinationis* (Linnaeus)	（ハエ目、ヒメコバエ科）旧北区
grass anthelid		*Pterolocera amplicornis* Walker	（チョウ目、Anthelidae）豪州区
grass aphid		*Metopolophium festucae* (Theobald)	（カメムシ目、アブラムシ科）旧北区

英名	和名	学名	所属、分布、ほか
grass armyworm			rice armyworm を見よ
grass bagworm		*Brachycyttarus griseus* de Joannis	(チョウ目、ミノガ科) 大洋区
grass blue butterfly			common grass-blue (1) を見よ
grass bob		*Suada swerge* (de Nicéville)	(チョウ目、セセリチョウ科) 東洋区
grass bug		*Corizus sidae* (Fabricius)	(カメムシ目、Corizidae) 新北区
grassbug		*Stenodema laevigatum* (Linnaeus)	(カメムシ目、カスミカメムシ科) 旧北区
grass bugs		Corizidae	(カメムシ目) の昆虫の総称　Coreidae に入れること多し
grass-carrying termites			rotten wood termites を見よ
grass-crown mealybug			Rhodes grass scale を見よ
grass caterpillar			grass webworm (2) を見よ　豪州での英名
grass chequered white			bath white を見よ
grass coccid		*Symonicoccus australis* (Maskell)	(カメムシ目、カタカイガラムシ科) 豪州区
grass cricket	クサヒバリ	*Paratrigonidium bifasciatum* Shiraki	(バッタ目、クサヒバリ科) 日本
grass cutworm	ムギヤガ	*Euxoa oberthueri* (Leech)	(チョウ目、ヤガ科) 日本、旧北区
grass darts		*Taractrocera*	(チョウ目、セセリチョウ科) の昆虫の総称
grass delphacid		*Sogatella kolophon* (Kirkaldy)	(カメムシ目、ウンカ科) 大洋区
grass demon	オオシロモンセセリ	*Udaspes folus* (Cramer)	(チョウ目、セセリチョウ科) 日本、東洋区
grass eggar		*Lasiocampa trifolii* (Denis et Schiffermüller)	(チョウ目、カレハガ科) 旧北区
grass emerald		*Pseudoterpna pruinata* (Hufnagel)	(チョウ目、シャクガ科) 旧北区
grass-feeding ruby-eye		*Perichares poaceaphaga* Burns	(チョウ目、セセリチョウ科) 新熱帯区
grass fleahopper		*Halticus chrysolepis* Kirkaldy	(カメムシ目、カスミカメムシ科). 新北区
grass flies			frit flies を見よ　豪州での英名
grass flies			opomyzid flies を見よ
grass fly	ムギキモグリバエ	*Meromyza nigriventris* Macquart	(ハエ目、キモグリバエ科) 日本、旧北区
grass grub beetle		*Costelytra zealandica* (White)	(コウチュウ目、コガネムシ科) 豪州区
grass grubs			May beetles (3) を見よ　幼虫の英名
grass jewel (1)	タイワンヒメシジミ	*Chilades trochylus* (Freyer)	(チョウ目、シジミチョウ科) 旧北区、東洋区、エチオピア区
grass jewel (2)		*Freyeria putli* (Kollar)	(チョウ目、シジミチョウ科) 豪州区
grass jewel			eastern grass blue を見よ
grass jewel-blue			grass jewel (2) を見よ

英名	和名	学名	所属、分布、ほか
grass leaf roller			rice leafroller を見よ
grass leafhopper	ウスアカヒメヨコバイ	*Zygina circumscripta* Matsumura	（カメムシ目、オオヨコバイ科）日本、東洋区
grass leafminer moths			grass miner moths を見よ
grass looper			yellow Mocis moth を見よ
grass looper			South American semilooper を見よ
grass mantis		*Thesprotia*	（カマキリ目、カマキリ科）の昆虫の総称
grass miner moths	クサモグリガ科	Elachistidae	（チョウ目）の昆虫の総称
grass miners			grass miner moths を見よ
grass moth (1)		*Chrysoteuchia culmella* Linnaeus	（チョウ目、メイガ科）全北区
grass moth (2)		*Agriphila straminella* Denis et Schiffermüller	（チョウ目、メイガ科）旧北区
grass moths (1)	ツトガ亜科	Crambinae	（チョウ目、メイガ科）の昆虫の総称
grass moths (2)	メイガ科	Pyralidae	（チョウ目）の昆虫の総称
grass nymph			eyed brown を見よ
grass-pear Bryobia		*Bryobia cristata* Duges	（アザミウマ目、アザミウマ科）旧北区
grass planthopper		*Chloriona kolophron* (Kirkaldy)	（カメムシ目、ウンカ科）大洋区
grass rivulet		*Perizoma albulata* (Denis et Schiffermüller)	（チョウ目、シャクガ科）旧北区
grass root Antonina		*Antoninoides parrotti* (Cockerell)	（カメムシ目、コナカイガラムシ科）新北区
grass-root beetle		*Brathinus nitidus* LeConte	（コウチュウ目、ハネカクシ科）新北区。水近くの草根付近に生息
grass-root beetles		Brathinidae	（コウチュウ目）の昆虫の総称
grassroot mealybug		*Rhizoecus rumicis* (Maskell)	（カメムシ目、コナカイガラムシ科）豪州区
grass sawfly		*Pachynematus extensicoris* (Norton)	（ハチ目、ハバチ科）新北区
grass sawfly		*Selandria serva* (Fabricius)	（ハチ目、ハバチ科）旧北区
grass scale			Rhodes grass scale を見よ
grass scales		Acleridae	（カメムシ目）の昆虫の総称
grass scolytid		*Hypothenemus pubescens* Hopkins	（コウチュウ目、キクイムシ科）新北区
grass sharpshooter		*Draeculacephala minerva* Ball	（カメムシ目、オオヨコバイ科）新北区
grass sheath miner		*Cerodontha (Cerodontha) dorsalis* (Loew)	（ハエ目、ハモグリバエ科）新北区
grass sheath miner			barley leafminer (2) を見よ
grass skippers			intermediate skippers を見よ
grass stem maggots			frit flies を見よ
grass thrips		*Aptinothrips rufus* (Haliday)	（アザミウマ目、アザミウマ科）旧北区
grass thrips		*Aptinothrips stylifer* Trybom	（アザミウマ目、アザミウマ科）旧北区

英名	和名	学名	所属、分布、ほか
grass thrips			American grass thrips を見よ　カナダでの英名
grass tubeworm			Texas grass tubeworm moth を見よ
grass tubeworm moth		*Acrolophus arcanella* (Clemens)	（チョウ目、Acrolophidae）新北区
grass tunnel moth		*Philobota chinoptera* Meyrick	（チョウ目、マルハキバガ科）豪州区
grass-veneers			grass moths (2) を見よ
grass wave		*Perconia strigillaria* (Hübner)	（チョウ目、シャクガ科）旧北区
grass-waved			grass wave を見よ
grass webworm		*Pediasia luteolella* Clemens	（チョウ目、メイガ科）新北区
grass webworm		*Calamotropha leptogrammella* (Meyrick)	（チョウ目、メイガ科）豪州区
grass webworm (1)		*Crambus agitatellus* Clemens	（チョウ目、メイガ科）新北区
grass webworm (2)	クロオビクロノメイガ	*Herpetogramma licarsisalis* (Walker)	（チョウ目、メイガ科）日本、全北区、東洋区、大洋区、豪州区。grass webworm moth, tropical grass webworm ともいう
grass webworm			grass moth (2) を見よ
grass webworms			grass moths (1)(2) を見よ　米国での英名
grass worm			black armyworm を見よ
grass worm			armyworm (1)(2) を見よ
grass yellow	キチョウ	*Eurema hecabe* (Linnaeus)	（チョウ目、シロチョウ科）日本、旧北区、東洋区、豪州区。沖縄以北のキチョウは *E. mandarina* (de l'Orza) キタキチョウという
grass yellows	キチョウ属	*Eurema*	（チョウ目、シロチョウ科）の昆虫の総称
grasshopper			Bombay locust を見よ
grasshopper			Javanese grasshopper を見よ
grasshopper bee fly		*Systoechus vulgaris* Loew	（ハエ目、ツリアブ科）新北区
grasshopper egg parasites		*Scelio*	（ハチ目、タマゴクロバチ科）の昆虫の総称
grasshopper maggots		*Blaesoxipha*	（ハエ目、ニクバエ科）の昆虫の総称
grasshoppers (1)	バッタ科	Acrididae	（バッタ目）の昆虫の総称　世界に約 7,000 種、日本に約 60 種
grasshoppers (2)	バッタ上科	Acridoidea	（バッタ目）の昆虫の総称　acridoids ともいう。
grasshoppers (3)	バッタ亜目	Caelifera	（バッタ目）の昆虫の総称
grasshoppers			locusts を見よ
grasshoppers			crickets (2) を見よ
grasslawn armyworm			dark mottled willow を見よ
grassveld sandman		*Spialia agylla* (Trimen)	（チョウ目、セセリチョウ科）エチオピア区
grassveld sylph		*Metisella malgacha* (Boisduval)	（チョウ目、セセリチョウ科）エチオピア区
grass-waved			grass wave を見よ
grassy cutworm			heart and dart を見よ　幼虫の英名

英名	和名	学名	所属、分布、ほか
grassy weevil			granary weevil を見よ
grateful midget moth		*Elaphria grata* Hübner	(チョウ目、ヤガ科) 新北区
grave digger	シデムシ		common burying beetle を見よ
gravedigger beetle		*Necrophorus germanicus* Linnaeus	(コウチュウ目、シデムシ科) 旧北区
gravel grasshopper		*Chorthippus pullus* (Phillipi)	(バッタ目、バッタ科) 旧北区
gray ant		*Solenopsis xyloni maniosa* Wheeler	(ハチ目、アリ科) 新北区。カンキツ害虫。southern fire ant を参照
gray ant		*Formica cinerea* Mayr	(ハチ目、アリ科) 旧北区。亜種 *F. c. neocinerea* Wheeler は grey ant
gray Archips moth		*Archips grisea* (Robinson)	(チョウ目、ハマキガ科) 新北区
gray bamboo longicorn beetle	ハイイロヤハズカミキリ	*Niphona furcata* (Bates)	(コウチュウ目、カミキリムシ科) 日本、東洋区。タケ害虫。gray bamboo longicorn ともいう
graybanded leafroller		*Argyrotaenia mariana* (Fernald)	(チョウ目、ハマキガ科) 新北区。gray-banded leafroller moth ともいう
gray-banded Zale moth		*Zale squamularis* (Drury)	(チョウ目、ヤガ科) 新北区。gray-banded Zale ともいう
gray-based crescent		*Castilia griseobasalis* (Rober)	(チョウ目、タテハチョウ科) 新熱帯区
gray bird grasshopper		*Schistocerca nitens* (Thunberg)	(バッタ目、バッタ科) 新北区。gray bird ともいう
gray blister beetle		*Epicauta cinerea* (Forster)	(コウチュウ目、ツチハンミョウ科) 新北区。ジャガイモ害虫
gray-blotched Epiblema moth		*Epiblema carolinana* (Walsingham)	(チョウ目、ハマキガ科) 新北区
gray blowfly			flesh fly (1) を見よ 米国での英名
gray citrus scale		*Coccus citricola* Campbell	(カメムシ目、カタカイガラムシ科) 旧北区
gray comma		*Nymphalis progne* (Cramer)	(チョウ目、タテハチョウ科) 新北区
gray comma			dark grey anglewing を見よ
gray copper	アイオアベニシジミ	*Lycaena dione* (Scudder)	(チョウ目、シジミチョウ科) 新北区
gray cracker		*Hamadryas februa* (Hübner)	(チョウ目、タテハチョウ科) 新熱帯区。grey cracker ともいう
gray dagger moth		*Acronicta grisea* Walker	(チョウ目、ヤガ科) 新北区。gray dagger ともいう
gray-edged Bomolocha moth		*Hypena madefactalis* Guenée	(チョウ目、ヤガ科) 新北区
gray Ethmia moth		*Ethmia monticola* (Walsingham)	(チョウ目、スエヒロキバガ科) 新北区
gray flesh flies			flesh flies (1) を見よ 南アフリカでの英名
gray fly			sheep bot fly を見よ
gray Furcula moth		*Furcula cinerea* (Walker)	(チョウ目、シャチホコガ科) 新北区
gray-green clubtail		*Arigomphus pallidus* (Rambur)	(トンボ目、サナエトンボ科) 新北区
gray ground cricket		*Allonemobius griseus* (Walker)	(バッタ目、コオロギ科) 新北区

英名	和名	学名	所属、分布、ほか
gray hairstreak		*Strymon clarionensis* Heid	（チョウ目、シジミチョウ科）新熱帯区
gray hairstreak (1)		*Strymon melinus* (Hübner)	（チョウ目、シジミチョウ科）新北区。gray hairstreak butterfly ともいう
gray half-spot moth		*Nedra ramosula* (Guenée)	（チョウ目、ヤガ科）新北区。gray half-spot ともいう
gray larch moth			spotted shoot を見よ
gray larch tortricid			larch tortrix (2) を見よ
gray Lasaia		*Lasaia sessilis* Schaus	（チョウ目、シジミタテハ科）新熱帯区
gray lawn leafhopper		*Exitianus exitiosus* (Uhler)	（カメムシ目、オオヨコバイ科）新北区
gray leafroller moth		*Syndemis afflictana* (Walker)	（チョウ目、ハマキガ科）新北区
gray looper moth		*Rachiplusia ou* (Guenée)	（チョウ目、ヤガ科）新北区、新熱帯区
gray marble	ハイイロツマキチョウ	*Anthocharis lanceolata* Lucas	（チョウ目、シロチョウ科）新北区
gray-marked mugwort leafhopper	ヨモギシロテンヨコバイ	*Mileewa margheritae* (Distant)	（カメムシ目、オオヨコバイ科）日本、旧北区、東洋区
gray-marked mugwort leafhopper			green willow leafhopper を見よ
gray marvel moth		*Anterastria teratophora* Herrich-Schäffer	（チョウ目、ヤガ科）新北区。gray marvel, grey marvel ともいう
gray metalmark		*Apodemia palmerii* (Edwards)	（チョウ目、シジミタテハ科）新北区
gray ministreak		*Ministrymon azia* (Hewitson)	（チョウ目、シジミチョウ科）新北区
gray patched prominent moth		*Dasylophia thyatiroides* (Walker)	（チョウ目、シャチホコガ科）新北区
gray pear scale			Italian pear scale を見よ
gray petaltail		*Tachopteryx thoreyi* (Hagen)	（トンボ目、ムカシヤンマ科）新北区
gray pineapple mealybug		*Dysmicoccus neobrevipes* Beardsley	（カメムシ目、コナカイガラムシ科）新北区、大洋区。grey pineapple mealybug とも記す
gray quaker moth		*Orthosia alurina* (Smith)	（チョウ目、ヤガ科）新北区
gray quaker moth		*Morrisonia mucens* (Hübner)	（チョウ目、ヤガ科）新北区
gray rough-wing moth		*Acleris scabrana* (Denis et Schiffermüller)	（チョウ目、ハマキガ科）新北区
gray sallow moth		*Psaphida grandis* (J. B. Smith)	（チョウ目、ヤガ科）新北区
gray sanddragon		*Progomphus borealis* McLachlan	（トンボ目、サナエトンボ科）新北区
gray scale			gray citrus scale を見よ
gray scoopwing moth		*Callizzia amorata* Packard	（チョウ目、ツバメガ科）新北区
gray shieldback		*Atlanticus dorsalis* (Burmeister)	（バッタ目、キリギリス科）新北区
gray silverfish		*Ctenolepisma longicaudata* Escherich	（シミ目、シミ科）汎世界

英名	和名	学名	所属、分布、ほか
gray skipper		*Mnasicles hicetaon* Godman	(チョウ目、セセリチョウ科) 新熱帯区
gray Sparganothis moth			aproned Cenopis moth を見よ
gray-spotted scrub-hairstreak		*Strymon astiocha* (Prittwitz)	(チョウ目、シジミチョウ科) 新熱帯区
gray spring moth		*Lomographa glomeraria* (Grote)	(チョウ目、シャクガ科) 新北区
gray spruce looper moth		*Caripeta divisata* Walker	(チョウ目、シャクガ科) 新北区
gray sugarcane mealybug			grey sugarcane mealybug を見よ
gray sunflower seed weevil		*Smicronyx sordidus* LeConte	(コウチュウ目、ゾウムシ科) 新北区。grey sunflower seed weevil とも記す
gray swordgrass moth		*Xylena cineritia* (Grote)	(チョウ目、ヤガ科) 新北区
gray tiger longicorn	ムネマダラトラカミキリ	*Xylotrechus grayii grayii* (White)	(コウチュウ目、カミキリムシ科) 日本、旧北区。アオギリ、キリ害虫。grey tiger longicorn ともいわれる
gray Tinsel		*Catapaecilma major emas* (Tinsel)	(チョウ目、シジミチョウ科) 東洋区
gray twig prunner			twig prunner を見よ
gray-waisted skimmer		*Cannaphila insularis* Kirby	(トンボ目、トンボ科) 新北区、新熱帯区
gray walking stick		*Pseudosermyle straminea* (Scudder)	(ナナフシ目、Heteromeniidae) 新北区
gray willow leaf beetle		*Tricholochmaea decora decora* (Say)	(コウチュウ目、ハムシ科) 新北区
gray-winged owlet moth		*Lesmone griseipennis* (Grote)	(チョウ目、ヤガ科) 新北区
gray-winged Pareuchaetes		*Pygarctia eglenensis* (Clemens)	(チョウ目、ヒトリガ科) 新北区
grayback			body louse を見よ
graybacks	ムカシヤンマ科	Petaluridae	(トンボ目) の昆虫の総称
grayish Zanclognatha moth		*Zanclognatha pedipilalis* (Guenée)	(チョウ目、ヤガ科) 新北区。grayish Zanclognatha ともいう
grayling			ジャノメチョウ科のチョウ。Satyrs を見よ
grayring	キマダラタカネジャノメ	*Hipparchia semele* (Linnaeus)	(チョウ目、タテハチョウ科) 旧北区。grayling butterfly ともいう
Gray's cattle-heart		*Parides aglaope* (Gray)	(チョウ目、アゲハチョウ科) 新熱帯区
Gray's polyptychus		*Polyptychoides grayi* (Walker)	(チョウ目、スズメガ科) エチオピア区
grease moth		*Aglossa cuprina* (Zeller)	(チョウ目、メイガ科) 全北区、豪州区、東洋区、新熱帯区
grease moth			small tabby を見よ
grease moth			large tabby を見よ
greasewood moth		*Agapema galbina* Clemens	(チョウ目、ヤママユガ科) 新北区
greasy cutworm			black cutworm を見よ 米国での幼虫の英名

英名	和名	学名	所属、分布、ほか
greasy cutworm moth			black cutworm を見よ
greasy Epitola		*Cephetola sublustris* (Bethune-Baker)	(チョウ目、シジミチョウ科) エチオピア区
greasy frit			marsh fritillary を見よ
greasy fritillary	ヒョウモンモドキ	*Melitaea artemis* Denis et Schiffermüller	(チョウ目、タテハチョウ科) 旧北区
great Agrias		*Agrias aedon rodriguezi* Schaus	(チョウ目、タテハチョウ科) 新熱帯区
great archduke		*Lexias cyanipardus* (Butler)	(チョウ目、タテハチョウ科) 東洋区。*L. c. sandakana* (Fruhstorfer) も同英名
great arctic			Nevada arctic を見よ
great ash sphinx		*Sphinx chersis* (Hübner)	(チョウ目、スズメガ科) 新北区。great ash sphinx moth ともいう
great banded grayling		*Brintesia circe* Fabricius	(チョウ目、タテハメチョウ科) 旧北区
Great Basin click beetle			Great Britain wireworm を見よ
Great Basin fritillary			Egleis fritillary を見よ
Great Basin ringlet			northwest ringlet を見よ
Great Basin snaketail		*Ophiogomphus morrisoni* Selys	(トンボ目、サナエトンボ科) 新北区
Great Basin sootywing		*Pholisora libya* (Scudder)	(チョウ目、セセリチョウ科) 新北区
Great Basin sootywing			Mojave sootywing を見よ
Great Basin tent caterpillar		*Malacosoma fragilis* (Stretch)	(チョウ目、カレハガ科) 新北区
Great Basin white			Becker's white を見よ
Great Basin wireworm			Great Britain wireworm を見よ
Great Basin wood nymph		*Cercyonis sthenele* (Boisduval)	(チョウ目、タテハチョウ科) 新北区
great black cutworm			cutworm (3) を見よ
great black-vein	タカムクシロチョウ	*Aporia agathon* (Gray)	(チョウ目、シロチョウ科) 旧北区、東洋区
great black wasp		*Sphex pennsylvanicus* Linnaeus	(ハチ目、アナバチ科) 新北区
great black water beetle		*Hydrous piceus* (Linnaeus)	(コウチュウ目、ガムシ科) 旧北区
great blue hairstreak			great purple hairstreak を見よ
great blue hookwing	オオルリモンタテハ	*Napeocles jucunda* Hübner	(チョウ目、タテハチョウ科) 新熱帯区
great blue mime		*Chilasa paradoxa aenigma* Zincken	(チョウ目、アゲハチョウ科) 東洋区。great mime を参照
great blue Papilio			giant blue swallowtail を見よ
great blue skimmer		*Libellula vibrans* (Fabricius)	(トンボ目、トンボ科) 新北区
great Borneo cicada			giant cicada を見よ

英名	和名	学名	所属、分布、ほか
Great Britain wireworm		*Ctenicera pruinina* (Horn)	（コウチュウ目、コメツキムシ科）新北区
great brocade	オオシラホシヤガ	*Eurois occulta* (Linnaeus)	（チョウ目、ヤガ科）日本、旧北区。great brocade moth ともいう
great brocaded rustic			great brocade を見よ
great brown twist		*Archips podana* (Scopoli)	（チョウ目、ハマキガ科）全北区
great brownie		*Miletus symethus* (Cramer)	（チョウ目、シジミチョウ科）東洋区
great capricorn beetle		*Cerambyx cerdo* Linnaeus	（コウチュウ目、カミキリムシ科）旧北区
great carpenter bee			eastern carpenter bee を見よ
great case		*Coleophora wockeella* Zeller	（チョウ目、ツツミノガ科）旧北区
great copper	オオベニシジミ	*Lycaena xanthoides* (Boisduval)	（チョウ目、シジミチョウ科）新北区
great cycadian		*Eumaeus childrenae* (Gray)	（チョウ目、シジミチョウ科）新熱帯区
great darkie		*Allotinus multistrigatus* de Nicéville	（チョウ目、シジミチョウ科）東洋区。
great darkie			crenulate darkie を見よ
great dart		*Agrotis crassa* Stephens	（チョウ目、ヤガ科）旧北区
great diving beetle			water beetle (1) を見よ
great diving beetles			diving beetles (2) を見よ
great dragonfly			brown hawker を見よ
great duffer		*Discophora timora* Westwood	（チョウ目、タテハチョウ科）東洋区
great eggfly			common eggfly を見よ
great Emesis		*Emesis mandana furor* Butler et Druce	（チョウ目、シジミタテハ科）新熱帯区。variable Emesis を参照
great Eurybia	アオメシジミタテハ	*Eurybia patrona* Weymer	（チョウ目、シジミタテハ科）新熱帯区。*E. p. persona* Staudinger も同英名
great evening brown	ウスベニコノマチョウ	*Melanitis zitenius* (Herbst)	（チョウ目、タテハチョウ科）東洋区
great figtree borer		*Batocera boisduvali* (Hope)	（コウチュウ目、カミキリムシ科）豪州区
great five-ring		*Ypthima dohertyi* (Moore)	（チョウ目、タテハチョウ科）東洋区
great gad fly		*Tabanus bovinus* Linnaeus	（ハエ目、アブ科）全北区
great golden digger wasp			golden digger wasp を見よ　great golden digger ともいう
great gray copper			great copper を見よ
great green bush-cricket	チョウセンヤブキリ	*Tettigonia viridissima* (Linnaeus)	（バッタ目、キリギリス科）日本、旧北区
great green grasshopper			great green bush-cricket を見よ
great green moth			green arches を見よ
great grey mountain moth		*Aletia mellifera* Walker	（チョウ目、ヤガ科）豪州区
great grig		*Cyphoderris monstrosa* Uhler	（バッタ目、Prophalangopsidae）新北区

英名	和名	学名	所属、分布、ほか
great Helen	イスワラモンキアゲハ	*Papilio iswara* White	（チョウ目、アゲハチョウ科）東洋区
great hockeystick sailer		*Phaedyma aspasia* (Leech)	（チョウ目、タテハチョウ科）東洋区
great hooktail		*Paragomphus magnus* Fraser	（トンボ目、サナエトンボ科）エチオピア区
great imperial		*Jacoona anasuja anasuja* (Felder)	（チョウ目、シジミチョウ科）東洋区
great Jay			pale triangle を見よ
great jewelmark		*Anteros kupris* Hewitson	（チョウ目、シジミタテハ科）新熱帯区
great kidney		*Drasteria grandirena* (Haworth)	（チョウ目、ヤガ科）旧北区
great leafwing		*Memphis schuasiana* (Godman et Salvin)	（チョウ目、タテハチョウ科）新熱帯区
great leopard moth		*Ecpantheria scribonia* (Stoll)	（チョウ目、ヒトリガ科）新北区。great leopard ともいう
great Marquis		*Bassarona dunya dunya* (Doubleday)	（チョウ目、タテハチョウ科）東洋区
great millet-stem fly			sorghum shootfly を見よ
great millet-stem maggot			sorghum shootfly を見よ　幼虫の英名
great mime	オオムラサキアゲハ	*Chilasa paradoxa* Zincken	（チョウ目、アゲハチョウ科）東洋区
great Mormon		*Papilio memnon* Linnaeus	（チョウ目、アゲハチョウ科）東洋区、豪州区。日本亜種は *P. m. thunbergii* von Siebold ナガサキアゲハ。*P. m. agenor* Linnaeus　も同英名
great nawab	フタオチョウ	*Polyura eudamippus* (Doubleday)	（チョウ目、タテハチョウ科）旧北区、東洋区。*P. e. formosana* (Rothschild), *P. e. weismanni* (Fritze) も同英名
great Nephele		*Nephele aequivalens* (Walker)	（チョウ目、スズメガ科）エチオピア区
great northern sulphur			giant sulphur を見よ
great oak beauty		*Hypomecis roboraria* (Denis et Schiffermüller)	（チョウ目、シャクガ科）旧北区。日本亜種は *H. r. displicens* (Butler) ハミスジエダシャク
great orange awlet		*Burara etelka* Hewitson	（チョウ目、セセリチョウ科）東洋区
great orange tip	ツマベニヤマキチョウ	*Anteos menippe* (Hübner)	（チョウ目、シロチョウ科）新熱帯区
great orange tip	ツマベニチョウ属	*Hebomoia*	（チョウ目、シロチョウ科）の昆虫の総称
great orange-tip		*Hebomoia glaucippe* (Linnaeus)	（チョウ目、シロチョウ科）旧北区、東洋区。*H. g. aturia* Fruhstorfer, *H. g. formosana* Fruhstorfer、日本のツマベニチョウ *H. g. liukiuyensis* Fruhstorfer も同英名
Great Pains cricket			Mormon cricket を見よ
great peacock			giant emperor を見よ
great peacock moth			giant emperor を見よ　米国での英名
great plains			Mormon cricket を見よ

英名	和名	学名	所属、分布、ほか
great polyptychus		*Pseudoclanis molitor* (Rothschild et Jordan)	(チョウ目、スズメガ科) エチオピア区
great pondhawk		*Erythemis vesiculosa* (Fabricius)	(トンボ目、トンボ科) 新北区、新熱帯区
great prominent		*Peridea anceps* (Goeze)	(チョウ目、シャチホコガ科) 旧北区
great pumpkin fly			pumpkin fly を見よ
great purple hairstreak		*Atlides halesus* (Cramer)	(チョウ目、シジミチョウ科) 新北区
great red sedge		*Phryganea grandis* Linnaeus	(トビケラ目、トビケラ科) 旧北区
great satyr		*Aulocera padma* (Kollar)	(チョウ目、タテハチョウ科) 東洋区
great satyr			Mermeria wood nymph を見よ
great scrub-hairstreak		*Strymon gabatha* (Hewitson)	(チョウ目、シジミチョウ科) 新熱帯区
great sergeant	ラリムナミスジ	*Athyma larymna* (Doubleday)	(チョウ目、タテハチョウ科) 東洋区
great silver beetle			great black water beetle を見よ
great silver diving beetle			great black water beetle を見よ
great silver water beetle			great black water beetle を見よ
great sooty satyr	フェルラジャノメ	*Satyrus ferula* (Fabricius)	(チョウ目、タテハチョウ科) 旧北区
great southern white	ガルフオオシロチョウ	*Ascia monuste* (Linnaeus)	(チョウ目、シロチョウ科) 新北区、新熱帯区
great spangled fritillary	オオギンボシヒョウモン (キベレギンボシヒョウモン)	*Speyeria cybele* (Fabricius)	(チョウ目、タテハチョウ科) 新北区
great spotted blue		*Phengaris atroguttata* (Oberthür)	(チョウ目、シジミチョウ科) 東洋区
great spreadwing			giant spreadwing を見よ
great spreadwing damselfly			giant spreadwing を見よ
great striped moth			common yellow underwing moth を見よ
great stripestreak		*Laothus gibberosa* Hewitson	(チョウ目、シジミチョウ科) 新熱帯区
great swift		*Pelopidas assamensis* (de Nicéville)	(チョウ目、セセリチョウ科) 東洋区
great tiger			black woolly-bear を見よ
great tiger-mimic		*Papilio zagreus* Doubleday	(チョウ目、アゲハチョウ科) 新熱帯区
great tiger moth			black woolly-bear を見よ　米国での英名
great Wallacean		*Zethera incerta* (Hewitson)	(チョウ目、タテハチョウ科) 東洋区
great waspel			hornet (1) を見よ
great white skipper			large white skipper を見よ
great windmill	ダサラダベニモンアゲハ	*Byasa dasarada* (Moore)	(チョウ目、アゲハチョウ科) 東洋区
great windmill (1)	オオベニモンアゲハ	*Byasa polyeuctes* (Doubleday)	(チョウ目、アゲハチョウ科) 旧北区、東洋区

英名	和名	学名	所属、分布、ほか
great winter moth			mottled umber を見よ
great yellow humble bee		*Bombus distinguendus* Morawitz	(ハチ目、ミツバチ科) 全北区
great yellow sailer		*Neptis radha* Moore	(チョウ目、タテハチョウ科) 東洋区
great zebra		*Graphium xenocles* (Doubleday)	(チョウ目、アゲハチョウ科) 東洋区
greater angle-wing			angular-winged katydid (1) を見よ
greater arctic			Nevada arctic を見よ
greater arid-land katydid		*Neobarrettia spinosa* (Caudell)	(バッタ目、キリギリス科) 新北区
greater arid-land predaceous katydid			greater arid-land katydid を見よ
greater banded hornet		*Vespa tropica* Linnaeus	(ハチ目、スズメバチ科) 東洋区、エチオピア区
greater bee fly			bee fly (1) を見よ
greater black-letter dart moth			spotted cutworm moth (2) を見よ
greater brown-cloaked bell		*Epiblema roborana* Denis et Schiffermüller	(チョウ目、ハマキガ科) 旧北区
greater Calephelis		*Calephelis sixola* McAlpine	(チョウ目、シジミタテハ科) 新熱帯区
greater cloaked marble		*Apotomis turbidana* (Treitschke)	(チョウ目、ハマキガ科) 旧北区
greater coconut spike moth			coconut-spike moth (1) を見よ
greater cotton-grass dwarf		*Biselachista albidella* (Nylander)	(チョウ目、クサモグリガ科) 旧北区
greater cream wave		*Scopula lactata* (Haworth)	(チョウ目、シャクガ科) 旧北区
greater date moth		*Arenipses sabella* Hampson	(チョウ目、メイガ科) 旧北区、東洋区
greater emperor			giant emperor を見よ
greater European spruce beetle			European spruce beetle を見よ
greater forceptails		*Aphylla*	(トンボ目、サナエトンボ科) の昆虫の総称
greater grain weevil			maize weevil を見よ
greater grapevine looper		*Eulithis gracilineata* Guenée	(チョウ目、シャクガ科) 新北区。幼虫はブドウその他につく。成虫は greater grapevine looper moth
greater grass emerald			grass emerald を見よ
greater grey shade		*Cnephasia chrysantheana* (Duponchel)	(チョウ目、ハマキガ科) 旧北区
greater hemlock flat-body		*Depressaria weirella* Stainton	(チョウ目、マルハキバガ科) 旧北区
greater horntail			horntail (1) を見よ
greater Idia moth		*Idia majoralis* (Smith)	(チョウ目、ヤガ科) 新北区。greater Idia ともいう
greater large blue			large blue (1) を見よ
greater metalmark		*Synargis nymphidioides septentrionalis* Callaghan, Llorente et Luis	(チョウ目、シジミタテハ科) 新熱帯区

英名	和名	学名	所属、分布、ほか
greater oak dagger moth		*Acronicta lobeliae* Guenée	（チョウ目、ヤガ科）新北区
greater peacock awl		*Allora major* (Rothschild)	（チョウ目、セセリチョウ科）豪州区
greater pear sucker			pear sucker を見よ
greater purple		*Agonopterix kaekeritziana* (Linnaeus)	（チョウ目、マルハキバガ科）旧北区
greater red dart moth		*Abagrotis alternata* Grote	（チョウ目、ヤガ科）新北区
greater rice weevil			maize weevil を見よ
greater sandy dung beetle		*Euoniticellus africanus* (Harold)	（コウチュウ目、コガネムシ科）豪州区
greater satin lutestring		*Tethea fluctuosa* Hübner	（チョウ目、トガリバガ科）旧北区
greater shieldback		*Idiostatus major* Caudell	（バッタ目、キリギリス科）新北区
greater silver water-beetle		*Hydrophilus piceus* (Linnaeus)	（コウチュウ目、ガムシ科）旧北区。イギリスの最大型甲虫
greater sod webworm			sod webworm moth を見よ
greater swallow prominent			swallow prominent を見よ
greater sweetpotato webworm moth		*Herpetogramma fluctuosalis* (Lederer)	（チョウ目、メイガ科）新北区
greater thorn-tipped longhorn beetle		*Pogonocherus hispidulus* (Piller et Mitterpacher)	（コウチュウ目、カミキリムシ科）旧北区
greater water boatmen			backswimmers を見よ
greater wax moth	ハチノスツヅリガ	*Galleria mellonella* (Linnaeus)	（チョウ目、メイガ科）日本、全北区、東洋区、豪州区。幼虫はミツバチの巣に生息
greater wax moth parasite	ハチミツガサムライコマユバチ	*Apanteles galleriae* Wilkinson	（ハチ目、コマユバチ科）日本。ハチノスツヅリガに寄生
greater wax moths			wax moths を見よ
Grecian anomalous blue		*Polyommatus aroaniensis* (Brown)	（チョウ目、シジミチョウ科）旧北区
Grecian Argus		*Kretania eurypilus* (Freyer)	（チョウ目、シジミチョウ科）旧北区
Grecian copper		*Lycaena ottomanus* (Lefebvre)	（チョウ目、シジミチョウ科）旧北区
Grecian grayling		*Pseudochazara graeca* (Staudinger)	（チョウ目、タテハチョウ科）旧北区
Grecian shoemaker			blue-spotted firewing を見よ
Greck Mazanne blue		*Polyommatus bellis* (Freyer)	（チョウ目、シジミチョウ科）旧北区
greedy scale		*Aspidiotus camelliae* Signoret	（カメムシ目、マルカイガラムシ科）新北区。温室害虫
greedy scale		*Hemiberlesia latestei* (Cockerell)	（カメムシ目、マルカイガラムシ科）豪州区
greedy scale (1)	ツバキマルカイガラムシ	*Hemiberlesia rapax* (Comstock)	（カメムシ目、マルカイガラムシ科）日本、汎世界。ツバキ、カンキツ、チャなどの害虫
Greek clouded yellow		*Colias libanotica heldreichii* Staudinger	（チョウ目、シロチョウ科）旧北区

英名	和名	学名	所属、分布、ほか
Greek honey bee		*Apis mellifica cecropia* Kiesenwetter	（ハチ目、ミツバチ科）旧北区
Greek-lettered argent		*Argyresthia goedartella* (Linnaeus)	（チョウ目、メムシガ科）旧北区
Greek mazarine blue		*Cyaniris helena* Staudinger	（チョウ目、シジミチョウ科）旧北区
green abietis aphid	トドミドリオオアブラムシ	*Cinara matsumurana* Hille Ris Lambers	（カメムシ目、アブラムシ科）日本
green-and-orange playboy		*Pilodeudorix mimeta* (Karsch)	（チョウ目、シジミチョウ科）エチオピア区
green and purple carpet moth		*Hydriomena similata* Walker	（チョウ目、シャクガ科）豪州区
green anglewing		*Polygonia faunus rusticus* (W. H. Edwards)	（チョウ目、タテハチョウ科）新北区
green ant			red ant (1) を見よ
green Antinephele		*Antinephele muscosa* Holland	（チョウ目、スズメガ科）エチオピア区
green apple aphid			apple aphid (1) を見よ
green apple aphis			apple aphid (1) を見よ
green apple leaf weevil			green leaf weevil (1) を見よ
green apple weevil			green leaf weevil (1) を見よ
green arches	アオバヤガ	*Anapleotoides prasina* (Denis et Schiffermüller)	（チョウ目、ヤガ科）日本、全北区。green arches moth ともいう
green aspen leafroller moth		*Apotomis removana* (Kearfott)	（チョウ目、ハマキガ科）新北区
green aspen leaftier moth		*Pandemis canadana* Kearfott	（チョウ目、ハマキガ科）新北区。green aspen leaftier ともいう
green awl		*Hasora discolor* (Felder et Felder)	（チョウ目、セセリチョウ科）豪州区
green awl		*Hasora salanga* (Plötz)	（チョウ目、セセリチョウ科）東洋区
green awlet		*Burara vasutana* Moore	（チョウ目、セセリチョウ科）東洋区
green-backed ruby-eye		*Perichares adela* (Hewitson)	（チョウ目、セセリチョウ科）新熱帯区
green-backed ruby-eye			Caribbean ruby-eye を見よ
green-backed ruby-eye			African mallow を見よ
green-backed skipper			Caribbean ruby-eye を見よ
green-banded jewel	カムサベニシキシジミ	*Hypochrysops theon* (C. et R. Felder)	（チョウ目、シジミチョウ科）豪州区
green-banded line-blue		*Nacaduba cyanea* (Cramer)	（チョウ目、シジミチョウ科）豪州区。*N. c. arinia* (Oberthür) は tailed green-banded blue という
green-banded ruby-eye		*Alera haworthiana* (Swainson)	（チョウ目、セセリチョウ科）新熱帯区
green-banded swallowtail	ニレウスルリアゲハ	*Papilio nireus* Linnaeus	（チョウ目、アゲハチョウ科）エチオピア区。*P. n. lyaeus* Doubleday も同英名
green-banded swallowtail			apple-green swallowtail を見よ

英名	和名	学名	所属、分布、ほか
green banded white		*Pieris krueperi devta* (de Nicéville)	（チョウ目、シロチョウ科）東洋区。Krueper's small white を参照
green baron		*Psaltoda magnifica* Moulds	（カメムシ目、セミ科）豪州区
green baron		*Euthalia adonia pinwilli* Pendlebury et Corbet	（チョウ目、タテハチョウ科）東洋区
green beetle			rose chafer (1) を見よ
green beetle			manuka beetle を見よ
green bird grasshopper			green valley grasshopper を見よ
green blister beetle	ミドリツチハンミョウ	*Lytta caraganae* Pallas	（コウチュウ目、ツチハンミョウ科）日本、旧北区
green blister beetle		*Lytta chloris* (Fall)	（コウチュウ目、ツチハンミョウ科）新北区
green blotched moth			spangled green moth を見よ
green blow fly	ミドリキンバエ	*Lucilia illustris* Meigen	（ハエ目、クロバエ科）日本、全北区
green blow fly (1)	ヒロズキンバエ	*Lucilia sericata* (Meigen)	（ハエ目、クロバエ科）日本、全北区、豪州区。ヒツジキンバエに似た普通種
green blowflies			greenbottle flies (1) を見よ
green-body matchstick		*Carnarvonella* sp.	（バッタ目、Eumastacidae）豪州区
green borer		*Anerastia ablutella* (Zeller)	（チョウ目、メイガ科）旧北区
green botfly	キンバエ	*Lucilia caesar* Linnaeus	（ハエ目、クロバエ科）日本、旧北区。幼虫は動物死体、糞などに発生
green botfly			green blow fly (1) を見よ
green bottle			green blow fly (1) を見よ　greenbottle とも記す
greenbottle fly		*Phaenicia sericata* (Meigen)	（ハエ目、クロバエ科）大洋区
green bottle fly			green blow fly (1) を見よ　greenbottle fly とも記す
green bottle fly			green botfly を見よ
greenbottle fly			bronze bottle fly を見よ
greenbottle fly			blow fly を見よ
green-brindled crescent		*Allophyes oxyacanthae* (Linnaeus)	（チョウ目、ヤガ科）旧北区
green-brindled crescent moth			green-brindled crescent を見よ
green brindled dot		*Valeria oleagina* (Denis et Schiffermüller)	（チョウ目、ヤガ科）旧北区
green broad-winged planthopper			green flatted planthopper を見よ
green broomweed looper		*Narraga fimetaria* Grote et Robinson	（チョウ目、シャクガ科）新北区
green bud moth			common long-cloaked marble を見よ
green budworm			grey bud moth を見よ　green budworm moth ともいう
green budworm			marbled orchard tortrix を見よ
green bunyip		*Tamasa rainbowi* Ashton	（カメムシ目、セミ科）豪州区

英名	和名	学名	所属、分布、ほか
green buprestid	アオマダラタマムシ	*Nipponbuprestis amabilis* (Snellen van Vollenhoven)	(コウチュウ目、タマムシ科) 日本
green cantharid			blister beetle を見よ
green capsid bug			common green capsid (1) を見よ
green carab beetle		*Calosoma schayeri* Erichson	(コウチュウ目、オサムシ科) 豪州区
green carpenter bees		*Lestis*	(ハチ目、アシブトハナバチ科) の昆虫の総称
green carpet		*Colostygia pectinataria* (Knoch)	(チョウ目、シャクガ科) 旧北区
green Carphoides moth		*Carphoides setigera* Rindge	(チョウ目、シャクガ科) 新北区
green carposinid moth		*Carposina viridis* (Walsingham)	(チョウ目、シンクイガ科) 大洋区
green-celled cattleheart	オオマエモンジャコウアゲハ	*Parides childrenae childrenae* (Gray)	(チョウ目、アゲハチョウ科) 新北区、新熱帯区
green-clouded swallowtail			spicebush swallowtail を見よ
green chafer	アオドウガネ	*Anomala albopilosa* (Hope)	(コウチュウ目、コガネムシ科) 日本、旧北区。多くの作物、果樹、樹木の害虫
green chafers		*Dichelonyx*	(コウチュウ目、コガネムシ科) の昆虫の総称
green chrysanthemum aphid			pale chrysanthemum aphid を見よ　北米での英名
green cicada	アオネゼミ	*Tregganua sibylla* (Stål)	(カメムシ目、セミ科) 豪州区
green citrus aphid			spirea aphid を見よ
green citrus longhorn		*Chelidonium gibbicolle* (White)	(コウチュウ目、カミキリムシ科) 豪州区。カンキツ害虫
green clover worm			green cloverworm moth を見よ
green cloverworm moth		*Plathypena scabra* (Fabricius)	(チョウ目、ヤガ科) 新北区
green club-tailed dragonfly			green snaketail を見よ
green cluster fly		*Dasyphora cyanella* (Meigen)	(ハエ目、イエバエ科) 旧北区
green cochlid	アオイラガ	*Parasa consocia* (Walker)	(チョウ目、イラガ科) 日本、旧北区、東洋区。ナシ、カキ、クリなどの害虫
green coconut bug		*Amblypelta cocophaga* China	(カメムシ目、ヘリカメムシ科) 東洋区。ヤシ害虫
green coffee caterpillar		*Eacles magnifica* Walker	(チョウ目、ヤママユガ科) 新熱帯区
green coffee scale			green scale を見よ
green comma		*Polygonia faunus* (Edwards)	(チョウ目、タテハチョウ科) 新北区
green commodore		*Sumalia daraxa* (Doubleday)	(チョウ目、タテハチョウ科) 東洋区
green copper		*Lycaena kasyapa* (Moore)	(チョウ目、シジミチョウ科) 東洋区
green corn aphid			corn leaf aphid を見よ
green cotton aphid			cotton aphid を見よ

英名	和名	学名	所属、分布、ほか
green cranberry span-worm			green span-worm を見よ
green cricket hunter			blue mud dauber を見よ
green currant worm			small gooseberry sawfly (1) を見よ　幼虫の英名
green cutworm		*Neumichtis saliaris* (Guenée)	(チョウ目、ヤガ科) 豪州区
green cutworm		*Anicla infecta* (Ochsenheimer)	(チョウ目、ヤガ科) 新北区、新熱帯区。green cutworm moth ともいう
green dancer			dragonflies (1) を見よ
green dark mayfly		*Ephemera danica* Müller	(カゲロウ目、モンカゲロウ科) 旧北区
green darner			common green darner を見よ
green demon Charaxes		*Charaxes cedreatis* Hewitson	(チョウ目、タテハチョウ科) エチオピア区
green dock beetle		*Gastroidea cyanea* Melsh	(コウチュウ目、ハムシ科) 新北区
green dock beetle (1)		*Gastrophysa viridula* (De Geer)	(コウチュウ目、ハムシ科) 旧北区
green drab moth		*Ophiusa tirhaca* (Cramer)	(チョウ目、ヤガ科) 旧北区、東洋区、豪州区、エチオピア区
green dragontail	スソビキアゲハ	*Lamproptera meges* (Zinken)	(チョウ目、アゲハチョウ科) 東洋区
green drake			英国でカゲロウを指す。green dark mayfly を見よ
green drake			brown mayfly を見よ　亜成虫の英名
greendrakes		*Ephemera*	(カゲロウ目、モンカゲロウ科) の昆虫の総称
green duke		*Euthalia sahadeva* (Moore)	(チョウ目、タテハチョウ科) 東洋区
green emerald damselfaly			willow emerald damselfly を見よ
green emperor		*Anax gibbosulus* Rambur	(トンボ目、ヤンマ科) 豪州区
green-eyed hook-tailed dragonfly			small pincertail を見よ
green-eyed monster			Buquet's vagrant を見よ
green-eyed owl-butterfly		*Dynastor macrosiris strix* Bates	(チョウ目、タテハチョウ科) 新熱帯区。black dynastor を参照
green-eyed skimmers	エゾトンボ科	Corduliidae	(トンボ目) の昆虫の総称　多くの種の翅に斑紋あり
green-faced clubtail		*Gomphus viridifrons* Hine	(トンボ目、サナエトンボ科) 新北区
green flasher		*Astraptes talus* (Cramer)	(チョウ目、セセリチョウ科) 新熱帯区
green flatted planthopper	アオバハゴロモ	*Geisha distinctissima* (Walker)	(カメムシ目、アオバハゴロモ科) 日本、旧北区、東洋区。ダイズ、果樹、樹木の害虫
green flattened chafer	ヒラタアオコガネ	*Anomala octiescostata* (Burmeister)	(コウチュウ目、コガネムシ科) 日本
green flies			aphids (1) (2) を見よ　greenflies とも記す
green flies			horse flies (1) を見よ
green flower beetle		*Anomala cupripes* (Hope)	(コウチュウ目、コガネムシ科) 東洋区
green flower chafer	アオハナムグリ	*Eucetonia roelofsi* (Harold)	(コウチュウ目、コガネムシ科) 日本
green fly			green leafhopper (1) を見よ　インドでの英名

英名	和名	学名	所属、分布、ほか
green fly			（緑色の）アブラムシ。英国での英名
green fly			small green leafhopper を見よ
greenfly			green peach aphid を見よ
greenfly			rose aphid (1) を見よ
green foresters			foresters (2) を見よ
green frogfly			green leafhopper (1)(2) を見よ
green fruit beetle		*Cotinis mutabilis* Gory et Percheron	（コウチュウ目、コガネムシ科）新北区、新熱帯区
green fruitworm (1)		*Lithophane antennata* (Brace)	（チョウ目、ヤガ科）新北区
green fruitworm (2)		*Himella intractata* Morrison	（チョウ目、ヤガ科）新北区
green fruitworm moth			green fruitworm (2) を見よ
green fruitworms		*Orthosia*	（チョウ目、ヤガ科）の昆虫の総称
green garden looper			silver Y moth (2) を見よ
green gem		*Microchrysa flavicornis* (Meigen)	（ハエ目、ミズアブ科）旧北区
green gem		*Poritia pleurata* Hewitson	（チョウ目、シジミチョウ科）東洋区
green gomphid			green snaketail を見よ
green grape capsid	ツマグロアオカスミカメムシ	*Lygocoris spinolae* (Meyer-Dur)	（カメムシ目、カスミカメムシ科）日本、旧北区
green grass bug			southern green stink bug を見よ
green grass Pyrgomorph		*Atractomorpha similis* Bolivar	（バッタ目、オンブバッタ科）豪州区、大洋区
green grasshopper			great green bush-cricket を見よ
green grasshoppers			long-horned grasshoppers (1) を見よ
greengrocer		*Cyclochila australasiae* (Donovan)	（カメムシ目、セミ科）豪州区。greengroce cicada ともいう
green ground beetle		*Chlaenius sericeus* Forster	（コウチュウ目、オサムシ科）新北区
green guava mealybug			green shield scale を見よ
green hairstreak	ミドリコツバメ	*Callophrys rubi* (Linnaeus)	（チョウ目、シジミチョウ科）旧北区。green hairstreak butterfly ともいう
green hairy caterpillar		*Rivula atimeta* (Swinhoe)	（チョウ目、ヤガ科）東洋区
green hairy caterpillar		*Rivula basalis* Hampson	（チョウ目、ヤガ科）東洋区
green Hawaiian blue			Hawaiian blue を見よ
green hawker		*Aeschna viridis* Eversmann	（トンボ目、ヤンマ科）旧北区
greenhead		*Tabanus nigrovittatus* Macquart	（ハエ目、アブ科）新北区
greenhead			horse flies (2) を見よ
greenhead			American horse-fly を見よ
green-head ant		*Rhytidoponera metallica* (Smith)	（ハチ目、アリ科）豪州区。green ant ともいう

英名	和名	学名	所属、分布、ほか
greenhead ants		*Rhytidoponera*	(ハチ目、アリ科) の昆虫の総称
green-headed flasher		*Astraptes latimargo bifascia* (Herrich-Schäffer)	(チョウ目、セセリチョウ科) 新熱帯区
green-headed sootywing		*Gorgopas chlorocephala* (Herrich-Schäffer)	(チョウ目、セセリチョウ科) 新熱帯区
green-headed sootywing		*Staphylus chlora* Evans	(チョウ目、セセリチョウ科) 新熱帯区
greenheaded spruce sawfly		*Pikonema dimmockii* (Cresson)	(ハチ目、ハバチ科) 新北区
green-heads			horse flies (1) を見よ　米国での英名
green-jacket skimmer			eastern pondhawk を見よ
green jay			spotted Jay を見よ
green June beetle		*Cotinis nitida* (Linnaeus)	(コウチュウ目、コガネムシ科) 新北区
green June beetle			green fruit beetle を見よ
green lacewing		*Mallada basalis* (Walker)	(アミメカゲロウ目、クサカゲロウ科) 日本、東洋区、大洋区、豪州区
green lacewing	ヨツボシクサカゲロウ	*Chrysopa septempunctata* Wesmael	(アミメカゲロウ目、クサカゲロウ科) 日本、旧北区
green lacewing		*Chrysopa phyllochroma* Wesmael	(アミメカゲロウ目、クサカゲロウ科) 日本、旧北区
green lacewing (1)		*Chrysopa californica* Coquillett	(アミメカゲロウ目、クサカゲロウ科) 新北区
green lacewing			common green lacewing を見よ
green lacewing fly			green lacewing (1) を見よ
green lacewings (1)	クサカゲロウ科	Chrysopidae	(アミメカゲロウ目) の昆虫の総称　成虫と一部幼虫は捕食性で、生物的防除に使用
green lacewings (2)		*Chrysopa, Mallada*	(アミメカゲロウ目、クサカゲロウ科) の昆虫の総称
green lappet moth			moss-green lappet を見よ
green larch looper			six-spotted angle moth を見よ　larch looper ともいう
green leaf beetle		*Cryptocephalus hypochaeridis* (Linnaeus)	(コウチュウ目、ハムシ科) 旧北区
green leaf bug	アカヒゲホソミドリカスミカメムシ	*Trigonotylus ruficornis* (Geoffroy)	(カメムシ目、カスミカメムシ科) 日本、全北区
green leaf weevil		*Polydrusus sericeus* (Schaller)	(コウチュウ目、ゾウムシ科) 旧北区
green leaf weevil (1)		*Phyllobius maculicornis* Germar	(コウチュウ目、ゾウムシ科) 旧北区
green leafhopper		*Nephotettix nigromaculatus* (Motschulsky)	(カメムシ目、オオヨコバイ科) 豪州区
green leafhopper	オオヨコバイ	*Cicadella viridis* (Linnaeus)	(カメムシ目、オオヨコバイ科) 日本、旧北区、東洋区。イネ、ムギ、ダイズ、果樹、サトウキビなどの害虫
green leafhopper (1)		*Empoasca flavescens* Fabricius	(カメムシ目、オオヨコバイ科) 東洋区

英名	和名	学名	所属、分布、ほか
green leafhopper (2)		*Empoasca decipiens* Paoli	（カメムシ目、オオヨコバイ科）旧北区
green leafhopper (3)		*Siphanta acuta* (Walker)	（カメムシ目、アオバハゴロモ科）豪州区
green leafhopper			West Indian canefly を見よ
green leafhopper			green rice leafhopper (3)(4) を見よ
green leafhoppers		*Nephotettix*	（カメムシ目、オオヨコバイ科）の昆虫の総称
green-legged matchstick		*Callitala major* Sjöstedt	（バッタ目、Morabidae）豪州区
green lestes			emerald damselfly を見よ
green Leuconycta moth		*Leuconycta diphteroides* (Guenée)	（チョウ目、ヤガ科）新北区
green lichen moth		*Izatha peroneanella* Walker	（チョウ目、マルハキバガ科）豪州区
green long-horn		*Adela viridella* Scopoli	（チョウ目、マガリガ科）旧北区
green longhorn moth	ミドリヒゲナガ	*Adela reaumurella* (Linnaeus)	（チョウ目、マガリガ科）日本、旧北区
green long-horned weevil	アオヒゲボソゾウムシ	*Phyllobius prolongatus* Motschulsky	（コウチュウ目、ゾウムシ科）日本
green longlegged fly			long-legged fly (1) を見よ
green looper		*Plusia signata* Fabricius	（チョウ目、ヤガ科）エチオピア区、東洋区、豪州区
green looper			wavy-lined emerald moth を見よ
green-maculated noctuid	アオバハガタヨトウ	*Antivaleria viridimacula* (Graeser)	（チョウ目、ヤガ科）日本、旧北区
green mantid		*Orthodera ministralis* (Fabricius)	（カマキリ目、カマキリ科）豪州区
green mantle	イブシキンシジミタテハ	*Caria mantinea* (C. et R. Felder)	（チョウ目、シジミタテハ科）新熱帯区
green maple longicorn beetle			maple twig borer を見よ
green-marbled sandman			African mallow を見よ
green-marbled sandman			marbled skipper を見よ
green-marked cutworm moth		*Melanchra insignis* Walker	（チョウ目、ヤガ科）豪州区
green marvel moth		*Acronicta fallax* (Herrich-Schäffer)	（チョウ目、ヤガ科）新北区。green marvel ともいう
green meadow grasshoppers			long-horned grasshoppers (1) を見よ
green mealy scale			green shield scale を見よ
green Mellana		*Quasimellana servilius* (Moschler)	（チョウ目、セセリチョウ科）新熱帯区
green midges		*Tanytarsus*	（ハエ目、ユスリカ科）の昆虫の総称
green milkweed beetle			dogbane beetle を見よ
green mirid		*Creontiades dilutus* (Stål)	（カメムシ目、カスミカメムシ科）豪州区
green mottled planthopper			common green planthopper を見よ

英名	和名	学名	所属、分布、ほか
green mountain grasshopper		*Miramella alpina* (Kollar)	（バッタ目、バッタ科）旧北区
green mugwort leafhopper	ミドリフトヨコバイ	*Laborrus impictifrons* (Boheman)	（カメムシ目、オオヨコバイ科）日本、旧北区
green needleworm			white-triangle Tortrix moth を見よ
green nettle weevil			nettle weevil を見よ
green oak leaf roller			green oak roller を見よ　green oak leafroller とも記す
green oak moth			green oak roller を見よ
green oak roller		*Tortrix viridana* (Linnaeus)	（チョウ目、ハマキガ科）旧北区
green oak tortrix			green oak roller を見よ
green oak tortrix moth			green oak roller を見よ
green oakblue		*Arhopala eumolphus* (Cramer)	（チョウ目、シジミチョウ科）東洋区。*A. e. maxwelli* Distant も同英名
green Oslaria moth		*Oslaria viridifera* (Grote)	（チョウ目、ヤガ科）新北区。green Oslaria ともいう
green ostomatid		*Temnochila chloridia* Mannerheim	（コウチュウ目、Ostomatidae）新北区
green owlet moth			common Merveille-du-jour を見よ　米国での英名
green paddy leafhopper			rice leafhopper (2) を見よ
green page moth		*Urania fulgens* (Walker)	（チョウ目、ツバメガ科）新熱帯区
green-patch swallowtail		*Battus laodamas copanae* (Reakirt)	（チョウ目、アゲハチョウ科）新熱帯区。*B. l. iopas* (Godman et Salvin) も同英名
green-patched looper moth		*Diachrysia balluca* Geyer	（チョウ目、ヤガ科）新北区
green-patched stink bug			birch shield bug を見よ
green pea louse			pea aphid (1) を見よ
green pea moth		*Laspeyresia dorsana* Fabricius	（チョウ目、ハマキガ科）旧北区
green peach aphid	モモアカアブラムシ	*Myzus persicae* (Sulzer)	（カメムシ目、アブラムシ科）日本、汎世界。多くの作物、果樹を加害する著名害虫
green pine chafer			green rose chafer (2) を見よ
green plant bug			southern green stink bug を見よ
green planthopper			green leafhopper (3) を見よ
green plum aphid			mealy plum aphid を見よ
green polished chafer	アオカナブン	*Rhomborrhina unicolor* Motschulsky	（コウチュウ目、コガネムシ科）日本
green polyptychus		*Chloroclanis virescens* (Butler)	（チョウ目、スズメガ科）エチオピア区
green potato aphid		*Aulacorthum pseudosolanii* Theobald	（カメムシ目、アブラムシ科）旧北区
green potato bug		*Cuspicona simplex* Walker	（カメムシ目、カメムシ科）豪州区

英名	和名	学名	所属、分布、ほか
green pubescent ground beetle			green ground beetle を見よ
green pug			green pug moth を見よ
green pug moth	リンゴアオナミシャク	*Pasiphila rectangulata* (Linnaeus)	(チョウ目、シャクガ科) 日本、旧北区。リンゴ害虫
green Rajah		*Charaxes nitebis* (Hewitson)	(チョウ目、タテハチョウ科) 東洋区
green rice bug			southern green stink bug を見よ
green rice caterpillar	フタオビコヤガ	*Naranga aenescens* Moore	(チョウ目、ヤガ科) 日本、東洋区。イネ害虫
green rice leafhopper (1)	クロスジツマグロヨコバイ	*Nephotettix nigropictus* (Stål)	(カメムシ目、オオヨコバイ科) 日本、東洋区。イネ害虫。rice green leafhopper が適切といわれる
green rice leafhopper (2)	タイワンツマグロヨコバイ	*Nephotettix virescens* (Distant)	(カメムシ目、オオヨコバイ科) 日本、東洋区。イネ害虫
green rice leafhopper (3)	ツマグロヨコバイ	*Nephotettix cincticeps* (Uhler)	(カメムシ目、オオヨコバイ科) 日本、旧北区、東洋区。イネ害虫
green rice leafhopper (4)	マラヤツマグロヨコバイ	*Nephotettix malayanus* Ishihara et Kawase	(カメムシ目、オオヨコバイ科) 日本、東洋区。イネ害虫
green rice moth	ミドリツヅリガ	*Doloessa viridis* Zeller	(チョウ目、メイガ科) 日本、東洋区。サツマイモ害虫
green rose and potato aphid			potato aphid (1) を見よ
green rose aphid	バラミドリアブラムシ	*Rhodobium porosum* (Sanderson)	(カメムシ目、アブラムシ科) 日本、汎世界。バラ害虫
green rose chafer (1)		*Epicometis hirta* Poda	(コウチュウ目、コガネムシ科) 旧北区
green rose chafer (2)		*Dichelonyx backi* (Kirby)	(コウチュウ目、コガネムシ科) 新北区
green rose chafer			rose chafer (1) を見よ
green Saliana		*Saliana hewitsoni* (Riley)	(チョウ目、セセリチョウ科) 新熱帯区
green sapphire	アオウラフチベニシジミ	*Heliophorus androcles* (Westwood)	(チョウ目、シジミチョウ科) 東洋区
green sawfly		*Rhogogaster viridis* (Linnaeus)	(ハチ目、ハバチ科) 旧北区
green scale	ミドリカタカイガラムシ	*Coccus viridis* (Green)	(カメムシ目、カタカイガラムシ科) 日本 (小笠原)、汎熱帯。カンキツ害虫
green scarab beetle		*Diphuncephala colaspidoides* (Gyllenhal)	(コウチュウ目、コガネムシ科) 豪州区
green semilooper			green looper を見よ
green-shaded honey			bee moth を見よ
green shieldback		*Idiostatus viridis* Rentz	(バッタ目、キリギリス科) 新北区
green shield bug (1)		*Piezodorus lituratus* (Fabricius)	(カメムシ目、カメムシ科) 旧北区
green shield bug			common green shieldbug を見よ
green shield bug			green stink bug (3) を見よ
green shieldbug			southern green stink bug を見よ
green shieldbug			common green shieldbug を見よ
green shield scale	ミドリワタカイガラムシ	*Pulvinaria psidii* Maskell	(カメムシ目、カタカイガラムシ科) 日本 (小笠原)、新北区、汎熱帯。柑橘害虫。米国での英名

英名	和名	学名	所属、分布、ほか
green shield scale			green scale を見よ
green shoot borer			apple leaf-curling moth を見よ
green-shouldered sootywing		*Gorgopas trochilus* (Hopffer)	(チョウ目、セセリチョウ科) 新熱帯区
green silver-lines		*Pseudoips prasinana britannica* (Warren)	(チョウ目、ヤガ科) 旧北区
green silver-lines			scarce silver-lines (1) を見よ
green silver-spangled shark	アオモンギンセダカモクメ	*Cucullia argentea* (Hufnagel)	(チョウ目、ヤガ科) 旧北区
green skimmer		*Orthetrum sabina* (Drury)	(トンボ目、トンボ科) 東洋区
green skipper		*Hesperia viridis* Edwards	(チョウ目、セセリチョウ科) 新北区、新熱帯区
green slug moth		*Parasa chloris* (Herrich-Schäffer)	(チョウ目、イラガ科) 新北区
green snaketail		*Ophiogomphus cecilia* (Geoffroy)	(トンボ目、サナエトンボ科) 旧北区。green gomphid ともいう
green snout beetle			gold-dust weevil を見よ
green soldier beetle		*Chauliognathus pulchellus* W. S. Macleay	(コウチュウ目、ジョウカイボン科) 豪州区
green soldier bug			green stink bug (2) を見よ
green sow-thistle aphid			currant-lettuce aphid を見よ　北米での英名
green span-worm		*Itame sulphurea* Packard	(チョウ目、シャクガ科) 新北区
green sphinx		*Tinostoma smaragditis* (Meyrick)	(チョウ目、スズメガ科)。大洋区。ハワイのカウアイ島山地に生息する緑色のスズメガ
green-spotted banner	オオベニスジシロモンタテハ	*Pyrrhogyra edocla edocla* Doubleday	(チョウ目、タテハチョウ科) 新熱帯区
green-spotted Brangas		*Brangas carthaea* (Hewitson)	(チョウ目、シジミチョウ科) 新熱帯区
green-spotted swallowtail	ウスミドリタイマイ (キミドリマダラタイマイ)	*Graphium tynderaeus* (Fabricius)	(チョウ目、アゲハチョウ科) エチオピア区
green-spotted triangle			tailed jay を見よ　豪州での英名。green-spotted triangle butterfly ともいう
green spring beetle		*Diphuncephala edwardsii* Waterhouse	(コウチュウ目、コガネムシ科) 豪州区
green springtail			varied springtail を見よ
green spruce aphid		*Cinara fornacula* Hottes	(カメムシ目、アブラムシ科) 新北区
green spruce aphid			spruce aphid を見よ
green spruce leafminer			dwarf pine bell を見よ
green stigma		*Bittacus chlorostigma* McLachlan	(シリアゲムシ目、ガガンボモドキ科) 新北区
green stink bug (1)	アオクサカメムシ	*Nezara antennata* Scott	(カメムシ目、カメムシ科) 日本、東洋区。多くの作物、果樹害虫
green stink bug (2)		*Acrosternum hilare* (Say)	(カメムシ目、カメムシ科) 新北区。green soldier bug ともいう
green stink bug (3)		*Plautia affinis* (Dallas)	(カメムシ目、カメムシ科) 豪州区
green stink bug			southern green stink bug を見よ

英名	和名	学名	所属、分布、ほか
green stink bugs		*Chlorochroa*	(カメムシ目、カメムシ科) の昆虫の総称
green stoneflies	ミドリカワゲラ科	Chloroperlidae	(カワゲラ目) の昆虫の総称
green-striped dappled white			green-striped white を見よ
green-striped darner		*Aeschna verticalis* Hagen	(トンボ目、ヤンマ科) 新北区
green-striped darner		*Austroaeschna forcipata* (Tillyard)	(トンボ目、ヤンマ科) 豪州区
green-striped forest looper		*Melanolophia imitata* Walker	(チョウ目、シャクガ科) 新北区
green-striped grasshopper		*Chortophaga viridifasciata* (De Geer)	(バッタ目、バッタ科) 新北区
green-striped green moth			green silver-lines を見よ
green striped hawk		*Theretra jugurtha* Boisduval	(チョウ目、スズメガ科) エチオピア区
green-striped mapleworm			striped maple worm を見よ
green-striped slim			emerald-striped slim を見よ
green-striped white	トラフシロチョウ	*Euchloe belemia* (Esper)	(チョウ目、シロチョウ科) 旧北区
green sugarcane aphid			cane aphid を見よ
green swallowtail			pipevine swallowtail を見よ
green sword-tail crickets		*Cyrtoxipha*	(バッタ目、クサヒバリ科) の昆虫の総称
green tiger beetle		*Cicindela campestris* Linnaeus	(コウチュウ目、ハンミョウ科) 旧北区
green tobacco aphid			green peach aphid を見よ
green tomato bug			southern green stink bug を見よ　米国での英名
green tortoise beetle	ミドリカメノコハムシ	*Cassida erudita* Baly	(コウチュウ目、ハムシ科) 日本
green tortoise beetle (1)		*Cassida viridis* Linnaeus	(コウチュウ目、ハムシ科) 旧北区
green tortoise beetle			thistle tortoise beetle を見よ
green tree ant		*Oecophilla virescens* Fabricius	(ハチ目、アリ科) エチオピア区、東洋区
green tree ant			red ant (1) を見よ
green tree cricket	アオマツムシ	*Truljalia hibinonis* (Matsumura)	(バッタ目、マツムシ科) 日本、旧北区
green treehopper		*Sextius virescens* (Fairmaire)	(カメムシ目、ツノゼミ科) 豪州区
green triangle		*Graphium macfarlanei* (Butler)	(チョウ目、アゲハチョウ科) 豪州区。green triangle-butterfly ともいう
green-underside blue	アレクシスカバイロシジミ	*Glaucopsyche alexis* (Poda)	(チョウ目、シジミチョウ科) 旧北区
green valley grasshopper		*Schistocerca shoshone* (Thomas)	(バッタ目、バッタ科) 新北区
green vegetable bug			southern green stink bug を見よ

英名	和名	学名	所属、分布、ほか
green vegetable bug egg parasite			stink bug egg parasite を見よ
green-veined cabbage white			mustard white を見よ
green-veined Charaxes	オナガフタオチョウ	*Charaxes candiope* (Godart)	（チョウ目、タテハチョウ科）エチオピア区
green-veined emperor			green-veined Charaxes を見よ
green-veined white	エゾスジグロシロチョウ	*Pieris napi japonica* Shirozu	（チョウ目、シロチョウ科）日本。*P. napi* Linnaeus の英国での英名。green-veined white butterfly ともいう
green velvet bud gall wasp		*Cynips longiventris* Hartig	（ハチ目、タマバチ科）旧北区
green velvet skipper	アオバセセリ	*Choaspes benjamini formosana* (Fruhstorfer)	（チョウ目、セセリチョウ科）日本、東洋区
greenweed flat-body moth		*Agonopterix atomella* (Denis et Schiffermüller)	（チョウ目、マルハキバガ科）旧北区
green whizzer		*Macrotristria intersecta* (Walker)	（カメムシ目、セミ科）豪州区
green willow leafhopper	ヤナギハトムネヨコバイ	*Macropsis prasina* (Boheman)	（カメムシ目、オオヨコバイ科）日本、旧北区。ヤナギ害虫
green willow sawfly		*Nematus virescens* Hartig	（ハチ目、ハバチ科）旧北区
green-winged stoneflies	アミメカワゲラ科	Isoperlidae	（カワゲラ目）の昆虫の総称
green-winged stoneflies		*Isoperla*	（カワゲラ目、アミメカゲロウ科）の昆虫の総称
green-winged stonefly		*Isoperla confusa* Frison	（カワゲラ目、アミメカゲロウ科）新北区
greenwings		*Neurobasis*	（トンボ目、イトトンボ科）の昆虫の総称
green worm of cabbage			diamondback moth を見よ　幼虫の英名
greenbottle flies		Luciliinae	（ハエ目、クロバエ科）の昆虫の総称
greenbottle flies		*Phaenicia*	（ハエ目、クロバエ科）の昆虫の総称
greenbottle flies (1)		*Lucilia*	（ハエ目、クロバエ科）の昆虫の総称
greenbottle flies			blow flies を見よ
greenbottles			greenbottle flies (1) を見よ
greenbottles			blow flies を見よ
greenbug	ムギミドリアブラムシ	*Schizaphis graminum* (Rondani)	（カメムシ目、アブラムシ科）日本、全北区。コムギ、イネ害虫
greenhouse aphid			green peach aphid を見よ
greenhouse camel cricket	クラズミウマ	*Tachycines asynamorus* (Adelung)	（バッタ目、カマドウマ科）日本、全北区。葉を巻いた巣を作る。温室害虫
greenhouse cockroach			Surinam cockroach を見よ
greenhouse grasshopper			greenhouse camel cricket を見よ
greenhouse leaf tier			celery leaf tier (2) を見よ

英名	和名	学名	所属、分布、ほか
greenhouse leaftier			rusty-dot pearl (1) を見よ
greenhouse leaftier			celery leaftier (1) を見よ
greenhouse orthezia			Jacarand bug を見よ
greenhouse scale			chaff scale (1) を見よ
greenhouse stone cricket			greenhouse camel cricket を見よ　北米での英名
greenhouse thrips	クロトンアザミウマ	*Heliothrips haemorrhoidalis* (Bouche)	(アザミウマ目、アザミウマ科) 日本、全北区、豪州区。クロトン、園芸植物、果樹の害虫
greenhouse whitefly	オンシツコナジラミ	*Trialeurodes vaporariorum* (Westwood)	(カメムシ目、コナジラミ科) 日本、汎世界。多くの作物、温室園芸植物などの害虫
greenhouse whitefly parasite	オンシツツヤコバチ	*Encarsia formosa* Gahan	(ハチ目、ツヤコバチ科) 日本、旧北区。豪州での英名
greenish black-tip	ウグイスシロチョウ	*Euchloe charlonia* (Donzel)	(チョウ目、シロチョウ科) 旧北区、エチオピア区
greenish blue	セイジコシジミ	*Aricia saepiolus* (Boisduval)	(チョウ目、シジミチョウ科) 新北区
greenish burrowing bee		*Andrena cuneilabris* Viereck	(ハチ目、ヒメハナバチ科) 新北区
greenish chestnut moth	クリミドリシンクイガ	*Eucoenogenes aestuosa* (Meyrick)	(チョウ目、ハマキガ科) 日本、東洋区。クリ害虫
greenish delicate geometrid	キトガリヒメシャク	*Scopula emissaria* (Walker)	(チョウ目、シャクガ科) 日本、東洋区、豪州区
greenish delicate geometrid			common emerald を見よ
greenish half mourner			bath white を見よ
greenish Mellana		*Quasimellana myron* (Godman)	(チョウ目、セセリチョウ科) 新熱帯区
greenish mountain blue		*Lycaena orbitulus* (Esper)	(チョウ目、シジミチョウ科) 東洋区
greenish narrow-winged noctuid	ホソアオバヤガ	*Ochropleura praecox flavomaculata* (Graeser)	(チョウ目、ヤガ科) 日本、旧北区
greenish noctuid			green arches を見よ
greenish oedemerid	アオカミキリモドキ	*Xanthochroa waterhousei* Harold	(コウチュウ目、カミキリモドキ科) 日本、旧北区
greenish prominent	アオシャチホコ	*Quadricalcarifera japonica* Nakatomi	(チョウ目、シャチホコガ科) 日本
greenish tailed looper moth	チズモンアオシャク	*Agathia carissima* Butler	(チョウ目、シャクガ科) 日本、旧北区
greenish yellow-brown hawk moth	トビイロスズメ	*Clanis bilineata tsinglauica* Mell	(チョウ目、スズメガ科) 日本、旧北区
greenish-yellow Sitochroa moth			sulphur-pearl を見よ
Greenland sulphur	ツンドラモンキチョウ	*Colias hecla* Lefebvre	(チョウ目、シロチョウ科) 新北区
greens			emeralds を見よ
Green's gem		*Libellago greeni* (Laidlaw)	(トンボ目、Chlorocyphidae) 東洋区
Green's mealybug			citrus mealybug (2) を見よ

英名	和名	学名	所属、分布、ほか
gregarious gall weevil		*Strongylorhinus clarki* Marshall	（コウチュウ目、ゾウムシ科）豪州区
gregarious gall weevil		*Strongylorhinus ochraceus* Schönherr	（コウチュウ目、ゾウムシ科）豪州区
gregarious iris leaf miner		*Cerodontha iridis* (Hendel)	（ハエ目、ハモグリバエ科）旧北区
gregarious leaf miner		*Cameraria cincinnatiella* (Chambers)	（チョウ目、ホソガ科）新北区
gregarious oak leafminer			gregarious leaf miner を見よ　gregarious oak leafminer moth ともいう
gregarious poplar sawfly		*Nematus melanaspis* Hartig	（ハチ目、ハバチ科）旧北区
gregarious spruce sawfly		*Pristiphora abietina* (Christ)	（ハチ目、ハバチ科）旧北区
Gregson's dart		*Agrotis spinifera* Hübner	（チョウ目、ヤガ科）旧北区、エチオピア区、東洋区
grevillea looper			pink bellied moth を見よ
grevillea mealybug		*Australicoccus grevilleae* (Fuller)	（カメムシ目、コナカイガラムシ科）豪州区
Greville's marble			dark long-cloaked marble を見よ
grey		*Hadena caesia* (Denis et Schiffermüller)	（チョウ目、ヤガ科）旧北区。亜種 *H. c. mananii* (Gregson) も同英名
grey albatross		*Appias melania* (Fabricius)	（チョウ目、シロチョウ科）豪州区
grey Amphipyra			smooth Amphipyra moth を見よ
grey angle-wing moth		*Hydriomena triphragma* Meyrick	（チョウ目、シャクガ科）豪州区
grey ant-blue		*Acrodipsas melania* (Sands)	（チョウ目、シジミチョウ科）豪州区
grey apple pigmy		*Stigmella incognitella* (Herrich-Schäffer)	（チョウ目、モグリチビガ科）旧北区
grey apple pigmy		*Nepticula pomella* Vaughan	（チョウ目、モグリチビガ科）旧北区
grey arches	オオシラホシヨトウ	*Polia nebulosa* (Hufnagel)	（チョウ目、ヤガ科）日本、旧北区
grey Asian grayling		*Pseudochazara geyeri* Herrich-Schäffer	（チョウ目、タテハチョウ科）旧北区
greyback cane beetle		*Dermolepida albohirtum* (Waterhouse)	（コウチュウ目、コガネムシ科）豪州区
greyback canegrub			greyback cane beetle を見よ
greybacks			human lice を見よ
greybanded leaf weevil		*Ethemaia sellata* Pascoe	（コウチュウ目、ゾウムシ科）豪州区
grey baron	コケムシナズマ	*Euthalia anosia* (Moore)	（チョウ目、タテハチョウ科）東洋区
grey birch		*Aethalura punctulata* (Denis et Schiffermüller)	（チョウ目、シャクガ科）旧北区
grey birch beauty			grey birch を見よ
grey blister beetle		*Epicauta aethiops* (Latreille)	（コウチュウ目、ツチハンミョウ科）エチオピア区

英名	和名	学名	所属、分布、ほか
grey-brand redeye		*Matapa druna* (Moore)	（チョウ目、セセリチョウ科）東洋区
grey-brown cutworm moth		*Melanchra mutans* Walker	（チョウ目、ヤガ科）豪州区
grey bud moth		*Hedya nubiferana* (Haworth)	（チョウ目、ハマキガ科）旧北区
grey bush brown		*Bicyclus taenias* (Hewitson)	（チョウ目、タテハチョウ科）エチオピア区
grey bush cricket		*Platycleis albopunctata* (Goeze)	（バッタ目、キリギリス科）旧北区
grey bush-cricket		*Platycleis denticulata* (Panz)	（バッタ目、キリギリス科）旧北区
grey bush-cricket		*Platycleis grisea* (Fabricius)	（バッタ目、キリギリス科）旧北区
grey cabbage aphid			cabbage aphid を見よ
grey carpet			pale grey carpet を見よ
grey chi			chi を見よ
grey Christmas beetle		*Trioplognathus griseopilosus* (Ohaus)	（コウチュウ目、コガネムシ科）豪州区
grey cluster bug		*Nysius clevedandensis* Evans	（カメムシ目、ナガカメムシ科）豪州区
grey cockroach			Madeira cockroach を見よ　南アフリカでの英名
grey cockroach			cinereous cockroach を見よ
grey commodore	ヒカゲタテハ	*Bhagadatta austenia* (Moore)	（チョウ目、タテハチョウ科）東洋区
grey coronet			grey を見よ
grey cossid			reed leopard を見よ
grey count		*Tanaecia lepidea* Butler	（チョウ目、タテハチョウ科）東洋区
grey dagger		*Acronicta psi* (Linnaeus)	（チョウ目、ヤガ科）旧北区。gray dagger ともいう
grey dagger moth			grey dagger を見よ
grey damsel bug		*Stalia major* Costa	（カメムシ目、マキバサシガメ科）旧北区
grey desert robber fly		*Promachus vertebrates* (Say)	（ハエ目、ムシヒキアブ科）新北区。アリゾナの岩絵の一つココペッサが本種に似ているとの話もある
grey diamond-backed smudge			diamondback moth を見よ
grey double-crescent bell		*Epinotia ramella* (Linnaeus)	（チョウ目、ハマキガ科）旧北区
grey drake			brown mayfly を見よ
grey dungball roller		*Sisyphus spinipes* (Thunberg)	（コウチュウ目、コガネムシ科）豪州区
grey early			early grey を見よ
grey Emesis		*Emesis eurydice* Godman	（チョウ目、シジミタテハ科）新熱帯区
grey evening moth		*Selidosema panagrata* Walker	（チョウ目、シャクガ科）豪州区
grey flesh fly		*Wohlfahrtia vigil* Walker	（ハエ目、ニクバエ科）新北区
grey forester		*Bebearia demetra* (Godart)	（チョウ目、タテハチョウ科）エチオピア区

英名	和名	学名	所属、分布、ほか
grey fruit-tree case			grey fruit-tree case moth を見よ
grey fruit-tree case moth		*Coleophora hemerobiella* (Scopoli)	(チョウ目、ツツミノガ科) 旧北区
greyfurrowed rose chafer		*Trichaulax philipsii* (Schreibers)	(コウチュウ目、コガネムシ科) 豪州区
grey glassy tiger	ユベンタヒメゴマダラ	*Ideopsis juventa sitah* Fruhstorfer	(チョウ目、タテハチョウ科) 東洋区
grey grain moth			large nutwing を見よ
grey-green moth		*Selidosema indistincta* Butler	(チョウ目、シャクガ科) 豪州区
grey hairstreak			gray hairstreak (1) を見よ
grey handkerchief		*Mestra hypermestra cana* Erichson	(チョウ目、タテハチョウ科) 新熱帯区
grey hill plume			grey wood plume を見よ
grey Hippotion		*Hippotion rosae* (Butler)	(チョウ目、スズメガ科) エチオピア区
grey larch moth			larch tortrix (2) を見よ
grey larch tortricid			larch tortrix (2) を見よ
grey larch tortrix			larch tortrix (2) を見よ
grey lawn leafhopper			gray lawn leafhopper を見よ
grey-lined lascar	オオキンミスジ	*Pantoporia dindinga* (Butler)	(チョウ目、タテハチョウ科) 東洋区
grey looper		*Cleora inflexaria* Snellen	(チョウ目、シャクガ科) 豪州区
grey moth			grey を見よ
grey mountain carpet		*Entephria caesiata* (Denis et Schiffermüller)	(チョウ目、シャクガ科) 旧北区。grey mountain moth、gray mountain carpet ともいう。日本亜種は *E. c. nebulosa* Inoue サザナミナミシャク
grey O moth			fringe-tree sallow moth を見よ
grey olearia moth		*Apatetris melanombra* Meyrick	(チョウ目、キバガ科) 豪州区
grey pansy	ハイイロタテハモドキ (ウスムラサキタテハモドキ)	*Junonia atlites* (Linnaeus)	(チョウ目、タテハチョウ科) 日本、東洋区
grey pine carpet		*Thera obeliscata* (Hübner)	(チョウ目、シャクガ科) 旧北区
grey pine needle aphid		*Schizolachnus pineti* (Fabricius)	(カメムシ目、Lachnidae) 旧北区
grey planthopper		*Anzora unicolor* (Walker)	(カメムシ目、アオバハゴロモ科) 豪州区
grey pug		*Eupithecia subfuscata* (Haworth)	(チョウ目、シャクガ科) 全北区。grey pug moth ともいう
grey pug		*Eupithecia castigata* Hübner	(チョウ目、シャクガ科) 旧北区
grey pyralid			tropical meal moth を見よ
grey red-barred twist		*Argyrotaenia pulchellana* Haworth	(チョウ目、ハマキガ科) 旧北区
grey ringlet			dingy ringlet を見よ
grey rustic		*Xestia castanea* (Esper)	(チョウ目、ヤガ科) 旧北区
grey sailor		*Neptis leucoporos cresina* Fruhstorfer	(チョウ目、タテハチョウ科) 東洋区

英名	和名	学名	所属、分布、ほか
grey scale			Ross's black scale (1)(2) を見よ　豪州での英名
grey scalloped bar		*Dyscia fagaria* (Thunberg)	（チョウ目、シャクガ科）旧北区
grey scrub hopper		*Aeromachus jhora* (de Nicéville)	（チョウ目、セセリチョウ科）東洋区
grey shoulder-knot		*Lithophane ornitopus* (Hufnagel)	（チョウ目、ヤガ科）旧北区
grey silver-barred piercer		*Pammene splendidulana* (Guenée)	（チョウ目、ハマキガ科）旧北区
grey silverline		*Aphnaeus gabriel* Swinhoe	（チョウ目、シジミチョウ科）東洋区
grey similar-wing			pine false looper moth を見よ
grey snout beetle		*Eremnus setulosus* Boheman	（コウチュウ目、ゾウムシ科）エチオピア区
grey spruce carpet			spruce carpet (1) を見よ
grey square		*Pardasena virgulana* (Mabille)	（チョウ目、ヤガ科）旧北区、エチオピア区
grey squirrel flea		*Orchopaeas howardi* (Baker)	（ノミ目、ナガノミ科）旧北区
grey stalk borer			sugarcane shoot borer を見よ
grey-streaked smudge		*Plutella porrectella* (Linnaeus)	（チョウ目、スガ科）旧北区
grey-striped fly		*Sarcophaga aurifrons* Macquart	（ハエ目、ニクバエ科）豪州区
grey-striped mosquito		*Aedes vittiger* (Skuse)	（ハエ目、カ科）豪州区
grey sugar-cane borer			sugarcane shoot borer を見よ
grey sugarcane mealybug	オガサワラコナカイガラムシ	*Dysmicoccus boninsis* (Kuwana)	（カメムシ目、コナカイガラムシ科）日本、東洋区、新北区。トウモロコシ、サトウキビの害虫
grey sugarcane mealybug			pink sugarcane mealybug を見よ
grey Temnora		*Temnora funebris* (Holland)	（チョウ目、スズメガ科）エチオピア区
grey tiger longicorn	クワヤマトラカミキリ	*Xylotrechus rusticus* (Linnaeus)	（コウチュウ目、カミキリムシ科）日本、旧北区
grey tortrix		*Cnephasia sitephensiana* (Doubleday)	（チョウ目、ハマキガ科）全北区。gray tortrix moth ともいう
grey trident			grey dagger を見よ
grey tussock moth		*Dasychira vagans* Barnes et McDunnough	（チョウ目、ドクガ科）新北区
grey willow leaf beetle		*Pyrrhalta decora* (Say)	（コウチュウ目、ハムシ科）新北区
grey wood plume		*Stenoptilia bipunctidactyla* Scopoli	（チョウ目、トリバガ科）旧北区
greyish black aphid		*Aphis laburni* Kaltenbach	（カメムシ目、アブラムシ科）東洋区
greyish longicorn beetle	ハイイロハナカミキリ	*Rhagium japonicus* Bates	（コウチュウ目、カミキリムシ科）日本、旧北区
greyish white granulated weevil	シロコブゾウムシ	*Episemus turritus* (Gyllenhal)	（コウチュウ目、ゾウムシ科）日本、旧北区

英名	和名	学名	所属、分布、ほか
greyish yellow-hindwinged noctuid	ハイイロキシタヤガ	*Xestia semiherbida decorata* (Butler)	（チョウ目、ヤガ科）日本、東洋区
Griffin's sheepmoth		*Hemileuca griffini* Tuskes	（チョウ目、ヤママユガ科）新北区。北米での英名
grig			コオロギ、キリギリス、バッタの俗称。小さなウナギにも使う
Grisatra underwing moth		*Catocala grisatra* Brower	（チョウ目、ヤガ科）新北区。Grisatra underwing ともいう
Grisea canegrub		*Lepidiota grisea* Britton	（コウチュウ目、コガネムシ科）豪州区
Grisons fritillary		*Mellicta varia* (Meyer-Dur)	（チョウ目、タテハチョウ科）旧北区
grizzled blue		*Orachrysops subravus* Henning et Henning	（チョウ目、シジミチョウ科）エチオピア区
grizzled bush brown		*Bicyclus ena* (Hewitson)	（チョウ目、タテハチョウ科）エチオピア区
grizzled mantis			lichen mantid を見よ
grizzled owlet		*Scythris picaepennis* (Haworth)	（チョウ目、キヌバコガ科）旧北区
grizzled pintail		*Acisoma variegatum* Kirby	（トンボ目、トンボ科）エチオピア区
grizzled skipper		*Pyrgus centaureae freija* (Warren)	（チョウ目、セセリチョウ科）新北区
grizzled skipper (1)	ヒメチャマダラセセリ	*Pyrgus malvae* (Linnaeus)	（チョウ目、セセリチョウ科）日本、旧北区。grizzled skipper butterfly ともいう
grizzled skippers		*Spialia*	（チョウ目、セセリチョウ科）の昆虫の総称
grizzled sootywing		*Staphylus tepeca* (Bell)	（チョウ目、セセリチョウ科）新熱帯区
Grose-Smith's flash		*Hypophytala henleyi* (Kirby)	（チョウ目、シジミチョウ科）エチオピア区
Grose-Smith's nymph		*Euriphene grosesmithi* (Staudinger)	（チョウ目、タテハチョウ科）エチオピア区
Grose-Smith's swallowtail		*Papilio grosesmithi* Rothschild	（チョウ目、アゲハチョウ科）エチオピア区
Grote's Bertholdia		*Bertholdia trigona* (Grote)	（チョウ目、ヒトリガ科）新北区。Grote's Bertholdia moth ともいう
Grote's buck moth		*Hemileuca grotei* Grote et Robinson	（チョウ目、ヤママユガ科）新北区
Grote's sallow moth		*Copivaleria grotei* (Smith)	（チョウ目、ヤガ科）新北区。Grote's sallow ともいう
Grote's sphinx moth		*Cautethia grotei* H. Edwards	（チョウ目、スズメガ科）新北区、新熱帯区。Grote's sphinx ともいう
ground and camel crickets			cave crickets (1) を見よ
ground beetle			strawberry ground beetle (1) を見よ
ground beetles	オサムシ科	Carabidae	（コウチュウ目）の昆虫の総称
ground beetles			bombardier beetles (1) を見よ
ground beetles			darkling beetles (1) を見よ　幼虫の英名
ground bugs	ツチカメムシ科	Cydnidae	（カメムシ目）の昆虫の総称
ground bugs			jumping ground bugs を見よ
ground bugs			chinch bugs (1) を見よ　英国での英名
ground crickets		*Allonemobius*	（バッタ目、コオロギ科）の昆虫の総称

英名	和名	学名	所属、分布、ほか
ground crickets			wood crickets を見よ
ground crickets			cave crickets (1) を見よ
ground fleas			springtails を見よ
ground fleas			blind springtails を見よ
ground fleas			white blind springtails を見よ
ground ivy		*Dasineura glechomae* (Kieffer)	（ハエ目、タマバエ科）がツタの葉の上面につくる虫えいでピンク色で有毛。旧北区
ground lackey		*Malacosoma castrensis* (Linnaeus)	（チョウ目、カレハガ科）旧北区
ground mantis		*Bolbe pygmaea* Saussure	（カマキリ目、Iridopterygidae）豪州区
ground mantis (1)		*Litaneutria minor* Scudder	（カマキリ目、カマキリ科）新北区
ground mealybug		*Rhizoecus terrestris* (Newstead)	（カメムシ目、コナカイガラムシ科）全北区
ground mealybug (1)		*Rhizoecus falcifer* Kunckel d'Herculais	（カメムシ目、コナカイガラムシ科）新北区、豪州区
ground mealybug			citrus ground mealybug を見よ
ground nesting wasps		Psammocharidae	（ハチ目）の昆虫の総称
ground nesting wasps			spider wasps を見よ
ground pearl		*Margarodes formicarum* Guilding	（カメムシ目、ワタフキカイガラムシ科）新熱帯区。西インド諸島のカイガラムシの雌。雌が作る貝殻に糸を通して首飾りとする
ground pearl (1)		*Margarodes vitis* (Philippi)	（カメムシ目、ワタフキカイガラムシ科）新熱帯区
ground pearl (2)		*Porphyrophora polonica* (Linnaeus)	（カメムシ目、ワタフキカイガラムシ科）旧北区
ground pearls			giant coccids を見よ　豪州での英名
ground-pearls		*Margarodes*	（カメムシ目、ワタフキカイガラムシ科）の昆虫の総称
ground puppy			changa を見よ
ground puppy changa			changa を見よ
ground shield bug		*Choreocoris paganus* (Fabricius)	（カメムシ目、キンカメムシ科）豪州区
ground skimmer			chalky percher を見よ
ground squirrel flea		*Diamanus montanus* (Baker)	（ノミ目、トゲノミ科）新北区
ground wasp			common wasp を見よ
ground weevil		*Barynotus obscurus* (Fabricius)	（コウチュウ目、ゾウムシ科）旧北区
ground weta		*Hemiandrus*	（バッタ目、Stenopelmatidae）の昆虫の総称
ground weta		*Hemiandrus furcifer* Ander	（バッタ目、Stenopelmatidae）豪州区
groundhoppers		*Tetrix*	（バッタ目、ヒシバッタ科）の昆虫の総称
groundhoppers			grouse locusts を見よ　英国での英名
groundhoppers			grasshoppers (3) を見よ

英名	和名	学名	所属、分布、ほか
groundlet fanner		*Glyphipterix equitella* (Scopoli)	(チョウ目、ホソハマキモドキガ科) 旧北区
groundlings		*Brachythemis*	(トンボ目、トンボ科) の昆虫の総称
groundnut aphid			cowpea aphid を見よ
groundnut beetle			groundnut bruchid を見よ
groundnut borer		*Caryedon fuscus* (Goeze)	(コウチュウ目、ハムシ科) 旧北区
groundnut borer			groundnut bruchid を見よ
groundnut bruchid	モモブトジマメゾウムシ	*Caryedon serratus* (Olivier)	(コウチュウ目、ハムシ科) 旧北区、大洋区、エチオピア区。ground nut seed beetle ともいう
groundnut bruchid			beet tortoise beetle (2) を見よ
groundnut bud borer		*Anarsia ephippias* Meyrick	(チョウ目、キバガ科) 東洋区
groundnut hopper		*Hilda patruelis* Stål	(カメムシ目、Tettigometridae) エチオピア区
groundnut leaf miner		*Aproaerema modicella* Deventer	(チョウ目、キバガ科) 東洋区
groundnut leaf webber			groundnut leaf miner を見よ
groundsel borer moth		*Papaipema pertincta* Dyar	(チョウ目、ヤガ科) 新北区。groundsel borer ともいう
groundselbush gall fly		*Rhopalomyia californica* Felt	(ハエ目、タマバエ科) 豪州区
groundselbush leaf beetle		*Trirhabada baccharidis* (Weber)	(コウチュウ目、ハムシ科) 豪州区
groundselbush leafwebbing caterpillar		*Aristotelia* sp.	(チョウ目、キバガ科) 豪州区
groundselbush stemborer		*Megacyllene mellyi* (Chevrolat)	(コウチュウ目、カミキリムシ科) 豪州区
grouse and pigmy locusts			grouse locusts を見よ
grouse locusts	ヒシバッタ科	Tettigidae	(バッタ目) の昆虫の総称
grouse locusts			long-horned grasshoppers (1) を見よ　米国での英名
grouse locusts			grasshoppers (2) を見よ
grousewinged backswimmer			common backswimmer を見よ
grub			昆虫 (とくに甲虫) の幼虫
grubs			May beetles (3) を見よ　幼虫の英名
Gruner's orange-tip	ヒメクモマツマキチョウ	*Anthocharis gruneri* Herrich-Schäffer	(チョウ目、シロチョウ科) 旧北区
gryllacridoids	コロギス上科	Gryllacridoidea	(バッタ目) の昆虫の総称
grylline insects		*Gryllinea*	(バッタ目、コオロギ科) の昆虫の総称
grylloblattids			rock crawlers (1)(2) を見よ
grylloid crickets	コオロギ上科	Grylloidea	(バッタ目) の昆虫の総称
gsika blue			tiny grass blue (2) を見よ　南米での英名
Guadalupe Island juniper hairstreak			Nelson's hairstreak を見よ
guar midge		*Contarinia texana* (Felt)	(ハエ目、タマバエ科) 新北区。guar は植物名

英名	和名	学名	所属、分布、ほか
guardpost skipper		*Euphyes peneia* (Godman)	（チョウ目、セセリチョウ科）新熱帯区
Guatemalan Actinote		*Actinote guatemalena guatemalena* (Bates)	（チョウ目、タテハチョウ科）新熱帯区
Guatemalan calico	ガテマラカスリタテハ	*Hamadryas guatemalena* (Bates)	（チョウ目、タテハチョウ科）新北区、新熱帯区
Guatemalan catone			East-Mexican banner を見よ
Guatemalan copper		*Iophanus pyrrhias* (Godman et Salvin)	（チョウ目、シジミチョウ科）新熱帯区
Guatemalan cracker			Guatemalan calico を見よ
Guatemalan crescent		*Tegosa guatemalena* (Bates)	（チョウ目、タテハチョウ科）新熱帯区
Guatemalan gemmed-satyr		*Cyllopsis guatemalena* Miller	（チョウ目、タテハチョウ科）新熱帯区
Guatemalan giant-skipper		*Agathymus indecisa* (Butler et Druce)	（チョウ目、セセリチョウ科）新熱帯区
Guatemalan hairstreak		*Callophrys guatemalena* Clench	（チョウ目、シジミチョウ科）新熱帯区
Guatemalan kelep		*Ectatomma tuberculatum* (Olivier)	（ハチ目、アリ科）新熱帯区、新北区。中米のアリで、boll weevil 防除のため米国に導入
Guatemalan leafwing		*Anaea forreri* (Godman et Salvin)	（チョウ目、タテハチョウ科）新北区、新熱帯区
Guatemalan metalmark		*Mesene croceella* H. Bates	（チョウ目、シジミタテハ科）新熱帯区
Guatemalan patch			Erodyle checkerspot を見よ
Guatemalan Remella		*Remella duena* (Evans)	（チョウ目、セセリチョウ科）新北区、新熱帯区
Guatemalan rubyspot		*Hetaerina rudis* Calvert	（トンボ目、イトトンボ科）新熱帯区
Guatemalan Sarota		*Sarota psaros psaros* Godman et Salvin	（チョウ目、シジミタテハ科）新熱帯区
Guatemalan scarlet-eye		*Cephise guatemalaensis* (Freeman)	（チョウ目、セセリチョウ科）新熱帯区
Guatemalan skipper		*Cogia mala* Evans	（チョウ目、セセリチョウ科）新熱帯区
Guatemalan' striped blue-skipper		*Quadrus contubernalis anicius* (Godman et Salvin)	（チョウ目、セセリチョウ科）新熱帯区。striped blue skipper を参照
Guatemalan sulphur		*Colias philodice guatemalena* Rober	（チョウ目、シロチョウ科）新熱帯区。clouded sulphur を参照
Guatemalan Tegosa			Guatemalan crescent を見よ
guava caterpillar moth		*Prostomeus brunneus* Busck	（チョウ目、キバガ科）新北区
guava fruit moth		*Argyresthia eugeniella* Busck	（チョウ目、メムシガ科）新北区。guava moth ともいう
guava leaf roller			eucalyptus leafroller を見よ
guava mealy scale		*Chloropulvinaria psidii* (Maskell)	（カメムシ目、カタカイガラムシ科）旧北区、東洋区、エチオピア区、新熱帯区
guava mealybug			striped mealybug を見よ
guava moth		*Coscinoptycha improbana* Meyrick	（チョウ目、シンクイガ科）豪州区

英名	和名	学名	所属、分布、ほか
guava scale			green shield scale を見よ　南アフリカでの英名
guava skipper		*Phocides palemon* (Cramer)	（チョウ目、セセリチョウ科）新北区
guava skipper			mangrove skipper (1) を見よ
guava whitefly			Florida whitefly を見よ
Guaymas skipper		*Polites norae* MacNeill	（チョウ目、セセリチョウ科）新熱帯区
Guderian's Charaxes			blue spangled Charaxes を見よ
guelder rose leaf beetle			cranberry tree leaf beetle を見よ
guelder-rose midget		*Phyllonorycter lantanella* (Schrank)	（チョウ目、ホソガ科）旧北区
Guenee's emerald		*Chlorocoma dichloraria* Guenée	（チョウ目、シャクガ科）豪州区
Guenee's Perisama		*Perisama lanice picteti* (Guenée)	（チョウ目、タテハチョウ科）新熱帯区
Guernsey carpet beetle		*Anthrenus sarnicus* Mroczkowski	（コウチュウ目、カツオブシムシ科）旧北区
Guerrero giant-skipper		*Turnerina hazelae* (Stallings et Turner)	（チョウ目、セセリチョウ科）新熱帯区
Guerrero mimic-white		*Lieinix neblina* J. et R. Maza	（チョウ目、シロチョウ科）新熱帯区
Guerrero pine-satyr		*Paramacera copiosa* Miller	（チョウ目、タテハチョウ科）新熱帯区
Guerrero skipperling		*Dalla mentor* Evans	（チョウ目、セセリチョウ科）新熱帯区
Guerrero sootywing		*Bolla guerra* Evans	（チョウ目、セセリチョウ科）新熱帯区
Guessfeldt's white blue		*Oboronia guessfeldti* (Dewitz)	（チョウ目、シジミチョウ科）エチオピア区
guest-flies			inquiline gall wasps を見よ
Guinea ant	オオシワアリ	*Tetramorium bicarinatum* (Nylander)	（ハチ目、アリ科）日本、旧北区、汎亜熱帯、汎熱帯。アフリカのギニアに由来
Guinea ant		*Tetramorium guineense* (Fabricius)	（ハチ目、アリ科）大洋区
Guinea cypress katydid		*Inscudderia strigata* (Scudder)	（バッタ目、キリギリス科）新北区
Guinea feather louse		*Goniodes numidae* Mjöberg	（ハジラミ目、チョウカクハジラミ科）新北区
Guinea pig lice			rodent chewing lice を見よ
Guinea pig louse		*Trimenopon hispidum* (Burmeister)	（ハジラミ目、ケモノタンカクハジラミ科）新北区
Guinea pig louse (1)	カビアハジラミ	*Gliricola porcelli* (Schrank)	（ハジラミ目、ナガケモノハジラミ科）日本、汎世界。モルモットに寄生
Guinea skimmer		*Orthetrum guineense* Ris	（トンボ目、トンボ科）エチオピア区
Guineafowl	シロボシタテハ	*Hamanumida daedalus* Fabricius	（チョウ目、タテハチョウ科）エチオピア区。Guineafowl butterfly ともいう
guitar beetle			fiddle beetle を見よ
gulf coast clubtail		*Gomphus modestus* Needham	（トンボ目、サナエトンボ科）新北区
gulf coast gray moth		*Anavitrinella atristrigaria* (Barnes et McDunnough)	（チョウ目、シャクガ科）新北区

英名	和名	学名	所属、分布、ほか
gulf fritillary	ヒョウモンドクチョウ	*Agraulis vanillae* (Linnaeus)	（チョウ目、タテハチョウ科）新北区、新熱帯区。*A. v. incarnata* (Riley) も同英名
gulf meadow katydid		*Conocephalus hygrophilus* Rehn et Hebard	（バッタ目、キリギリス科）新北区
gulf pine sphinx			Brou's sphinx moth を見よ
gulf wireworm		*Conoderus amphicollis* (Gyllenhal)	（コウチュウ目、コメツキムシ科）新北区
gull-feather case	シモフリツツミノガ	*Coleophora laripenella* (Zetterstedt)	（チョウ目、ツツミノガ科）日本、旧北区
gull-feather case			crabweed case を見よ
gulls	マルバネシロチョウ属	*Cepora*	（チョウ目、シロチョウ科）の昆虫の総称
gum emperor			emperor (1) を見よ　豪州区での英名
gum emperor moth			emperor (1) を見よ
gum leaf katydid		*Torbia viridissima* (Brunner von Wattenwyl)	（バッタ目、キリギリス科）豪州区
gum snout moth		*Pinara fervens* Walker	（チョウ目、シャクガ科）豪州区
gum tree thrips		*Isoneurothrips australis* Bagnall	（アザミウマ目、アザミウマ科）日本、旧北区、エチオピア区、大洋区、豪州区、新熱帯区
gum tree weevil			eucalyptus weevil を見よ
gumleaf grasshopper		*Goniaea australasiae* Leach	（バッタ目、バッタ科）豪州区
gumleaf skeletonizer		*Uraba lugens* Walker	（チョウ目、ヤガ科）豪州区
gumtree hoppers		*Eurymela*	（カメムシ目、Eurymelidae) の昆虫の総称
gumtree scale		*Eriococcus coriaceus* Maskell	（カメムシ目、フクロカイガラムシ科）豪州区
gumtree scale ladybird			black lady beetle を見よ
gum-tree weevil			eucalyptus weevil を見よ
Gundlach's trig		*Cyrtoxipha gundlachi* Saussure	（バッタ目、クサヒバリ科）新北区
Gurdaspur borer		*Bissetia steniella* (Hampson)	（チョウ目、メイガ科）東洋区。インドの地名に由来
Gurney's shieldback		*Aglaothorax gurneyi* (Rentz et Birchim)	（バッタ目、キリギリス科）新北区
Gurney's shieldback		*Idiostatus gurneyi* Rentz	（バッタ目、キリギリス科）新北区
gurt fly		*Chlorops taeniops* (Wiedemann)	（ハエ目、キモグリバエ科）旧北区
gurt-fly of barley			gurt fly を見よ
gut-mine pigmy		*Stigmella viscerella* (Stainton)	（チョウ目、モグリチビガ科）旧北区
Gutierrezia mealybug		*Sirococcus gutierreziae* (Cockerell)	（カメムシ目、コナカイガラムシ科）新北区
Guyana spotted cockroach			South American cockroach を見よ
Guyanan Sarota			Gyas jewelmark を見よ
Gyas jewelmark		*Sarota gyas* (Cramer)	（チョウ目、シジミタテハ科）新熱帯区

英名	和名	学名	所属、分布、ほか
Gydo blue		*Lepidochrysops gydoae* Dickson et Wykeham	（チョウ目、シジミチョウ科）エチオピア区
gyne			雌アリ、女王アリを指す
gynoparae	産性虫、産性型		アブラムシ類で alienicolae が生れる胎生で、単性生殖雌 sexupara ともいう
gypsy ant			spider-like spine-waisted ant を見よ
gypsy cuckoo bee		*Bombus bohemicus* Seidl	（ハチ目、ミツバチ科）旧北区
gypsy moth	マイマイガ	*Lymantria dispar* (Linnaeus)	（チョウ目、ドクガ科）日本、全北区。多くの作物、樹木の著名害虫。成虫の飛翔が一定方向でないことに由来
gypsy moth parasitoid		*Chelonus melanoscelus* (Ratzeburg)	（ハチ目、コマユバチ科）日本、全北区、東洋区

英名	和名	学名	所属、分布、ほか
H			
Habilis underwing moth		*Catocala habilis* Grote	(チョウ目、ヤガ科) 新北区。Habilis underwing ともいう
hackberry butterflies			emperors (1) を見よ
hackberry butterfly		*Asterocampa celtis* (Boisduval et LeConte)	(チョウ目、タテハチョウ科) 新北区
hackberry emperor	エノキタテハ	*Asterocampa celtis antonia* (W. H. Edwards)	(チョウ目、タテハチョウ科) 新熱帯区
hackberry engraver		*Scolytus muticus* Say	(コウチュウ目、キクイムシ科) 新北区
hackberry lace bug		*Corythucha celtides* Osborn et Drake	(カメムシ目、グンバイムシ科) 新北区
hackberry nipplegall maker		*Pachypsylla celtidismamma* (Fletcher)	(カメムシ目、キジラミ科) 新北区。ハックベリーの葉に虫えいを作る
hacklegill mayflies	カクカゲロウ科	Potamanthidae	(カゲロウ目) の昆虫の総称
hacklegill mayflies (1)		Behningidae	(カゲロウ目) の昆虫の総称
Hadassa firetip		*Pyrrhopyge hadassa* Hewitson	(チョウ目、セセリチョウ科) 新熱帯区
Hadrian's white Charaxes		*Charaxes hadrianus* Ward	(チョウ目、タテハチョウ科) エチオピア区
hag			clothes moth (1) を見よ
hag monkey slug moth			hag moth を見よ
hag moth		*Phobetron pithecium* (Smith et Abbott)	(チョウ目、イラガ科) 新北区
Hagen's bluet		*Enallagma hageni* (Walsh)	(トンボ目、イトトンボ科) 新北区
Hagen's sphinx		*Ceratomia hageni* Grote	(チョウ目、スズメガ科) 新北区。Hagen's sphinx moth ともいう
Hagen's sprite		*Pseudagrion hageni* Karsch	(トンボ目、イトトンボ科) エチオピア区
hairless flower thrips		*Pseudanaphothrips achaetus* (Bagnall)	(アザミウマ目、アザミウマ科) 豪州区
hairstreak butterflies	ムラサキミドリシジミ属	*Quercusia*	(チョウ目、シジミチョウ科) の昆虫の総称
hairstreaks	ミドリシジミ亜科	Theclinae	(チョウ目、シジミチョウ科) の昆虫の総称
hairstreaks			tits (1) を見よ
hairy ammophila		*Sphex hirsuta* Scopoli	(ハチ目、アナバチ科) 全北区
hairy angle		*Darpa hanria* (Moore)	(チョウ目、セセリチョウ科) 東洋区
hairy ant	ヒゲナガアメイロアリ	*Paratrechina longicornis* (Latreille)	(ハチ目、アリ科) 日本、新北区、豪州区、汎亜熱帯、汎熱帯
hairy bear beetle		*Paracotalpa granicollis* (Haldeman)	(コウチュウ目、コガネムシ科) 新北区
hairy beetle			green rose chafer (1) を見よ
hairy black springtail			glacier flea を見よ
hairy borer		*Ipochus fasciatus* LeConte	(コウチュウ目、カミキリムシ科) 新北区。クルミ害虫
hairy cellar beetle		*Mycetaea hirta* (Marsham)	(コウチュウ目、テントウムシ科) 旧北区

英名	和名	学名	所属、分布、ほか
hairy chinch bug		*Blissus leucopterus hirtus* Montandon	（カメムシ目、ナガカメムシ科）新北区
hairy cicada		*Tettigarcta crinita* Distant	（カメムシ目、Tettigarctidae）豪州区。山地に生息
hairy cicadas		Tettigarctidae	（カメムシ目）の昆虫の総称
hairy colletid bees		*Leioproctus*	（ハチ目、ムカシハナバチ科）の昆虫の総称
hairy desert cockroach		*Eremoblatta subdiaphana* Scudder	（ゴキブリ目、ムカシゴキブリ科）新北区
hairy dragonfly			hairy hawker を見よ
hairy dragonfly			brown hawker を見よ
hairy dry-wooden longicorn beetle	マルクビケマダラカミキリ	*Hesperophanes campestris* (Faldermann)	（コウチュウ目、カミキリムシ科）日本、旧北区。hairy dry-wooden longicorn ともいう
hairy dusk-hawker		*Gynacantha villosa* Grunberg	（トンボ目、ヤンマ科）エチオピア区
hairy dusky wing			persius duskywing を見よ
hairy flies			tangle-veined flies を見よ
hairy flower bees			hairy-footed bees を見よ
hairy flower wasps			scoliid wasps を見よ　豪州での英名
hairy-footed bees		*Anthophora*	（ハチ目、コシブトハナバチ科）の昆虫の総称
hairy-footed bees			long-tongued bees を見よ
hairy fungus beetle	チャイロコキノコムシ	*Typhaea stercorea* (Linnaeus)	（コウチュウ目、コキノコムシ科）日本、汎世界。貯穀害虫
hairy fungus beetle		*Berginus pumilus* LeConte	（コウチュウ目、コキノコムシ科）新北区
hairy fungus beetles	コキノコムシ科	Mycetophagidae	（コウチュウ目）の昆虫の総称
hairy green leaf-roller weevil	ヒメケブカチョッキリ	*Involvulus pilosus* (Roelofs)	（コウチュウ目、オトシブミ科）日本
hairy ground beetles		Panagaeini	（コウチュウ目、オサムシ科）の昆虫の総称
hairy hawker		*Brachytron pratense* (Müller)	（トンボ目、ヤンマ科）旧北区。hairy dragonfly ともいう。
hairy juniper longicorn	ケブカマルクビカミキリ	*Atimia okayamensis* Hayashi	（コウチュウ目、カミキリムシ科）日本
hairy kelp fly		*Chaetocoelopa littoralis* Robineau-Desvoidy	（ハエ目、ハマベバエ科）旧北区
hairy leafeating caterpillar		*Xanthodes ongenita* (Hampson)	（チョウ目、ヤガ科）豪州区
hairy-legged flies		*Trichopoda*	（ハエ目、ヤドリバエ科）の昆虫の総称
hairy-legged fly		*Trichopoda (Trichopoda) indivisa* Townsend	（ハエ目、ヤドリバエ科）新北区
hairy-legged horsefly	オオキボシアブ	*Hybomitra bimaculata* (Macquart)	（ハエ目、アブ科）日本、旧北区
hairy-legged mining bee		*Dasypoda altercator* (Harris)	（ハチ目、ケアシハナバチ科）旧北区
hairy-legged mining bee		*Dasypoda hirtipes* (Fabricius)	（ハチ目、ケアシハナバチ科）旧北区
hairy-legged mining bee		*Dasypoda plumipes* Panzer	（ハチ目、ケアシハナバチ科）旧北区

英名	和名	学名	所属、分布、ほか
hairy line blue		*Erysichton lineata* (Murray)	(チョウ目、シジミチョウ科) 豪州区。hairy line-blue butterfly ともいう
hairy maggot blowfly	ホホジロオビキンバエ	*Chrysomya rufifacies* (Macquart)	(ハエ目、クロバエ科) 日本、全北区、豪州区
hairy mothflies			moth flies を見よ
hairy poplar sawfly			poplar sawfly (2) を見よ
hairy powder-post beetle		*Lyctus pubescens* Panzer	(コウチュウ目、ナガシンクイムシ科) 旧北区
hairy powderpost beetle (1)	ケブトヒラタキクイムシ	*Minthea rugicollis* (Walker)	(コウチュウ目、ナガシンクイムシ科) 日本、汎世界。乾材害虫
hairy rice caterpillar	クロスジシロコブガ	*Nola taeniata* Snellen	(チョウ目、コブガ科) 日本、旧北区、東洋区、豪州区。マメ科牧草害虫
hairy rice stemborer		*Niphadoses palleucus* Common	(チョウ目、メイガ科) 豪州区
hairy rove beetle	オオハネカクシ	*Creophilus maxillosus* (Linnaeus)	(コウチュウ目、ハネカクシ科) 日本、全北区。東洋区。動物死体に普通
hairy sand wasp		*Podalonia hirsuta* (Scopoli)	(ハチ目、アナバチ科) 旧北区
hairy scarab		*Saulostomus villosus* Waterhouse	(コウチュウ目、コガネムシ科) 豪州区
hairy small weevil	ケブカヒメカタゾウムシ	*Arrhaphogaster pilosa* Roelfs	(コウチュウ目、ゾウムシ科) 日本
hairy spider beetle		*Ptinus villiger* (Reitter)	(コウチュウ目、ヒョウホンムシ科) 新北区
hairy-veined barklice		Ptiloneuridae	(チャタテムシ目) の昆虫の総称
hairy vetch bruchid			vetch weevil を見よ
hairy whirligig	エゾコオナガミズスマシ	*Orectochilus villosus* (Muller)	(コウチュウ目、ミズスマシ科) 日本、旧北区
hairy-winged barklice	ケブカチャタテ科	Amphipsocidae	(チャタテムシ目) の昆虫の総称
hairy wood ant		*Formica lugubris* Zetterstedt	(ハチ目、アリ科) 旧北区
hairymary caterpillar			urticating anthelid を見よ
Haitian cockroach			discoid cockroach を見よ
Haitian mimic		*Dismorphia spio* (Godart)	(チョウ目、シロチョウ科) 新熱帯区
Haitian skipper		*Choranthus haitensis* Skinner	(チョウ目、セセリチョウ科) 新北区
Hakea moth		*Oenochroma vinosa* (Warren)	(チョウ目、シャクガ科) 旧北区
Hakea seed-moth		*Carposina autologa* Meyrick	(チョウ目、シンクイガ科) 豪州区、エチオピア区
Hakone globular treehopper	ハコネマルツノゼミ	*Gargara doenitzi* Matsumura	(カメムシ目、ツノゼミ科) 日本、旧北区
Hakone pine sawfly	ハコネマツハバチ	*Gilpinia hakonensis* (Matsumura)	(ハチ目、マツハバチ科) 日本
Hakone sawfly			Hakone pine sawfly を見よ
Halciones hairstreak			royal hairstreak を見よ
Haldeman's green potato beetle		*Leptinotarsa haldemani* Rogers	(コウチュウ目、ハムシ科) 新北区

英名	和名	学名	所属、分布、ほか
Haldeman's shieldback		*Pediodectes haldemani* (Girard)	（バッタ目、キリギリス科）新北区
half-blind sphinx moth		*Perigonia lusca* (Fabricius)	（チョウ目、スズメガ科）新北区、新熱帯区。half-blind sphinx ともいう
half-blue hairstreak		*Contrafacia ahola* (Hewitson)	（チョウ目、シジミチョウ科）新熱帯区
half-edged roadside-skipper		*Amblyscirtes anubis* (Godman)	（チョウ目、セセリチョウ科）新熱帯区
half-loopers			underwing moths を見よ　幼虫の英名
half-tailed skipper		*Niconiades incomptus* Austin	（チョウ目、セセリチョウ科）新熱帯区
half-winged conehead		*Belocephalus subapterus* Scudder	（バッタ目、キリギリス科）新北区
half-winged geometer		*Phigalia titea* Cramer	（チョウ目、シャクガ科）新北区。half-wing moth ともいう
half-yellow moth		*Ponometia semiflava* (Guenée)	（チョウ目、ヤガ科）新北区。half-yellow ともいう
halfordia fruit fly		*Dacus halfordiae* Tryon	（ハエ目、ミバエ科）豪州区
halictid bees			sweat bees (1) を見よ
halictophagids	クシヒゲネジレバネ科	Halictophagidae	（ネジレバネ目）の昆虫の総称
Halimede Eurybia		*Eurybia halimede* (Hübner)	（チョウ目、シジミタテハ科）新熱帯区
Haliphron birdwing	フチグロキシタアゲハ	*Troides haliphron* (Boisduval)	（チョウ目、アゲハチョウ科）東洋区
haliplid beetles			crawling water beetles を見よ
Hall scale		*Nilotaspis halli* (Green)	（カメムシ目、マルカイガラムシ科）全北区
halloween lady beetle			Asian multicolored lady beetle を見よ
halloween pennant		*Celithemis eponina* (Drury)	（トンボ目、トンボ科）新北区
Hall's celery aphid			honeysuckle aphid を見よ
Hall's Theope		*Theope pseudopedias* Hall	（チョウ目、シジミタテハ科）新熱帯区
Halobates sea skaters		*Halobates*	（カメムシ目、アメンボ科）の昆虫の総称
Halyzia skipper		*Metisella halyzia* (Hewitson)	（チョウ目、セセリチョウ科）豪州区
ham beetle			red-legged ham beetle を見よ
ham beetle			larder beetle を見よ
ham beetles		Corynetidae	（コウチュウ目）の昆虫の総称
ham skipper			cheese skipper を見よ　幼虫の英名
Hamadryad			Cairns Hamadryad を見よ
Hamakua pamakami plume moth		*Oidaematophorus beneficus* Yano et Heppner	（チョウ目、トリバガ科）新北区、大洋区。ハワイの雑草の生物的防除に使用、英名はその雑草のハワイ名による
hamamelis gall aphid	イガフシマンサクアブラムシ	*Hamamelistes betulinus miyabei* (Matsumura)	（カメムシ目、アブラムシ科）日本
Hamon's sprite		*Pseudagrion hamoni* Fraser	（トンボ目、イトトンボ科）エチオピア区

英名	和名	学名	所属、分布、ほか
Hampson's hedge blue		*Acytolepis lilacea* (Hampson)	(チョウ目、シジミチョウ科) 東洋区
Hampson's hedge hopper		*Baracus hampsoni* Elwes et Edwards	(チョウ目、セセリチョウ科) 東洋区
Hampson's hedge hopper			hedge hopper を見よ
Hamra beetle			red pumpkin beetle を見よ
Handley's small brown			dingy skipper を見よ
handmaid		*Dysauxes ancilla* (Linnaeus)	(チョウ目、ヒトリガ科) 旧北区
handmaidens			wasp moths を見よ　南米での英名
handsome bush cricket		*Phyllopalpus pulchellus* (Uhler)	(バッタ目、コオロギ科) 新北区
handsome clubtail		*Gomphus crassus* Hagen	(トンボ目、サナエトンボ科) 新北区
handsome earwig		*Prolabia pulchella* (Serville)	(バッタ目、コオロギ科) 新北区。体長 7 mm。
handsome fungus beetles	テントウムシダマシ科	Endomychidae	(コウチュウ目) の昆虫の総称
handsome Macrotona grasshopper		*Macrotona mjobergi* Sjöstedt	(バッタ目、バッタ科) 豪州区。handsome Macrotona ともいう。
handsome meadow katydid		*Orchelimum pulchellum* Davis	(バッタ目、キリギリス科) 新北区
handsome trip			handsome bush cricket を見よ
hang flies			scorpion flies (1) を見よ
hanging clubtails		*Stylurus*	(トンボ目、サナエトンボ科) の昆虫の総称
hanging flies	ガガンボモドキ科	Bittacidae	(シリアゲムシ目) の昆虫の総称
hanging flies			scorpion flies (1) を見よ
hanging scorpionflies			hanging flies を見よ
hangingfly		*Bittacus italicus* (Müller)	(シリアゲムシ目、ガガンボモドキ科) 旧北区
Hanham's owlet moth		*Phalaenostola hanhami* (Smith)	(チョウ目、ヤガ科) 新北区。Hanham's owlet, Hanham's snout moth ともいう
Hannington's fritillary		*Issoria hanningtoni* Elwes	(チョウ目、タテハチョウ科) エチオピア区
Hanno blue		*Hemiargus hanno* (Stoll)	(チョウ目、シジミチョウ科) 新熱帯区
Hannyngton's Apollo	チベットチビウスバアゲハ	*Parnassius hannyngtoni* Avinoff	(チョウ目、アゲハチョウ科) 東洋区
Hantamsberg skolly		*Thestor calviniae* Riley	(チョウ目、シジミチョウ科) エチオピア区
hardback			yellow mealworm を見よ
hard-back beetle		*Pentodon idiota* (Herbst)	(コウチュウ目、コガネムシ科) 旧北区
hard-back beetles			Hercules beetles を見よ
hardbacks			May beetles (2) を見よ
hard maple budminer		*Ectoedemia ochrefasciella* (Chambers)	(チョウ目、モグリチビガ科) 新北区。hard maple budminer moth ともいう
hard wax scale			Chinese wax scale (2) を見よ

英名	和名	学名	所属、分布、ほか
hardwood stump borer beetle		*Mallodon dasytomus* Say	（コウチュウ目、カミキリムシ科）新北区
hardwood tussock moth		*Dasychira dorsipennata* (Barnes et McDunnough)	（チョウ目、ドクガ科）新北区
hare louse		*Haemodipsus lyriocephalus* (Burmeister)	（シラミ目、フトゲジラミ科）旧北区、豪州区
Harford's sulphur	ハーフォードモンキチョウ	*Colias harfordii* Hy. Edwards	（チョウ目、シロチョウ科）新熱帯区
Harkness's dancer		*Argia harknessi* Calvert	（トンボ目、イトトンボ科）新熱帯区
harlequin		*Praetaxila segecia punctaria* (Fruhstorfer)	（チョウ目、シジミタテハ科）豪州区。harlequin metalmark を参照
harlequin (1)		*Taxila haquinus* (Fabricius)	（チョウ目、シジミタテハ科）東洋区
harlequin beetle	テナガカミキリ	*Acrocinus longimanus* Linnaeus	（コウチュウ目、カミキリムシ科）新熱帯区。道化師ハーレキンに由来
harlequin beetles			hister beetles を見よ　中南米での英名
harlequin bug		*Dindymus versicolor* (Herrich-Schäffer)	（カメムシ目、ホシカメムシ科）豪州区
harlequin bug (1)		*Murgantia histrionica* (Hahn)	（カメムシ目、カメムシ科）新北区。体長約10 mm。翅に黒赤の斑紋のあるキャベツ害虫。calico bug, fire bug ともいう
harlequin bug (2)		*Tectocoris diophthalmus* (Thunberg)	（カメムシ目、キンカメムシ科）豪州区
harlequin bugs		*Bagrada*	（カメムシ目、カメムシ科）の昆虫の総称
harlequin cabbage-bug			harlequin bug (1) を見よ
harlequin cockroach	イエゴキブリ	*Neostylopyga rhombifolia* (Stoll)	（ゴキブリ目、ゴキブリ科）日本、東洋区、新北区、新熱帯区。屋内で食品を加害
harlequin darner		*Gomphaeschna furcillata* (Say)	（トンボ目、ヤンマ科）新北区
harlequin flies			march flies を見よ　南アフリカでの英名
harlequin hawk		*Batocnema africanus* (Distant)	（チョウ目、スズメガ科）エチオピア区
harlequin ladybird			Asian multicolored lady beetle を見よ
harlequin metalmark		*Praetaxila segecia* (Hewitson)	（チョウ目、シジミタテハ科）豪州区
harlequin moth		*Diphthera ludifica* Linnaeus	（チョウ目、ヤガ科）旧北区
harlequin moth			magpie を見よ
harlequin skipper			fulvous ranger を見よ
harlequin sprite			Newton's sprite を見よ
harlequin webworm moth		*Diathrausta harlequinalis* Dyar	（チョウ目、メイガ科）新北区
harmonia mantle		*Caria rhacotis* Godman et Salvin	（チョウ目、シジミタテハ科）新熱帯区
harmonia ringlet		*Hermeuptychia harmonia* (Butler)	（チョウ目、タテハチョウ科）新熱帯区
harmonia tiger			tiger butterfly を見よ

英名	和名	学名	所属、分布、ほか
harmonia tigerwing		*Tithorea harmonia hippothous* Godman et Salvin	(チョウ目、タテハチョウ科) 新熱帯区。*T. h. salvadoris* Staudinger も同英名
harmonia tigerwing			tiger butterfly を見よ
harnessed moth		*Apantesis phalerata* (Harris)	(チョウ目、ヒトリガ科) 新北区
harnessed tiger moth			harnessed moth を見よ
Harper scale	シロナガカキカイガラムシ	*Neopinnaspis harperi* McKenzie	(カメムシ目、マルカイガラムシ科) 日本、東洋区。マンゴー、マサキ、モクセイなどを加害
Harpoon clubtail		*Gomphus descriptus* Banks	(トンボ目、サナエトンボ科) 新北区
Harris' checkerspot	ハリスシロホシタテハ	*Chlosyne harrisii* (Scudder)	(チョウ目、タテハチョウ科) 新北区
harrisia cactus longicorn		*Alcidion cereicola* Fisher	(コウチュウ目、カミキリムシ科) 豪州区
harrisia cactus mealybug		*Hypogeococcus pungens* (Granara de Willink)	(カメムシ目、コナカイガラムシ科) 豪州区
harrisia cactus weevil		*Eriocereophaga humeridens* O'Brien	(コウチュウ目、ゾウムシ科) 豪州区
Harris's three-spot moth		*Harrisimemna trisignata* (Walker)	(チョウ目、ヤガ科) 新北区。Harris's three-spot ともいう
Hartig's brahmaea		*Acanthobrahmaea europaea* (Hartig)	(チョウ目、イボタガ科) 旧北区
Hartmann's talontail		*Crenigomphus hartmanni* (Förster)	(トンボ目、サナエトンボ科) エチオピア区
Hartweg's sphinx		*Dolbogene hartwegii* (Butler)	(チョウ目、スズメガ科) 新北区
harvest flies			cicadas を見よ
harvest flies			dog-day cicadas を見よ
harvest fly		*Cicada tibicen* Linnaeus	(カメムシ目、セミ科) 新北区、新熱帯区
harvest men			crane flies (1) を見よ　ザトウムシ目の動物の総称でもある
harvested ants		*Pogonomyrmex*	(ハチ目、アリ科) の昆虫の総称
harvester	アメリカアリマキシジミ	*Feniseca tarquinius* Fabricius	(チョウ目、シジミチョウ科) 新北区
harvester ant		*Pogonomyrmex comanche* Wheeler	(ハチ目、アリ科) 新北区
harvester ant		*Messor structor* (Latreille)	(ハチ目、アリ科) 旧北区
harvester ant (1)		*Veromessor pergandei* (Mayr)	(ハチ目、アリ科) 新北区
harvester ant (2)		*Messor barbarus* Linnaeus	(ハチ目、アリ科) 旧北区、エチオピア区。harvested ant ともいう
harvester ant (3)		*Pogonomyrmex rugosus* Emery	(ハチ目、アリ科) 新北区
harvester ant (4)	クロナガアリ	*Messor aciculatus* (F. Smith)	(ハチ目、アリ科) 日本、旧北区。草の種子を巣に貯蔵し餌とする
harvester ant (5)		*Pogonomyrmex salinus* Olsen	(ハチ目、アリ科) 新北区
harvester ants		*Messor*	(ハチ目、アリ科) の昆虫の総称

英名	和名	学名	所属、分布、ほか
harvester ants		*Aphenogaster*	（ハチ目、アリ科）の昆虫の総称
harvester ants			agricultural ants を見よ
harvester moth			harvester を見よ
harvester termite		*Hodotermes mossambicus* Hagen	（シロアリ目、シュウカクシロアリ科）エチオピア区
harvester termites		*Drepanotermes*	（シロアリ目、シロアリ科）の昆虫の総称
harvester termites			rotten-wood termites を見よ
harvesters		Gerydinae	（チョウ目、シジミチョウ科）の昆虫の総称
harvesters	アシナガシジミ亜科	Miletinae	（チョウ目、シジミチョウ科）の昆虫の総称
Harvey's eyed-metalmark		*Mesosemia harveyi* DeVries et Hall	（チョウ目、シジミタテハ科）新熱帯区
Hassan greenstreak		*Chalybs hassan* (Stoll)	（チョウ目、シジミチョウ科）新熱帯区
Hast's button		*Acleris hastiana* (Linnaeus)	（チョウ目、ハマキガ科）旧北区
hatchet wasps			ensign wasps を見よ　豪州での英名
hau leafminer		*Philodoria hauicola* (Swezey)	（チョウ目、ホソガ科）新北区
Hawaiian aalii long-horned beetle		*Plagithmysus dodonaeae* (Swezey)	（コウチュウ目、カミキリムシ科）大洋区
Hawaiian alani long-horned beetle		*Plagithmysus bishopi* Sharp	（コウチュウ目、カミキリムシ科）大洋区
Hawaiian antlion		*Eidoleon wilsoni* (McLachlan)	（アミメカゲロウ目、ウスバカゲロウ科）大洋区、新北区
Hawaiian aweoweo long-horned beetle		*Plagithmysus chenopodii* (Perkins)	（コウチュウ目、カミキリムシ科）大洋区
Hawaiian beet webmoth			beet webworm を見よ　Hawaiian beet webworm moth ともいう
Hawaiian black wattle long-horned beetle		*Plagithmysus decurrensae* Gressitt et Davis	（コウチュウ目、カミキリムシ科）大洋区
Hawaiian blue	ハワイシジミ	*Udara blackburni* (Tuely)	（チョウ目、シジミチョウ科）大洋区
Hawaiian bud moth		*Helicoverpa hawaiiensis* (Quaintance et Brues)	（チョウ目、ヤガ科）新北区、大洋区
Hawaiian carpenter ant		*Camponotus variegatus* (F. Smith)	（ハチ目、アリ科）新北区、大洋区
Hawaiian ebony leaf miner		*Caloptilia mabaella* (Swezey)	（チョウ目、ホソガ科）大洋区
Hawaiian flower thrips			flower thrips (1) を見よ　米国での英名
Hawaiian grass bug		*Oronomiris hawaiiensis* Kirkaldy	（カメムシ目、オオヨコバイ科）大洋区
Hawaiian grass thrips		*Anaphothrips swezeyi* Moulton	（アザミウマ目、アザミウマ科）新北区、大洋区
Hawaiian koa long-horned beetle		*Plagithmysus acaciae* Gressitt et Davis	（コウチュウ目、カミキリムシ科）大洋区。ハワイの植物 koa に由来
Hawaiian long-horned		*Plagithmysus microgaster* (Sharp)	（コウチュウ目、カミキリムシ科）大洋区

英名	和名	学名	所属、分布、ほか
Hawaiian mamane long-horned beetle		*Plagithmysus blackburni* (Sharp)	（コウチュウ目、カミキリムシ科）大洋区。ハワイの植物 mamane に由来
Hawaiian ohia long-horned beetle		*Plagithmysus bilineatus* Sharp	（コウチュウ目、カミキリムシ科）大洋区。ハワイの植物 ohia に由来
Hawaiian pelagic water strider		*Halobates hawaiiensis* Usinger	（カメムシ目、アメンボ科）大洋区
Hawaiian pug-moth		*Eupithecia orichloris* Meyrick	（チョウ目、シャクガ科）大洋区
Hawaiian scale	ハワイカキカイガラムシ	*Acanthomytilus hawaiiensis* (Maskell)	（カメムシ目、マルカイガラムシ科）日本、汎熱帯。カンキツ、バラ、サルスベリの害虫
Hawaiian serpentine leafminer		*Liriomyza* sp.	（ハエ目、ハモグリバエ科）大洋区
Hawaiian sphinx		*Hyles calida* (Butler)	（チョウ目、スズメガ科）大洋区、新北区
Hawaiian sweetpotato leaf miner			sweetpotato leafminer を見よ
Hawaiian sweetpotato leafroller		*Udea despecta* (Butler)	（チョウ目、メイガ科）大洋区
Hawaiian tobacco hornworm			Blackburn's sphinx moth を見よ
Hawaiian tomato hornworm			Blackburn's sphinx moth を見よ
Hawaiian upland damselfly		*Megalagrion hawaiiensis* (McLachlan)	（トンボ目、イトトンボ科）大洋区
Hawaiian vegetable leafminer			vegetable leafminer を見よ
Hawaiian wood gnat		*Sylvicola cinctus* (Fabricius)	（ハエ目、カバエ科）大洋区
Hawequas widow		*Torynesis hawequas* Dickson	（チョウ目、タテハチョウ科）エチオピア区
hawk flies			robber flies (1) を見よ
hawk flies			flower flies を見よ　英国での英名
hawk lice	オオハジラミ科	Laemobothriidae	（ハジラミ目）の昆虫の総称
hawk louse		*Laemobothrion maximum* Scopoli	（ハジラミ目、オオハジラミ科）新北区。タカ、ミサゴに外部寄生
hawk moth		*Daphnis hypothous* (Cramer)	（チョウ目、スズメガ科）旧北区、東洋区、豪州区
hawk moth		*Nephele subvaria* (Walker)	（チョウ目、スズメガ科）豪州区
hawk moths	スズメガ科	Sphingidae	（チョウ目）の昆虫の総称　世界に 1,100 種
hawker dragonflies			darners を見よ
hawker dragonflies			dragonflies (2) を見よ
hawkers			darners を見よ
hawklet			savanna hawkmoth を見よ
hawkmoths		*Hyles*	（チョウ目、スズメガ科）の昆虫の総称
Haworth's crescent			Haworth's minor を見よ
Haworth's fanner		*Glyphipterix haworthana* Stainton	（チョウ目、ホソハマキモドキガ科）旧北区

英名	和名	学名	所属、分布、ほか
Haworth's glyphipterid moth			Haworth's fanner を見よ
Haworth's minor		*Celaena haworthii* (Curtis)	(チョウ目、ヤガ科) 旧北区
Haworth's pug		*Eupithecia haworthiata* Doubleday	(チョウ目、シャクガ科) 旧北区
Haworth's purple		*Eriocrania haworthi* Bradley	(チョウ目、スイコバネガ科) 旧北区
hawthorn aphid (1)		*Nearctaphis crataegifoliae* Fitch	(カメムシ目、アブラムシ科) 新北区
hawthorn aphid (2)		*Dysaphis apiifolia* (Theobald)	(カメムシ目、アブラムシ科) 全北区、エチオピア区、新熱帯区
hawthorn button-top midge		*Dasineura crataegi* (Winnertz)	(ハエ目、タマバエ科) 旧北区
hawthorn-carrot aphid		*Dysaphis crataegi* (Kaltenbach)	(カメムシ目、アブラムシ科) 全北区、東洋区
hawthorn cosmet		*Blastodacna hellerella* (Duponchel)	(チョウ目、カザリバガ科) 旧北区
hawthorn fruit miner		*Batrachedra curvilineella* Chambers	(チョウ目、カザリバガ科) 新北区
hawthorn lace bug		*Corythucha cydoniae* (Fitch)	(カメムシ目、グンバイムシ科) 新北区
hawthorn leaf beetle		*Lochmaea crataegi* (Forster)	(コウチュウ目、ハムシ科) 旧北区
hawthorn moth			hawthorn webber moth を見よ
hawthorn orange midget		*Phyllonorycter oxyacanthae* (Frey)	(チョウ目、ホソガ科) 旧北区
hawthorn parsley aphid		*Dysaphis petroselini* (Borner)	(カメムシ目、アブラムシ科) 旧北区
hawthorn-parsley aphid			hawthorn aphid (2) を見よ
hawthorn red midget			apple leafminer (2) を見よ
hawthorn red midget moth			apple leafminer (2) を見よ
hawthorn sawfly	コヒラクチハバチ	*Trichiosoma tibiale* Stephens	(ハチ目、コンボウハバチ科) 日本。旧北区
hawthorn shieldbug		*Acanthosoma haemorrhoidale* (Linnaeus)	(カメムシ目、ツノカメムシ科) 旧北区
hawthorn stem midge		*Thomasiniana crataegi* Barnes	(ハエ目、タマバエ科) 旧北区
hawthorn sucker		*Psylla melanoneura* Forster	(カメムシ目、キジラミ科) 旧北区
hawthorn underwing moth		*Catocala crataegi* Saunders	(チョウ目、ヤガ科) 新北区。hawthorn underwing ともいう
hawthorn webber			hawthorn webber moth を見よ　幼虫の英名
hawthorn webber moth		*Scythropia crataegella* (Linnaeus)	(チョウ目、スガ科) 旧北区
hawthorn white			black-veined white (1) を見よ
hay-stack bug			corn stack bug を見よ

英名	和名	学名	所属、分布、ほか
Hayden grasshopper		*Derotmema haydeni* (Thomas)	（バッタ目、バッタ科）新北区
Hayden's ringlet		*Coenonympha haydenii* (Edwards)	（チョウ目、タテハチョウ科）新北区
Hayhurst's scallopwing			southern sootywing を見よ
hayworm			gold fringe を見よ　幼虫の英名
hazel aphid		*Corylobium avellanae* (Schrank)	（カメムシ目、アブラムシ科）旧北区
hazel aphid			filbert aphid を見よ
hazel flat-body		*Semioscopis avellanella* (Hübner)	（チョウ目、マルハキバガ科）旧北区
hazel flea beetle		*Haltica brevicollis* Foudras	（コウチュウ目、ハムシ科）旧北区
hazel leaf roller			pear leaf roller (1) を見よ
hazel leaf roller weevil			hazel weevil を見よ
hazel leaf-roller weevil			pear leaf roller (1) を見よ
hazel leaftier moth		*Nites grotella* Robinson	（チョウ目、マルハキバガ科）新北区
hazel pot beetle		*Cryptocephalus coryli* (Linnaeus)	（コウチュウ目、ハムシ科）旧北区
hazel sawfly		*Croesus septentrionalis* (Linnaeus)	（ハチ目、ハバチ科）旧北区
hazel tortrix moth			chequared fruit-tree tortrix を見よ
hazel twist		*Archips sorbiana* Hübner	（チョウ目、ハマキガ科）旧北区
hazel weevil		*Apoderus coryli* (Linnaeus)	（コウチュウ目、オトシブミ科）旧北区
hazelnut capricorn beetle		*Oberea linearis* (Linnaeus)	（コウチュウ目、カミキリムシ科）旧北区
hazelnut flea beetle			hazel weevil を見よ
hazelnut lace bug		*Corythucha coryli* Osborn et Drake	（カメムシ目、グンバイムシ科）新北区
hazelnut weevil		*Curculio neocorylus* Gibson	（コウチュウ目、ゾウムシ科）全北区
hazelnut weevil		*Curculio obtusus* (Blanchard)	（コウチュウ目、ゾウムシ科）新北区
hazelnut weevil			acorn weevil (1) を見よ
head and body louse			body louse を見よ
head louse			body louse を見よ
head louse of chickens			chicken head louse を見よ
head maggot			sheep bot fly を見よ　幼虫の英名
head-standing moths			shiny head-standing moths を見よ
head worm			sheep bot fly を見よ　幼虫の英名
heart and club		*Agrotis clavis* Hufnagel	（チョウ目、ヤガ科）旧北区

英名	和名	学名	所属、分布、ほか
heart and dart		*Agrotis exclamationis* (Linnaeus)	(チョウ目、ヤガ科) 旧北区。日本亜種 *A. e. informis* Leech センモンヤガはムギ、トウモロコシ、ダイズ、アブラナ科野菜などを加害。heart and dart moth ともいう。
heart moth		*Dicycla oo* (Linnaeus)	(チョウ目、ヤガ科) 旧北区
heart pinion			heart moth を見よ
hearth cricket			house cricket を見よ
heath assassin bug		*Coranus subapterus* (De Geer)	(カメムシ目、サシガメ科) 旧北区
heath bee-fly		*Bombylius minor* Linnaeus	(ハエ目、ツリアブ科) 旧北区
heath bumblebee		*Bombus jonellus* (Kirby)	(ハチ目、ミツバチ科) 全北区
heath bush cricket		*Gampsocleis glabra* (Herbst)	(バッタ目、キリギリス科) 旧北区
heath case		*Coleophora pyrrhulipennella* Zeller	(チョウ目、ツツミノガ科) 旧北区
heath damsel bug		*Nabis ericetorum* Scholtz	(カメムシ目、マキバサシガメ科) 旧北区
heath fritillary	キタコヒョウモンモドキ	*Mellicta athalia* (Rottemburg)	(チョウ目、タテハチョウ科) 旧北区
heath fritillary			Assmann's fritillary を見よ
heath grass veneer		*Crambus ericellus* Hübner	(チョウ目、メイガ科) 旧北区
heath grasshopper		*Chorthippus vagans* (Eversmann)	(バッタ目、バッタ科) 旧北区
heath humble bee			heath bumble bee を見よ
heath humble bee			red-tailed bumble bee (1) を見よ
heath lattice			latticed heath を見よ
heath ochre			Phigalia skipper を見よ
heath potter		*Eumenes coarctatus* (Linnaeus)	(ハチ目、スズメバチ科) 旧北区
heath Probole moth		*Probole nepiasaria* (Walker)	(チョウ目、シャクガ科) 新北区
heath rivulet		*Perizoma minorata* (Treitschke)	(チョウ目、シャクガ科) 旧北区
heath rustic		*Xestia agathina* (Duponchel)	(チョウ目、ヤガ科) 旧北区
heath silverfish		*Dilta littoralis* Womersley	(シミ目、イシノミ科) 旧北区
heath spittle bug		*Clastoptera saintcyri* Provancher	(カメムシ目、コガシラアワフキ科) 新北区
heath villa		*Villa venusta* (Meigen)	(ハエ目、ツリアブ科) 旧北区
heather beetle		*Lochmaea suturalis* (Thomson)	(コウチュウ目、ハムシ科) 旧北区
heather-bill			dragonflies (1) を見よ
heather blue		*Agriades cassiope* Emmel et Emmel	(チョウ目、シジミチョウ科) 新北区
heather bug		*Rhacognathus punctatus* (Linnaeus)	(カメムシ目、カメムシ科) 旧北区
heather felt scale		*Eriococcus devoniensis* (Green)	(カメムシ目、コナカイガラムシ科) 旧北区

英名	和名	学名	所属、分布、ほか
heather flea beetle		*Altica ericeti* (Allard)	（コウチュウ目、ハムシ科）旧北区
heatherflee			dragonflies (1) を見よ
heather groundling		*Neofaculta ericetella* (Geyer)	（チョウ目、キバガ科）旧北区
heather ladybird		*Chilocorus bipunctulatus* (Linnaeus)	（コウチュウ目、テントウムシ科）旧北区
heather moth			common heath を見よ
heather neb		*Aristotelia ericinella* (Zeller)	（チョウ目、キバガ科）旧北区
heather scale		*Acanthococcus devoniensis* (Green)	（カメムシ目、コナカイガラムシ科）旧北区
heather weevil		*Strophosomus lateralis* (Paykull)	（コウチュウ目、ゾウムシ科）旧北区
heavy dragoons			bed bug (1) を見よ
Hebard's trig		*Falcicula hebardi* Rehn	（バッタ目、クサヒバリ科）新北区
hebe tiger moth		*Arctia hebe* Linnaeus	（チョウ目、ヒトリガ科）旧北区
Hebrew character		*Orthosia gothica* (Linnaeus)	（チョウ目、ヤガ科）旧北区。日本亜種は *O. g. askoldensis* (Staudinger) カシワキリガ。ヘブライ文字に由来
Hebrew character moth			Hebrew character を見よ
Hebrew moth		*Polygrammate hebraeicum* Hübner	（チョウ目、ヤガ科）新北区。Hebrew ともいう
Hebrew Sonia moth		*Sonia paraplesiana* Blanchard	（チョウ目、ハマキガ科）新北区
Hebrus metalmark		*Menander hebrus* (Cramer)	（チョウ目、シジミタテハ科）新熱帯区
Hecale longwing		*Heliconius hecale fornarina* Hewitson	（チョウ目、タテハチョウ科）新熱帯区。*H. h. zuleika* Hewitson も同英名。polymorphic longwing を参照
Hecale longwing			tiger longwing を見よ
Hecate dropwing		*Trithemis hecate* Ris	（トンボ目、トンボ科）エチオピア区
Hecate hairstreak		*Michaelus hecate* (Godman et Salvin)	（チョウ目、シジミチョウ科）新熱帯区
Hecebolus skipper		*Panoquina hecebola* Scudder	（チョウ目、セセリチョウ科）新北区、新熱帯区
Hecla sulphur			Greenland sulphur を見よ
Hecuba longwing		*Heliconius hecuba* (Hewitson)	（チョウ目、タテハチョウ科）新熱帯区
Hedemann's leafwing		*Memphis hedemanni* (R. Felder)	（チョウ目、タテハチョウ科）新熱帯区
hedge blues		*Lycaenopsis*	（チョウ目、シジミチョウ科）の昆虫の総称
hedge brown		*Maniola tithonus* (Linnaeus)	（チョウ目、タテハチョウ科）旧北区
hedge-chafer			June beetle (1) を見よ
hedge cupid		*Bothrinia chennellii* (de Nicéville)	（チョウ目、シジミチョウ科）東洋区
hedge eye			hedge brown を見よ

英名	和名	学名	所属、分布、ほか
hedge gothic			hedge rustic を見よ
hedge grasshopper			giant grasshopper を見よ
hedgehog caterpillars			tiger moths (1) を見よ　幼虫の英名
hedgehog flea		*Archaeopsylla erinacei* (Bouché)	（ノミ目、ヒトノミ科）旧北区
hedge hopper		*Baracus vittatus* (Felder)	（チョウ目、セセリチョウ科）東洋区
hedge-hoppers		*Baracus*	（チョウ目、セセリチョウ科）の昆虫の総称
hedge-parsley aphid		*Semiaphis anthrisci* (Kaltenbach)	（カメムシ目、アブラムシ科）旧北区
hedgerow hairstreak	サビイロカラスシジミ	*Satyrium saepium* (Boisduval)	（チョウ目、シジミチョウ科）新北区、新熱帯区。*S. s. chalcis* (W. H. Edwards) も同英名
hedge rustic		*Tholera cespitis* (Denis et Schiffermüller)	（チョウ目、ヤガ科）旧北区
Heeger's midget		*Phyllonorycter heegeriella* (Zeller)	（チョウ目、ホソガ科）旧北区
Heeger's whittlesea piercer		*Laspeyresia corollana* Hübner	（チョウ目、ハマキガ科）旧北区
heel flies			warble flies (1)(3) を見よ
heel fly			cattle warble fly (2) を見よ
heelwalkers	カカトアルキ目	Mantophasma	の昆虫の総称　2002年に記載された目。カマキリとナナフシに似る
Hegemon dartwhite		*Catasticta prioneris hegemon* Godman et Salvin	（チョウ目、シロチョウ科）新熱帯区
Heidelberg copper		*Chrysoritis aureus* van Son	（チョウ目、シジミチョウ科）エチオピア区
Heitzman's dagger moth		*Acronicta heitzmani* Covell et Metzler	（チョウ目、ヤガ科）新北区
helcomyzid flies		Helcomyzidae	（ハエ目）の昆虫の総称
helcomyzids			helcomyzid flies を見よ
Helena blue Morpho		*Morpho helena* Staudinger	（チョウ目、タテハチョウ科）新熱帯区
Helena brown		*Tisiphone helena* (Olliff)	（チョウ目、タテハチョウ科）豪州区
Helena gum moth			Helena moth を見よ
Helena moth		*Opodiphthera helena* (White)	（チョウ目、ヤママユガ科）豪州区
Helena redeye		*Semanga helena* (Rober)	（チョウ目、シジミチョウ科）東洋区
Helenor Morpho			common Morpho を見よ
Helfer's dune grasshopper		*Trimerotropis helferi* Strohecker	（バッタ目、バッタ科）新北区。体長3cm。海浜性の希種
heliconians	ドクチョウ亜科	Heliconiinae	（チョウ目、タテハチョウ科）の昆虫の総称
Heliconides metalmark		*Pheles heliconides* Herrich-Schäffer	（チョウ目、シジミタテハ科）新熱帯区
Heliconius mimic		*Eueides isabella* (Stoll)	（チョウ目、タテハチョウ科）新北区、新熱帯区
helicopter damselfly	ハビロイトトンボ	*Megaloprepus caerulatus* (Drury)	（トンボ目、Pseudostigmatidae）新熱帯区。開帳19cmの最大級のトンボ
Helicta satyr		*Neonympha helicta* (Hübner)	（チョウ目、タテハチョウ科）新北区
heliodinid moths			sun moths (2) を見よ

英名	和名	学名	所属、分布、ほか
Helios blue ringlet		*Caeruleuptychia helios* Weymer	（チョウ目、タテハチョウ科）新熱帯区
heliotrope flea beetle		*Longitarsus albinens* (Fondras)	（コウチュウ目、ハムシ科）豪州区
heliotrope moth	ベニゴマダラヒトリ	*Utetheisa pulchelloides* Hampson	（チョウ目、ヒトリガ科）日本、東洋区、豪州区、大洋区
hellgrammite			dobsonfly を見よ
hellgrammites			dobsonflies (1)(2)(3) を見よ
helmet scale			hemispherical scale (2) を見よ
helmeted squash bug		*Euthochtha galeator* (Fabricius)	（カメムシ目、ヘリカメムシ科）新北区、大洋区
helodid beetles			marsh beetles を見よ
helomyzid flies	トゲハネバエ科	Helomyzidae	（ハエ目）の昆虫の総称
helorid wasps	クシヅメクロバチ科	Heloridae	（ハチ目）の昆虫の総称
helorids			helorid wasps を見よ
Helvina lady slipper		*Pierella helvina* (Hewitson)	（チョウ目、タテハチョウ科）新北区、新熱帯区
Helvina satyr			Helvina lady slipper を見よ
hemercallis aphid	カンゾウコブアブラムシ	*Myzus hemerocallis* Takahashi	（カメムシ目、アブラムシ科）日本、旧北区、東洋区、豪州区
hemerobiids			brown lacewings (1) を見よ
hemimerids		Diploglossata	ハサミムシ目として扱われる。
hemipeplid beetles		Hemipeplidae	（コウチュウ目）の昆虫の総称　ヒラタムシ科の異名
Hemiphlebia damselfly		*Hemiphlebia mirabilis* Selys	（トンボ目、ムカシトンボ科）豪州区。絶滅危惧種
hemipteroids		Hemipteroidea	の昆虫の総称
hemispherical scale (1)		*Saissetia hemisphaerica* (Targioni)	（カメムシ目、カタカイガラムシ科）新北区、新熱帯区。カンキツなど各種樹木害虫
hemispherical scale (2)	ハンエンカタカイガラムシ	*Saissetia coffeae* (Walker)	（カメムシ目、カタカイガラムシ科）日本、東洋区、新北区、汎熱帯。カンキツ、マンゴー、チャなどの害虫
hemlock angle moth		*Macaria fissinotata* (Walker)	（チョウ目、シャクガ科）新北区。hemlock angle ともいう
hemlock borer		*Melanaphila fulvoguttata* (Harris)	（コウチュウ目、タマムシ科）新北区
hemlock caterpillar	ツガカレハ	*Dendrolimus superans* (Butler)	（チョウ目、カレハガ科）日本、旧北区。モミ、ツガ、エゾマツなどの害虫
hemlock looper (1)		*Lambdina fiscellaria* (Guenée)	（チョウ目、シャクガ科）新北区。hemlock looper moth ともいう。亜種 *L. f. lugubrosa* (Hulst) は western hemlock looper, *L. f. somniaria* (Hulst) は western oak looper という
hemlock looper (2)		*Lambdina fervidaria* Hübner	（チョウ目、シャクガ科）新北区
hemlock moth			Alstroemer's flat-body を見よ
hemlock needleminer moth		*Epinotia tsugana* Freeman	（チョウ目、ハマキガ科）新北区

英名	和名	学名	所属、分布、ほか
hemlock sawfly		*Neodiprion tsugae* Middleton	(ハチ目、マツハバチ科) 新北区
hemlock scale		*Abgrallaspis ithacae* (Ferris)	(カメムシ目、マルカイガラムシ科) 新北区
hemlock scale		*Dynaspidiotus abietis* (Schrank)	(カメムシ目、マルカイガラムシ科) 旧北区
hemlock scale		*Physokermes hemicryphus* Dalman	(カメムシ目、カタカイガラムシ科) 旧北区
hemlock scale (1)	ツガコノハカイガラムシ	*Fiorinia externa* Ferris	(カメムシ目、マルカイガラムシ科) 日本、新北区。ツガ害虫
hemlock scale (2)	ツガマルカイガラムシ	*Nuculaspis tsugae* (Marshall)	(カメムシ目、マルカイガラムシ科) 日本。モミ、ツガ害虫
hemlock spruce woolly aphid	ハリモミヒメカサアブラムシ	*Aphrastasia tsugae* (Annand)	(カメムシ目、カサアブラムシ科) 日本、東洋区、新北区。トウヒ類害虫
hemlock woolly adelgid			hemlock spruce woolly aphid を見よ
hemlock xyloryctid	ツガヒロバキバガ	*Metathrinca tsugensis* (Kearfott)	(チョウ目、ヒロバキバガ科) 日本。ツガ害虫
Hemon blue hairstreak		*Theritas hemon* (Cramer)	(チョウ目、シジミチョウ科) 新熱帯区。Hemon hairstreak ともいう
hemp-agrimony case		*Coleophora follicularis* (Vallot)	(チョウ目、ツツミノガ科) 旧北区
hemp-agrimony plume	ウスキヒメトリバ	*Adaina microdactyla* Hübner	(チョウ目、トリバガ科) 日本、旧北区
hemp aphid	アサイボアブラムシ	*Phorodon cannabis* (Passerini)	(カメムシ目、アブラムシ科) 日本、旧北区。アサ、ホップ害虫
hemp borer	クロヒメハナノミ	*Mordellistena comes* Marseul	(コウチュウ目、ハナノミ科) 日本
hemp dagger moth	アサケンモン	*Plataplecta pruinosa consanguis* (Butler)	(チョウ目、ヤガ科) 日本、東洋区
hemp flea			hemp flea beetle を見よ
hemp flea beetle	アサトビハムシ	*Psylloides attenuata* (Koch)	(コウチュウ目、ハムシ科) 日本、旧北区
hemp flea beetle			hop flea beetle を見よ
hemp helodid	トビイロマルハナノミ	*Scirtes japonicus* Kiesenwetter	(コウチュウ目、マルハナノミ科) 日本、東洋区
hemp leafminer	アサハモグリバエ	*Liriomyza cannabis* Hendel	(ハエ目、ハモグリバエ科) 日本、旧北区。アサ害虫
hemp longicorn			hemp longicorn beetle を見よ
hemp longicorn beetle	アサカミキリ	*Thyestilla gebleri* (Faldermann)	(コウチュウ目、カミキリムシ科) 日本、旧北区。アサ、カラムシ害虫
hemp moth			hemp stemborer を見よ
hemp palm scale			cyanophyllum scale を見よ
hemp sawfly		*Trichiocampus cannabis* Xiao et Huang	(ハチ目、ハバチ科) 旧北区
hemp stemborer	ヨツスジヒメシンクイ	*Grapholita delineana* Walker	(チョウ目、ハマキガ科) 日本、旧北区。ホップ害虫
hemp weevil	アササルゾウムシ	*Ceutorhynchus rubripes* Hustache	(コウチュウ目、ゾウムシ科) 日本。アサ害虫

英名	和名	学名	所属、分布、ほか
hemp weevil	オオクチブトサルゾウムシ	*Rhinoncus pericarpius* (Linnaeus)	（コウチュウ目、ゾウムシ科）日本、旧北区
hen flea			chicken flea を見よ
hen flea			sticktight flea を見よ
henbane flea beetle		*Psylliodes hyoscyami* (Linnaeus)	（コウチュウ目、ハムシ科）旧北区
henbane fly			beet leaf miner (1) を見よ
Henning's black-eye		*Leptomyrina henningi* Dickson	（チョウ目、シジミチョウ科）エチオピア区
Henning's copper		*Aloeides henningi* Tite et Dickson	（チョウ目、シジミチョウ科）エチオピア区
Henry's elfin	ヘンリックコツバメ	*Callophrys henrici* (Grote et Robinson)	（チョウ目、シジミチョウ科）新北区
Henry's marsh moth		*Simyra insularis* (Herrich-Schäffer)	（チョウ目、ヤガ科）新北区。幼虫は cattail caterpillar
Henshaw's satyr			Sonoran satyr を見よ
Hepburn's metalmark		*Apodemia hepburni* Godman et Salvin	（チョウ目、シジミタテハ科）新北区
hepialid moths			ghost moths を見よ
hepialids			ghost moths を見よ
Hera buckmoth			Hera moth を見よ
Hera moth		*Hemileuca hera* (Harris)	（チョウ目、ヤママユガ科）新北区。Hera buckmoth ともいう
Heraclea sister		*Adelpha heraclea* (C. et R. Felder)	（チョウ目、タテハチョウ科）新熱帯区
Heracleum stem borer moth		*Papaipema harrisii* (Grote)	（チョウ目、ヤガ科）新北区
Heraclides hairstreak		*Temecla heraclides* (Godman et Salvin)	（チョウ目、シジミチョウ科）新熱帯区
herald			spectacle (2) を見よ
herald moth			spectacle (2) を見よ
Heraldica clearwing		*Ithomia heraldica* Bates	（チョウ目、タテハチョウ科）新熱帯区
Heraldica hairstreak		*Nicolaea heraldica* (Dyar)	（チョウ目、シジミチョウ科）新熱帯区
Heras skipper		*Librita heras* Godman	（チョウ目、セセリチョウ科）新熱帯区
herbacea leafwing		*Memphis herbacea* (Butler et Druce)	（チョウ目、タテハチョウ科）新熱帯区
herbarium beetle	オオヒメマキムシ	*Cartodere filum* (Aube)	（コウチュウ目、ヒメマキムシ科）旧北区
herbarium beetle (1)		*Tricorynus herbarium* Gorham	（コウチュウ目、シバンムシ科）大洋区
herbarium moth		*Eois ptelearia* Riley	（チョウ目、シャクガ科）新北区
herbivorous Pleuroptya moth		*Pleuroptya silicalis* (Guenée)	（チョウ目、メイガ科）新北区
herculean ant (1)		*Camponotus ligniperda* (Latreille)	（ハチ目、アリ科）新北区
herculean ant (2)		*Camponotus herculeanus* (Linnaeus)	（ハチ目、アリ科）旧北区

英名	和名	学名	所属、分布、ほか
herculean carpenter ant		*Camponotus herculeanus pennsylvanicus* De Geer	（ハチ目、アリ科）新北区
herculean scale		*Clavaspis herculeana* (Doane et Hadden)	（カメムシ目、マルカイガラムシ科）新北区、豪州区
Hercules beetle	ヘラクレスオオカブトムシ	*Dynastes hercules* (Linnaeus)	（コウチュウ目、コガネムシ科）新熱帯区。体長15 cmの大型甲虫。ギリシャ神話のヘラクレスに由来
Hercules beetles	カブトムシ亜科	Dynastinae	（コウチュウ目、コガネムシ科）の昆虫の総称
Hercules Morpho		*Morpho hercules* (Dalman)	（チョウ目、タテハチョウ科）新熱帯区
Hercules moth		*Coscinocera hercules* (Miskin)	（チョウ目、ヤママユガ科）豪州区。豪州に生息する世界最大級の蛾
herd-grass stem fly			rice stem fly (1)(3) を見よ
Hereford case		*Coleophora sylvaticella* Wood	（チョウ目、ツツミノガ科）旧北区。イングランドの都市名に由来
Hereford midget		*Phyllonorycter distendella* (Zeller)	（チョウ目、ホソガ科）旧北区
Hereford's Eupithecia moth		*Eupithecia herefordaria* Cassino et Swett	（チョウ目、シャクガ科）新北区。Hereford's Eupithecia ともいう
Herekopare weta		*Deinacrida carinata* Salmon	（バッタ目、Stenopelmatidae）豪州区
Herman's shieldback		*Idiostatus hermanii* (Thomas)	（バッタ目、キリギリス科）新北区
Hermathena longwing		*Heliconius hermathena* (Hewitson)	（チョウ目、タテハチョウ科）新熱帯区
Hermes copper	エルメスシジミ	*Lycaena hermes* (Edwards)	（チョウ目、シジミチョウ科）新北区、新熱帯区
Hermes ringlet		*Hermeuptychia hermes* (Fabricius)	（チョウ目、タテハチョウ科）新熱帯区。
Hermes satyr			Hermes ringlet を見よ
Hermesia skipper		*Vacerra hermesia* (Hewitson)	（チョウ目、セセリチョウ科）新熱帯区
Herminia glider	ナカグロキイロタテハ	*Cymothoe herminia* (Grose-Smith)	（チョウ、タテハチョウ科）エチオピア区
Hermione skipper		*Lento hermione hermione* (Schaus)	（チョウ目、セセリチョウ科）新熱帯区
hermit		*Grais stigmaticus* Mabille	（チョウ目、セセリチョウ科）新北区、新熱帯区。hermit skipper ともいう
hermit		*Chazara briseis* (Linnaeus)	（チョウ目、タテハチョウ科）旧北区
hermit flower beetle		*Osmoderma eremicola* Knoch	（コウチュウ目、コガネムシ科）新北区
hermit sphinx		*Sphinx eremitus* (Hübner)	（チョウ目、スズメガ科）新北区。hermit sphinx moth ともいう
Hermodora metalmark		*Detritivora hermodora* (C. et R. Felder)	（チョウ目、シジミタテハ科）新熱帯区
Herodias underwing moth		*Catocala herodias* Strecker	（チョウ目、ヤガ科）新北区。Herodias underwing ともいう
herolids			helorid wasps を見よ
heron louse		*Ciconiphilus decimfasciatus* (Boisduval et Lacordaire)	（ハジラミ目、タンカクハジラミ科）旧北区

英名	和名	学名	所属、分布、ほか
Herpetogramma moth		*Herpetogramma thestealis* (Walker)	（チョウ目、メイガ科）新北区
hesitant dagger moth		*Acronicta haesitata* Grote	（チョウ目、ヤガ科）新北区
Hesperina eyed-metalmark		*Mesosemia hesperina hesperina* Butler	（チョウ目、シジミタテハ科）新熱帯区
Hesperis eighty-eight			Hesperis numberwing を見よ
Hesperis fritillary	ウスグロギンボシヒョウモン	*Speyeria atlantis hesperis* (Edwards)	（チョウ目、タテハチョウ科）新北区。Atlantic fritillary を参照
Hesperis numberwing		*Callicore hesperis* (Guérin-Méneville)	（チョウ目、タテハチョウ科）新熱帯区
Hesperitis groundstreak		*Ziegleria hesperitis* (Butler et Druce)	（チョウ目、シジミチョウ科）新熱帯区
Hessel's hairstreak	ヘッセルスギカラスシジミ	*Callophrys hesseli* (Rawson et Ziegler)	（チョウ目、シジミチョウ科）新北区
Hessian fly		*Mayetiola destructor* (Say)	（ハエ目、タマバエ科）全北区。小麦、大麦、ライの害虫。
Hessian fly parasite		*Platygaster hiemalis* Forbes	（ハチ目、ハラビロクロバチ科）新北区
Hessian gall midge			Hessian fly を見よ
heterocerid beetles			variegated mud-loving beetles を見よ
heterocerid moths			moths を見よ
heteromerans			heteromerous beetles を見よ
heteromerous beetles	異節群	Heteromera	（コウチュウ目）の昆虫の総称
heterometabolic insects	不完全変態類	Heterometabola	の昆虫の総称
heterothripids			tree thrips を見よ
Hewitson's Aguna		*Aguna aurunce* (Hewitson)	（チョウ目、セセリチョウ科）新熱帯区
Hewitson's blackstreak			black hairstreak (1) を見よ
Hewitson's blue hairstreak		*Thecla coronata* Hewitson	（チョウ目、シジミチョウ科）新熱帯区
Hewitson's Clito		*Clito zelotes* (Hewitson)	（チョウ目、セセリチョウ科）新熱帯区
Hewitson's commander		*Euryphura plautilla* (Hewitson)	（チョウ目、タテハチョウ科）エチオピア区
Hewitson's dwarf crow	ヒュウイットソンルリマダラ	*Euploea hewitsonii* C. et R. Felder	（チョウ目、タテハチョウ科）東洋区
Hewitson's egg white		*Dismorphia lua* (Hewitson)	（チョウ目、シロチョウ科）新熱帯区
Hewitson's forest blue	ウラスジモルフォシジミ	*Hewitsonia boisduvali* (Hewitson)	（チョウ目、シジミチョウ科）エチオピア区
Hewitson's glory		*Asterope degandii* Hewitson	（チョウ目、タテハチョウ科）新熱帯区
Hewitson's metalmark		*Nymphidium onaeum* Hewitson	（チョウ目、シジミタテハ科）新熱帯区
Hewitson's mimic		*Eresia datis* Hewitson	（チョウ目、タテハチョウ科）新熱帯区
Hewitson's oakblue		*Arhopala oenea* (Hewitson)	（チョウ目、シジミチョウ科）東洋区

英名	和名	学名	所属、分布、ほか
Hewitson's olivewing	ミズイロタテハ	*Nessaea hewitsoni* (C. et R. Felder)	(チョウ目、タテハチョウ科) 新熱帯区
Hewitson's palmfly	ヘリボシルリモンジャノメ	*Elymnias hewitsoni* Wallace	(チョウ目、タテハチョウ科) 東洋区
Hewitson's Pierella			golden lady slipper を見よ
Hewitson's pigmy		*Baeotis bacaenis* Hewitson	(チョウ目、シジミタテハ科) 新熱帯区
Hewitson's pink forester		*Euphaedra hewitsoni* Hecq	(チョウ目、タテハチョウ科) エチオピア区
Hewitson's Raven		*Pereute telthusa* (Hewitson)	(チョウ目、シロチョウ科) 新熱帯区
Hewitson's Ridens		*Ridens mephitis* (Hewitson)	(チョウ目、セセリチョウ科) 新熱帯区
Hewitson's sabre-wing		*Jemadia hewitsonii* (Mabille)	(チョウ目、セセリチョウ科) 新熱帯区
Hewitson's sailor		*Dynamine sosthenes* Hewitson	(チョウ目、タテハチョウ科) 新熱帯区
Hewitson's small tree-nymph		*Ideopsis hewitsonii* Kirsch	(チョウ目、タテハチョウ科) 東洋区
Hewitson's tiger		*Melinaea menophilus* (Hewitson)	(チョウ目、タテハチョウ科) 新熱帯区
Hewitson's Uraneis		*Uraneis ucubis* Hewitson	(チョウ目、シジミチョウ科) 新熱帯区
Hexapod			insects を見よ
hexapodans			insects を見よ
hexapods			insects を見よ
Hexham grey			Scotch grey を見よ
hey worm	アカマエアツバ	*Simplicia rectalis* (Eversmann)	(チョウ目、ヤガ科) 日本、旧北区
Heyden's strawberry pigmy		*Nepticula fragariella* Heinemann	(チョウ目、モグリチビガ科) 旧北区
Heyden's strawberry pigmy			golden pigmy を見よ
Hezia clearwing		*Callithomia hezia hedila* Godman et Salvin	(チョウ目、タテハチョウ科) 新熱帯区。*C. h. wellingi* Fox も同英名
hiba torynid			thuja torymid を見よ
hibiscus caterpillar	フタトガリコヤガ	*Xanthodes transversus* Guenée	(チョウ目、ヤガ科) 日本、東洋区、大洋区、豪州区。ワタ、ハイビスカスなどの害虫
hibiscus caterpillar			cotton semi-looper を見よ
hibiscus flower beetle		*Aethina concolor* (Macleay)	(コウチュウ目、ケシキスイ科) 豪州区
hibiscus greenish noctuid	ミドリリンガ	*Clethrophora distincta* (Leech)	(チョウ目、ヤガ科) 日本、旧北区、東洋区
hibiscus ground mealybug	ハイビスカスネコナカイガラムシ	*Rhizoecus hibisci* Kawai et Takagi	(カメムシ目、コナカイガラムシ科) 日本。ハイビスカス害虫
hibiscus leaf caterpillar	オオアカキリバ	*Anomis privata* (Walker)	(チョウ目、ヤガ科) 日本、新北区。リンゴなど果樹、ハイビスカス害虫。成虫は hibiscus-leaf caterpillar moth
hibiscus leaf caterpillar			hibiscus looper を見よ

英名	和名	学名	所属、分布、ほか
hibiscus leafminer		*Prectopa hibiscella* (Swezey)	（チョウ目、ホソガ科）新北区
hibiscus long-horned beetle			Gelonaetha long-horned beetle を見よ
hibiscus looper	アカキリバ	*Anomis mesogona* (Walker)	（チョウ目、ヤガ科）日本、東洋区。トマト、ナシなど果樹害虫
hibiscus mealybug	タマコナカイガラムシ	*Nipaecoccus vastator* (Maskell)	（カメムシ目、コナカイガラムシ科）日本、東洋区、大洋区、新熱帯区。カンキツ、ビワ、ハイビスカスなどの害虫
hibiscus mealybug (1)	ワタコナカイガラムシ	*Maconellicoccus hirsutus* (Green)	（カメムシ目、コナカイガラムシ科）日本、東洋区、大洋区、豪州区、エチオピア区。カンキツ、ハイビスカスなどの害虫
hibiscus mealybug			spherical mealybug を見よ
hibiscus psylla	ヤマアサキジラミ	*Mesohomotoma camphorae* Kuwayama	（カメムシ目、ネッタイキジラミ科）日本、東洋区。ハイビスカス害虫
hibiscus snow scale			lesser snow scale を見よ
hibiscus thrips		*Liothrips varicornis* Hood	（アザミウマ目、クダアザミウマ科）大洋区
hibiscus whitefly			azalea whitefly (1) を見よ
Hicetas hairstreak		*Aubergina hicetas* (Godman et Salvin)	（チョウ目、シジミチョウ科）新熱帯区
hickory bark beetle		*Scolytus quadrispinosus* Say	（コウチュウ目、キクイムシ科）新北区
hickory bullet gall midge		*Caryomyia tubicola* (Osten Sacken)	（ハエ目、タマバエ科）新北区
hickory hairstreak	トネリコカラスシジミ	*Satyrium caryaevorus* (McDunnough)	（チョウ目、シジミチョウ科）新北区
hickory Halisidota			hickory tiger moth を見よ
hickory horned devil			royal walnut moth を見よ
hickory leafroller		*Argyrotaenia juglandana* (Fernald)	（チョウ目、ハマキガ科）新北区。hickory leafroller moth ともいう
hickory leafstem borer moth		*Acrobasis angusella* Grote	（チョウ目、メイガ科）新北区
hickory phylloxeran		*Phylloxera caryaecaulis* (Fitch)	（カメムシ目、ネアブラムシ科）新北区
hickory plant bug		*Lygocoris caryae* (Knight)	（カメムシ目、カスミカメムシ科）新北区
hickory shoot borer moth			pecan nut case-bearer (1) を見よ　hickory shoot moth ともいう
hickory shuckworm		*Cydia caryana* (Fitch)	（チョウ目、ハマキガ科）新北区。ペカン害虫。hickory shuckworm moth ともいう
hickory tiger moth		*Halisidota caryae* (Harris)	（チョウ目、ヒトリガ科）新北区。hickory tiger ともいう
hickory tussock moth			hickory tiger moth を見よ　北米での英名
hidden-ray skipper		*Conga chydaea* (Butler)	（チョウ目、セセリチョウ科）新北区、新熱帯区
hidden snout weevil	クチカクシゾウムシ亜科	Cryptorhynchinae	（コウチュウ目、ゾウムシ科）の昆虫の総称
hidden-yellow skipper		*Ouleus cyrna* (Mabille)	（チョウ目、セセリチョウ科）新熱帯区

英名	和名	学名	所属、分布、ほか
hide beetle	ハラジロカツオブシムシ	*Dermestes maculatus* De Geer	(コウチュウ目、カツオブシムシ科) 日本、汎世界。貯穀害虫
hide beetles (1)	コブスジコガネムシ科	Trogidae	(コウチュウ目) の昆虫の総称
hide beetles (2)		*Dermestes*	(コウチュウ目、カツオブシムシ科) の昆虫の総称
hide beetles			larder beetles を見よ
hiding eyed-metalmark		*Mesosemia hypermegala* Stichel	(チョウ目、シジミタテハ科) 新熱帯区
hieroglyphic flat		*Odina hieroglyphica ortina* Evans	(チョウ目、セセリチョウ科) 東洋区
hieroglyphic moth		*Diphthera festiva* (Fabricius)	(チョウ目、ヤガ科) 新北区、新熱帯区
higejiro-ring-legged earwig	ヒゲジロハサミムシ	*Gonolabis marginalis* (Dohrn)	(ハサミムシ目、ハサミムシ科) 日本、旧北区、東洋区
Higgin's anomalous blue		*Polyommatus nephohiptamenos* (Brown)	(チョウ目、シジミチョウ科) 旧北区
Higgin's anomalous blue			Gallo's anomalous blue を見よ
high-back prominent	セダカシャチホコ	*Rabtala cristata* (Butler)	(チョウ目、シャチホコガ科) 日本、旧北区、東洋区
high brown friitillary		*Fabriciana adippe* (Denis et Schiffermüller)	(チョウ目、タテハチョウ科) 旧北区。日本亜種のウラギンヒョウモン *F. a. pallescens* (Butler) も同英名
high brown fritillary		*Argynnis cydippe* (Linnaeus)	(チョウ目、タテハチョウ科) 旧北区
high flier			purple emperor (2) を見よ
high mountain blue		*Agriades franklinii* (Curtis)	(チョウ目、シジミチョウ科) 新北区
high plains grasshopper		*Dissosteira longipennis* (Thomas)	(バッタ目、バッタ科) 新北区
highland darter		*Sympetrum nigrescens* Lucas	(トンボ目、トンボ科) 旧北区
highland sympetrum			highland darter を見よ
highlands skipper		*Aethilla chiapa* Freeman	(チョウ目、セセリチョウ科) 新熱帯区
highveld blue		*Lepidochrysops praeterita* Swanepoel	(チョウ目、シジミチョウ科) エチオピア区
Hikosan bamboo scale	ヒコサンナガカイガラムシ	*Kuwanaspis hikosani* (Kuwana)	(カメムシ目、マルカイガラムシ科) 日本
hilarimorphid flies		Hilarimorphidae	(ハエ目) の昆虫の総称
Hildebrandt's Charaxes		*Charaxes hildebrandti* (Dewitz)	(チョウ目、タテハチョウ科) エチオピア区
Hildegard's buff		*Baliochila hildegarda* (Kirby)	(チョウ目、シジミチョウ科) エチオピア区
hill ant			common red ant を見よ
hill bush brown		*Bicyclus campus* (Karsch)	(チョウ目、タテハチョウ科) エチオピア区
hill cuckoo bee		*Psithyrus rupestris* (Fabricius)	(ハチ目、ミツバチ科) 旧北区

英名	和名	学名	所属、分布、ほか
hill hedge blue		*Celastrina argiolus kollari* (Westwood)	（チョウ目、シジミチョウ科）東洋区。*C. a. sikkima* (Moore) も同英名。holly blue を参照
hill hedge blue			holly blue を見よ
hill Jezebel	ベラドンナカザリシロチョウ	*Delias belladonna* (Fabricius)	（チョウ目、シロチョウ科）東洋区
hill soldier		*Oxycera pardalina* Meigen	（ハエ目、ミズアブ科）旧北区
Hill's brown blowfly		*Calliphora hilli* Patton	（ハエ目、クロバエ科）．豪州区
Hill's wave moth		*Idaea hilliata* (Hulst)	（チョウ目、シャクガ科）新北区
hillside brown		*Stygionympha vigilans* (Trimen)	（チョウ目、タテハチョウ科）エチオピア区
hilltop hopper		*Platylesches dolomitica* Henning et Henning	（チョウ目、セセリチョウ科）エチオピア区
Hilo noctuid moth		*Hypena newelli* (Swezey)	（チョウ目、ヤガ科）大洋区
Himalayan black-vein	ヒマラヤシロチョウ	*Aporia leucodice* (Eversmann)	（チョウ目、シロチョウ科）旧北区、東洋区
Himalayan dark dart			Himalayan grass dart を見よ
Himalayan dart		*Potanthus dara* (Kollar)	（チョウ目、セセリチョウ科）東洋区
Himalayan five-ring		*Ypthima sakra* Moore	（チョウ目、タテハチョウ科）東洋区
Himalayan grain moth			grain worm を見よ
Himalayan grass dart		*Taractrocera danna* (Moore)	（チョウ目、セセリチョウ科）東洋区
Himalayan jester		*Symbrenthia brabira* Moore	（チョウ目、タテハチョウ科）東洋区
Himalayan jester (1)		*Symbrenthia hypselis* (Godart)	（チョウ目、タテハチョウ科）東洋区
Himalayan oak silkworm		*Antheraea roylei* Moore	（チョウ目、ヤママユガ科）東洋区
Himalayan pierrot			veined pierrot を見よ
Himalayan red flash		*Rapala selira* (Moore)	（チョウ目、シジミチョウ科）東洋区
Himalayan relict dragonfly		*Epiophlebia laidlawi* Tillyard	（トンボ目、ムカシトンボ科）東洋区
Himalayan sailer		*Neptis mahendra* Moore	（チョウ目、タテハチョウ科）東洋区
Himalayan sergeant	ヒマラヤミスジ	*Athyma opalina* (Kollar)	（チョウ目、タテハチョウ科）旧北区、東洋区
Himalayan spotted flat		*Celaenorrhinus munda* (Moore)	（チョウ目、セセリチョウ科）東洋区
Himalayan swift		*Polytremis discreta* (Elwes et Edwards)	（チョウ目、セセリチョウ科）東洋区
Himalayan tailed-cupid			Chapman's cupid を見よ
Himalayan white flat		*Seseria dohertyi* Watson	（チョウ目、セセリチョウ科）東洋区
Himalayan yellow flat		*Celaenorrhinus dhanada* (Moore)	（チョウ目、セセリチョウ科）東洋区
Himmelman's plume moth		*Geina tenuidactyla* (Fitch)	（チョウ目、トリバガ科）新北区
hinoki cypress borer			black-spotted buprestid を見よ

英名	和名	学名	所属、分布、ほか
Hintza blue		*Castalius hintza* Trimen	（チョウ目、シジミチョウ科）エチオピア区
Hintza blue			blue pied pierrot を見よ
Hintz's skimmer		*Orthetrum hintzi* Schmidt	（トンボ目、トンボ科）エチオピア区
hippo bell		*Epiblema turbidana* Treitschke	（チョウ目、ハマキガ科）旧北区
hippo flies			horse flies (1) を見よ
hippoboscid fly			sheep ked を見よ
Hirayama fruit fly	ヒラヤマアミメケブカミバエ	*Campiglossa hirayamae* (Matsumura)	（ハエ目、ミバエ科）日本、旧北区
Hirlanda white		*Cunizza hirlanda* (Stoll)	（チョウ目、シロチョウ科）新熱帯区
Hispaniolan sanddragon		*Progomphus serenus* Hagen	（トンボ目、サナエトンボ科）新熱帯区
hister beetle		*Hister nomas* Erichson	（コウチュウ目、エンマムシ科）豪州区
hister beetles	エンマムシ科	Histeridae	（コウチュウ目）の昆虫の総称　屍体、菌、樹皮下、アリの巣に生息
histerids			hister beetles を見よ　中南米での英名
hive bee			honey bee を見よ
hive bees			honey bees を見よ
hoary anglewing			hoary comma を見よ
hoary beetle			summer chafer を見よ
hoary comma		*Polygonia gracilis* (Grote et Robinson)	（チョウ目、タテハチョウ科）新北区
hoary double-crescent bell		*Epinotia bilunana* (Haworth)	（チョウ目、ハマキガ科）旧北区
hoary edge			hoary-edged skipper を見よ
hoary-edged skipper		*Achalarus lyciades* Geyer	（チョウ目、セセリチョウ科）新北区
hoary elfin		*Callophrys polios* (Cook et Watson)	（チョウ目、シジミチョウ科）新北区
hoary footman		*Eilema caniola* (Hübner)	（チョウ目、ヒトリガ科）旧北区
hoary grey		*Scoparia pyralella* Hübner	（チョウ目、メイガ科）旧北区
hoary knot-horn		*Gymnancyla canella* Hübner	（チョウ目、メイガ科）旧北区
hoary palmer		*Unkana ambasa batara* Distant	（チョウ目、セセリチョウ科）東洋区
hoary plume		*Platyptilia isodactyla* Zeller	（チョウ目、トリバガ科）旧北区
hoary plume moth			common plume を見よ
hoary sealed bell	アザミスソモンヒメハマキ	*Eucosma cana* (Haworth)	（チョウ目、ハマキガ科）日本、旧北区
hoary skipper		*Carrhenes canescens* (Felder)	（チョウ目、セセリチョウ科）新北区、新熱帯区
hoary spreadwing			hoary skipper を見よ
Hobart brown			Tasmanian brown を見よ
Hobart's red glider		*Cymothoe hobarti* Butler	（チョウ目、タテハチョウ科）エチオピア区
hobby-horse			dragonflies (1) を見よ
Hobley's ciliate blue		*Anthene hobleyi* (Neave)	（チョウ目、シジミチョウ科）エチオピア区

英名	和名	学名	所属、分布、ほか
Hobomok skipper		*Poanes hobomok* Harris	（チョウ目、セセリチョウ科）新北区
hockeystick sailer		*Neptis nycteus* de Nicéville	（チョウ目、タテハチョウ科）東洋区
Hodges's clubtail		*Gomphus hodgesi* Needham	（トンボ目、サナエトンボ科）新北区
Hodgkinson's emerald		*Chlorissa cloraria* (Hübner)	（チョウ目、シャクガ科）旧北区
Hodson's ciliate blue		*Anthene hodsoni* (Talbot)	（チョウ目、シジミチョウ科）エチオピア区
Hoffmann's checkerspot			aster checkerspot を見よ
Hoffmann's giant-skipper		*Agathymus hoffmanni* (Freeman)	（チョウ目、セセリチョウ科）新熱帯区
Hoffmann's groundstreak		*Ziegleria hoffmani* Johnson	（チョウ目、シジミチョウ科）新熱帯区
Hoffmann's skipper		*Melanopyge hoffmanni* (Freeman)	（チョウ目、セセリチョウ科）新熱帯区
Hoffmann's checkerspot		*Chlosyne hoffmanni* (Behr)	（チョウ目、タテハチョウ科）新北区、新熱帯区
Hoffman's cochlid moth		*Cochylis hoffmanana* (Kearfott)	（チョウ目、ホソハマキガ科）全北区
hog caterpillar			Virginia creeper sphinx moth を見よ
hog louse	ブタジラミ	*Haematopinus suis* (Linnaeus)	（シラミ目、ケモノジラミ科）日本、汎世界。ブタ、ウシに寄生する 6 mm の大型シラミ
hog sphinx			Virginia creeper sphinx moth を見よ　北米での英名
hogweed aphid		*Anuraphis subterranea* (Walker)	（カメムシ目、アブラムシ科）旧北区
Holarctic grass skipper			silver-spotted skipper (2) を見よ
holcocera		*Holcocera iceryaella* (Riley)	（チョウ目、ネマルハガ科）新北区。カンキツ害虫
Holland's castor		*Ariadne merionoides* (Holland)	（チョウ目、タテハチョウ科）東洋区
Holland's cochylid moth		*Thyraylia hollandana* (Kearfott)	（チョウ目、ハマキガ科）新北区
Holland's flash		*Hypophytala benitensis* (Holland)	（チョウ目、シジミチョウ科）エチオピア区
Holland's sylph		*Ceratrichia hollandi* Bethune-Baker	（チョウ目、セセリチョウ科）エチオピア区
Holland's white spots		*Osmodes hollandi* Evans	（チョウ目、セセリチョウ科）エチオピア区
hollow-spotted angle moth		*Digrammia gnophosaria* (Guenée)	（チョウ目、シャクガ科）新北区
hollow-spotted Blepharomastix moth		*Blepharomastix ranalis* (Guenée)	（チョウ目、メイガ科）新北区
hollow-spotted Plagodis moth		*Plagodis alcoolaria* (Guenée)	（チョウ目、シャクガ科）新北区。hollow-spotted Plagodis ともいう
holly aphid		*Aphis ilicis* Kaltenbach	（カメムシ目、アブラムシ科）旧北区
holly blue		*Celastrina argiolus* Linnaeus	（チョウ目、シジミチョウ科）旧北区。日本亜種は *C. a. ladonides* (de l'Orza) ルリシジミ。holly blue butterfly ともいう

英名	和名	学名	所属、分布、ほか
holly borer moth		*Synanthedon kathyae* (Duckworth et Eichlin)	（チョウ目、スカシバガ科）新北区
holly butterfly			green hairstreak を見よ
holly leaf tier			holly tortrix を見よ
holly leafminer		*Phytomyza ilicis* (Curtis)	（ハエ目、ハモグリバエ科）全北区。幼虫がヒイラギの葉に作る虫えいを holly leaf-mine という
holly scale			bay-tree scale を見よ
holly tortrix		*Rhopobota unipunctana* Haworth	（チョウ目、ハマキガ科）全北区
holly tortrix			black-headed fireworm を見よ
holly tortrix moth			holly tortrix を見よ
holly whitefly	イヌツゲクビレコナジラミ	*Aleurotuberculatus hikosanensis* Takahashi	（カメムシ目、コナジラミ科）日本。ツゲ、トベラなどの害虫
hollyhock aphid	アオヒメヒゲナガアブラムシ	*Macrosiphoniella yomogifoliae* (Matsumura)	（カメムシ目、アブラムシ科）日本、旧北区、東洋区
hollyhock groundling		*Pexicopia malvella* (Hübner)	（チョウ目、キバガ科）旧北区
hollyhock leak skeletonizer			ribbed cocoon maker moth を見よ
hollyhock moth		*Crocidosema plebejana* Zeller	（チョウ目、ハマキガ科）東洋区、新北区、大洋区
hollyhock plant bug		*Brooksetta althaeae* (Hussey)	（カメムシ目、カスミカメムシ科）新北区
hollyhock pyralid	ウスオビキノメイガ	*Microstega jessica* (Butler)	（チョウ目、メイガ科）日本
hollyhock seed moth			hollyhock groundling を見よ
hollyhock weevil		*Apion longirostre* Olivier	（コウチュウ目、ホソクチゾウムシ科）新北区
Holmes's skolly		*Thestor holmesi* van Son	（チョウ目、シジミチョウ科）エチオピア区
holomerous insects		Holomerentoma	の昆虫の総称
holometabolic insects	完全変態類	Holometabola	の昆虫の総称
homelsss bees			cuckoo bees (1) を見よ
Homerus swallowtail	ホメルスアゲハ	*Papilio homerus* Fabricius	（チョウ目、アゲハチョウ科）新熱帯区
Homeyer's silverline		*Cigaritis homeyeri* (Dewitz)	（チョウ目、シジミチョウ科）エチオピア区
homopteran bugs			homopterous insects を見よ
homopterans			homopterous insects を見よ
homopterous insects	ヨコバイ亜目	Homoptera	（カメムシ目）の昆虫の総称　世界に約 32,000 種
Hondo spruce sawfly			spruce sawfly (1) を見よ
Honduran metalmark			harmonia mantle を見よ
Honduran ruby-eye		*Carystoides hondura* Evans	（チョウ目、セセリチョウ科）新熱帯区
hone-wort sawfly			carrot sawfly を見よ
honest Pero moth		*Pero honestraria* (Walker)	（チョウ目、シャクガ科）新北区

英名	和名	学名	所属、分布、ほか
honey ant			honeypot ant ともいわれる。腹部に蜜など液体食料をためるカーストをもつ。下記の honey ant (1)(2) の属のほか *Melphorus, Tapinolepis, Myrmecocystus* などの属に見られる
honey ant		*Myrmecocystus mexicanus* Wesmael	(ハチ目、アリ科) 新熱帯区
honey ant (1)	ミツアリ	*Camponotus inflatus* Lubbock	(ハチ目、アリ科) 豪州区。ミツツボアリともいう
honey ant (2)		*Myrmecocystus melliger* Forel	(ハチ目、アリ科) 新北区
honey ants		*Myrmecocystus*	(ハチ目、アリ科) の昆虫の総称　蜜をためるアリ
honey bee	セイヨウミツバチ	*Apis mellifera* Linnaeus	(ハチ目、ミツバチ科) 日本、汎世界。採蜜用に各地で飼養
honey-bee parasite			bee lice を見よ
honey bees		Apinae	(ハチ目、ミツバチ科) の昆虫の総称
honey bees			bees (1) を見よ
honey beetle		*Heteronyx flavus* Blackburn	(コウチュウ目、コガネムシ科) 豪州区
honeybrown beetle		*Echnolagria grandis* (Gyllenhal)	(コウチュウ目、ゴミムシダマシ科) 豪州区
honeylocust gall midge		*Dasineura gleditchiae* (Osten Sacken)	(ハエ目、タマバエ科) 全北区
honey locust moth		*Sphingicampa bicolor* (Harris)	(チョウ目、ヤママユガ科) 新北区
honeylocust plant bug		*Diaphnocoris chlorionis* (Say)	(カメムシ目、カスミカメムシ科) 新北区。サイカチにつく
honeylocust pod gall midge			honeylocust gall midge を見よ
honey moth			lesser wax moth を見よ　英国での英名
honey wasp		*Brachygastra mellifica* (Say)	(ハチ目、スズメバチ科) 新北区
honeycomb moth			greater wax moth を見よ　honeycomb ともいう
honeydew moth		*Cryptoblabes gnidiella* Milliere	(チョウ目、メイガ科) 旧北区、東洋区、新熱帯区、エチオピア区、大洋区。honey locust サイカチに由来
honeypot ant			Australian desert ant を見よ
honeypot ant			honey ant (1)(2) を見よ
honeys			cereal moths を見よ
honeys			wax moths を見よ
honeysuckle aphid		*Hyadaphis foeniculi* (Passerini)	(カメムシ目、アブラムシ科) 旧北区
honeysuckle clearwing	クロスキバホウジャク	*Hemaris affinis* (Bremer)	(チョウ目、スズメガ科) 日本、旧北区
honeysuckle leafminer	スイカズラモグリガ	*Perittia lonicerae* Zimmerman et Bradley	(チョウ目、クサモグリガ科) 日本、新北区
honeysuckle leafminer	スイカズラハモグリバエ	*Chromatomyia suikazurae* Sasakawa	(ハエ目、ハモグリバエ科) 日本、旧北区

英名	和名	学名	所属、分布、ほか
honeysuckle moth		*Ypsolopha dentella* (Fabricius)	(チョウ目、スガ科) 全北区
honeysuckle sawfly		*Zaraea inflata* Norton	(ハチ目、コンボウハバチ科) 新北区
honeysuckle striped sawfly		*Zaraea fasciata* (Linnaeus)	(ハチ目、コンボウハバチ科) 日本、旧北区
honeysuckle thrips	キイロハナアザミウマ	*Thrips flavus* Schrank	(アザミウマ目、アザミウマ科) 日本、旧北区、東洋区。ダイズ、果樹、バラなどの害虫
honeysuckle whitefly		*Aleyrodes lonicerae* Walker	(カメムシ目、コナジラミ科) 旧北区
Honrath's white-lady			Almansor white-lady swordtail を見よ
hood casemake caddisflies	ホソバトビケラ科	Molannidae	(トビケラ目) の昆虫の総称
hood-shape-head leafhopper	ズキンヨコバイ	*Idiocerus vitticollis* Matsumura	(カメムシ目、オオヨコバイ科) 日本、旧北区
hooded grouse locust		*Paratettix cucullatus* (Burmeister)	(バッタ目、ヒシバッタ科) 新北区
hooded mantis			tropical shield mantis を見よ
hook-faced conehead		*Pyrgocorypha uncinata* (Harris)	(バッタ目、キリギリス科) 新北区
hook-lined satyr		*Taygetis uncinata* Weymer	(チョウ目、タテハチョウ科) 新熱帯区
hook-marked bell moth		*Cnephasia jactatana* Walker	(チョウ目、ハマキガ科) 豪州区
hook marked conch		*Agapata hamana* Linnaeus	(チョウ目、ハマキガ科) 旧北区
hook-streak grass veneer		*Crambus pratella* Linnaeus	(チョウ目、メイガ科) 旧北区
hook-tip fern looper			large hook-tip moth を見よ
hook-tip moth			arched hooktip を見よ
hook-tip moths	カギバガ科	Drepanidae	(チョウ目) の昆虫の総称
hook-tipped Amyna moth		*Amyna bullula* (Grote)	(チョウ目、ヤガ科) 新北区
hooktipped brown lacewing		*Drepanacra binocula* (Newman)	(アミメカゲロウ目、ヒメカゲロウ科) 豪州区
hook-tipped roller		*Ancylis apicella* (Denis et Schiffermüller)	(チョウ目、ハマキガ科) 旧北区
hook-tipped rubyspot		*Hetaerina curvicauda* Garrison	(トンボ目、イトトンボ科) 新熱帯区
hook-tipped sootywing		*Bolla orsines* (Godman et Salvin)	(チョウ目、セセリチョウ科) 新熱帯区
hook tips			hook-tip moths を見よ　hooktips とも記す
hook-winged geometrid	カギバアオシャク	*Tanaorhinus reciprocata confuciaria* (Walker)	(チョウ目、シャクガ科) 日本、旧北区
hook-winged Tortrix moth		*Acleris effractana* (Hübner)	(チョウ目、ハマキガ科) 日本、全北区
hooked awlking		*Choaspes furcata* Evans	(チョウ目、セセリチョウ科) 東洋区
hooked blotch roller		*Ancylis unculana* (Haworth)	(チョウ目、ハマキガ科) 旧北区
hooked marbled roller		*Ancylis laetana* Staudinger	(チョウ目、ハマキガ科) 旧北区
hooked oakblue			comic oakblue (1) を見よ

英名	和名	学名	所属、分布、ほか
hooked platypodid	ソトハナガキクイムシ	*Crossotarsus externedentatus* (Fairmaire)	（コウチュウ目、ナガキクイムシ科）日本、東洋区、大洋区、エチオピア区。材木害虫
hooked redwing		*Phylloxiphia vicina* (Rothschild et Jordan)	（チョウ目、スズメガ科）エチオピア区
hoop-pine bark beetle		*Hylurdrectonus piniarius* Schedl	（コウチュウ目、ゾウムシ科）豪州区
hoop-pine borer		*Pachycotes australis* Schedl	（コウチュウ目、ゾウムシ科）豪州区
hoop-pine borer		*Pachycotes clavatus* Schedl	（コウチュウ目、ゾウムシ科）豪州区
hoop-pine branchcutter		*Strongylurus decoratus* (McKeown)	（コウチュウ目、カミキリムシ科）豪州区
hoop-pine jewel beetle		*Prospheres aurantiopictus* (Laporte et Gory)	（コウチュウ目、タマムシ科）豪州区
hoop-pine longicorn		*Diotimana undulata* (Pascoe)	（コウチュウ目、カミキリムシ科）豪州区
hoop pine moth		*Milionia isodoxa* Prout	（チョウ目、シャクガ科）東洋区
hoop-pine seed moth		*Hieromantis ephodophora* Meyrick	（チョウ目、マルハキバガ科）豪州区
hoop-pine stitch beetle		*Hyleops glabratus* Schedl	（コウチュウ目、ゾウムシ科）豪州区
hop aphid			damson-hop aphid を見よ
hop borer		*Hydraecia immanis* Guenée	（チョウ目、ヤガ科）新北区
hop capsid		*Calocoris fulvomaculatus* (De Geer)	（カメムシ目、カスミカメムシ科）旧北区
hop cone flea			hop flea beetle を見よ
hop-damson aphid			damson-hop aphid を見よ
hop dog			pale tussock を見よ　幼虫の英名
hop flea			brassy flea beetle (1) を見よ
hop flea beetle		*Psylliodes punctulatus* Melsheimer	（コウチュウ目、ハムシ科）新北区
hop flea beetle			hemp flea beetle を見よ
hop flea beetle			brassy flea beetle (1) を見よ
hop fly			ホップにつくアブラムシ
hop frog-fly			hop jumper を見よ
hop froghopper			hop jumper を見よ
hop-garden earwig			short-winged earwig (2) を見よ
hop jumper	キスジカンムリヨコバイ	*Evacanthus interruptus* (Linnaeus)	（カメムシ目、オオヨコバイ科）日本、旧北区。ホップ、キク害虫
hop leaf miner		*Agromyza flaviceps* Fallén	（ハエ目、ハモグリバエ科）旧北区
hop leafhopper			hop jumper を見よ
hop louse			damson-hop aphid を見よ
hop merchant			comma (2) を見よ
hop merchant			question mark を見よ
hop moth			ghost moth を見よ
hop plant bug		*Taedia hawleyi* (Knight)	（カメムシ目、カスミカメムシ科）新北区

英名	和名	学名	所属、分布、ほか
hop plant-louse			damson-hop aphid を見よ　北米での英名
hop root borer			ghost moth を見よ
hop root weevil		*Epipolaeus caliginosus* (Fabricius)	（コウチュウ目、ゾウムシ科）旧北区
hop snout moth		*Hypena humuli* Harris	（チョウ目、ヤガ科）新北区。幼虫は hop looper というホップ害虫
hop stalk borer moth		*Papaipema circumlucens* (Smith)	（チョウ目、ヤガ科）新北区。hops stalk borer ともいう
hop string maggot			hop string midge を見よ　幼虫の英名
hop string midge		*Contarinia humuli* Theobald	（ハエ目、タマバエ科）旧北区
hop-tree ermine moth		*Yponomeuta atomocella* (Dyar)	（チョウ目、スガ科）新北区
hop vine borer			hop borer を見よ
hop vine moth			hop borer を見よ　hop vine borer moth ともいう
Hopffer's Andean white			Nereina white を見よ
Hopffer's dartwhite		*Catasticta pieris* (Hopffer)	（チョウ目、シロチョウ科）新熱帯区
Hopffer's firetip		*Oxynetra hopfferi* Staudinger	（チョウ目、セセリチョウ科）新熱帯区
Hopffer's skiperling		*Dalla plancus* (Hopffer)	（チョウ目、セセリチョウ科）新熱帯区
Hopffer's tailed owlet		*Opoptera arsippe* (Hopffer)	（チョウ目、タテハチョウ科）新熱帯区
hopline chafers		Hopliinae	（コウチュウ目、コガネムシ科）の昆虫の総称
hopper			バッタ、ヨコバイなどの飛ぶ虫
hopper			Asiatic migratory locust を見よ
hoppers		*Platylesches*	（チョウ目、セセリチョウ科）の昆虫の総称
hops azure		*Celastrina ladon humulus* Scott et D. Wright	（チョウ目、シジミチョウ科）新北区。spring azure を参照
Horace's duskywing	ホレースミヤマセセリ	*Erynnis horatius* (Scudder et Burgess)	（チョウ目、セセリチョウ科）新北区、新熱帯区
horehound bug		*Agonoscelis rutila* (Fabricius)	（カメムシ目、カメムシ科）豪州区
horehound long-horn			copper-japan longhorn を見よ
horehound plume		*Pterophorus spilodactylus* Curtis	（チョウ目、トリバガ科）旧北区
Hori scale	ホリイコノハカイガラムシ	*Fiorinia horii* Kuwana	（カメムシ目、マルカイガラムシ科）日本
horizontal borer		*Austroplatypus incompertus* (Schedl)	（コウチュウ目、ナガキクイムシ科）豪州区
horn blower			tobacco hornworm を見よ
horn compressed vegetable moth			Indian meal moth (1) を見よ
horn flies			stable flies を見よ
horn fly		*Haematobia stimulans* (Meigen)	（ハエ目、イエバエ科）全北区

英名	和名	学名	所属、分布、ほか
horn fly (1)	ノサシバエ	*Haematobia irritans* (Linnaeus)	(ハエ目、イエバエ科) 日本、大洋区、新北区。ウシなどから吸血。幼虫は牛糞に発生
horn moth		*Ceratophaga vastella* (Zeller)	(チョウ目、ヒロズコガ科) エチオピア区
horn worm			tobacco worm を見よ
hornbeam aphid		*Myzocallis carpini* Koch	(カメムシ目、アブラムシ科) 旧北区
hornbeam leaf gall midge		*Zygiobia carpini* (Low)	(ハエ目、タマバエ科) 旧北区
hornbeam leafhopper		*Typhlocyba bifasciata* Boheman	(カメムシ目、オオヨコバイ科) 旧北区
hornbeam whitefly		*Asterobemisia carpini* (Koch)	(カメムシ目、コナジラミ科) 旧北区
horned bee			red Osmia を見よ
horned beetle			Japanese horned beetle を見よ
horned caterpillar			evening brown を見よ
horned clubtail		*Arigomphus cornutus* (Tough)	(トンボ目、サナエトンボ科) 新北区
horned fungus beetle			forked fungus beetle を見よ
horned longhorn		*Necydalis cavipennis* LeConte	(コウチュウ目、カミキリムシ科) 新北区
horned moth			hornet moth を見よ
horned Passalus			bess-beetle を見よ
horned Passalus beetle			bess-beetle を見よ
horned Passalus beetles			peg beetles (1) を見よ
horned pigeon louse		*Coloceras damicorne* (Nitzsch)	(ハジラミ目、チョウカクハジラミ科) 旧北区
horned powder-post beetles			false powderpost beetles を見よ
horned spanworm			filament borer を見よ
horned squash bug		*Anasa armigera* (Say)	(カメムシ目、ヘリカメムシ科) 新北区
horned talontail		*Crenigomphus cornutus* Pinhey	(トンボ目、サナエトンボ科) エチオピア区
horned treehopper		*Centrotus cornutus* (Linnaeus)	(カメムシ目、ツノゼミ科) 旧北区
horned treehopper	ツノゼミ	*Orthobelus flavipes* Uhler	(カメムシ目、ツノゼミ科) 日本、旧北区
horned treehopper (1)		*Umbonia crossicornis* Amyot et Serville	(カメムシ目、ツノゼミ科) 新北区。体長 5 mm
horned treehoppers			treehoppers (1) を見よ
horned wax scale			Indian wax scale を見よ
hornet		*Vespa crabro* Linnaeus	(ハチ目、スズメバチ科) 旧北区。日本亜種は *V. c. flavofascia* Cameron モンスズメバチ。giant hornet を本種に使用するのは稀
hornet			Japanese yellow hornet を見よ
hornet			baldfaced hornet を見よ

英名	和名	学名	所属、分布、ほか
hornet clearwing		*Sesia bombeciformis* Hübner	（チョウ目、スカシバガ科）全北区
hornet clearwing			hornet moth を見よ
hornet clearwing			red oak clearwing moth を見よ
hornet clearwing moth			hornet moth を見よ
hornet-fly			hornet を見よ
hornet moth		*Sesia apiformis* (Clerck)	（チョウ目、スカシバガ科）旧北区。スズメバチに似ることから
hornet robberfly		*Asilus crabroniformis* Linnaeus	（ハエ目、ムシヒキアブ科）旧北区
hornets (1)		Vespinae	（ハチ目、スズメバチ科）の昆虫の総称
hornets (2)	スズメバチ科	Vespidae	（ハチ目）の昆虫の総称
hornets			common wasps (1) を見よ　hornet は広く使用されるが、*Vespa* 属の種に用いるのが本来
hornets			wasps (1) を見よ
hornets			crabronid wasps を見よ
hornicle			hornet を見よ
Horniman's green-banded swallowtail	オナガルリアゲハ	*Papilio hornimani* Distant	（チョウ目、アゲハチョウ科）エチオピア区
Horniman's swallowtail			Horniman's green-banded swallowtail を見よ
Horn's checkered beetle		*Trichodes peninsularis* Horn	（コウチュウ目、カッコウムシ科）新北区
horntail (1)		*Urocerus gigas* (Linnaeus)	（ハチ目、キバチ科）旧北区。日本亜種は *U. g. orientalis* Maa モミノオオキバチ
horntail (2)		*Sirex noctilio* Fabricius	（ハチ目、キバチ科）旧北区、豪州区
horntail wasps	キバチ科	Siricidae	（ハチ目）の昆虫の総称
horntails (1)	ハバチ亜目（広腰亜目）	Symphyta	（ハチ目）の昆虫の総称
horntails		*Urocerus*	（ハチ目、キバチ科）の昆虫の総称
horntails			horntail wasps を見よ
hornworm moths			hawk moths を見よ
hornworms			hawk moths を見よ
horrid Zale moth		*Zale horrida* Hübner	（チョウ目、ヤガ科）新北区
horse adder			dragonflies (1) を見よ
horse ant			common red ant を見よ
horse ant			herculean ant (2) を見よ
horsebane angle flat-body		*Depressaria ultimella* Stainton	（チョウ目、マルハキバガ科）旧北区
horse-bean longhorn		*Dendrobias mandibularis* Serville	（コウチュウ目、カミキリムシ科）新北区。カンキツ害虫。horse-bean longhorn beetle ともいう
horsebiting louse		*Werneckiella equi* (Denny)	（シラミ目、Trichodectidae）豪州区
horse biting louse	ウマハジラミ	*Bovicola equi* (Denny)	（ハジラミ目、ケモノハジラミ科）日本、汎世界。ウマに寄生

英名	和名	学名	所属、分布、ほか
horse bloodsucking louse			horse sucking louse を見よ
horse bot flies	ウマバエ科	Gasterophilidae	（ハエ目）の昆虫の総称　horse botflies とも記す。
horse botflies			horse bot flies を見よ
horse bot fly	ウマバエ	*Gasterophilus intestinalis* (De Geer)	（ハエ目、ウマバエ科）日本、汎世界。幼虫はウマの胃に寄生
horse bots			horse bot flies を見よ
horse-chestnut borer			leopard moth を見よ
horse chestnut cockchafer		*Melolontha hippocastani* Fabricius	（コウチュウ目、コガネムシ科）旧北区
horse chestnut leaf-miner		*Cameraria ohridella* (Deschka et Dimic)	（チョウ目、ホソガ科）旧北区
horse-chestnut longwing			horse-chestnut moth を見よ
horse-chestnut moth		*Pachycnemia hippocastanaria* (Hübner)	（チョウ目、シャクガ科）旧北区。horse chestnut ともいう
horse chestnut scale		*Pulvinaria regalis* Canard	（カメムシ目、カタカイガラムシ科）旧北区
horse chewing louse			horse biting louse を見よ
horse flies (1)	アブ科	Tabanidae	（ハエ目）の昆虫の総称　greenheads, purple eyes とも呼ばれる
horse flies (2)		*Tabanus*	（ハエ目、アブ科）の昆虫の総称
horse flies (3)		*Chrysops*	（ハエ目、アブ科）の昆虫の総称
horse-fly (1)		*Chrysops coecutiens* Linnaeus	（ハエ目、アブ科）旧北区
horse fly (2)		*Tabanus sudeticus* Zeller	（ハエ目、アブ科）全北区
horse fly			forest fly を見よ
horse-fly			twin-lobed deerfly を見よ
horse-fly			black deerfly を見よ
horse-fly			small horse-fly を見よ
horse-fly			western horsefly を見よ
horse ked			forest fly を見よ
horse-long-cripple			dragonflies (1) を見よ
horse-nettle beetle		*Leptinotarsa juncta* (Germar)	（コウチュウ目、ハムシ科）新北区
horse-nettle borer			Riley's clearwing moth を見よ
horse-nettle leafminer			eggplant leaf miner (2) を見よ
horse louse			horse sucking louse を見よ
horse louse-fly			forest fly を見よ　豪州での英名
horse lubber		*Taeniopoda eques* (Burmeister)	（バッタ目、バッタ科）新北区
horse lubber grasshopper			horse lubber を見よ
horse nit fly			horse bot fly を見よ　米国での英名

英名	和名	学名	所属、分布、ほか
horse nostril fly		*Rhinoestrus purpureus* (Brauer)	(ハエ目、ヒツジバエ科) 旧北区
horse-radish flea beetle		*Phyllotreta armoraciae* (Koch)	(コウチュウ目、ハムシ科) 全北区。野菜害虫
horseradish webworm		*Plutella armoraciae* Busck	(チョウ目、スガ科) 新北区
horseshoe crab beetles			feather-winged beetles を見よ
horse-shoe groundling		*Teleiodes scriptella* (Hübner)	(チョウ目、キバガ科) 旧北区
horse-stang			dragonflies (1) を見よ
horse-stinger			dragonflies (1) を見よ
horse sucking louse	ウマジラミ	*Haematopinus asini* (Linnaeus)	(シラミ目、ケモノジラミ科) 日本、汎世界。ウマに寄生
horsetail flea beetle		*Hippuriphila modeeri* (Linnaeus)	(コウチュウ目、ハムシ科) 旧北区
horsetail gemmed satyr			Chermock's gemmed-satyr を見よ
horse tick			forest fly を見よ　米国での英名
Horsfield's baron	ルリヘリコイナズマ	*Tanaecia iapis puseda* Moore	(チョウ目、タテハチョウ科) 東洋区
Horsfield's bush brown	ウスオビコジャノメ	*Mycalesis horsfieldi* Moore	(チョウ目、タテハチョウ科) 東洋区
Horvath katydid	シブイロカヤキリモドキ	*Xestophrys horvathi* Bolivar	(バッタ目、キリギリス科) 日本、東洋区、大洋区
horvathian smaller water-strider	ホルバートケシカタビロアメンボ	*Microvelia horvathi* Lundblad	(カメムシ目、カタビロアメンボ科) 日本、東洋区
Hosta skipper		*Zera phila hosta* Evans	(チョウ目、セセリチョウ科) 新熱帯区
Hosta spreadwing		*Zera hosta* Evans	(チョウ目、セセリチョウ科) 新熱帯区
hot-bed bug		*Xylocoris galactinus* (Fieber)	(カメムシ目、ハナカメムシ科) 旧北区
hot-house thrips			banded glasshouse thrips を見よ
hot pepper gall midge			citrus gall midge を見よ
Hottentot bug (1)		*Eurygaster maurus* Linnaeus	(カメムシ目、キンカメムシ科) 旧北区
Hottentot bug (2)		*Eurygaster hottentotta* (Fabricius)	(カメムシ目、キンカメムシ科) 旧北区
Hottentot bug			cereal bug を見よ
Hottentot god			mantids (2) を見よ　南アフリカ、豪州での英名。Hottentot's god とも記される
Hottentot mantis		*Bolbena hottentotta* Karny	(カマキリ目、Iridopterygidae) エチオピア区
Hottentot skipper			Latreille's skipper を見よ
Hottes' green goldenglow aphid		*Macrosiphum cockerelli* Hottes	(カメムシ目、アブラムシ科) 旧北区
hour-glass skipperling		*Piruna penaea* (Dyer)	(チョウ目、セセリチョウ科) 新北区、新熱帯区
house ant	ムネボソアリ	*Leptothrax congruus* Smith	(ハチ目、アリ科) 日本

英名	和名	学名	所属、分布、ほか
house ant			thief ant を見よ
house-bock			house longhorn を見よ
house-bug			bed bug (1) を見よ
house cricket	イエコオロギ	*Acheta domesticus* (Linnaeus)	（バッタ目、コオロギ科）全北区、豪州区。欧州から北米に入った種
house crickets			field crickets (1) を見よ
house flies		Muscinae	（ハエ目、イエバエ科）の昆虫の総称
house flies	ヒメイエバエ科	Fanniidae	（ハエ目）の昆虫の総称
house flies (1)	イエバエ科	Muscidae	（ハエ目）の昆虫の総称
house fly	イエバエ	*Musca domestica* Linnaeus	（ハエ目、イエバエ科）日本、汎世界。世界中に普通で人家や家畜小屋に多し
house fly pupal parasite		*Spalangia endius* Walker	（ハチ目、コガネコバチ科）新北区
house gnat			common mosquito を見よ　house-gnat とも記す
house longhorn		*Hylotrupes bajulus* (Linnaeus)	（コウチュウ目、カミキリムシ科）全北区
house longhorn beetle			house longhorn を見よ
house longicorn beetle			house longhorn を見よ
house mosquito			common mosquito を見よ
house mosquitoes		*Culex*	（ハエ目、カ科）の昆虫の総称
house moth			small tabby を見よ
house psocid		*Ectopsocus maindroni* Badonnel	（チャタテムシ目、Ectopsocidae）大洋区
house windowfly		*Scenopinus fenestralis* (Linnaeus)	（ハエ目、マドギワアブ科）旧北区
household casebearer		*Phereoeca walsinghami* Busck	（チョウ目、ヒロズコガ科）新北区
household casebearer		*Tineola uterella* Walsingham	（チョウ目、ヒロズコガ科）新北区。household casebearer moth ともいう
household casebearer moth		*Phereoeca praecox* Gozzmany et Vári	（チョウ目、ヒロズコガ科）新北区
household case-bearing moth		*Phereoeca allutella* (Rebel)	（チョウ目、ヒロズコガ科）東洋区、大洋区、エチオピア区
hover flies			flower flies を見よ　英国での英名
hover-flies			wasp flies を見よ
hover fly		*Volucella bombylans* (Linnaeus)	（ハエ目、ハナアブ科）新北区
hover-fly (1)	ホソヒラタアブ	*Episyrphus balteatus* (De Geer)	（ハエ目、ハナアブ科）日本、東洋区、豪州区
hover-fly (2)	オオフタホシヒラタアブ	*Syrphus ribesii* (Linnaeus)	（ハエ目、ハナアブ科）日本、全北区
hover-fly		*Cheilosia intonsa* Loew	（ハエ目、ハナアブ科）旧北区
hover-fly		*Chrysotoxum bicinctum* (Linnaeus)	（ハエ目、ハナアブ科）旧北区

英名	和名	学名	所属、分布、ほか
hover-fly		*Pelecocera tricincta* Meigen	（ハエ目、ハナアブ科）旧北区
hover-fly		*Eristalis intricarius* (Linnaeus)	（ハエ目、ハナアブ科）旧北区
hover-fly			large syrphid fly を見よ
hover wasps	ハラホソバチ亜科	Stenogastrinae	（ハチ目、スズメバチ科）の昆虫の総称
hovering hawkmoth			hummingbird hawkmoth を見よ
Howard margarodid scale	ハワードワラジカイガラムシ	*Drosicha howardi* (Kuwana)	（カメムシ目、ワタフキカイガラムシ科）日本、旧北区
Howard warajicoccus			Howard margarodid scale を見よ
Howard's whitefly		*Aleurothrixus howardii* (Quaintance)	（カメムシ目、コナジラミ科）新熱帯区。カンキツ害虫
Howarth's cave cricket		*Caconemobius howarthi* Gurney et Rentz	（バッタ目、コオロギ科）大洋区。ハワイ固有種
Howarth's white		*Ganyra howarthi* (Dixey)	（チョウ目、シロチョウ科）新熱帯区
Huachuca giant skipper			brigadier を見よ
Hualapai buckmoth		*Hemileuca hualapai* (Neumoegen)	（チョウ目、ヤママユガ科）新北区
huanacina dancer		*Argia huanacina* Förster	（トンボ目、イトトンボ科）新熱帯区
Huancabamba crescent		*Higginsius fasciata* (Hopffer)	（チョウ目、タテハチョウ科）新熱帯区
Huastecan Calephelis		*Calephelis huasteca* McAlpine	（チョウ目、シジミタテハ科）新熱帯区
Huastecan skipper		*Decinea decinea huasteca* (Freeman)	（チョウ目、セセリチョウ科）新熱帯区
Hubbard's angel insect		*Zoratypus hubbardi* Caudell	（ジュズヒゲムシ目、ジュズヒゲムシ科）新北区
Hubbell's ground cricket		*Pictonemobius hubbelli* Walker et Mays	（バッタ目、コオロギ科）新北区
Hubner's blue ringlet		*Chloreuptychia chlorimene* Hübner	（チョウ目、タテハチョウ科）新熱帯区
Hubner's glad-eye		*Calisto herophile* Hübner	（チョウ目、タテハチョウ科）新熱帯区
Hubner's glasswing		*Scada reckia* (Hübner)	（チョウ目、タテハチョウ科）新熱帯区
Hubner's green-veined white			dark-veined white を見よ
Hubner's owlet		*Opsiphanes invirae* (Hübner)	（チョウ目、タテハチョウ科）新熱帯区
Hubner's Pero moth		*Pero ancetaria* (Hübner)	（チョウ目、シャクガ科）新北区
Hubner's shoemaker			one-spotted Prepona を見よ
Hubner's sister		*Adelpha plesaure* Hübner	（チョウ目、タテハチョウ科）新熱帯区
Hubner's skipper		*Papias phaeomelas* (Hübner)	（チョウ目、セセリチョウ科）新熱帯区
Hubner's Telemiades		*Telemiades epicalus* Hübner	（チョウ目、セセリチョウ科）新熱帯区
huckleberry planthopper		*Oliarus cinnamomeus* Provancher	（カメムシ目、ヒシウンカ科）新北区。湿地に生息。希な種
huckleberry sphinx		*Paonias astylus* (Drury)	（チョウ目、スズメガ科）新北区。huckleberry sphinx moth ともいう

英名	和名	学名	所属、分布、ほか
Huebner's metalmark		*Adelotypa huebneri* Butler	（チョウ目、シジミタテハ科）新熱帯区
huge-comma moth			wavy huge-comma を見よ
Hugo's Diopetes		*Pilodeudorix hugoi* Libert	（チョウ目、シジミチョウ科）エチオピア区
huhu beetle		*Prionoplus reticularis* White	（コウチュウ目、カミキリムシ科）豪州区
Hulst's underwing moth		*Catocala luctuosa* Hulst	（チョウ目、ヤガ科）新北区
hum-buz			June beetle (1) を見よ
human body louse			body louse を見よ
human bot fly		*Dermatobia hominis* (Linnaeus)	（ハエ目、ヒツジバエ科）新北区、新熱帯区。カに産卵し、幼虫はカの吸血時に人の皮膚に入る
human flea	ヒトノミ	*Pulex irritans* Linnaeus	（ノミ目、ヒトノミ科）日本、汎世界。ヒトはじめ多くの動物に寄生
human head louse			body louse を見よ
human lice	ヒトジラミ科	Pediculidae	（シラミ目）の昆虫の総称
human louse			body louse を見よ
human warble fly			human bot fly を見よ
humble bee flies			bee flies を見よ
humble bees			bumble bees (1)(2)(3) を見よ
Humboldt shieldback		*Idiostatus inermoides* Rentz	（バッタ目、キリギリス科）新北区
Humboldt's Elzunia		*Elzunia humboldt* (Latreille)	（チョウ目、タテハチョウ科）新熱帯区
Humboldt's Perisama		*Perisama humboldtii* Guérin-Méneville	（チョウ目、タテハチョウ科）新熱帯区
hummingbird			hummingbird hawkmoth を見よ
humming bird flies			tangle-veined flies を見よ　南アフリカでの英名
hummingbird clearwing			hummingbird moth を見よ　hummingbird clearwing moth ともいう
humming-bird hawk			hummingbird hawkmoth を見よ
hummingbird hawkmoth	ホウジャク	*Macroglossum stellatarum* (Linnaeus)	（チョウ目、スズメガ科）日本、旧北区
hummingbird hawkmoths	ホウジャクガ亜科	Macroglossinae	（チョウ目、スズメガ科）の昆虫の総称
hummingbird lice			bird lice (2) を見よ
hummingbird moth		*Hemaris thysbe* (Fabricius)	（チョウ目、スズメガ科）新北区
hummingbird moth			striped hawk を見よ
hummingbird moths			bee hawk moths を見よ　ハチドリに似ることから
hummingbird moths			hawk moths を見よ
hummock skipper		*Polygonus leo* (Gmelin)	（チョウ目、セセリチョウ科）新北区、新熱帯区
hump beetle			museum beetle (1) を見よ
hump beetle			spider beetle (1) を見よ

英名	和名	学名	所属、分布、ほか
hump mite			museum beetle (1) を見よ
hump-winged cricket			great grig を見よ
hump-winged crickets			primitive katydids を見よ
hump-winged grigs			primitive katydids を見よ
humpbacked flies	ノミバエ科	Phoridae	（ハエ目）の昆虫の総称　成虫胸部の形状に由来
humpbacked flies			alderflies (1) を見よ
humpbacked flies			small-headed flies を見よ
humpbacked fly		*Megaselia aletiae* (Comstock)	（ハエ目、ノミバエ科）新北区
humped claw		*Psoricoptera gibbosella* (Zeller)	（チョウ目、キバガ科）旧北区
humped sphecodes		*Sphecodes gibbus* Linnaeus	（ハチ目、コハナバチ科）旧北区
humpless casemake caddisflies	カクスイトビケラ科	Brachycentridae	（トビケラ目）の昆虫の総称
hunch-backed beetle		*Omophron americanum* Dejean	（コウチュウ目、オサムシ科）新北区
hunch-backed beetle		*Omophron obliteratum* Horn	（コウチュウ目、オサムシ科）新北区
Hungarian glider		*Neptis rivularis* (Scopoli)	（チョウ目、タテハチョウ科）旧北区、東洋区。日本亜種は *N. r. insularum* Fruhstorfer フタスジチョウ
Hungarian skipper		*Spialia orbifer* (Hübner)	（チョウ目、セセリチョウ科）旧北区
Hungerford's crawling water beetle		*Brychius hungerfordi* Spangler	（コウチュウ目、コガシラミズムシ科）新北区
hunter fly		*Coenosia attenuata* Stein	（ハエ目、イエバエ科）全北区
Hunter's butterfly			painted beauty を見よ
hunting billbug	シバオサゾウムシ	*Sphenophorus venatus vestitus* Chittenden	（コウチュウ目、オサゾウムシ科）日本、新北区。シバ害虫
Huntington's blue		*Echinargus huntingtoni hannoides* (Clench)	（チョウ目、シジミチョウ科）新熱帯区
Husian's whitefly		*Aleurocanthus husaini* Corbett	（カメムシ目、コナジラミ科）東洋区。カンキツ害虫
Hutchinson's highflyer		*Aphnaeus hutchinsonii* Trimen	（チョウ目、シジミチョウ科）エチオピア区
Hutchinson's silver spot			Hutchinson's highflyer を見よ
hyacinth glider		*Miathyria marcella* (Selys)	（トンボ目、トンボ科）新熱帯区
Hyagnis skipper		*Moeris hyagnis hyagnis* Godman	（チョウ目、セセリチョウ科）新熱帯区
hyaline grass bug	スカシヒメヘリカメムシ	*Liorhyssus hyalinus* (Fabricius)	（カメムシ目、ヒメヘリカメムシ科）日本、汎世界。ニンジン、ワタ害虫
hyaline maple leafhopper	ミスジトガリヨコバイ	*Japananus hyalinus* (Osborn)	（カメムシ目、オオヨコバイ科）日本、全北区。カエデ類害虫
hyaline-spotted clothes moths			monopis moths を見よ
hyaline swift		*Parnara amalia* (Semper)	（チョウ目、セセリチョウ科）豪州区

英名	和名	学名	所属、分布、ほか
hyblaeid moths			hawk moths を見よ
hybrid Argus		*Callerebia hybrida* Butler	(チョウ目、タテハチョウ科) 東洋区
hybrid sapphire		*Heliophorus hybrida* (Tytler)	(チョウ目、シジミチョウ科) 東洋区
hybrid wood white		*Leptosia hybrida* Bernardi	(チョウ目、シロチョウ科) エチオピア区
hybrizontid parasites		Hybrizontidae	(ハチ目) の昆虫の総称
Hydaspe fritillary	ニセチャマダラギンボシヒョウモン	*Speyeria hydaspe* (Boisduval)	(チョウ目、タテハチョウ科) 新北区
Hydaspes eighty-eight	コベニモンウズマキタテハ	*Callicore hydaspes* (Drury)	(チョウ目、タテハチョウ科) 新熱帯区
hydraenid water beetles			minute moss beetles を見よ
hydrangea aphid	サビタトックリアブラムシ	*Rhopalosiphoninus hydrangeae* (Matsumura)	(カメムシ目、アブラムシ科) 日本、旧北区
hydrangea leaftier moth		*Olethreutes ferriferana* (Walker)	(チョウ目、ハマキガ科) 新北区
hydrangea sawfly	アジサイハバチ	*Perineura okutanii* Takeuchi	(ハチ目、ハバチ科) 日本。アジサイ害虫
hydrangea scale	アジサイワタカイガラムシ	*Pulvinaria hydrangeae* Steinweden	(カメムシ目、カタカイガラムシ科) 日本、全北区、豪州区
hydrangea scale			cottony mulberry scale を見よ
hydrangea sphinx		*Darapsa versicolor* (Harris)	(チョウ目、スズメガ科) 新北区。hydrangea shinx moth ともいう
hydrilla leafcutter moth		*Parapoynx diminutalis* (Snellen)	(チョウ目、メイガ科) 新北区
hydriomene moth		*Perizoma custodiata* (Guenée)	(チョウ目、シャクガ科) 新北区
hydriomene moth		*Perizoma epictata* (Barnes et McDunnough)	(チョウ目、シャクガ科) 新北区
hydrometrid bugs			water measures を見よ
hygrobiid water beetles		Hygrobiidae	(コウチュウ目) の昆虫の総称
hylophilid beetles		Hylophilidae	(コウチュウ目) の昆虫の総称 クビボソムシ科 Aderidae の異名
hymenopterans			wasps (2) を見よ
hymenopteroid insects		Hymenopteroidea	の昆虫の総称
hymenopteroids			hymenopteroid insects を見よ
hymenopteron			wasps (2) を見よ
Hypargyra skipper		*Paracarystus hypargyra* (Herrich-Schäffer)	(チョウ目、セセリチョウ科) 新熱帯区
Hypena moth		*Hypena laceratalis* Walker	(チョウ目、ヤガ科) 豪州区
hyperoscelidid gnats		Hyperoscelididae	(ハエ目) の昆虫の総称
hypnotist		*Nirvanopsis hypnus* (Tsukada et Nishiyama)	(チョウ目、タテハチョウ科) 東洋区
Hypocala moth		*Hypocala andremona* (Stoll)	(チョウ目、ヤガ科) 新北区
hypogaeic army ant		*Dorylus laevigatus* Smith	(ハチ目、アリ科) 東洋区

英名	和名	学名	所属、分布、ほか
hypogastrurid springtails	ヒメトビムシ科	Hypogastruridae	（トビムシ目）の昆虫の総称
Hystaspes eighty-eight			Zelphanta numberwing を見よ
hyster skipper		*Codatractus hyster* (Dyar)	（チョウ目、セセリチョウ科）新熱帯区

英名	和名	学名	所属、分布、ほか
I			
iacchoides skipper		*Trapezites iacchoides* Waterhouse	（チョウ目、セセリチョウ科）豪州区
iapygids			earwiglike entotrophs を見よ
ibalid wasps	ヒラタタマバチ科	Ibalidae	（ハチ目）の昆虫の総称　キバチ科の種に寄生
ibalids			ibalid wasps を見よ
Iberian bluetail		*Ischnura graellsii* (Rambur)	（トンボ目、イトトンボ科）旧北区
Iberian marbled white			southern marbled white を見よ
Iberian sooty copper		*Lycaena bleusei* Oberthür	（チョウ目、シジミチョウ科）旧北区
ibis-feather case		*Coleophora ibipennella* Zeller	（チョウ目、ツツミノガ科）旧北区
ibis fly			yellow-legged water-snipefly を見よ
ibis skipper		*Vertica ibis* Evans	（チョウ目、セセリチョウ科）新熱帯区
ice crawlers			rock crawlers (2) を見よ
Iceland dart		*Euxoa islandica islandica* Staudinger	（チョウ目、ヤガ科）旧北区
Icelandic plume		*Stenoptilia islandica* Staudinger	（チョウ目、トリバガ科）旧北区
ichneumon		*Rhyssa lineolata* Kirby	（ハチ目、ヒメバチ科）新北区
ichneumon flies			ichneumons を見よ
ichneumon fly	ヒメバチ		（ハチ目、ヒメバチ科）の昆虫。ichneumons を見よ
ichneumon fly			yellow Ophion を見よ
ichneumon fly			cabbage white butterfly parasite を見よ
ichneumon wasps			ichneumons を見よ
ichneumon wasps			ichneumonoid wasps を見よ
ichneumonid wasps			ichneumons を見よ　豪州での英名
ichneumonids			ichneumons を見よ
ichneumonoid flies			ichneumons を見よ
ichneumonoid flies			ichneumonoid wasps を見よ
ichneumonoid wasps	ヒメバチ上科	Ichneumonoidea	（ハチ目）の昆虫の総称
ichneumons	ヒメバチ科	Ichneumonidae	（ハチ目）の昆虫の総称
Icilius blue	イシリウスヒスイシジミ	*Jalmenus icilius* Hewitson	（チョウ目、シジミチョウ科）豪州区。Icilius blue butterfly ともいう
Ictinus blue	ウスグロヒスイシジミ	*Jalmenus ictinus* Hewitson	（チョウ目、シジミチョウ科）豪州区
Idas blue		*Plebejus idas* (Linnaeus)	（チョウ目、シジミチョウ科）全北区
Idas blue			northern blue (1) を見よ
Idomeneus giant owl	イドメウスフクロチョウ	*Caligo idomeneus* (Linnaeus)	（チョウ目、タテハチョウ科）新熱帯区
Ignoble bush brown		*Bicyclus ignobilis* (Butler)	（チョウ目、タテハチョウ科）エチオピア区
ignorant Apamea moth		*Apamea indocilis* (Walker)	（チョウ目、ヤガ科）新北区。ignorant Apamea ともいう

英名	和名	学名	所属、分布、ほか
Iguala sootywing		*Staphylus iguala* (Williams et Bell)	(チョウ目、セセリチョウ科) 新熱帯区
Ilavia hairstreak		*Satyrium ilavia* (Beutenmüller)	(チョウ目、シジミチョウ科) 新北区
Ilex hairstreak	ヨーロッパベニモンカラスシジミ	*Satyrium ilicis* (Esper)	(チョウ目、シジミチョウ科) 旧北区
Ilex sucker	ネグロキジラミ	*Petalolyma bicolor* (Kuwayama)	(カメムシ目、キジラミ科) 日本、東洋区
Ilfracombe humble bee			broken-belted bumblebee を見よ Ilfracombe bumblebee とも記す
ilia underwing		*Catocala ilia* (Cramer)	(チョウ目、ヤガ科) 新北区。ilia underwing moth ともいう
ilima leafminer		*Philodoria marginestrigata* (Walsingham)	(チョウ目、ホソガ科) 新北区
ilima moth		*Amyna natalis* (Walker)	(チョウ目、ヤガ科) 東洋区、豪州区、大洋区、新北区
Illidge's ant-blue		*Acrodipsas illidgei* Waterhouse et Lyell	(チョウ目、シジミチョウ科) 豪州区
Illinissa glasswing		*Hyposcada illinissa* (Hewitson)	(チョウ目、タテハチョウ科) 新熱帯区
Illinois river cruiser			swift river cruiser を見よ
Illioneus giant owl		*Caligo illioneus* (Cramer)	(チョウ目、タテハチョウ科) 新熱帯区
illustrious sprite		*Celaenorrhinus illustris* (Mabille)	(チョウ目、セセリチョウ科) エチオピア区
Ilus swallowtail		*Mimoides ilus branchus* (Doubleday)	(チョウ目、アゲハチョウ科) 新熱帯区。*M. i. occiduus* (Vazquez) も同英名
imago			昆虫の成虫を指す
imbricated snout beetle		*Epicaerus imbricatus* (Say)	(コウチュウ目、ゾウムシ科) 新北区
imitator mole cricket		*Scapteriscus imitatus* Nickle et Castner	(バッタ目、ケラ科) 新熱帯区
Imma hairstreak		*Contrafacia imma* (Prittwitz)	(チョウ目、シジミチョウ科) 新熱帯区
Imma moths		Immidae	(チョウ目) の昆虫の総称
immaculata skipper		*Enosis immaculata immaculata* (Hewitson)	(チョウ目、セセリチョウ科) 新熱帯区
immaculate emperor	ウラギンアメリカコムラサキ	*Doxocopa laure griseldis* (C. et R. Felder)	(チョウ目、タテハチョウ科) 新熱帯区。Laure を参照
immaculate grass-veneer moth			hook-streak grass veneer を見よ
immaculate green hairstreak	ムモンウラアオシジミ	*Callophrys affinis* (Edwards)	(チョウ目、シジミチョウ科) 新北区
immaculate Holomelina		*Holomelina immaculata* (Reakirt)	(チョウ目、ヒトリガ科) 新北区。immaculate Holomelina moth ともいう
immaculate roadside-skipper		*Amblyscirtes novimmaculatus* Warren	(チョウ目、セセリチョウ科) 新熱帯区
immaculate skipper		*Anthoptus insignis* (Plötz)	(チョウ目、セセリチョウ科) 新熱帯区
immaculate snow flat			immaculate suffused snow flat を見よ
immaculate spirit			immaculate wood white を見よ

英名	和名	学名	所属、分布、ほか
immaculate suffused snow flat		*Tagiades gana* (Moore)	（チョウ目、セセリチョウ科）東洋区
immaculate tufted-skipper		*Pachyneuria licisca* (Plötz)	（チョウ目、セセリチョウ科）新熱帯区
immaculate wood white		*Leptosia nupta* (Butler)	（チョウ目、シロチョウ科）エチオピア区
immigrant acacia weevil			vine weevil を見よ
immigrant pinion moth		*Lithophane oriunda* (Grote)	（チョウ目、ヤガ科）新北区
impatiens aphid	ホウセンカコブアブラムシ	*Eumyzus impatiensae* (Shinji)	（カメムシ目、アブラムシ科）日本
impatiens hawk moth			vine hawk moth (2) を見よ
imperial Apollo	ミカドウスバアゲハ	*Parnassius imperator augustus* Oberthür	（チョウ目、アゲハチョウ科）東洋区
imperial Arcas	クジャクシジミ	*Arcas imperialis* (Cramer)	（チョウ目、シジミチョウ科）新熱帯区
imperial awlet		*Bibasis imperialis* (Plötz)	（チョウ目、セセリチョウ科）東洋区
imperial banner		*Epiphile imperator* Attal	（チョウ目、タテハチョウ科）新熱帯区
imperial blue	ヒスイシジミ	*Jalmenus evagoras* (Donovan)	（チョウ目、シジミチョウ科）東洋区。imperial blue butterfly ともいう
imperial blue Charaxes		*Charaxes imperialis* Butler	（チョウ目、タテハチョウ科）エチオピア区
imperial hairstreak			common imperial blue を見よ
imperial Jezebel			imperial white butterfly を見よ
imperial moth		*Eacles imperialis* (Drury)	（チョウ目、ヤママユガ科）新北区。幼虫は種々の木につく。亜種 *E. i. pini* Mitchener も同英名
imperial white butterfly		*Delias harpalyce* (Donovan)	（チョウ目、シロチョウ科）豪州区
imperials		*Cheritrella*	（チョウ目、シジミチョウ科）の昆虫の総称
imported birch leaf-miner			birch leaf-mining sawfly (2) を見よ　米国での英名
imported cabbage butterfy			common white を見よ　*P. rapae* Linnaeus の米国での英名。幼虫は imported cabbage worm
imported cabbage leaf-miner		*Scaptomyza graminum* (Fallén)	（ハエ目、ショウジョウバエ科）旧北区
imported cabbage maggot			cabbage root fly (2) を見よ　米国での英名
imported cabbage webworm			cabbage webworm (1) を見よ
imported cabbage worm			cabbage webworm (1) を見よ
imported clover weevil			clover seed weevil (3) を見よ
imported crucifer weevil		*Baris lepidii* Germar	（コウチュウ目、ゾウムシ科）新北区
imported currant borer			currant clearwing を見よ　米国での英名

英名	和名	学名	所属、分布、ほか
imported currant worm			gooseberry sawfly (1) を見よ　imported currantworm とも記す
imported fire ant		*Solenopsis saevissima richteri* Forel	（ハチ目、アリ科）新北区
imported fire ant (1)		*Solenopsis invicta* Buren	（ハチ目、アリ科）新北区、豪州区
imported fire ant			fire ant (1) を見よ
imported Klamathweed beetle			alligatorweed flea beetle を見よ
imported longhorned weevil	チビメナガゾウムシ	*Calomycterus setarius* Roelofs	（コウチュウ目、ゾウムシ科）日本、全北区
imported mealybug		*Pseudococcus importus* McKenzie	（カメムシ目、コナカイガラムシ科）新北区
imported onion maggot			onion maggot を見よ　米国での英名
imported pea moth			pea moth を見よ　米国での英名
imported poplar and willow leaf beetle			willow leaf beetle (1) を見よ
imported rose stem girdler			rose stem girdler を見よ
imported turnip leaf-miner		*Scaptomyza flaveola* (Meigen)	（ハエ目、ショウジョウバエ科）旧北区
imported willow and poplar borer			willow beetle を見よ
imported willow leaf beetle			willow leaf beetle (1) を見よ
impostor duskywing			blue-banded skipper を見よ
impressed dagger moth		*Acronicta impressa* Walker	（チョウ目、ヤガ科）新北区
impure ringlet		*Ypthima impura* Elwes et Edwards	（チョウ目、タテハチョウ科）エチオピア区
Ina grass dart		*Taractrocera ina* Waterhouse	（チョウ目、セセリチョウ科）豪州区
Ina skipper		*Methionopsis ina* (Plötz)	（チョウ目、セセリチョウ科）新熱帯区
Inca doctor			Inca metalmark (1) を見よ
Inca metalmark		*Aricoris incana* (Stichel)	（チョウ目、シジミタテハ科）新熱帯区
Inca metalmark (1)	キオビシジミタテハ	*Ancyluris inca inca* (Saunders)	（チョウ目、シジミタテハ科）新熱帯区
Inca skipper		*Vehilius inca* (Scudder)	（チョウ目、セセリチョウ科）新熱帯区
incense cedar sphinx		*Sphinx libocedrus* H. Edwards	（チョウ目、スズメガ科）新北区。incense cedar sphinx moth ともいう
incense cedar hairstreak			Nelson's hairstreak を見よ
incense-cedar wasp		*Syntexis libocedrii* Rohwer	（ハチ目、Anaxyelidae）新北区
inchman ant			bulldog ant (2) を見よ
inchmen			bulldog ants を見よ
inchmen jumper ants			bulldog ants を見よ
inchworm moths			geometer moths を見よ

英名	和名	学名	所属、分布、ほか
inchworms			geometer moths を見よ
included cordgrass borer moth		*Photedes includens* (Walker)	（チョウ目、ヤガ科）新北区
inconsolable underwing moth		*Catocala insolabilis* Guenée	（チョウ目、ヤガ科）新北区。insolable underwing ともいう
inconspicuos satyr		*Taygetis inconspicua* Draudt	（チョウ目、タテハチョウ科）新熱帯区
inconspicuous Euselasia		*Euselasia inconspicua* (Godman et Salvin)	（チョウ目、シジミタテハ科）新熱帯区
inconspicuous Pentila		*Pentila inconspicua* Druce	（チョウ目、シジミチョウ科）エチオピア区
inculta skipper		*Anthoptus inculta* (Dyar)	（チョウ目、セセリチョウ科）新熱帯区
incurvalids			yucca moths (1) を見よ
incurved shieldback		*Idionotus incurvus* Rentz et Birchim	（バッタ目、キリギリス科）新北区
Indamora glider		*Cymothoe indamora* (Hewitson)	（チョウ、タテハチョウ科）エチオピア区
indecorous eggar		*Bombycopsis indecora* (Walker)	（チョウ目、シャクガ科）エチオピア区
indefinite marble		*Argyroploce lacunana* (Denis et Schiffermüller)	（チョウ目、ハマキガ科）旧北区
indented Dichomeris moth		*Dichomeris inserrata* (Walsingham)	（チョウ目、キバガ科）新北区。indented Dichomeris ともいう
Indian ace		*Helpe homolea* (Hewitson)	（チョウ目、セセリチョウ科）東洋区
Indian aphid	ゴンズイフクレアブラムシ	*Indomegoura indica* (Van der Goot)	（カメムシ目、アブラムシ科）日本、旧北区、東洋区
Indian awlking		*Choaspes benjaminii* Guérin-Méneville	（チョウ目、セセリチョウ科）東洋区。*C. b. flavens* Eliot も同英名
Indian awlking			green velvet skipper を見よ
Indian Babul blue		*Azanus uranus* Butler	（チョウ目、シジミチョウ科）東洋区
Indian bee			giant honey bee を見よ
Indian bee			Indian honey bee を見よ
Indian borer			coffee white borer を見よ
Indian cabbage white	タイワンモンシロチョウ	*Pieris canidia* (Linnaeus)	（チョウ目、シロチョウ科）旧北区、東洋区
Indian clothes moth		*Trichophaga abruptella* Wollaston	（チョウ目、ヒロズコガ科）東洋区
Indian clothes moth		*Lobesia abscisana* Doubleday	（チョウ目、ハマキガ科）旧北区
Indian cotton leafhopper			okra leafhopper を見よ
Indian cupid		*Chilades parrhasius* (Fabricius)	（チョウ目、シジミチョウ科）東洋区
Indian cupid (1)	タイワンツバメシジミ	*Everes lacturnus kawaii* Matsumura	（チョウ目、シジミチョウ科）日本、東洋区。*E. l. rileyi* (Godfrey)、*E. l. assamica* (Tytler) も同英名。tailed cupid (1) を参照
Indian dart		*Potanthus pseudomaesa* (Moore)	（チョウ目、セセリチョウ科）東洋区

英名	和名	学名	所属、分布、ほか
Indian dingiest sailer			chocolate sailor を見よ
Indian duskhawker		*Gynacantha dravida* Lieftinck	(トンボ目、ヤンマ科) 東洋区
Indian flour moth			Mediterranean flour moth を見よ
Indian flower manntis		*Creobroter pictipennis* Wood-Mason	(カマキリ目、カマキリ科) 東洋区
Indian fritillary	ツマグロヒョウモン	*Argyreus hyperbius hyperbius* (Linnaeus)	(チョウ目、タテハチョウ科) 日本、旧北区、東洋区、豪州区、エチオピア区。*A. h. sumatrensis* Fruhstorfer も同英名
Indian fruit fly			Mediterranean fruit fly を見よ 米国での英名
Indian grizzled skipper			Indian skipper (1) を見よ
Indian gypsy moth			apple hairy caterpillar を見よ
Indian honey bee	インドミツバチ	*Apis cerana indica* Fabricius	(ハチ目、ミツバチ科) 東洋区
Indian house cricket	カマドコオロギ	*Gryllodes sigillatus* (Walker)	(バッタ目、コオロギ科) 日本、豪州区、大洋区
Indian jumping ant		*Harpegnathos saltator* (Jerdon)	(ハチ目、アリ科) 東洋区
Indian lac insect			lac insect を見よ
Indian leaf butterfly			orange oakleaf を見よ
Indian leafwing	ムラサキコノハチョウ	*Kallima paralekta* (Horsfield)	(チョウ目、タテハチョウ科) 東洋区
Indian mallow bug			hyaline grass bug を見よ
Indian map butterfly			common map を見よ
Indian meal moth (1)	ノシメマダラメイガ	*Plodia interpunctella* (Hübner)	(チョウ目、メイガ科) 日本、汎世界。シイタケ、貯穀害虫
Indian meal moth (2)		*Euzophera osseatella* Treitschke	(チョウ目、メイガ科) 汎世界
Indian monarch			common tiger (3) を見よ
Indian moon moth		*Actias selene* Hübner	(チョウ目、ヤママユガ科) 旧北区、東洋区。Indian luna moth ともいう。moon moth も参照
Indian oakblue		*Arhopala alax* (Evans)	(チョウ目、シジミチョウ科) 東洋区
Indian oakblue		*Arhopala alemon* (de Nicéville)	(チョウ目、シジミチョウ科) 東洋区
Indian orange tips	メスシロキチョウ属	*Ixias*	(チョウ目、シロチョウ科) の昆虫の総称
Indian painted lady	アカタテハ	*Vanessa indica* (Herbst)	(チョウ目、タテハチョウ科) 日本、東洋区
Indian palm bob	クロボシセセリ	*Suastus gremius gremius* (Fabricius)	(チョウ目、セセリチョウ科) 日本、東洋区
Indian plum Judy		*Abisara echerius suffusa* Moore	(チョウ目、シジミタテハ科) 東洋区
Indian purple emperor	シロコムラサキ	*Mimathyma ambica* (Kollar)	(チョウ目、タテハチョウ科) 東洋区
Indian purple hairstreak	ヒメムラサキミドリシジミ	*Esakiozephyrus bieti* (Oberthür)	(チョウ目、シジミチョウ科) 東洋区
Indian rat flea			Oriental rat flea を見よ

英名	和名	学名	所属、分布、ほか
Indian red admiral			Indian painted lady を見よ
Indian red flash		*Baspa melampus* (Stoll)	（チョウ目、シジミチョウ科）東洋区
Indian rockdweller		*Bradinopyga geminata* (Rambur)	（トンボ目、トンボ科）東洋区
Indian silkmoth			gorse silkworm を見よ
Indian skipper		*Hesperia sassacus sassacus* Harris	（チョウ目、セセリチョウ科）新北区
Indian skipper (1)		*Spialia galba* (Fabricius)	（チョウ目、セセリチョウ科）東洋区
Indian stick-insect		*Carausius morosus* Sinety	（ナナフシ目、コブナナフシ科）旧北区
Indian sugarcane leafhopper		*Pyrilla perpusilla* Walker	（カメムシ目、Lophopidae）東洋区
Indian sun hemp moth			crimson speckled を見よ
Indian sunbeam	マレーウラギンシジミ	*Curetis thetis* (Drury)	（チョウ目、シジミチョウ科）東洋区
Indian sweetpotato weevil			West Indian sweetpotato weevil を見よ
Indian toe-biters			giant water bugs (3) を見よ
Indian tortoiseshell	カシミールコヒオドシ	*Aglais cashmirensis* (Kollar)	（チョウ目、タテハチョウ科）東洋区
Indian tortoiseshell		*Nymphalis kaschmirensis* (Kollar)	（チョウ目、タテハチョウ科）東洋区
Indian wax scale	ツノロウムシ	*Ceroplastes ceriferus* (Fabricius)	（カメムシ目、カタカイガラムシ科）日本、汎世界。多くの果樹、園芸植物の害虫
Indian weed caterpillar		*Heliothis rubrescens* (Walker)	（チョウ目、ヤガ科）豪州区
Indian white admiral		*Limenitis trivena* Moore	（チョウ目、タテハチョウ科）東洋区
Indian white wax scale			Indian wax scale を見よ
Indian yellow-lined gem		*Libellago lineata indica* (Fraser)	（トンボ目、Chlorocyphidae）東洋区
Indies bush cricket		*Antillicharis oriobates* Otte et Perez Gelabert	（バッタ目、コオロギ科）新北区
Indies short-tailed cricket		*Anurogryllus celerinictus* Walker	（バッタ目、コオロギ科）新北区
indigo dropwing		*Trithemis festiva* (Rambur)	（トンボ目、トンボ科）東洋区
indigo elongate weevil	アイノカツオゾウムシ	*Lixus maculatus* Roelofs	（コウチュウ目、ゾウムシ科）日本、旧北区
indigo flash	クラルシジミ	*Rapala varuna* (Horsfield)	（チョウ目、シジミチョウ科）東洋区、豪州区。*R. v. simsoni* (Miskin), *R. v. orseis* Hewitson も同英名
indigo flea beetle	ミドリトビハムシ	*Crepidodera japonica* Baly	（コウチュウ目、ハムシ科）日本
indigo playboy		*Deudorix odana* Druce	（チョウ目、シジミチョウ科）エチオピア区
indigo stem borer moth		*Papaipema boptisiae* (Bird)	（チョウ目、ヤガ科）新北区
indigo weevil	タデサルゾウムシ	*Homorosoma asperum* (Roelofs)	（コウチュウ目、ゾウムシ科）日本、旧北区

英名	和名	学名	所属、分布、ほか
indigobush twig borer moth		*Hystrichophora taleuna* (Grote)	（チョウ目、ハマキガ科）新北区
indiscrete Cissusa moth		*Cissusa indiscreta* (Edwards)	（チョウ目、ヤガ科）新北区
indistinct angel moth		*Olceclostera indistincta* (H. Edwards)	（チョウ目、カイコガ科）新北区
Indochina mantis			giant Asian mantis (1) を見よ
Indo-Chinese labyrinth		*Neope pulahoides* (Moore)	（チョウ目、タテハチョウ科）東洋区
indomitable Melipotis moth		*Melipotis indomita* (Walker)	（チョウ目、ヤガ科）新北区、新熱帯区、大洋区
Indra swallowtail	タカネキアゲハ	*Papilio indra* Reakirt	（チョウ目、アゲハチョウ科）新北区。Indra swallowtail butterfly ともいう
Induna Acraea		*Telchinia induna* (Trimen)	（チョウ目、タテハチョウ科）エチオピア区
inept Drasteria moth		*Drasteria inepta* (H. Edwards)	（チョウ目、ヤガ科）新北区。inept Drasteria ともいう
infant		*Archiearis infans* (Möschler)	（チョウ目、シャクガ科）新北区。infant moth ともいう
ingrailed clay		*Diarsia mendica* (Fabricius)	（チョウ目、ヤガ科）旧北区
inimical borer moth		*Pseudogalleria inimicella* (Zeller)	（チョウ目、ハマキガ科）新北区
inkblot Palpita moth		*Palpita illibalis* (Hübner)	（チョウ目、メイガ科）新北区
inky skipper	ホソバマダラセセリ	*Erynnis marloyi* (Boisduval)	（チョウ目、セセリチョウ科）旧北区
inland armyworm		*Persectania dyscrita* Common	（チョウ目、ヤガ科）豪州区
inland field cricket		*Lepidogryllus parvalus* (Walker)	（バッタ目、コオロギ科）豪州区
inland floodwater mosquito	キンイロヤブカ	*Aedes vexans* (Meigen)	（ハエ目、カ科）日本、旧北区
inland green tree hopper			common garden katidid を見よ
inland hairstreak		*Jalmenus aridus* Graham et Moulds	（チョウ目、シジミチョウ科）豪州区
inland hunter		*Austrogomphus australis* Dale	（トンボ目、サナエトンボ科）豪州区
inland katydid			common garden katydid を見よ
inocellid snakeflies			snake-flies (2) を見よ
inocellids			snake-flies (2) を見よ
inornate moth			bella moth を見よ
inornate Olethreutes moth		*Olethreutes inornatana* (Clemens)	（チョウ目、ハマキガ科）新北区
inornate Pyrausta moth		*Pyrausta inornatalis* (Fernald)	（チョウ目、メイガ科）新北区、新熱帯区
inornate ringlet		*Coenonympha tullia inornata* W. H. Edwards	（チョウ目、タテハチョウ科）新北区。large heath を参照
inornate ringlet			prairie ringlet を見よ

英名	和名	学名	所属、分布、ほか
inornate scale		*Aonidiella inornata* McKenzie	（カメムシ目、マルカイガラムシ科）新北区
Inous blue		*Jalmenus inous* Hewitson	（チョウ目、シジミチョウ科）豪州区。Inous blue butterfly ともいう
inquiline gall wasps		Synerginae	（ハチ目、タマバチ科）の昆虫の総称
inquisitive monkey		*Janomima westwoodi* Aurivillius	（チョウ目、オビガ科）エチオピア区
insects	昆虫類	Insecta	六脚類 Hexapoda との術語もある
insidious flower bug		*Orius insidiosus* (Say)	（カメムシ目、ハナカメムシ科）新北区
inspector		*Chalcostephia flavifrons* Kirby	（トンボ目、トンボ科）エチオピア区
interior least clubtail		*Stylogomphus sigmastylus* Cook et Laudermilk	（トンボ目、サナエトンボ科）新北区
intermediate bushbrown		*Mycalesis intermedia* (Moore)	（チョウ目、タテハチョウ科）東洋区
intermediate maplet		*Chersonesia intermedia* Martin	（チョウ目、タテハチョウ科）東洋区
intermediate skippers		Hesperiinae	（チョウ目、セセリチョウ科）の昆虫の総称
intermediate sphinx moth		*Eumorpha intermedia* (Clark)	（チョウ目、スズメガ科）新北区。intermediate sphinx ともいう
international swallowtail			common yellow swallowtail を見よ　*P. machaon* L. の英名
internodal borer		*Chilo sacchariphagus indicus* (Kapur)	（チョウ目、メイガ科）東洋区。イネ害虫
interoceanic ear moth		*Amphipoea interoceanica* Smith	（チョウ目、ヤガ科）新北区
interrupted dagger moth		*Acronicta interrupta* Guenée	（チョウ目、ヤガ科）新北区
intertidal dwarf bugs	サンゴカメムシ科	Omaniidae	（カメムシ目）の昆虫の総称
introduced basswood thrips		*Thrips calcaratus* Uzel	（アザミウマ目、アザミウマ科）旧北区
introduced coffee bean weevil			coffee weevil を見よ
introduced pine sawfly		*Diprion similis* (Hartig)	（ハチ目、マツハバチ科）全北区
invasive flea beetle		*Luperomorpha xanthodera* (Fairmaire)	（コウチュウ目、ハムシ科）旧北区
invasive garden ant		*Lasius neglectus* Van Loon, Boomsma et Andrasfalvy	（ハチ目、アリ科）旧北区
Invermay bug		*Nysius turneri* Evans	（カメムシ目、ナガカメムシ科）豪州区。オーストラリアの地名に由来
inverness gold-dot bentwing	ポプラシロハモグリ	*Proleucoptera sinuella* (Reutti)	（チョウ目、ハモグリガ科）旧北区
inverted Y slug moth			yellow-collared slug moth を見よ
inyo shieldback		*Idiostatus inyo* Rehn et Hebard	（バッタ目、キリギリス科）新北区
Io			Io moth を見よ　ギリシャ神話のイオに由来
Io moth		*Automeris io* (Fabricius)	（チョウ目、ヤママユガ科）新北区

英名	和名	学名	所属、分布、ほか
Iolas blue		*Iolana iolas* (Ochsenheimer)	（チョウ目、シジミチョウ科）旧北区
Iphicleola sister		*Adelpha iphicleola iphicleola* (Bates)	（チョウ目、タテハチョウ科）新熱帯区
Iphiclus sister		*Adelpha iphiclus* (Linnaeus)	（チョウ目、タテハチョウ科）新熱帯区
Iphidamas cattleheart	シロモンジャコウアゲハ	*Parides iphidamas iphidamas* (Fabricius)	（チョウ目、アゲハチョウ科）新熱帯区
ipsilon dart			black cutworm を見よ　ipsilon dart moth ともいう。
Ira hairstreak		*Michaelus ira* Hewitson	（チョウ目、シジミチョウ科）新熱帯区
irapuan		*Melipona ruficrus* Latreille	（ハチ目、ミツバチ科）新熱帯区
Irenea metalmark		*Thisbe irenea* Stoll	（チョウ目、シジミタテハ科）新熱帯区
Irene's opal		*Chrysoritis irene* (Pennington)	（チョウ目、シジミチョウ科）エチオピア区
iridescent centurion		*Sargus iridatus* (Scopoli)	（ハエ目、ミズアブ科）旧北区
iridescent club-horned sawfly	アカガネコンボウハバチ	*Abia iridescens* Marlatt	（ハチ目、コンボウハバチ科）日本
iridescent Strobisia moth		*Strobisia iridipennella* Clemens	（チョウ目、キバガ科）新北区
iris aphid		*Aphis newtoni* Theobald	（カメムシ目、アブラムシ科）旧北区
iris aphid			tulip aphid を見よ
iris borer		*Macronoctua onusta* Grote	（チョウ目、ヤガ科）新北区。成虫は　iris borer moth
iris-eyed silkmoth		*Automeris iris* (Walker)	（チョウ目、ヤママユガ科）新北区
iris flea beetle		*Aphthona nonstriata* (Goeze)	（コウチュウ目、ハムシ科）旧北区
iris gelechiid	アヤメキバガ	*Monochroa* sp.	（チョウ目、キバガ科）日本
iris leaf miner		*Cerodontha ireos* (Goureau)	（ハエ目、ハモグリバエ科）旧北区
iris leaf miner		*Dizygomyza iridis* Hendel	（ハエ目、ハモグリバエ科）旧北区
iris leafminer	ヒオウギハモグリバエ	*Amauromyza belamcandae* Sasakawa	（ハエ目、ハモグリバエ科）日本
iris leafminer		*Cerodontha iraeos* (Robineau-Desvoidy)	（ハエ目、ハモグリバエ科）日本
iris leafminer	ヒメアヤメハモグリバエ	*Cerodontha iridicola* (Koizumi)	（ハエ目、ハモグリバエ科）日本。アヤメ害虫
iris sawfly		*Rhadinoceraca micans* (Klug)	（ハチ目、ハバチ科）旧北区
iris skipper		*Arrhenes dshillus iris* (Waterhouse)	（チョウ目、セセリチョウ科）豪州区
iris thrips		*Iridothrips iridis* (Watson)	（アザミウマ目、アザミウマ科）全北区
iris weevil		*Mononychus vulpeculus* (Fabricius)	（コウチュウ目、ゾウムシ科）新北区
iris whitefly		*Aleyrodes spiraeoides* Quaintance	（カメムシ目、コナジラミ科）新北区、新熱帯区
Irish bluet			crescent bluet を見よ

英名	和名	学名	所属、分布、ほか
Irish burnet		*Zygaena minos* (Denis et Schiffermüller)	（チョウ目、マダラガ科）旧北区
Irish damselfly			crescent bluet を見よ
Irish major		*Oxycera fallenii* Staeger	（ハエ目、ミズアブ科）旧北区
Irish winter gnat		*Trichocera fuscata* Meigen	（ハエ目、ガガンボダマシ科）旧北区
Irmina sister		*Adelpha irmina* Doubleday	（チョウ目、タテハチョウ科）新熱帯区
iron blue		*Alainites muticus* (Linnaeus)	（カゲロウ目、コカゲロウ科）旧北区
iron prominent		*Notodonta dromodarius* (Linnaeus)	（チョウ目、シャチホコガ科）旧北区
iron Rajah	マルスフタオチョウ	*Charaxes mars* Staudinger	（チョウ目、タテハチョウ科）東洋区
ironbark beetle		*Zopherosis georgei* White	（コウチュウ目、Zopheridae）豪州区。ironbark borer ともいう
ironbark lace lerp		*Cardiaspina vittaformis* (Froggatt)	（カメムシ目、キジラミ科）豪州区
ironbark sawfly		*Lophyrotoma analis* (Costa)	（ハチ目、Pergidae）豪州区
ironclad beetle		*Phloeodes diabolicus* LeConte	（コウチュウ目、ゴミムシダマシ科）新北区
ironclad beetle		*Phloeodes pustulosus* (LeConte)	（コウチュウ目、ゴミムシダマシ科）新北区
ironclad beetle		*Zopherus haldemani* Horn	（コウチュウ目、Zopheridae）新北区
ironclads		*Phloeodes*	（コウチュウ目、ゴミムシダマシ科）の昆虫の総称
ironweed borer moth		*Papaipema cerussata* (Grote)	（チョウ目、ヤガ科）新北区
ironweed clearwing moth		*Carmenta bassiformis* Walker	（チョウ目、スカシバガ科）新北区
ironweed root moth		*Polygrammodes flavidalis* (Guenée)	（チョウ目、メイガ科）新北区
ironwood leafminer moth		*Stilbosis ostryaeella* (Chambers)	（チョウ目、カザリバガ科）新北区
ironwood tubemaker moth		*Acrobasis sylviella* Ely	（チョウ目、メイガ科）新北区
irregular-marked noctuid	ウンモンクチバ	*Mocis annetta* (Walker)	（チョウ目、ヤガ科）日本、旧北区、東洋区
irregular striped hawk		*Hippotion irregularis* (Walker)	（チョウ目、スズメガ科）エチオピア区
irregularity marked leaf bug	トビマダラカスミカメムシ	*Phytocoris nowickyi* Fieber	（カメムシ目、カスミカメムシ科）日本、旧北区
irrorated white neb		*Psamathocrita osseella* (Stainton)	（チョウ目、キバガ科）旧北区
Irumu ciliate blue		*Anthene irumu* (Stempffer)	（チョウ目、シジミチョウ科）エチオピア区
Irving's blue		*Lepidochrysops irvingi* (Swanepoel)	（チョウ目、シジミチョウ科）エチオピア区
Isabella moth			banded woollybear を見よ
Isabella tiger-moth			banded woollybear を見よ

英名	和名	学名	所属、分布、ほか
Isabella's heliconian			Heliconius mimic を見よ
Isabella's longwing		*Eueides isabella eva* (Fabricius)	(チョウ目、タテハチョウ科) 新熱帯区。*E. i. nigricornis* Maza も同英名
Isabella's longwing			Heliconius mimic を見よ
Isabelline tiger			banded woollybear を見よ
ischnocerans			ischnocerous lice を見よ
ischnocerous lice		*Ischnocera*	(ハジラミ目) の昆虫の総称
Ishigaki whitefly	イシガキコナジラミ	*Aleurotrachelus ishigakiensis* (Takahashi)	(カメムシ目、コナジラミ科) 日本
Isidora leafwing		*Zaretis isidora* (Cramer)	(チョウ目、タテハチョウ科) 新熱帯区
Isis blue		*Azanus isis* (Drury)	(チョウ目、シジミチョウ科) エチオピア区
island bluetail		*Ischnura genei* (Rambur)	(トンボ目、イトトンボ科) 旧北区
island fruit fly		*Rioxa musae* Froggatt	(ハエ目、ミバエ科) 豪州区。カンキツ害虫
island fruit fly (1)		*Diroxa pornia* (Walker)	(ハエ目、ミバエ科) 東洋区、豪州区
island pinhole borer	フィリッピンザイノキクイムシ	*Xyleborus perforaus* (Wollaston)	(コウチュウ目、キクイムシ科) 日本、東洋区、豪州区
island stinkbug		*Oebalus insularis* (Stål)	(カメムシ目、カメムシ科) 新熱帯区
island swallowtail			Schaus swallowtail を見よ
Isle of Wight neb		*Metzneria littorella* (Douglas)	(チョウ目、キバガ科) 旧北区
Isle of Wight wainscot			large wainscot を見よ
Isle of Wight wave		*Idaea humiliata* (Hufnagel)	(チョウ目、シャクガ科) 旧北区
Ismare tiger	トガリシロオビマダラ	*Danaus ismare* Cramer	(チョウ目、タテハチョウ科) 東洋区
isotomid springtails			smooth springtails を見よ
Isshiki orneodid	ニジュウシトリバ	*Alucita spilodesma* (Meyrick)	(チョウ目、ニジュウシトリバガ科) 日本。一色博士は Issiki を使用
issid planthoppers	マルウンカ科	*Issidae*	(カメムシ目) の昆虫の総称
Issiki xylorictid	イッシキフタホシコガ	*Odites isshikii* (Takahashi)	(チョウ目、ヒゲナガキバガ科) 日本、旧北区
Istar sphinx		*Sphinx istar* (Rothschild et Jordan)	(チョウ目、スズメガ科) 新北区、新熱帯区。Istar sphinx moth ともいう
Italian bee			Italian honey bee を見よ
Italian beetle			house longhorn を見よ
Italian cricket			tree cricket (1) を見よ
Italian goldenring		*Cordulegaster trinacriae* Waterston	(トンボ目、オニヤンマ科) 旧北区
Italian honey bee		*Apis mellifica ligustica* Spinola	(ハチ目、ミツバチ科) 旧北区
Italian locust		*Calliptamus italicus* (Linnaeus)	(バッタ目、バッタ科) 旧北区
Italian marbled white		*Melanargia arge* (Sulzer)	(チョウ目、タテハチョウ科) 旧北区
Italian pear scale		*Epidiaspis leperii* (Signoret)	(カメムシ目、マルカイガラムシ科) 全北区。クルミ害虫

英名	和名	学名	所属、分布、ほか
Itatiaia owlet		*Dasyophthalma rusina* Latreille et Godart	（チョウ目、タテハチョウ科）新熱帯区
Ithomioides crescent			variable mimic (1) を見よ
ithonid lacewings			moth lacewings を見よ
Ituri fairy playboy		*Paradeudorix ituri* (Bethune-Baker)	（チョウ目、シジミチョウ科）エチオピア区
Itys bush brown		*Mycalesis itys* C. et R. Felder	（チョウ目、タテハチョウ科）東洋区
Itys leafwing		*Zaretis itys itys* (Cramer)	（チョウ目、タテハチョウ科）新熱帯区
ivana leafroller moth		*Argyrotaenia ivana* (Fernald)	（チョウ目、ハマキガ科）新北区
ivory marked beetle		*Eburia quadrigeminata* (Say)	（コウチュウ目、カミキリムシ科）新北区、大洋区。ivory marked borer ともいう
ivory pintail		*Acisoma trifidum* Kirby	（トンボ目、トンボ科）新北区
ivy and hop weevil		*Liophloeus tessulatus* (Müller)	（コウチュウ目、ゾウムシ科）旧北区
ivy aphid		*Aphis hederae* Kaltenbach	（カメムシ目、アブラムシ科）全北区、豪州区
ivy bark beetle		*Kissophagus hederae* (Schmitt)	（コウチュウ目、キクイムシ科）旧北区
ivy boring beetle		*Ochina ptinoides* (Marsham)	（コウチュウ目、シバンムシ科）旧北区
ivy lacebug		*Derephysia foliacea* (Fallén)	（カメムシ目、グンバイムシ科）旧北区
ivy leafroller		*Cryptoptila immersana* (Walker)	（チョウ目、ハマキガ科）豪州区
ivy scale		*Aspidiotus nerii* (Bouché)	（カメムシ目、マルカイガラムシ科）新北区、豪州区
ivy scale			oleander scale を見よ
ixora whitefly		*Aleurotuberculatus minutus* (Singh)	（カメムシ目、コナジラミ科）東洋区、大洋区

英名	和名	学名	所属、分布、ほか

J

英名	和名	学名	所属、分布、ほか
Jacarand bug		*Orthezia insignis* Browne	(カメムシ目、ハカマカイガラムシ科) 東洋区、全北区、豪州区、エチオピア区。米国に移入
jack-borer			June beetle (1) を見よ
jack jumper ant		*Myrmecia pilosula* (F. Smith)	(ハチ目、アリ科) 豪州区
Jack pine budworm		*Choristoneura pinus* Freeman	(チョウ目、ハマキガ科) 新北区。Jack pine budworm moth ともいう
Jack pine false looper			washed-out Zale moth を見よ
Jack pine resin midge		*Cecidomyia resinicola* (Osten Sacken)	(ハエ目、タマバエ科) 新北区。Jack pine midge ともいう
Jack pine sawfly		*Neodiprion pratti banksianae* Rohwer	(ハチ目、マツハバチ科) 新北区
Jack pine tip beetle		*Conophthrous banksianae* McPherson	(コウチュウ目、キクイムシ科) 新北区
Jack pine tip moth		*Rhyacionia granti* Miller	(チョウ目、ハマキガ科) 新北区
Jack pine tube moth		*Argyrotaenia tabulala* Freeman	(チョウ目、ハマキガ科) 新北区
jackal flies	クロコバエ科	Milichiidae	(ハエ目) の昆虫の総称
jackfruit fly		*Dacus umbrosus* Linnaeus	(ハエ目、ミバエ科) 東洋区
jackfruit fruit borer		*Glyphodes caesalis* Walker	(チョウ目、メイガ科) 東洋区
jackfruit longhorn beetle		*Apriona germarii* Hope	(コウチュウ目、カミキリムシ科) 東洋区
Jackquin's drill		*Dichrorampha alpinana* (Treitschke)	(チョウ目、ハマキガ科) 旧北区
Jackson's cupid		*Euchrysops reducta* Hulstaert	(チョウ目、シジミチョウ科) エチオピア区
Jackson's dotted border		*Mylothris jacksoni* Sharpe	(チョウ目、シロチョウ科) エチオピア区
Jackson's leaf butterfly		*Mallika jacksoni* (Sharpe)	(チョウ目、タテハチョウ科) エチオピア区
Jackson's playboy		*Deudorix jacksoni* Talbot	(チョウ目、シジミチョウ科) エチオピア区
Jackson's sapphire		*Iolaus jacksoni* (Stempffer)	(チョウ目、シジミチョウ科) エチオピア区
Jackson's silver spot		*Aphnaeus jacksoni* Stempffer	(チョウ目、シジミチョウ科) エチオピア区
Jackson's swallowtail		*Papilio jacksoni* Sharpe	(チョウ目、アゲハチョウ科) エチオピア区
jacky breeze			dragonflies (1) を見よ
Jacqueline's gemmed-satyr		*Cyllopsis jacquelineae* Miller	(チョウ目、タテハチョウ科) 新熱帯区
Jacques' beetle		*Subcoccinella vigintiquatuorpunctata* (Linnaeus)	(コウチュウ目、テントウムシ科) 新北区
jade-blue hairstreak		*Arawacus jada* (Hewitson)	(チョウ目、シジミチョウ科) 新北区
jade clubtail		*Arigomphus submedianus* (Williamson)	(トンボ目、サナエトンボ科) 新北区
jade flies			dog-day cicadas を見よ
jade hunter		*Austrogomphus ochraceus* (Selys)	(トンボ目、サナエトンボ科) 豪州区

英名	和名	学名	所属、分布、ほか
jagged border skipper			Juba skipper を見よ
jagged-edged yellow		*Eurema gratiosa* Doubleday	（チョウ目、シロチョウ科）新熱帯区
jaguar flower moth		*Schinia jaguarina* Guenée	（チョウ目、ヤガ科）新北区、新熱帯区
Jair underwing moth		*Catocala jair* Strecker	（チョウ目、ヤガ科）新北区。Jair underwing ともいう
Jakama hairstreak	ジャカマミドリシジミ	*Chrysozephyrus jakamensis* (Tytler)	（チョウ目、シジミチョウ科）東洋区
Jalapus cloudywing		*Thessia jalapus* (Plötz)	（チョウ目、セセリチョウ科）新北区
Jamaican field cricket		*Gryllus assimilis* (Fabricius)	（バッタ目、コオロギ科）新北区
Jamaican monarch		*Danaus cleophile* (Godart)	（チョウ目、タテハチョウ科）新熱帯区
Jamaican sister		*Adelpha abyla* (Hewitson)	（チョウ目、タテハチョウ科）新熱帯区
Jameson's large sailer		*Neptis jamesoni* Godman et Salvin	（チョウ目、タテハチョウ科）エチオピア区
Janae's Ymeldia moth		*Ymeldia janae* Hodges	（チョウ目、キバガ科）新北区
Janais patch			crimson patch を見よ
Janardana bush brown			common bush brown (1) を見よ
Janetta Themis forester		*Euphaedra janetta* (Butler)	（チョウ目、タテハチョウ科）エチオピア区
Janias greenstreak		*Chalybs janias* (Cramer)	（チョウ目、シジミチョウ科）新熱帯区
Janse's giant cupid		*Lepidochrysops jansei* van Someren	（チョウ目、シジミチョウ科）エチオピア区
Janse's sphinx		*Xenosphingia jansei* Jordan	（チョウ目、スズメガ科）エチオピア区。Janse's sphinx hawk ともいう
Janse's widow		*Dira jansei* (Swierstra)	（チョウ目、タテハチョウ科）エチオピア区
Janthina hairstreak		*Janthecla janthina* (Hewitson)	（チョウ目、シジミチョウ科）新熱帯区
Janthodonia hairstreak		*Janthecla janthodonia* (Dyar)	（チョウ目、シジミチョウ科）新熱帯区
January Melipotis moth		*Melipotis januaris* Guenée	（チョウ目、ヤガ科）新北区、新熱帯区
Japan bark beetle			Japanese bark beetle を見よ
Japan club-horned sawfly	ルリコンボウハバチ	*Orientabia japonica* (Cameron)	（ハチ目、コンボウハバチ科）日本、旧北区
Japan leaf-cutting bee	ヤマトハキリバチ	*Megachile japonica* Alfken	（ハチ目、ハキリバチ科）日本
Japan soldier fly	ミズアブ	*Stratiomys japonica* (van der Wulp)	（ハエ目、ミズアブ科）日本、旧北区
Japanese alder broad-head leafhopper	マエキヒロズヨコバイ	*Oncopsis flavicollis* (Linnaeus)	（カメムシ目、オオヨコバイ科）日本、旧北区
Japanese alder elongate bark beetle	ハンノナガコキクイムシ	*Ernoporus longus* Eggers	（コウチュウ目、キクイムシ科）日本
Japanese alder elongate leafhopper			alder leafhopper（1）を見よ

英名	和名	学名	所属、分布、ほか
Japanese alder geometrid			large thorn を見よ
Japanese alder horntail	ハンノクビナガキバチ	*Xiphydria alnivora* Matsumura	(ハチ目、クビナガキバチ科) 日本、旧北区
Japanese alder leafhopper			fruit-tree leafhopper を見よ
Japanese alder maculated aphid	ハンノブチアブラムシ	*Symydobius alniaria* (Matsumura)	(カメムシ目、アブラムシ科) 日本、旧北区
Japanese alder pyralid	ヨスジノメイガ	*Pagyda quadrilineata* Butler	(チョウ目、メイガ科) 日本、東洋区
Japanese alder round scale			trilobe scale を見よ
Japanese alder sawfly			alder sawfly (1)(2) を見よ
Japanese alder scale	ハンノキシロカイガラムシ	*Chionaspis alnus* Kuwana	(カメムシ目、マルカイガラムシ科) 日本、旧北区
Japanese alder spanworm	ハンノトビスジエダシャク	*Aethalura ignobilis* (Butler)	(チョウ目、シャクガ科) 日本
Japanese alder sucker			alder psylla を見よ
Japanese alder tetocallis	ケヤキブチアブラムシ	*Tinocallis zelkowae* (Takahashi)	(カメムシ目、アブラムシ科) 日本、旧北区、東洋区
Japanese alder xylococcus	ハンノモグリカイガラムシ	*Xylococcus japonicus* Oguma	(カメムシ目、ワタフキカイガラムシ科) 日本、旧北区
Japanese alpine	ベニヒカゲ	*Erebia niphonica niphonica* Janson	(チョウ目、タテハメチョウ科) 日本
Japanese apple leaf miner			Northants bentwing を見よ
Japanese awlking			green velvet skipper を見よ
Japanese bagberry whitefly			myrica whitefly を見よ　米国での英名
Japanese bamboo shot-hole beetle			Japanese shothole borer を見よ
Japanese bark beetle	ニホンキクイムシ	*Scolytus japonicus* Chapuis	(コウチュウ目、キクイムシ科) 日本、旧北区、東洋区。果樹、サクラ、チャの害虫
Japanese basket osier lacebug	コリヤナギグンバイ	*Cysteochia salicorum* (Baba)	(カメムシ目、グンバイムシ科) 日本
Japanese beech cottony aphid			woolly beech aphid を見よ
Japanese beech gall midge	ブナハスジトガリタマバエ	*Mikiola fagi* (Hartig)	(ハエ目、タマバエ科) 日本、旧北区
Japanese beetle	マメコガネ	*Popillia japonica* Newman	(コウチュウ目、コガネムシ科) 日本、全北区。1913年に日本から米国に入った害虫
Japanese bibionid	メスアカケバエ	*Bibio rufiventris* (Duda)	(ハエ目、ケバエ科) 日本、旧北区
Japanese black rice bug			black rice bug を見よ
Japanese broad-winged katydid	クダマキモドキ	*Holochlora japonica* Brunner von Wattenwyl	(バッタ目、キリギリス科)、日本、新北区
Japanese broad-winged planthopper	ベッコウハゴロモ	*Orosanga japonicus* (Melichar)	(カメムシ目、ハゴロモ科) 日本、旧北区、東洋区
Japanese burrowing cricket			mikado okame cricket を見よ

英名	和名	学名	所属、分布、ほか
Japanese butterbur aphid	フキアブラムシ	*Aphis fukii* Shinji	(カメムシ目、アブラムシ科) 日本、旧北区、東洋区
Japanese butterbur lacebug	エグリグンバイ	*Cochiochila conchata* (Matsumura)	(カメムシ目、グンバイムシ科) 日本、旧北区
Japanese cabbage white			common white を見よ
Japanese camel cricket	マダラカマドウマ	*Diestrammena marmorata* de Haan	(バッタ目、カマドウマ科) 日本、全北区、東洋区
Japanese camellia scale			greedy scale (1) を見よ
Japanese carrion beetle	オオヒラタシデムシ	*Eusilpha japonica* (Motschulsky)	(コウチュウ目、シデムシ科) 日本
Japanese cedar bark midge			cryptomeria bark midge を見よ
Japanese cedar gall midge			cryptomeria needle gall midge を見よ
Japanese cedar mealybug			pear mealybug を見よ
Japanese cedar scale	スギシロホシカイガラムシ	*Cryptoparlatorea leucaspis* Linnaeus	(カメムシ目、マルカイガラムシ科) 日本
Japanese cedar torymid			cryptomeria torymid を見よ
Japanese cherry fruit fly	オウトウハマダラミバエ	*Rhacochlaena japonica* Ito	(ハエ目、ミバエ科) 日本、旧北区。オウトウ害虫
Japanese cherry tree borer			cherry tree borer (1) を見よ
Japanese chrysanthemum aphid			mugwort long-horned aphid を見よ
Japanese chrysanthemum gall midge	キクヒメタマバエ	*Rhopalomyia chrysanthemum* Monzen	(ハエ目、タマバエ科) 日本。キク害虫
Japanese citrus flower-bud midge	ミカンツボミタマバエ	*Contarinia okadai* (Miyoshi)	(ハエ目、タマバエ科) 日本。カンキツ害虫
Japanese citrus thrips	トゲナシクダアザミウマ	*Ecacanthothrips inarmatus* Kurosawa	(アザミウマ目、クダアザミウマ科) 日本
Japanese clouded apollo	ウスバシロチョウ	*Parnassius glacialis* Butler	(チョウ目、アゲハチョウ科) 日本、旧北区
Japanese cockchafer	コフキコガネ	*Melolontha japonica* Burmeister	(コウチュウ目、コガネムシ科) 日本。果樹、スギなどの害虫
Japanese cockroach	ヤマトゴキブリ	*Periplaneta japonica* Karny	(ゴキブリ目、ゴキブリ科) 日本
Japanese conifer scale			conifer scale を見よ
Japanese cottonwood aphid	ドロハケアブラムシ	*Chaitophorus dorocolus* Matsumura	(カメムシ目、アブラムシ科) 日本、旧北区
Japanese cottonwood looper	オオハガタナミシャク	*Ecliptopera umbrosaria* (Motschulsky)	(チョウ目、シャクガ科) 日本、旧北区
Japanese cottonwood pyralid	ヒメアカマダラメイガ	*Nephopterix adelphella* (Fischer von Röslerstamm)	(チョウ目、メイガ科) 日本、旧北区

英名	和名	学名	所属、分布、ほか
Japanese cucumber tree aphid	ホオノキヒゲナガアブラムシ	*Neocalaphis magnolicolens* (Takahashi)	(カメムシ目、アブラムシ科) 日本
Japanese death's head			death's head を見よ
Japanese dogwood aphid			dogwood aphid を見よ
Japanese dor beetle			Japanese dung beetle (1) を見よ
Japanese drepanid	ヤマトカギバ	*Nordstromia japonica* (Moore)	(チョウ目、カギバガ科) 日本、旧北区
Japanese dung beetle	ミヤマダイコクコガネ	*Copris pecuarius* Lewis	(コウチュウ目、コガネムシ科) 日本、旧北区
Japanese dung beetle (1)	オオセンチコガネ	*Geotrupes auratus* Motschulsky	(コウチュウ目、センチコガネ科) 日本、全北区
Japanese elder aphid	モクレンヒゲナガマダラアブラムシ	*Neocalaphis magnoliae* (Essig et Kuwana)	(カメムシ目、アブラムシ科) 日本、旧北区
Japanese elm bark beetle			elm bark beetle (3) を見よ
Japanese elm cutworm			lesser spotted pinion を見よ
Japanese elm fall canker worm	ホソウスバフユシャク	*Inurois tenuis* Butler	(チョウ目、シャクガ科) 日本、旧北区
Japanese elm leafroller	ニレハマキ	*Acleris boscana ulmicola* (Meyrick)	(チョウ目、ハマキガ科) 日本、旧北区。リンゴ、バラ害虫
Japanese elm moth	アトベリクチブサガ	*Ypsolopha vittelus* (Linnaeus)	(チョウ目、スガ科) 日本
Japanese elm sawfly	ニレチュウレンジ	*Arge captiva* (Smith)	(ハチ目、ミフシハバチ科) 日本、旧北区
Japanese elm woolly aphid	ニレワタムシ	*Eriosoma japonicum* (Matsumura)	(カメムシ目、アブラムシ科) 日本、旧北区
Japanese emperor	オオムラサキ	*Sasakia charonda charonda* (Hewitson)	(チョウ目、タテハチョウ科) 日本
Japanese eriococcus	キフクロカイガラムシ	*Eriococcus japonicus* Kuwana	(カメムシ目、フクロカイガラムシ科) 日本
Japanese exoristine tachinid			Japanese tachina fly を見よ
Japanese ficus agaonid	イヌビワコバチ	*Blastophaga nipponica* Grandi	(ハチ目、イチジクコバチ科) 日本
Japanese firefly			Genji firefly を見よ
Japanese flatheaded borer			Japanese jewel beetle を見よ
Japanese flea cricket			pygmy mole cricket を見よ
Japanese flower beetle			Japanese beetle を見よ．
Japanese fruit scale			white peach scale を見よ　アジアでの英名
Japanese garden fleahopper	オオクロトビカスミカメムシ	*Ectometopterus micantulus* (Horvath)	(カメムシ目、カスミカメムシ科) 日本
Japanese genji firefly			Genji firefly を見よ
Japanese giant hornet			Asian giant hornet を見よ
Japanese giant silk moth			giant silk moth を見よ

英名	和名	学名	所属、分布、ほか
Japanese giant weevil	オオゾウムシ	*Sipalinus gigas* (Fabricius)	(コウチュウ目、オサゾウムシ科) 日本、旧北区、東洋区
Japanese grain moth			stored nut moth を見よ　米国での英名
Japanese grasshopper			rice grasshopper (1) を見よ
Japanese grasshopper			greenhouse camel cricket を見よ
Japanese grasshoppers		*Oxya*	(バッタ目、バッタ科) 米国での英名
Japanese green-lined albizzia longicorn			two-lined albizzia longhorn を見よ
Japanese ground beetle	マイマイカブリ	*Damaster blaptoides* Kollar	(コウチュウ目、オサムシ科) 日本。貝を捕食する
Japanese ground grasshopper	ツチイナゴ	*Patanga japonica* (Bolivar)	(バッタ目、バッタ科) 日本、旧北区、東洋区
Japanese grouse locust	ヒシバッタ	*Tetrix japonica* (Bolivar)	(バッタ目、ヒシバッタ科) 日本、旧北区
Japanese hackberry cimbex	ホシアシブトハバチ	*Agenocimbex jucunda* (Mocsary)	(ハチ目、コンボウハバチ科) 日本
Japanese hackberry oystershell scale	エノキカキカイガラムシ	*Lepidosaphes celtis* Kuwana	(カメムシ目、マルカイガラムシ科) 日本
Japanese hazelnut leaf roller			chequered fruit-tree tortrix を見よ
Japanese hazelnut looper			June highflier を見よ
Japanese hemlock scale			hemlock scale (2) を見よ
Japanese holly gall midge	イヌツゲタマバエ	*Asteralobia sasakii* (Monzen)	(ハエ目、タマバエ科) 日本
Japanese holly looper	マエキトビエダシャク	*Nothomiza formosa* (Butler)	(チョウ目、シャクガ科) 日本
Japanese honey bee	ニホンミツバチ	*Apis cerana japonica* Radoszkowski	(ハチ目、ミツバチ科) 日本。野生のミツバチ
Japanese honey suckle cimbex	ネジロコンボウハバチ	*Zaraea metallica* Mocsary	(ハチ目、コンボウハバチ科) 日本、旧北区
Japanese horned aphid	ヤマトツノアブラムシ	*Ceratovacuna japonica* (Takahashi)	(カメムシ目、アブラムシ科) 日本
Japanese horned beetle	カブトムシ	*Allomyrina dichotoma dichotoma* Linnaeus	(コウチュウ目、コガネムシ科) 日本、旧北区。日本を代表するコガネムシ
Japanese horntail	ニホンキバチ	*Urocerus japonicus* (Smith)	(ハチ目、キバチ科) 日本、旧北区。モミ、マツ類害虫
Japanese jewel beetle	ヤマトタマムシ	*Chrysochroa fulgidissima* (Schoenherr)	(コウチュウ目、タマムシ科) 日本、旧北区、東洋区
Japanese katydid	キリギリス	*Gampsocleis buergeri* (de Haan)	(バッタ目、キリギリス科)、日本。草原に普通
Japanese larch aphid	エゾマツカサアブラムシ	*Adelges japonicus* (Monzen)	(カメムシ目、カサアブラムシ科) 日本、旧北区
Japanese larch bark borer	カラマツカミキリ	*Tetropium morishimaorum* Kusama et Takakura	(コウチュウ目、カミキリムシ科) 日本、旧北区。カラマツ害虫
Japanese larch big aphid			larch aphid (1) を見よ

英名	和名	学名	所属、分布、ほか
Japanese larch eurytomid			larch seed chalcid (1) を見よ
Japanese larch flattened sawfly	カラマツヒラタハバチ	*Cephalcia koebelei* Rohwer	(ハチ目、ヒラタハバチ科) 日本
Japanese larch gelechid	ウスグロコガ	*Gelechia tragicella* (Heyden)	(チョウ目、キバガ科) 日本、旧北区
Japanese larch sawfly			larch sawfly (1) を見よ
Japanese laurel whitefly			aucuba whitefly を見よ
Japanese lawn grass cutworm			lawn grass cutworm を見よ
Japanese leaf-cut weevil	ヒメゴマダラオトシブミ	*Paraplapoderus vanvolxemi* (Roelofs)	(コウチュウ目、オトシブミ科) 日本
Japanese linden aphid	シナノキハネマダラアブラムシ	*Tiliaphis shinae* (Shinji)	(カメムシ目、アブラムシ科) 日本、旧北区
Japanese linden bark beetle	シナノキクイムシ	*Hylesinus tiliae* (Niijima)	(コウチュウ目、キクイムシ科) 日本
Japanese linden geometer			square-spot を見よ
Japanese linden globose aphid	メナモミトックリアブラムシ	*Rhopalosiphoninus tiliae* (Matsumura)	(カメムシ目、アブラムシ科) 日本、旧北区
Japanese linden hornworm	ヒメクチバスズメ	*Marumba jankowskii* Oberthür	(チョウ目、スズメガ科) 日本、旧北区
Japanese linden leafhopper	シナノキハトムネヨコバイ	*Macropsis tiliae* (Germar)	(カメムシ目、オオヨコバイ科) 日本、旧北区
Japanese long scale			pear white scale を見よ
Japanese maple aphid	クロニタイケアブラムシ	*Periphyllus kuwanai* (Takahashi)	(カメムシ目、アブラムシ科) 日本。全北区。カエデ類害虫
Japanese maple aphid	ヒラヤマカマガタアブラムシ	*Yamatocallis hirayamae* Matsumura	(カメムシ目、アブラムシ科) 日本、旧北区
Japanese maple cottony scale			cottony maple scale (2) を見よ
Japanese maple prominent	クロエグリシャチホコ	*Ptilodon okanoi* (Inoue)	(チョウ目、シャチホコガ科) 日本
Japanese maple scale			pear white scale を見よ
Japanese mason bee	マメコバチ	*Osmia cornifrons* (Radoszkowski)	(ハチ目、ハキリバチ科) 日本、旧北区
Japanese mealybug	フジコナカイガラムシ	*Planococcus kraunhiae* (Kuwana)	(カメムシ目、コナカイガラムシ科) 日本、全北区。カンキツ、ナシなど果樹、樹木害虫
Japanese mountain alder aphid			Japanese alder tetocallis を見よ
Japanese mugwort leafworm	ヨモギキリガ	*Orthosia ella* (Butler)	(チョウ目、ヤガ科) 日本、旧北区
Japanese nine-spotted moth			pear amatid を見よ
Japanese oak leafroller	ウスアミメキハマキ	*Tortrix sinapina* (Butler)	(チョウ目、ハマキガ科) 日本、旧北区。カシ類害虫
Japanese oak silkmoth			Japanese silkmoth を見よ

英名	和名	学名	所属、分布、ほか
Japanese oak silkworm			Japanese silkmoth を見よ
Japanese oakworm moth			Japanese silkmoth を見よ
Japanese orange fly	ミカンバエ	*Bactrocera tsuneonis* (Miyake)	(ハエ目、ミバエ科) 日本、東洋区。カンキツ害虫。black fly ともいわれる
Japanese orange fruit fly			Japanese orange fly を見よ　米国での英名
Japanese paper mulberry skeletonizer	コウゾハマキモドキ	*Choreutis hyligenes* (Butler)	(チョウ目、ハマキモドキガ科) 日本、東洋区
Japanese paper wasp	キイロフタモンアシナガバチ	*Polistes chinensis antennalis* Pérez	(ハチ目、スズメバチ科) 日本
Japanese paper wasp	セグロアシナガバチ	*Polistes jadwigae* Dalla Torre	(ハチ目、スズメバチ科) 日本
Japanese pepper longicorn beetle	タイワンメダカカミキリ	*Stenhomalus taiwanus* Matsushita	(コウチュウ目、カミキリムシ科) 日本、東洋区。乾材害虫。Japanese pepper longicorn ともいう
Japanese pine needle gall midge			pine needle gall midge を見よ
Japanese pine sawyer	マツノマダラカミキリ	*Monochamus alternatus* Hope	(コウチュウ目、カミキリムシ科) 日本、旧北区、東洋区。マツ害虫
Japanese pine sawyer beetle			Japanese pine sawyer を見よ
Japanese pine shoot moth	マツヅアカシンムシ	*Petrova cristata* Walsingham	(チョウ目、ハマキガ科) 日本、旧北区、東洋区
Japanese pine spittlebug			pine froghopper (1) を見よ
Japanese pirate ant			slave-maker ant を見よ
Japanese platypodid	ヤチダモノナガキクイムシ	*Crossotarsus niponicus* Blandford	(コウチュウ目、ナガキクイムシ科) 日本、東洋区。クリ、材木害虫
Japanese potato aphid			potato aphid (2) を見よ
Japanese prominent			lobster moth を見よ
Japanese red pine aphid	マツエダオオアブラムシ	*Cinara pinidensiflorae* (Essig et Kuwana)	(カメムシ目、アブラムシ科) 日本、旧北区、東洋区
Japanese red pine bark beetle			fir bark beetle (2) を見よ
Japanese red scale			Asiatic red scale を見よ
Japanese rhinoceros beetle			Japanese horned beetle を見よ
Japanese rice leaf miner			rice leafminer (1) を見よ
Japanese rice leafroller	イネタテハマキ	*Cnaphalocrocis exigua* (Butler)	(チョウ目、メイガ科) 日本。イネ害虫
Japanese rice planthopper	サッポロウンカ	*Changeondelphax velitchkovskyi* (Melichar)	(カメムシ目、ウンカ科) 日本、旧北区、東洋区。イネ科牧草害虫
Japanese rice planthopper	サッポロトビウンカ	*Unkanodes sapporonus* (Matsumura)	(カメムシ目、ウンカ科) 日本、旧北区、東洋区。トウモロコシ害虫。アジアでの英名
Japanese rice root aphid			rice root aphid を見よ　アジアでの英名

英名	和名	学名	所属、分布、ほか
Japanese salix aphid	ヤナギケアブラムシ	*Chaitophorus salijaponicus* Essig et Kuwana	(カメムシ目、アブラムシ科) 日本、旧北区
Japanese scale			pear white scale を見よ
Japanese shothole borer	ニホンタケナガシンクイ	*Dinodercus japonicus* Lesne	(コウチュウ目、ナガシンクイムシ科) 日本、東洋区
Japanese silkmoth	ヤママユ	*Anthraea yamamai* (Guérin-Méneville)	(チョウ目、ヤママユガ科) 日本、旧北区。クリ、サクラ、カシ類害虫
Japanese silver fir aphid	モミイボアブラムシ	*Elatobium momii* (Shinji)	(カメムシ目、アブラムシ科) 日本
Japanese silver fir cottony scale	モミシロカイガラムシ	*Pseudaulacaspis momi* (Kuwana)	(カメムシ目、マルカイガラムシ科) 日本
Japanese silver fir scale	モミカキカイガラムシ	*Lepidosaphes okitsuensis* Kuwana	(カメムシ目、マルカイガラムシ科) 日本
Japanese silver fir torymid			abies torymid を見よ
Japanese skimmer	シオヤトンボ	*Orthetrum japonicum* (Uhler)	(トンボ目、トンボ科) 日本。
Japanese snowflower aphid	ウツギアブラムシ	*Micromyzus dicrvillae* (Matsumura)	(カメムシ目、アブラムシ科) 日本、旧北区
Japanese snowflower sawfly	ウツギハバチ	*Asiemphytus deutziae* (Takeuchi)	(ハチ目、ハバチ科) 日本
Japanese spined tree looper	ウスムラサキエダシャク	*Selenia adustaria* Leech	(チョウ目、シャクガ科) 日本
Japanese spring orange whitefly	エゴノキコナジラミ	*Rhachisphora styraci* (Takahashi)	(カメムシ目、コナジラミ科) 日本
Japanese stink bug	ツノアオカメムシ	*Pentatoma japonica* (Distant)	(カメムシ目、カメムシ科) 日本、旧北区
Japanese swift moth			swift moth を見よ
Japanese tachina fly	ブランコヤドリバエ	*Exorista japonica* (Townsend)	(ハエ目、ヤドリバエ科) 日本、旧北区、東洋区
Japanese tarnished plant bug			pasture leaf bug を見よ
Japanese termite	ヤマトシロアリ	*Reticulitermes speratus* (Kolbe)	(シロアリ目、ミゾガシラシロアリ科) 日本、旧北区
Japanese thyridid	マドガ	*Thyris usitata* Butler	(チョウ目、マドガ科) 日本、旧北区
Japanese tree cricket			tree cricket (2)(3) を見よ
Japanese tussock moth	ヒメシロモンドクガ	*Orgyia thyellina* Butler	(チョウ目、ドクガ科) 日本、旧北区、東洋区。ダイズ、果樹、チャ、園芸植物などの害虫
Japanese ulmus gall aphid	ニレタマフシ	*Kaltenbachiella japonica* (Matsumura)	(カメムシ目、アブラムシ科) 日本、旧北区
Japanese velvety longicorn			velvety longicorn beetle を見よ
Japanese walking-stick	ナナフシ	*Phraortes elongatus* (Thunberg)	(ナナフシ目、ナナフシ科) 日本
Japanese walnut aphid	サワグルミミツアブラムシ	*Kurisakia onigurumii* (Shinji)	(カメムシ目、アブラムシ科) 日本
Japanese water bug	コオイムシ	*Diplonychus japonicus* Vuillefroy	(カメムシ目、コオイムシ科) 日本、旧北区

英名	和名	学名	所属、分布、ほか
Japanese water scorpion	タイコウチ	*Laccotrephes japonensis* Scott	(カメムシ目、タイコウチ科) 日本、旧北区
Japanese wax scale	カメノコロウムシ	*Ceroplastes japonicus* Green	(カメムシ目、カタカイガラムシ科) 日本、旧北区。カンキツ、チャ、ツバキなどの害虫
Japanese waxtree cottony scale	ナガワタカイガラムシ	*Pulvinaria hazeae* Kuwana	(カメムシ目、カタカイガラムシ科) 日本
Japanese waxtree scale	ビワコカタカイガラモドキ	*Nipponaclerda biwakoensis* (Kuwana)	(カメムシ目、カタカイガラモドキ科) 日本、旧北区
Japanese weevil			gooseberry weevil (1) を見よ
Japanese wheat stem sawfly	キベリクキバチ	*Hartigia viator* (Smith)	(ハチ目、クキバチ科) 日本、旧北区
Japanese white ant			Japanese termite を見よ
Japanese white birch aphid	カバハマキヒラタアブラムシ	*Hamamelistes betulinus* (Horvath)	(カメムシ目、アブラムシ科) 日本、全北区
Japanese white birch gelechid	カバオオフサキバガ	*Dichomeris ustalella* (Fabricius)	(チョウ目、キバガ科) 日本、旧北区
Japanese white birch leaf worm	カバキリガ	*Orthosia evanida* (Butler)	(チョウ目、ヤガ科) 日本、旧北区、東洋区
Japanese white birch prominent	シーベリスシャチホコ	*Odontosia sieversi japonica* Matsumura	(チョウ目、シャチホコガ科) 日本
Japanese white birch worm			beautiful-bordered pearl を見よ
Japanese white oak spined aphid			oak spined aphid を見よ
Japanese white pine aphid	ゴヨウマツオオアブラムシ	*Cinara shinjii* Inouye	(カメムシ目、アブラムシ科) 日本
Japanese white-spotted longicorn			white-spotted longicorn beetle (1) を見よ
Japanese wild silk moth			Japanese silkmoth を見よ
Japanese winged-walking-stick	トビナナフシ	*Micadina phluctaenoides* (Rehn)	(ナナフシ目、ナナフシ科) 日本
Japanese wisteria bark beetle	フジノキクイムシ	*Xyleborus kraunhiae* Niijima	(コウチュウ目、キクイムシ科) 日本
Japanese wisteria cottony scale			wisteria scurfy scale を見よ
Japanese wisteria gelechid	フジフサキバガ	*Dichomeris oceanis* Meyrick	(チョウ目、キバガ科) 日本、東洋区
Japanese wisteria mealybug			Japanese mealybug を見よ
Japanese wisteria scale			wisteria scurfy scale を見よ
Japanese yellow hornet	キイロスズメバチ	*Vespa simillima xanthoptera* Cameron	(ハチ目、スズメバチ科) 日本。Japanese hornet ともいう
Japetus beauty			narrow-lined beauty を見よ
japygids			earwiglike entotrophs を見よ
japygids			entotrophs を見よ
Jaque's beetle			Jacques' beetle を見よ
jar flies			dog-day cicadas を見よ

英名	和名	学名	所属、分布、ほか
jarr worm			mole cricket (3) を見よ
jarrah leafminer		*Perthida glyphopa* Common	(チョウ目、マガリガ科) 豪州区
Jarvis's fruit fly		*Dacus jarvisi* (Tryon)	(ハエ目、ミバエ科) 豪州区
jasione pug		*Eupithecia denotata jasionata* Crewe	(チョウ目、シャクガ科) 旧北区
jasmine bud worm			Arabian jasmine pyralid を見よ
jasmine flower borer			Arabian jasmine pyralid を見よ
jasmine leaf webworm		*Nausinoe geometralis* Guenée	(チョウ目、メイガ科) 東洋区
jasmine sphinx			rustic sphinx を見よ
Jason's Mylon		*Mylon jason* (Ehrmann)	(チョウ目、セセリチョウ科) 新熱帯区
Jatropha leafminer moth		*Neurobathra curcassi* Busck	(チョウ目、ホソガ科) 新北区
jaundice fly			wheat midge を見よ
Jaunty dropwing		*Trithemis stictica* (Burmeister)	(トンボ目、トンボ科) エチオピア区
Javan crow		*Euploea gamelia* (Hübner)	(チョウ目、タテハチョウ科) 東洋区
Javan tiger		*Parantica pseudomelaneus* (Moore)	(チョウ目、タテハチョウ科) 東洋区
Javanese grasshopper		*Valanga nigricornis* Burmeister	(バッタ目、バッタ科) 東洋区、豪州区。イネ害虫
Javanese leaf insect		*Phyllium bioculatum* Gray	(ナナフシ目、Phyllidae) 東洋区
Javelin moth		*Bactra verutana* Zeller	(チョウ目、ハマキガ科) 新北区、新熱帯区
Jeannel's blue		*Leptotes jeanneli* (Stempffer)	(チョウ目、シジミチョウ科) エチオピア区
Jeffery's blue		*Lepidochrysops jefferyi* (Swierstra)	(チョウ目、シジミチョウ科) エチオピア区
Jeffery's bush brown		*Bicyclus jefferyi* Fox	(チョウ目、タテハチョウ科) エチオピア区
Jeffery's silver spot		*Aphnaeus jefferyi* Hawker-Smith	(チョウ目、シジミチョウ科) エチオピア区
jeffry-cock			June beetle (1) を見よ
jehol bark beetle	ネッカコキクイムシ	*Cryphalus jehorensis* Murayama	(コウチュウ目、キクイムシ科) 日本、旧北区
jelly grub		*Niphadolepis alianta* Karsch	(チョウ目、イラガ科) エチオピア区
Jemima Eurybia		*Eurybia jemima* Hewitson	(チョウ目、シジミタテハ科) 新熱帯区
Jemina clearwing		*Dircenna jemina* (Geyer)	(チョウ目、タテハチョウ科) 新熱帯区
Jenny-spinners			crane flies (1) を見よ
Jera skipper		*Monca jera* (Godman)	(チョウ目、セセリチョウ科) 新熱帯区
Jerdon's jumping ant			Indian jumping ant を見よ
Jerdons silverspot		*Clossiana jerdoni* (Lang)	(チョウ目、タテハチョウ科) 東洋区
Jerine's widow		*Dingana jerinae* Henning et Henning	(チョウ目、タテハチョウ科) エチオピア区
Jersey black arches		*Nola chlamytulalis* (Hübner)	(チョウ目、コブガ科) 旧北区

英名	和名	学名	所属、分布、ほか
Jersey emerald		*Pseudoterpna coronillaria* (Hübner)	(チョウ目、シャクガ科) 旧北区
Jersey grasshopper		*Euchorthippus elegantulus* Zeuner	(バッタ目、バッタ科) 旧北区
Jersey grasshopper			sharp-tailed grasshopper を見よ
jersey tiger (1)		*Euplagia hera* Linnaeus	(チョウ目、ヒトリガ科) 旧北区
Jersey tiger (2)		*Euplagia quadripunctaria* (Poda)	(チョウ目、ヒトリガ科) 旧北区
Jerusalem artichoke tuber aphid		*Trama troglodytes* von Heyden	(カメムシ目、Lachnidae) 旧北区。Jerusalem artichoke キクイモ に由来
Jerusalem cricket		*Stenopelmatus fuscus* Haldeman	(バッタ目、コオロギ科) 新北区。体長 5 cm の普通種
Jerusalem crickets			cave crickets (2) を見よ
Jesia ringlet		*Euptychia jesia* Butler	(チョウ目、タテハチョウ科) 新熱帯区
Jesia satyr			Jesia ringlet を見よ
Jessica underwing moth		*Catocala jessica* Hy. Edwards	(チョウ目、ヤガ科) 新北区。Jessica underwing ともいう
jesters	キミスジ属	*Symbrenthia*	(チョウ目、タテハチョウ科) の昆虫の総称
Jesus bug			common water strider を見よ　北米での英名
jet ant			field ant を見よ
jet-black ant			field ant を見よ
Jethys mimic-white		*Enantia jethys* (Boisduval)	(チョウ目、シロチョウ科) 新熱帯区
jewel beetle (1)		*Chrysochroa fulminans* (Fabricius)	(コウチュウ目、タマムシ科) 東洋区。カンキツ害虫
jewel beetle		*Stigmodera raei* Hope	(コウチュウ目、タマムシ科) 豪州区
jewel beetle		*Stigmodera gratiosa* Chevrolat	(コウチュウ目、タマムシ科) 豪州区
jewel beetle			European oak borer を見よ
jewel beetles			flat-headed beetles を見よ
jewel blues		*Chilades*	(チョウ目、シジミチョウ科) の昆虫の総称
jewel bugs			shield-backed bugs を見よ　豪州での英名
jewel four-line blue		*Nacaduba sanaya elioti* Corbet	(チョウ目、シジミチョウ科) 東洋区
jewel fourring		*Ypthima avanta* Moore	(チョウ目、タテハチョウ科) 東洋区
jewel-studded skipper		*Thracides phidon* (Cramer)	(チョウ目、セセリチョウ科) 新熱帯区
jewel wasps	コガネコバチ科	Pteromalidae	(ハチ目) の昆虫の総称
jewel wasps			cuckoo wasps (1) を見よ
jeweled hairstreak		*Atlides carpasia* (Hewitson)	(チョウ目、シジミチョウ科) 新熱帯区
jeweled satyr moth		*Phrygionis paradoxata* (Guenée)	(チョウ目、シャクガ科) 新北区、新熱帯区
jeweled skipper		*Phanes aletes* (Geyer)	(チョウ目、セセリチョウ科) 新北区、新熱帯区
jeweled tailed slug moth		*Packardia geminata* (Packard)	(チョウ目、イラガ科) 新北区
jewelled grass-blue			grass jewel (1) を見よ

英名	和名	学名	所属、分布、ほか
jewelled Nawab	ホウセキフタオチョウ	*Polyura delphis* (Doubleday)	（チョウ目、タテハチョウ科）東洋区
jewels	ニシキシジミ属	*Hypochrysops*	（チョウ目、シジミチョウ科）の昆虫の総称
jezebel nymph	クロカザリタテハ	*Mynes geoffroyi* (Guérin-Méneville)	（チョウ目、タテハチョウ科）豪州区
Jezebel palmfly	アカネルリモンジャノメ	*Elymnias vasudeva* Moore	（チョウ目、タテハチョウ科）東洋区
Jezebels	カザリシロチョウ属	*Delias*	（チョウ目、シロチョウ科）の昆虫の総称 Jezebel は邪悪な女の意
jigger			chigoe flea を見よ
jigger flea			chigoe flea を見よ
Jitterbug daisy copper		*Chrysoritis zeuxo* (Linnaeus)	（チョウ目、シジミチョウ科）エチオピア区
Joan swallowtail			Ozark swallowtail を見よ
Joanna's skipper		*Joanna joanna* Evans	（チョウ目、セセリチョウ科）新熱帯区
Joaquin shieldback		*Ateloplus joaquin* Rentz	（バッタ目、キリギリス科）新北区
Jocose sallow			joker moth を見よ
Jodutta Acraea		*Acraea jodutta* (Fabricius)	（チョウ目、タテハチョウ科）エチオピア区
Jodutta glider	ヘリグロキイロタテハ	*Cymothoe jodutta* (Westwood)	（チョウ、タテハチョウ科）エチオピア区
Joe Bassett			June beetle (1) を見よ　幼虫の英名
Joe Pye plume moth			eupatorium plume moth を見よ
Johanna's Zulu		*Alaena johanna* Sharpe	（チョウ目、シジミチョウ科）エチオピア区
John-and-Avery's skipper		*Halotus jonaveriorum* Burns	（チョウ目、セセリチョウ科）新熱帯区
Johnny white-face		*Leucorrhinia intacta* Hagen	（トンボ目、トンボ科）新北区
Johnson's Euchlaena moth		*Euchlaena johnsonaria* (Fitch)	（チョウ目、シャクガ科）新北区
Johnson's hairstreak		*Callophrys johnsoni* (Skinner)	（チョウ目、シジミチョウ科）新北区
Johnson's problematic hairstreak		*Rhamma familiaris* (Johnson)	（チョウ目、シジミチョウ科）新熱帯区
Johnston's Acraea		*Acraea johnstoni* Godman	（チョウ目、タテハチョウ科）エチオピア区
Johnswort pigmy		*Fomoria septembrella* (Stainton)	（チョウ目、モグリチビガ科）旧北区
joined underwing moth		*Catocala junctura* Walker	（チョウ目、ヤガ科）新北区。joined underwing ともいう
joint			jointworms (2) を見よ　幼虫の英名
joint-worms		*Harmolita*	（ハチ目、カタビロコバチ科）の昆虫の総称
jointworms (1)		*Tetramesa*	（ハチ目、カタビロコバチ科）の昆虫の総称
jointworms (2)	カタビロコバチ科	Eurytomidae	（ハチ目）の昆虫の総称
joker		*Byblia*	（チョウ目、タテハチョウ科）の昆虫の総称
joker (1)	ヒメキマダラタテハ	*Byblia ilithyia* (Drury)	（チョウ目、タテハチョウ科）東洋区、エチオピア区
joker			common joker (1) を見よ

英名	和名	学名	所属、分布、ほか
joker butterfly			common joker (1) を見よ
joker moth		*Feralia jocosa* Guenée	（チョウ目、ヤガ科）新北区
Jola grasshopper		*Colemania sphenarioides* Bolivar	（バッタ目、バッタ科）東洋区
Jonas sawfly	ウンモンチュウレンジ	*Arge jonasi* (Kirby)	（ハチ目、ミフシハバチ科）日本、旧北区
Jone's forestwatcher			forest-watcher を見よ
Jone's mantis		*Pyrgomantis jonesi* Kirby	（カマキリ目、Tarachodidae）エチオピア区
Jordan's Altinote		*Altinote hilaris* Jordan	（チョウ目、タテハチョウ科）新熱帯区
Jordan's grass yellow		*Eurema andersoni jordani* Corbet et Pendlebury	（チョウ目、シロチョウ科）東洋区。one-spot grass yellow を参照
Jordan's Mormon	ジョルダンアゲハ	*Papilio jordani* Fruhstorfer	（チョウ目、アゲハチョウ科）東洋区
Jordan's pretty satyr		*Paralasa jordana* (Staudinger)	（チョウ目、タテハチョウ科）旧北区
Jordan's sailer		*Neptis jordani* Neave	（チョウ目、タテハチョウ科）エチオピア区
Jordan's sister	ヒメキオビイチモンジ	*Adelpha jordani* Fruhstorfer	（チョウ目、タテハチョウ科）新熱帯区
Jordan's swallowtail			Jordan's Mormon を見よ
Joseph's coat moth		*Agarista agricola* Donovan	（チョウ目、ヤガ科）豪州区
Joshua bug			bed bug (1) を見よ　モーゼの後継者ヨシュアに由来
Joshua tree yucca moth		*Tegeticula paradoxa* Riley	（チョウ目、マガリガ科）新北区
joyful Holomelina		*Holomelina laeta* (Guérin-Méneville)	（チョウ目、ヒトリガ科）新北区。joyful Holomelina moth ともいう
Juana copper		*Aloeides juana* Tite et Dickson	（チョウ目、シジミチョウ科）エチオピア区
Juanita sphinx moth		*Proserpinus juanita* (Strecker)	（チョウ目、スズメガ科）新北区。Juanita sphinx ともいう
Juanita's hairtail		*Anthene juanitae* Henning et Henning	（チョウ目、シジミチョウ科）エチオピア区
Juba skipper		*Hesperia juba* (Scudder)	（チョウ目、セセリチョウ科）新北区、新熱帯区
jubilee fan-foot		*Zanclognatha lunalis* (Scopoli)	（チョウ目、ヤガ科）旧北区
Jucunda page			great blue hookwing を見よ
Judaic tortoise beetle		*Agroiconota judaica* (Fabricius)	（コウチュウ目、ハムシ科）新熱帯区
Judies		*Abisara*	（チョウ目、シジミタテハ科）の昆虫の総称
Judith's agave borer			Neumoegen's giant-skipper を見よ
Judith's underwing moth		*Catocala judith* Strecker	（チョウ目、ヤガ科）新北区。Judith's underwing ともいう
Judys			metalmarks (1) を見よ
jugate moths	同翅亜目	Jugatae	（チョウ目）の昆虫の総称
juglans aphid			Japanese walnut aphid を見よ
jugular-horned beetles	デバヒラタムシ科	Prostomidae	（コウチュウ目）の昆虫の総称　朽木に生息。

英名	和名	学名	所属、分布、ほか
juice mydas fly		*Opomydas limbatus* (Williston)	（ハエ目、ムシヒキアブモドキ科）新北区
Julia	チャイロドクチョウ	*Dryas iulia* (Fabricius)	（チョウ目、タテハチョウ科）新北区。*D. i. moderata* (Riley) も同英名。Julia butterfly ともいう
Julia heliconian			Julia を見よ
Julia longwing butterfly			Julia を見よ
Julia skimmer		*Orthetrum julia* Kirby	（トンボ目、トンボ科）エチオピア区
Julia skipper		*Nastra julia* (Freeman)	（チョウ目、セセリチョウ科）新北区、新熱帯区。Julia's skipper ともいう
Julia's Dicymolomia moth		*Dicymolomia julianalis* (Walker)	（チョウ目、メイガ科）新北区、新熱帯区
Julia's giant-skipper		*Agathymus juliae* (Stallings et Turner)	（チョウ目、セセリチョウ科）新熱帯区
Julia's Protea butterfly		*Capys juliae* Henning et Henning	（チョウ目、シジミチョウ科）エチオピア区
Juliette longwing		*Eueides aliphera* (Godart)	（チョウ目、タテハチョウ科）新熱帯区。Juliette ともいう
July acorn piercer		*Pammene fasciana* (Linnaeus)	（チョウ目、ハマキガ科）旧北区
July belle		*Scotopteryx luridata plumbaria* (Fabricius)	（チョウ目、シャクガ科）旧北区
July high flier moth			June highflier を見よ　July highflier moth ともいう。flyer も使用される
July highflier			June highflier を見よ
July lead-belle			July belle を見よ
jumper			hop jumper を見よ
jumper			cheese skipper を見よ　幼虫の英名
jumper ants			bulldog ants を見よ
jumper ants			keleps を見よ　豪州での英名
jumping ants			keleps を見よ
jumping bean moth		*Emporia melanobasis* Janse	（チョウ目、メイガ科）新熱帯区。南米での英名
jumping bean moth (1)		*Cydia deshaisiana* (Lucas)	（チョウ目、ハマキガ科）大洋区、新熱帯区
jumping borer			lesser cornstalk borer を見よ
jumping bristletails			machilids を見よ
jumping bugs			shore bugs を見よ
jumping bush cricket		*Orocharis saltator* Uhler	（バッタ目、コオロギ科）新北区
jumping ground bugs	ムクゲカメムシ科	Dipsocoridae	（カメムシ目）の昆虫の総称　体長 2 mm 以内の小型カメムシ
jumping ground bugs			jumping soil bugs を見よ
jumping plant lice and suckers			jumping plantlice を見よ
jumping plantlice	キジラミ科	Psyllidae	（カメムシ目）の昆虫の総称
jumping soil bugs	ノミカメムシ科	Schizopteridae	（カメムシ目）の昆虫の総称　体長 2 mm 以下で主に熱帯湿地に生息

英名	和名	学名	所属、分布、ほか
jumping springtails	イシノミ目	Microcoryphia	の昆虫の総称
jumping springtails			machilids を見よ
jumping thrips	メロアザミウマ科	Merothripidae	（アザミウマ目）の昆虫の総称
jumping tree bugs	ダルマカメムシ科	Isometopidae	（カメムシ目）の昆虫の総称
June beetle (1)		*Melolontha melolontha* (Linnaeus)	（コウチュウ目、コガネムシ科）旧北区。英国の普通種
June beetle		*Phyllophaga fusca* (Fröhlich)	（コウチュウ目、コガネムシ科）新北区
June beetle		*Phyllophaga (Lachnosterna) citri* (Smith)	（コウチュウ目、コガネムシ科）新熱帯区。カンキツ害虫
June beetle		*Phyllophaga (Lachnosterna) puberula* (J. Duval)	（コウチュウ目、コガネムシ科）新熱帯区。カンキツ害虫
June beetle		*Phyllophaga (Lachnosterna) subssericans* (J. Duval)	（コウチュウ目、コガネムシ科）新熱帯区。カンキツ害虫
June beetle		*Serica alternata* LeConte	（コウチュウ目、コガネムシ科）新北区。アボガド害虫
June beetle		*Serica fimbriata* LeConte	（コウチュウ目、コガネムシ科）新北区。アボガド害虫
June beetle		*Coenonycha testacea* (Cazier)	（コウチュウ目、コガネムシ科）新北区。アボカド害虫
June beetle			（コウチュウ目、コガネムシ科）米国北部では *Phyllophaga* 属の種、南部では figeater (*Cotinis nitida*) にも使われる
June beetles			May beetles (2)(3) を見よ
June beetles			white grubs を見よ
June beetles			chafer beetles を見よ
June beetles			flower beetles (3) を見よ
June bug			June beetle (1) を見よ
June bug			rose chafer (1) を見よ
June bug			summer chafer を見よ
June bug			green fruit beetle を見よ
June bugs			chafer beetles を見よ
June bugs			May beetles (1)(2)(3) を見よ
June highflier		*Hydriomena furcata* (Thunberg)	（チョウ目、シャクガ科）旧北区。日本亜種は *H. f. nexifasciata* Butler ヤナギナミシャク
jungle glory	ルリモンワモンチョウ（マルバネワモンチョウ）	*Thaumantis diores* Doubleday	（チョウ目、タテハチョウ科）東洋区
jungle nymph			spiny stick insect を見よ
jungle skipper		*Papias subcostulata* (Herrich-Schäffer)	（チョウ目、セセリチョウ科）新熱帯区
jungle threadtail		*Elattoneura caesia* (Hagen)	（トンボ目、Protoneuridae）東洋区
jungleking		*Thauria lathyi* Fruhstorfer	（チョウ目、タテハチョウ科）東洋区
junglewatchers		*Hylaeothemis*	（トンボ目、トンボ科）の昆虫の総称

英名	和名	学名	所属、分布、ほか
juniper aphid			thuya aphid を見よ
juniper bark borer	ビャクシンカミキリ	*Semanotus bifasciatus* (Motschulsky)	(コウチュウ目、カミキリムシ科) 日本、旧北区、東洋区
juniper budworm moth		*Cudonigera houstonana* (Grote)	(チョウ目、ハマキガ科) 新北区
juniper bush katydid		*Insara juniperi* Hebard	(バッタ目、キリギリス科) 新北区
juniper carpet		*Thera juniperata* (Linnaeus)	(チョウ目、シャクガ科) 旧北区。juniper carpet moth ともいう
juniper crest		*Dichomeris juniperella* (Linnaeus)	(チョウ目、キバガ科) 旧北区
juniper gelechiid moth		*Gelechia sabinella* (Zeller)	(チョウ目、キバガ科) 新北区
juniper geometer moth		*Patalene olyzonaria* (Walker)	(チョウ目、シャクガ科) 新北区
juniper hairstreak		*Mitoura siva* Edwards	(チョウ目、シジミチョウ科) 新北区
juniper hairstreak (1)	オリーブスギカラスシジミ	*Callophrys gyrneus* (Hübner)	(チョウ目、シジミチョウ科) 新北区
juniper hawk		*Oligographa juniperi* (Boisduval)	(チョウ目、スズメガ科) エチオピア区
juniper leafminer	ビャクシンハモグリガ	*Argyresthia sabinae* Moriuti	(チョウ目、メムシガ科) 日本。イブキ害虫
juniper looper moth		*Eupithecia interruptofasciata* Packard	(チョウ目、シャクガ科) 新北区
juniper midge		*Contarinia juniperina* Felt	(ハエ目、タマバエ科) 新北区
juniper pug		*Eupithecia pusillata* (Denis et Schiffermüller)	(チョウ目、シャクガ科) 旧北区
juniper pug		*Eupithecia sobrinata* Hübner	(チョウ目、シャクガ科) 旧北区
juniper quadrangular bud gall midge	タマイブキノタマバエ	*Aschistonyx eppoi* Inouye	(ハエ目、タマバエ科) 日本。イブキ害虫
juniper sawfly		*Monoctenus juniperi* (Linnaeus)	(ハチ目、マツハバチ科) 旧北区
juniper scale		*Carulaspis juniperi* (Bouché)	(カメムシ目、マルカイガラムシ科) 日本、汎世界
juniper scale	ビャクシンコノハカイガラムシ	*Fiorinia pinicola* Maskell	(カメムシ目、マルカイガラムシ科) 日本、東洋区。イヌマキ、トベラなどの害虫
juniper scale			minute cypress scale を見よ
juniper seed chalcid		*Megastigmus kuntzei* Kapuscinski	(ハチ目、オナガコバチ科) 旧北区
juniper shieldbug		*Cyphostethus tristriatus* (Fabricius)	(カメムシ目、ツノカメムシ科) 旧北区
juniper shoot moth			violet juniper argent を見よ
juniper tip midge		*Oligotrophus betheli* Felt	(ハエ目、タマバエ科) 新北区
juniper tip moth		*Glyphidocera juniperella* Adamski	(チョウ目、Glyphidoceridae) 新北区
juniper tussock moth	ウチジロマイマイ	*Parocneria furva* (Leech)	(チョウ目、ドクガ科) 日本、旧北区。イブキ、ヒノキ害虫
juniper webber			white-bordered crest を見よ

英名	和名	学名	所属、分布、ほか
juniper webber moth			white-bordered crest を見よ
juniper webworm			white-bordered crest を見よ　juniper webworm moth ともいう
juniperus scale	ネズヒメシロカイガラムシ	*Pinnaspis juniperi* Takahashi	(カメムシ目、マルカイガラムシ科) 日本
Juno buck moth		*Hemileuca juno* Packard	(チョウ目、ヤママユガ科) 新北区
Juno heliconian			scarce silver-spotted flambeau を見よ
Juno longwing		*Dione juno huascuma* (Reakirt)	(チョウ目、タテハチョウ科) 新熱帯区。scarce silver-spotted flambeau を参照
Junod's swordtail		*Graphium junodi* (Trimen)	(チョウ目、アゲハチョウ科) エチオピア区
Jurgensen's doctor			costa-spotted metalmark を見よ
Justina sister		*Adelpha justina* (C. et R. Felder)	(チョウ目、タテハチョウ科) 新熱帯区
jute hairy caterpillar		*Spilarctia obliqua* Walker	(チョウ目、ヒトリガ科) 東洋区
jute looper	ヒメアカキリバ	*Anomis involuta* (Walker)	(チョウ目、ヤガ科) 日本、東洋区、豪州区
jute semilooper		*Cosmophila sabulifera* Guenée	(チョウ目、ヤガ科) 旧北区、東洋区、豪州区、大洋区、エチオピア区。ジュート害虫
jute stem weevil		*Apion corchori* Marshall	(コウチュウ目、ホソクチゾウムシ科) 東洋区
Jutta arctic			Baltic grayling を見よ
Juturna underleaf		*Eurybia juturna* C. et R. Felder	(チョウ目、シジミタテハ科) 新熱帯区
Juvenal's duskywing		*Erynnis juvenalis juvenalis* (Fabricius)	(チョウ目、セセリチョウ科) 新北区、新熱帯区
Juventus skipper		*Callimormus juventus* Scudder	(チョウ目、セセリチョウ科) 新熱帯区
Jyntea hedge blue		*Celastrina argiolus jynteana* (de Nicéville)	(チョウ目、シジミチョウ科) 東洋区。holly blue を参照

英名	和名	学名	所属、分布、ほか
K			
Kabru hairstreak	カノミドリシジミ	*Chrysozephyrus kabrua* (Tytler)	（チョウ目、シジミチョウ科）東洋区
kadondong beetle		*Podontia quatuordecimpunctata* Linnaeus	（コウチュウ目、ハムシ科）東洋区
Kadoyama xyleborus	カドヤマキクイムシ	*Xyleborus kadoyamensis* Murayama	（コウチュウ目、キクイムシ科）日本、東洋区
Kaempfer cicada	ニイニイゼミ	*Platypleura kaempferi* (Fabricius)	（カメムシ目、セミ科）日本、旧北区、東洋区
Kahli swallowtail		*Papilio kahli* Chermock et Chermock	（チョウ目、アゲハチョウ科）新北区
Kahului aweoweo long-horned beetle		*Plaithmysus kahului* Samuelson	（コウチュウ目、カミキリムシ科）大洋区
Kaikoura giant weta		*Deinacrida parva* Buller	（バッタ目、Stenopelmatidae）豪州区。Kaikoura weta ともいう
Kaiser-i-Hind	テングアゲハ	*Teinopalpus imperialis* Hope	（チョウ目、アゲハチョウ科）東洋区
Kakadu vicetail		*Hemigomphus magela* Watson	（トンボ目、サナエトンボ科）豪州区
Kakamega swift		*Borbo kaka* (Evans)	（チョウ目、セセリチョウ科）エチオピア区
Kakamega sylph		*Metisella kakamega* de Jong	（チョウ目、セセリチョウ科）エチオピア区
Kakum Diopetes		*Pilodeudorix corruscans* (Aurivillius)	（チョウ目、シジミチョウ科）エチオピア区
Kalahari copper		*Aloeides simplex* (Trimen)	（チョウ目、シジミチョウ科）エチオピア区
Kalahari orange tip		*Colotis lais* (Butler)	（チョウ目、シロチョウ科）エチオピア区
Kalinzu leaf sitter		*Gorgyra kalinzu* Evans	（チョウ目、セセリチョウ科）エチオピア区
Kaltenbach's purple		*Eriocrania chrysolepidella* (Zeller)	（チョウ目、スイコバネガ科）旧北区
Kamakura oystershell scale	カマクラカキカイガラムシ	*Lepidosaphes kamakurensis* Kuwana	（カメムシ目、マルカイガラムシ科）日本
Kamani psyllid		*Leptynoptera sulfurea* Crawford	（カメムシ目、キジラミ科）東洋区、大洋区
Kamehameha butterfly	カメハメハアカタテハ	*Vanessa tameamea* (Eshscholtz)	（チョウ目、タテハチョウ科）大洋区。ハワイ王朝カメハメハ一世に由来。Kamehameha ともいう
Kamehameha lady			Kamehameha butterfly を見よ
Kammanassie widow		*Dingana kammanassiensis* Henning et Henning	（チョウ目、タテハチョウ科）エチオピア区
Kanara oakblue		*Arhopala alea* (Hewitson)	（チョウ目、シジミチョウ科）東洋区
Kanara oakblue		*Panchala canaraica* (Moore)	（チョウ目、シジミチョウ科）東洋区
Kanara swift		*Caltoris canaraica* (Moore)	（チョウ目、セセリチョウ科）東洋区
kangaroo beetle		*Sagra papuana* Jacoby	（コウチュウ目、ハムシ科）豪州区
kangaroo bot fly		*Tracheomyia macropi* (Froggatt)	（ハエ目、ヒツジバエ科）豪州区
kangaroo louse		*Heterodoxus longitarsus* (Piaget)	（シラミ目、ミナミケモノハジラミ科）豪州区

英名	和名	学名	所属、分布、ほか
Kanshul longtail		*Polythrix kanshul* Shuey	(チョウ目、セセリチョウ科) 新熱帯区
Kaplan's copper		*Aloeides kaplani* Tite et Dickson	(チョウ目、シジミチョウ科) エチオピア区
Kaplan's Thestor		*Thestor kaplani* Dickson et Stephen	(チョウ目、シジミチョウ科) エチオピア区
kapra beetle			Khapra beetle を見よ
Karakoram banded Apollo		*Parnassius staudingeri hunza* Grumm-Grshimailo	(チョウ目、アゲハチョウ科) 東洋区
karbi		*Trigona carbonaria* Smith	(ハチ目、ミツバチ科) 豪州区
Karkloof blue		*Orachrysops ariadne* (Butler)	(チョウ目、シジミチョウ科) エチオピア区
Karkloof emperor			pearl Charaxes を見よ
Karkloof emperor		*Charaxes karkloof* van Someren et Jackson	(チョウ目、タテハチョウ科) エチオピア区
Karner blue		*Lycaeides melissa samuelis* Nobokov	(チョウ目、シジミチョウ科) 新北区。orange-bordered blue を参照
Karoo brown		*Stygionympha irrorata* (Trimen)	(チョウ目、タテハチョウ科) エチオピア区
Karoo caterpillar		*Loxostege frustalis* Zeller	(チョウ目、メイガ科) エチオピア区
Karoo copper		*Chrysoritis chrysantas* (Trimen)	(チョウ目、シジミチョウ科) エチオピア区
Karoo daisy copper			Natal copper を見よ
Karoo daisy copper			Karoo copper を見よ
Karoo dancer		*Alenia sandaster* (Trimen)	(チョウ目、セセリチョウ科) エチオピア区
Karoo sandman			Karoo dancer を見よ
Karoo tent caterpillar		*Phiala patagiata* Aurivillius	(チョウ目、オビガ科) エチオピア区
Karoo Thestor		*Thestor protumnus aridus* van Son	(チョウ目、シジミチョウ科) エチオピア区
Karoo widow		*Tarsocera fulvina* Vári	(チョウ目、タテハチョウ科) エチオピア区
Karry-long-legs			crane flies (1) を見よ
Karwar lascar		*Pantoporia karwara* Fruhstorfer	(チョウ目、タテハチョウ科) 東洋区
Karwinski's beauty		*Smyrna karwinskii* (Hübner)	(チョウ目、タテハチョウ科) 新北区、新熱帯区
Kashmir wall		*Lasiommata maerula* C. et R. Felder	(チョウ目、タテハチョウ科) 東洋区
Kashmir white		*Pieris deota* (de Nicéville)	(チョウ目、シロチョウ科) 東洋区
Kashmir willow defoliator			apple hairy caterpillar を見よ
Kathleen's shieldback		*Idiostatus kathleenae* Rentz	(バッタ目、キリギリス科) 新北区
Katie's springtail		*Katianna australis* Womersley	(トビムシ目、マルトビムシ科) 豪州区
katsura long-horned aphid	カツラフクレアブラムシ	*Acyrthosiphon cercidiphylli* (Matsumura)	(カメムシ目、アブラムシ科) 日本
katydid		*Cyrtophyllus concavus* Scudder	(バッタ目、バッタ科) 新北区。カンキツ害虫

英名	和名	学名	所属、分布、ほか
katydid		*Microcentrum lanceolatum* Burmeister	（バッタ目、キリギリス科）新熱帯区。カンキツ害虫
katydid			ウマオイムシに似た米国産キリギリス科の昆虫。雄の鳴声が katy did, katy didn't と聞こえるという
katydids			locusts を見よ
katydids			long-horned grasshoppers (1)(3) を見よ
katydids			crickets (3) を見よ
Kauai bog damselfly		*Megalagrion paludicola* (Maciorek et Howarth)	（トンボ目、イトトンボ科）大洋区。ハワイの Kauai 島に由来
Kauai flightless stag beetle		*Apterocyclus honoluluensis* Waterhouse	（コウチュウ目、コガネムシ科）大洋区
Kauai 'le'ie damselfly		*Megalagrion kauaiensis* Perkins	（トンボ目、イトトンボ科）大洋区
Kauai mountain damselfly		*Megalagrion heterogamias* Perkins	（トンボ目、イトトンボ科）大洋区
Kaumana cave cricket		*Caconemobius varius* Gurney et Rentz	（バッタ目、コオロギ科）大洋区。ハワイ固有種
kauri coccid		*Conifericoccus agathidis* Brimblecombe	（カメムシ目、ワタフキカイガラムシ科）豪州区。英名はカウリマツに由来
kauri moths		Agathiphagidae	（チョウ目）の昆虫の総称
kauri thrips		*Oxythrips agathidis* Morison	（アザミウマ目、アザミウマ科）豪州区
Kawasaki horntail			alder wood wasp を見よ
kaya spined beetle	クロルリトゲハムシ	*Rhdinosa nigrocyanea* (Motschulsky)	（コウチュウ目、ハムシ科）日本、旧北区、東洋区
Kearfott's Acrobasis moth		*Acrobasis kearfottella* Dyar	（チョウ目、メイガ科）新北区
Kearfott's Rolandylis moth		*Rolandylis maiana* (Kearfott)	（チョウ目、ハマキガ科）新北区
ked			sheep ked を見よ
Kedong cupid		*Chilades naidina kedonga* Grose-Smith	（チョウ目、シジミチョウ科）エチオピア区
keds			pupiparous flies を見よ
keds			louse flies を見よ
keeled Apollo	マルバネウスバ	*Parnassius jacquemonti* Boisduval	（チョウ目、アゲハチョウ科）旧北区、東洋区
keeled orthetrum		*Orthetrum coerulescens* (Fabricius)	（トンボ目、トンボ科）旧北区
keeled shieldback		*Neduba carinata* Walker	（バッタ目、キリギリス科）新北区
keeled skimmer			keeled orthetrum を見よ
keeler grasshopper		*Melanoplus keeleri* (Thomas)	（バッタ目、バッタ科）新北区
Keferstein's mapwing		*Hypanartia kefersteini* (Doubleday)	（チョウ目、タテハチョウ科）新熱帯区
Keifer mealybug		*Spirococcus keiferi* McKenzie	（カメムシ目、コナカイガラムシ科）新北区
Keila hairstreak		*Ostrinotes keila* (Hewitson)	（チョウ目、シジミチョウ科）新北区、新熱帯区

英名	和名	学名	所属、分布、ほか
Keiser's forktail		*Macrogomphus annulatus keiseri* Lieftinck	（トンボ目、サナエトンボ科）東洋区
kelep			Guatemalan kelep を見よ
keleps	ハリアリ亜科	Ponerinae	（ハチ目、アリ科）の昆虫の総称
kelp flies			seaweed flies を見よ
kelp fly			seaweed fly を見よ
kelp muscid		*Fucellia costalis* Stein	（ハエ目、イエバエ科）新北区
Kelso shieldback		*Eremopedes kelsoensis* Tinkham	（バッタ目、キリギリス科）新北区
Kemner's skipper		*Paratrytone kemneri* Steinhauser	（チョウ目、セセリチョウ科）新熱帯区
Kemner's skipperling		*Piruna kemneri* Freeman	（チョウ目、セセリチョウ科）新熱帯区
Kemp's Hellula moth		*Helllula kempae* Munroe	（チョウ目、メイガ科）新北区、新熱帯区
Kendall's checkerspot		*Chlosyne kendallorum* Opler	（チョウ目、タテハチョウ科）新熱帯区
Kendall's satyr		*Splendeuptychia kendalli* Miller	（チョウ目、タテハチョウ科）新熱帯区
Kennicott grasshopper		*Melanoplus kennicotti* Scudder	（バッタ目、バッタ科）新北区
Kent black arches		*Meganola albula* (Denis et Schiffermüller)	（チョウ目、コブガ科）旧北区。亜種 *M. a. formosana* (Wileman et West) トビモンシロコブガは日本（沖縄）、台湾。亜種 *M. a. pacifica* (Inoue) は日本（本、四、九）。イングランドの州名に由来
Kent ermel		*Yponomeuta plumbella* (Denis et Schiffermüller)	（チョウ目、スガ科）旧北区
Kent seathorn groundling		*Gelechia hippophaella* (Schrank)	（チョウ目、キバガ科）旧北区
Kentish glory		*Endromis versicolora* (Linnaeus)	（チョウ目、Endromidae）旧北区、エチオピア区。Kentish glory moth ともいう
Kentish piercer		*Cydia caecana* (Schlager)	（チョウ目、ハマキガ科）旧北区
Kent's geometer moth		*Selenia kentaria* (Grote et Robinson)	（チョウ目、シャクガ科）新北区
Kentucky lichen moth		*Cisthene kentuckiensis* (Dyar)	（チョウ目、ヒトリガ科）新北区
Kenya mealybug			coffee mealybug (1) を見よ
Kenyan bush brown		*Bicyclus kenia* (Rogenhofer)	（チョウ目、タテハチョウ科）エチオピア区
Kenyan dung beetle		*Copris fallaciosus* Gillet	（コウチュウ目、コガネムシ科）豪州区
Kenyan fiery Acraea		*Acraea pudorina* Staudinger	（チョウ目、タテハチョウ科）エチオピア区
Kenyan white cloaked skipper		*Leucochitonea hindei* Druce	（チョウ目、セセリチョウ科）エチオピア区
Kerea satyr		*Taygetis kerea* Butler	（チョウ目、タテハチョウ科）新熱帯区
kermes		*Kermes ilicis* Linnaeus	（カメムシ目、タマカイガラムシ科）旧北区。エンジムシ（カイガラムシ類）の一種で、雌を乾燥して染料をとる
kermes		*Kermes vermilio* (Planchon)	（カメムシ目、タマカイガラムシ科）旧北区

英名	和名	学名	所属、分布、ほか
kermes (1)	タマカイガラムシ科	Kermidae	（カメムシ目）の昆虫の総称　Kermestidae は異名
kermes (2)		*Kermes quercus* Linnaeus	（カメムシ目、タマカイガラムシ科）旧北区
kermes berry			kermes (1)(2) を見よ
kermes coccids			kermes (1) を見よ
kermes scale	ナラタマカイガラムシ	*Kermes nakagawae* Kuwana	（カメムシ目、タマカイガラムシ科）日本、旧北区。カシ類害虫
kermes scale moth		*Euclemensia bassettella* (Clemens)	（チョウ目、カザリバガ科）新北区
kernel grub		*Assara seminivale* (Turner)	（チョウ目、メイガ科）豪州区
Kershaw's brown		*Oreixenica kershawi* (Miskin)	（チョウ目、タテハチョウ科）豪州区
Kersten's ciliate blue		*Anthene kersteni* (Gerstaecker)	（チョウ目、シジミチョウ科）エチオピア区
Kersten's sprite		*Pseudagrion kersteni* (Gerstacker)	（トンボ目、イトトンボ科）エチオピア区
Ketsi blue		*Lepidochrysops ketsi* Cottrell	（チョウ目、シジミチョウ科）エチオピア区
Kew arches			lily borer を見よ
keyhole glider		*Tramea basilaris* (Palisot de Beauvois)	（トンボ目、トンボ科）エチオピア区、東洋区
keyhole wasp		*Pachodynerus nasidens* (Latreille)	（ハチ目、スズメバチ科）新北区
Keys bush cricket		*Orocharis diplastes* Walker	（バッタ目、コオロギ科）新北区
Keys conehead		*Belocephalus sleighti* Davis	（バッタ目、キリギリス科）新北区
Keys green June beetle		*Cotinis aliena* Woodruff	（コウチュウ目、コガネムシ科）新北区
Keys scaly cricket		*Cycloptilum irregularis* Love et Walker	（バッタ目、コオロギ科）新北区
Keys wood cricket		*Gryllus cayensis* Walker	（バッタ目、コオロギ科）新北区
khaki silverline		*Cigaritis rukmini* de Nicéville	（チョウ目、シジミチョウ科）東洋区
Khamti oakblue		*Arhopala khamti* Doherty	（チョウ目、シジミチョウ科）東洋区
Khapra beetle	ヒメアカツオブシムシ	*Trogoderma granarium* (Everts)	（コウチュウ目、カツオブシムシ科）日本、汎世界。貯蔵食品の大害虫
Kheil's albatross		*Appias mata* (Kheil)	（チョウ目、シロチョウ科）東洋区
kiawe bean weevil		*Algarobius bottimeri* Kingsolver	（コウチュウ目、ハムシ科）新北区、大洋区。ハワイのマメ科の木 kiawe に由来。
kiawe flower moth		*Ithome concolorella* (Chambers)	（チョウ目、カザリキバガ科）新北区、大洋区
kiawe moth			fragile gray を見よ
kiawe roundheaded borer		*Placosternus crinicornis* (Chevrolat)	（コウチュウ目、カミキリムシ科）大洋区、新熱帯区
kiawe scolytid	ナガサキコキクイムシ	*Hypothenemus birmanus* (Eichhoff)	（コウチュウ目、キクイムシ科）日本、東洋区、大洋区、新北区、新熱帯区
kidney-spotted miner		*Lacinipolia renigera* Stephens	（チョウ目、ヤガ科）新北区

英名	和名	学名	所属、分布、ほか
kidney-vetch sober	クロチビキバガ	*Aproaerema anthyllidella* (Hübner)	(チョウ目、キバガ科) 日本、旧北区。マメ科牧草害虫
Kikuyu grass bug			grass leafhopper を見よ
Kikuyu sailer		*Neptis kikuyuensis* Jackson	(チョウ目、タテハチョウ科) エチオピア区
Kilimanjaro swallowtail		*Papilio sjoestedti* Aurivillius	(チョウ目、アゲハチョウ科) エチオピア区
killer bee	アメリカナイズドミツバチ		(ハチ目、ミツバチ科) 新北区。*Apis mellifera scutellata* Lepeletier アメリカミツバチと *A. mellifera* Linnaeus セイヨウミツバチの交雑種。攻撃性が強い。
Kimball's leafroller moth		*Argyrotaenia kimballi* Obraztsov	(チョウ目、ハマキガ科) 新北区
Kimball's Palpita moth		*Palpita kimballi* Munroe	(チョウ目、メイガ科) 新北区
Kimberley emperor		*Anax georgius* Selys	(トンボ目、ヤンマ科) 豪州区
Kimberley hunter		*Austrogomphus mouldsorum* Theischinger	(トンボ目、サナエトンボ科) 豪州区
Kimberley spotted opal		*Nesolycaena caesia* Apice et Miller	(チョウ目、シジミチョウ科) 豪州区
kindred button		*Acleris lipstana* (Denis et Schiffermüller)	(チョウ目、ハマキガ科) 旧北区
king billy			monarch butterfly を見よ　カナダで使用
king blue		*Lepidochrysops tantalus* (Trimen)	(チョウ目、シジミチョウ科) エチオピア区
King Cerulean			dark Cerulean (1) を見よ
King Christmas beetle		*Anoplognathus viridiaeneus* (Donovan)	(コウチュウ目、コガネムシ科) 豪州区
king copper		*Tylopaedia sardonyx* (Trimen)	(チョウ目、シジミチョウ科) エチオピア区
King cracker			red cracker を見よ
king crickets			cave crickets (2) を見よ　豪州での英名
king crickets			gryllacridoids を見よ　南アフリカでの英名
king crow		*Euploea phaenareta castelnaui* C. et R. Felder	(チョウ目、タテハチョウ科) 東洋区。giant crow を参照
king fisher			dragonflies (1) を見よ
King George			demoiselle Agrion を見よ
king leafwing		*Polygrapha tyrianthina* (Salvin et Godman)	(チョウ目、タテハチョウ科) 新熱帯区
King page	タスキアゲハ	*Papilio thoas* Linnaeus	(チョウ目、アゲハチョウ科) 新北区、新熱帯区
King page swallowtail			King page を見よ
King Papilio			regal swallowtail を見よ
King shoemaker		*Prepona demophon* (Linnaeus)	(チョウ目、タテハチョウ科) 新熱帯区
King shoemaker (1)		*Prepona meander* (Cramer)	(チョウ目、タテハチョウ科) 新熱帯区
king stag beetle		*Phalacrognathus muelleri* (Macleay)	(コウチュウ目、クワガタムシ科) 豪州区

英名	和名	学名	所属、分布、ほか
King swallowtail			King page を見よ
kingfisher wasp		*Trielis alcione* (Banks)	(ハチ目、ツチバチ科) 新北区
kingfisher			demoiselle Agrion を見よ
King's bee-hawk		*Cephonodes kingi* Macleay	(チョウ目、スズメガ科) 旧北区
King's hairstreak		*Satyrium kingi* (Klots et Clench)	(チョウ目、シジミチョウ科) 新北区
kinnarid planthoppers		Kinnaridae	(カメムシ目) の昆虫の総称　体長 3 〜 4 mm。熱帯に多い
Kiowa dancer		*Argia immunda* (Hagen)	(トンボ目、イトトンボ科) 新北区
Kirbari hairstreak	キルバリミドリシジミ	*Chrysozephyrus kirbariensis* (Tytler)	(チョウ目、シジミチョウ科) 東洋区
Kirby's backswimmer		*Notonecta kirbyi* Hungerford	(カメムシ目、マツモムシ科) 新北区
Kirby's dropwing		*Trithemis kirbyi* Selys	(トンボ目、トンボ科) 旧北区、東洋区、エチオピア区
Kirby's swordtail		*Graphium kirbyi* (Hewitson)	(チョウ目、アゲハチョウ科) エチオピア区
Kiriakoff's sailer		*Neptis kiriakoffi* Overlaet	(チョウ目、タテハチョウ科) エチオピア区
Kirkaldy's whitefly		*Dialeurodes kirkaldyi* (Kotinsky)	(カメムシ目、コナジラミ科) 新北区。Kirkaldy whitefly ともいう
Kirk's Charaxes		*Charaxes kirki* Butler	(チョウ目、タテハチョウ科) エチオピア区
Kiso bark beetle	キソキクイムシ	*Polygraphus kisoensis* Niijima	(コウチュウ目、キクイムシ科) 日本
kissing bug		*Triatoma infestans* Klug	(カメムシ目、サシガメ科) 新熱帯区
kissing bug			assassin bug (1) を見よ
kissing bug			masked hunter bug を見よ
kissing bugs			assassin bugs (1) を見よ
Kitale giant cupid		*Lepidochrysops kitale* (Stempffer)	(チョウ目、シジミチョウ科) エチオピア区
kitchen cockroach			common cockroach を見よ
kite-feather case	カンバマエジロツツミノガ	*Coleophora milvipennis* Zeller	(チョウ目、ツツミノガ科) 日本、旧北区
kiteflee			dragonflies (1) を見よ
kite swallowtail			scarce swallowtail を見よ
kite swallowtails	アオスジアゲハ属	*Graphium*	(チョウ目、アゲハチョウ科) の昆虫の総称
kite-tailed robberfly		*Machimus atricapillus* (Fallén)	(ハエ目、ムシヒキアブ科) 旧北区
kitten moth			alder kitten を見よ
kitten moth			sallow kitten (1) を見よ
kitten moth			poplar kitten (1) を見よ
Kitty-witch			June beetle (1) を見よ
Kitui grizzled skipper		*Spialia kituina* (Karsch)	(チョウ目、セセリチョウ科) エチオピア区
Klamathweed beetle (1)		*Chrysolina quadrigemina* (Suffrian)	(コウチュウ目、ハムシ科) 全北区、豪州区。オトギリソウの 1 種 Klamath weed (St. Johnswort ともいう) の生物的防除に次種ほかと一緒に使用された

英名	和名	学名	所属、分布、ほか
Klamath weed beetle (2)		*Chrysolina hyperici* (Forster)	(コウチュウ目、ハムシ科) 全北区、豪州区
Klamath weed midge			St. John'swort midge を見よ
Kleemann's midget		*Phyllonorycter kleemannella* (Fabricius)	(チョウ目、ホソガ科) 旧北区
Klu psyllid		*Heteropsylla fusca* Crawford	(カメムシ目、キジラミ科) 大洋区
Klug's clearwing		*Dircenna klugii* (Geyer)	(チョウ目、タテハチョウ科) 新北区、新熱帯区
Klug's Xenica			marbled Xenica を見よ
knapweed aphid		*Dactynotus jaceae* Linnaeus	(カメムシ目、アブラムシ科) 旧北区
knapweed carder bee		*Bombus ruderarius* (Müller)	(ハチ目、ミツバチ科) 旧北区
knapweed carder bee			shrill carder-bee を見よ
knapweed fritillary		*Melitaea phoebe* (Goeze)	(チョウ目、タテハチョウ科) 旧北区
knapweed gall fly		*Urophora jaceana* (Hering)	(ハエ目、ミバエ科) 旧北区
knapweed green case		*Coleophora paripannella* Zeller	(チョウ目、ツツミノガ科) 旧北区
knapweed green case (1)		*Coleophora alcyonipennella* Kollar	(チョウ目、ツツミノガ科) 旧北区、豪州区
knapweed root-borer moth		*Apapeta zoegana* (Linnaeus)	(チョウ目、ハマキガ科) 新北区
knapweed Surray case		*Coleophora conspicuella* Zeller	(チョウ目、ツツミノガ科) 旧北区
Knaus' saxinis		*Saxinis knausi* Schaeffer	(コウチュウ目、ハムシ科) 新北区
knife mole cricket			western mole cricket を見よ
knight		*Lebadea martha* (Fabricius)	(チョウ目、タテハチョウ科) 東洋区。*L. m. parkeri* Eliot も同英名
knob-tipped ringtail		*Erpetogomphus constrictor* Ris	(トンボ目、サナエトンボ科) 新熱帯区
knock beetle			furniture beetle (1) を見よ
knock beetle			common furniture beetle を見よ
knock beetle			death watch beetle (2) を見よ　カナダでの英名
knoister			bed bug (1) を見よ
knopper gall wasp			marble gall wasp を見よ
knopper gall wasp			acorn cup gall cynipid を見よ
knotgrass	ナシケンモン	*Viminia rumicis* (Linnaeus)	(チョウ目、ヤガ科) 日本、旧北区。果樹、バラ、サクラ、ポプラなどの害虫
knotgrass moth			knotgrass を見よ
knot-horn		*Scrobipalpa artemisiella* (Treitschke)	(チョウ目、キバガ科) 旧北区
knot-horn moths			cereal moths を見よ　knot-horns ともいう。
Knotweed leaf beetle		*Gastroidea polygoni* (Linnaeus)	(コウチュウ目、ハムシ科) 旧北区
koa bug		*Coleotichus blackburniae* White	(カメムシ目、キンカメムシ科) 大洋区、新北区。マメ科の木 koa に由来

英名	和名	学名	所属、分布、ほか
koa haole looper			fragile gray を見よ
koa haole moth		*Semiothisa santaremaria* (Walker)	(チョウ目、シャクガ科)。大洋区
koa haole seed weevil		*Araecerus levipennis* Jordan	(コウチュウ目、ヒゲナガゾウムシ科) 新北区、大洋区。koa haole seed beetle ともいう
koa long-horned beetle			Megopis long-horned beetle を見よ
koa moth		*Scotorythra paludicola* (Butler)	(チョウ目、シャクガ科) 新北区
koa seedworm		*Cryptophlebia illepida* (Butler)	(チョウ目、ハマキガ科) 新北区
Koch's sandgroper		*Cylindraustralia kochii* (Saussure)	(バッタ目、Cylindrachetidae) 豪州区
Kodiak ringlet		*Coenonympha kodiak* Edwards	(チョウ目、タテハチョウ科) 全北区
Koele mountain damselfly		*Megalagrion koelense* (Blackburn)	(トンボ目、イトトンボ科) 大洋区
Koh-i-noor butterfly		*Amathuxidia amythaon* Doubleday	(チョウ目、タテハチョウ科) 東洋区。東インド会社から英女王に献上されたダイアモンドに由来。*A. a. dilucida* Honrath も同英名
Kojima xyleborus	コジマキクイムシ	*Xyleborus kojimai* Murayama	(コウチュウ目、キクイムシ科) 日本
kokeshi longicorn			kokeshi longicorn beetle を見よ
kokeshi longicorn beetle	クロトラカミキリ	*Chlorophorus diadema inhirsutus* (Motschulsky)	(コウチュウ目、カミキリムシ科) 日本。カンキツ、乾材害虫
Kola nut weevil		*Sophrorhinus insperatus* Faust	(コウチュウ目、ゾウムシ科) エチオピア区
Kola nut weevil		*Batanogastris kolae* (Deshrochers)	(コウチュウ目、ゾウムシ科) エチオピア区
Kollar's wanderer		*Leptophobia penthica* (Kollar)	(チョウ目、シロチョウ科) 新熱帯区
Kolumbacs' fly		*Simulium columbaschense* (Fabricius)	(ハエ目、ブユ科) 旧北区
komani psyllid		*Leptinotarsa sulfurea* Crawford	(カメムシ目、キジラミ科) 大洋区
Kondo ground mealybug			citrus ground mealybug を見よ 米国での英名
Kondo mealybug			citrus ground mealybug を見よ
koo-tsabe fly			alkali fly を見よ
Koochahble fly			alkali fly を見よ
Koppie blue		*Lepidochrysops ortygia* (Trimen)	(チョウ目、シジミチョウ科) エチオピア区
Koppie Charaxes			foxy Charaxes を見よ
Koppie emperor			foxy Charaxes を見よ
Korean aphid	ヤドリギアブラムシ	*Tuberaphis coreanus* Takahashi	(カメムシ目、アブラムシ科) 日本、旧北区
Korean pine bark beetle	モミノコキクイムシ	*Cryphalus abietis* (Retzeburg)	(コウチュウ目、キクイムシ科) 日本、旧北区

英名	和名	学名	所属、分布、ほか
kou leafworm		*Ethmia nigroapicella* (Saalmüller)	（チョウ目、マルハキバガ科）日本、東洋区、豪州区、大洋区、新北区、エチオピア区。ムラサキ科のkouに由来
kou leafworm		*Ethmia colonella* Walsingham	（チョウ目、スエヒロキバガ科）新北区
Krause's Acraea mimic	ホソチョウシジミ	*Mimacraea krausei* Dewitz	（チョウ目、シジミチョウ科）エチオピア区
Kretschmarr cave mold beetle		*Texamaurops reddelli* Barr et Steeves	（コウチュウ目、ハネカクシ科）新北区
Kriemhild fritillary		*Clossiana kriemhild* (Strecker)	（チョウ目、タテハチョウ科）新北区
Krishna peacock	タカネクジャクアゲハ	*Papilio krishna* Moore	（チョウ目、アゲハチョウ科）東洋区
Kroon's flat		*Calleagris krooni* Vári	（チョウ目、セセリチョウ科）エチオピア区
Kroon's skipper			Kroon's flatを見よ
Krueper's small white		*Pieris krueperi* Staudinger	（チョウ目、シロチョウ科）旧北区
Kubusi streamjack		*Metacnemis valida* (Hagen)	（トンボ目、モノサシトンボ科）エチオピア区
kudzu bug			rice bug (3)を見よ
Kuehn's Jezebel	クーニイカザリシロチョウ	*Delias kuhni* Honrath	（チョウ目、シロチョウ科）東洋区
Kuehn's plushblue		*Flos kuehni* (Rober)	（チョウ目、シジミチョウ科）東洋区
Kuekenthal's tiger	キュウケンタリイアサギマダラ	*Parantica kuekenthali* (Pagenstecher)	（チョウ目、タテハチョウ科）東洋区
Kuekenthal's yellow tiger			Kuekenthal's tigerを見よ
Kumamoto xyleborus	クマモトキクイムシ	*Xyleborus kumamotoensis* Murayama	（コウチュウ目、キクイムシ科）日本
Kungu fly		*Chaoborus* sp.	（ハエ目、ケヨソイカ科）新北区。雲のように群飛する
Kuno scale	タマカタカイガラムシ	*Eulecanium kunoense* (Kuwana)	（カメムシ目、カタカイガラムシ科）日本、全北区。リンゴ、モモ、サクラなどの害虫
kurrajong leaf-tier		*Lygropia dytusalis* (Walker)	（チョウ目、メイガ科）豪州区。豪州の樹の現地名
kurrajong pod beetle		*Idaethina froggatti* Kirejtshuk et Lawrence	（コウチュウ目、ケシキスイ科）新北区、豪州区
kurrajong psyllid			kurrajong star psyllidを見よ
kurrajong seed weevil		*Tepperia sterculiae* Lea	（コウチュウ目、ゾウムシ科）豪州区
kurrajong star psyllid		*Protyora sterculiae* (Froggatt)	（カメムシ目、ネッタイキジラミ科）豪州区
kurrajong twig psyllid			kurrajong star psyllidを見よ
kurrajong wattle psyllid		*Aconopsylla sterculiae* (Froggatt)	（カメムシ目、キジラミ科）豪州区
kurrajong weevil		*Axionicus insignis* Pascoe	（コウチュウ目、ゾウムシ科）豪州区
kusamaki scale	ニッポンカキカイガラムシ	*Lepidosaphes japonica* (Kuwana)	（カメムシ目、マルカイガラムシ科）日本、旧北区

英名	和名	学名	所属、分布、ほか
Kuwana coccus	クワナカタカイガラムシ	*Lecanium kuwanai* (Kanda)	(カメムシ目、カタカイガラムシ科) 日本
Kuwana globular scale	クワナタマカイガラムシ	*Kermes kuwanae* Kanda	(カメムシ目、タマカイガラムシ科) 日本
Kuwana pear aphid	ナシハマキワタムシ	*Prociphilus kuwanai* Monzen	(カメムシ目、アブラムシ科) 日本
Kuwayama aphid	クワヤマアブラムシ	*Neorhopalomyzus lonicericola* (Takahashi)	(カメムシ目、アブラムシ科) 日本、旧北区
Kuwayama long-winged planthopper	クワヤマハネナガウンカ	*Zoraida kuwayamae* (Matsumura)	(カメムシ目、ハネナガウンカ科) 日本
Kyabobo green sapphire		*Iolaus kyabobo* Larsen	(チョウ目、シジミチョウ科) エチオピア区
kyoso fly			silkworm tachina fly (2) を見よ　蟹蛆（キョウソ）といわれることに由来
Kyoto moth	ヒマラヤスギキバガ	*Autosticha kyotensis* (Matsumura)	(チョウ目、マルハキバガ科) 日本、新北区

英名	和名	学名	所属、分布、ほか
L			
l-album wainscot		*Mythimna l-album* (Linnaeus)	(チョウ目、ヤガ科) 旧北区
L-marked tussock moth	エルモンドクガ	*Arctornis lingrum ussuricum* Bytinski-Salz	(チョウ目、ドクガ科) 日本、旧北区
La Palma grayling		*Hipparchia tilosi* Manil	(チョウ目、タテハチョウ科) 旧北区
La Plata weevil		*Sphenophorus brunnipennis* (Germar)	(コウチュウ目、ゾウムシ科) 豪州区
Labe satyr		*Cissia labe* (Butler)	(チョウ目、タテハチョウ科) 新熱帯区
labidurid earwigs			long-horned earwigs を見よ
labiid earwigs			little earwigs (1) を見よ
laboratory fruit fly			pomace fly を見よ
laboratory stick-insect			Indian stick-insect を見よ
Labrador carpet moth		*Xanthorhoe labradorensis* (Packard)	(チョウ目、シャクガ科) 新北区
Labrador skipper		*Hesperia comma borealis* Lindsey	(チョウ目、セセリチョウ科) 新北区。silver-spotted skipper (2) を参照
Labrador sulphur		*Colias nastes nastes* Boisduval	(チョウ目、シロチョウ科) 全北区
Labrador tiger moth	ダイセツヒトリ	*Apantesis quenselii* (Paykull)	(チョウ目、ヒトリガ科) 新北区。日本亜種 *A. q. daisetsuzana* Matsumura。カナダ東部の半島名に由来
laburnum aphid		*Aphis cytisorum* Hartig	(カメムシ目、アブラムシ科) 旧北区
laburnum leaf miner		*Leucoptera laburmella* (Stainton)	(チョウ目、ハモグリガ科) 全北区。laburnum leaf miner moth ともいう
labyrinth moth		*Phaecasiophora niveiguttana* Grote	(チョウ目、ハマキガ科) 新北区
lac insect	ラックカイガラムシ	*Laccifer lacca* Linnaeus	(カメムシ目、ラックカイガラムシ科) 東洋区。熱帯でイチジクの枝につき、雌の貝殻からシェラックを作る
lac insects	ラックカイガラムシ科	Kerriidae	(カメムシ目) の昆虫の総称
lac insects			scale insects を見よ
lac scale insects			lac insects を見よ
lac scales			lac insects を見よ
lace border	フチグロシロヒメシャク	*Scopula ornata* (Scopoli)	(チョウ目、シャクガ科) 日本、旧北区
lace-border wave			lace border を見よ
lace-bug		*Corythucha monacha* (Stål)	(カメムシ目、グンバイムシ科) 旧北区
lace bug			pear lace bug (1) を見よ
lace bugs	グンバイムシ科	Tingidae	(カメムシ目) の昆虫の総称
lace-capped caterpillar			white-streaked prominent moth を見よ
lace-winged roadside skipper		*Amblyscirtes aesculapius* (Fabricius)	(チョウ目、セセリチョウ科) 新北区
laced fritillary			Indian fritillary を見よ

英名	和名	学名	所属、分布、ほか
lacewing flies			brown lacewings (1) を見よ
lacewing flies			lacewings (1) を見よ
lacewings	クサカゲロウ	*Chrysopa*	(アミメカゲロウ目、クサカゲロウ科) の昆虫の総称
lacewings	ハレギチョウ属	*Cethosia*	(チョウ目、タテハチョウ科) の昆虫の総称
lacewings	ヒメカゲロウ上科	Hemerobioidea	(アミメカゲロウ目) の昆虫の総称
lacewings (1)	アミメカゲロウ目	Neuroptera	の昆虫の総称　世界に約 5,000 種。タテハチョウ科の *Cethosia* 属の種も lacewing と呼ばれる。red lacewing 参照
lacewings			green lacewings (1)(2) を見よ
Lacey's scrub-hairstreak			Alea hairstreak を見よ
lachnine plantlice	オオアブラムシ亜科	Lachninae	(カメムシ目、アブラムシ科) の昆虫の総称
lackey		*Malacosoma neustria* Linnaeus	(チョウ目、カレハガ科) 全北区。日本亜種　*M. n. testacea* (Motschulsky) オビカレハは tent caterpillar
lackey caterpillar			lackey を見よ　幼虫の英名
lackey moth			lackey を見よ
lackey moth			American tent caterpillar を見よ　北米での英名
lackey moth			forest tent caterpillar を見よ　北米での英名
lackey moths			tent caterpillar moths (1) を見よ
lacquer psylla	セグロヒメキジラミ	*Calophya nigridorsalis* Kuwayama	(カメムシ目、ヒメキジラミ科) 日本。ハゼ害虫
Lacrines metalmark		*Emesis lacrines* Hewitson	(チョウ目、シジミタテハ科) 新熱帯区
lactuca aphid	ワダンコブアブラムシ	*Myzus lactucicola* Takahashi	(カメムシ目、アブラムシ科) 日本
Lacuna moth			dark strawberry tortrix moth を見よ
Lacustra sleepy duskywing		*Erynnis brizo lacustra* (Wright)	(チョウ目、セセリチョウ科) 新熱帯区。sleepy duskywing を参照
Ladak tortoiseshell	マルバネコヒオドシ	*Aglais ladakensis* (Moore)	(チョウ目、タテハチョウ科) 東洋区
Ladakh banded Apollo	タカネウスバアゲハ	*Parnassius stoliczkanus* C. et R. Felder	(チョウ目、アゲハチョウ科) 東洋区
Ladakh clouded yellow	ラダックモンキチョウ	*Colias ladakensis* C. Felder	(チョウ目、シロチョウ科) 東洋区
ladder-backed Ethmia moth		*Ethmia delliella* (Fernald)	(チョウ目、スエヒロキバガ科) 新北区、新熱帯区
lady beetle			seven-spot ladybird を見よ
lady beetles			ladybird beetles を見よ
lady bug			seven-spot ladybird を見よ
ladybird			seven-spot ladybird を見よ
ladybird			variable ladybird を見よ
ladybird			vedalia を見よ
ladybird beetles	テントウムシ科	Coccinellidae	(コウチュウ目) の昆虫の総称

英名	和名	学名	所属、分布、ほか
ladybird parasitoid	テントウハラボソコマユバチ	*Dinocampus coccinellae* (Shrank)	(ハチ目、コマユバチ科) 日本、汎世界
ladybirds			ladybird beetles を見よ
ladybugs			ladybird beetles を見よ
ladycow			seven-spot ladybird を見よ
lady fly			dragonflies (1) を見よ
ladyfly			seven-spot ladybird を見よ
lady slipper	フトオハカマジャノメ	*Pierella hyalinus* Gmelin	(チョウ目、タテハチョウ科) 新熱帯区
lady's maid			black and orange を見よ
Laetitia's forester		*Bebearia laetitia* (Plötz)	(チョウ目、タテハチョウ科) エチオピア区
Lagora eyemark			Lagora metalmark を見よ
Lagora metalmark		*Leucochimona lagora* (Herrich-Schäffer)	(チョウ目、シジミタテハ科) 新熱帯区
lagriid beetles			long-joined beetles を見よ
laguna marine springtail		*Entomobrya laguna* Bacon	(トビムシ目、アヤトビムシ科) 新北区
lake darner		*Aeschna eremita* Kirby	(トンボ目、ヤンマ科) 新北区
lake-flies			mayflies を見よ
lake olive		*Cloeon simile* Eaton	(カゲロウ目、コカゲロウ科) 旧北区
Lake Tahoe benthic stonefly			Tahoe wingless stonefly を見よ　アメリカ西部の湖の名に由来
lakin grasshopper		*Melanoplus lakinus* (Scudder)	(バッタ目、バッタ科) 新北区
Lamas blue		*Leptotes lamasi* Balint et Johnson	(チョウ目、シジミチョウ科) 新熱帯区
lamellicorn beetles		Lamellicornia	(コウチュウ目) の昆虫の総称
lamellicorn beetles			chafer beetles を見よ
lamellicorns			chafer beetles を見よ
lamiine longhorns			flat-faced long-horns を見よ
Lampethusa banner	アカムラサキタテハ	*Epiphile lampethusa* Doubleday	(チョウ目、タテハチョウ科) 新熱帯区
Lampetia hairstreak		*Erora lampetia* Godman et Salvin	(チョウ目、シジミチョウ科) 新熱帯区
Lampeto metalmark		*Caria mantinea lampeto* Godman et Salvin	(チョウ目、シジミタテハ科) 新熱帯区。green mantle を参照
lamplight Altinote		*Altinote ozomene* (Godart)	(チョウ目、タテハチョウ科) 新熱帯区。*A. o. nox* (Bates) も同英名
lance flies			lonchaeid flies を見よ　豪州での英名
lance sergeant		*Athyma pravara helma* Fruhstorfer	(チョウ目、タテハチョウ科) 東洋区。unbroken sergeant を参照
lance-wing moths		Pterolonchidae	(チョウ目) の昆虫の総称
lance-winged moth fly		*Maruina lanceolata* (Kincaid)	(ハエ目、チョウバエ科) 新北区、新熱帯区
lanceolate Helcystogramma moth		*Helcystogramma hystricella* Braun	(チョウ目、キバガ科) 新北区。lanceolate moth ともいう

英名	和名	学名	所属、分布、ほか
lancet clubtail		*Gomphus exilis* Selys	（トンボ目、サナエトンボ科）新北区
land bugs		Geocorisae	（カメムシ目）の昆虫の総称
land bugs			true bugs を見よ
land midges			堆肥やごみに発生する小型ハエ類の俗称
Landbeck's scarce flat		*Calleagris landbecki* (Druce)	（チョウ目、セセリチョウ科）エチオピア区
lange grizzled skipper		*Muschampia proto* Ochsenheimer	（チョウ目、セセリチョウ科）旧北区
Langeberg skolly		*Thestor pictus* van Son	（チョウ目、シジミチョウ科）エチオピア区
Lange's metalmark			Mormon metalmark を見よ　Lange metalmark とも記す
Langmaid's yellow underwing			lesser broad-border を見よ
Lang's short-tailed blue		*Leptotes pirithous* (Linnaeus)	（チョウ目、シジミチョウ科）旧北区、エチオピア区
Langton's forester moth			fireweed caterpillar を見よ
languriid beetles			lizard beetles を見よ
Lannin's Nephele		*Nephele lannini* Jordan	（チョウ目、スズメガ科）エチオピア区
Lansdorf's crescent		*Eresia lansdorfi* (Godart)	（チョウ目、タテハチョウ科）新熱帯区
lantana bug			Jacarand bug を見よ　東アフリカでの英名
lantana cerambycid		*Plagiohammus spinipennis* Thomson	（コウチュウ目、カミキリムシ科）新北区、大洋区
lantana defoliator			Hypena moth を見よ
lantana defoliator caterpillar		*Hypena strigata* (Fabricius)	（チョウ目、ヤガ科）新北区
lantana flower caterpillar		*Epinotia lantana* (Busck)	（チョウ目、ハマキガ科）豪州区
lantana gall fly		*Eutreta xanthochaeta* Aldrich	（ハエ目、ミバエ科）新北区
lantana hispid			lantana leafminer (2) を見よ
lantana lace bug		*Teleonemia scrupulosa* Stål	（カメムシ目、グンバイムシ科）新北区、豪州区
lantana lace bug		*Teleonemia elata* Drake	（カメムシ目、グンバイムシ科）東洋区、豪州区、新熱帯区
lantana lace bug		*Leptobyrsa decora* Drake	（カメムシ目、グンバイムシ科）豪州区
lantana leaf beetle		*Octotoma scabriculus* Sahlberg	（コウチュウ目、ハムシ。新北区
lantana leaf beetle			lantana leafminer (1)(2) を見よ
lantana leaf-folding caterpillar			lantana leaftier を見よ
lantana leafminer		*Cremastobombycia lantanella* (Schrank)	（チョウ目、ホソガ科）新北区、新熱帯区、大洋区
lantana leafminer (1)		*Octotoma scabripennis* Guérin-Méneville	（コウチュウ目、ハムシ科）豪州区、新熱帯区
lantana leafminer (2)		*Uroplata girardi* Pic	（コウチュウ目、ハムシ科）豪州区、新北区。lantana leaf-mining beetle ともいう

英名	和名	学名	所属、分布、ほか
lantana leafmining fly		*Calycomyza lantanae* (Frick)	(ハエ目、ハモグリバエ科) 豪州区
lantana leaftier		*Salbia haemorrhoidalis* Guenée	(チョウ目、メイガ科) 新北区、新熱帯区、豪州区、大洋区。lantana leaftier moth ともいう
lantana moth		*Diastema tigris* Guenée	(チョウ目、ヤガ科) 新北区、東洋区、大洋区、豪州区、エチオピア区。lantana control moth ともいう。北米からアフリカ、豪州などにランタナ防除のため導入された
lantana plume moth		*Lantanophaga pusillidactyla* (Walker)	(チョウ目、トリバガ科) 新北区、新熱帯区、豪州区、大洋区
lantana scrub-hairstreak			smaller lantana butterfly を見よ
lantana seed fly		*Ophiomyia lantanae* Froggatt	(ハエ目、ハモグリバエ科) 新北区、大洋区、豪州区。ランタナ防除のためハワイに導入された
lantana stick moth		*Neogalea sunia* (Guenée)	(チョウ目、ヤガ科) 新北区、大洋区。幼虫は lantana stick caterpillar
lantern bug		*Enchophora sanguinea* Distant	(カメムシ目、ビワハゴロモ科) 新熱帯区
lantern bugs	ウンカ上科	Fulgoroidea	(カメムシ目、ビワハゴロモ科) の昆虫の総称
lantern bugs			lantern flies を見よ
lantern click beetle		*Pyrophorus schotti* LeConte	(コウチュウ目、コメツキムシ科) 新北区。成虫、幼虫とも光る
lantern click beetles			click beetles (2) を見よ
lantern flies	ビワハゴロモ科	Fulgoridae	(カメムシ目) の昆虫の総称 発光すると考えられていたため
Laodice untailed Charaxes		*Charaxes lycurgus* (Fabricius)	(チョウ目、タテハチョウ科) エチオピア区
Laoma spreadwing		*Potamanaxas laoma* (Hewitson)	(チョウ目、セセリチョウ科) 新熱帯区
laphriine flies		Laphriinae	(ハエ目、ムシヒキアブ科) の昆虫の総称
Lapland carrion beetle			common carrion beetle を見よ
Lapland fritillary		*Euphydryas iduna* (Dalman)	(チョウ目、タテハチョウ科) 旧北区
Lapland ringlet		*Erebia embla* (Thunberg)	(チョウ目、タテハチョウ科) 旧北区
Lapland springfly		*Diura bicaudata* (Linnaeus)	(カワゲラ目、アミメカワゲラ科) 旧北区
lappet		*Gastropacha quercifolia* Linnaeus	(チョウ目、カレハガ科) 旧北区。日本亜種は *G. q. cerridifolia* Felder et Felder ヒロバカレハ
lappet			lappet moth を見よ
lappet caterpillars			tent caterpillar moths (1) を見よ
lappet moth		*Phyllodesma americana* (Harris)	(チョウ目、カレハガ科) 新北区
lappet moth			lappet を見よ
lappet moths			tent caterpillar moths (1) を見よ
lappet moths			yarn moths を見よ
larch adelges			larch aphid (1) を見よ

英名	和名	学名	所属、分布、ほか
larch aphid		*Cinara cuneomaculata* (del Guercio)	(カメムシ目、アブラムシ科) 旧北区
larch aphid (1)	カラマツオオアブラムシ	*Cinara laricis* (Hartig)	(カメムシ目、アブラムシ科) 日本、全北区。カラマツ害虫
larch bark beetle	カラマツコキクイムシ	*Cryphalus laricis* Niijima	(コウチュウ目、キクイムシ科) 日本、旧北区。カラマツ、マツ、モミなどの害虫
larch bark beetle			larch ips を見よ
larch-boring argent		*Blastotere laevigatella* Heydenreich	(チョウ目、メムシガ科) 旧北区
larch bud midge	カラマツメタマバエ	*Dasineura nipponica* (Inouye)	(ハエ目、タマバエ科) 日本。カラマツ害虫
larch bud midge		*Dasineura laricis* (Loew)	(ハエ目、タマバエ科) 旧北区
larch bud moth			larch needleworm moth を見よ
larch bud moth			larch tortrix (2) を見よ
larch casebearer		*Coleophora lamella* Hübner	(チョウ目、ツツミノガ科) 新北区
larch casebearer	カラマツツツミノガ	*Coleophora obducta* (Meyrick)	(チョウ目、ツツミノガ科) 日本。カラマツ害虫
larch casebearer (1)		*Coleophora laricella* (Hübner)	(チョウ目、ツツミノガ科) 全北区。larch casebearer moth ともいう
larch chermes			larch aphid (1) を見よ
larch chermesid			larch aphid (1) を見よ
larch-cone gall cynipid		*Andricus fecundator* (Hartig)	(ハチ目、タマバチ科) 旧北区。artichoke gall を参照
larch cone maggot	カラマツタネバエ	*Lasiomma laricicola* (Karl)	(ハエ目、ハナバエ科) 日本、旧北区。カラマツ害虫
larch-gall piercer		*Laspeyresia zebeana* Ratzeburg	(チョウ目、ハマキガ科) 旧北区
larch geometrid moth	ミスジツマキリエダシャク	*Zethenia rufescentaria* (Motschulsky)	(チョウ目、シャクガ科) 日本。カラマツ、マツ、スギ害虫
larch ips	カラマツヤツバキクイムシ	*Ips cembrae* (Heer)	(コウチュウ目、キクイムシ科) 日本、旧北区、東洋区。カラマツ害虫
larch ladybird		*Aphidecta obliterata* (Linnaeus)	(コウチュウ目、テントウムシ科) 旧北区、エチオピア区
larch lappet moth		*Tolype laricis* (Fitch)	(チョウ目、カレハガ科) 新北区
larch leaf miner			larch casebearer (1) を見よ
larch leafroller	カラマツヒメハマキ	*Spilonota eremitana* Moriuti	(チョウ目、ハマキガ科) 日本、旧北区。カラマツ害虫
larch longhorn beetle		*Tetropium gabrieli* Weise	(コウチュウ目、カミキリムシ科) 全北区。larch longicorn beetle ともいう
larch longicorn beetle			Japanese larch bark borer を見よ
larch mining-case			larch casebearer (1) を見よ　larch-mining case とも記す
larch moth			larch pug を見よ
larch needleworm moth		*Zeiraphera improbana* Walker	(チョウ目、ハマキガ科) 新北区
larch pug	ホソカバスジナミシャク	*Eupithecia lariciata* (Freyer)	(チョウ目、シャクガ科) 日本、全北区。larch pug moth ともいう

英名	和名	学名	所属、分布、ほか
larch pug moth		*Eupithecia annulata* (Hulst)	（チョウ目、シャクガ科）新北区
larch pyralid	カラマツマダラメイガ	*Cryptoblabes loxiella* Ragonot	（チョウ目、メイガ科）日本、東洋区。カラマツ、カシ類害虫
larch ringlet	カラマツベニヒカゲ	*Erebia scipio* Boisduval	（チョウ目、タテハチョウ科）旧北区
larch sawfly	カラマツアカハバチ	*Pachynematus itoi* Okutani	（ハチ目、ハバチ科）日本。カラマツ害虫
larch sawfly	カラマツキハラハバチ	*Pristiphora wesmaeli* (Tischbein)	（ハチ目、ハバチ科）日本、旧北区。カラマツ害虫
larch sawfly	ヒメカラマツハバチ	*Anoplonyx destructor* Benson	（ハチ目、ハバチイ科）日本、旧北区
larch sawfly (1)	カラマツコハバチ	*Pristiphora politivaginata* (Takeuchi)	（ハチ目、ハバチ科）日本。カラマツ害虫
larch sawfly (2)	カラマツアカハラハバチ	*Pristiphora erichsoni* (Hartig)	（ハチ目、ハバチ科）日本、全北区
larch seed chalcid		*Megastigmus seitneri* Hoffmeyer	（ハチ目、オナガコバチ科）旧北区
larch seed chalcid (1)	カラマツノミカタビロコバチ	*Eurytoma laricis* Yano	（ハチ目、カタビロコバチ科）日本。カラマツ害虫
larch shoot borer			larch shoot moth (1) を見よ　幼虫の英名
larch shoot moth (1)	カラマツエダモグリガ	*Argyresthia laevigatella* (Herrich-Schäffer)	（チョウ目、メムシガ科）日本、全北区。カラマツ害虫
larch shoot moth		*Argyresthia laricella* Kearfott	（チョウ目、メムシガ科）新北区
larch shoot tortricid moth		*Spilonota lariciana* (Heinemann)	（チョウ目、ハマキガ科）旧北区
larch silkworm		*Hyalophora columbia* (Smith)	（チョウ目、ヤママユガ科）新北区
larch thrips		*Taeniothrips laricivorus* Kratochvil et Farsky	（アザミウマ目、アザミウマ科）旧北区
larch Tolype moth			larch lappet moth を見よ
larch tortrix (1)	ハイイロアミメヒメハマキ	*Zeiraphera griseana* (Hübner)	（チョウ目、ハマキガ科）日本、旧北区。カラマツ害虫
larch tortrix (2)		*Zeiraphera diniana* (Guenée)	（チョウ目、ハマキガ科）旧北区
larch tortrix moth			larch tortrix (1) を見よ
larch torymid	カラマツモンオナガコバチ	*Megastigmus inamurae* Yano	（ハチ目、オナガコバチ科）日本。カラマツ害虫
larch-tree moth			larch casebearer (1) を見よ
larch-tree tortrix			larch tortrix (1) を見よ
larch twist			larch webworm を見よ
larch web-spinning sawfly		*Cephalcia alpina* Klug	（ハチ目、ヒラタハバチ科）旧北区
larch webworm	カラマツイトヒキハマキ	*Ptycholomoides aeriferana* (Herrich-Schäffer)	（チョウ目、ハマキガ科）日本、旧北区。カラマツ害虫
larch weevil		*Hylobius pinastri* (Gyllenhal)	（コウチュウ目、ゾウムシ科）旧北区

英名	和名	学名	所属、分布、ほか
larder beetle	オビカツオブシムシ	*Dermestes lardarius* Linnaeus	(コウチュウ目、カツオブシムシ科) 日本、汎世界。貯蔵食品害虫
larder beetles	カツオブシムシ科	Dermestidae	(コウチュウ目) の昆虫の総称　ほとんどの種の成虫幼虫は乾燥動植物を食う
large Acraea skipper		*Fresna cojo* (Karsch)	(チョウ目、セセリチョウ科) エチオピア区
large agarista			Joseph's coat moth を見よ
large ambrosia beetle		*Platypus froggatti* Sampson	(コウチュウ目、ナガキクイムシ科) 豪州区
large American raspberry aphid			rubus aphid (2) を見よ
large angle-shades		*Trigonophora meticulosa* Linnaeus	(チョウ目、ヤガ科) 旧北区
large ant-blue		*Acrodipsas brisbanensis* (Miskin)	(チョウ目、シジミチョウ科) 豪州区
large apple fruit moth	シロヒメシンクイ	*Spilonota albicana* (Motschulsky)	(チョウ目、ハマキガ科) 日本、旧北区
large apple leaf roller	オオフタスジハマキ	*Hoshinoa adumbratana* (Walsingham)	(チョウ目、ハマキガ科) 日本
large apple tortrix	オオアトキハマキ	*Archips ingentanus* (Chrystoph)	(チョウ目、ハマキガ科) 日本、旧北区
large argent-and-sable			argent and sable を見よ
large ash bark beetle		*Hylesinus crenatus* (Fabricius)	(コウチュウ目、キクイムシ科) 旧北区
large aspen pigmy		*Nepticula argyropeza* Zeller	(チョウ目、モグリチビガ科) 旧北区
large aspen tortrix		*Choristoneura conflictana* (Walker)	(チョウ目、ハマキガ科) 新北区。large aspen tortrix moth ともいう
large auger beetle		*Bostrychopsis jesuita* (Fabricius)	(コウチュウ目、ナガシンクイムシ科) 豪州区。カンキツ害虫
large autumnal carpet		*Oporinia autumnata* Borkhausen	(チョウ目、シャクガ科) 旧北区
large backswimmer		*Notonecta insulata* Kirby	(カメムシ目、マツモムシ科) 新北区
large backswimmers		*Notonecta*	(カメムシ目、マツモムシ科) の昆虫の総称
large bagworm			Saunders's case moth を見よ
large banded awl		*Hasora khoda* (Mabille)	(チョウ目、セセリチョウ科) 豪州区
large banded grasshopper		*Arcyptera fusca* (Pallas)	(バッタ目、バッタ科) 旧北区
large banded-wing grasshopper			banded wing grasshopper を見よ
large Batrachedra moth			giant Batrachedra moth を見よ
large bee flies		*Bombylius*	(ハエ目、ツリアブ科) の昆虫の総称
large beefly			bee fly (1) を見よ
large bigeyed bug			bigeyed bug (1) を見よ
large birch pigmy		*Ectoedemia occultella* (Linnaeus)	(チョウ目、モグリチビガ科) 旧北区
large birch sawfly			birch sawfly (1) を見よ

英名	和名	学名	所属、分布、ほか
large black blossom longicorn	オオクロハナカミキリ	*Leptura thoracica* (Creutzer)	（コウチュウ目、カミキリムシ科）日本、旧北区
large black chafer	オオクロコガネ	*Holotrichia parallela* (Motschulsky)	（コウチュウ目、コガネムシ科）日本、旧北区。ダイズ、ナシ、サトイモなどの害虫
large black conifer longicorn			large black longicorn beetle を見よ
large black hunting wasp		*Priocnemis monachus* (Smith)	（ハチ目、ベッコウバチ科）豪州区
large black longhorn beetle		*Stictoleptura scutellata* (Fabricius)	（コウチュウ目、カミキリムシ科）旧北区
large black longicorn beetle	オオクロカミキリ	*Megasemum quadricostulatum* Kraatz	（コウチュウ目、カミキリムシ科）日本、旧北区。マツ、材木害虫
large black rove-beetle		*Ocypus olens* (Müller)	（コウチュウ目、ハネカクシ科）旧北区
large black-spotted geometrid	オオゴマダラエダシャク	*Parapercnia giraffata* (Guenée)	（チョウ目、シャクガ科）日本、旧北区、東洋区
large blackberry aphid			rubus aphid (2) を見よ
large blood-vein		*Calothysanis amata* Linnaeus	（チョウ目、シャクガ科）旧北区
large blue (1)		*Maculinea arionides* (Staudinger)	（チョウ目、シジミチョウ科）旧北区。日本亜種は *M. a. takamukui* (Matsumura) オオゴマシジミ
large blue	コウザンゴマシジミ	*Maculinea arion* (Linnaeus)	（チョウ目、シジミチョウ科）旧北区。large blue butterfly ともいう
large blue butterflies	ゴマシジミ属	*Maculinea*	（チョウ目、シジミチョウ科）の昆虫の総称
large blue Charaxes			blue Charaxes を見よ
large blue cuckoo wasp		*Chrysis coerulans* Fabricius	（ハチ目、セイボウ科）新北区
large blue emperor			blue Charaxes を見よ
large blue flea beetle		*Altica lythri* (Aube)	（コウチュウ目、ハムシ科）旧北区
large Bomolocha moth		*Hypena edictalis* Walker	（チョウ目、ヤガ科）新北区
large bostrichid	オオナガシンクイ	*Heterobostrychus hamatipennis* (Lesne)	（コウチュウ目、ナガシンクイムシ科）日本、東洋区、エチオピア区。乾材害虫
large branded swift		*Pelopidas subochracea* (Moore)	（チョウ目、セセリチョウ科）豪州区
large branded swift			Chinese swift を見よ
large Brazilian skipper			Brazilian skipper を見よ
large broad-mouth weevil	オオクチブトゾウムシ	*Macrocorynus variabilis* (Roelofs)	（コウチュウ目、ゾウムシ科）日本
large broadwings		*Brachycercus harrisellus* Curtis	（カゲロウ目、ヒメカゲロウ科）旧北区
large bronze azure			large brown azure を見よ
large brook dun		*Ecdyonurus torrentis* Kimmins	（カゲロウ目、ヒラタカゲロウ科）旧北区
large brown American cockroach			American cockroach を見よ

英名	和名	学名	所属、分布、ほか
large brown aphid			balsam twig aphid を見よ
large brown azure		*Ogyris idmo* Hewitson	（チョウ目、シジミチョウ科）豪州区
large brown carpet moth		*Chloroclystis dryas* Meyrick	（チョウ目、シャクガ科）豪州区
large brown cicada	アブラゼミ	*Graptopsaltria nigrofusca* (Motschulsky)	（カメムシ目、セミ科）日本、旧北区
large brown click beetle	オオカバイロコメツキ	*Ectinus dahuricus persimilis* (Lewis)	（コウチュウ目、コメツキムシ科）日本、旧北区
large brown cricket		*Brachytrupes portentosus* Lichtenstein	（バッタ目、コオロギ科）東洋区。カンキツ害虫
large brown house moth			southern old lady moth を見よ
large brown mantids		*Archimantis*	（カマキリ目、カマキリ科）の昆虫の総称
large brown mantis			stick mantis を見よ
large brown pine weevil			pine weevil (1) を見よ
large brown skipper			Araxes skipper を見よ
large brown-striped geometrid	オオトビスジエダシャク	*Ectropis excellens* (Butler)	（チョウ目、シャクガ科）日本、旧北区
large brown wainscot moth		*Leucania purdii* Fereday	（チョウ目、ヤガ科）豪州区
large bulb fly			narcissus bulb fly を見よ
large bush brown		*Bicyclus italus* (Hewitson)	（チョウ目、タテハチョウ科）エチオピア区
large cabbage-heart caterpillar			cabbage head caterpillar を見よ　large cabbage-heart caterpillar moth ともいう
large cabbage white			large white を見よ
large caddisflies	トビケラ科	Phryganeidae	（トビケラ目）の昆虫の総称
large California spanworm		*Procherodes truxaliata* (Guenée)	（チョウ目、シャクガ科）新北区
large carder bee		*Bombus muscorum* Linnaeus	（ハチ目、ミツバチ科）旧北区
large carpenter ant	ムネアカオオアリ	*Camponotus obscuripes* Mayr	（ハチ目、アリ科）日本、旧北区。木材、立木に営巣
large carpenter bees			carpenter bees を見よ　米国での英名
large carpenter bees			small carpenter bees (1) を見よ
large carrot flat-body		*Agonopterix ciliella* (Stainton)	（チョウ目、マルハキバガ科）旧北区
large case moth		*Psyche unicolor* Butler	（チョウ目、ミノガ科）旧北区
large chequered skipper	チョウセンギンボシセセリ	*Heteropterus morpheus* Pallas	（チョウ目、セセリチョウ科）旧北区
large chestnut aphid	クリオオアブラムシ	*Lachnus tropicalis* (Van der Goot)	（カメムシ目、アブラムシ科）日本、旧北区、東洋区
large chestnut weevil		*Curculio caryatrypes* (Boheman)	（コウチュウ目、ゾウムシ科）新北区
large chicken louse	マルハジラミ	*Goniodes gigas* (Taschenberg)	（ハジラミ目、チョウカクハジラミ科）日本、汎世界
large chocolate tip			chocolate-tip を見よ

英名	和名	学名	所属、分布、ほか
large ciliate blue		*Anthene lemnos* (Hewitson)	（チョウ目、シジミチョウ科）エチオピア区
large citrus butterfly	メスアカモンアゲハ	*Papilio aegeus* Donovan	（チョウ目、アゲハチョウ科）豪州区
large-clawed scarab			western rose chafer を見よ
large clouded brindle		*Apamea characterea* Hübner	（チョウ目、ヤガ科）旧北区
large clouded knot-horn			sunflower moth (1) を見よ
large clover case		*Coleophora trifolii* (Curtis)	（チョウ目、ツツミノガ科）全北区。large clover case-bearer ともいう
large coconut spathe-boring moth		*Acritocera negligens* Butler	（チョウ目、ボクトウガ科）大洋区
large common crane fly			common cranefly (2) を見よ
large common tubic		*Hoffmannophila pseudospretella* (Stainton)	（チョウ目、マルハキバガ科）全北区、豪州区
large cone-head		*Ruspolia nitidula* (Scopoli)	（バッタ目、キリギリス科）旧北区、エチオピア区
large copper	チョウセンベニシジミ（オオベニシジミ）	*Lycaena dispar* (Haworth)	（チョウ目、シジミチョウ科）旧北区。large copper butterfly ともいう
large cottony scale		*Pulvinaria mammeae* Maskell	（カメムシ目、カタカイガラムシ科）大洋区
large cottony scale (1)		*Pulvinaria amygdali* Cockerell	（カメムシ目、カタカイガラムシ科）新北区
large dagger		*Apatele cuspis* Hübner	（チョウ目、ヤガ科）旧北区
large dagger moth			broom caterpillar を見よ
large daimyo oak aphid	カシオオアブラムシ	*Lachnus roboris* (Linnaeus)	（カメムシ目、アブラムシ科）日本、旧北区
large dark olive		*Baetis rhodani* (Pictet)	（カゲロウ目、コカゲロウ科）旧北区
large dark prominent	トビマダラシャチホコ	*Notodonta torva* (Hübner)	（チョウ目、シャチホコガ科）日本、全北区
large darter		*Telicota anisodesma* Lower	（チョウ目、セセリチョウ科）豪州区
large dingy skipper		*Toxidia peron* (Latreille)	（チョウ目、セセリチョウ科）豪州区
large dingy skippy			large dingy skipper を見よ
large dragonflies			darners を見よ
large dragonfly			common blue darner を見よ
large duck louse	カモハジラミ	*Trinoton querquedulae* (Linnaeus)	（ハジラミ目、タンカクハジラミ科）日本、汎世界。カモ類に寄生
large ear	エゾショウブヨトウ	*Amphipoea lucens* (Freyer)	（チョウ目、ヤガ科）日本、旧北区
large earth humble bee			buff-tailed bumble bee を見よ
large elephant hawk			elephant hawk moth を見よ
large elephant hawk-moth			elephant hawk moth を見よ　原名亜種の英名
large elliptic water beetle		*Cybister ellipticus* LeConte	（コウチュウ目、ゲンゴロウ科）新北区

英名	和名	学名	所属、分布、ほか
large elliptic water beetle		*Cybister explanatus* (LeConte)	(コウチュウ目、ゲンゴロウ科) 新北区
large elm bark beetle		*Hylastinus crenatus* (Fabricius)	(コウチュウ目、キクイムシ科) 旧北区
large elm bark beetle			elm bark beetle (1) を見よ
large elm leaf beetle		*Monocesta coryli* (Say)	(コウチュウ目、ハムシ科) 新北区
large emerald		*Geometra papilionaria* (Linnaeus)	(チョウ目、シャクガ科) 旧北区。日本亜種は *G. p. subrigua* (Prout) オオシロオビアオシャク
large emperor			giant emperor を見よ
large emperor moth			giant emperor を見よ
large ermine knot-horn		*Myelois cribrumella* (Hübner)	(チョウ目、メイガ科) 旧北区
large fairy hairstreak		*Hypolycaena antifaunus* (Westwood)	(チョウ目、シジミチョウ科) エチオピア区
large false wireworm			striate false wireworm を見よ
large faum		*Discophora sondaica* (Boisduval)	(チョウ目、タテハチョウ科) 東洋区
large fern weevil			fern weevil (1) を見よ
large fig hornworm			fig sphinx moth を見よ
large flat			large sprite を見よ
large flat-headed pine heartwood borer		*Chalcophora virginiensis* (Drury)	(コウチュウ目、タマムシ科) 新北区、大洋区。large flat-headed pine heartwood borer beetle ともいう
large flax flea beetle		*Aphthona euphorbiae* (Schrank)	(コウチュウ目、ハムシ科) 旧北区
large fleck-winged snipefly		*Rhagio notatus* (Meigen)	(ハエ目、シギアブ科) 旧北区
large flesh fly			friendly fly を見よ
large flour beetle			dark flour beetle を見よ　カナダでの英名
large footman			four-spotted footman を見よ
large forest bob		*Scobura cephaloides* (de Nicéville)	(チョウ目、セセリチョウ科) 東洋区
large 4-lineblue		*Nacaduba pactolus* (Felder)	(チョウ目、シジミチョウ科) 東洋区、豪州区。*N. p. odon* Fruhstorfer も同英名
large fruit bark beetle			shot-hole borer (3) を見よ
large fruit-flies			fruit flies を見よ
large fruit-tree tortrix			great brown twist を見よ　large fruit-tree Tortrix moth ともいう
large garden bumble bee		*Bombus ruderatus* (Fabricius)	(ハチ目、ミツバチ科) 旧北区
large garden bumblebee		*Bombus hortorum* (Linnaeus)	(ハチ目、ミツバチ科) 旧北区
large glasswing	クロテンスカシシジミ	*Ornipholidotos peucetia* (Hewitson)	(チョウ目、シジミチョウ科) エチオピア区
large globular scale	オオタマカイガラムシ	*Kermes vastus* Kuwana	(カメムシ目、タマカイガラムシ科) 日本、旧北区
large gold grasshopper		*Chrysochraon dispar* (Germar)	(バッタ目、バッタ科) 旧北区

英名	和名	学名	所属、分布、ほか
large golden phytometra			scarce burnished brass を見よ
large goldenfort	ゴールパラヤマヒカゲ	*Lethe goalpara* (Moore)	（チョウ目、タテハチョウ科）東洋区
large grape plume moth	ブドウオオトリバ	*Platyptilia ignifera* Meyrick	（チョウ目、トリバガ科）日本、東洋区。ブドウ害虫
large grass-yellow			grass yellow を見よ
large green aporandria		*Aporandria specularia* Guenée	（チョウ目、シャクガ科）東洋区
large green-banded blue	タスキシジミ	*Danis danis* (Cramer)	（チョウ目、シジミチョウ科）豪州区
large green dun		*Ecdyonurus insignis* (Eaton)	（カゲロウ目、ヒラタカゲロウ科）旧北区
large green forester		*Bebearia barombina* (Staudinger)	（チョウ目、タテハチョウ科）エチオピア区
large green jassid		*Batracomorphus angustatus* (Osborn)	（カメムシ目、オオヨコバイ科）豪州区
large green sapphire		*Iolaus calisto* (Westwood)	（チョウ目、シジミチョウ科）エチオピア区
large green sawfly		*Perga affinis insularis* Riek	（ハチ目、Pergidae）豪州区
large green stoneflies		*Stenoperla*	（カワゲラ目、Eusteniidae）の昆虫の総称
large green underwing		*Lycaena galathea* Blanchard	（チョウ目、シジミチョウ科）東洋区
large green weevil			common leaf weevil を見よ
large greenish silk moth			lunar moth (2) を見よ
large grey		*Scoparia subfusca* Haworth	（チョウ目、メイガ科）旧北区
large grey cutworm moth		*Melanchra ustistrigata* Walker	（チョウ目、ヤガ科）豪州区
large grizzled skipper (1)		*Pyrgus albescens* Plötz	（チョウ目、セセリチョウ科）新北区、新熱帯区
large grizzled skipper (2)		*Pyrgus alveus* Hübner	（チョウ目、セセリチョウ科）旧北区
large grizzled skipper			forest sandman を見よ
large groundstreak		*Lamprospilus sethon* (Godman et Salvin)	（チョウ目、シジミチョウ科）新熱帯区
large guava blue		*Deudorix perse* Hewitson	（チョウ目、シジミチョウ科）東洋区
large hairtail			large ciliate blue を見よ
large harvester		*Megalopalpus metaleucus* Karsch	（チョウ目、シジミチョウ科）エチオピア区
large hawthorn sawfly			hawthorn sawfly を見よ
large-headed grasshopper		*Phoetaliotes nebrascensis* (Thomas)	（バッタ目、バッタ科）新北区。長翅と短翅あり
large heath	オオヒメヒカゲ（タイリクヒメヒカゲ）	*Coenonympha tullia* (Müller)	（チョウ目、タテハチョウ科）全北区。large heath butterfly ともいう
large heath			gatekeeper を見よ

英名	和名	学名	所属、分布、ほか
large hedge blue		*Celastrina huegelii* (Moore)	（チョウ目、シジミチョウ科）東洋区
large hen louse			large chicken louse を見よ　米国、豪州での英名
large hickory lecanium		*Eulecanium caryae* (Fitch)	（カメムシ目、カタカイガラムシ科）新北区
large hidden-mouth weevil	オオクチカクシゾウムシ	*Syrotelus septentrionalis* (Roelofs)	（コウチュウ目、ゾウムシ科）日本、旧北区
large hook-tip moth		*Sarisa muriferata* (Walker)	（チョウ目、シャクガ科）豪州区
large hopper			robust hopper を見よ
large horsefly			horse fly (2) を見よ
large horse-fly			great gad fly を見よ
large hydrophilid	ガムシ	*Hydrophilus acuminatus* Motschulsky	（コウチュウ目、ガムシ科）日本、旧北区、東洋区
large Hypenodes moth		*Hypenodes caducus* Walker	（チョウ目、ヤガ科）新北区
large Indian bee			giant honey bee を見よ
large Japanese alder blue longicorn	ミナミハンノオオカミキリ	*Eutetrapha chrysochloris chrysargrea* Bates	（コウチュウ目、カミキリムシ科）日本
large japyx		*Japyx diversiunguis* Silvestri	（コムシ目、ハサミコムシ科）。新北区
large jewel blue		*Polyommatus loewii* Zeller	（チョウ目、シジミチョウ科）旧北区、東洋区
large keeled Apollo		*Parnassius tianschianicus* (Oberthür)	（チョウ目、アゲハチョウ科）東洋区
large kissing bug	オオサシガメ	*Triatoma rubrofasciatus* (De Geer)	（カメムシ目、サシガメ科）日本、東洋区、新北区、汎熱帯
large lace border		*Scopula limboundata* (Haworth)	（チョウ目、シャクガ科）全北区。成虫は large lace-border moth
large lancewing		*Epermenia illigerella* (Hübner)	（チョウ目、ササベリガ科）旧北区
large lacewings			giant lacewings を見よ
large lantana butterfly	ランタナカラスシジミ	*Tmolus echion* (Linnaeus)	（チョウ目、シジミチョウ科）大洋区。ランタナ防除のためテキサスからハワイに導入された。large lantana ともいう
large lappet			lappet を見よ
large larch bark beetle			larch ips を見よ
large larch sawfly			larch sawfly (2) を見よ
large larix bark beetle			larix engraver を見よ
large leaf-cutting bee	オオハキリバチ	*Chalicodoma sculpturalis* (Smith)	（ハチ目、ハキリバチ科）日本
large leafeating ladybird		*Epilachna guttatopustulata* (Fabricius)	（コウチュウ目、テントウムシ科）豪州区
large leaf sitter		*Gorgyra afikpo* Druce	（チョウ目、セセリチョウ科）エチオピア区
large-legged thrips			jumping thrips を見よ
large long-necked leaf-cut weevil	オオツルクビオトシブミ	*Paracycnotrachelus longiceps* (Motschulsky)	（コウチュウ目、オトシブミ科）日本、旧北区
large looper moth		*Autographa ampla* Walker	（チョウ目、ヤガ科）新北区

英名	和名	学名	所属、分布、ほか
large lurid glider	ウラグロキイロタテハ	*Cymothoe hypatha* (Hewitson)	（チョウ、タテハチョウ科）エチオピア区
large Madagascar babul blue		*Azanus sitalces* (Mabille)	（チョウ目、シジミチョウ科）エチオピア区
large mango tipborer			mango shoot caterpillar (1) を見よ
large maple spanworm		*Prochoerodes transversata* Drury	（チョウ目、シャクガ科）新北区。large maple spanworm moth ともいう
large marble			creamy marblewing を見よ
large marbled bush brown		*Bicyclus mandanes* Hewitson	（チョウ目、タテハチョウ科）エチオピア区
large marbled tortrix			oak Nycteoline moth を見よ
large marsh grasshopper		*Stethophyma grossum* (Linnaeus)	（バッタ目、バッタ科）旧北区
large marsh horsefly			autumnal breeze fly を見よ
large marsh neb		*Monochroa pandalis* (Hübner)	（チョウ目、キバガ科）旧北区
large metallic oakblue		*Arhopala aedias agnis* C. et R. Felder	（チョウ目、シジミチョウ科）東洋区
large milkweed bug		*Oncopeltus fasciatus* (Dallas)	（カメムシ目、ナガカメムシ科）新北区。トウワタに多い
large moonbeam			Diana moonbeam を見よ
large mossy Lithacodia moth		*Protodeltote muscosula* (Guenée)	（チョウ目、ヤガ科）新北区。large mossy Lithacodia ともいう
large moth borer			giant moth borer を見よ
large moth borer			giant sugarcane borer を見よ
large mountain grasshopper		*Stauroderus scalaris* (Fischer-Waedheim)	（バッタ目、バッタ科）旧北区
large narcissus bulb fly			narcissus bulb fly を見よ
large narcissus fly			narcissus bulb fly を見よ
large necklace moth		*Hypsoropha monilis* (Fabricius)	（チョウ目、ヤガ科）新北区
large nutmeg		*Apamea infesta* Ochsenheimar	（チョウ目、ヤガ科）旧北区
large nutmeg			large nutwing を見よ
large nutwing		*Apamea anceps* (Denis et Schiffermüller)	（チョウ目、ヤガ科）旧北区
large oak blue		*Arhopala amantes* (Hewitson)	（チョウ目、シジミチョウ科）東洋区
large orange Acraea		*Telchinia anacreon* Trimen	（チョウ目、タテハチョウ科）エチオピア区
large orange sprite		*Celaenorrhinus meditrina* (Hewitson)	（チョウ目、セセリチョウ科）エチオピア区
large orange sulphur			orange giant sulphur を見よ
large orange tip	マルバネツマアカシロチョウ	*Colotis antevippe* (Boisduval)	（チョウ目、シロチョウ科）エチオピア区
large Paectes moth		*Paectes abrostoloides* (Guenée)	（チョウ目、ヤガ科）新北区、新熱帯区

英名	和名	学名	所属、分布、ほか
large pale clothes moth		*Tinea pallescentella* Stainton	（チョウ目、ヒロズコガ科）旧北区、豪州区、新熱帯区
large pale clothes moth		*Tinea straminella* Chambers	（チョウ目、ヒロズコガ科）新北区
large pale stonefly			stonefly を見よ
large pale-tipped black moth	オオウスヅマカラスヨトウ	*Amphipyra erebina* Butler	（チョウ目、ヤガ科）日本、旧北区
large palm forester		*Bebearia cocalioides* Hecq	（チョウ目、タテハチョウ科）エチオピア区
large parnassian	ミヤマウスバアゲハ	*Parnassius phoebus* (Fabricius)	（チョウ目、アゲハチョウ科）旧北区
large pathfinder		*Catuna angustatum* (Felder et Felder)	（チョウ目、タテハチョウ科）エチオピア区
large pathfinder skipper		*Pardaleodes tibullus* (Fabricius)	（チョウ目、セセリチョウ科）エチオピア区
large peach aphid	カワリコブアブラムシ	*Myzus varians* Davidson	（カメムシ目、アブラムシ科）日本
large pear fruitborer			pear fruit moth を見よ
large pear sucker			pear sucker を見よ
large phoenix			phoenix を見よ
large pigeon louse		*Menopon giganteum* Denny	（ハジラミ目、タンカクトリハジラミ科）旧北区
large pincertail		*Onychogomphus uncatus* (Charpentier)	（トンボ目、サナエトンボ科）旧北区
large pine aphid		*Cinara pinea* (Mordvilko)	（カメムシ目、アブラムシ科）旧北区
large pine weevil			pine weevil (1) を見よ
large pintail beetle		*Mordella antarctica* White	（コウチュウ目、ハナノミ科）豪州区
large plain stiletto		*Thereva cinifera* Meigen	（ハエ目、ツルギアブ科）旧北区
large poplar and willow borer			poplar longhorn を見よ
large poplar borer			poplar longhorn を見よ
large poplar longhorn			poplar longhorn を見よ　large poplar longhorn beetle ともいう
large poultry louse			chicken body louse を見よ　米国での英名
large powderpost beetle			capuchin beetle を見よ
large purple line-blue			six-line blue を見よ
large purplewing			Florida purplewing を見よ
large purplish gray moth		*Iridopsis vellivolata* (Hulst)	（チョウ目、シャクガ科）新北区
large ranunculus		*Polymixis flavicincta* (Denis et Schiffermüller)	（チョウ目、ヤガ科）旧北区
large raspberry aphid		*Amphorophora idaei* (Börner)	（カメムシ目、アブラムシ科）旧北区
large recluse		*Leona binoevatus* (Mabille)	（チョウ目、セセリチョウ科）エチオピア区
large red-belted clearwing		*Synanthedon culiciformis* Linnaeus	（チョウ目、スカシバガ科）全北区。large red-belted clearwing moth ともいう

英名	和名	学名	所属、分布、ほか
large red-black sawfly			wheat sawfly (1)(2) を見よ
large red damselfly		*Pyrrhosoma nymphula* (Sulzer)	(トンボ目、イトトンボ科) 旧北区
large red-spot ciliate blue		*Anthene lusones* (Hewitson)	(チョウ目、シジミチョウ科) エチオピア区
large red-tailed bumblebee			red-tailed bumble bee (1) を見よ
large reddish blunt-tipped moth			hibiscus leaf caterpillar を見よ
large rice grasshoppers		*Hieroglyphus*	(バッタ目、バッタ科) の昆虫の総称
large ringlet	ニセクモマベニヒカゲ	*Erebia euryale* (Esper)	(チョウ目、タテハチョウ科) 旧北区
large rivulet			rivulet を見よ
large roadside skipper		*Amblyscirtes exoteria* (Herrich-Schäffer)	(チョウ目、セセリチョウ科) 新北区、新熱帯区
large rose sawfly		*Arge ochropus* (Gmelin)	(ハチ目、ミフシハバチ科) 旧北区
large rove beetles	ハネカクシ亜科	Staphylininae	(コウチュウ目、ハネカクシ科) の昆虫の総称
large ruby tiger moth		*Phragmatobia assimilans* Walker	(チョウ目、ヒトリガ科) 新北区
large ruby tiger moth			line ruby tiger moth を見よ
large sabred grasshopper		*Homocoryphus nitidulus* (Scopoli)	(バッタ目、キリギリス科) 旧北区
large salmon Arab	トキイロチョウ	*Colotis fausta* (Olivier)	(チョウ目、シロチョウ科) 旧北区、東洋区、エチオピア区
large sand cricket		*Neotridactylus apicalis* (Say)	(バッタ目、ノミバッタ科) 新北区
large sand scarab		*Pericoptus truncatus* (Fabricius)	(コウチュウ目、コガネムシ科) 豪州区
large seraphim			seraphim (1) を見よ
large shoot weevil		*Rhinaria concavirostris* Lea	(コウチュウ目、ゾウムシ科) 豪州区
large shot-hole borer		*Scolytus mali* (Bechstein)	(コウチュウ目、キクイムシ科) 旧北区
large silver-spotted copper		*Argyrospodes argyraspis* (Trimen)	(チョウ目、シジミチョウ科) エチオピア区
large silver-spotted copper		*Trimenia argyroplaga* (Dickson)	(チョウ目、シジミチョウ科) エチオピア区
large silverstripe	オオヤマミドリヒョウモン	*Childrena childreni* (Gray)	(チョウ目、タテハチョウ科) 東洋区
large skipper		*Ochlodes sylvanus* (Esper)	(チョウ目、セセリチョウ科) 旧北区
large skipper (1)	コキマダラセセリ	*Ochlodes venatus* (Bremer et Grey)	(チョウ目、セセリチョウ科) 日本、東洋区、旧北区。large skipper butterfly ともいう
large skippers		Coedialinae	(チョウ目、セセリチョウ科) の昆虫の総称
large slate hairstreak		*Barangas caranus* (Stoll)	(チョウ目、シジミチョウ科) 新熱帯区
large snow flat			immaculate suffused snow flat を見よ
large sod webworm			sod webworm moth を見よ
large speckled bush cricket		*Isophya pyrenea* (Serville)	(バッタ目、キリギリス科) 旧北区

英名	和名	学名	所属、分布、ほか
large spiny stick insect		*Argosarchus spiniger* (White)	（ナナフシ目、ナナフシ科）豪州区
large-spot ermine moth			Kent ermel を見よ
large spotted Acraea	ゼーテースホソチョウ	*Acraea zetes* (Linnaeus)	（チョウ目、タテハチョウ科）エチオピア区
large-spotted Evergestis moth		*Evergestis unimacula* (Grote et Robinson)	（チョウ目、メイガ科）新北区
large spotted flat		*Celaenorrhinus patula* de Nicéville	（チョウ目、セセリチョウ科）東洋区
large spring tiger	ギフチョウ	*Luehdorfia japonica* Leech	（チョウ目、アゲハチョウ科）日本
large sprite		*Celaenorrhinus rutilans* (Mabille)	（チョウ目、セセリチョウ科）エチオピア区
large sprite			Christmas forester を見よ
large spurwing		*Procloeon pennulatum* (Eaton)	（カゲロウ目、コカゲロウ科）旧北区
large spurwing			marbled spurwing を見よ
large squaregill mayflies		Neoephemeridae	（カゲロウ目）の昆虫の総称
large stable tabby			large tabby を見よ
large stoneflies			common stoneflies (1) を見よ
large stonefly		*Perlodes mortoni* (Klapalek)	（カワゲラ目、アミメカワゲラ科）旧北区
large streaked flat		*Celaenorrhinus aspersa* Leech	（チョウ目、セセリチョウ科）東洋区
large streaked flat-body		*Agonopterix ulicetella* Stainton	（チョウ目、マルハキバガ科）旧北区
large striped blue		*Atlides polybe* (Linnaeus)	（チョウ目、シジミチョウ科）新熱帯区
large striped flea beetle		*Phyllotreta nemorum* (Linnaeus)	（コウチュウ目、ハムシ科）旧北区
large-striped grass-veneer moth		*Crambus quinquareatus* Zeller	（チョウ目、メイガ科）新北区
large striped hawk-moth		*Hippotion osiris* (Dalman)	（チョウ目、スズメガ科）エチオピア区
large striped looper moth		*Xanthorhoe clarata* Walker	（チョウ目、シャクガ科）豪州区
large striped swordtail	オナガコモンタイマイ	*Graphium antheus* (Cramer)	（チョウ目、アゲハチョウ科）エチオピア区
large suffused snow flat			immaculate suffused snow flat を見よ
large sugarcane moth borer			cane-moth borer を見よ
large sulphur			orange-barred sulphur を見よ
large summer dun		*Siphlonurus*	（カゲロウ目、フタオカゲロウ科）の昆虫の総称
large swift			ghost moth を見よ
large sword tailed bush cricket		*Polysarcus denticauda* (Charpentier)	（バッタ目、キリギリス科）旧北区
large syrphid			large syrphid fly を見よ

英名	和名	学名	所属、分布、ほか
large syrphid fly		*Scaeva pyrastri* (Linnaeus)	(ハエ目、ハナアブ科) 旧北区、新北区
large tabby		*Aglossa pinguinalis* (Linnaeus)	(チョウ目、メイガ科) 旧北区、豪州区、エチオピア区。large tabby moth ともいう
large tawny wall	アカマダラジャノメ	*Rhaphicera satricus* (Doubleday)	(チョウ目、タテハチョウ科) 東洋区
large teak borer		*Duomitus ceramicus* Walker	(チョウ目、ボクトウガ科) 東洋区
large thatch groundling		*Platyedra vilella* Zeller	(チョウ目、キバガ科) 旧北区
large thatch groundling (1)		*Platyedra subcinerea* (Haworth)	(チョウ目、キバガ科) 全北区
large thorn		*Ennomos autumnaria* (Werneburg)	(チョウ目、シャクガ科) 日本 (沖縄)、旧北区、東洋区。日本の2亜種 *E. a. intermedia* Inoue キリバエダシャク日本 (北)、サハリン。*E. a. nephotropa* Prout 日本 (本、四、九)
large thorn			canary-shouldered thorn を見よ
large three-ring		*Ypthima nareda* (Kollar)	(チョウ目、タテハチョウ科) 東洋区
large three-ring		*Ypthima motschulskyi* (Bremer et Grey)	(チョウ目、タテハチョウ科) 東洋区。日本亜種は *Y. m. niphonica* Murayama ウラナミジャノメ
large tiger			tiger mimic-queen (1) を見よ
large tiger blue			Hewitson's forest blue を見よ
large tiger longicorn			large tiger longicorn beetle を見よ
large tiger longicorn beetle	オオトラカミキリ	*Xylotrechus villioni* Villard	(コウチュウ目、カミキリムシ科) 日本、旧北区
large Tolype		*Tolype velleda* (Stoll)	(チョウ目、カレハガ科) 新北区。large Tolype moth ともいう
large tortoiseshell	ヨーロッパヒオドシチョウ	*Nymphalis polychloros* Linnaeus	(チョウ目、タテハチョウ科) 旧北区。large tortoiseshel butterfly ともいう
large tree nymph			Siam tree nymph を見よ
large turkey louse	ツノハジラミ	*Chelopistes meleagridis* (Linnaeus)	(ハジラミ目、チョウカクハジラミ科) 日本、全北区
large twentyeight-spotted lady beetle			potato ladybird を見よ
large twin-spot carpet		*Xanthorhoe quadrifasciata* (Clerck)	(チョウ目、シャクガ科) 旧北区。日本亜種は *X. q. ignobilis* (Butler) ヨスジナミシャク
large two-spotted leafhopper			green rice leafhopper (2) を見よ
large vagrant	ワタリウスアオシロチョウ	*Nepheronia argia varia* (Trimen)	(チョウ目、シロチョウ科) エチオピア区。*N. argia* (Fabricius) も同英名
large variable diadem	スルスミムラサキ	*Hypolimnas dinarcha* (Hewitson)	(チョウ目、タテハチョウ科) エチオピア区
large variable eggfly			large variable diadem を見よ
large velvet bush brown		*Bicyclus sophrosyne* (Plotz)	(チョウ目、タテハチョウ科) エチオピア区
large vine hawk-moth			elephant hawk moth を見よ
large wainscot	ヨシヨトウ	*Rhizedra lutosa* (Hübner)	(チョウ目、ヤガ科) 日本、旧北区。large wainscot moth ともいう

英名	和名	学名	所属、分布、ほか
large wall brown	マイラツマジロウラジャノメ	*Lasiommata maera* (Linnaeus)	（チョウ目、タテハチョウ科）旧北区。large wall ともいう
large wallaby louse fly		*Ortholfersia macleayi* (Leach)	（ハエ目、シラミバエ科）豪州区
large walnut aphid		*Callaphis juglandis* (Goeze)	（カメムシ目、アブラムシ科）旧北区、東洋区
large water boatman		*Hesperocorixa laevigata* (Uhler)	（カメムシ目、ミズムシ科）新北区
large wax moth			greater wax moth を見よ
large weevil			Japanese giant weevil を見よ
large western speckle-wing		*Callibaetis californicus* Banks	（カゲロウ目、コカゲロウ科）新北区
large wheat shoot fly			late wheat shoot fly を見よ
large whirligig beetles		*Dineutus*	（コウチュウ目、ミズスマシ科）の昆虫の総称
large white	オオモンシロチョウ	*Pieris brassicae* (Linnaeus)	（チョウ目、シロチョウ科）日本、旧北区。large white butterfly ともいう
large white-banded blackish geometrid	オオシロオビクロナミシャク	*Rheumaptera hastata rikovskensis* (Matsumura)	（チョウ目、シャクガ科）日本、旧北区
large white-faced darter			yellow-spotted whiteface を見よ
large white flat		*Satarupa gopala* (Moore)	（チョウ目、セセリチョウ科）東洋区
large white-headed leafhopper	シロズオオヨコバイ	*Oniella leucocephala* Matsumura	（カメムシ目、カンムリヨコバイ科）日本
large white plume			white plume moth (1) を見よ
large white skipper		*Heliopetes ericetorum* Boisduval	（チョウ目、セセリチョウ科）新北区、新熱帯区
large white-spots		*Osmodes laronia* (Hewitson)	（チョウ目、セセリチョウ科）エチオピア区
large widow		*Torynesis magna* (van Son)	（チョウ目、タテハチョウ科）エチオピア区
large willow aphid	ヤナギコブオオアブラムシ	*Tuberolachnus salignus* (Gmelin)	（カメムシ目、アブラムシ科）日本、全北区。ヤナギ害虫
large wood nymph	ベニジャノメ	*Cercyonis pegala* (Fabricius)	（チョウ目、タテハチョウ科）新北区。*C. p. nephele* (Kirby) は wood nymph
large yellow		*Citrinophila erastus* (Hewitson)	（チョウ目、シジミチョウ科）。エチオピア区
large yellow tussock caterpillar		*Calliteara horsfieldii* Saunders	（チョウ目、ドクガ科）東洋区
large yellow underwing moth			common yellow underwing moth を見よ large yellow underwing ともいう
large Yeomen	ウスチャタテハ	*Cirrochroa aoris* Doubleday	（チョウ目、タテハチョウ科）東洋区
larger apple curculio			western apple curculio を見よ
larger azalea lacebug	サツマグンバイ	*Stephanitis propinqua* Horvath	（カメムシ目、グンバイムシ科）日本、旧北区
larger black flour beetle		*Cynaeus angustus* (LeConte)	（コウチュウ目、ゴミムシダマシ科）新北区
larger black flour beetle (1)	コゴメゴミムシダマシ	*Latheticus oryzae* Waterhouse	（コウチュウ目、ゴミムシダマシ科）日本、汎世界。貯穀害虫

英名	和名	学名	所属、分布、ほか
larger blotch-marked bell		*Epiblema scutulana* (Denis et Schiffermüller)	(チョウ目、ハマキガ科) 旧北区
larger boxelder leafroller moth		*Archips negundana* (Dyar)	(チョウ目、ハマキガ科) 新北区。larger boxelder leafroller ともいう
larger brown bush crickets			bush crickets (2) を見よ
larger bud moth			common long-cloaked marble を見よ
larger burdock weevil	オオゴボウゾウムシ	*Larinus meleagris* Petri	(コウチュウ目、ゾウムシ科) 日本
larger cabbage cutworm			cutworm (3) を見よ
larger cabinet beetle		*Trogoderma versicolor* (Creutzer)	(コウチュウ目、カツオブシムシ科) 旧北区
larger cabinet beetle			warehouse beetle (1) を見よ
larger chestnut weevil		*Curculio probuscideus* Fabricius	(コウチュウ目、ゾウムシ科) 新北区
larger dingy marble		*Endothenia oblongana* (Haworth)	(チョウ目、ハマキガ科) 旧北区
larger elm bark beetle			elm bark beetle (1) を見よ
larger elm leaf beetle			large elm leaf beetle を見よ
larger elongate weevil	オオカツオゾウムシ	*Lixus divaricatus* Motschulsky	(コウチュウ目、ゾウムシ科) 日本、旧北区
larger ensign wasp		*Evania appendigaster* Linnaeus	(ハチ目、ヤセバチ科) 大洋区
larger European bark beetle			elm bark beetle (1) を見よ
larger European bark beetle			large elm bark beetle を見よ　米国での英名
larger frosted chafer	オオコフキコガネ	*Melolontha frater* Arrow	(コウチュウ目、コガネムシ科) 日本
larger grain borer		*Prostephanus truncatus* (Horn)	(コウチュウ目、ナガシンクイムシ科) 新北区
larger grained moth	オオモクメシャチホコ	*Cerura menciana* Moore	(チョウ目、シャチホコガ科) 日本
larger green damselfly			emerald damselfly を見よ
larger green weevil	オオアオゾウムシ	*Chlorophanus grandis* Roelofs	(コウチュウ目、ゾウムシ科) 日本
larger greenish narrow-winged noctuid	オオホソアオバヤガ	*Ochropleura praecurrens* (Staudinger)	(チョウ目、ヤガ科) 日本、旧北区
larger Hawaiian cutworm		*Agrotis crinigera* (Butler)	(チョウ目、ヤガ科) 新北区
larger horned citrus bug			spined citrus bug を見よ
larger irregular-marked noctuid	オオウンモンクチバ	*Mocis undata* (Fabricius)	(チョウ目、ヤガ科) 日本、東洋区、豪州区、大洋区
larger Japanese earwig	オオハサミムシ	*Labidura riparia japonica* (de Haan)	(ハサミムシ目、オオハサミムシ科) 日本、汎世界

英名	和名	学名	所属、分布、ほか
larger lantana butterfly			large lantana butterfly を見よ
larger manchurian ash bark beetle	ヤチダモノオオキクイムシ	*Hylesinus nobilis* Blandford	(コウチュウ目、キクイムシ科) 日本、旧北区
larger meadow katydids		*Orchelimum*	(バッタ目、キリギリス科) の昆虫の総称
larger mugwort leaf beetle	オオヨモギハムシ	*Chrysolina angusticollis* (Motschulsky)	(コウチュウ目、ハムシ科) 日本、旧北区
larger oraesia			fruit piercing moth (6) を見よ
larger pale booklouse			booklouse (2) を見よ
larger pale trogiid			booklouse (2) を見よ
larger pellucid hawk moth	オオスカシバ	*Cephonedes hylas* Linnaeus	(チョウ目、スズメガ科) 日本、旧北区、東洋区、豪州区、エチオピア区。クチナシ害虫
larger peppermint small weevil	オオハッカヒメゾウムシ	*Baris pilosa* Roelofs	(コウチュウ目、ゾウムシ科) 日本
larger pine looper			satin beauty を見よ
larger pine scolytid			larch ips を見よ
larger pine shoot borer		*Hylesinus piniperda* Linnaeus	(コウチュウ目、キクイムシ科) 旧北区
larger pine weevil		*Pissodes pini* (Linnaeus)	(コウチュウ目、ゾウムシ科) 旧北区
larger pitch borer			(コウチュウ目、キクイムシ科) の昆虫
larger pith borer			pine shoot beetle (1) を見よ
larger potato lady beetle			potato ladybird を見よ
larger refuse beetle	オオゴモクムシ	*Harpalus capito* Morawitz	(コウチュウ目、オサムシ科) 日本、旧北区
larger rice bug	オオクモヘリカメムシ	*Anacanthocoris striicornis* (Scott)	(カメムシ目、ヘリカメムシ科) 日本、旧北区、東洋区。カンキツ、カキなどの害虫
larger rice crane fly	ドウボソガガンボ	*Tipula longicauda* Matsumura	(ハエ目、ガガンボ科) 日本
larger roadside-skipper		*Amblyscirtes folia* Godman	(チョウ目、セセリチョウ科) 新熱帯区
larger round-necked flattened longicorn			opaque sawyer を見よ
larger Sakhalin fir bark beetle	トドマツオオキクイムシ	*Xyleborus validus* Eichhoff	(コウチュウ目、キクイムシ科) 日本、東洋区
larger saltmarsh case		*Coleophora maritimella* Newman	(チョウ目、ツツミノガ科) 旧北区
larger shot-hole borer			shot-hole borer (3) を見よ
larger squash bug			larger rice bug を見よ
larger striated chafer	オオスジコガネ	*Mimela costata* (Hope)	(コウチュウ目、コガネムシ科) 日本
larger wax moth			greater wax moth を見よ
larger white-banded greenish geometrid			large emerald を見よ
larger white-hindwinged catocala	オオシロシタバ	*Catocala lara* Bremer	(チョウ目、ヤガ科) 日本、旧北区
larger white longicorn beetle	オオシロカミキリ	*Olenecamptus eretaceus* Bates	(コウチュウ目、カミキリムシ科) 日本

英名	和名	学名	所属、分布、ほか
larger yamfly		*Loxura cassiopeia cassiopeia* Distant	（チョウ目、シジミチョウ科）東洋区
larger yellow ant		*Acanthomyops interjectus* (Mayr)	（ハチ目、アリ科）新北区
larger zigzag-marked leafhopper	オオイナズマヨコバイ	*Metalimnus formosus* (Boheman)	（カメムシ目、オオヨコバイ科）日本、旧北区
larix engraver	カラマツキクイムシ	*Orthotomiicus laricis* (Fabricius)	（コウチュウ目、キクイムシ科）日本、旧北区
larrid wasps			sandloving wasps を見よ
Larsen's Epitolina		*Epitolina larseni* Libert	（チョウ目、シジミチョウ科）エチオピア区
larva			昆虫、とくに完全変態類の幼虫の英名
lash-faced barklouse		*Trichopsocus clarus* (Banks)	（チャタテムシ目、Trichopsocidae）旧北区、豪州区
Lassaux's sphinx moth		*Erinnyis lassauxii* (Boisduval)	（チョウ目、スズメガ科）新北区、新熱帯区。Lassaux's sphinx ともいう。
Last's albatross		*Appias lasti* (Grose-Smith)	（チョウ目、シロチョウ科）エチオピア区
Last's ciliate blue		*Anthene lasti* (Grose-Smith et Kirby)	（チョウ目、シジミチョウ科）エチオピア区
Lasus metalmark		*Perophthalma lasus* (Westwood)	（チョウ目、シジミタテハ科）新熱帯区
Latagus hairstreak		*Lathecla latagus* (Godman et Salvin)	（チョウ目、シジミチョウ科）新熱帯区
Latania scale	ヤシシロマルカイガラムシ	*Hemiberlesia lataniae* (Signoret)	（カメムシ目、マルカイガラムシ科）日本、東洋区、新北区、汎熱帯。カンキツ、クワ、チャ、ハイビスカスなどの害虫
late cattle grub			cattle warble fly (1) を見よ　幼虫の英名
late wheat shoot fly		*Phorbia securis* Tiensuu	（ハエ目、ハナバエ科）旧北区
lateral lady beetle		*Hyperaspis lateralis* Mulsant	（コウチュウ目、テントウムシ科）新北区
lateral ladybird beetle			lateral lady beetle を見よ
lateral lined armyworm			velvet armyworm moth を見よ
laterite ochre		*Trapezites waterhouse* Mayo et Atkins	（チョウ目、セセリチョウ科）豪州区
Latham's Eucosma moth		*Eucosma lathami* Forbes	（チョウ目、ハマキガ科）新北区
lathridiids			plaster beetles を見よ
Lathy's great satyr		*Pronophila unifasciata* (Lathy)	（チョウ目、タテハチョウ科）新熱帯区
latin	ムラサキツマキリヨトウ	*Callopistria juventina* (Stoll)	（チョウ目、ヤガ科）日本、旧北区
latin moth			latin を見よ
Latreille's Altinote			orange-disked Altinote を見よ
Latreille's 89 butterfly		*Diaethria euclides* (Latreille)	（チョウ目、タテハチョウ科）新熱帯区
Latreille's latin		*Callopistria latreillei* (Duponchel)	（チョウ目、ヤガ科）旧北区、東洋区、エチオピア区

英名	和名	学名	所属、分布、ほか
Latreille's skipper		*Gegenes hottentota* (Latreille)	(チョウ目、セセリチョウ科) エチオピア区
latrine fly	コブアシヒメイエバエ	*Fannia scalaris* (Fabricius)	(ハエ目、ヒメイエバエ科) 日本、汎世界
lattice brown	ロクセラナキマダラモドキ	*Kirinia roxelana* Cramer	(チョウ目、タテハチョウ科) 旧北区
lattice moths		Arrhenophanidae	(チョウ目) の昆虫の総称
latticed heath	ヒメアミメエダシャク	*Chiasmia clathrata* (Linnaeus)	(チョウ目、シャクガ科) 日本、旧北区
laugher			tree-lichen beauty を見よ
laugher moth		*Charadra deridens* (Guenée)	(チョウ目、ヤガ科) 新北区。laugher ともいう
Laura's clubtail		*Stylurus laurae* Williamson	(トンボ目、サナエトンボ科) 新北区
Laura's leafwing		*Memphis laura balboa* (Hall)	(チョウ目、タテハチョウ科) 新熱帯区。Laura leafwing ともいう
Laure	シロオビアメリカコムラサキ	*Doxocopa laure* (Drury)	(チョウ目、タテハチョウ科) 新北区、新熱帯区。Laure butterfly ともいう
laurel psyllid			bay sucker を見よ
laurel scale			laurel-tree scale を見よ
laurel sphinx		*Sphinx kalmiae* J. E. Smith	(チョウ目、スズメガ科) 新北区。laurel sphinx moth ともいう
laurel swallowtail			Palamedes swallowtail を見よ
laurel-tree scale		*Aonidia lauri* Bouché	(カメムシ目、マルカイガラムシ科) 旧北区
Laurentian skipper		*Hesperia comma laurentina* (Lyman)	(チョウ目、セセリチョウ科) 新北区。silver-spotted skipper (2) を参照
Laurentian tiger beetle		*Cicindela denikei* Brown	(コウチュウ目、ハンミョウ科) 新北区
Laureolus skipper		*Cymaenes laurelolus laureolus* (Schaus)	(チョウ目、セセリチョウ科) 新熱帯区
Lausus hairstreak		*Thereus lausus* (Cramer)	(チョウ目、シジミチョウ科) 新熱帯区
lauxanid fly		*Minettia lupulina* (Fabricius)	(ハエ目、シマバエ科) 日本、新北区
lauxaniid flies	シマバエ科	Lauxaniidae	(ハエ目) の昆虫の総称
lauxaniids			lauxaniid flies を見よ
lavender count	コキタスコイナズマ	*Tanaecia cocytus* (Fabricius)	(チョウ目、タテハチョウ科) 東洋区
lavender dancer		*Argia hinei* Kennedy	(トンボ目、イトトンボ科) 新北区
Laverna Calephelis		*Calephelis laverna laverna* (Godman et Salvin)	(チョウ目、シジミタテハ科) 新熱帯区
Laverna emperor		*Apatura laverna* Leech	(チョウ目、タテハチョウ科) 旧北区
Laviana skipper		*Heliopetes laviana* (Hewitson)	(チョウ目、セセリチョウ科) 新北区、新熱帯区
Laviana white skipper			Laviana skipper を見よ
Lavinia clearwing			Lavinia glasswing を見よ
Lavinia emperor	アカモンルリオビコムラサキ	*Doxocopa lavinia* (Butler)	(チョウ目、タテハチョウ科) 新熱帯区

英名	和名	学名	所属、分布、ほか
Lavinia glasswing		*Hypoleria lavinia* Hewitson	（チョウ目、タテハチョウ科）新熱帯区
lawn armyworm			rice armyworm を見よ
lawn armyworm parasite		*Apanteles marginiventris* (Cresson)	（ハチ目、コマユバチ科）大洋区
lawn caterpillar			dark mottled willow を見よ
lawn chinch bug		*Blissus insularis* Barber	（カメムシ目、ナガカメムシ科）新北区。イネ害虫
lawn fly		*Hydrellia tritici* Coquillett	（ハエ目、ミギワバエ科）豪州区
lawn grass bagworm	シバミノガ	*Nipponopsyche fuscescens* Yazaki	（チョウ目、ミノガ科）日本。シバ害虫
lawn grass cutworm	スジキリヨトウ	*Spodoptera depravata* (Butler)	（チョウ目、ヤガ科）日本、旧北区。イネ科牧草、シバの害虫
lawn ground cricket	ネッタイシバスズ	*Dianemobius taprobanensis* (Walker)	（バッタ目、コオロギ科）日本、東洋区
lawn leafhopper		*Recila hospes* (Kirkaldy)	（カメムシ目、オオヨコバイ科）新北区、大洋区
lawn moths			grass moths (1) (2) を見よ
lawn springtail		*Bourletiella arvalis* (Fitch)	（トビムシ目、マルトビムシ科）新北区
lawn webworm	ツトガ	*Ancylolomia japonica* Zeller	（チョウ目、メイガ科）日本、旧北区、東洋区。イネ、シバ害虫
lax beetles			blister beetles (3) を見よ
layman		*Amauris albimaculata* Butler	（チョウ目、タテハチョウ科）エチオピア区
Lazy mountain satyr		*Eretris subrufescens* (Smith et Kirby)	（チョウ目、タテハチョウ科）新熱帯区
Le Cerf's grayling		*Hipparchia miguelensis* (Le Cerf)	（チョウ目、タテハチョウ科）旧北区
Le Cerf's white Charaxes		*Charaxes lecerfi* Lathy	（チョウ目、タテハチョウ科）エチオピア区
Le Doux's glassy Acraea		*Acraea endoscota* Le Doux	（チョウ目、タテハチョウ科）エチオピア区
Le Gras pierrot		*Tarucus legrasi* Stempffer	（チョウ目、シジミチョウ科）エチオピア区
Leach's grass-veneer moth		*Crambus leachellus* (Zincken)	（チョウ目、メイガ科）新北区。Leach's grass-veneer ともいう
lead belle		*Scotopteryx mucronata umbrifera* (Heydeman)	（チョウ目、シャクガ科）旧北区
lead cable borer		*Scobicia declivis* (LeConte)	（コウチュウ目、ナガシンクイムシ科）新北区。電線、カシ材を加害
lead-colored lichen moth		*Cisthene plumbea* Stretch	（チョウ目、ヒトリガ科）新北区
lead-coloured case		*Coleophora murinipennella* (Duponchel)	（チョウ目、ツツミノガ科）旧北区
lead-coloured drab		*Orthosia populeti* (Fabricius)	（チョウ目、ヤガ科）旧北区
lead-coloured pug		*Eupithecia plumbeolata* (Haworth)	（チョウ目、シャクガ科）旧北区
lead-streaked piercer		*Cydia gemmiferana* Treitschke	（チョウ目、ハマキガ科）旧北区

英名	和名	学名	所属、分布、ほか
Leada spreadwing		*Carrhenes leada* Butler	(チョウ目、セセリチョウ科) 新熱帯区
leaden ciliate blue			black-striped hairtail を見よ
leaden-fringed drill		*Dichrorampha plumbagana* (Treitschke)	(チョウ目、ハマキガ科) 旧北区
leaden hairtail			black-striped hairtail を見よ
leaden spider wasp		*Pompilus cinereus* Fabricius	(ハチ目、ベッコウバチ科) 日本、旧北区、エチオピア区、東洋区、豪州区
leaf and blossom aphis			oat bird-cherry aphid を見よ
leaf-and-bud weevils	チョッキリゾウムシ亜科	Rhynchitinae	(コウチュウ目、オトシブミ科) の昆虫の総称
leaf and planthoppers			true leaf hoppers を見よ
leaf bagworm			orange case moth を見よ
leafbeet pyralid	ヒメクロミスジノメイガ	*Hedylepta misera* (Butler)	(チョウ目、メイガ科) 日本、旧北区
leaf beetle	ポプラハムシ	*Chrysomela tremula* Fabricius	(コウチュウ目、ハムシ科) 日本、旧北区
leaf beetles (1)	ハムシ科	Chrysomelidae	(コウチュウ目) の昆虫の総称
leaf beetles (2)	ハムシ亜科	Chrysomelinae	(コウチュウ目、ハムシ科) の昆虫の総称
leaf beetles (3)		*Chrysomela*	(コウチュウ目、ハムシ科) の昆虫の総称
leaf blotch miner moth			madrone skin miner を見よ
leaf blotch miner moths			leaf blotch-miners を見よ
leafblotch miners (1)	ホソガ科	Gracillariidae	(チョウ目) の昆虫の総称
leaf blotch-miners		*Phyllonorycter*	(チョウ目、ホソガ科) の昆虫の総称
leaf blue			purple leaf blue を見よ
leaf bugs			plant bugs (1) を見よ
leaf-bunching currant aphid			gooseberry aphid (1) を見よ
leaf butterflies	コノハチョウ属	*Kallima*	(チョウ目、タテハチョウ科) の昆虫の総称
leaf butterflies			leafwings を見よ
leaf butterfly	マレーコノハチョウ	*Kallima limborgii amplirufa* Fruhstorfer	(チョウ目、タテハチョウ科) 東洋区
leaf butterfly			orange oakleaf を見よ
leaf case moth			orange case moth を見よ
leaf commodore			dry leaf を見よ
leaf crumpler			apple leaf crumpler を見よ leaf crumpler moth ともいう
leaf-curl plum aphid	ムギワラギクオマルアブラムシ	*Brachycaudus helichrysi* (Kaltenbach)	(カメムシ目、アブラムシ科) 日本、汎世界。ナシ、モモ、ウメなどの害虫
leaf-curling apple midge			apple leaf midge を見よ
leaf-curling currant aphid			currant-lettuce aphid を見よ
leaf-curling pear midge			pear leaf roller (2) を見よ

英名	和名	学名	所属、分布、ほか
leaf curling plum aphid			leaf-curl plum aphid を見よ
leaf curling plum aphid			oat bird-cherry aphid を見よ
leaf cutter ant			leaf cutting ant (1) を見よ
leafcutter ants		*Atta, Acromyrmex*	（ハチ目、アリ科）の昆虫の総称　植物の葉を切って巣にもち帰り、唾液で菌を培養して餌とする
leafcutter bee			common leaf-cutter bee を見よ　leaf-cutter bee とも記す。
leafcutter bees			leaf-cutting bees (1)(2) を見よ　leaf-cutter bees とも記す
leafcutter moths			yucca moths (1) を見よ
leaf cutters			agricultural ants を見よ
leaf cutting ant		*Atta octospinosa* Reich	（ハチ目、アリ科）新熱帯区。カンキツ害虫
leaf cutting ant		*Atta sexdens* Linnaeus	（ハチ目、アリ科）新熱帯区。カンキツ害虫
leaf cutting ant (1)		*Atta cephalotes* Linnaeus	（ハチ目、アリ科）新熱帯区。カンキツ害虫
leaf cutting ant			Texas leaf cutting ant を見よ
leaf-cutting ants		*Acromyrmex*	（ハチ目、アリ科）の昆虫の総称　leafcutter ants を参照
leaf-cutting ants			leafcutter ants を見よ
leafcutting bee		*Megachile pascoensis* Mitchell	（ハチ目、ハキリバチ科）新北区
leaf-cutting bee		*Melipona testacea* Klug	（ハチ目、ミツバチ科）新熱帯区
leaf-cutting bee			short leafcutter bee を見よ
leaf-cutting bees (1)	ハキリバチ科	Megachilidae	（ハチ目）の昆虫の総称
leaf-cutting bees (2)		*Megachile*	（ハチ目、ハキリバチ科）の昆虫の総称　leafcutting bees とも記す
leafcuttig bees			leaf-cutting bees (2) を見よ
leaf-eating beetles			leaf beetles (2) を見よ
leaf-eating crane fly		*Cylindrotoma splendens* Doane	（ハエ目、ガガンボ科）新北区
leaf-eating lady beetles		Epilachnini	（コウチュウ目、テントウムシ科）の昆虫の総称
leaf eating weevil			brown leaf weevil を見よ
leaf eating weevil			citrus leafeating weevil を見よ
leaf flies			leafminer flies を見よ
leaffolder			rice leafroller を見よ　豪州での英名
leaf-footed bug (1)		*Leptoglossus phyllopus* (Linnaeus)	（カメムシ目、ヘリカメムシ科）新北区
leaf-footed bug		*Acanthocephala terminalis* (Dallas)	（カメムシ目、ヘリカメムシ科）新北区
leaf-footed bugs		*Acanthocephala*	（カメムシ目、ヘリカメムシ科）の昆虫の総称
leaf-footed bugs			coreid bugs を見よ
leaf-footed flies			coreid bugs を見よ

英名	和名	学名	所属、分布、ほか
leaffooted pine seed bug		*Leptoglossus corculus* (Say)	（カメムシ目、ヘリカメムシ科）新北区
leaf-footed plant bug (1)	アシビロヘリカメムシ	*Fabrictilis gonagra* (Fabricius)	（カメムシ目、ヘリカメムシ科）日本、東洋区、大洋区、豪州区
leaf-footed plant bug (2)		*Leptoglossus australis* (Fabricius)	（カメムシ目、ヘリカメムシ科）エチオピア区、東洋区、豪州区
leaf-gall midge			mango gall fly を見よ
leaf gall thrips		*Liothrips karnyi* (Bagnall)	（アザミウマ目、クダアザミウマ科）東洋区
leaf-horned beetles			lamellicorn beetles を見よ
leaf insect		*Phobaeticus philippinicus* (Hennemann et Conle)	（ナナフシ目、コノハムシ科）東洋区
leaf insects			stick insects を見よ　*Phyllium* 属など。walking leaf ともいう。木の葉虫
leaf insects	コノハムシ科	Phylliidae	（ナナフシ目）の昆虫の総称
leaf litter barklice	スカシチャタテ科	Hemipsocidae	（チャタテムシ目）の昆虫の総称
leaf litter barklice			hemipeplid beetles を見よ
leaf litter moth			four-spotted yellowneck moth を見よ
leaf-locust		*Pseudophyllus prasinus* (Karny)	（バッタ目、キリギリス科）東洋区
leaf longitudinal roller		*Laspeyresia leucostoma* Meyrick	（チョウ目、ハマキガ科）旧北区、東洋区。チャ害虫
leaf mantis		*Choeradodis strumaria* (Linnaeus)	（カマキリ目、カマキリ科）新熱帯区
leaf mantis			tropical shield mantis を見よ
leaf midges		*Dasineura*	（ハエ目、タマバエ科）の昆虫の総称
leaf miner		*Promecotheca papuana* Cziki	（コウチュウ目、ハムシ科）豪州区
leaf miner		*Acrocercops cathedraea* Meyrick	（チョウ目、ホソガ科）日本、東洋区
leaf miner			citrus peel miner を見よ
leaf miner			simple-dot slender を見よ
leaf miner			sweetpotato leafminer を見よ
leaf miner-and roller			citrus leafroller (1) を見よ
leafminer flies	ハモグリバエ科	Agromyzidae	（ハエ目）の昆虫の総称　世界に 1,800 種
leaf miner fly			vegetable leafminer を見よ
leafminer moths			leafblotch miners を見よ
leafminer parasite		*Apanteles bedelliae* Viereck	（ハチ目、コマユバチ科）大洋区
leafminers		*Liriomyza*	（ハエ目、ハモグリバエ科）の昆虫の総称
leaf miners			leafminer flies を見よ
leaf miners			shield bearers を見よ
leaf miners			Douglas moths を見よ
leaf miners			midget moths を見よ
leaf miners			yucca moths (1) を見よ
leaf miners			leafblotch miners を見よ
leaf-mining beetle			spiny leaf beetle を見よ

英名	和名	学名	所属、分布、ほか
leaf-mining flies			leafminer flies を見よ
leaf-mining fly		*Melanagromyza theae* Bigot	(ハエ目、ハモグリバエ科) 東洋区。チャ害虫
leaf-mining leaf beetles	トゲハムシ亜科	Hispinae	(コウチュウ目、ハムシ科) の昆虫の総称
leaf-mining moth			white blotch oak leaf miner を見よ
leaf moths			window-winged moths を見よ
leaf roller (1)	テングハマキ	*Sparganothis (Platynota) pilleriana* Schiffermüller	(チョウ目、ハマキガ科) 日本。旧北区。ダイズ、イチゴ、カンキツ害虫
leaf roller		*Depressaria culcitella* Herrich-Schäffer	(チョウ目、マルハキバガ科) 日本。カンキツ害虫
leaf roller			common orchard moth を見よ
leaf roller			black-shaped Platynota moth を見よ
leaf roller			oblique tortrix を見よ
leafroller moths	ハマキガ科	Tortricidae	(チョウ目) の昆虫の総称　世界に 4,000 種以上。幼虫は leaf roller といわれる
leaf rollers			olethreutid moths を見よ
leaf rollers			leafroller moths を見よ　幼虫の英名。Leafrollers とも記す
leaf-rolling cricket		*Camptonotus carolinensis* (Gerstaecker)	(バッタ目、コロギス科) 新北区。葉を巻いた巣を作る。夜行性でアブラムシを捕食
leaf-rolling crickets		Gryllacridinae	(バッタ目、コロギス科) の昆虫の総称
leaf-rolling crickets			tree crickets (1) を見よ
leaf-rolling grasshopper			leaf-rolling cricket を見よ
leaf-rolling grasshoppers			leaf-rolling crickets を見よ
leafrolling peach sawfly		*Pamphilius persicus* MacGillivray	(ハチ目、ヒラタハバチ科) 新北区
leaf-rolling pear aphid		*Dysaphis reaumuri* (Mordvilko)	(カメムシ目、アブラムシ科) 旧北区
leaf-rolling rose sawfly		*Blennocampa pusilla* (Klug)	(ハチ目、ハバチ科) 旧北区
leaf-rolling sawflies			web-spinning sawflies を見よ
leaf-rolling thrips		*Hoplandrothrips marshalli* Karny	(アザミウマ目、クダアザミウマ科) エチオピア区
leaf-rolling weevils	オトシブミ亜科	Attelabinae	(コウチュウ目、オトシブミ科) の昆虫の総称
leaf-rolling weevils			attelabid weevils を見よ
leaf skeletonizer moths (1)	ハモグリガ科	Lyonetiidae	(チョウ目) の昆虫の総称　、
leaf skeletonizer moths (2)	マダラガ科	Zygaenidae	(チョウ目) の昆虫の総称
leaf skeletonizer moths			smoky moths (1) を見よ
leaf-stalk aphid			lettuce root aphid を見よ
leaf webber		*Dudua aprobola* Meyrick	(チョウ目、ハマキガ科) 東洋区

英名	和名	学名	所属、分布、ほか
leaf weevils		*Phyllobius*	（コウチュウ目、ゾウムシ科）の昆虫の総称
leaf weevils		*Phytonomus*	（コウチュウ目、ゾウムシ科）の昆虫の総称
leafblister sawfly		*Phylacteophaga eucalypti* Froggatt	（ハチ目、Pergidae）豪州区
leafblister sawfly		*Phylacteophaga froggatti* Riek	（ハチ目、Pergidae）豪州区
leafhopper assassin bug		*Zelus renardii* Kolenati	（カメムシ目、サシガメ科）新北区、大洋区。生物的防除のためハワイに導入された
leafhopper egg-predator		*Cyrtorhinus mundulus* (Breddin)	（カメムシ目、カスミカメムシ科）大洋区、豪州区
leafhoppers	ヨコバイ亜科	Deltocephalinae	（カメムシ目、オオヨコバイ科）の昆虫の総称
leafhoppers (1)	ツノゼミ上科	Membracoidea	（カメムシ目）の昆虫の総称
leafhoppers (2)	オオヨコバイ科	Cicadellidae	（カメムシ目）の昆虫の総称　世界に20,000種以上。sharpshooter, doggers ともいわれる
leafhoppers			lantern bugs を見よ
leaftiers	ノメイガ亜科	Pyraustinae	（チョウ目、メイガ科）の昆虫の総称
leafwing		*Anaea cubana* Salvin	（チョウ目、タテハチョウ科）新北区
leafwing butterfly			Australian leafwing (1) を見よ
leafwings		Charaxinae	（チョウ目、タテハチョウ科）の昆虫の総称
leafy spurge gall midge		*Spurgia capitigena* (Bremi)	（ハエ目、タマバエ科）新北区。*Euphorbia esula* の防除に使用された
leafy spurge gall midge		*Spurgia esulae* Gagné	（ハエ目、タマバエ科）新北区
leafy spurge hawk moth			spurge hawkmoth を見よ
Leanira checkerspot		*Chlosyne leanira* (C. et R. Felder)	（チョウ目、タテハチョウ科）新北区
leapers			crickets (2) を見よ
leaping beetles			click beetles (1) を見よ
leaproach		*Saltoblatella montistabularis* Bohn, Picker, Klass et Colville	（ゴキブリ目、チャバネゴキブリ科）エチオピア区
least black arches	ヒメコブガ	*Nola confusalis* (Herrich-Schäffer)	（チョウ目、コブガ科）日本、旧北区
least carpet		*Idaea vulpinaria* (Herrich-Schäffer)	（チョウ目、シャクガ科）旧北区。*I. v. atrosignaria* Lempke も同英名
least case		*Coleophora juncicolella* Stainton	（チョウ目、ツツミノガ科）旧北区
least Dichomeris moth		*Dichomeris siren* Hodges	（チョウ目、キバガ科）新北区
least hawthorn pigmy		*Stigmella perpygmaeella* (Doubleday)	（チョウ目、モグリチビガ科）旧北区
least heliconian			Juliette longwing を見よ
least Holomelina			ruddy Holomelina を見よ
least Hypocrisias		*Hypocrisias minima* (Neumoegen)	（チョウ目、ヒトリガ科）新北区
least katydid		*Brachyinsara hemiptera* Hebard	（バッタ目、キリギリス科）新北区

英名	和名	学名	所属、分布、ほか
least-marked Euchlaena moth		*Euchlaena irraria* (Barnes et McDunnough)	（チョウ目、シャクガ科）新北区
least meadow brown			small heath を見よ
least minor		*Photedes captiuncula* (Treitschke)	（チョウ目、ヤガ科）旧北区
least owlet		*Scythris siccella* (Zeller)	（チョウ目、キヌバコガ科）旧北区
least Prepona		*Prepona dexamenus medinai* Beutelspacher	（チョウ目、タテハチョウ科）新熱帯区
least satin lutestring		*Tethea duplaris* Linnaeus	（チョウ目、トガリバガ科）旧北区
least scarlet-eye		*Nascus paulliniae* (Sepp)	（チョウ目、セセリチョウ科）新熱帯区
least shieldback		*Atlanticus monticola* Davis	（バッタ目、キリギリス科）新北区
least skipper		*Ancyloxypha numitor* (Fabricius)	（チョウ目、セセリチョウ科）新北区、新熱帯区
least skipperling			least skipper を見よ
least water-snipefly		*Atrichops crassipes* (Meigen)	（ハエ目、ナガレアブ科）旧北区
least yellow underwing		*Noctua interjecta* Hübner	（チョウ目、ヤガ科）旧北区
leather beetle			Frisch's carrion beetle を見よ
leather beetle			hide beetle を見よ
leather-coloured sod webworm		*Crambus trisectus* Walker	（チョウ目、メイガ科）新北区
leather jacket grubs			crane flies (1) を見よ　幼虫の英名
leatherjacket			common crane fly (1)(2) を見よ　幼虫の英名
leatherjackets			crane flies (1)(2) を見よ　幼虫の英名
leatherleaf fern borer moth		*Undulambia polystichalis* (Capps)	（チョウ目、メイガ科）新北区。leatherleaf fern borer ともいう
leather wing			dragonflies (1) を見よ
leather-winged beetle	ジョウカイボン	*Athemus suturellus* (Motschulsky)	（コウチュウ目、ジョウカイボン科）日本、旧北区
leather-winged beetles			soldier beetles (1) を見よ
Lebbeck mealybug			spherical mealybug を見よ
Lebbeck mealybug		*Nipaecoccus filamentosus* (Cockerell)	（カメムシ目、コナカイガラムシ科）東洋区、エチオピア区
Lebeau's silkmoth		*Rothschildia lebeau* (Guérin-Méneville)	（チョウ目、ヤママユガ科）新北区
Leche's twist	オオギンスジアカハマキ	*Ptycholoma lecheana* (Linnaeus)	（チョウ目、ハマキガ科）日本、旧北区
Leche's twist moth			Leche's twist を見よ
LeConte's Haploa		*Haploa lecontei* (Guérin-Méneville)	（チョウ目、ヒトリガ科）新北区。LeConte's Haploa moth ともいう
Leda hairstreak	レダカラスシジミ	*Ministrymon leda* (Edwards)	（チョウ目、シジミチョウ科）新北区
Ledum piercer		*Cydia janthinana* (Duponchel)	（チョウ目、ハマキガ科）旧北区

英名	和名	学名	所属、分布、ほか
Leech's sailor		*Neptis arachne* Leech	（チョウ目、タテハチョウ科）旧北区
leek grasshopper	イナゴモドキ	*Mecostethus alliaceus* (Germar)	（バッタ目、バッタ科）日本。ダイズ害虫
leek moth		*Acrolepia assectella* (Zeller)	（チョウ目、アトヒゲコガ科）全北区
leek smudge			leek moth を見よ
leersia aphid	カンスゲワタムシ	*Colopha kansugei* (Uye)	（カメムシ目、アブラムシ科）日本
Lefebvre's ringlet	ピレネークロベニヒカゲ	*Erebia lefebvrei* Boisduval	（チョウ目、タテハチョウ科）旧北区
Legge's snow horned skipper		*Chondrolepis leggei* (Heron)	（チョウ目、セセリチョウ科）エチオピア区
legionary ant		*Neivamyrmex nigrescens* (Cresson)	（ハチ目、アリ科）新北区
legionary ant			driver ant を見よ
legionary ants			army ants (1) を見よ　*Eciton* 属のアリを指すこともある
legionary ants			safari ants を見よ
legume lady bug			Jacque's beetle を見よ
legume leafminer	マメハモグリバエ	*Liriomyza trifolii* (Burgess)	（ハエ目、ハモグリバエ科）日本、全北区、大洋区。エンドウ、各種野菜、キクなどの害虫
legume leafminer			vegetable leafminer を見よ
legume pod borer			bean pod borer (1) を見よ
legume-pod moth			lima-bean pod borer を見よ
legume webspinner		*Lamprosema alstitalis* (Walker)	（チョウ目、メイガ科）豪州区
legume weevils			seed weevils (1) を見よ
legume weevils			pea weevils を見よ
leguminosae rivula	テンクロアツバ	*Rivula sericealis* (Scopoli)	（チョウ目、ヤガ科）日本、旧北区。イネ科牧草の害虫
leguminous seed weevils	コガタゾウムシ亜科	Tychiinae	（コウチュウ目、ゾウムシ科）の昆虫の総称
Leichhardt's grasshopper		*Petasida ephippigera* White	（バッタ目、オンブバッタ科）豪州区
leiodids			round fungus beetles を見よ
Lelex egg white		*Dismorphia lelex* (Hewitson)	（チョウ目、シロチョウ科）新熱帯区
Lember's ghost moth		*Hepialus lembertii* Dyar	（チョウ目、コウモリガ科）新北区
lemon bell		*Thiodia citrana* (Hübner)	（チョウ目、ハマキガ科）旧北区
lemon bird			brimstone を見よ
lemon bud moth		*Prays parilis* Turner	（チョウ目、スガ科）豪州区、東洋区
lemon butterfly			chequered swallowtail を見よ
lemon clouded yellow		*Colias thrasibulus* Fabricius	（チョウ目、シロチョウ科）東洋区
lemon emigrant			lemon migrant を見よ
lemon migrant	ウスキシロチョウ	*Catopsilia pomona* (Fabricius)	（チョウ目、シロチョウ科）日本、東洋区、豪州区

英名	和名	学名	所属、分布、ほか
lemon pansy	ジャノメタテハモドキ	*Junonia lemonias lemonias* (Linnaeus)	(チョウ目、タテハチョウ科) 日本、東洋区、旧北区
lemon Plagodis moth		*Plagodis serinaria* Herrich-Schäffer	(チョウ目、シャクが科) 新北区
lemon tipped hunter		*Austrogomphus prasinus* Tillyard	(トンボ目、サナエトンボ科) 豪州区
lemon traveller	トガリツマアカシロチョウ	*Colotis subfasciatus subfasciatus* (Swainson)	(チョウ目、シロチョウ科) エチオピア区
lemon tree borer		*Oemona hirta* (Fabricius)	(コウチュウ目、カミキリムシ科) 豪州区
lemon-tree borer parasite		*Xanthocryptus novozealandicus* (Dalla Torre)	(ハチ目、ヒメバチ科) 豪州区
lemon wheat blossom midge			wheat blossom midge を見よ
lemon white			greenish black-tip を見よ　東洋区の亜種 *E. c. lucilla* Butler の英名
lemoniid moths			autumn silkworm moths を見よ
Lempke's gold spot			Lempke's spot を見よ
Lempke's gold spot			Putnam's looper moth を見よ
Lempke's spot		*Plusia putnami gracilis* (Lempke)	(チョウ目、ヤガ科) 旧北区
Lena sooty wing			Mojave sootywing を見よ
Lenea clearwing		*Callithomia lenea* (Cramer)	(チョウ目、タテハチョウ科) 新熱帯区
Lenis sootywing		*Staphylus lenis* Steinhauser	(チョウ目、セセリチョウ科) 新熱帯区
lenticular gall			lenticular gall wasp の虫えいの英名
lenticular gall wasp		*Numismalis quercusbaccarum* (Linnaeus)	(ハチ目、タマバチ科) 旧北区
Lentiginos moth		*Sparganothoides lentiginosana* (Walsingham)	(チョウ目、ハマキガ科) 新北区、新熱帯区
lentil weevil		*Bruchus ervi* Frölich	(コウチュウ目、ハムシ科) 旧北区。lentil ソラマメに由来
lentil weevil		*Bruchus signaticornis* Gyllenhal	(コウチュウ目、ハムシ科) 旧北区
lentil weevil	ヒラマメゾウムシ	*Bruchus lentis* Frölich	(コウチュウ目、ハムシ科) 東洋区
Leonard's skipper			Leonardus skipper を見よ
Leonardus skipper		*Hesperia leonardus* Harris	(チョウ目、セセリチョウ科) 新北区
Leonora's dancer		*Argia leonorae* Garrison	(トンボ目、イトトンボ科) 新北区
leopard	テンジクゴマダラ	*Neurosigma siva* (Westwood)	(チョウ目、タテハチョウ科) 東洋区
leopard			common leopard を見よ
leopard			triangle (1) を見よ　英国での英名
leopard butterflies	ヒョウモンツバメ属	*Cigaritis*	(チョウ目、シジミチョウ科) の昆虫の総称
leopard butterfly			tawny silverline を見よ
leopard lacewing		*Cethosia cyane* (Drury)	(チョウ目、タテハチョウ科) 東洋区

英名	和名	学名	所属、分布、ほか
leopard moth	ゴマフボクトウ	*Zeuzera pyrina* (Linnaeus)	(チョウ目、ボクトウガ科) 日本、全北区。カキ、チャ、サクラ、ツバキ、カシ類などの害虫
leopard moths		*Zeuzera*	(チョウ目、ボクトウガ科) の昆虫の総称
leopard moths			carpenter moths (1) を見よ
leopard spot			eighty-eight butterfly (1) を見よ
leopard-spotted beauty		*Baeotus deucalion* (Felder)	(チョウ目、タテハチョウ科) 新熱帯区
leopards	ウラベニヒョウモン属	*Phalanta*	(チョウ目、タテハチョウ科) の昆虫の総称
Lepcha bushbrown		*Mycalesis lepcha* (Moore)	(チョウ目、タテハチョウ科) 東洋区
lepidopterans			butterflies and moths を見よ
lepidopteroids			butterflies and moths を見よ
lepidopteron			butterflies and moths を見よ
lepidopterous insects			butterflies and moths を見よ
lepidostomatids			bizarre caddisflies を見よ
Leplastrier's piercer		*Selania leplastriana* (Curtis)	(チョウ目、ハマキガ科) 旧北区
leposcelid booklice			booklice (1) を見よ
Leprea brown		*Nesoxenica leprea* (Hewitson)	(チョウ目、タテハチョウ科) 豪州区
Leprieur's glory		*Asterope leprieuri* (Feisthamel)	(チョウ目、タテハチョウ科) 新熱帯区
Leptines ciliate blue		*Anthene leptines* (Hewitson)	(チョウ目、シジミチョウ科) エチオピア区
leptocoris bug		*Leptocoris mitellata* Bergroth	(カメムシ目、ヒメヘリカメムシ科) 豪州区
leptogasterine flies		Leptogasterinae	(ハエ目、ムシヒキアブ科) の昆虫の総称
leptopine weevils		*Gymnopholus*	(コウチュウ目、ゾウムシ科) の昆虫の総称
Lerna sister			Erotia sister を見よ
lerp insects			jumping plantlice を見よ　豪州での英名
lerp psylla	エノキカイガラキジラミ	*Celtisaspis japonica* (Miyatake)	(カメムシ目、キジラミ科) 日本、旧北区。エノキ害虫
Lesbia clouded yellow		*Colias lesbia* (Fabricius)	(チョウ目、シロチョウ科) 新熱帯区
Lesne's earwig		*Forficula lesnei* Finot	(ハサミムシ目、クギヌキハサミムシ科) 旧北区
Lesotho blue		*Lepidochrysops lerothodi* (Trimen)	(チョウ目、シジミチョウ科) エチオピア区
lespedeza plume moth	ナカノホソトリバ	*Stenoptilia emarginata* (Snellen)	(チョウ目、トリバガ科) 日本、旧北区
lespedeza webworm		*Tetralopha scortealis* (Lederer)	(チョウ目、メイガ科) 新北区。成虫は lespedeza webworm moth
lespedeza worm			grape colaspis (2) を見よ
less rich sailer		*Neptis nashona* Swinhoe	(チョウ目、タテハチョウ科) 東洋区
lesser albatross	ナミエシロチョウ	*Appias paulina distanti* (Moore)	(チョウ目、シロチョウ科) 東洋区
lesser albatross			Ward's albatross を見よ

英名	和名	学名	所属、分布、ほか
lesser angle-wing			angular-winged katydid (2) を見よ
lesser antler sawfly			bristly rose slug sawfly を見よ
lesser apple leaf folder			yellow headed fireworm を見よ
lesser apple leaf-roller			yellow headed fireworm を見よ
lesser appleworm	アメリカリンゴシンクイ	*Grapholita prunivora* (Walsh)	(チョウ目、ハマキガ科) 新北区。成虫は lesser appleworm moth
lesser arid-land katydid		*Neobarrettia victoriae* (Caudell)	(バッタ目、キリギリス科) 新北区
lesser army worm			beet armyworm を見よ　lesser armyworm とも記される
lesser armyworm parasite fly		*Lespesia archippivora* (Riley)	(ハエ目、ヤドリバエ科) 大洋区、新北区
lesser ash bark beetle			fig tree bark borer を見よ
lesser aspen webworm moth		*Meroptera pravella* (Grote)	(チョウ目、メイガ科) 新北区
lesser auger beetle		*Heterobostrychus aequalis* (Waterhouse)	(コウチュウ目、ナガシンクイムシ科) 豪州区
lesser backswimmers			pygmy backswimmer bugs を見よ
lesser-banded hornet	ツマグロスズメバチ	*Vespa affinis* (Linnaeus)	(ハチ目、スズメバチ科) 日本、東洋区
lesser bath white			small bath white を見よ
lesser batwing		*Atrophaneura aidoneus* (Doubleday)	(チョウ目、アゲハチョウ科) 東洋区
lesser bee moth			lesser wax moth を見よ　英国での英名
lesser belle			salix noctuid を見よ
lesser black-letter dart moth			spotted cutworm moth (1) を見よ
lesser black-patches		*Anthene lamprocles* (Hewitson)	(チョウ目、シジミチョウ科) エチオピア区
lesser bladder cicada		*Cystosoma schmeltzi* Distant	(カメムシ目、セミ科) 豪州区
lesser bloody nosed beetle		*Timarcha goettingensis* (Linnaeus)	(コウチュウ目、ハムシ科) 旧北区
lesser blue Charaxes		*Charaxes numenes* (Hewitson)	(チョウ目、タテハチョウ科) エチオピア区
lesser blue-dusted elfin		*Sarangesa majorella* (Mabille)	(チョウ目、セセリチョウ科) エチオピア区
lesser bombardier beetle		*Brachynus explodens* Duftschmid	(コウチュウ目、ホソクビゴミムシ科) 旧北区
lesser-bordered yellow-underwing			lesser broad-border を見よ
lesser bottle cicada			small bottle cicada を見よ
lesser brimstone		*Gonepteryx aspasia* Ménétriès	(チョウ目、シロチョウ科) 東洋区。日本亜種は *G. a. niphonica* Verity スジボソヤマキチョウ
lesser brimstone		*Gonepteryx mahaguru* Gistel	(チョウ目、シロチョウ科) 旧北区、東洋区

英名	和名	学名	所属、分布、ほか
lesser broad-border		*Noctua janthina* (Denis et Schiffermüller)	（チョウ目、ヤガ科）旧北区
lesser broad-bordered yellow underwing			lesser broad-border を見よ
lesser brown blowfly (1)		*Calliphora dubia* (Macquart)	（ハエ目、クロバエ科）豪州区
lesser brown blowfly (2)		*Calliphora augur* (Fabricius)	（ハエ目、クロバエ科）豪州区
lesser brown budworm			beautiful brindled groundling を見よ
lesser brown evening moth		*Selidosema pannularia* Guenée	（チョウ目、シャクガ科）豪州区
lesser brown lacewings		Sympherobiidae	（アミメカゲロウ目）の昆虫の総称
lesser brown striped hawk		*Basiothia charis* (Boisduval)	（チョウ目、スズメガ科）エチオピア区
lesser bud moth			beautiful brindled groundling を見よ
lesser bulb fly	コブアシハイジマハナアブ	*Eumerus tuberculatus* Rondani	（ハエ目、ハナアブ科）日本
lesser bulb fly			onion bulb fly を見よ
lesser cabbage moth			diamondback moth を見よ
lesser canna leafroller		*Geshna cannalis* (Quaintance)	（チョウ目、メイガ科）新北区。lesser canna leafroller moth ともいう
lesser chicken louse	ヒメニワトリハジラミ	*Goniocotes hologaster* Nitzsch	（ハジラミ目、チョウカクハジラミ科）新北区、大洋区
lesser cloaked marble		*Hedya pruniana* (Hübner)	（チョウ目、ハマキガ科）旧北区
lesser clouded yellow	クリステーメモンキチョウ	*Colias chrysotheme* (Esper)	（チョウ目、シロチョウ科）旧北区
lesser clover leaf weevil			clover leaf weevil（2）を見よ
lesser clover weevil			clover leaf weevil（2）を見よ
lesser cockroach		*Ectobius panzeri* Stephens	（ゴキブリ目、チャバネゴキブリ科）旧北区
lesser coconut spike moth			coconut moth（2）を見よ
lesser common rustic		*Mesapamea secalella* Remm	（チョウ目、ヤガ科）旧北区
lesser common rustic		*Mesapamea didyma* Esper	（チョウ目、ヤガ科）旧北区
lesser cornborer			lesser grain borer を見よ
lesser cornstalk borer	モロコシマダラメイガ	*Elasmopalpus lignosellus* (Zeller)	（チョウ目、メイガ科）新北区、新熱帯区。イネ、トウモロコシなどの害虫。lesser cornstalk borer moth ともいう
lesser cotton worm			beet armyworm を見よ　北アフリカでの幼虫の英名
lesser cream striped hawk		*Theretra monteironis* (Butler)	（チョウ目、スズメガ科）エチオピア区
lesser cream wave		*Scopula immutata* (Linnaeus)	（チョウ目、シャクガ科）旧北区

英名	和名	学名	所属、分布、ほか
lesser crepuscular skipper		*Gretna cylinda* (Hewitson)	(チョウ目、セセリチョウ科) エチオピア区
lesser darkie		*Allotinus unicolor* C. et R. Felder	(チョウ目、シジミチョウ科) 東洋区
lesser dart		*Potanthus omaha omaha* (Edwards)	(チョウ目、セセリチョウ科) 東洋区
lesser date moth		*Batrachedra amydraula* Meyrick	(チョウ目、カザリバガ科) 旧北区。油ヤシ害虫
lesser death watch			booklouse (2) を見よ
lesser dried-fish beetle	ホシカカツオブシムシ	*Dermestes carnivorus* Fabricius	(コウチュウ目、カツオブシムシ科) 日本
lesser dung flies			small dung flies を見よ
lesser earwig	ミジンハサミムシ	*Labia minor* (Linnaeus)	(ハサミムシ目、クロハサミムシ科) 日本、旧北区、新北区
lesser earwigs			little earwigs (1) を見よ
lesser emerald		*Hemistola immaculata* (Thunberg)	(チョウ目、シャクガ科) 旧北区
lesser emperor		*Anax parthenope* (Selys)	(トンボ目、ヤンマ科) 旧北区、東洋区、エチオピア区。yellow-ringed emeperor ともいう。日本亜種は *A. p. julius* Brauer ギンヤンマ
lesser emperor dragonfly			lesser emperor を見よ
lesser ensign wasp		*Szepligetella sericia* (Cameron)	(ハチ目、ヤセバチ科) 新北区
lesser European bark beetle			elm bark beetle (2) を見よ
lesser field crickets		*Miogryllus*	(バッタ目、コオロギ科) の昆虫の総称
lesser field grasshopper		*Chorthippus mollis* (Charpentier)	(バッタ目、バッタ科) 旧北区
lesser fiery copper	クジャクシジミ	*Thersamonia thersamon* (Esper)	(チョウ目、シジミチョウ科) 旧北区、東洋区
lesser fig-tree blue		*Myrina dermaptera* (Wallengren)	(チョウ目、シジミチョウ科) エチオピア区
lesser flour beetle	ガイマイゴミムシダマシ	*Alphitobius diaperinus* (Panzer)	(コウチュウ目、ゴミムシダマシ科) 日本、汎世界。貯穀害虫
lesser frit fly		*Oscinella pusilla* (Meigen)	(ハエ目、キモグリバエ科) 旧北区
lesser ghost moth		*Fraus simulans* Walker	(チョウ目、コウモリガ科) 豪州区
lesser giant hunting ant			bullet ant を見よ
lesser glow-worm		*Phosphaenus hemipterus* (Goeze)	(コウチュウ目、ホタル科) 旧北区
lesser grain borer	コナナガシンクイ	*Rhyzopertha dominica* (Fabricius)	(コウチュウ目、ナガシンクイムシ科) 日本、汎世界。貯穀害虫
lesser grain weevil			rice weevil を見よ
lesser grape root borer moth		*Vitacea scepsiformis* (Edwards)	(チョウ目、スカシバガ科) 新北区
lesser grapevine looper moth		*Eulithis diversilineata* (Hübner)	(チョウ目、シャクガ科) 新北区。幼虫は lesser grapevine looper

英名	和名	学名	所属、分布、ほか
lesser grass blue		*Zizina otis* (Fabricius)	（チョウ目、シジミチョウ科）東洋区。日本亜種は *Z. o. emelina* (de l'Orza) シルビアシジミ。*Z. o. lampa* (Corbet) も同英名
lesser grass blues	シルビアシジミ属	*Zizina*	（チョウ目、シジミチョウ科）の昆虫の総称
lesser gray Chytolita			morbid owlet moth を見よ
lesser green emperor			pale spotted emperor を見よ
lesser green leafhopper			green leafhopper (1) を見よ
lesser grey groundling		*Teleiodes paripunctella* (Thunberg)	（チョウ目、キバガ科）旧北区
lesser grey shade		*Cnephasia virgaureana* Treitschke	（チョウ目、ハマキガ科）旧北区
lesser gull	ウスムラサキシロチョウ	*Cepora nadina* (Lucas)	（チョウ目、シロチョウ科）東洋区。*C. n. andersoni* Distant も同英名
lesser harlequin		*Laxita thuisto thuisto* (Hewitson)	（チョウ目、シジミタテハ科）東洋区
lesser honey bee	コミツバチ	*Apis florea* (Fabricius)	（ハチ目、ミツバチ科）東洋区
lesser horned citrus bug		*Vitellus antemna* Breddin	（カメムシ目、カメムシ科）豪州区
lesser horned swift		*Borbo lugens* (Hopffer)	（チョウ目、セセリチョウ科）エチオピア区
lesser house fly	ヒメイエバエ	*Fannia canicularis* (Linnaeus)	（ハエ目、ヒメイエバエ科）日本、汎世界
lesser Jay	エベモンミカドアゲハ	*Graphium evemon* (Boisduval)	（チョウ目、アゲハチョウ科）東洋区
lesser lattice brown		*Kirinia climene* (Esper)	（チョウ目、タテハチョウ科）旧北区
lesser lawn leafhopper		*Graminella sonorus* (Ball)	（カメムシ目、オオヨコバイ科）新北区
lesser lichen case bearer			spring grey smoke を見よ
lesser loblolly pineconeworm moth		*Dioryctria taedivorella* Neunzig et Leidy	（チョウ目、メイガ科）新北区
lesser luteous snout			morbid owlet moth を見よ
lesser lutestring			oak lutestring を見よ
lesser maple leaf blotch miner moth		*Phyllonorycter lucidicostella* (Clemens)	（チョウ目、ホソガ科）新北区
lesser maple leafroller moth		*Acleris chalybeana* (Fernald)	（チョウ目、ハマキガ科）新北区
lesser maple spanworm moth		*Speranza pustularia* Guenée	（チョウ目、シャクガ科）新北区
lesser marbled fritillary		*Brenthis ino* Rottemburg	（チョウ目、タテハチョウ科）旧北区。日本亜種は *B. i. tigroides* (Fruhstorfer) コヒョウモン
lesser marsh grasshopper		*Chorthippus albomarginatus* (De Geer)	（バッタ目、バッタ科）旧北区
lesser mealworm			lesser flour beetle を見よ
lesser mealworm beetle			lesser flour beetle を見よ
lesser mealworm beetle			black fungus beetle を見よ

英名	和名	学名	所属、分布、ほか
lesser migratory grasshopper		*Melanoplus mexicanus* (Saussure)	（バッタ目、バッタ科）新北区、新熱帯区。カンキツ害虫。亜種 *M. m. atlanis* (Riley) は lesser migratory locust という
lesser millet skipper			small branded swift を見よ
lesser mime	キボシアゲハ	*Chilasa epycides* (Hewitson)	（チョウ目、アゲハチョウ科）東洋区
lesser mottled grasshopper		*Stenobothrus stigmaticus* (Rambur)	（バッタ目、バッタ科）旧北区
lesser mountain ringlet	チビモンベニヒカゲ	*Erebia melampus* (Fuessly)	（チョウ目、タテハチョウ科）旧北区
lesser mulberry pyralid			mulberry pyralid を見よ
lesser oak carpenter worm			little carpenterworm を見よ
lesser ochreous dart		*Andronymus helles* Evans	（チョウ目、セセリチョウ科）エチオピア区
lesser orchid weevil		*Orchidophilus pereginator* Buchanan	（コウチュウ目、ゾウムシ科）新北区
lesser ox warble fly			cattle warble fly (2) を見よ
lesser peach borer			lesser peach tree borer を見よ
lesser peach tree borer		*Synanthedon pictipes* (Grote et Robinson)	（チョウ目、スカシバガ科）新北区。成虫は lesser peach tree borer moth
lesser pecan tree borer		*Aegeria geliformis* Walker	（チョウ目、スカシバガ科）新北区
lesser pine borer		*Asemum moestum* Haldeman	（コウチュウ目、カミキリムシ科）旧北区
lesser pine katydid		*Orchelimum minor* Bruner	（バッタ目、キリギリス科）新北区
lesser pine sawfly			fox colored sawfly を見よ
lesser pine shoot beetle		*Myelophilus minor* Hartig	（コウチュウ目、キクイムシ科）旧北区
lesser pumpkin fly			Ethiopian fruit fly を見よ
lesser punch		*Dodona dipoea* Hewitson	（チョウ目、シジミタテハ科）東洋区
lesser purple emperor	タイリクコムラサキ	*Apatura ilia* Schiffermüller	（チョウ目、タテハチョウ科）旧北区
lesser purple emperor			purple emperor (1) を見よ
lesser purple flat-body		*Agonopterix purpurea* (Haworth)	（チョウ目、マルハキバガ科）旧北区
lesser Queensland fruit fly		*Dacus neohumeralis* Hardy	（ハエ目、ミバエ科）豪州区
lesser rice leafroller		*Cnaphalocrocis poeyalis* (Boisduval)	（チョウ目、メイガ科）エチオピア区、東洋区、豪州区
lesser rice weevil			rice weevil を見よ
lesser rose aphid		*Myzaphis rosarum* (Kaltenbach)	（カメムシ目、アブラムシ科）日本、全北区、豪州区、新熱帯区、エチオピア区。温室害虫
lesser satin moth			common lutestring を見よ
lesser shieldback		*Ateloplus minor* Caudell	（バッタ目、キリギリス科）新北区
lesser shoot weevil		*Rhachiodes dentifer* Boheman	（コウチュウ目、ゾウムシ科）豪州区
lesser shot-hole borer			Saxena ambrosia beetle を見よ

英名	和名	学名	所属、分布、ほか
lesser shot-hole borer			shot-hole borer (3) を見よ
lesser silver-spotted frit			Queen of Spain fritillary を見よ
lesser snow scale	コンマカイガラムシ	*Pinnaspis strachani* (Cooley)	（カメムシ目、マルカイガラムシ科）日本、東洋区、大洋区、新北区、新熱帯区。カンキツ害虫
lesser spotted fritillary		*Melitaea trivia* (Schiffermüller)	（チョウ目、タテハチョウ科）旧北区
lesser spotted pinion	ニレキリガ	*Cosmia affinis* (Linnaeus)	（チョウ目、ヤガ科）日本、旧北区
lesser stable flies			stable flies を見よ　南アフリカでの英名
lesser stag beetle		*Dorcus parallelipipedus* (Linnaeus)	（コウチュウ目、クワガタムシ科）旧北区
lesser strawberry weevil		*Otiorhynchus rugusostriatus* (Goeze)	（コウチュウ目、ゾウムシ科）旧北区
lesser striped flea beetle			striped flea beetle (2) を見よ
lesser swallow prominent		*Pheosia gnoma* (Fabricius)	（チョウ目、シャチホコガ科）旧北区
lesser Tasmanian darner		*Austroaeschna hardyi* Tillyard	（トンボ目、ヤンマ科）豪州区
lesser tawny tubic		*Batia lunaris* (Haworth)	（チョウ目　マルハキバガ科）旧北区
lesser thorn-tipped longhorn beetle		*Pogonocherus hispidus* (Linnaeus)	（コウチュウ目、カミキリムシ科）旧北区
lesser three-ring		*Ypthima inica* Hewitson	（チョウ目、タテハチョウ科）東洋区
lesser treble-bar		*Aplocera efformata* (Guenée)	（チョウ目、シャクガ科）旧北区
lesser vagabond sod webworm		*Agriphila ruricolellus* (Zeller)	（チョウ目、メイガ科）新北区
lesser viburnum clearwing moth		*Synanthedon fatifera* Hodges	（チョウ目、スカシバガ科）新北区
lesser vine sphinx			banded sphinx moth を見よ
lesser wainscot		*Depressaria chaerophyllella* Zeller	（チョウ目、マルハキバガ科）旧北区
lesser wainscot flat-body			lesser wainscot を見よ
lesser wanderer		*Danaus petilia* (Stoll)	（チョウ目、タテハチョウ科）豪州区
lesser wanderer			African monarch を見よ
lesser wanderer butterfly		*Danaus chrysippus petilia* (Stoll)	（チョウ目、タテハチョウ科）豪州区。African monarch を参照
lesser wasp moth		*Pseudocharis minima* (Grote)	（チョウ目、ヒトリガ科）新北区、新熱帯区
lesser water boatman		*Plea atomaria* (Pallas)	（カメムシ目、マルミズムシ科）旧北区
lesser water boatman		*Micronecta scholtzi* (Fieber)	（カメムシ目、ミズムシ科）旧北区
lesser water boatman			common corixa を見よ
lesser water boatmen			water boatman bugs を見よ
lesser water boatmen			pygmy backswimmer bugs を見よ
lesser water weevil			rice water weevil をみよ

英名	和名	学名	所属、分布、ほか
lesser water-measurer		*Hydrometra gracilenta* Horvath	（カメムシ目、イトアメンボ科）旧北区
lesser wax moth	コハチノスツヅリガ	*Achroia grisella* (Fabricius)	（チョウ目、メイガ科）日本、全北区、東洋区、豪州区。幼虫は乾燥食品害虫
lesser waxworm			lesser wax moth を見よ　英国での英名
lesser whirligig		*Gyrinus minutus* Linnaeus	（コウチュウ目、ミズスマシ科）旧北区
lesser willow sawfly		*Nematus pavidus* Lepeletier	（ハチ目、ハバチ科）旧北区
lesser yellow underwing		*Noctua comes* Hübner	（チョウ目、ヤガ科）全北区。lesser yellow underwing moth ともう
lesser zebra		*Graphium macareus* (Godart)	（チョウ目、アゲハチョウ科）東洋区。*G. m. perakensis* (Fruhstorfer) も同英名
Lesson's snout-nose		*Pseudorhyncus lessonii* Serville	（バッタ目、キリギリス科）豪州区
less-small midget		*Phyllonorycter froelichiella* (Zeller)	（チョウ目、ホソガ科）旧北区
lestid damselflies			emerald damselflies を見よ
lethal Pyrausta moth		*Pyrausta lethalis* (Grote)	（チョウ目、メイガ科）新北区
Leto fritillary		*Speyeria leto* (Behr)	（チョウ目、タテハチョウ科）新北区
lettered China-mark	シロアヤヒメノメイガ	*Diasemia litterata* (Scopoli)	（チョウ目、メイガ科）旧北区
lettered Habrosyne		*Habrosyne scripta* (Gosse)	（チョウ目、トガリバガ科）新北区。lettered Hybrosyne moth ともいう
lettered sphinx		*Deidamia inscripta* Harris	（チョウ目、スズメガ科）新北区。lettered sphinx moth ともいう
lettered Zanclognatha moth		*Zanclognatha lituralis* (Hübner)	（チョウ目、ヤガ科）新北区。lettered Zanclognatha ともいう
lettuce aphid		*Acyrthosiphon scariolae* Nevsky	（カメムシ目、アブラムシ科）旧北区
lettuce aphid	ノゲシフクレアブラムシ	*Hyperomyzus cardnellinus* (Theobald)	（カメムシ目、アブラムシ科）日本、東洋区、豪州区、エチオピア区
lettuce aphid (1)		*Nasonovia ribisnigri* Mosley	（カメムシ目、アブラムシ科）旧北区
lettuce aphid			lettuce root aphid を見よ
lettuce fruit fly	ツママダラミバエ	*Trupanea amoena* (Frauenfeld)	（ハエ目、ミバエ科）日本、旧北区。レタス害虫
lettuce looper		*Rachiplusia nu* Guenée	（チョウ目、ヤガ科）新熱帯区
lettuce root aphid		*Pemphigus bursarius* (Linnaeus)	（カメムシ目、アブラムシ科）全北区
lettuce root fly			chrysanthemum stool fly を見よ
lettuce seed fly		*Pegohylemyia gnava* (Meigen)	（ハエ目、ハナバエ科）旧北区
lettuce shark		*Cucullia lactucae* (Denis et Schiffermüller)	（チョウ目、ヤガ科）旧北区
Leucadia longwing		*Heliconius leucadia* (Bates)	（チョウ目、タテハチョウ科）新熱帯区
leucaena psyllid	ギンネムキジラミ	*Heteropsylla cubana* Crawford	（カメムシ目、キジラミ科）日本、大洋区、豪州区
Leucas sister		*Adelpha barnesia leucas* Fruhstorfer	（チョウ目、タテハチョウ科）新熱帯区

英名	和名	学名	所属、分布、ほか
Leuce yellow		*Pyrisitia leuce* (Boisduval)	（チョウ目、シロチョウ科）新北区
Leucodesma crescent			whitened crescent を見よ
Leucogyna stripestreak			narrow-lined hairstreak を見よ
Leucone skipper		*Nastra leucone leucone* (Godman)	（チョウ目、セセリチョウ科）新熱帯区
leucospidid wasps	シリアゲコバチ科	Leucospididae	（ハチ目）の昆虫の総称
leucospidids			leucospidid wasps を見よ
leuctrids			rolled-winged stoneflies を見よ　オナシカワゲラ科の異名 Leuctridae から
levant blackneck		*Tathorhynchus exsiccata* (Lederer)	（チョウ目、ヤガ科）旧北区、エチオピア区
Levant clubtail		*Gomphus davidi* Selys	（トンボ目、サナエトンボ科）旧北区
Levantine skipper		*Thymelicus hyrax* (Lederer)	（チョウ目、セセリチョウ科）旧北区
levels cleg		*Haematopota subcylindrica* Pandelle	（ハエ目、アブ科）旧北区
levels yellow-horned horsefly		*Hybomitra ciureai* (Séguy)	（ハエ目、アブ科）旧北区
Levona sister		*Adelpha levona* Steinhauser et Miller	（チョウ目、タテハチョウ科）新熱帯区
Levona's Zestusa		*Zestusa levona* Steinhauser	（チョウ目、セセリチョウ科）新熱帯区
Lewenhoek's signal		*Pancalia leuwenhoekella* (Linnaeus)	（チョウ目、カザリバガ科）旧北区
Lewes wave		*Scopula immorata* (Linnaeus)	（チョウ目、シャクガ科）旧北区
Lewin's bag-shelter moth		*Panacela lewinae* (Lewin)	（チョウ目、オビガ科）豪州区
Lewis ambrosia beetle	ルイスザイノキクイムシ	*Ambrosiodmus lewisi* (Blandford)	（コウチュウ目、キクイムシ科）日本、旧北区、東洋区。木材害虫
Lewis diving beetle	マルコガタノゲンゴロウ	*Cybister lewisianus* Sharp	（コウチュウ目、ゲンゴロウ科）日本、東洋区
Lewis earwig	コブハサミムシ	*Anechura harmandi* (Burr)	（ハサミムシ目、クギヌキハサミムシ科）日本、旧北区
Lewis elongate scolytid			Lewis platypodid を見よ
Lewis gourd-shaped weevil	ワモンヒョウタンゾウムシ	*Sympiezominas lewisi* (Roelofs)	（コウチュウ目、ゾウムシ科）日本
Lewis large bark beetle	ルイスオオキクイムシ	*Hyorrhynchus lewisi* Blandford	（コウチュウ目、キクイムシ科）日本、東洋区
Lewis platypodid	ルイスナガキクイムシ	*Platypus lewisi* Blandford	（コウチュウ目、ナガキクイムシ科）日本、東洋区。クリ、木材害虫
Lewis round bark beetle			Lewis ambrosia beetle を見よ
Libbekh mealybug		*Pseudococcus perniciosus* Newstead	（カメムシ目、コナカイガラムシ科）旧北区。カンキツ害虫
libelluloid dragonflies			skimmers (1) を見よ
Libert's giant Epitola		*Epitola uranioides uranioides* Libert	（チョウ目、シジミチョウ科）エチオピア区

英名	和名	学名	所属、分布、ほか
Libert's orange playboy		*Hypomyrina mimetica* Libert	（チョウ目、シジミチョウ科）エチオピア区
library beetle			herbarium beetle (1) を見よ
library beetle			biscuit beetle を見よ
Librita skipper		*Librita librita* (Plötz)	（チョウ目、セセリチョウ科）新熱帯区
Libye ringlet		*Magneuptychia libye* (Linnaeus)	（チョウ目、タテハチョウ科）新熱帯区
Lical butterfly		*Cigaritis lilacinus* (Moore)	（チョウ目、シジミチョウ科）東洋区
lice			phthirapterous insects を見よ
lice			human lice を見よ
lice			damson-hop aphid を見よ　幼虫の英名
lice			pubic lice を見よ
lice			chewing lice を見よ
lichee stinkbug		*Tessaratoma papillosa* Drury	（カメムシ目、カメムシ科）東洋区
lichen bag moth		*Cebysa leucotelus* Walker	（チョウ目、ミノガ科）豪州区
lichen case-bearer			Linnaeus's virgin case を見よ
lichen case moth			lichen bag moth を見よ
lichen forest moth		*Elvia glaucata* Walker	（チョウ目、シャクガ科）豪州区
lichen mantid		*Gonatista grisea* (Fabricius)	（カマキリ目、カマキリ科）新北区
lichen mimic			lichen mantid を見よ　lichen mimic mantis ともいう
lichen moth (1)		*Declana atronivea* Walker	（チョウ目、シャクガ科）新北区、豪州区。北米での英名
lichen moth (2)		*Lycomorpha pholus* (Drury)	（チョウ目、ヒトリガ科）新北区
lichen moths		*Cisthene*	（チョウ目、ヒトリガ科）の昆虫の総称　米国での英名
lichen sober		*Acanthophila alacella* (Zeller)	（チョウ目、キバガ科）旧北区
lichen-eating caterpillar		*Manulea replana* (Lewin)	（チョウ目、ヒトリガ科）豪州区
lichmen moths			footmen を見よ
Lidderdale's dawnfly		*Capila lidderdali* (Elwes)	（チョウ目、セセリチョウ科）東洋区
Lieftinck's sprite		*Archibasis lieftincki* Coniff et Bedjanic	（トンボ目、イトトンボ科）東洋区
lieutenants		*Brachydiplax*	（トンボ目、トンボ科）の昆虫の総称
lift moths			shield bearer moths を見よ
lifts			shield bearer moths を見よ
light arches		*Apamea lithoxylaea* (Denis et Schiffermüller)	（チョウ目、ヤガ科）旧北区
light birch pigmy		*Stigmella lapponica* (Wocke)	（チョウ目、モグリチビガ科）旧北区
light brindled-brown groundling		*Calyocolum proximum* Haworth	（チョウ目、キバガ科）旧北区

英名	和名	学名	所属、分布、ほか
light brocade		*Lacanobia w-latinum* (Hufnagel)	（チョウ目、ヤガ科）旧北区
light brown apple moth		*Epiphyas postvittana* (Walker)	（チョウ目、ハマキガ科）全北区、豪州区。カンキツ害虫
light brown carpet moth		*Chloroclystis sphragitis* Meyrick	（チョウ目、シャクガ科）豪州区
light-brown flea beetle		*Longitarsus victoriensis* Blackburn	（コウチュウ目、ハムシ科）豪州区
light brown forester		*Bebearia zonara* (Bulter)	（チョウ目、タテハチョウ科）エチオピア区
light carpet moth		*Hydrelia lucata* (Guenée)	（チョウ目、シャクガ科）新北区
light crimson underwing		*Catocala promissa* (Denis et Schiffermüller)	（チョウ目、ヤガ科）旧北区
light emerald		*Campaea margaritata* (Linnaeus)	（チョウ目、シャクガ科）旧北区
light emerald moth			light emerald を見よ
light ermine moth			dark-spotted tiger moth を見よ
light feathered rustic		*Agrotis cinerea* (Denis et Schiffermüller)	（チョウ目、ヤガ科）旧北区
light flies	デガシラバエ科	Pyrgotidae	（ハエ目）の昆虫の総称　成虫は灯火にくる
light grey tortrix			allied shade moth を見よ
light knot-grass		*Acronicta menyanthidis* (Esper)	（チョウ目、ヤガ科）旧北区
light knot-grass dagger			light knot-grass を見よ
light knot-grass moth			light knot-grass を見よ
light-loving noctuid moth		*Agrotis photophila* (Butler)	（チョウ目、ヤガ科）大洋区
light magpie		*Abraxas pantaria* Linnaeus	（チョウ目、シャクガ科）旧北区
light Marathyssa moth		*Marathyssa basalis* Walker	（チョウ目、ヤガ科）新北区
light marbled brown			marbled prominent を見よ
light orange underwing	クロフカバシャク	*Archiearis notha* (Hübner)	（チョウ目、シャクガ科）日本、旧北区
light red Acraea		*Acraea nohara* Boisduval	（チョウ目、タテハチョウ科）エチオピア区
light Scottish stiletto		*Thereva inornata* Verrall	（ハエ目、ツルギアブ科）旧北区
light silver-striped piercer			spruce seed moth (1) を見よ
light spectacle			dark spectacle を見よ
light straw ace		*Pithauria stramineipennis* Wood-Mason et de Nicéville	（チョウ目、セセリチョウ科）東洋区
light streak		*Pleurota bicostella* (Clerck)	（チョウ目、マルハキバガ科）旧北区
light-tipped demon		*Indothemis carnatica* (Fabricius)	（トンボ目、トンボ科）東洋区
lighthouse gall			ground ivy を見よ
lighting beetle			glow-worm (2) を見よ

英名	和名	学名	所属、分布、ほか
lighting beetles			fireflies (1) を見よ
lightingbugs			fireflies (2) を見よ
lightning beetles			fireflies (1) を見よ
lightningbugs			fireflies (1) を見よ
Ligilla skipperling		*Dalla ligilla* (Hewitson)	（チョウ目、セセリチョウ科）新熱帯区
ligularia fruit fly	ツワブキケブカミバエ	*Paratephritis fukaii* Shiraki	（ハエ目、ミバエ科）日本
Ligurian bee			Italian honey bee を見よ
Ligurian leafhopper		*Eupteryx decemnotata* Rey	（カメムシ目、オオヨコバイ科）全北区
Ligurina hairstreak		*Kolana ligurina* (Hewitson)	（チョウ目、シジミチョウ科）新熱帯区
ligustrum dagger moth			coronet を見よ
ligustrum fruit gall midge	イボタミタマバエ	*Asphondylia sphaera* Monzen	（ハエ目、タマバエ科）日本
ligustrum globular treehopper	イボタマルツノゼミ	*Gargara ligustri* (Matsumura)	（カメムシ目、ツノゼミ科）日本、旧北区
ligustrum horn worm	コエビガラスズメ	*Sphinx costricta* Butler	（チョウ目、スズメガ科）日本
ligustrum longicorn	イボタサビカミキリ	*Sophronica obrioides* (Bates)	（コウチュウ目、カミキリムシ科）日本、東洋区
ligustrum moth		*Brahmaea wallichii* (Gray)	（チョウ目、イボタガ科）旧北区。日本亜種は *B. w. japonica* Butler イボタガ
ligustrum planthopper	イボタヒシウンカ	*Kuvera ligustri* Matsumura	（カメムシ目、ヒシウンカ科）日本、旧北区
Lila ruby-eye		*Carystoides lila* Evans	（チョウ目、セセリチョウ科）新熱帯区
lilac-banded Euselasia		*Euselasia perisama* Hall et Lamas	（チョウ目、シジミタテハ科）新熱帯区
lilac-banded longtail		*Urbanus dorantes* (Stoll)	（チョウ目、セセリチョウ科）新北区
lilac beauty (1)	イチモジエダシャク	*Apeira syringaria* (Linnaeus)	（チョウ目、シャクガ科）日本、旧北区
lilac beauty (2)	カクタシンジュタテハ	*Salamis cacta* (Fabricius)	（チョウ目、タテハチョウ科）エチオピア区
lilac beauty moth			lilac beauty (1) を見よ
lilac-bordered copper			Nivalis copper を見よ
lilac borer			ash borer を見よ
lilac grass-skipper			Doubleday's skipper を見よ
lilac hunter		*Zephyrogomphus lateralis* (Selys)	（トンボ目、サナエトンボ科）豪州区
lilac leaf miner		*Caloptilia syringella* (Fabricius)	（チョウ目、ホソガ科）全北区。lilac leafminer moth ともいう
lilac leaf miner			confluent-barred slender を見よ
lilac mother-of-pearl			lilac beauty (2) を見よ
lilac nymph		*Sallya rosa* (Hewitson)	（チョウ目、タテハチョウ科）エチオピア区
lilac oakblue		*Arhopala camdeo* (Moore)	（チョウ目、シジミチョウ科）東洋区
lilac pyralid	マエアカスカシノメイガ	*Palpita nigropunctalis* (Bremer)	（チョウ目、メイガ科）日本、旧北区。オリーブ、モクセイなどの害虫

英名	和名	学名	所属、分布、ほか
lilac silverline		*Aphnaeus lilacinus* Moore	（チョウ目、シジミチョウ科）東洋区
lilac thorn			lilac beauty (1) を見よ
lilac tip		*Colotis celimene amina* (Hewitson)	（チョウ目、シロチョウ科）エチオピア区。magenta tip を参照
lilac tip			magenta tip を見よ
lilac tree nymph			lilac nymph を見よ
lilacine bushbrown	コジャノメ	*Mycalesis francisca perdiccas* Hewitson	（チョウ目、タテハチョウ科）日本、旧北区。*M. francisca* (Stoll) 東洋区の英名
liliaceous leafminer	ササカワフンバエ	*Parallelomma sasakawae* Hering	（ハエ目、フンバエ科）日本
lily aphid			mottled arum aphid を見よ　豪州での英名
lily beetle		*Crioceris lilii* (Scopoli)	（コウチュウ目、ハムシ科）旧北区
lily black thrips			lily thrips を見よ
lily borer		*Brithys pancratii* (Cyrillo)	（チョウ目、ヤガ科）旧北区
lily bulb thrips	ユリクダアザミウマ	*Liothrips vaneecki* Priesner	（アザミウマ目、クダアザミウマ科）日本、全北区、東洋区。ユリ害虫
lily bush-cricket		*Tylopsis liliifolia* Fabricius	（バッタ目、ツユムシ科）旧北区
lily caterpillar		*Spodoptera picta* (Guérin-Méneville)	（チョウ目、ヤガ科）豪州区
lily leaf beetle		*Lilioceris nigripes* (Fabricius)	（コウチュウ目、ハムシ科）豪州区
lily leaf beetle	カタクリハムシ	*Sangariola punctatostriata* (Motschulsky)	（コウチュウ目、ハムシ科）日本、旧北区。ユリ害虫
lily leaf beetle			lily beetle を見よ
lily leaf beetle			scarlet lily beetle を見よ
lily leaf miner		*Liriomyza urophorina* Mik	（ハエ目、ハモグリバエ科）旧北区
lily moth	ハマオモトヨトウ	*Brithys crini* (Fabricius)	（チョウ目、ヤガ科）日本、旧北区、エチオピア区、豪州区。幼虫は lily borer
lily moth		*Polytela gloriosae* Fabricius	（チョウ目、ヤガ科）東洋区
lily-of-the-valley aphid		*Aulacorthum speyeri* Börner	（カメムシ目、アブラムシ科）旧北区
lily pilly psyllid			eugenia psyllid を見よ
lily reddish leaf beetle	ユリクビナガハムシ	*Lilioceris merdigera* (Linnaeus)	（コウチュウ目、ハムシ科）日本、旧北区、東洋区、新熱帯区
lily thrips	ユリキイロアザミウマ	*Frankliniella lilivora* Kurosawa	（アザミウマ目、アザミウマ科）日本、旧北区
lily thrips			lily bulb thrips を見よ
lily weevil		*Agasphaerops nigra* Horn	（コウチュウ目、ゾウムシ科）新北区
lily weevils			bulb snout beetles を見よ
lilypad clubtail		*Arigomphus furcifer* (Hagen)	（トンボ目、サナエトンボ科）新北区
lilypad forktail		*Ischnura kellicotti* Williamson	（トンボ目、イトトンボ科）新北区
lilypad whiteface		*Leucorrhinia caudalis* (Charpentier)	（トンボ目、トンボ科）旧北区
lilysquatters		*Paracercion*	（トンボ目、イトトンボ科）の昆虫の総称

英名	和名	学名	所属、分布、ほか
lima bean borer			lima-bean pod borer を見よ
lima-bean pod borer	シロイチモンジマダラメイガ	*Etiella zinckenella* (Treitschke)	(チョウ目、メイガ科) 日本、全北区、東洋区、豪州区、エチオピア区、新熱帯区。ダイズなどマメ類害虫
lima-bean vine borer		*Monoptilota pergratialis* (Hulst)	(チョウ目、メイガ科) 新北区。成虫は lima bean vine borer moth
limacodids			slug caterpillar moths を見よ
limantriids			tussock moths (1) を見よ
lime beetle		*Stenostola dubia* (Laicharting)	(コウチュウ目、カミキリムシ科) 旧北区
lime blue		*Chilades lajus* (Cramer)	(チョウ目、シジミチョウ科) 東洋区。*C. l. tavoyanus* Evans も同英名
lime butterfly			chequered swallowtail を見よ *P. d. malayanus* Wallace も同英名
lime hawk		*Mimas tiliae* (Linnaeus)	(チョウ目、スズメガ科)旧北区
lime hawk moth			lime hawk を見よ
lime leaf aphid		*Eucallipterus tiliae* (Linnaeus)	(カメムシ目、アブラムシ科) 全北区、豪州区。lime aphid ともいう
lime leaf gall midge		*Didymomyia tiliacea* (Bremi)	(ハエ目、タマバエ科) 旧北区
lime leaf-mining sawfly		*Parna tenella* (Klug)	(ハチ目、ハバチ科) 日本、旧北区
lime leaf-petiole gall midge			linden bud gall midge を見よ
lime leaf-roll gall midge		*Dasineura tiliamvolvens* (Rubsaamen)	(ハエ目、タマバエ科) 旧北区
lime-speck pug		*Eupithecia centaureata* Denis et Schiffermüller	(チョウ目、シャクガ科) 旧北区
lime swallowtail			chequered swallowtail を見よ
lime tree borer		*Chelidonium cinctum* Guérin	(コウチュウ目、カミキリムシ科) 東洋区。カンキツ害虫
Limenia hairstreak		*Strymon limenia* (Hewitson)	(チョウ目、シジミチョウ科) 新北区
limnophilid flies			northern caddisflies を見よ
limpet caterpillar		*Cathopsyche reidi* Watt	(チョウ目、ミノガ科) 東洋区
limpets			ant flies を見よ 幼虫の英名
Lincoln underwing moth		*Catocala lincolnana* Brower	(チョウ目、ヤガ科) 新北区。Lincoln underwing ともいう
Linda's emperor		*Doxocopa linda* (Felder)	(チョウ目、タテハチョウ科) 新熱帯区
Linda's hairtail		*Anthene lindae* Henning et Henning	(チョウ目、シジミチョウ科) エチオピア区
Linda's roadside skipper			Arkansas roadside skipper を見よ
linden ambrosia beetle	カナクギノキクイムシ	*Indocryphalus pubipennis* (Blandford)	(コウチュウ目、キクイムシ科) 日本、旧北区、東洋区。木材害虫
linden aphid			lima leaf aphid を見よ
linden bark beetle			linden ambrosia beetle を見よ
linden bark borer			Linnaeus's cosmet を見よ

英名	和名	学名	所属、分布、ほか
linden borer		*Saperda vestita* Say	(コウチュウ目、カミキリムシ科) 新北区
linden bud gall midge		*Contarinia tiliarum* (Kieffer)	(ハエ目、タマバエ科) 旧北区
linden burncow		*Lampra rutilans* (Fabricius)	(コウチュウ目、タマムシ科) 旧北区
linden gall midge		*Cecidomyia tiliaria* Harris	(ハエ目、タマバエ科) 旧北区
linden lace bug			basswood lace bug を見よ
linden looper		*Erannis tiliaria* (Harris)	(チョウ目、シャクガ科) 新北区。linden looper moth ともいう
linden prominent moth		*Ellida caniplaga* (Walker)	(チョウ目、シャチホコガ科) 新北区
linden scale		*Lepidosaphes tiliae* Linnaeus	(カメムシ目、マルカイガラムシ科) 旧北区
Lindsey's branded skipper			Lindsey's skipper を見よ
Lindsey's skipper		*Hesperia lindseyi* (Holland)	(チョウ目、セセリチョウ科) 新北区
line grass yellow		*Eurema laeta lineata* (Miskin)	(チョウ目、シロチョウ科) 豪州区。lined grass-yellow とも記す
line-grass yellow			spotless grass yellow を見よ　豪州での英名
line ruby tiger moth		*Phragmatobia lineata* Newman et Donahue	(チョウ目、ヒトリガ科) 新北区。lined ruby tiger moth ともいう。
linear bug		*Phaenacantha australiae* Kirkaldy	(カメムシ目、Colobathristidae) 豪州区
linear-winged grasshopper		*Aptenopedes sphenarioides* Scudder	(バッタ目、バッタ科) 新北区
lineate bark beetle			ambrosia beetle (1) を見よ
lineate bark-beetle			striped ambrosia beetle を見よ
lineate chafer	スジコガネ	*Mimela testaceipes* (Motschulsky)	(コウチュウ目、コガネムシ科) 日本、旧北区。多くの作物、樹木害虫
lined acrobat ant			acrobat ant を見よ
lined acrobatic ant			acrobat ant を見よ
lined click beetle			common click beetle (1) を見よ　北米での英名
lined copper looper			dark-spotted looper moth を見よ
lined corn-borer			lined stalk borer moth を見よ
lined corn-borer			broken-banded leafroller moth を見よ
lined grasshopper		*Stenobothrus lineatus* (Panzer)	(バッタ目、バッタ科) 旧北区
lined hawk		*Theretra orpheus* (Herrich-Schäffer)	(チョウ目、スズメガ科) エチオピア区
lined quaker moth		*Apamea inficita* Walker	(チョウ目、ヤガ科) 新北区。lined quaker という
lined spittle bug		*Philaenus lineatus* Linnaeus	(カメムシ目、コガシラアワフキ科) 新北区
lined spittlebug		*Neophilaeus lineatus* (Linnaeus)	(カメムシ目、アワフキムシ科) 全北区
lined spittle insect			lined spittle bug を見よ

英名	和名	学名	所属、分布、ほか
lined stalk borer			lined stalk borer moth を見よ
lined stalk borer moth		*Oligia fractilinea* (Grote)	(チョウ目、ヤガ科) 全北区
lineella underwing			little lined underwing moth を見よ
ling pug		*Eupithecia goossensiata* Mabille	(チョウ目、シャクガ科) 旧北区
ling tubic		*Amphisbatis incongruella* (Stainton)	(チョウ目、キバガ科) 旧北区
link moth			orange-spotted castniid を見よ
Linna palm dart		*Telicota linna* Evans	(チョウ目、セセリチョウ科) 東洋区
Linnaeus' horse bot fly			throat bot fly を見よ
Linnaeus' Idea	クロオビオオゴマダラ	*Idea idea* (Linnaeus)	(チョウ目、タテハチョウ科) 東洋区
Linnaeus' 17-year cicada			seventeen-year locust (1) を見よ
Linnaeus's cosmet		*Chrysoclista linneella* (Clerck)	(チョウ目、カザリバガ科) 旧北区
Linnaeus's leaf-insect		*Phyllium siccifolium* (Linnaeus)	(ナナフシ目、コノハムシ科) 東洋区
Linnaeus's virgin case		*Dahlica lichenella* (Linnaeus)	(チョウ目、ミノガ科) 旧北区
Lintner's Gluphisia moth		*Gluphisia lintneri* (Grote)	(チョウ目、シャチホコガ科) 新北区
Liodes Emesis		*Emesis liodes* Godman et Salvin	(チョウ目、シジミタテハ科) 新熱帯区
Liodes hairtail		*Anthene liodes* (Hewitson)	(チョウ目、シジミチョウ科) エチオピア区
liodid beetles		Liodidae	(コウチュウ目) の昆虫の総称
lion ants			pyramid ants を見よ
lion beetle		*Ulochaetes leoninus* LeConte	(コウチュウ目、カミキリムシ科) 新北区
lion swordtail	オオオナガタイマイ	*Graphium androcles* (Boisduval)	(チョウ目、アゲハチョウ科) 東洋区
liopterid wasps		Liopteridae	(ハチ目) の昆虫の総称
liopterids			liopterid wasps を見よ
lip botfly			nose bot fly を見よ
Lipara buff		*Baliochila lipara* Stempffer et Bennett	(チョウ目、シジミチョウ科) エチオピア区
liposcelid booklice			booklice (1) を見よ
Lirides banner		*Ectima lirides* Staudinger	(チョウ目、タテハチョウ科) 新熱帯区
Liris skipper		*Lerema liris* Evans	(チョウ目、セセリチョウ科) 新北区、新熱帯区
Lissa recluse		*Leona lissa* Evans	(チョウ目、セセリチョウ科) エチオピア区
Lister's hairstreak		*Pamela dudgeonii* (de Nicéville)	(チョウ目、シジミチョウ科) 東洋区
listrosceline grasshoppers	ウマオイ亜科	Listroscelinae	(バッタ目、キリギリス科) の昆虫の総称
Lisus hairstreak		*Theritas lisus* (Stoll)	(チョウ目、シジミチョウ科) 新熱帯区

英名	和名	学名	所属、分布、ほか
Litana skipper		*Vacerra litana* (Hewitson)	（チョウ目、セセリチョウ科）新熱帯区
litchi bud moth		*Crocidosema litchivora* Almela	（チョウ目、ハマキガ科）新北区
litchi fruit moth			macadamia nutborer を見よ
litchi lantern bug		*Pyrops candelaria* (Linnaeus)	（カメムシ目、ビワハゴロモ科）東洋区
litchi stink bug		*Lyramorpha rosea* Westwood	（カメムシ目、Tessaratomidae）豪州区
litchi stink bug (1)		*Tesseratoma papillosa* (Drury)	（カメムシ目、カメムシ科）東洋区
lithocarpus bark beetle	マテバシイキクイムシ	*Xyleborus glabratus* Eichhoff	（コウチュウ目、キクイムシ科）日本、東洋区
lithocarpus scale	ムツレタマカイガラムシ	*Kermes mutsurensis* Kuwana	（カメムシ目、タマカイガラムシ科）日本
Lithochroa blue		*Jalmenus lithochroa* Waterhouse	（チョウ目、シジミチョウ科）豪州区。Lithochroa blue butterfly ともいう
Litocala moth		*Litocala sexsignata* (Harvey)	（チョウ目、ヤガ科）新北区
litsea longicorn beetle	ヒメリンゴカミキリ	*Oberea hebescens* Bates	（コウチュウ目、カミキリムシ科）日本。フジ、クスノキ害虫
litter moth			dried leaf moth を見よ　北米での英名
little Acraea		*Acraea axina* Westwood	（チョウ目、タテハチョウ科）エチオピア区
little ant	ヒメアリ	*Monomorium intrudens* F. Smith	（ハチ目、アリ科）日本、旧北区
little arches moth		*Drasteria petricola* (Walker)	（チョウ目、ヤガ科）新北区
little banded swift			Bengal swift を見よ
little banded Yeoman		*Paduca fasciata fasciata* (C. et R. Felder)	（チョウ目、タテハチョウ科）東洋区
little banner	ツマグロヒメカバタテハ	*Nica flavilla* Godart	（チョウ目、タテハチョウ科）新熱帯区
little bark beetle		*Crypturgus pusillus* (Gyllenhal)	（コウチュウ目、キクイムシ科）旧北区
Little Barrier Island giant weta			weta-punga を見よ
Little Barrier Island weta			weta-punga を見よ
little bear		*Pocalta ursina* (Horn)	（コウチュウ目、コガネムシ科）新北区。ブドウ害虫
little bear moths			bear moths を見よ
little beggar moth		*Eubaphe meridiana* (Slosson)	（チョウ目、シャクガ科）新北区
little black ant		*Monomorium minimum* (Buckley)	（ハチ目、アリ科）新北区
little black ant		*Prenolepis imparis nitens* (Mayr)	（ハチ目、アリ科）旧北区
little black-knees		*Methiolopsis geniculata* (Stål)	（バッタ目、バッタ科）豪州区
little blue			azure dartlet を見よ

英名	和名	学名	所属、分布、ほか
little blue			small blue butterfly を見よ
little blue cattle louse			small blue cattle louse を見よ
little blue dragonlet		*Erythrodiplax minuscula* (Rambur)	(トンボ目、トンボ科) 新北区、新熱帯区
little bluet		*Enallagma minusculum* Morse	(トンボ目、イトトンボ科) 新北区
little-bull clothes		*Ochsenheimeria urella* Fischer von Röslerstamm	(チョウ目、Ochsenheimeriidae) 旧北区
little Carol's wasp moth			Florida Euceron を見よ
little carpenter bee		*Ceratina (Zadontomerus) dupla* (Say)	(ハチ目、ミツバチ科) 新北区
little carpenterworm		*Prionoxystus macmurtrei* (Guérin-Méneville)	(チョウ目、ボクトウガ科) 新北区。成虫は little carpenterworm moth.
little cloud Ancylis moth			apple leaf sewer を見よ
little dusk-hawker		*Gynacantha manderica* Grunberg	(トンボ目、ヤンマ科) エチオピア区
little earwig			lesser earwig を見よ　米国での英名
little earwigs	クロハサミムシ科	Spongiphoridae	(ハサミムシ目) の昆虫の総称
little earwigs (1)	チビハサミムシ科	Labiidae	(ハサミムシ目) の昆虫の総称
little emerald	ナミガタウスキアオシャク	*Jodis lactearia* (Linnaeus)	(チョウ目、シャクガ科) 日本、旧北区
little Epitola		*Stempfferia zelza* (Hewitson)	(チョウ目、シジミチョウ科) エチオピア区
little fire ant		*Ochetomyrmex auropunctata* (Roger)	(ハチ目、アリ科) 新北区、新熱帯区
little fire ant		*Ochetomyrmex*	(ハチ目、アリ科) の昆虫の総称
little fritillary		*Mellicta asteria* (Freyer)	(チョウ目、タテハチョウ科) 旧北区
little glassywing		*Pompeius verna* (Edwards)	(チョウ目、セセリチョウ科) 新北区。新熱帯区の *P. v. sequoyah* (Freeman) も同英名
little grass cicada		*Kikihia muta* (Fabricius)	(カメムシ目、セミ科) 豪州区
little green leafhopper		*Balclutha incisa hospes* (Kirkaldy)	(カメムシ目、オオヨコバイ科) 新北区、大洋区
little hairtail		*Anthene minima* (Trimen)	(チョウ目、シジミチョウ科) エチオピア区
little hickory aphid		*Monellia caryella* (Fitch)	(カメムシ目、アブラムシ科) 全北区
little horn bug			lesser stag beetle を見よ
little house fly			lesser house fly を見よ
little house fly			latrine fly を見よ
little Janne			tailed orange を見よ
little lancewing		*Cataplectica profugella* (Stainton)	(チョウ目、ササベリガ科) 旧北区
little lined underwing moth		*Catocala lineella* Grote	(チョウ目、ヤガ科) 新北区。little lined underwing ともいう
little maplet		*Chersonesia peraka peraka* Distant	(チョウ目、タテハチョウ科) 東洋区

英名	和名	学名	所属、分布、ほか
little metalmark		*Calephelis virginiensis* (Guérin-Méneville)	（チョウ目、シジミタテハ科）新北区
little nymph underwing moth			tiny nymph underwing moth を見よ
little orange tip		*Colotis etrida* (Boisduval)	（チョウ目、シロチョウ科）東洋区
little Oregon skipper			Mardon skipper を見よ
little pasture cockchafer		*Aphodius frenchi* Blackburn	（コウチュウ目、コガネムシ科）豪州区
little pine-tree beetle			two-toothed pine beetle を見よ
little pondskater		*Gerris argentatus* Schummel	（カメムシ目、アメンボ科）旧北区
little red ant			Pharaoh's ant を見よ
little red louse			cattle chewing louse を見よ
little sailor		*Dynamine ate* (Godman et Salvin)	（チョウ目、タテハチョウ科）新熱帯区
little scarlet		*Crocothemis sanguinolenta* (Burmeister)	（トンボ目、トンボ科）エチオピア区。small scarlet ともいう
little 17-year cicada		*Magicicada septendecula* Alexander et Moore	（カメムシ目、セミ科）新北区
little snipefly		*Chrysopilus asiliformis* (Preyssler)	（ハエ目、シギアブ科）旧北区
little spurthroated grasshopper		*Melanoplus infanitilis* Scudder	（バッタ目、バッタ科）新北区
little stout crawer mayflies		Tricorythidae	（カゲロウ目）の昆虫の総称
little sulphur			little yellow を見よ
little sulphurs			grass yellows を見よ
littte 13-year cicada		*Magicicada tredecula* Alexander et Moore	（カメムシ目、セミ科）新北区
little thorn	アトボシエダシャク	*Cepphis advenaria* (Hübner)	（チョウ目、シャクガ科）日本、旧北区
little tiger blue		*Tarucus balkanicus* (Freyer)	（チョウ目、シジミチョウ科）旧北区、エチオピア区
little toe biter		*Belostoma flumineum* Say	（カメムシ目、コオイムシ科）新北区
little underwing moth		*Catocala minuta* Edwards	（チョウ目、ヤガ科）新北区。little underwing ともいう
little virgin moth		*Grammia anna* (Grote)	（チョウ目、ヒトリガ科）新北区
little virgin tiger moth		*Apantesis virguncula* Saunders	（チョウ目、ヒトリガ科）新北区
little whisp		*Agriocnemis exilis* Selys	（トンボ目、イトトンボ科）エチオピア区
little white lichen moth		*Clemensia albata* (Packard)	（チョウ目、ヒトリガ科）新北区
little whites	ツマグロシロチョウ属	*Euchloe*	（チョウ目、シロチョウ科）の昆虫の総称
little wife underwing moth		*Catocala muliercula* Guenée	（チョウ目、ヤガ科）新北区。little wife underwing ともいう
little wingless grasshopper		*Gymnoscirtetes pusillus* Scudder	（バッタ目、バッタ科）新北区

英名	和名	学名	所属、分布、ほか
little wood satyr			wood satyr を見よ
little wood satyr		*Megisto cymela cymela* (Cramer)	（チョウ目、タテハチョウ科）新北区
little yellow		*Eurema lisa* (Boisduval et LeConte)	（チョウ目、シロチョウ科）新北区、新熱帯区
little yellow ant			vagrant ant を見よ
little yellow ant			Pharaoh's ant を見よ
little Yukatan mantis		*Mantoida maya* Saussure et Zehntner	（カマキリ目、Mantoididae）新北区、新熱帯区
live oak bagworm		*Oiketicus abbotii* Grote	（チョウ目、ミノガ科）新北区、新熱帯区。カンキツ害虫
live oak cluster beetle		*Cibdelis blaschkei* Mannerheim	（コウチュウ目、ゴミムシダマシ科）新北区
live oak leaf blotchminer			oak leaf blotchminer を見よ
live oak Metria moth		*Metria amella* (Guenée)	（チョウ目、ヤガ科）新北区
live oak ribbed casemaker		*Bucculatrix albertiella* Busck	（チョウ目、チビガ科）新北区
lively ant-guest beetles		Cossyphodinae	（コウチュウ目、ゴミムシダマシ科）の昆虫の総称
liver argent	イボタコスガ	*Zelleria hepariella* Stainton	（チョウ目、スガ科）日本、旧北区
liverfluke snail predator-fly		*Sepedomerus macropus* (Walker)	（ハエ目、ヤチバエ科）新熱帯区、新北区、大洋区
Liverpool feather-horn			middle feather clothes を見よ
livestock biting lice		*Bovicola*	（ハジラミ目、ケモノハジラミ科）の昆虫の総称
livestock sucking lice		*Haematopinus*	（シラミ目、ケモノジラミ科）の昆虫の総称
livestock sucking lice		*Linognathus*	（シラミ目、ケモノホソジラミ科）の昆虫の総称
livid crescent plume		*Marasmarcha lunadactyla* Haworth	（チョウ目、トリバガ科）旧北区
living bamboo longicorn	サビアヤカミキリ	*Abryna obscura* Schwarzer	（コウチュウ目、カミキリムシ科）日本、東洋区。タケ害虫。living bamboo longicorn beetle ともいう
lizard barklice	ケチャタテ科	Caeciliidae	（チャタテムシ目）の昆虫の総称
lizard barklouse		*Valenzuela flavidus* (Stephens)	（チャタテムシ目、チャタテムシ科）旧北区
lizard beetles	コメツキモドキ科	Languriidae	（コウチュウ目）の昆虫の総称
lizard weevil		*Cathormiocerus britannicus* Blair	（コウチュウ目、ゾウムシ科）旧北区
llight southeastern drywood termite		*Incisitermes snyderi* (Light)	（シロアリ目、レイビシロアリ科）新北区、新熱帯区
loaiasis vector		*Chrysops dimidiata* Wulp	（ハエ目、アブ科）エチオピア区
Loammi skipper			southern dusted skipper を見よ
lobed evening brown			dusky evening brown を見よ
lobed skipper		*Osphantes ogowena* (Mabille)	（チョウ目、セセリチョウ科）エチオピア区
lobed Temnora		*Temnora pseudopylas* (Rothschild)	（チョウ目、スズメガ科）エチオピア区

英名	和名	学名	所属、分布、ほか
Lobelia blue ringlet		*Caeruleuptychia lobelia* Butler	（チョウ目、タテハチョウ科）新熱帯区
Lobelia dagger moth			greater oak dagger moth を見よ
loblolly pine sawfly		*Neodiprion taedae linearis* Ross	（ハチ目、マツハバチ科）新北区
loblolly pineconeworm moth		*Dioryctria merkeli* Mutuura et Munroe	（チョウ目、メイガ科）新北区
lobster			lobster moth を見よ　米国での英名
lobster caterpillar			longan prominent moth を見よ
lobster cockroach			cinereous cockroach を見よ　米国での英名
lobster moth		*Stauropus fagi* (Linnaeus)	（チョウ目、シャチホコガ科）旧北区。日本亜種は *S. f. persimilis* Butler シャチホコガ。幼虫の形態に由来
lobster prominent			lobster moth を見よ
local chalk carpet		*Scotopteryx bipunctaria* (Denis et Schiffermüller)	（チョウ目、シャクガ科）旧北区
local hawthorn ermel			hawthorn webber moth を見よ
local long-tailed satin		*Trichocercus sparshalli* (Curtis)	（チョウ目、シャチホコガ科）旧北区、豪州区
locust			イナゴ、バッタなどバッタ科の昆虫をいうが、とくに群集して大害をなすトノサマバッタの類を指す
locust			米国でセミを指す
locust			Asiatic migratory locust を見よ
locust			dragonflies (1) を見よ
locust bean moth	イナゴマメマダラメイガ	*Ectomyelois ceratoniae* (Zeller)	（チョウ目、メイガ科）全北区、豪州区、エチオピア区。トベラ、貯穀害虫
locust borer		*Megacyllene robiniae* (Forster)	（コウチュウ目、カミキリムシ科）新北区。locust borer beetle ともいう
locust clearwing			western poplar clearwing を見よ
locust digitate leafminer moth		*Parectopa robiniella* Clemens	（チョウ目、ホソガ科）全北区。幼虫は locust digitate leafminer
locust egg parasites			scelionid wasps を見よ　南アフリカでの英名
locust gall midge	ハリエンジュハベリタマバエ	*Obolodiplosis robiniae* (Haldeman)	（ハエ目、タマバエ科）日本、全北区。ニセアカシア害虫
locust leaf miner		*Odontota dorsalis* (Thunberg)	（コウチュウ目、ハムシ科）新北区
locust leaf roller		*Nephopterix subcaesiella* (Clemens)	（チョウ目、メイガ科）新北区。成虫は locust leafroller moth
locust looper			faint-spotted angle moth を見よ
locust thrips	ビワハナアザミウマ	*Thrips coloratus* Schmutz	（アザミウマ目、アザミウマ科）日本、東洋区
locust treehopper		*Thelia bimaculata* Fabricius	（カメムシ目、ツノゼミ科）新北区。体長 10 〜 11 mm
locust twig borer		*Ecdytolopha insiticiana* Zeller	（チョウ目、ハマキガ科）新北区。locust twig borer noth ともいう
locust underwing moth		*Euparthenos nubilis* (Hübner)	（チョウ目、ヤガ科）新北区。locust underwing ともいう

英名	和名	学名	所属、分布、ほか
locusts	バッタ目	Orthoptera	の昆虫の総称．世界に約 20,000 種
locusts			grasshoppers (1) (2) を見よ
locusts			dog-day cicadas を見よ
lodgepole-cone beetle			ponderosa pine cone beetle を見よ
lodgepole needle miner (1)		*Coleotechnites starki* Freeman	（チョウ目、キバガ科）新北区
lodgepole needle-miner		*Coleotechnites milleri* (Busck)	（チョウ目、キバガ科）新北区。lodgepole needleminer moth ともいう
lodgepole needletier moth			Jack pine tube moth を見よ
lodgepole pine beetle		*Dendroctonus murrayanae* Hopkins	（コウチュウ目、キクイムシ科）新北区
lodgepole pine needle-miner			lodgepole needle miner (1) を見よ
lodgepole pinecone borer moth			cone moth を見よ
lodgepole sawfly		*Neodiprion burkei* Middleton	（ハチ目、マツハバチ科）新北区
lodgepole terminal weevil		*Pissodes terminalis* Hopping	（コウチュウ目、ゾウムシ科）新北区
Loefling's curl		*Aleimma loeflingiana* (Linnaeus)	（チョウ目、ハマキガ科）旧北区
Loewenstein's blue		*Lepidochrysops loewensteini* (Swanepoel)	（チョウ目、シジミチョウ科）エチオピア区
Loew's blue	シロモンヒメシジミ	*Plebejus loewii* (Zeller)	（チョウ目、シジミチョウ科）旧北区
loft fly			cluster fly を見よ
lofty bath white			peak white を見よ
log cabin bagworm			common bagworm を見よ
log cabin caddisworms		*Oecetis*	（トビケラ目、ヒゲナガトビケラ科）の昆虫の総称
loganberry beetle			raspberry beetle (1)(2) を見よ
loganberry cane fly	バラモグリハナバエ	*Pegomyia rubivora* (Coquillett)	（ハエ目、ハナバエ科）日本、全北区。バラ害虫
Logan's button	ウスジロハマキ	*Acleris logiana* (Clerck)	（チョウ目、ハマキガ科）日本、旧北区
Logan's button (1)		*Acleris schalleriana* (Linnaeus)	（チョウ目、ハマキガ科）旧北区
Loki hairstreak		*Callophrys gryneus loki* (Skinner)	（チョウ目、シジミチョウ科）新北区、新熱帯区
Loki hairstreak			Skinner's hairstreak を見よ
Loki juniper hairstreak			Loki hairstreak を見よ juniper hairstreak (1) を参照
Lollia hairstreak		*Dicya lollia* (Godman et Salvin)	（チョウ目、シジミチョウ科）新熱帯区
Lompobatang lady		*Vanessa buana* (Fruhstorfer)	（チョウ目、タテハチョウ科）東洋区
lonchaeid flies	クロツヤバエ科	Lonchaeidae	（ハエ目）の昆虫の総称
lonchaeids			lonchaeid flies を見よ 豪州での英名

英名	和名	学名	所属、分布、ほか
London brindled-beauty			brindled beauty を見よ
long-banded silverline		*Cigaritis lohita* (Horsfield)	（チョウ目、シジミチョウ科）東洋区。*C. l. senama* Seitz も同英名
long-beaked clover aphid			hawthorn aphid (1) を見よ
long-beaked Donacaula moth		*Donacaula longirostrallus* Clemens	（チョウ目、メイガ科）新北区
long-bodied water scorpion			water stick insect を見よ
long-brand bushbrown		*Mycalesis visala* Moore	（チョウ目、タテハチョウ科）東洋区。*M. v. phamis* Talbot et Corbet も同英名。long brand bushbrown とも記す
long-branded crow	アルゲアルリマダラ	*Euploea algea* (Godart)	（チョウ目、タテハチョウ科）東洋区、豪州区
long brown scale		*Coccus elongatus* (Signoret)	（カメムシ目、カタカイガラムシ科）全北区。樹木害虫
long brown scale (1)	ナガカタカイガラムシ	*Coccus longulus* (Douglas)	（カメムシ目、カタカイガラムシ科）日本、東洋区、汎熱帯。カンキツ、パパイア、ラン、クロトンなどの害虫
long-chirp field cricket		*Gryllus multipulsator* Weissman	（バッタ目、コオロギ科）新北区
long-clasped midget		*Phyllonorycter cerasicolella* (Herrich-Schäffer)	（チョウ目、ホソガ科）旧北区
long dash		*Polites mystic* (Edwards)	（チョウ目、セセリチョウ科）新北区
long-headed flies			long-legged flies (1) を見よ
long-headed flour beetle			larger black flour beetle (1) を見よ
long-headed grasshopper			snouted grasshopper を見よ
long-headed leafhopper	ホソサジヨコバイ	*Nirvana pallida* Melichar	（カメムシ目、オオヨコバイ科）日本、東洋区
long-headed wasps		*Dolichovespula*	（ハチ目、スズメバチ科）の昆虫の総称
long horn beetles			long-horned beetles を見よ
longhorn beetles			long-horned beetles を見よ
longhorn crazy ant			hairy ant を見よ
longhorn grasshoppers			bush crickets (1) を見よ
longhorn grasshoppers			long-horned grasshoppers (3) を見よ
long-horn grasshoppers			long-horned grasshoppers（1）を見よ
longhorn moths	ヒゲナガガ亜科	Adelinae	（チョウ目、マガリガ科）の昆虫の総称
longhorn moths			yucca moths (1) を見よ
long-horned beetles			long-horned leaf beetles を見よ
long-horned ant lions			owlflies を見よ
long-horned bee		*Eucera longicornis* (Linnaeus)	（ハチ目、ミツバチ科）旧北区

英名	和名	学名	所属、分布、ほか
long-horned beetle		*Sybra alternans* (Wiedemann)	（コウチュウ目、カミキリムシ科）大洋区
long-horned beetles	カミキリムシ科	Cerambycidae	（コウチュウ目）の昆虫の総称　世界に 35,000 種
long-horned black legionnaire		*Beris geniculata* Haliday	（ハエ目、ミズアブ科）旧北区
long-horned bugs	顕角群亜目	Gymnocerata	（カメムシ目）の昆虫の総称
longhorned caddis		*Triplectides australis* Navas	（トビケラ目、ヒゲナガトビケラ科）豪州区
long-horned caddisflies	ヒゲナガトビケラ科	Leptoceridae	（トビケラ目）の昆虫の総称　世界に 800 種
longhorned casemaker caddisflies			long-horned caddisflies を見よ
long-horned cleg		*Haematopota grandis* Meigen	（ハエ目、アブ科）旧北区
long-horned earwigs	オオハサミムシ科	Labiduridae	（ハサミムシ目）の昆虫の総称　世界に約 60 種
long-horned Eucera			long-horned bee を見よ
longhorned fairy moths			yucca moths (1) を見よ
long-horned flies			thread-horned flies を見よ
long-horned general		*Stratiomys longicornis* (Scopoli)	（ハエ目、ミズアブ科）旧北区
longhorned grasshopper		*Conocephalus saltator* (Saussure)	（バッタ目、キリギリス科）大洋区
long-horned grasshoppers			crickets (3) を見よ
long-horned grasshoppers (1)	キリギリス科	Tettigoniidae	（バッタ目）の昆虫の総称　世界に 約 400 種
long-horned grasshoppers (2)		*Conocephalus*	（バッタ目、キリギリス科）の昆虫の総称
long-horned grasshoppers (3)	キリギリス上科	Tettigonioidea	（バッタ目）の昆虫の総称
long-horned groundhopper		*Tetrix tenuicornis* (Sahlberg)	（バッタ目、ヒシバッタ科）旧北区
long-horned leaf beetles	ネクイハムシ亜科	Donaciinae	（コウチュウ目、ハムシ科）の昆虫の総称
long-horned leaf-cut weevil	ヒゲナガオトシブミ	*Paracycnotrachelus longicornis* (Roelofs)	（コウチュウ目、オトシブミ科）日本
long-horned locusts			long-horned grasshoppers (1) を見よ
longhorned moths		Lecithoceridae	（チョウ目）の昆虫の総称
long-horned owlet moth		*Hypenula cacuminalis* (Walker)	（チョウ目、ヤガ科）新北区
long-horned rice bug	ヒゲナガカメムシ	*Pachygrontha antennata* (Uhler)	（カメムシ目、ナガカメムシ科）日本、旧北区
long-horned soldier		*Vanoyia tenuicornis* (Macquart)	（ハエ目、ミズアブ科）旧北区
long horned swift			foolish swift を見よ

英名	和名	学名	所属、分布、ほか
long-horned walking-stick	エダナナフシ	*Phraortes illepidus* (Brunner von Wattenwyl)	(ナナフシ目、ナナフシ科) 日本、東洋区
long-horned weevil	ミカンセマルヒゲナガゾウムシ	*Phloeobius alternatus* (Wiedemann)	(コウチュウ目、ヒゲナガゾウムシ科) 日本、東洋区
long-horned white-spotted longicorn	ヒゲナガシラホシカミキリ	*Eumecocera argyrosticta* (Fischer)	(コウチュウ目、カミキリムシ科) 日本
long-horned wood borers			long-horned beetles を見よ
long-horned wood-boring beetles			long-horned beetles を見よ
longhorns			longhorn moths を見よ
longhorns			long-horned beetles を見よ
long-horns			yucca moths (1) を見よ
long-horns			thread-horned flies を見よ
long-joined beetles	ハムシダマシ科	Lagriidae	(コウチュウ目) の昆虫の総称
long-jointed bark beetles			long-joined beetles を見よ
longleaf pine seedworm		*Cydia ingens* Heinrich	(チョウ目、ハマキガ科) 新北区。longleaf pine seedworm moth ともいう
long-legged anabrus		*Anabrus longipes* Caudell	(バッタ目、キリギリス科) 新北区
long-legged ant	アシナガキアリ	*Anoplolepis gracilipes* (Jerdon)	(ハチ目、アリ科) 東洋区、エチオピア区
long-legged armoured katydid		*Acanthoplus longipes* (Charpentier)	(バッタ目、キリギリス科) エチオピア区
long-legged assassin bugs		*Empicoris*	(カメムシ目、サシガメ科) の昆虫の総称
long-legged bandwing		*Heteroptemis obscurella* (Blanchard)	(バッタ目、バッタ科) 豪州区
long-legged chafer	アシナガコガネ	*Hoplia communis* Waterhouse	(コウチュウ目、コガネムシ科) 日本
long-legged china-mark		*Dolicarthria punctalis* (Denis et Schiffermüller)	(チョウ目、メイガ科) 旧北区
long-legged darkling beetle		*Stenocra dentata* Herbst	(コウチュウ目、ゴミムシダマシ科) エチオピア区
long-legged flies		*Condylostylus*	(ハエ目、アシナガバエ科) の昆虫の総称
long-legged flies		*Dolichopoda*	(ハエ目、アシナガバエ科) の昆虫の総称
long-legged flies (1)	アシナガバエ科	Dolichopodidae	(ハエ目) の昆虫の総称　世界に 6,000 種
long-legged flies (2)	オドリバエ科	Empididae	(ハエ目) の昆虫の総称　世界に 3,000 種
long-legged fly		*Condylostylus occidentalis* (Bigot)	(ハエ目、アシナガバエ科) 東洋区、新北区
long-legged fly (1)		*Austrosciapus connexus* (Walker)	(ハエ目、アシナガバエ科) 豪州区
long-legged leaf-cut weevil	アシナガオトシブミ	*Phialodes rufipennis* Roelofs	(コウチュウ目、オトシブミ科) 日本
long-legged tabby		*Synaphe angustalis* (Denis et Schiffermüller)	(チョウ目、メイガ科) 旧北区
long-legged tabby		*Synaphe punctalis* (Fabricius)	(チョウ目、メイガ科) 旧北区

英名	和名	学名	所属、分布、ほか
long-legged taylers			crane flies (1) を見よ
long-legged wasp	キアシナガバチ	*Polistes rothneyi iwatai* van der Vecht	(ハチ目、スズメバチ科) 日本
long-lipped beetles		Telegeusidae	(コウチュウ目) の昆虫の総称
long mussel scale			Glover scale を見よ　南アフリカでの英名
long-necked ground beetles		Odocanthini	(コウチュウ目、オサムシ科) の昆虫の総称
long-necked snakeflies			snake-flies (3) を見よ
long-nosed cattle louse	ウシホソジラミ	*Linognathus vituli* (Linnaeus)	(シラミ目、ケモノホソジラミ科) 日本、汎世界。ウシに寄生
long-nosed ox louse			long-nosed cattle louse を見よ
long-nosed planthopper	テングスケバ	*Dictyophara patruelis* (Stål)	(カメムシ目、テングスケバ科) 日本、旧北区、東洋区
long-nosed weevils			attelabid weevils を見よ
long-palped Mycterophora moth		*Mycterophora longipalpata* Hulst	(チョウ目、ヤガ科) 新北区
long-palpi cochilid	テングイラガ	*Microleon longipalpis* Butler	(チョウ目、イラガ科) 日本、旧北区
long palpi tortrix			vine leaf-roller moth を見よ
long-palps Peoria moth		*Peoria longipalpella* (Ragonot)	(チョウ目、メイガ科) 新北区
long-proboscis scale			black thread scale を見よ
long scale			long brown scale (1) を見よ
long scale			Glover scale を見よ
long skimmer		*Orthetrum trinacria* (Selys)	(トンボ目、トンボ科) エチオピア区
long soft scale			long brown scale (1) を見よ
long-spined mealybug			long-tailed mealybug (1) を見よ
long-spotted silverdrop		*Epargyreus deleoni* Freeman	(チョウ目、セセリチョウ科) 新熱帯区
long-spurred meadow katydid		*Orchelimum silvaticum* McNeill	(バッタ目、キリギリス科) 新北区
long-spurred shieldback		*Atlanticus calcaratus* Rehn et Hebard	(バッタ目、キリギリス科) 新北区
long-sting		*Megarhyssa lunator* Fabricius	(ハチ目、ヒメバチ科) 新北区
long-sting			sirex parasite (1) を見よ
long-streak sailor		*Neptis philyra* Ménétriès	(チョウ目、タテハチョウ科) 旧北区
long-streaked Halisidota			long-streaked tussock moth を見よ
long-streaked sallow midget	ヤナギギンモンホソガ	*Phyllonorycter salicicolella* (Sircom)	(チョウ目、ホソガ科) 日本、旧北区
long-streaked tussock moth		*Leucanopsis longa* (Grote)	(チョウ目、ヒトリガ科) 新北区
long-tailed fruitfly parasite			long-tailed fruit fly egg parasite を見よ

英名	和名	学名	所属、分布、ほか
long-tailed admiral		*Antanartia schaeneia* (Trimen)	（チョウ目、タテハチョウ科）エチオピア区。long-tail admiral ともいう
long-tailed Aguna			tailed Aguna を見よ
long-tailed blue			pea blue を見よ　long-tailed blue butterfly ともいう
long-tailed burnet moths		Himantopteridae	（チョウ目）の昆虫の総称
long-tailed citrus mealybug			citrus mealybug (2) を見よ
long-tailed flasher		*Astraptes megalurus* (Mabille)	（チョウ目、セセリチョウ科）新熱帯区
long-tailed fruit fly egg parasite		*Biosteres longicaudatus* Ashmead	（ハチ目、コマユバチ科）新北区、大洋区
long-tailed greenish silk moth			lunar moth (1) を見よ
long-tailed hairstreak			Martial hairstreak を見よ
long-tailed line-blue			common lineblue を見よ
long-tailed mayfly		*Palingenia longicauda* Olivier	（カゲロウ目、Palingeniidae）旧北区
long-tailed meadow katydid		*Conocephalus attenuatus* (Scudder)	（バッタ目、キリギリス科）新北区
long-tailed mealybug	ナガオコナカイガラムシ	*Pseudococcus longispinus* (Targioni-Tozetti)	（カメムシ目、コナカイガラムシ科）日本、新北区、汎熱帯。カンキツ、マンゴー害虫
long-tailed mealybug (1)		*Pseudococcus adonidum* (Linnaeus)	（カメムシ目、コナカイガラムシ科）旧北区、エチオピア区、東洋区、豪州区、新熱帯区。カンキツ害虫
long-tailed metalmark	オナガツバメシジミタテハ	*Rhetus arcius* (Linnaeus)	（チョウ目、シジミタテハ科）新北区。新熱帯区の *R. a. beutelspacheri* Llorente, *R. a. thia* (Morisse) も同英名
long-tailed mistletoe hairstreak		*Iolaphilus timon* Fabricius	（チョウ目、シジミチョウ科）エチオピア区
long-tailed parlatoria		*Parlatoria longispina* Morgan	（カメムシ目、マルカイガラムシ科）新熱帯区。カンキツ害虫
long-tailed pea-blue			pea blue を見よ
longtailed scarlet-eye		*Cephise aelius* (Plotz)	（チョウ目、セセリチョウ科）新熱帯区
longtailed silk moth			lunar moth (2) を見よ
long-tailed silverfish		*Ctenolopisma longicaudata* Escherich	（シミ目、シミ科）新北区
long-tailed skimmer			white tail を見よ
long-tailed skipper		*Urbanus proteus* (Linnaeus)	（チョウ目、セセリチョウ科）新北区
long-tailed striped swordtail		*Graphium liponesco* (Suffert)	（チョウ目、アゲハチョウ科）エチオピア区
long-tailed wasps			ichneumonoid wasps を見よ
long-toed water beetles	ドロムシ科	Dryopidae	（コウチュウ目）の昆虫の総称　幼虫は水生。世界に 200 種
long-tongued bees	コシブトハナバチ科	Anthophoridae	（ハチ目）の昆虫の総称　世界に 4,000 種以上

英名	和名	学名	所属、分布、ほか
longwing crescent	ホソバマネシミカズキタテハ	*Eresia phillyra phyllyra* (Hewitson)	（チョウ目、タテハチョウ科）新熱帯区
longwing fritillaries		*Agraulis*	（チョウ目、タテハチョウ科）の昆虫の総称
long-winged conehead		*Conocephalus discolor* Thunberg	（バッタ目、キリギリス科）旧北区
long-winged crest		*Dichomeris fasciella* (Hübner)	（チョウ目、キバガ科）旧北区
long-winged dagger moth		*Acronicta longa* Guenée	（チョウ目、ヤガ科）新北区
long-winged grasshopper		*Aiolopus thalassinus* (Fabricius)	（バッタ目、バッタ科）旧北区、豪州区
long-winged greenstreak			mountain greenstreak を見よ
long-winged grouse locust	ハネナガヒシバッタ	*Euparatettix insularis* Bei-Bienko	（バッタ目、ヒシバッタ科）日本、東洋区、豪州区
longwinged oak longicorn			longwinged oak longicorn beetle を見よ
longwinged oak longicorn beetle	ナガゴマフカミキリ	*Mesosa longipennis* Bates	（コウチュウ目、カミキリムシ科）日本。フジ、木材害虫
long-winged orange Acraea		*Telchinia alalonga* (Henning et Henning)	（チョウ目、タテハチョウ科）エチオピア区
long-winged pearl		*Perinephela lancealis* Schiffermüller	（チョウ目、メイガ科）旧北区
long-winged rice grasshopper			rice grasshopper (1) を見よ
long-winged scaly cricket		*Hoplosphyrum boreale* (Scudder)	（バッタ目、コオロギ科）新北区
long-winged skipper			Ocola skipper を見よ
long-winged vegetable grasshopper	ハネナガフキバッタ	*Eirenephilus longipennis* (Shiraki)	（バッタ目、バッタ科）日本、旧北区
longwings			heliconians を見よ
longan prominent moth		*Stauropus alternus* Walker	（チョウ目、シャチホコガ科）東洋区
longan scale	リュウガンコノハカイガラムシ	*Thysanofiorinia nephelii* (Maskell)	（カメムシ目、マルカイガラムシ科）日本、汎熱帯
longanberry crown-borer			raspberry clearwing moth を見よ
longicera long-horned aphid	タデケクダヒゲナガアブラムシ	*Trichosiphonaphis polygoniformosana* (Takahashi)	（カメムシ目、アブラムシ科）日本、旧北区、東洋区
longicorn			long-horned beetles を見よ
longicorn beetle			house longhorn を見よ
longicorn beetles			long-horned beetles を見よ
longicornis			long-horned beetles を見よ
longicorns			long-horned beetles を見よ　中南米での英名
longitudinal striped leaf beetle	フタスジツツハムシ	*Cryptocephalus bilineatus* (Linnaeus)	（コウチュウ目、ハムシ科）日本、旧北区

英名	和名	学名	所属、分布、ほか
Long's brownie		*Miletus longeana* (de Nicéville)	（チョウ目、シジミチョウ科）東洋区
lonicera aphid	ニッポンオマルアブラムシ	*Amphicercidus japonicus* (Hori)	（カメムシ目、アブラムシ科）日本、旧北区
looped three-ring		*Ypthima watsoni* (Moore)	（チョウ目、タテハチョウ科）東洋区
looper			South American semilooper を見よ
looper caterpillar		*Chrysodeixis subsidens* (Walker)	（チョウ目、ヤガ科）豪州区
looper caterpillar			tobacco looper を見よ
looper caterpillar			silver Y moth (2) を見よ
looper caterpillars			geometer moths を見よ 幼虫の英名
looper moths			gems (1) を見よ
loopers			geometer moths を見よ
loosestrife borer moth		*Papaipema lysimachiae* Bird	（チョウ目、ヤガ科）新北区
lophopid planthoppers		Lophopidae	（カメムシ目）の昆虫の総称
lophopteryx prominent	エグリシャチホコ	*Ptilodon robusta* (Matsumura)	（チョウ目、シャチホコガ科）日本
lopp			human flea を見よ
loquat leaf roller	クロスジキンノメイガ	*Pleuroptya balteata* (Fabricius)	（チョウ目、メイガ科）日本、旧北区、東洋区
loquat stink bug	アヤナミカメムシ	*Agonoscelidus nubila* (Fabricius)	（カメムシ目、カメムシ科）日本、旧北区、東洋区
Lora Aborn's moth			chrysanthemum flower borer (1) を見よ
Lord Howe Island cicada		*Psaltoda insularis* Ashton	（カメムシ目、セミ科）豪州区
Lord Howe Island stick insect		*Dryococelus australis* Montrouzier	（ナナフシ目、ナナフシ科）豪州区
Lord Howe Island wax scale		*Ceroplastes insulanus* De Lotto	（カメムシ目、カタカイガラムシ科）豪州区
Lorenzo red tab policeman		*Coeliades lorenzo* Evans	（チョウ目、セセリチョウ科）エチオピア区
Lorey leafworm	クサシロキヨトウ	*Mythimna loreyi* (Duponchel)	（チョウ目、ヤガ科）日本、旧北区、東洋区、エチオピア区。イネ、サトウキビ、トウモロコシ害虫
Lorimer's rustic		*Caradrina flavirena* (Guenée)	（チョウ目、ヤガ科）旧北区
Lorkovic's brassy ringlet		*Erebia calcaria* Lorkovic	（チョウ目、タテハチョウ科）旧北区
Lorquin's admiral	ツマアカイチモンジ（ロルキンイチモンジ）	*Limenitis lorquini* Boisduval	（チョウ目、タテハチョウ科）新北区。*L. l. powelli* Austin et Emmel も同英名
Lorquin's angle moth		*Speranza lorquinaria* (Guenée)	（チョウ目、シャクガ科）新北区
Lorquin's blue		*Cupido lorquini* (Herrich-Schäffer)	（チョウ目、シジミチョウ科）旧北区
Loruhama eyemark		*Mesosemia loruhama* Hewitson	（チョウ目、シジミタテハ科）新熱帯区

英名	和名	学名	所属、分布、ほか
Lorza's sister		*Adelpha cocala lorzae* (Boisduval)	（チョウ目、タテハチョウ科）新熱帯区
lost egg skipper			Lindsey's skipper を見よ
lost metalmark		*Calephelis perditalis* Barnes et McDunnough	（チョウ目、シジミタテハ科）新北区
lost owlet moth		*Ledaea perditalis* (Walker)	（チョウ目、ヤガ科）新北区
Lotana blue		*Lepidochrysops lotana* Swanepoel	（チョウ目、シジミチョウ科）エチオピア区
Loteni brown		*Neita lotenia* (van Son)	（チョウ目、タテハチョウ科）エチオピア区
Lotis blue		*Lycaeides argyrognomon lotis* Lintner	（チョウ目、シジミチョウ科）新北区。Reverdin's blue を参照
lotus borer			American lotus borer moth を見よ
lotus hairstreak		*Callophrys perplexa perplexa* W. Barnes et Benjamin	（チョウ目、シジミチョウ科）新熱帯区
lotus lily midge	ハスムグリユスリカ	*Stenochironomus nelumbus* Tokunaga et Kuroda	（ハエ目、ユスリカ科）日本
lotus ruby-eye		*Perichares lotus* (Butler)	（チョウ目、セセリチョウ科）新熱帯区
loud-singing bush crickets		*Orocharis*	（バッタ目、コオロギ科）の昆虫の総称
Louise's underwing moth		*Catocala louiseae* Bauer	（チョウ目、ヤガ科）新北区
Louisiana angle-wing		*Microcentrum louisianum* Hebard	（バッタ目、キリギリス科）新北区
Louisiana Macrochilo moth		*Macrochilo louisiana* (Forbes)	（チョウ目、ヤガ科）新北区。Louisiana Macrochilo, Louisiana snout-moth ともいう
louse	シラミ		複数形は lice。sucking lice を見よ
louse flies	シラミバエ科	Hippoboscidae	（ハエ目）の昆虫の総称　世界に330種
louse flies			pupiparous flies を見よ
louse fly			sheep ked を見よ
louse-like flies			louse flies を見よ
lousiness			sucking lice を見よ
lousy watchman			dor beetle (1) を見よ
lovage weevil			alfalfa snout beetle を見よ
love bug		*Plecia nearctica* Hardy	（ハエ目、ケバエ科）新北区。自動車の排気に集まり問題となる
loving barklice	クロフチャタテ科	Philotarsidae	（チャタテムシ目）の昆虫の総称
Lower's darter		*Telicota mesoptis* (Lower)	（チョウ目、セセリチョウ科）豪州区
lowland branded blue		*Uranothauma falkensteini* (Dewitz)	（チョウ目、シジミチョウ科）エチオピア区
lowland owl-butterfly		*Opsiphanes invirae relucens* Fruhstorfer	（チョウ目、タテハチョウ科）新熱帯区。Hubner's owlet を参照
lowland tree termite		*Incisitermes immigrans* (Snyder)	（シロアリ目、レイビシロアリ科）日本、新北区、大洋区
lubber grasshopper		*Brachystola magna* (Girard)	（バッタ目、Romaleidae）新北区。体長8 cmの大型種
lubber grasshoppers		Romaleidae	（バッタ目）の昆虫の総称

英名	和名	学名	所属、分布、ほか
lubberly band-winged grasshopper		*Agymnastus ingens* (Scudder)	(バッタ目、バッタ科) 新北区
lubberly grasshopper		*Romalea microptera* (Beauvais)	(バッタ目、Romaleidae) 新北区
lubberly grasshopper			lubber grasshopper を見よ
lubbers		Romaleinae	(バッタ目、Romaleidae) の昆虫の総称
Lucagus hairstreak		*Dicya lucagus* (Godman et Salvin)	(チョウ目、シジミチョウ科) 新熱帯区
lucanids			stag beetles (1) を見よ
lucerne aphid parasite		*Aphidius ervi* Haliday	(ハチ目、コマユバチ科) 日本、豪州区
lucerne blue			pea blue を見よ
lucerne blue			common grass-blue (1) を見よ
lucerne blue butterfly			pea blue を見よ
lucerne bug			alfalfa plant bug を見よ
lucerne butterfly			African clouded yellow を見よ
lucerne caterpillar			beet armyworm を見よ
lucerne crownborer		*Corrhenes stigmatica* (Pascoe)	(コウチュウ目、カミキリムシ科) 豪州区
lucerne crownborer		*Zygrita diva* Thomson	(コウチュウ目、カミキリムシ科) 豪州区
lucerne earth flea			lucerne flea を見よ 南アフリカでの英名
lucerne flea	キマルトビムシ	*Sminthurus viridis* (Linnaeus)	(トビムシ目、マルトビムシ科) 日本、旧北区、豪州区
lucerne flower midge			alfalfa leaf midge を見よ
lucerne grub			black alfalfa leaf-beetle を見よ 幼虫の英名
lucerne ladybird			Jacque's beetle を見よ
lucerne leaf midge			alfalfa leaf midge を見よ
lucerne leafeating beetle		*Colaspoides foveiventris* Lea	(コウチュウ目、ハムシ科) 豪州区
lucerne leafhopper		*Austroasca alfalfae* (Evans)	(カメムシ目、オオヨコバイ科) 豪州区
lucerne leafroller			lucerne webworm を見よ
lucerne looper		*Zermizinga indocilisaria* Walker	(チョウ目、シャクガ科) 豪州区
lucerne moth		*Cydia medicaginis* Kusnetsov	(チョウ目、ハマキガ科) 旧北区
lucerne moth			North American grass webworm を見よ
lucerne plant bug			alfalfa plant bug を見よ
lucerne seed chalcid			clover seed chalcid (1) を見よ
lucerne seed wasp		*Bruchophagus roddi* (Gussakovski)	(ハチ目、カタビロコバチ科) 豪州区、新北区
lucerne seed wasp			clover seed chalcid (1) を見よ
lucerne seed web moth	ヒメイチモジマダラメイガ	*Etiella behrii* (Zeller)	(チョウ目、メイガ科) 日本、東洋区、豪州区。ダイズ害虫
lucerne sprout midge			alfalfa sprout midge を見よ

英名	和名	学名	所属、分布、ほか
lucerne webworm		*Merophyas divulsana* Walker	(チョウ目、ハマキガ科) 豪州区
lucerne weevil		*Phytonomus variabilis* (Herbst)	(コウチュウ目、ゾウムシ科) 旧北区
lucerne weevil			alfalfa weevil を見よ
Lucia widow		*Palpopleura lucia* (Drury)	(トンボ目、トンボ科) エチオピア区
Luciana underwing moth		*Catocala luciana* Strecker	(チョウ目、ヤガ科) 新北区。Luciana underwing ともいう
lucifer skipper		*Decinea lucifer* (Hübner)	(チョウ目、セセリチョウ科) 新熱帯区
Lucius metalmark		*Metacharis lucius* (Fabricius)	(チョウ目、シジミタテハ科) 新熱帯区
lucky bugs			whirligig beetles を見よ
Luda skipper		*Neoxeniades luda* (Hewitson)	(チョウ目、セセリチョウ科) 新熱帯区
Ludo grass skipper			Ludo skipper を見よ
Ludo skipper		*Lento ludo* Evans	(チョウ目、セセリチョウ科) 新熱帯区
Lukenia giant cupid		*Lepidochrysops lukenia* van Someren	(チョウ目、シジミチョウ科) エチオピア区
Lulworth skipper butterfly		*Thymelicus acteon* (Rottemburg)	(チョウ目、セセリチョウ科) 旧北区。英デボンシャーの地名に由来。Lulworth skipper ともいう
luminous percher		*Diplacodes luminans* (Karsch)	(トンボ目、トンボ科) エチオピア区
luna hornet moth			hornet clearwing を見よ
luna moth	アメリカオオミズアオ	*Actias luna* (Linnaeus)	(チョウ目、ヤママユガ科) 新北区。開帳 12 cm
lunar double stripe		*Minucia lunaris* (Denis et Schiffermüller)	(チョウ目、ヤガ科) 旧北区
lunar Eudesmia		*Eudesmia menea* (Drury)	(チョウ目、ヒトリガ科) 新熱帯区
lunar giant ichneumon		*Megarhyssa macrurus icterosticta* Michener	(ハチ目、ヒメバチ科) 新北区
lunar hornet moth			hornet clearwing を見よ
lunar marbled brown		*Drymonia ruficornis* (Hufnagel)	(チョウ目、シャチホコガ科) 旧北区
lunar marbled brown moth			lunar marbled brown を見よ
lunar moth (1)	オナガミズアオ	*Actias gnoma* (Butler)	(チョウ目、ヤママユガ科) 日本。オオミズアオに似た大型種
lunar moth (2)	オオミズアオ	*Actias artemis aliena* Butler	(チョウ目、ヤママユガ科) 日本。青白色の大型種。幼虫はリンゴ、サクラ、カシ類などの害虫
lunar spotted pinion	ナシキリガ	*Cosmia pyralina* (Denis et Schiffermüller)	(チョウ目、ヤガ科) 日本
lunar thorn		*Selenia lunularia* (Hübner)	(チョウ目、シャクガ科) 旧北区
lunar underwing		*Omphaloscelis lunosa* (Haworth)	(チョウ目、ヤガ科) 旧北区
lunar underwing			lunar yellow underwing を見よ
lunar yellow underwing		*Noctua orbona* (Hufnagel)	(チョウ目、ヤガ科) 旧北区

英名	和名	学名	所属、分布、ほか
lunar yellow underwing			common yellow underwing moth を見よ
lunate zale moth			moon umber を見よ　lunate zale ともいう
Lundgren's shieldback		*Idionotus lundgreni* Rentz et Birchim	(バッタ目、キリギリス科) 新北区
lungwort slender		*Acrocercops imperialella* (Zeller)	(チョウ目、ホソガ科) 旧北区
lupin aphid		*Macrosiphum albifrons* Essig	(カメムシ目、アブラムシ科) 旧北区
Lupina Emesis			ochreous Emesis を見よ
lupine blue		*Aricia lupini* (Boisduval)	(チョウ目、シジミチョウ科) 新北区
lupine borer moth		*Comadia bertholdi* (Grote)	(チョウ目、ボクトウガ科) 新北区
lupine ghost moth		*Phymatopus californicus* (Boisduval)	(チョウ目、コウモリガ科) 新北区
Lupita's satyr		*Zischkaia lupita* (Reakirt)	(チョウ目、タテハチョウ科) 新熱帯区
lurcher	キオビコノハ	*Yoma sabina* (Cramer)	(チョウ目、タテハチョウ科) 東洋区、豪州区
Lurid glider	ルリダキイロタテハ	*Cymothoe lurida* (Butler)	(チョウ目、タテハチョウ科) エチオピア区
Lusk's pine moth		*Coloradia luski* Barnes et Benjamin	(チョウ目、ヤママユガ科) 新北区
lustrous copper	キララタカネシジミ	*Lycaena cupreus* (Edwards)	(チョウ目、シジミチョウ科) 新北区
lustrous copper			Snow's copper を見よ
luteous alderfly		*Sialis flarilatera* (Linnaeus)	(アミメカゲロウ目、センブリ科) 旧北区
Luzon peacock	ルソンカラスアゲハ	*Papilio chikae* Igarashi	(チョウ目、アゲハチョウ科) 東洋区
Lybia longwing		*Eueides lybia* (Fabricius)	(チョウ目、タテハチョウ科) 新熱帯区
Lycaste tigerwing		*Hypothyris lycaste* (Fabricius)	(チョウ目、タテハチョウ科) 新熱帯区。*H. l. dionaea* (Hewitson) も同英名
lychee bush cricket		*Xenogryllus* sp.	(バッタ目、コオロギ科) 新北区
lychee stink bug			litchi stink bug (1) を見よ
lychnis			lychnis coronet を見よ
lychnis coronet		*Hadena bicruris* (Hufnagel)	(チョウ目、ヤガ科) 旧北区
lychnis groundling		*Calyocolum viscariella* (Stainton)	(チョウ目、キバガ科) 旧北区
lychnis moth			lychnis　coronet を見よ
lycid beetles			net-winged beetles (1) を見よ
lycid-mimic longhorned beetle		*Tethlimmena aliena* Bates	(コウチュウ目、カミキリムシ科) 新熱帯区
lycid-mimic moth		*Correbia lycoides* (Walker)	(チョウ目、ヒトリガ科) 新熱帯区
lycid moth			lichen moth (2) を見よ　北米での英名
Lycidas swallowtail			Cramer's swallowtail を見よ
Lycisca metalmark			blue-winged Eurybia を見よ
Lycoa Acraea		*Acraea lycoa* Godart	(チョウ目、タテハチョウ科) エチオピア区
Lycorias metalmark			Banner metalmark を見よ

英名	和名	学名	所属、分布、ほか
Lycortas skipper		*Orthos lycortas* (Godman)	（チョウ目、セセリチョウ科）新北区、新熱帯区
lyctid powderpost beetle			southern lyctus beetle を見よ
lyctid powder-post beetles			lyctids を見よ
lyctids	ヒラタキクイムシ亜科	Lyctinae	（コウチュウ目、ナガシンクイムシ科）の昆虫の総称
Lydd beauty		*Peribatodes ilicaria* Geyer	（チョウ目、シャクガ科）旧北区
Lyde hairstreak		*Kolana lyde* (Godman et Salvin)	（チョウ目、シジミチョウ科）新熱帯区
Lydenburg opal		*Chrysoritis aethon* (Trimen)	（チョウ目、シジミチョウ科）エチオピア区
Lydia tiger moth		*Phoenicoprocta lydia* (Druce)	（チョウ目、ヒトリガ科）新北区、新熱帯区
Lyell's swift		*Pelopidas lyelli* (Rothschild)	（チョウ目、セセリチョウ科）豪州区
lygaeid bugs			chinch bugs (1) を見よ
Lygus Acraea		*Acraea lygus* Druce	（チョウ目、タテハチョウ科）エチオピア区
Lygus bugs		*Lygus*	（カメムシ目、カスミカメムシ科）の昆虫の総称
Lyman's Haploa		*Haploa confusa* (Lyman)	（チョウ目、ヒトリガ科）新北区
lyme grass	ホソバウスキヨトウ	*Photedes elymi* (Treitschke)	（チョウ目、ヤガ科）日本、旧北区
lyme grass moth			lyme grass を見よ
lyme-grass wainscot			lyme grass を見よ
Lympharis Morpho		*Morpho lympharis* Butler	（チョウ目、タテハチョウ科）新熱帯区
lyonetiid moths			leaf skeletonizer moths (1) を見よ
Lyra clearwing		*Greta andromica lyra* (Salvin)	（チョウ目、タテハチョウ科）新熱帯区
Lyra metalmark		*Lyropteryx lyra cleadas* Druce	（チョウ目、シジミタテハ科）新熱帯区
lyrate grappletail		*Heliogomphus lyratus* Fraser	（トンボ目、サナエトンボ科）東洋区
lyreman			dog-day cicada (2) を見よ
lyre-tipped spreadwing		*Lestes unguiculatus* (Hagen)	（トンボ目、アオイトトンボ科）新北区
Lysias skipper		*Corticea lysias lysias* (Plötz)	（チョウ目、セセリチョウ科）新熱帯区
Lyside		*Kricogonia lyside* (Godart)	（チョウ目、シロチョウ科）新北区、新熱帯区
Lyside sulphur			Lyside を見よ
Lysimnia tigerwing		*Mechanitis lysimnia* (Fabricius)	（チョウ目、タテハチョウ科）新熱帯区。*M. l. utemaia* Reakirt も同英名
Lysippus metalmark		*Riodinia lysippus* (Linnaeus)	（チョウ目、シジミタテハ科）新熱帯区

世界の昆虫英名辞典
vol. 1 A-L

発行日　2018年5月12日　初版　第1刷

編　集

矢野宏二
(やのこうじ)

発　行

櫂歌書房
(とうかしょぼう)

ISBN 978-4-434-24028-7

有限会社 櫂歌書房
〒811-1365 福岡市南区皿山4丁目14-2
TEL: 092-511-8111　FAX: 092-511-6641
E-mail: e@touka.com　http://www.touka.com

発売所　星雲社